W0090538

FORTSCHRITTE DER CHEMIE ORGANISCHER NATURSTOFFE

PROGRESS IN THE CHEMISTRY OF ORGANIC NATURAL PRODUCTS

PROGRÈS DANS LA CHIMIE DES SUBSTANCES ORGANIQUES NATURELLES

HERAUSGEGEBEN VON EDITED BY RÉDIGÉ PAR

L. ZECHMEISTER

CALIFORNIA INSTITUTE OF TECHNOLOGY, PASADENA

ZWÖLFTER BAND TWELFTH VOLUME DOUZIÈME VOLUME

VERFASSER AUTHORS AUTEURS

G. W. BEADLE · A. J. HAAGEN-SMIT · T. G. HALSALL · F. T. HAXO
E. R. H. JONES · R. MICHEL · J. ROCHE · K. SLOTTA · A. R. THOMPSON
E. O. P. THOMPSON · R. TSCHESCHE · F. L. WARREN

MIT 15 ABBILDUNGEN WITH 15 ILLUSTRATIONS AVEC 15 ILLUSTRATIONS

WIEN · SPRINGER-VERLAG · 1955

Softcover reprint of the hardcover 1st edition 1987

ISBN 978-3-7091-7168-4 ISBN 978-3-7091-7166-0 (eBook)
DOI 10.1007/978-3-7091-7166-0

Inhaltsverzeichnis.
Contents. — Table des matières.

Sesquiterpenes and Diterpenes.

By A. J. HAAGEN-SMIT, Pasadena, California.

With 2 Figures.

Contents.

Introduction.

The study of terpenes is intimately connected with the development of organic chemistry. In the early years of terpene research the peculiar structures of this type of compounds posed formidable problems for which methods of attack had to be invented; and the organic chemist of today owes a great deal to the pioneers in this field such as WALLACH, SEMMLER and many others.

Two dozen years after WALLACH started inventorizing the terpenes in 1890, the detailed structure of most of the hundred monoterpenes listed in the handbooks on essential oils and terpenes by GUENTHER (*89*) and by SIMONSEN (*188*) was well established. At the present time it is a rare event when new structures are added such as lavandulol (I) from oil of French lavander (*185*), the thujaplicins (II) from the heartwood of Coniferae (*6, 74, 147*), nepetalactone (III), the main constituent of catnip oil (*135*) and shonanic acid (IV) from *Libocedrus formosana* (*166a*).

At the present time the majority of the monoterpenes can be assigned their absolute stereochemical configurations (*28, 81, 81a, 137, 167, 168*). A few examples of these structures are presented in formulae (V), (VI), (VII), and (VIII). The field of monoterpenes is far from being exhausted and has in recent years been stimulated by investigations in the related fields of steroids and other natural products. Physical methods of

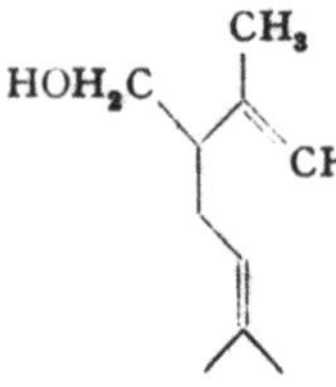

(I.) Lavandulol.

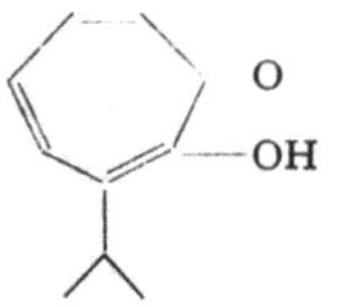

(II.) α-Thujaplicin.

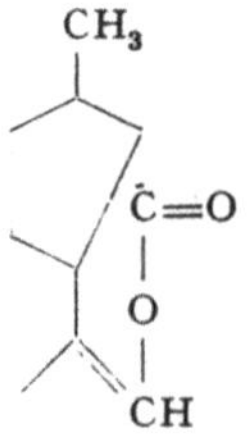

(III.) Nepetalactone.

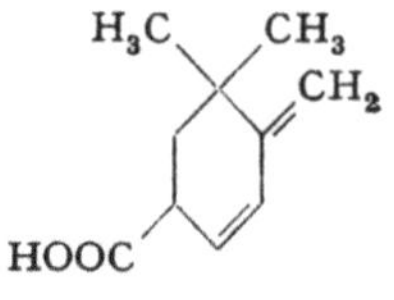

(IV.) Shonanic acid.

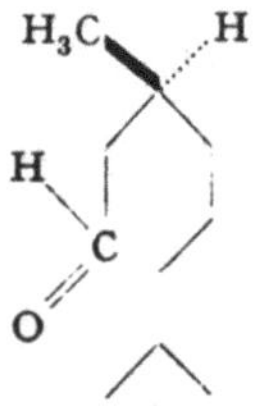

(V.) (+)-Citronellal.

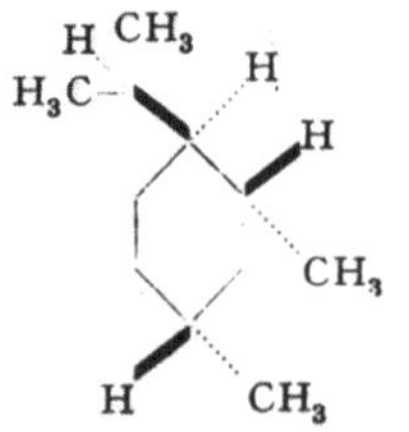

(VI.) (—)-Menthol.

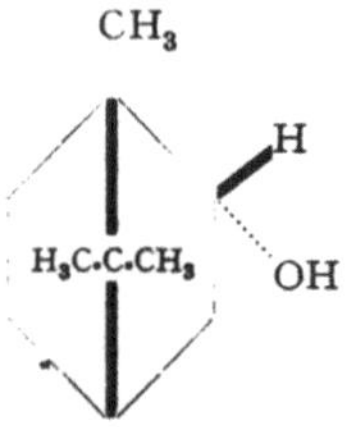

(VII.) (—)-Borneol.

(VIII.) (+)-α-Pinene.

structural analysis and chromatographic procedures in the purification of terpenes and their degradation products have become regular experimental tools for the terpene chemist. This influence is especially noticeable in the sesqui- and higher terpene groups, and is to a large extent responsible for the rapid expansion of our knowledge of a variety of complex structures.

I. Sesquiterpenes.

1. Acyclic Sesquiterpenes.

While the monoterpene chemistry was developing rapidly, it was realized that the building principle of the terpenes could probably be applied to compounds with multiples of five carbon atoms, the sesqui- and di-terpenes and their oxygen derivatives. The first member of the sesquiterpene family to which a correct structure could be assigned by KERSCHBAUM (*111*) in 1913 was the acyclic farnesol (IX). Nerolidol (X), another acyclic sesquiterpene, is related to farnesol in the same manner as linalool to geraniol in the monoterpenes (*174*).

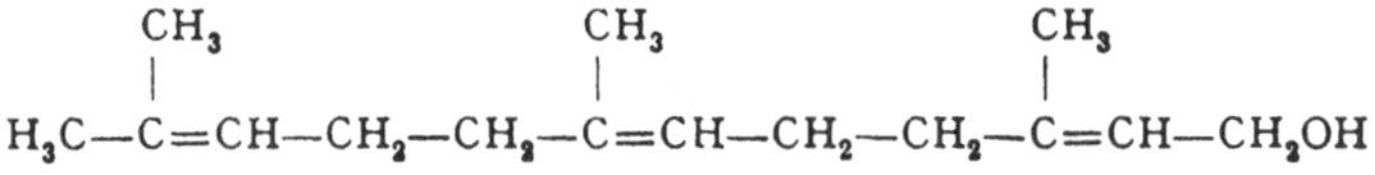

$$H_3C-\underset{\displaystyle CH_3}{\overset{|}{C}}=CH-CH_2-CH_2-\underset{\displaystyle CH_3}{\overset{|}{C}}=CH-CH_2-CH_2-\underset{\displaystyle CH_3}{\overset{|}{C}}=CH-CH_2OH$$

(IX.) Farnesol.

$$H_3C-\overset{\displaystyle CH_3}{\overset{|}{C}}=CH-CH_2-CH_2-\overset{\displaystyle CH_3}{\overset{|}{C}}=CH-CH_2-CH_2-\underset{\underset{\displaystyle OH}{|}}{\overset{\overset{\displaystyle CH_3}{|}}{C}}-CH_2=CH_2$$

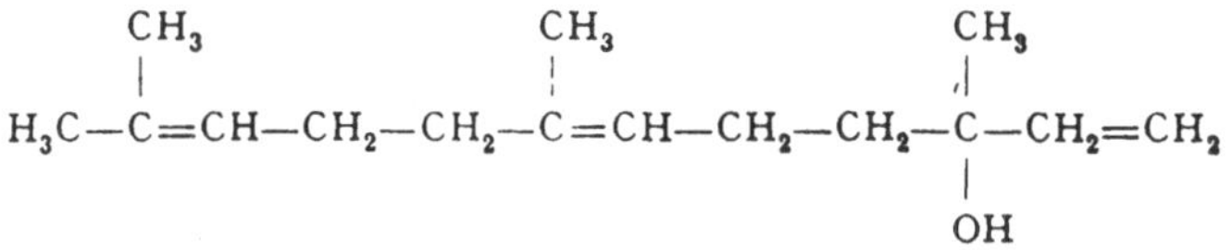

(X.) Nerolidol.

The isolation by chromatographic methods of farnesene, the dehydration product of farnesol, and a variety of other sesquiterpenes by ŠORM, HEROUT, and co-workers (*98*, *99*, *198*, *211*) illustrates the revolutionary effect which the introduction of this method will have on our knowledge of naturally occurring terpenes.

Recently, ipomeamarone and ngaione, volatile compounds from black-rotted sweet potatoes have been added to the farnesol type structure (*30*, *123*). They are formulated as *cis*- and *trans*-forms of (XI).

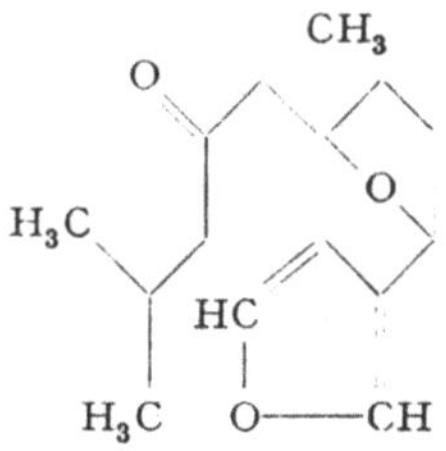

CH₃ O O H₃C HC H₃C O—CH

(XI.) Ipomeamarone (*cis*). Ngaione (*trans*).

2. Bisabolene and Cadinene Type Sesquiterpenes.

The application of the dehydrogenation method with sulfur (*237*) by RUZICKA and MEYER (*180*) in 1921 proved to be of major importance for the development of the sesqui- and higher terpenes. The recognition of the carbon skeleton of the cyclic terpenes has provided a firm basis for the interpretation of the oxidation experiments, and as a result, the detailed structure of a large number of sesquiterpenes and diterpenes has been settled.

The discovery of cadalene, 1,6-dimethyl-4-isopropyl-naphthalene (XV), among the dehydrogenation products of cadinene showed that sesquiterpenes may be constructed from isoprene units arranged as in farnesol (XII) and that the cyclization follows along the lines found in the monoterpene series. Bisabolene (XIII) and cadinene (XIV) can be constructed from a farnesol chain by cyclizations, comparable to the formation of dipentene from geraniol.

CH_2OH

(XII.) Farnesol. (XIII.) Bisabolene type. (XIV.) Cadinene type. (XV.) Cadalene.

In thirty years, many representatives of this type of ring closures have been found in mono-, bi- and tricyclic terpenes.

These structures vary in number and position of double bonds and the presence of carbonyl and hydroxyl groups. For example, the cadinenes found in nature are represented as eight structures by Herout et al. (*100, 101*), all yielding the same dihydrochloride. These isomers are shown in *Chart 1*, formulae (XVI)–(XX).

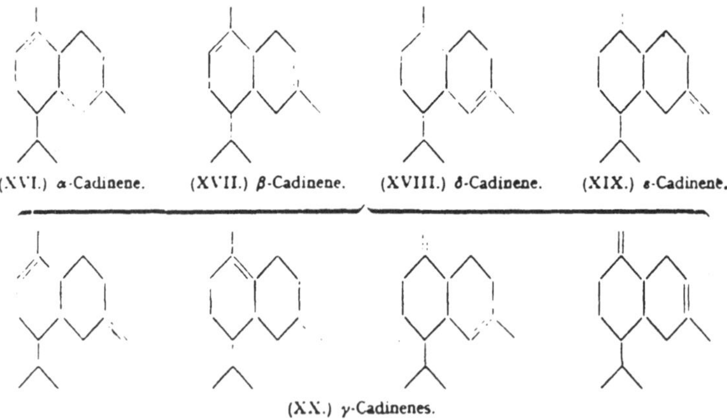

Chart 1. Isomeric Cadinenes.

The presently known structures of bisabolene and cadinene type sesquiterpenes described by Simonsen and Barton (*188*) are presented in *Chart 2*, formulae (XXI)-(XLIII). Included are, some recently investigated members of this group: bisabolol (XXVI) (*142, 212*), calamenene (XXXIV) (*210*), calamendiol (XXXIX) (*225*), lanceol

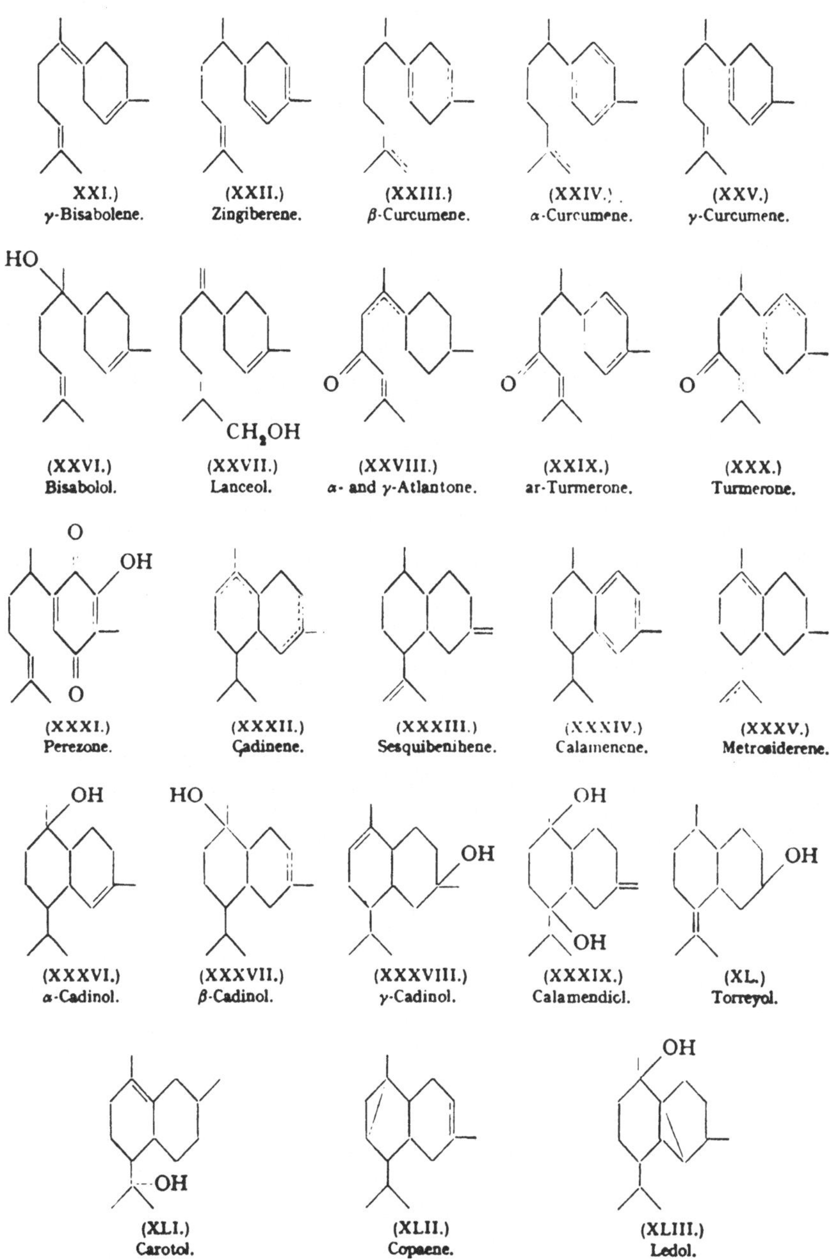

Chart 2. Sesquiterpenes of the Bisabolene and Cadinene Types.

(XXVII) (*77*), metrosiderene (XXXV) (*62*); furthermore, the "irregularly built" carotol (XLI) (*188*), and the tricyclic structures of copaene (*188*), and ledol (*119*,) tentatively presented as (XLII) and (XLIII).

The lively interest in the stereochemical relationships among the steroids has stimulated similar investigations in the sesquiterpene field. ARIGONI and JEGER (*12*) presented the absolute configuration of zingiberene and perezone as (XLIV) and (XLV), respectively, (*137*, *140*, *145*), while BRADFIELD, FRANCIS, and SIMONSEN (*36*) and MILLS (*137*) assigned structure (XLVI) to lanceol.

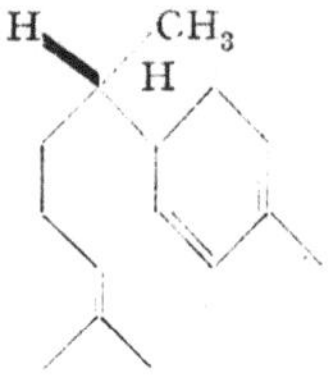

H CH$_3$ H

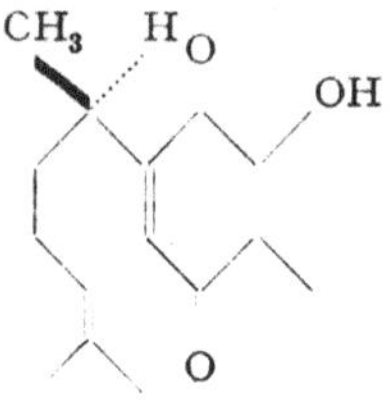

CH$_3$ H O OH O

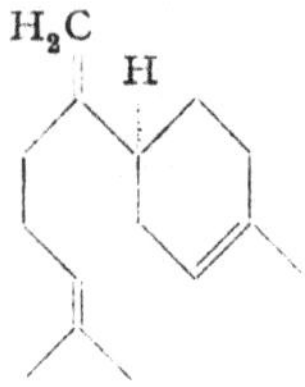

H$_2$C H

3. Eudesmol Type Sesquiterpenes.

A new type of cyclic sesquiterpene was added by RUZICKA (*181*) when eudesmol was dehydrogenated. The naphthalene hydrocarbon isolated contained one carbon less than cadalene, and proved to be 1-methyl-7-isopropyl-naphthalene (eudalene) (XLVII). Eudesmol had to be represented by (XLIX) and (L) (*179*) since only so could the structure be divided into isoprene units and explain the removal of a methyl group.

As in the cadalene group, we can form the eudesmol carbon skeleton from a farnesol chain. This time the type of ring formation is like that found in cyclogeraniol or ionone. The second cyclization is again like that of geraniol to dipentene.

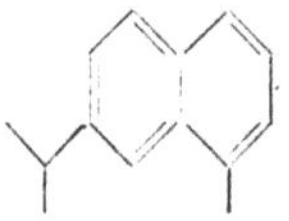

CH$_3$ HOOC HOOC CH$_3$

(XLVII.) Eudalene. 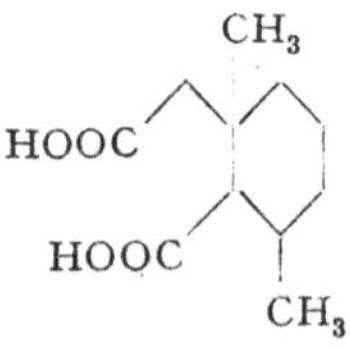 (XLVIII.) Dicarboxylic acid $C_{11}H_{18}O_4$.

The position of the methyl group has been established with certainty by PLATTNER et al. (*157*) by the isolation of the acid (XLVIII). The same position of the angular methyl group has been established for the cyperones by synthesis of β-cyperone (LII) (*2*, *133*, *173*).

Additional evidence for the position of the angular methyl in eudesmol type sesquiterpenes can be found in the investigations on the structure and synthesis of santonin and its derivatives (*1*, *1a*, *1b*, *53*,

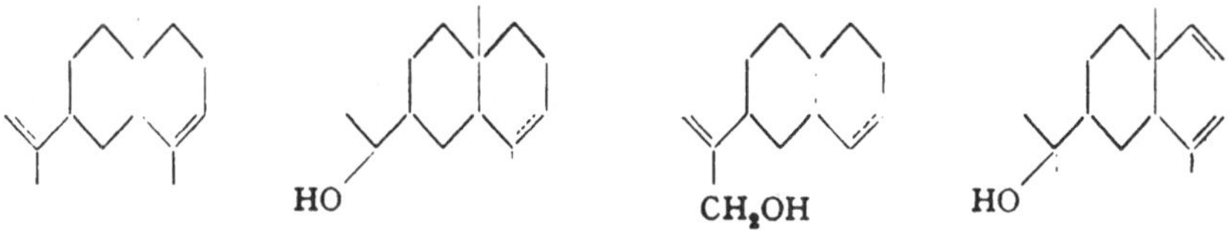

HO CH₂OH HO

(LIII.) α- and β-Selinene. (LIV.) α- and β-Eudesmol. (LV.) Sesquibenihiol. (LVI.) Elemol.

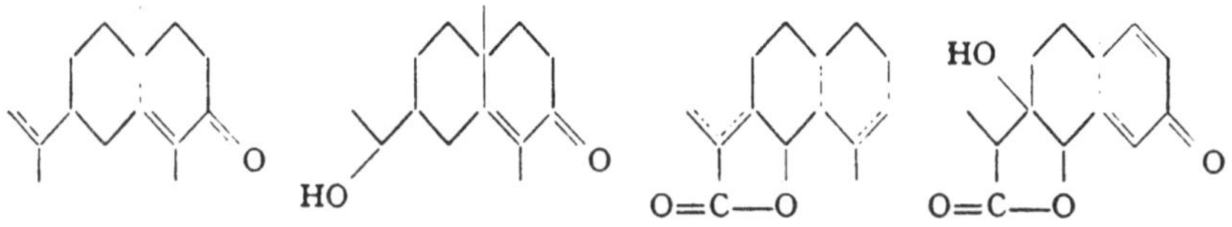

(LVII.) α-Cyperone. (LVIII.) Carissone. (LIX.) Alantolactones. (LX.) Artemisin.

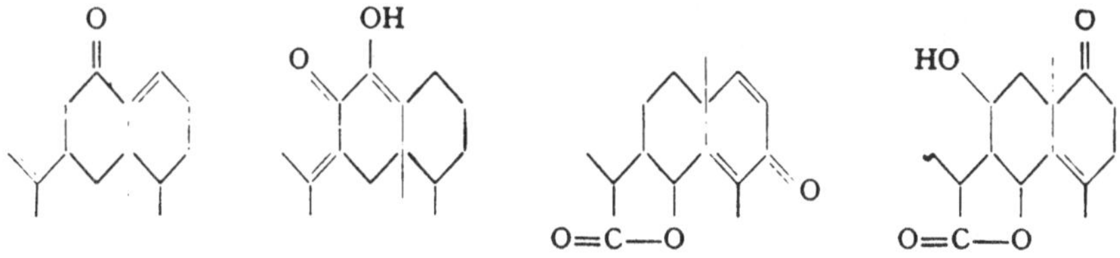

(LXI.) Eremophilone. (LXII.) Hydroxy-eremophilone. (LXIII.) α- and β-Santonin. (LXIV.) ψ-Santonin.

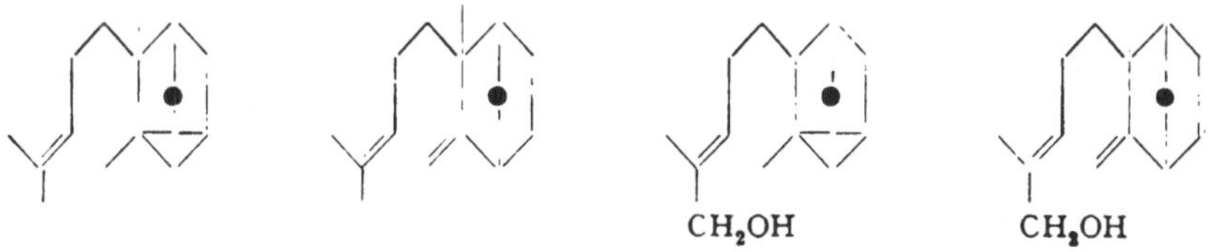

(LXV.) α-Santalene. (LXVI.) β-Santalene. (LXVII.) α-Santalol. (LXVIII.) β-Santalol.

Chart 3. Eudesmol Type Sesquiterpenes

54, 54b, 65, 86, 87, 131, 132, 239). The naturally occurring lactones may be assigned the complete configuration shown in (XLVIII a), (XLVIII b), and (XVLIII c) (*1c, 18, 60, 62a*). ψ-Santonin is formulated as (LXIV) in *Chart 3* (*54a, 58, 65, 65a*).

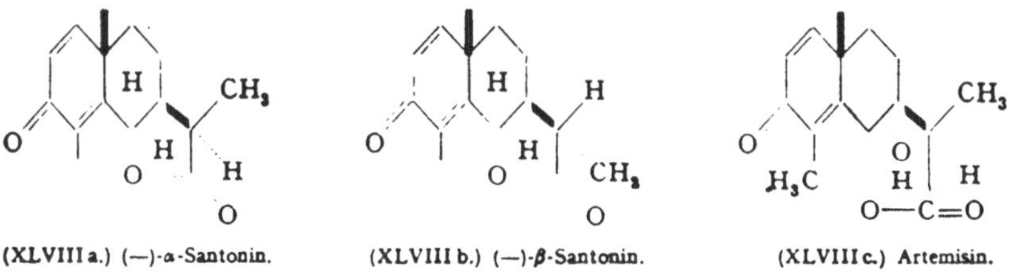

(XLVIII a.) (—)-α-Santonin. (XLVIII b.) (—)-β-Santonin. (XLVIII c.) Artemisin.

The method of molecular rotation differences has been used by KLYNE (*115, 116*) to correlate the stereochemistry of sesqui- and diterpenes with that of the triterpenes and steroids. Oxidative degradation as well as conformational and rotational analyses indicate that α- and β-eudesmol (XLIX; L) and α-selinene (LI) are *trans*-decalin derivatives (*19, 171, 173*). Natural α-cyperone has been assigned the absolute configuration (LII) by MCQUILLIN (*134a*).

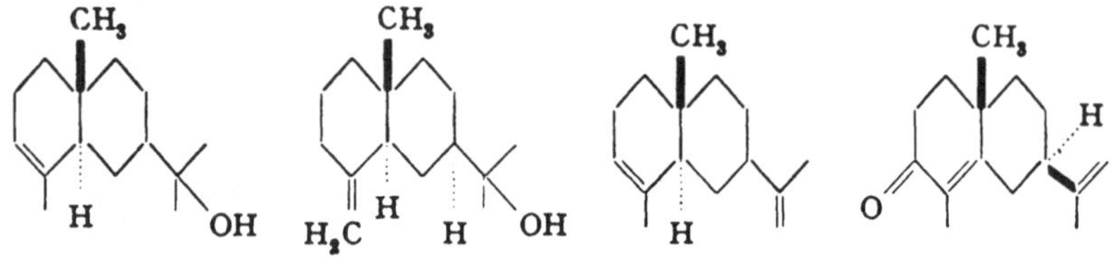

(XLIX.) α-Eudesmol. (L.) β-Eudesmol. (LI.) α-Selinene. (LII.) (+)-α-Cyperone.

The members of the eudesmol group and related tricyclic structures are shown in *Chart 3* including the recently established structure of carissone (LVIII) (*27*).

4. Elemol.

Included in the eudalene type structures in *Chart 3* are the "irregularly built" elemol (LVI) and eremophilone (LXI). Elemol, extensively studied by RUZICKA, was recently reinvestigated by ŠORM and coworkers (*217, 218*). Dehydrogenation of tetrahydro-elemane to 1-methyl-2,4-diisopropyl-benzene (LXIX) proved that elemol has the skeleton of 1-methyl-1-ethyl-2,4-diisopropyl-cyclohexane. On the basis of earlier degradation reactions elemol is represented by (LVI). The positions of the double bonds are, however, uncertain. Elemol, composed of three isoprene units in "irregular" combination, cannot be described as a

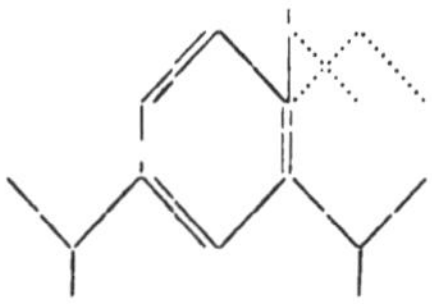

(LXIX.) 1-Methyl-2,4-diisopropyl-benzene.

cyclization product of farnesol unless we assume with RUZICKA (*175*) that a ring opening occurred in a eudaline type structure, giving rise to the vinyl and isopropylidene groups. Similar reactions are seen in the *in vitro* conversion of ocimene and myrcene from α- and β-pinene by way of biradicals.

5. Eremophilone.

Eremophilone does not follow the "farnesol" rule (p. 31), but gives eudalene upon dehydrogenation. Oxidation experiments by PENFOLD and SIMONSEN (*37, 152*) established that the angular methyl group is present in position 10, and not at $C_{(9)}$ as in eudesmol.

Because of the deviation of eremophilone from the normal isoprene type structure, investigations are still continuing on this subject, and recently GEISSMAN (*83*) presented confirmatory evidence regarding the structure of hydroxy-eremophilone by clearing up the constitution of a degradation product, $C_{15}H_{22}O_4$, formerly regarded as $C_{12}H_{18}O_3$ (*37*). The reinvestigation showed that this compound is formed by opening of the 6-membered ring at the double bond, followed by lactone formation. These observations support the "irregular" structure of hydroxy-eremophilone.

OH O H$_3$C CH$_3$ → O$_3$ → O C OH O H$_3$C O H$_3$C H$_3$C CH$_3$

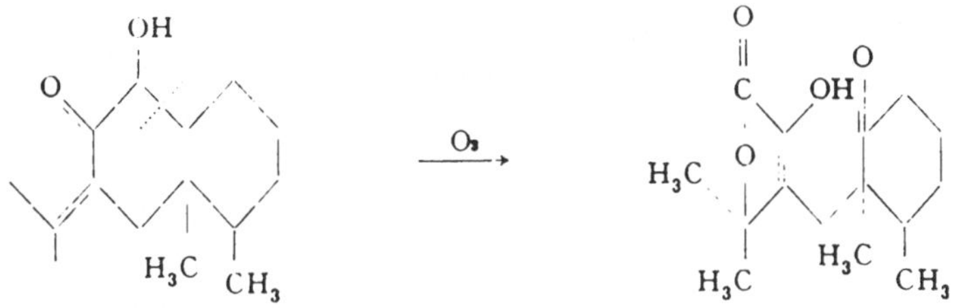

(LXX.) Hydroxy-eremophilone. (LXXI.) $C_{15}H_{22}O_4$.

ROBINSON (cf. *188*) postulated that a methyl shift might have occurred during the formation from the "regularly" built precursors (LXXII) or (LXXIII).

O CH$_3$ HO CH$_3$ — CH$_3$ O OH CH$_3$

(LXXII.) (LXXIII.)

Possible precursors of eremophilone.

6. Iresin.

Dipentene-like ring closures are seen in the cadinene group, and a dipentene and cyclogeraniol cyclization in the eudalene group. Recently, a sesquiterpene was discovered which can be constructed through two cyclogeraniol type ring closures. DJERASSI (*70, 70a*) has shown that iresin is probably represented by (LXXIV). It is interesting that this type of cyclization is most common in the di- and triterpenes and the steroids but is apparently rare among the sesquiterpenes. A degradation of higher terpenes could be considered in its biological formation. In the laboratory this type of cyclization of sesquiterpenes was accomplished using sesquilavandulol, farnesylic acid and farnesal (LXXV → LXXVI) by SCHINZ et al. (*43, 61, 124*) as well as by STOLL and COMMARMONT (*215*).

(LXXIV.) Iresin. (LXXV.) Farnesylic acid. (LXXI.) α-Bicyclo-farnesylic acid.

7. Azulene Type Sesquiterpenes.

While many of the sesquiterpenes could be classified as belonging either to the cadalene or to the eudalene type, in some cases blue and violet hydrocarbons, the azulenes, could be isolated instead of the naphthalene derivatives. The assignment of the structure of the sesquiterpene azulene precursors was possible after PFAU and PLATTNER (*153*) had shown that the azulenes consisted of a bicyclic system containing a five- and a seven-membered ring with five double bonds in conjugated position. The azulenes derived from sesquiterpenes

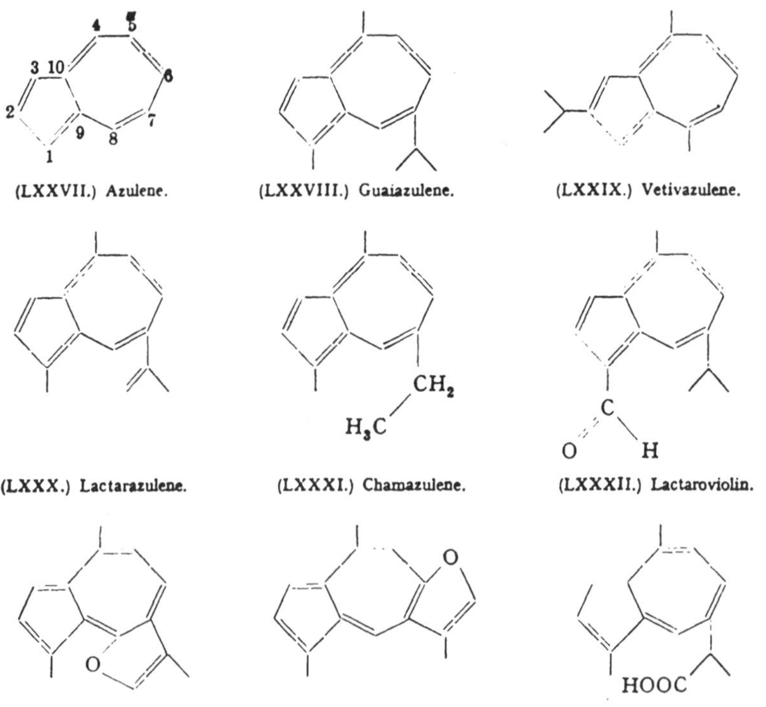

(LXXVII.) Azulene. (LXXVIII.) Guaiazulene. (LXXIX.) Vetivazulene.

(LXXX.) Lactarazulene. (LXXXI.) Chamazulene. (LXXXII.) Lactaroviolin.

(LXXXII a.) Artemazulene. (LXXXII b.) Linderazulene. (LXXXII c.) Chamazulene carboxylic acid

are substitution products of the parent hydrocarbon, azulene, $C_{10}H_8$, (LXXVII), an isomer of naphthalene. At present the structures of guaiazulene (LXXVIII) (*153*), vetivazulene (LXXIX) (*153*), lactarazulene (LXXX) (*191, 238*) and chamazulene (LXXXI) (*136, 197, 202, 219*) are known. Elemazulene is identical with vetivazulene (*190*). Red-violet lactaroviolin, obtained from the mushroom *Lactarius deliciosus* L., belongs to the guaiazulene type and contains an aldehyde group (LXXXII) (*110, 137, 160, 161, 191, 192*). Recently, an azulenoid acid (LXXXIIc) was isolated from *Achillea millefolium* L. by STAHL (*214, 214a*). Artemazulene (LXXXIIa) (*103, 103a*) and linderazulene (LXXXIIb) (*220*) have been recognized as furano-azulenes or furazulenes.

The azulenes indicate the nature of the carbon skeleton of the sesquiterpenes from which they originate. As in the case of cadalene and eudalene, we may expect these precursors to differ in position and number of double bonds, the occurrence of oxygen-containing groups and the number of rings which may have opened or formed during the dehydrogenation process.

The largest group of azulene type sesquiterpenes is derived from guaiazulene. Members of this group are, guaiol (LXXXIII); partheniol (LXXXIV); α-chigadmarene (LXXXV); δ-guajene (XCIa) (*193a*); the sesquiterpene oxides germacrol (LXXXVI) (*228a*), kessyl alcohol (LXXXVII) (*226*), patchouli alcohol (LXXXVIII) (*224*), and kessoglycol (XCIIa) (*236a*). Their formulae are shown in *Chart 4*. Aromadendrene (XC) and globulol (XCI), in which the position of the cyclopropane ring is uncertain, are tentatively presented in Chart 4 (*33, 230*). The precursors of chamazulene, lactaroviolin and lactarazulene also belong to this group.

Chamazulene, identical with lindazulene, was recognized as 1,4-dimethyl-7-ethyl-azulene (LXXXI) (*136, 197, 202, 219—221*). This azulene is not present in the chamomille flowers but is formed from a precursor during isolation. Similar pro-azulenes are found in *Artemisia absinthium* L. and are hydroxy- and acetoxy-substitution products of "guaianolide" (XCII) containing two double bonds (*47, 48, 102, 103*). These lactones are of physiological interest since their antiphlogistic activity is equal to that of chamazulene.

Dehydrogenations with sulfur or selenium are carried out at high temperatures, and changes in structure sometimes do occur. Thermal rearrangements of the substituents may take place, as seen in the formation of Se-guaiazulene from guaiol, whereby the methyl group at position 1 has shifted to position 2 (for numbering cf. p. 10). Similar shifts have been observed by HERZ (*105*) in the preparation of 4- and 5-alkyl substituted azulenes when the 4- and 5-methyl groups moved to positions 3 and 2. COCKER et al. (57) have shown that elimination of

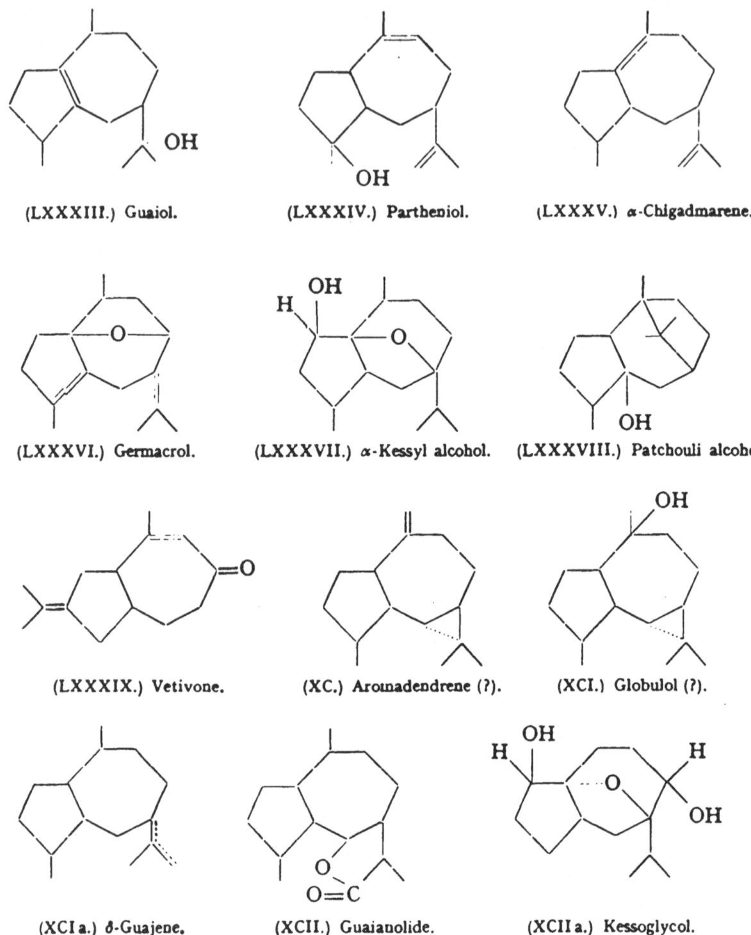

Chart 4. Azulene Type Sesquiterpenes.

non-angular alkyl groups may also occur in aromatization reactions. These results contain a warning that a source of error may exist in locating alkyl groups in azulenogens by dehydration and dehydrogenation of bicyclodecane derivatives.

Ring opening is evident in the formation of guaiazulene from the tricyclic structures aromadendrene (*33*, *230*), patchouli alcohol (*224*), ledol (*119*), and caryophyllene oxide (*223*). The position of the third ring in these compounds is therefore uncertain, and should be determined by less drastic methods than dehydrogenation.

Synthesis and Properties of Azulenes. The study of azulenes has been actively pursued in recent years, not only because of their importance in establishing the structure of the naturally occurring terpene precursors, but especially for their interesting physical properties. These investigations have been reviewed by several authors (*85a, 90, 165*) and these reviews were recently brought up to date in an extensive survey by Treibs, Kirchhof and Ziegenbein (*232a*).

The azulenes of natural origin, S- and Se-guaiazulene and chamazulene have been synthesized (*136, 146a, 195, 200, 203, 219*). Dehydrogenation methods with aluminium oxide and palladium charcoal were investigated by the Swiss workers (*88, 120, 122, 169*), and the dehydrogenation with halogens was introduced by Treibs (*223, 234*). A catalytic dehydrogenation of octahydroazulene to azulene has been described by Kováts, Günthard, and Plattner et al. (*121*) using carbon disulfide and a molybdenum-nickel sulfide catalyst on an alumina carrier. Maximum yields of 60% have thus been obtained. A new method of synthesis of the azulenes has been reported by Braude and Forbes (*39*) and by Pommer (*166*). Standard and improved procedures were used in preparing a variety of alkyl azulenes (*8, 14, 40, 71, 72, 118, 127a, 139, 149, 189, 196, 199, 209, 232a*) and their oxygen derivatives (*10, 159, 235*), furthermore, polycyclic azulenes (*11, 16, 34, 69, 227, 229, 232, 233*) and azulene analogues containing nitrogen as a ring member (*127, 148, 150, 231*).

The study of the chemical and physical properties of the azulenes has been actively pursued (*49, 95, 97, 105, 162, 213, 229a*).

Azulenes easily form addition products with acids, and this property has been extensively used in the purification process with H_3PO_4. When azulenes are dissolved in strong acids the blue color disappears and the solutions turn yellow, brown or reddish, depending on the presence of substituents and on the concentration. The absorption spectra show that a radical change in the electronic system has taken place; and the calculations of Heilbronner (*97*) provide an explanation for the shift in the bands of the spectra upon solution in acids, and are in agreement with the observed dipole moments of the azulenes.

The degree of basicity of the azulenes is comparable to that of the nitroanilines (*162*) and can be explained by the formation of azulenium

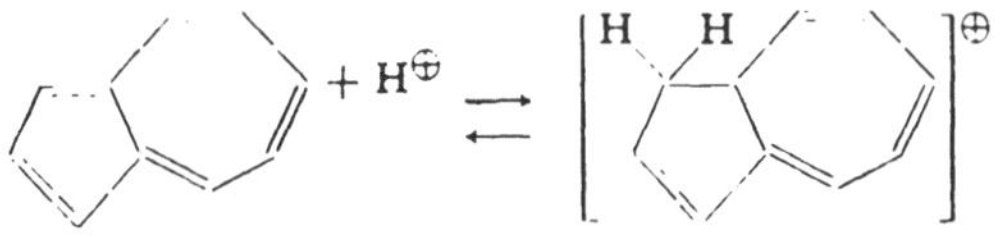

A + $H^{\oplus}$ ⇄ *B* (8 resonance structures)

(XCIII.) Azulenium ion.

ions (XCIII). Among the eight possible azulenium ions, structure *B* is shown to have the greatest resonance energy. The other seven ionic structures are present in negligible quantities in equilibrium with *B*.

On the basis of these considerations and calculations of BROWN (*41*), and other authors (*155, 166*) electrophilic attack of the azulene nucleus is predicted for position 1. Nucleophilic and radical attack will both take place at position 4. ANDERSON et al. (*7, 9*) have shown that FRIEDEL-CRAFTS acetylation of azulene results in the formation of 1-acetyl-azulene and a diacetyl-azulene. 1-Nitro-azulene has been obtained from the treatment of azulene with cupric nitrate and acetic anhydride. Benzene diazonium chloride with azulene gave 1-azulene-azobenzene. *n*-Bromo-succinimid gives rise to 1-bromo-azulenes and some dibromo-azulenes.

8. Longifolene.

For a long time the sesquiterpenes which could not be assigned a naphthalene or an azulene type structure offered great difficulties in their structure determination. However, in the last few years, the structures of longifolene, clovene, caryophyllene alcohol, caryophyllene, and humulene have been established.

Longifolene (XCV), occurring in the oleoresins of several species of the genus pine, contains a tricyclic system consisting of a fused 7- and two 5-membered rings. BRADFIELD, FRANCIS, and SIMONSEN suggested a structure with a cyclopentane and cyclohexane ring and a vinyl side-chain (*36*).

The presence of a vinyl group was concluded from the formation of the aldehyde and alcohol from iso-longifolic acid, which was held

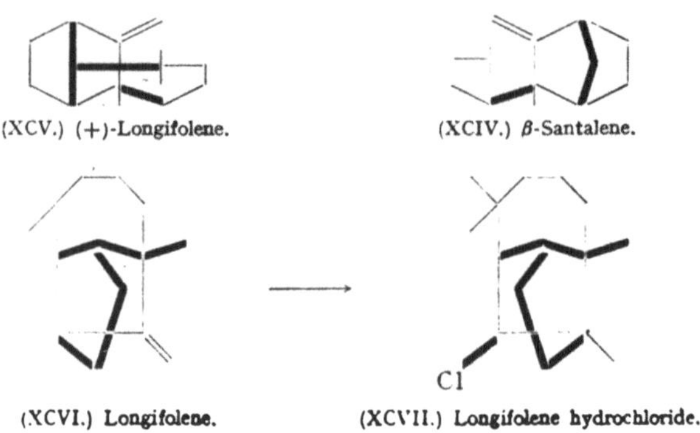

(XCV.) (+)-Longifolene. (XCIV.) β-Santalene.

(XCVI.) Longifolene. (XCVII.) Longifolene hydrochloride.

to be a C_{14}-acid. NAFFA and OURISSON (*141, 151*) have shown that longifolic acid and iso-longifolic acid are formed by terminal oxidation of a methylene group, and that these acids have 15 carbon atoms instead of 14. Their findings necessitated a revision of the earlier formula which contained a vinyl group. On the basis of X-ray studies by MOFFETT and ROGERS (*138*) on longifolene hydrochloride, longifolene itself is now considered to be an isoprene homologue of camphene. The third isoprene unit has its "tail" joined to the exo-member of the gem-dimethyl group of camphene. This structure is similar to that of bicyclic β-santalene

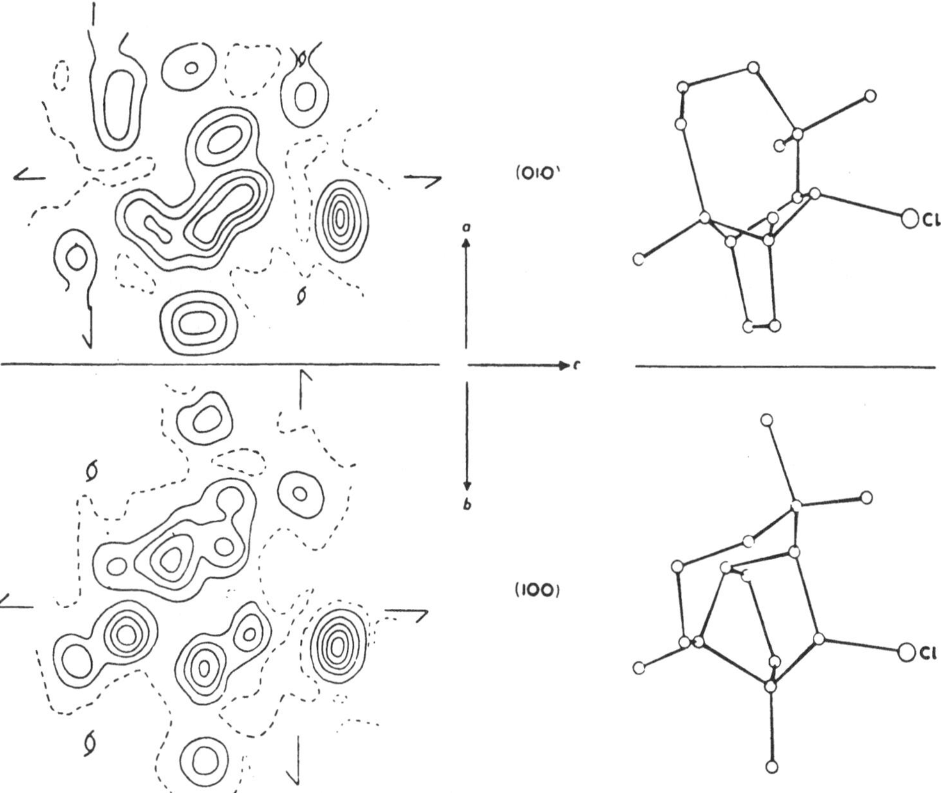

Fig. 1. Space formule of Longifolene hydrochloride as determined by X-ray studies, according to MOFFETT and ROGERS (*138*). Corresponding portions of the projections of the electron density on (010) and (100). Contours are drawn at intervals of 2e. A^{-3} on the carbon atoms and 4e. A^{-3} on the chlorine: the first is broken. Although this X-ray study cannot give the absolute configuration of the molecule, the model drawn here is based on OURISSON's correlation with FISCHER's convention, which has recently been shown to represent true configuration. [From: Chemistry and Industry *1953*, 916.]

(XCIV); however, in longifolene the isoprene "head" is cyclized to a seven-membered ring. Molecular rotation considerations show (+)-longifolene to derive from (+)-camphene (*151*). Formula (XCV) represents the absolute configuration of longifolene with respect to E. FISCHER's convention. A comparison of formulae (XCVI) and (XCVII) illustrates the WAGNER-MEERWEIN rearrangement in the formation of the hydrochloride, and *Figure 1* shows the space formula for the hydrochloride as determined by X-ray studies.

9. Caryophyllene.

Treibs (*223*) has shown that the hydrocarbons present in the oil of cloves are accompanied by an oxide which is formed by the action of oxygen or per-acids on β-caryophyllene. He concluded that the oxide was a secondary product formed during the processing of the flowers. Naves (*143*) found that caryophyllene is not present in the oil obtained by extraction with benzene, but that only a small amount of the epoxide could be isolated. He concludes, therefore, that the flowers do not contain the sesquiterpenes as such, but that they are present in complexes which are decomposed in the steam distillation process. Similar interesting observations have been made by Naves (*85*, *144*) on other flower oils.

Until 1934, caryophyllene was held to be an isoprene homologue of pinene. When the investigation of nor-caryophyllenic acid showed that the carboxyl groups are situated on adjacent carbon atoms, the pinene-like structure could no longer be defended. The structure of nor-caryophyllenic acid has since been confirmed by synthesis of *DL-trans*-nor-caryophyllenic acid (XCVIII), with which it is identical (*184*). Its homologue, (+)-*trans*-caryophyllenic acid (XCIX) (*44*) has also been synthesized and proved to be identical with the corresponding oxidation product of caryophyllene.

H_3C, H_3C / COOH / COOH

(XCVIII.) Nor-caryophyllenic acid.

H_3C, H_3C / CH_2—COOH / COOH

(XCIX.) Caryophyllenic acid.

Ruzicka et al. (*178*) in a systematic oxidative degradation of dihydro-caryophyllene gave evidence that the second ring contained at least eight carbon atoms. At the time this degradation was carried out, the idea of a ring system of more than six carbon atoms did not find general acceptance, and the results of Ruzicka were attributed to the lack of homogeneity of caryophyllene. In 1950, however, Šorm et al. (*194*), oxidizing caryophyllene oxide with permanganate, obtained a crystalline oxido-ketone, $C_{14}H_{22}O_2$, which in infrared analysis exhibited a carbonyl stretching frequency at 1698 cm^{-1}. This value was lower than that for aliphatic ketones or cyclohexanones and more in agreement with large-ring ketones.

From similar studies on dihydro-caryophyllene oxide Šorm *et al.* concluded that caryophyllene alcohol contained a four- and a nine-membered ring. This new type of bicyclic structure received support from Barton and Lindsey (*21*, *23*, *24*) as well as from Dawson, Ramage, and Wilson (*66*, *67*, *169a*, *169b*). The synthesis of this ring system, 2,8,8-trimethyl-bicyclo-(5,2,0)-nonane was accomplished by DOLEJŠ and ŠORM (*73*). The position of the exo- and endocyclic double bonds was determined by ATWATER and REID (*15*) in an oxidative degradation of dihydro-caryophyllene. This degradation established the presence of the grouping $—CH_2—CH=C(CH_3)—CH_2—$ which can only be accomodated in structure (C) that represents according to BARTON and NICKON (*24a*) the absolute configuration of caryophyllene.

In addition to caryophyllene, the oil of cloves contains related sesquiterpenes originally named α- and γ-caryophyllene. BARTON, BRUUN, and LINDSEY (*21, 22*) conclude that γ-caryophyllene should be considered as a stereochemically different compound, due to *cis-trans* isomerism at the endocyclic double bond; and it is therefore proposed by AEBI, BARTON, and LINDSAY (*4, 5*) that the name β-caryophyllene be replaced by caryophyllene and that of γ-caryophyllene by iso-caryophyllene. Caryophyllene is the *trans*-isomer, and it is represented as (C), while isocaryophyllene (CI) has *cis* configuration.

10. β-Caryophyllene Alcohol.

By the hydration of caryophyllene with the BERTRAM-WALBAUM reagent, crystalline β-caryophyllene alcohol is formed. This extremely stable alcohol is tricyclic, and its structure (CII) and that of its chloride has now been established (*20, 76, 129*). The X-ray studies by ROBERTSON and TODD (*172*) on caryophyllene hydrochloride confirmed the *trans*-fusion of the cyclobutane ring to the larger ring; and they also found that the H-atom attached to $C_{(5)}$, which corresponds to $C_{(1)}$ in β-caryophyllene, must lie on the same side of the molecule as the methylene bridge $C_{(12)}$. The shape of the chloride as determined by X-ray measurements is shown in *Figure 2*.

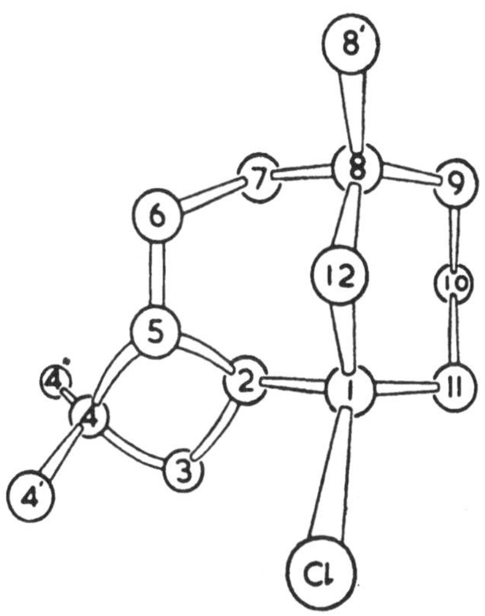

Fig. 2. Perspective drawing of the caryophyllene chloride molecule, according to ROBERTSON and TODD (*172*). [From: Chemistry and Industry *1953*, 437.]

When caryophyllene is treated with acid, the tricyclic sesquiterpenes clovene and iso-clovene are formed. Caryophyllene alcohol when dehydrated with P_2O_5 yields only pseudo-clovene. These results are explained by AEBI, BARTON, and LINDSEY (*3a*), if clovene is represented by (CIV), with a methylene bridge opposite to that of β-caryophyllene alcohol (CII). This structure has been confirmed by a stepwise degradation to *p*-cymene and to clovenic acid (*3*). Pseudo-clovene is tentatively presented as (CIII) (*128—130*).

11. Humulene.

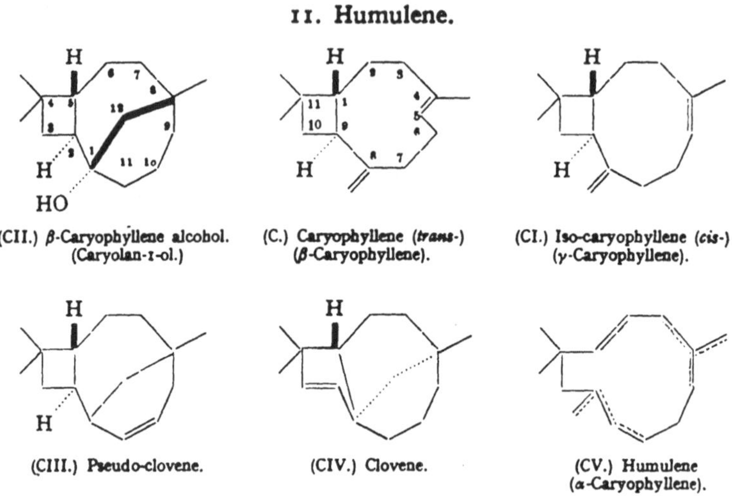

(CII.) β-Caryophyllene alcohol. (Caryolan-1-ol.)

(C.) Caryophyllene (*trans*-) (β-Caryophyllene).

(CI.) Iso-caryophyllene (*cis*-) (γ-Caryophyllene).

(CIII.) Pseudo-clovene.

(CIV.) Clovene.

(CV.) Humulene (α-Caryophyllene).

That α-caryophyllene is identical with humulene (CV) and contains an eleven membered ring structure with three double bonds was proven by Šorm et al. (*104, 193, 204, 205*) by synthesis of the parent hydrocarbon and comparison of the infrared spectra of the natural and synthetic products (*50, 50a, 51, 52, 68, 78—80, 94, 206*).

12. Cedrene.

The last few years have seen a rapid solution to the puzzling structure of cedrene and cedrol, and their tricyclic structures are now represented by (CVI) and (CVII) (*156, 216*). The early investigations of cedrene by Semmler and by Ruzicka and their co-workers led to the opening of ring *A* with the formation of cedrene dicarboxylic acid, $C_{14}H_{22}O_4$, and nor-cedrene dicarboxylic acid, $C_{13}H_{20}O_4$, (CVIII) (*182, 187*). Ring *B* was opened in nor-cedrene dicarboxylic acid by introducing a double bond in the α-position to the carboxyl group (*164*). Subsequent oxidation led to a monocyclic dicarboxylic acid, $C_{11}H_{18}O_4$ (CIX). The last ring, *C*, was opened in a similar manner by substitution with bromine, removal of HBr and oxidation of the newly formed double bond (*156*). The resulting open-chain methyl-keto-dicarboxylic acid, $C_{10}H_{16}O_5$, (CX) and its NaOBr oxidation product, α,α-dimethyl-β-carboxy-adipic acid, $C_9H_{14}O_6$ (CXI), were synthesized and proved to be identical with the oxidation products of cedrene (*158*).

The present formula of cedrene permits ready interpretation of the origin of previously obtained oxidation products. For example, α,α-dimethyl-tricarballylic acid (CXII), formed in the oxidation of cedrene or norcedrene-dicarboxylic acid, contains ring members from all three

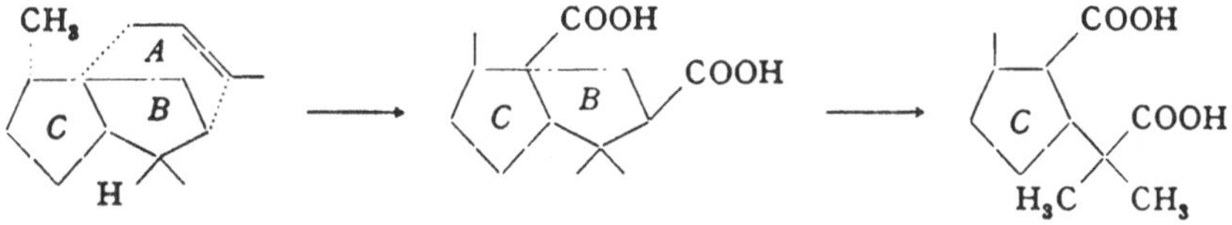

(CVI.) Cedrene.

(CVIII.) Nor-cedrene-dicarboxylic acid.

(CIX.) Dicarboxylic acid $C_{11}H_{18}O_4$.

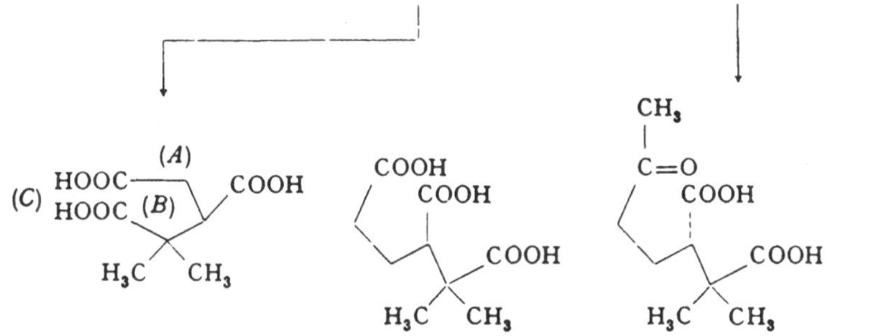

(CXII.) α,α-Dimethyl-tricarballylic acid.

(CXI.) α,α-Dimethyl-β-carboxy-adipic acid.

(CX.) 2,2-Dimethyl-3-carboxy-6-keto-heptylic acid.

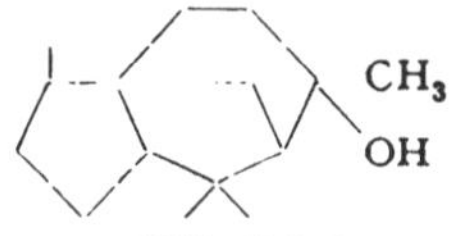

(CVII.) Cedrol.

rings, and is formed by the rupture of rings *A*, *B* and *C* (*163, 222*). From stereochemical considerations, a *cis*-fusion of rings *B* and *C* and

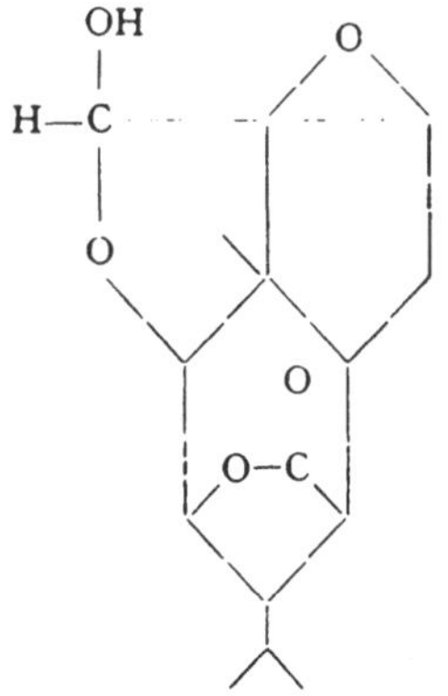

(CXIII.) Coriamyrtion.

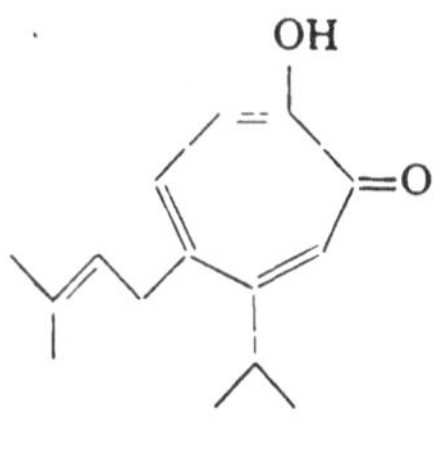

(CXIV.) Nootkatin.

a *trans*-position of the methyl group in ring *C* with regard to the gem-dimethylacetic acid residue in (CIX) is considered most probable.

Closely related to the sesquiterpenes are the toxic components of *Coriaria japonica* GR., coriamyrtion ($C_{15}H_{18}O_5$) and tutin ($C_{15}H_{18}O_6$). Coriamyrtion has been assigned formula (CXIII) by KARIYONE et al. (*108*, *109*). The tropolone, nootkatin, from heartwood of *Chamaecyparis nootkatensis* SPACH is an isoprene homologue of the thujaplicins and has been presented by ERDTMAN and HARVEY (*75*) as (CXIV). This structure has been confirmed by CAMPBELL and ROBERTSON (*45*) by X-ray analysis.

II. Diterpenes.

1. Structure.

The diterpenes contain five carbon atoms more than the sesquiterpenes, and we may therefore expect a greater complexity and a larger number of members in this group. The number of fully established structures is still relatively small, due mainly to the great experimental difficulties in their study. Recent thorough accounts of our knowledge in this field are given by SIMONSEN and BARTON (*188*) and by BARTON (*17*).

The diterpenes are composed of four isoprene units and, with the exception of camphorene and methyl vinhaticoate, abietic acid and related resin acids, they can be constructed from an isoprene homologue of farnesol-geranyl geraniol. Since phytol (CXV) is a naturally occurring open-chain diterpene built along the line of geranyl-geraniol, RUZICKA expressed this structural relationship among the diterpenes in the "phytol" rule (*175*).

$$\begin{array}{l} \quad\;\; CH_3 \qquad\qquad\qquad\qquad\quad CH_3 \qquad\qquad\qquad\qquad\quad CH_3 \\ \quad\;\;\; | \qquad\qquad\qquad\qquad\qquad\;\; | \qquad\qquad\qquad\qquad\qquad\;\; | \\ H_3C—CH—CH_2—CH_2(CH_2—CH—CH_2—CH_2)_2CH_2—C{=}CH—CH_2OH \end{array}$$

(CXV.) Phytol.

The common occurrence of the geraniol-diterpene type of ring closure in the lower terpenes would lead one to expect such structures in the diterpenes. However, it was only recently that this type was found by ŠORM et al. (*207*, *208*) in a systematic investigation of wormwood oil. The carbon skeletons of two of these diterpenes have been well

(CXVI.) Diterpene from wormwood.

CH_2OH

(CXVII.) Vitamin A_1

established by synthesis of their hydrogenation product, 2,6-dimethyl-10-β-tolyl-undecane. Formula (CXVI) is suggested for one of the wormwood diterpenes. All other presently known cyclic diterpenes contain a cyclogeraniol ring structure. One such ring is present in Vitamin A_1 (axerophthol) (CXVII) and its several *cis*-forms such as neo-vitamin A_1, and in the corresponding aldehyde, retinene (*186*).

The majority of the known diterpenes contain two or more rings, and their aromatization by dehydrogenation with sulfur or selenium has laid the foundation for our present knowledge of these systems. VESTERBERG (*237*) as early as 1903 obtained retene (CXVIII) from abietic acid by heating with sulfur. Twenty years later RUZICKA and BALAS (*176*) isolated a different hydrocarbon, "pimanthrene" (CXIX), from the dehydrogenation products of dextropimaric acid. A third type of phenanthrene derivative was obtained from agathene-dicarboxylic acid. This 1,7,8-trimethylphenanthrene (CXX) is accompanied by agathalene (CXXI) and 1,1,4,7-tetramethylphenalan ($C_{17}H_{20}$) (CXXII) (*42*).

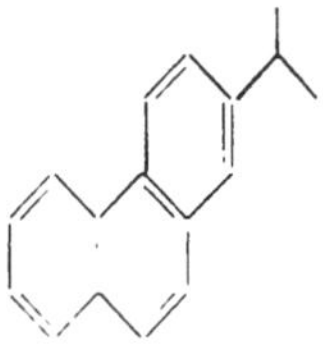

(CXVIII.) Retene.

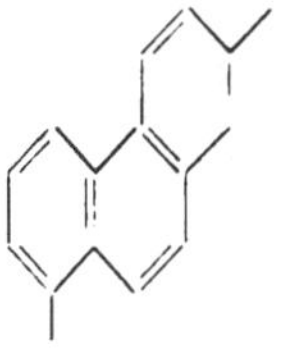

(CXIX.) Pimanthrene.

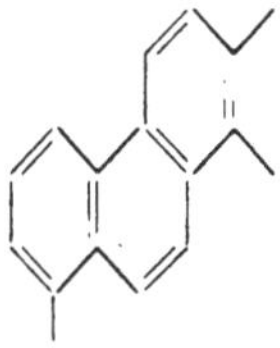

(CXX.) 1,7,8-Trimethylphenanthrene.

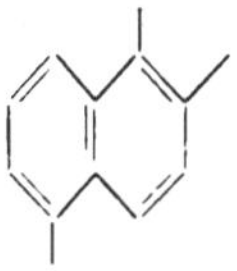

(CXXI.) Agathalene.

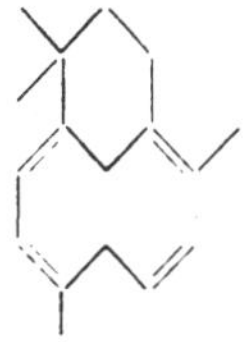

(CXXII.) 1,1,4,7-Tetrametylphenalan ($C_{17}H_{20}$).

The dehydrogenation products mentioned indicate the nature of the carbon skeleton of the original diterpene. Guided by these experiments further structural analysis established the constitution of several of the naturally occurring diterpenes. In *Chart* 5 are presented the formulae of the resin acids, abietic acid (CXXXII), dextropimaric acid (CXXVII), levopimaric acid (CXXXI), neo-abietic acid (CXXXIII) and agathene dicarboxylic acid (CXXIII); the resin alcohols, manool (CXXII b) and sclareol (CXXII a); the oxides, manoyl oxide (CXXIV) and keto-manoyl oxide (CXXV); the phenols, ferruginol (CXXXIV) and sugiol (CXXXV); and the hydrocarbon camphorene (CXXXVI).

The tetracyclic hydrocarbons phyllocladene and iso-phyllocladene are related to both abietic acid and dextropimaric acid, and recently structure (CXXIX) was proposed by BRANDT (*38*) for phyllocladene.

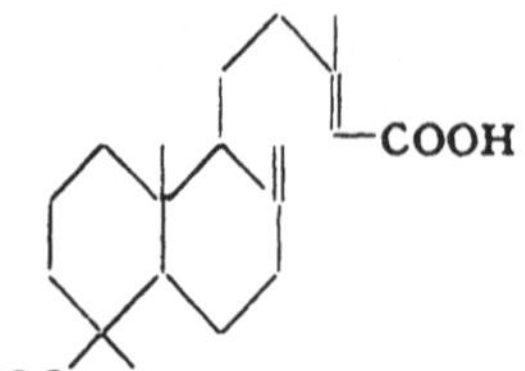

(CXXII a.) Sclareol. (CXXII b.) Manool. (CXXIII.) Agathene dicarboxylic acid

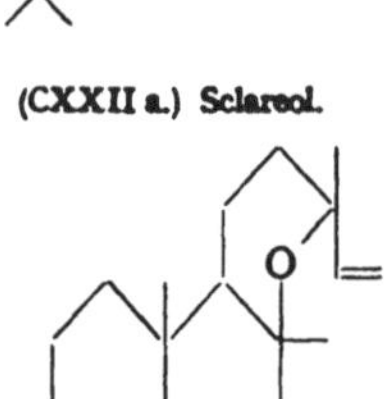

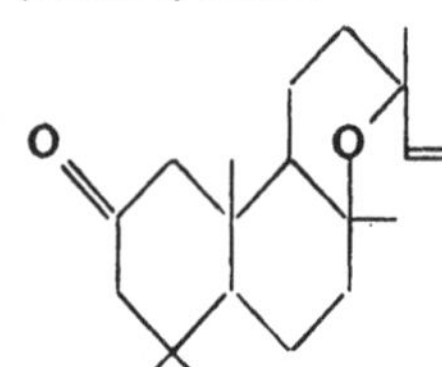

(CXXIV.) Manoyl oxide. (CXXV.) Keto-manoyl oxide. (CXXVI.) Marrubiin.

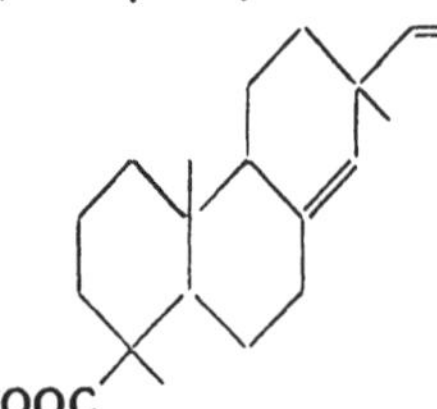

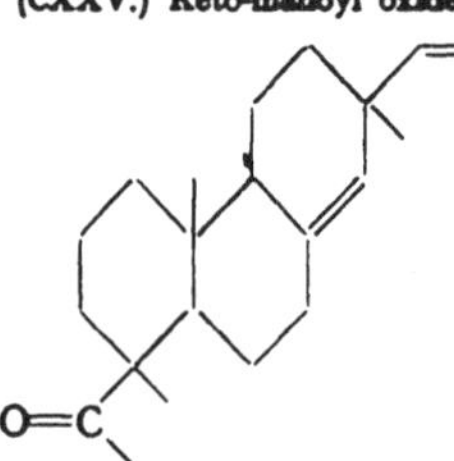

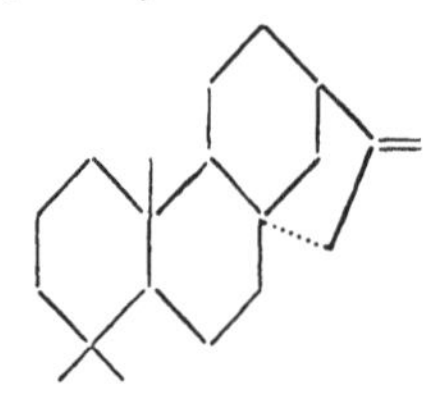

(CXXVII.) Dextropimaric acid. (CXXVIII.) Cryptopinone. (CXXIX.) Phyllocladene.

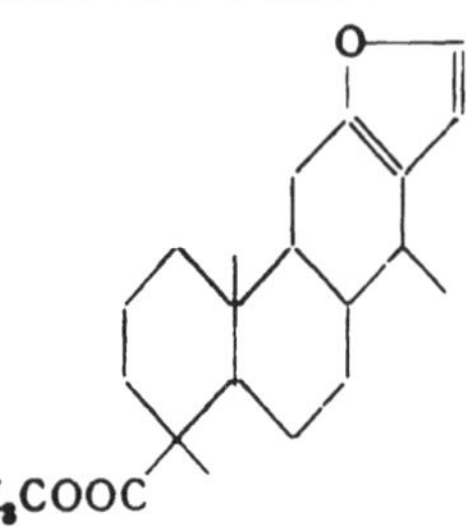

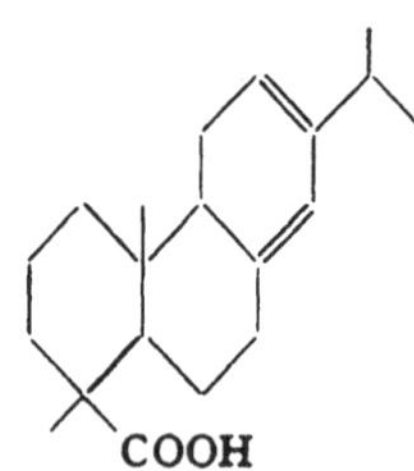

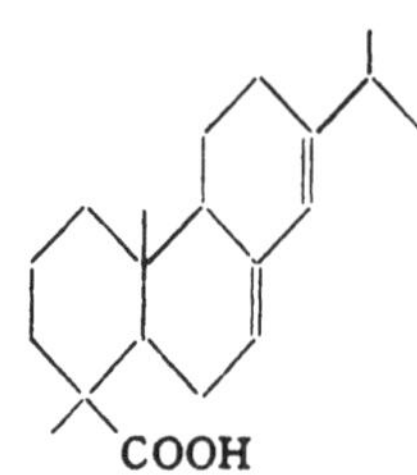

(CXXX.) Methyl vinhaticoate. (CXXXI.) Levopimaric acid. (CXXXII.) Abietic acid.

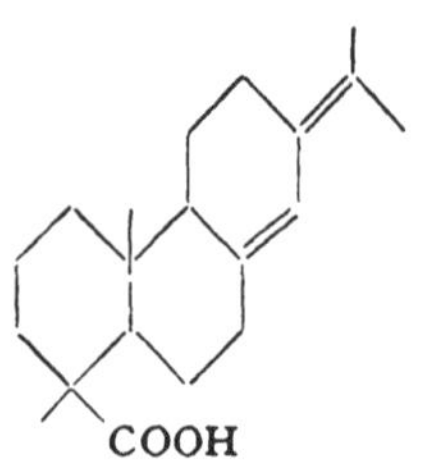

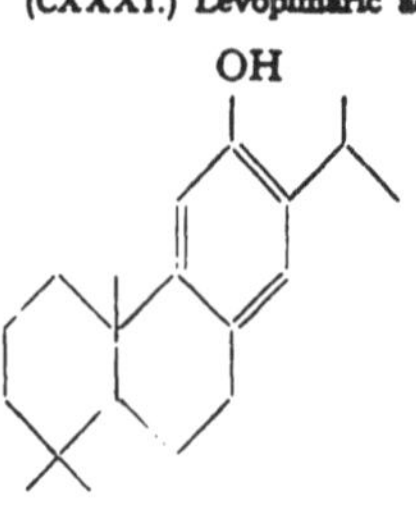

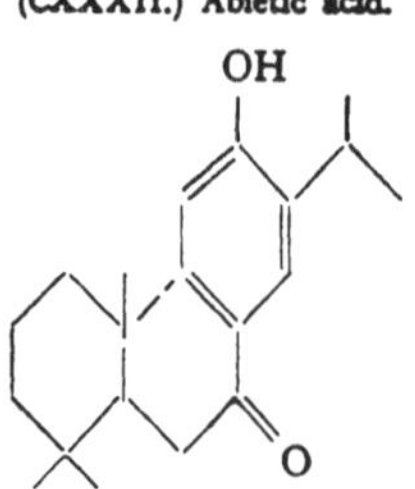

(CXXXIII.) Neo-abietic acid. (CXXXIV.) Ferruginol. (CXXXV.) Sugiol.

Chart 5. Diterpenes.

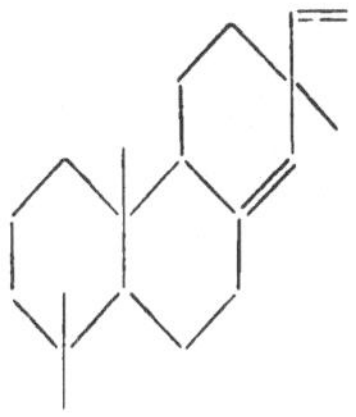

(CXXXV a.) Rimuene. 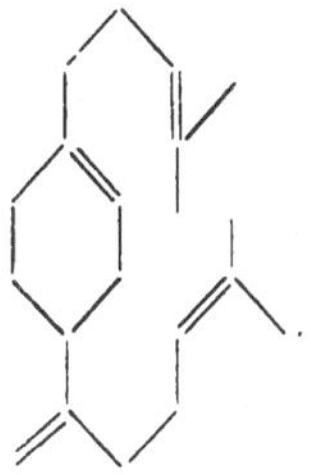 (CXXXVI.) α-Camphorene. 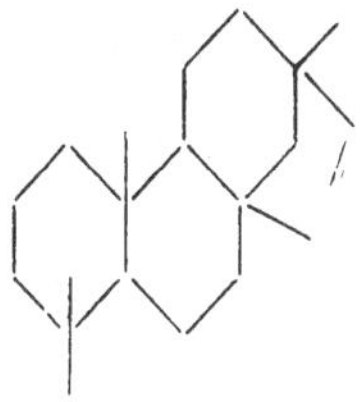 (CXXXVII a.) Mirene (?).

Chart 5. Diterpenes (continued).

Included in *Chart 5* is marrubiin (CXXVI), the bitter principle of horehound (*Marrubium vulgare* L.), which contains a bicyclic system and is related to sclareol and agathic acid. The ring system gives rise to 1,5,6-trimethyl-naphthalene upon dehydrogenation. This part of the structure of marrubiin was further studied by HARDY and RIGBY (*93*) who established the structure of a γ-lactone, $C_{14}H_{18}O_2$, H_2O (CXXXVII) obtained by GHIGI (*84*) in the oxidation of anhydro-marrubiin. Isolation of (CXXXVIII) and (CXXXVIII a) by COCKER et al. (*56, 59*) and BURN and RIGBY (*42 a*) proves the structure (CXXVI) of marrubiin; only the configuration and position of substitution of the furane ring remain uncertain.

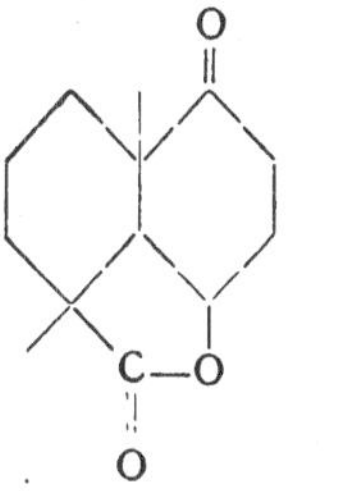

(CXXXVII.) γ-Lactone, $C_{14}H_{18}O_2$, H_2O. 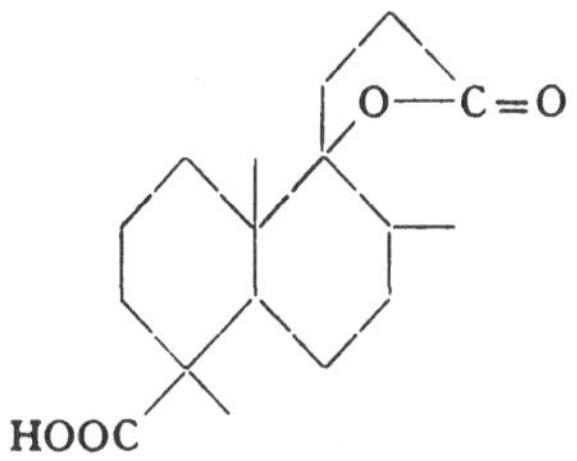 (CXXXVIII.) 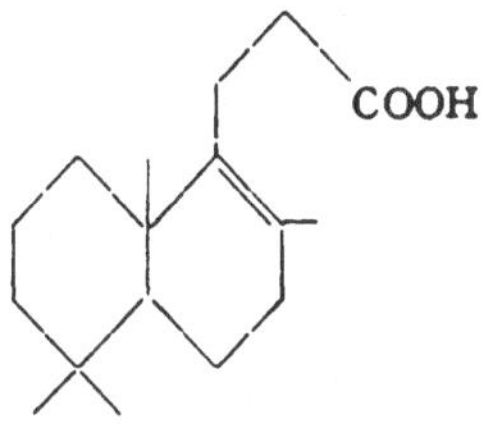 (CXXXVIII a.)

From the heartwood of *Plathymenia reticulata* BENTH, KING et al. (*112, 113*) isolated a diterpene methylester, termed methyl vinhaticoate. After opening the furane ring by ozonolysis and marking the resulting carbonyl group with a methyl group, dehydrogenation with selenium yielded a 1,3,8-trimethyl-phenanthrene. The ester is formulated as (CXXX) and does not follow the isoprene rule.

Vinhaticoic acid is related to the agathic acid type of resin acids, and it is possible that a methyl shift has occurred in the biochemical formation of the diterpene acid during the ring closure of a farnesol type precursor.

2. Stereochemistry of the Diterpenes.

The similarity in the building principle of the cyclic diterpenes to that of the triterpenes and steroids expresses itself also in some stereochemical relations. The *trans*-locking of the rings *A* and *B* of the diterpenes has been rigorously established by comparison with the degradation products of lanostadienol and ergosterol D. The relation between lanostadienol (CXXXIX) and manool (CXLI) was established through degradations leading to a bicyclic carboxylic acid (CXL) which still contained rings *A* and *B* of the starting products (*125*).

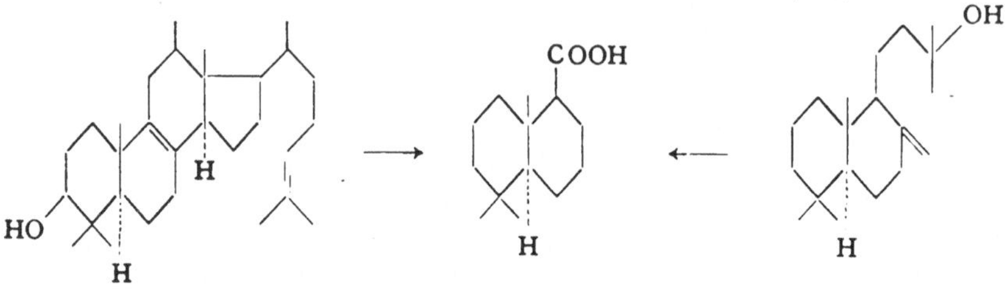

(CXXXIX.) Lanostadienol. (CXL.) 1,1,10-Trimethyldecalyl-(5)-carboxylic acid. (CXLI.) Manool.

The configurations of ergosterol D (CXLII) and abietic acid (CXLIV) were correlated through degradations leading to the dicarboxylic acid (CXLIII) (*106*). *Trans*-locking of the rings *A* and *B* of abietic and dextropimaric acids was concluded from the degradation to the tricarboxylic acid (CXLV). This acid is optically inactive, and only two configurations are therefore possible. By comparing the steric

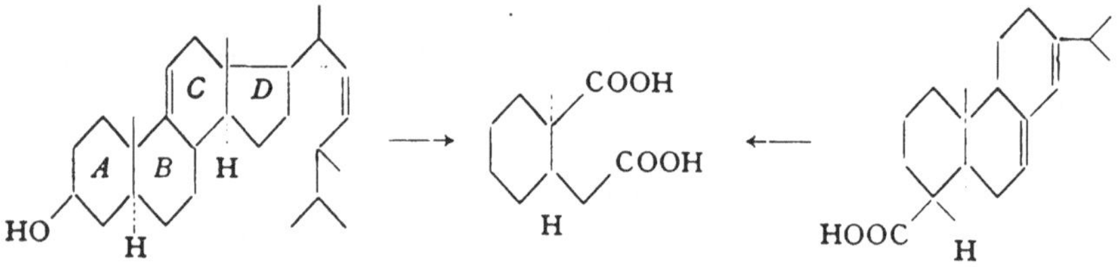

(CXLII.) Ergosterol D. (CXLIII.) Dicarboxylic acid. (CXLIV.) Abietic acid.

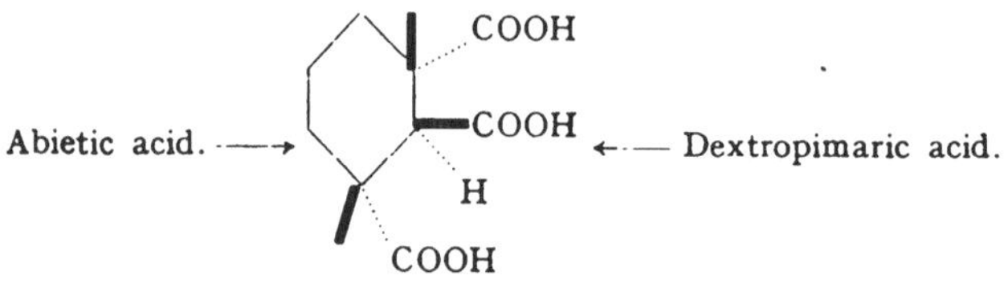

Abietic acid. → ← Dextropimaric acid.

(CXLV.) Tricarboxylic acid.

hindrance of the carboxyl groups in molecular models and the observed reactivity of these groups, CAMPBELL and TODD (*46*) concluded a *trans*-fusion of rings *A* and *B* in abietic, dehydro-abietic, levopimaric and podocarpic acids. BARTON (*25*, *26*) confirmed these conclusions and assigned configuration as shown in (CXLV) to the natural oxidation product on the basis of studies of the effect of configuration on the dissociation constants of polycarboxylic acids.

Manool is directly related to dehydroabietic acid, and therefore also to abietic acid through a conversion to a hydrocarbon C_{20} (CXLVI). This establishes the same configuration at carbon atoms $C_{(11)}$ and $C_{(12)}$ in these acids and also in the related manoyl oxide, keto-manoyl oxide and sclareol (*107*).

(CXLVI.) Hydrocarbon C_{20}.

RUZICKA, ZWICKY, and JEGER (*183*) have shown that agathene dicarboxylic acid has the same ring fusion *A*/*B* as abietic acid; and CAMPBELL and TODD (*46*) proved by partial synthesis this relation for ferruginol and dehydroabietic acid. An optically active form of acid (CXLV) was obtained by RUZICKA and BERNOLD (*177*) form agathic acid, which establishes the position of the $C_{(1)}$ carboxyl group as *trans*- to the H at $C_{(11)}$. Agathic acid and also podocarpic acid are therefore epimeric with abietic acid with regard to the carboxyl group at $C_{(1)}$.

These investigations furnish experimental proof of the *trans*-construction of rings *A* and *B* in all known di- and triterpenes and 5 α-steroids. They also confirm the conclusions based on molecular rotation shifts by KLYNE (*115*, *116*) and in the course of asymetric synthesis by DAUBEN and other authors (*63*, *65*, *167*, *168*). JEGER et al. (*13*, *170*) have shown that the usual projection formulae of the steroids and, therefore, also those of the terpenes represent the absolute configurations, supporting the conclusion reached by PRELOG (*167*, *168*) and MILLS (*137*). The correlations between di- and triterpenes and steroids have been discussed by KLYNE (*115*, *116*) and presented in schematic form by RUZICKA (*175*).

(CXLVII.) Abietane.

(CXLVIII.) Pimarane.

KLYNE (*117*) suggests a nomenclature which brings out more clearly the relationship of the diterpenes with the triterpenes and steroids. This system retains the numbering of the ring carbon atoms; however, the presently accepted nomenclature of rings *C* and *B* are changed

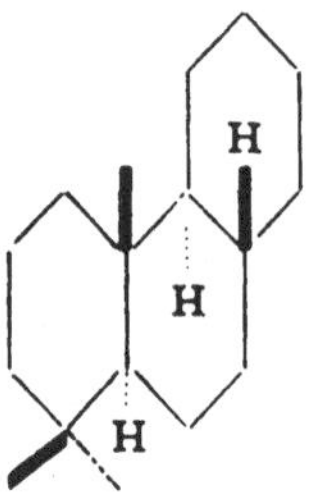

(CXLIX.) Podocarpane. 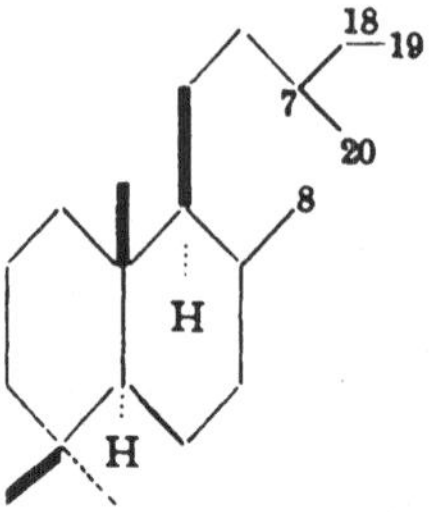 (CL.) 7,8-seco-Pimarane. 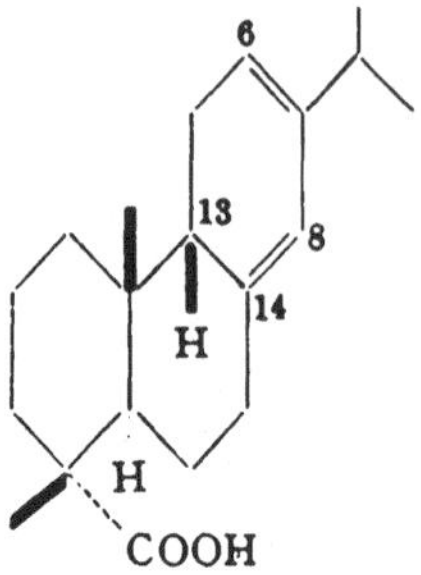 (CLI.) Levopimaric acid.

into *B* and *C*. The stereochemical configurations of the fundamental hydrocarbons are shown in formulae (CXLVII)–(CL). If inversion has taken place at a ring junction, this is indicated by a prefix giving the number of the C atom and the Greek letter α or β. Thus, levopimaric acid is named systematically 13 β-abieta-6,8(14)-diene-15-oic acid (CLI).

III. The Biogenesis of Sesqui- and Diterpenes.

The biogenesis of terpenes and steroids has been reviewed by RUZICKA (*175*), MONDON (*139*), BLOCH (*32*), BONNER (*35*), the writer (*91*), and on pp. 131–168 of the present Volume by TSCHESCHE (*230*). It has been shown that acetate groups are the fundamental building blocks of squalene, rubber and steroids, and it is likely that these findings can be extrapolated to the lower terpenes. If we represent the carbonyl end of acetic acid with a cross and the methyl group with a circle, the carbon chain formed in the biosynthesis is represented as (CLII). The

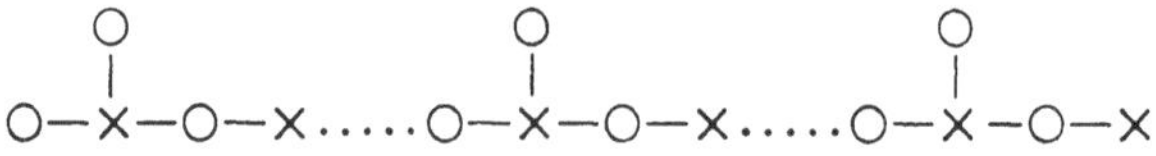

```
    O              O              O
    |              |              |
O—X—O—X.....O—X—O—X.....O—X—O—X
```

(CLII.)

intermediate in this synthesis, acetoacetyl-Coenzyme A ("Co A"), formed from two mols of acetyl-CoA, reacts with another acetyl-CoA to form β-hydroxy-β-methylglutaryl-CoA, from which the repeating units of β-methyl-isovaleryl-CoA or β-methyl-crotonic acid may be formed by decarboxylation and dehydration. This postulated mechanism is illustrated in the formation of a farnesol type sesquiterpene in (CLIII) [cf. LIPMANN (*126*)]. Condensation of several such units accounts for the isoprene type structure of the terpenes in regular "head-to-tail" arrangement as expressed in the geraniol, farnesol and phytol rules.

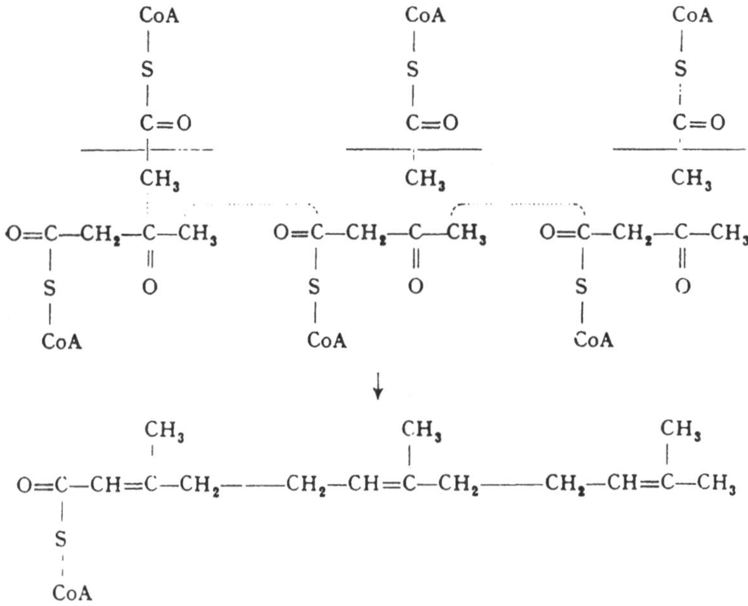

(CLIII.) Formation of a farnesol type sesquiterpene.

The precursors, possibly still connected with the enzyme system, undergo several changes such as oxidation, reduction and cyclization before they are finally deposited in the oil glands. Following RUZICKA's schematic presentation, the variations in ring types are shown in *Charts 6* and *7* (pp. 28–29).

The heavy broken line represents in these Charts the primary "regular" acyclic chain. Through cross-linking, indicated by a thin line, the majority of the terpenes can be constructed. RUZICKA has discussed the formation of these ring systems on the basis of ionic and radical mechanisms. He proposes that the "irregularly" built abietic acid, levopimaric acid and neo-abietic acid also adhere to this ring formation by postulating an ionic mechanism that involve a methyl shift in a dextropimaric acid type intermediate (CLXXVI) as indicated in *Chart 8*, p. 29.

WENKERT (*237a*) has proposed similar relationships proceeding by preferred protonation of the $\Delta^{8,14}$-instead of the $\Delta^{18,19}$-linkage as indicated in (CLXXIX a–c) (Chart 8), and has discussed the relationship between rimuene (CXXXV a), mirene (CXXXVII a), phyllocladene (CXXIX) and the interesting *Garrya* and *Aconite* diterpene alkaloids such as *veatchine* (*237c*) and *atisine* (*151b*).

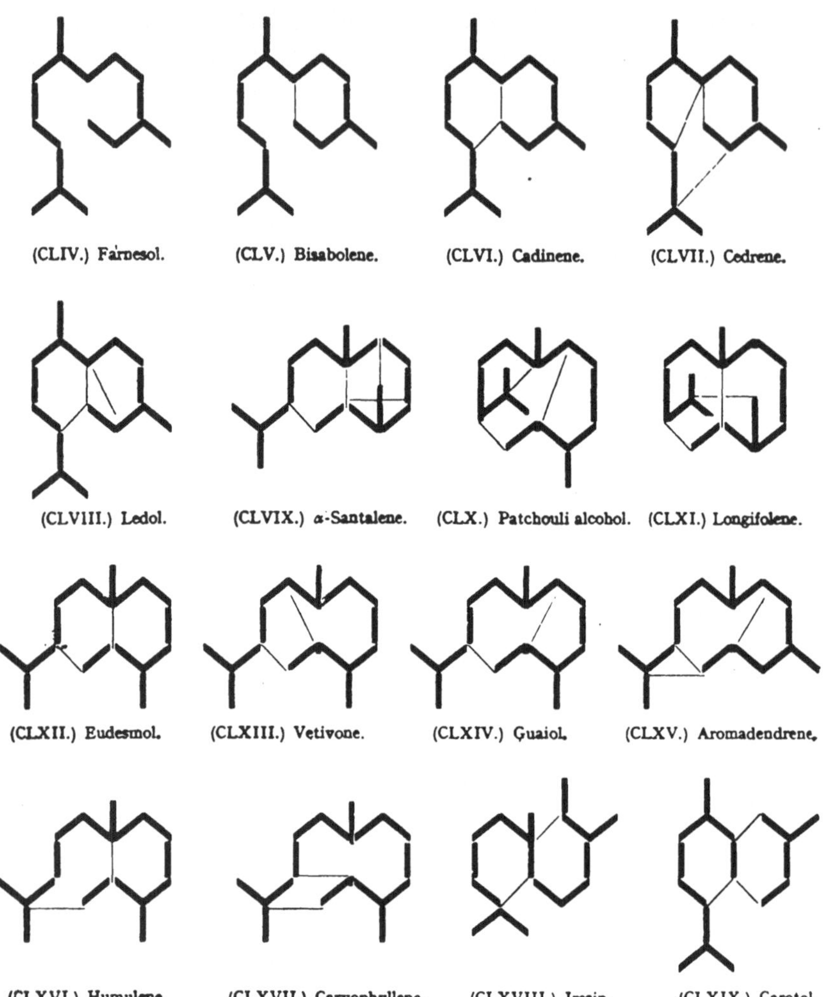

Chart 6. Carbon Skeletons of Sesquiterpenes and the Farnesol Rule.

Biogenesis and Nomenclature. A convenient way of describing the cyclization in the terpene field based on the isoprene building pattern was recently proposed by KLYNE (*117*). The individual atoms of an isoprene unit are lettered as in (CLXXX), the two "head" atoms being called *w* and the tail atoms *z*. The units of a linear isoprene structure with only head-to-tail (*w–z*) links are numbered 1, 2, 3, etc., starting from the head (free) end. The carbon skeleton of the terpenes can then be designated by numbers for units and letters for the linking carbon atoms. The linkages in the main chain are designated as (*z–w*) in geraniol, and $(z-w)_2$ in farnesol. The cross-linkages are indicated by the appropriate

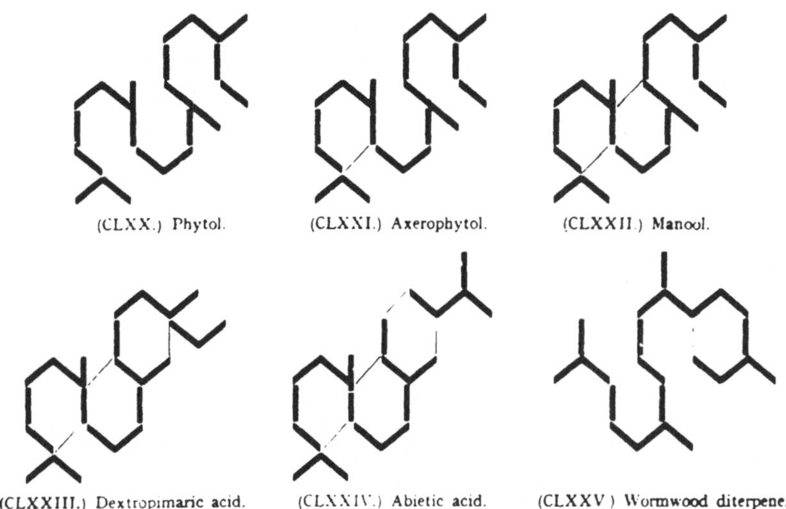

(CLXX.) Phytol. (CLXXI.) Axerophytol. (CLXXII.) Manool.

(CLXXIII.) Dextropimaric acid. (CLXXIV.) Abietic acid. (CLXXV.) Wormwood diterpene.

Chart 7. Carbon Skeletons of Diterpenes and the Phytol Rule.

C A B ⊕ HOOC → HOOC → HOOC

(CLXXVI.) (CLXXVII.) (CLXXVIII.) Levopimaric acid.

Me

18 19 8 14 H⁺ → ⊕ →

R *R* *R*

(CLXXIX a.) (CLXXIX b.) (CLXXIX c.) Abietic acid.

Chart 8. Ionic Mechanisms in the Biogenesis of Sesquiterpenes.

letters and numbers of the units. This numbering system is shown in the case of eudesmol (CLXXXI).

```
  w' 1                  w' 2                  w' 3
   C                     C                     C
   |                     |                     |
C—C—C—C    . . . .    C—C—C—C    . . . .    C—C—C—C
w1 x1 y1 z1          w2 x2 y2 z2          w3 x3 y3 z3
```

(CLXXX.) Farnesol $(z—w)_3$.

The letters indicate equivalent position of the carbon atoms in the biogenesis; and evidence of their common origin can be seen in the final product. For example, the carbon atoms x, situated at the methyl branch are often hydroxylated or connected by double bonds to neighboring atoms. In naturally occurring monoterpenes, ring formation always involves carbon atoms y and z; in the sesquiterpenes, the addition of one more isoprene unit permits the linking of 1 y–z 2, 1 y–z 3, and 2 y–z 3. All three combinations are frequently seen in this group. The x–y linkages, indicating a geraniol-cyclogeraniol cyclization, are present in one of the most frequently occurring monoterpenes, pinene, but are also seen in a larger variety of higher terpenes and become the sole building principle in tri- and tetraterpenes and in the steroids.

This system is more than a convenience in describing the different terpenes, since it facilitates the prediction of possible occurrence of new ring systems and their probable stereochemical configuration. It also poses some questions regarding the manner in which the cyclization of an isoprenic chain took place. For example, eudesmol can be described as $(z\text{–}w)_2$: 1 y–z 3, 2 x–y 3 or $(z\text{–}w)_2$: 1 y–z 3, 3 x–w 2, depending on which manner of synthesis from a possible open-chain precursor is selected. Biosynthetic studies with labelled compounds may help us decide which route is correct.

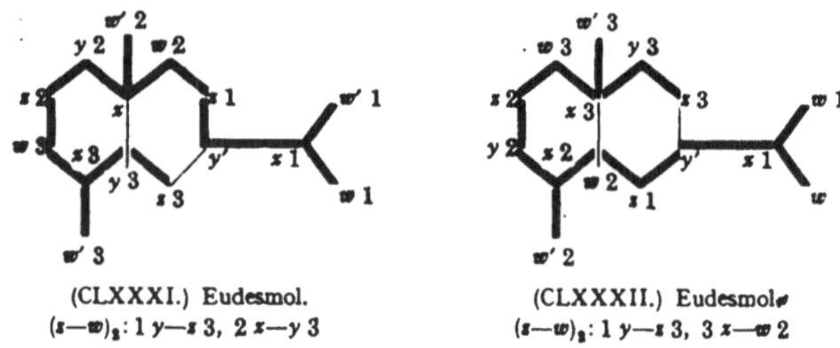

(CLXXXI.) Eudesmol.
$(z\text{—}w)_2$: 1 y—z 3, 2 x—y 3

(CLXXXII.) Eudesmol.
$(z\text{—}w)_2$: 1 y—z 3, 3 x—w 2

Similar studies could provide experimental evidence in support of the "head-to-tail" rule by eliminating exceptions such as eremophilone and abietic acid. If, for example, the methyl shift in the formation of abietic acid from a phytol type chain, as proposed by Ruzicka, is correct, the methyl groups in the isopropyl group in abietic acid originate from the carbonyl, as well as from the methyl group of acetate. In the usual rearrangement, both of these groups arise from the methyl groups of acetate.

Up to the present time, no pertinent work has been reported with isotopic precursors. In some studies on the formation of monoterpenes, precursors of the geraniol type have been injected into plants and the formation of ring compounds, limonene and terpineol, has been established, lending support to the geraniol rule (*81b*).

The terpene chemist has built up a tremendous amount of information on volatile plant products in the course of half a century. From this heterogeneous mass of data crystallized the *isoprene rule*, which recognizes as a common structural feature of the terpenes that their carbon skeletons are composed of isoprene units in regular or irregular arrangement. An extension of this rule ("*biogenetic isoprene rule*") proposed by RUZICKA, ESCHENMOSER, and HEUSSER (*175*) requires that the carbon skeleton of natural terpenic compounds can be deduced from postulated simple precursors of the geraniol, farnesol, geranyl-geraniol and squalene type structure by accepted reaction mechanisms. This rule expresses, therefore, that carbon-carbon rearrangements may have occurred in the formation of the terpene end products. Such rearrangements have been discussed for eremophilone and abietic acid (pp. 9,27). In these cases a methyl shift is postulated in order to construct these compounds from a "regular" isoprene chain.

These rules are deduced from regularities in the structures of the end products of the reactions leading to the terpenes. A real understanding of these rules, and, in general, the biochemical history of these compounds will come from a study of the intermediate stages in their formation. Recent progress in analytical and biochemical techniques promises a rapid development of this most interesting aspect of the terpene field.

References.

1. ABE, Y., T. HARUKAWA, H. ISHIKAWA, T. MIKI, M. SUMI and T. TOGA: Santonin. I. The Synthesis of Two Optically Inactive Stereoisomerides of Santonin. J. Amer. Chem. Soc. **75**, 2567 (1953).

1a. — — — — — — A New Racemic Stereoisomeride of Santonin. Proc. Japan Acad. **29**, 113 (1953) [Chem. Abstr. **48**, 10706 (1954)].

1b. ABE, Y., T. HARUKAWA, H. ISHIKAWA, T. MIKI and M. SUMI: Nomenclature of Santonin and Related Compounds. Chem. and Ind. **1955**, 91.

1c. ABE, Y. and M. SUMI: Configuration of the Santonins. Chem. a. Ind. **1955**, 253.

2. ADAMSON, P. S., F. I. McQUILLIN, R. ROBINSON and J. L. SIMONSEN: Synthetic Cyperones and their Comparison with α-and β-Cyperones. J. Chem. Soc. (London) **1937**, 1576.

3. AEBI, A., D. H. R. BARTON, A. W. BURGSTAHLER and A. S. LINDSEY: Sesquiterpenoids. V. The Stereochemistry of the Tricyclic Derivatives of Caryophyllene. J. Chem. Soc. (London) **1954**, 4659.

3a. AEBI, A., D. H. R. BARTON and A. S. LINDSEY: The Relationship between Clovene and β-Caryophyllene Alcohol. Chem. and Ind. **1953**, 748.

4. — — — Sesquiterpenoids. III. The Stereochemistry of Caryophyllene. J. Chem. Soc. (London) **1953**, 3124.

5. — — — The Nature of γ-Caryophyllene. Chem. and Ind. **1953**, 487.

6. ANDERSON, A. B. and E. C. SHERRARD: Dehydroperillic Acid, an Acid from Western Red Cedar (*Thuja Plicata* DON). J. Amer. Chem. Sc . **55**, 3813 (1933).

7. ANDERSON, A. G., Jr. and J. A. NELSON: Electrophilic Substitution of Azulene. J. Amer. Chem. Soc. **72**, 3824 (1950).

8. — — Azulene. I. An Improved Synthesis. J. Amer. Chem. Soc. **73**, 232 (1951).

9. ANDERSON, A. G., Jr., J. A. NELSON and J. J. TAZUMA: Azulene. III. Electrophilic Substitution. J. Amer. Chem. Soc. **75**, 4980 (1953).

10. ANDERSON, A. G., Jr. and J. J. TAZUMA: Azulene. II. Synthesis of Methyl 1-Azuloate. J. Amer. Chem. Soc. **75**, 4979 (1953).

11. ANDERSON, A. G., Jr. and S. Y. WANG: An Attempted Synthesis of 1,10-Cyclopentenoheptalene. 1,8-Tetramethylene-azulene. J. Organ. Chem. (USA) **19**, 277 (1954).

12. ARIGONI, D. und O. JEGER: Über Sesquiterpene und Azulene. 111. Mitt. Über die absolute Konfiguration des Zingiberens. Helv. Chim. Acta **37**, 881 (1954).

13. ARIGONI, D., B. RINIKER und O. JEGER: Über Steroide und Sexualhormone. 200. Mitt. Über die Bestimmung der relativen Konfiguration von C-20 bei Steroiden mit Hilfe der Konstellationsanalyse. Helv. Chim. Acta **37**, 878 (1954).

14. ARNOLD, H. und W. SPIELMANN: Darstellung des 5-Methyl-2-isopropyl- und des 5,7-Dimethyl-2-isopropyl-azulens. Weitere Untersuchungen der Beziehungen zwischen Farbe und Konstitution bei Azulenen. Ber. dtsch. chem. Ges. **83**, 28 (1950).

15. ATWATER, N. W. and E. B. REID: On the Nature of β-Caryophyllene. Chem. and Ind. **1953**, 688.

16. BAKER, W. and P. G. JONES: Attempts to Prepare New Aromatic Systems. III. *cyclo*Pent(*a*)indene. J. Chem. Soc. (London) **1951**, 787.

17. BARTON, D. H. R.: The Chemistry of the Diterpenoids. Quart. Rev. Chem. Soc. (London) **3**, 36 (1949).

18. — The Stereochemistry of Santonin, β-Santonin, and Artemisin. J. Organ. Chem. (USA) **15**, 466 (1950).

19. — Conformational Analysis of Substituted *cyclo*Hexanes. Chem. and Ind. **1953**, 664.

20. BARTON, D. H. R., T. BRUUN and A. S. LINDSEY: Tricyclic Derivatives of Caryophyllene. Chem. and Ind. **1951**, 910.

21. — — — Sesquiterpenoids. II. Tricyclic Derivatives of Caryophyllene. J. Chem. Soc. (London) **1952**, 2210.

22. — — — The Constitution of Caryophyllene. Chem. and Ind. **1952**, 691.

23. BARTON, D. H. R. and A. S. LINDSEY: The Constitution of Caryophyllene. Chem. and Ind. **1951**, 313.

24. — — Sesquiterpenoids. I. Evidence for a Nine-membered Ring in Caryophyllene. J. Chem. Soc. (London) **1951**, 2988.

24a. BARTON, D. H. R. and A. NICKON: Sesquiterpenoids. VI. The Absolute Configuration of Caryophyllene. J. Chem. Soc. (London) **1954**, 4665.

25. BARTON, D. H. R. and G. A. SCHMEIDLER: The Application of the Method of Electrostatic Energy Differences. I. Stereochemistry of the Diterpenoid Resin Acids. J. Chem. Soc. (London) **1948**, 1197.

26. — — Stereochemistry of the $C_{11}H_{16}O_6$ Tricarboxylic Acid from Abietic Acid. J. Chem. Soc. (London) **1949**, Suppl. Issue No. 1, S 232.

27. BARTON, D. H. R. and E. J. TARLTON: Sesquiterpenoids. IV. The Constitution of Carissone. J. Chem. Soc. (London) **1954**, 3492.

28. BIRCH, A. J.: Homocyclic Compounds. Annu. Rep. Chem. Soc. (London) **47**, 177 (1950).

29. BIRCH, A. J. and F. N. LAHEY: The Structure of Aromadendrene. I. Austral. J. Chem. **6**, 379 (1953).

30. BIRCH, A. J., R. A. MASSY-WESTROPP, S. E. WRIGHT, T. KUBOTA, T. MATSUURA and M. D. SUTHERLAND: Ipomeamarone and Ngaione. Chem. and Ind. **1954**, 902.

31. BIRCH, A. J. and A. R. MURRAY: The Constitution of Lanceol. J. Chem. Soc. (London) **1951**, 1888.

32. BLOCH, K.: Biosynthesis of Steroids. Sympos. Steroid Hormones, pp. 33—45. Madison, Wisc.: Univ. Wisconsin Press. 1950.

32a. — Biological Synthesis of Cholesterol. Harvey Lect. **48**, 68 (1952—53).

33. BLUMANN, A., A. R. H. COLE, K. J. L. THIEBERG and D. E. WHITE: The Constitution of Globulol. Chem. and Ind. **1954**, 1426.

34. BOEKELHEIDE, V., W. E. LANGELAND and C. T. LIU: A Study of the Synthesis and Some Properties of Acepleiadiene. J. Amer. Chem. Soc. **73**, 2432 (1951).

35. BONNER, J., M. W. PARKER and J. C. MONTERMOSO: Biosynthesis of Rubber. Science (Washington) **120**, 549 (1954).

36. BRADFIELD, A. E., E. M. FRANCIS and J. L. SIMONSEN: The Constituents of Indian Turpentine from *Pinus longifolia* ROXB., Part III (continued). J. Chem. Soc. (London) **1934**, 188.

37. BRADFIELD, A. E., N. HELLSTRÖM, A. R. PENFOLD and J. L. SIMONSEN: The Constitution of Eremophilone, Hydroxyeremophilone, and Hydroxydihydroeremophilone. II. J. Chem. Soc. (London) **1938**, 767.

38. BRANDT, C. W.: Constitution of Phyllocladene and Related Diterpenes. New Zealand J. Sci. Technol. **34** B, 46 (1952).

39. BRAUDE, E. A. and W. F. FORBES: Polycyclic Systems. V. A. New Route to Azulenes. J. Chem. Soc. (London) **1953**, 2208.

40. — — A New Route to Azulenes. Nature (London) **168**, 874 (1951).

41. BROWN, R. D.: A Quantum-mechanical Investigation of the Azulene Molecule. I. Trans. Faraday Soc. **44**, 984 (1948).

42. BÜCHI, G. and J. J. PAPPAS: Terpenes. I. Structure and Synthesis of the $C_{17}H_{20}$ Hydrocarbon Obtained by Dehydrogenation of Agathic Acid. J. Amer. Chem. Soc. **76**, 2963 (1954).

42a. BURN, D. and W. RIGBY: The Structure of Marrubiin. Chem. a. Ind. **1955**, 386.

43. CALIEZI, A. und H. SCHINZ: Zur Kenntnis der Sesquiterpene und Azulene. 102. Mitt. Über α-, β- und Allo-bicyclofarnesylsäure. Helv. Chim. Acta **35**, 1637 (1952).

44. CAMPBELL, A. and H. N. RYDON: The Synthesis of Caryophyllenic Acid. J. Chem. Soc. (London) **1953**, 3002.

45. CAMPBELL, R. B. and J. M. ROBERTSON: The Structure of Nootkatin. An X-Ray Determination. Chem. and Ind. **1952**, 1266.

46. CAMPBELL, W. P. and D. TODD: The Structure and Configuration of Resin Acids. Podocarpic Acid and Ferruginol. J. Amer. Chem. Soc. **64**, 928 (1942).

47. ČEKAN, Z., V. HEROUT and F. ŠORM: On Terpenes. LXII. Isolation and Properties of the Pro-chamazulene from *Matricaria chamomilla* L., a Further Compound of the Guaianolide Group. Coll. Czech. Chem. Comm. **19**, 798 (1954).

48. — — — A Chamazulene Precursor from Chamomile (*Matricaria chamomilla* L.). Chem. and Ind. **1954**, 604.

49. CHOPARD-DIT-JEAN, L. H. und E. HEILBRONNER: Zur Kenntnis der Sesquiterpene und Azulene. 103. Mitt. Die basischen Eigenschaften der Azulene (Teil III). Helv. Chim. Acta **35**, 2170 (1952).

50. CLARKE, P.: The Structure of Humulene. Chem. and Ind. **1954**, 661.

50a. CLARKE, P. and G. R. RAMAGE: The Caryophyllenes. XII. The Structure of Some Derivatives of Humulene. J. Chem. Soc. (London) **1954**, 4345.

51. CLEMO, G. R. and J. O. HARRIS: The Structure of Humulene. Chem. and Ind. **1951**, 799.

52. — — The Chemistry of Humulene. II. J. Chem. Soc. (London) **1952**, 665.

53. CLEMO, G. R. and F. J. McQUILLIN: Synthesis in the Santonin Series. I. Some Oxidation Products of α-(2-Keto-4-methyl*cyclo*hexyl)propionic Acid. J. Chem. Soc. (London) **1952**, 3835.

54. — — Synthesis in the Santonin Series. II. J. Chem. Soc. (London) **1952**, 3839.

54a. COCKER, W.: The Chemistry of ψ-Santonin. Chem. and Ind. **1955**, 19.

54b. COCKER, W. and R. S. CAHN: Nomenclature of the Santonins. Chem. and Ind. **1955**. 384.

55. COCKER, W., B. E. CROSS, S. R. DUFF, J. T. EDWARD and T. F. HOLLEY: The Constitution of Marrubiin. I. J. Chem. Soc. (London) **1953**, 2540.

56. COCKER, W., B. E. CROSS, S. R. DUFF and T. F. HOLLEY: The Constitution of Marrubiin. Chem. and Ind. **1952**, 827.

57. COCKER, W., B. E. CROSS, J. T. EDWARD, D. S. JENKINSON and J. McCORMICK: The Elimination of Non-angular Alkyl Groups in Aromatisation Reactions. II. J. Chem. Soc. (London) **1953**, 2355.

58. COCKER, W., B. E. CROSS and D. H. HAYES: The Structure of ψ-Santonin: Position of the Double Bond. Chem. and Ind. **1952**, 314.

59. COCKER, W., J. T. EDWARD and T. F. HOLLEY: The Chemistry of Marrubiin. Chem. and Ind. **1953**, 1227.

60. COCKER, W. and T. B. H. McMURRY: The Stereochemistry of Santonin. Chem. and Ind. **1954**, 1199.

61. COLOMBI, L. und H. SCHINZ: Zur Kenntnis der Sesquiterpene und Azulene. 101. Mitt. Synthese und Cyclisation von „Sesquilavandulol". Helv. Chim. Acta **35**, 1066 (1952).

62. CORBETT, R. E. and W. G. HANGER: Structure of Metrosiderene. J. Chem. Soc. (London) **1954**, 1179.

62a. COREY, E. J.: The Stereochemistry of Santonin, β-Santonin and Artemisin. J. Amer. Chem. Soc. **77**, 1044 (1955).

63. CRAM, D. J. and F. A. ABD ELHAFEZ: Studies in Stereochemistry. X. The Rule of "Steric Control of Asymmetric Induction" in the Synthesis of Acyclic Systems. J. Amer. Chem. Soc. **74**, 5828 (1952).

64. DAUBEN, W. G., D. F. DICKEL, O. JEGER und V. PRELOG: Untersuchungen über asymmetrische Synthesen. III. Über die Anwendung der asymmetrischen Synthese zur Konfigurationsbestimmung bei Triterpenen und Steroiden. Helv. Chim. Acta **36**, 325 (1953).

65. DAUBEN, W. G. and P. D. HANCE: The Position of the Double Bond in Pseudosantonin. J. Amer. Chem. Soc. **75**, 3352 (1953).

65a. — — The Structure of ψ-Santonin. J. Amer. Chem. Soc. **77**, 606 (1955).

66. DAWSON, T. L., G. R. RAMAGE and R. WHITEHEAD: The Structure of Caryophyllene. Chem. and Ind. **1952**, 450.

67. DAWSON, T. L., G. R. RAMAGE and B. WILSON: The Structure of Caryophyllene. Chem. and Ind. **1951**, 464.

68. DEV, S.: Structure of Humulene. Current Sci. (India) **20**, 296 (1951).

69. — 1:8-*cyclo*Hepteno-azulene. Chem. and Ind. **1954**, 1021.

70. DJERASSI, C., W. RITTEL, A. L. NUSSBAUM, F. W. DONOVAN and J. HERRAN: Terpenoids. XV. The Constitution of Iresin. A New Fundamental Sesquiterpene Skeleton. J. Amer. Chem. Soc. **76**, 6410 (1954).

70a. DJERASSI, C., P. SENGUPTA, J. HERRAN and F. WALLS: Terpenoids. V. The Isolation of Iresin, a New Sesquiterpene Lactone. J. Amer. Chem. Soc. **76**, 2966 (1954) (and private communication).

71. DOERING, W. v. E. and L. H. KNOX: Synthesis of Substituted Tropolones. J. Amer. Chem. Soc. **75**, 297 (1953).

72. DOERING, W. v. E., J. R. MAYER and C. H. DEPUY: Two-Step Synthesis of Azulene. J. Amer. Chem. Soc. **75**, 2386 (1953).

73. DOLEJŠ, L. and F. ŠORM: On Terpenes. LIII. Synthesis of 2,8,8-Trimethylbicyclo-(5,2,0)-nonane: a Proof of the Constitution of β-Caryophyllene. Collect. Czech. Chem. Communs. **19**, 559 (1954).

74. ERDTMAN, H. and J. GRIPENBERG: Antibiotic Substances from the Heart Wood of *Thuja plicata* DON. Nature (London) **161**, 719 (1948).

75. ERDTMAN, H. and W. E. HARVEY: The Chemistry of the Natural Order Cupressales. IX. Nootkatin. Chem. and Ind. **1952**, 1267.

76. ESCHENMOSER, A. und Hs. H. GÜNTHARD: Zur Kenntnis der Sesquiterpene und Azulene. 98. Mitt. Zur Cyclisation des β-Caryophyllens. Helv. Chim. Acta **34**, 2338 (1951).

77. ESCHENMOSER, A., J. SCHREIBER und W. KELLER: Zur Kenntnis der Sesquiterpene und Azulene. 97. Mitt. Zur Konstitution des Lanceols. Helv. Chim. Acta **34**, 1667 (1951).

78. FAWCETT, R. W. and J. O. HARRIS: Synthesis in the Humulene Series. Chem. and Ind. **1953**, 18, 67.

79. -- -- The Structure of Humulene. Chem. and Ind. **1954**, 405.

80. — -- The Chemistry of Humulene. IV—V. J. Chem. Soc. (London) **1954**, 2669, 2673.

81. FREUDENBERG, K. und W. HOHMANN: Die Konfiguration des tertiären Kohlenstoffatoms. V. Liebigs Ann. Chem. **584**, 54 (1953).

81a. FREUDENBERG, K. und W. LWOWSKI: Die Konfiguration des Camphers. Liebigs Ann. Chem. **587**, 213 (1954).

81b. FUJITA, Y.: Fundamental Studies of Essential Oils, 296—299; 203—412 (Ogawa Perfume Times, No. 202, 627 pp., Sept. 1951; Osaka and Tokyo: Ogawa and Co.) [in Japanese].

82. GALLOWAY, W. L., D. H. REID and W. H. STAFFORD: The Fine Structure of Azulene. Chem. and Ind. **1954**, 724.

83. GEISSMAN, T. A.: On the Structure of Eremophilone. J. Amer. Chem. Soc. **75**, 4008 (1953).

84. GHIGI, E.: The Constitution of Marrubiin. Gazz. chim. ital. **81**, 336 (1951).

85. GLICHITCH, L. S. and Y. R. NAVES: Composition of Oil of Ylang-Ylang.

85a. GORDON, M: The Azulenes. Chem. Rev. **50**, 127 (1952).
Parfums de France **10**, 7, 41 (1932) [Chem. Abstr. **26**, 2554 (1932)].

86. GUNSTONE, F. D. and R. M. HEGGIE: Experiments on the Synthesis of Santonin. I. The Preparation of the Lactone of α-(2-Hydroxy-3-keto*cyclo*hexyl)propionic Acid. J. Chem. Soc. (London) **1952**, 1354.

87. — -- Experiments on the Synthesis of Santonin. II. The Preparation of Compounds containing the Dienone System present in Santonin. J. Chem. Soc. (London) **1952**, 1437.

88. GÜNTHARD, Hs. H., R. SÜESS, L. MARTI, A. FÜRST und PL. A. PLATTNER: Zur Kenntnis der Sesquiterpene und Azulene. 95. Mitt. Dehydrierung von Decahydro-naphtalinen und Hydro-azulenen. Helv. Chim. Acta **34**, 959 (1951).

89. GUENTHER, E.: The Essential Oils. Vol. II. The Constituents of Essential Oils. New York: Van Nostrand Co. 1941.

90. HAAGEN-SMIT, A. J.: Azulenes. Fortschr. Chem. organ. Naturstoffe **5**, 40 (1948).

91. -- The Biogenesis of Terpenes. Annu. Rev. Plant Physiol. **4**, 305 (1953).

92. HALSALL, T. G., E. R. H. JONES and G. D. MEAKINS: The Chemistry of the Triterpenes. XI. The Conversion of Lupeol into Germanicol (*iso*Lupeol). The Structure of Lupeol Hydrochloride. J. Chem. Soc. (London) **1952**, 2862.

93. HARDY, D. G. and W. RIGBY: The Constitution of Marrubiin. Chem. and Ind. 1953, 1150.

94. HARRIS, J. O.: The Chemistry of Humulene. III. J. Chem. Soc. (London) 1953, 184.

95. HEILBRONNER, E. und A. ESCHENMOSER: Über das Benzo-cycloheptatrienylium-Kation. Helv. Chim. Acta **36**, 1101 (1953).

96. HEILBRONNER, E. und R. W. SCHMID: Zur Kenntnis der Sesquiterpene und Azulene. 113. Mitt. Azulenaldehyde und Azulenketone: Die Struktur des Lactaroviolins. Helv. Chim. Acta **37**, 2018 (1954).

97. HEILBRONNER, E. und M. SIMONETTA: Zur Kenntnis der Sesquiterpene und Azulene. 100. Mitt. Die basischen Eigenschaften der Azulene (Teil II). Helv. Chim. Acta **35**, 1049 (1952).

98. HEROUT, V., V. BENEŠOVÁ and J. PLÍVA: On Terpenes. XLI. The Sesquiterpenes of Ginger Oil. Collect. Czech. Chem. Communs. **18**, 248 (1953).

99. HEROUT, V. and D. J. DIMITROV: On Terpenes. XLIV. Sesquiterpenic Hydrocarbons of Ylang-Ylang Oil. Chem. Listy **46**, 432 (1952).

100. HEROUT, V., T. KOLOŠ and J. PLÍVA: On Terpenes. XLIX. Sesquiterpenes of the Cadinene Type in Javanese Citronellal. Chem. Listy **47**, 440 (1953).

101. HEROUT, V. and F. ŠANTAVÝ: On Terpenes. XLVIII. The Constitution of ε- and δ-Cadinene. Collect. Czech. Chem. Communs. **19**, 118 (1954).

102. HEROUT, V. and F. ŠORM: On Terpenes. LXI. Contribution to the Constitution of Pro-chamazulenogen, the Natural Precursor of Chamazulene in *Artemisia absinthium* L. Collect. Czech. Chem. Communs. **19**, 792 (1954).

103. — — On the Components of Wormwood (*Artemisia absinthium* L.) and the Isolation of a Crystalline Pro-chamazulenogen. Collect. Czech. Chem. Communs. **18**, 854 (1953).

103 a. — — O terpenech. LXI. Příspěvek ke konstituci pro-chamazulenogenu, pirozencho zdroje chamazulenu v *Artemisia absinthium* L. Chem. Listy **48**, (cf. *232 a*).

104. HEROUT, V., M. STREIBL, J. MLEZIVA and F. ŠORM: On Terpenes. XIV. On the Identity of Humulene and "α-Caryophyllene". Collect. Czech. Chem. Communs. **14**, 716 (1949).

105. HERZ, W.: A Novel Rearrangement in the Synthesis of Azulenes. J. Amer. Chem. Soc. **76**, 3349 (1954).

106. HEUSSER, H., E. BERIGER, R. ANLIKER, O. JEGER und L. RUZICKA: Über Steroide und Sexualhormone. 196. Mitt. Über die experimentelle Verknüpfung der Steroide mit Di- und Triterpenen. I. Abbau des Ergosterins zur trans(+)-1-Methyl-1-carboxy-cyclohexyl-(2)-essigsäure (Xa). Helv. Chim. Acta **36**, 1918 (1953).

107. JEGER, O., O. DÜRST und G. BÜCHI: Zur Kenntnis der Diterpene. 56. Mitt. Überführung des Manools in ein Umwandlungsprodukt der Abietinsäure. Helv. Chim. Acta **30**, 1853 (1947).

108. KARIYONE, T., N. KAWANO and M. TETSUÖ: Components of *Coriaria japonica*. Chemical Constitution of Coriamyrtin. J. pharmac. Soc. Japan **73**, 925 (1953).

109. KARIYONE, T. and T. OKUDA: The Components of Japanese Toxic Plants. I. The Toxic Components of *Coriaria japonica*. Bull. Inst. Chem. Res. Kyoto Univ. **31**, 387 (1953) [Chem. Abstr. **48**, 9971 (1954)].

110. KARRER, P., H. RUCKSTUHL und E. ZBINDEN: Über Lactaroviolin, einen Farbstoff aus *Lactarius deliciosus*. Helv. Chim. Acta **28**, 1176 (1945).

111. KERSCHBAUM, M.: Über den aliphatischen Sesquiterpen-Alkohol Farnesol. Ber. dtsch. chem. Ges. **46**, 1732 (1913).

112. KING, F. E. and T. J. KING: The Chemistry of Extractives from Hardwoods. XV. The Constitution of Methyl Vinhaticoate. J. Chem. Soc. (London) 1953, 4158.

113. KING, F. E., T. J. KING and K. G. NEILL: The Chemistry of Extractives from Hardwoods. XI. The Isolation of a Diterpene Ester (Methyl Vinhaticoate), and of 6:7:3':4'-Tetrahydroxyflavanone (Plathymenin), and 2:4:5:3':4'-Pentahydroxychalkone (*neo*Plathymenin), from *Plathymenia reticulata*. J. Chem. Soc. (London) **1953**, 1055.

114. KIRCHNER, J. G. and J. M. MILLER: Preparation of Terpeneless Essential Oils. A Chromatographic Process. Ind. Eng. Chem. **44**, 318 (1952).

115. KLYNE, W.: The Molecular Rotations of Polycyclic Compounds. I. General Principles and the Correlation of the Triterpenoids with the Steroids. J. Chem. Soc. (London) **1952**, 2916.

116. — The Molecular Rotations of Polycyclic Compounds. II. Diterpenoids and Sesquiterpenoids. J. Chem. Soc. (London) **1953**, 3072.

117. — Symbolic Representations of Natural Isoprene Structures. Chem. and Ind. **1954**, 725.

118. KNESSL, O. and A. VLASTIBOROVÁ: Terpenes. LVI. Paper Chromatography of Azulenes. Chem. Listy **48**, 212 (1954) [Chem. Abstr. **48**, 12708 (1954)].

119. KOMPPA, G. and G. A. NYMAN: Über die Konstitution des Ledumcamphers und des Ledens. C. R. Trav. Lab. Carlsberg **22**, 272 (1938).

120. KOVÁTS, E., A. FÜRST und Hs. H. GÜNTHARD: Zur Kenntnis der Sesquiterpene und Azulene. 109. Mitt. Über stereoisomere Bicyclo-(0,3,5)-decan-8-ole und deren Wasserabspaltungsprodukte. Helv. Chim. Acta **37**, 534 (1954).

121. KOVÁTS, E., Hs. H. GÜNTHARD und PL. A. PLATTNER: Katalytische Dehydrierungen mit Schwefelkohlenstoff. I. Dehydrierung von Oktahydro-azulen. Helv. Chim. Acta **37**, 2123 (1954).

122. KOVÁTS, E., PL. A. PLATTNER und Hs. H. GÜNTHARD: Zur Kenntnis der Sesquiterpene und Azulene. 110. Mitt. Dehydrierungen in der Azulenreihe. II. Helv. Chim. Acta **37**, 983 (1954).

123. KUBOTA, T. and N. ICHIKAWA: On the Chemical Constitution of Ipomeanine, a new Ketone from the Black-rotted Sweet Potato. Chem. and Ind. **1954**, 902.

124. KUHN, W. und H. SCHINZ: Zur Kenntnis der Sesquiterpene und Azulene. 104. Mitt. Die Sesquilavandulylsäure und ihre Umwandlungsprodukte. Helv. Chim. Acta **35**, 2395 (1952).

125. KYBURZ, E., B. RINIKER, H. R. SCHENK, H. HEUSSER und O. JEGER: Über Steroide und Sexualhormone. 193. Mitt. Über die konfigurative Verknüpfung des Lanostadienols mit cyclischen Diterpenen und Triterpenen. Helv. Chim. Acta **36**, 1891 (1953).

126. LIPMANN, F.: On Chemistry and Function of Coenzyme A. Bacteriol. Rev. **17**, 1 (1953).

127. LLOYD, D.: 1-Aza-2,3-benzazulene. Chem. and Ind. **1953**, 921.

127a. LLOYD, D. and F. ROWE: Azulenes. II. Application of the Pinacone Reduction to the Synthesis of 1-Substituted Azulenes. J. Chem. Soc. (London) **1954**, 4232.

128. LUTZ, A. W. and E. B. REID: On the Nature of β-Caryophyllene Alcohol, Clovene and Clovenic Acid. Chem. and Ind. **1953**, 278.

129. — — On the Relationship between β-Caryophyllene Alcohol and Clovene. Chem. and Ind. **1953**, 749.

130. — — Clovene and β-Caryophyllene Alcohol. J. Chem. Soc. (London) **1954**, 2265.

131. MATSUI, M., S. KITAMURA, Y. SUZUKI and M. HAMURO: Studies on Santonin. I. Experiments on the Synthesis of Santonin. I. Preparation of 2-Keto-1-10-dimethyldecalin. Bull. Chem. Soc. Japan **27**, 5 (1954).

132. MATSUI, M., K. TOKI, S. KITAMURA, Y. SUZUKI and M. HAMURO: Studies on Santonin. II. Experiments on the Synthesis of Santonin. II. Preparation of a Stereoisomeride of Santonin. Bull. Chem. Soc. Japan **27**, 7 (1954).

132a. MAYOT, M., G. BERTHIER et B. PULLMAN: Sur le calcul des charges électriques dans les hydrocarbures non alternants. J. chim. physique **50**, 170 (1953).

133. McQUILLIN, F. J.: Structure of β-Cyperone. J. Chem. Soc. (London) **1951**, 716.

134. — Synthesis in the Santonin Series. Chem. and Ind. **1954**, 311.

134a. — The Structure of Cyperone. III. Natural and Synthetic Cyperones. J. Chem. Soc. (London) **1955**, 528.

135. MEINWALD, J.: The Structure of Nepetalactone. Chem. and Ind. **1954**, 488.

136. MEISELS, A. and A. WEIZMANN: The Structure of Chamazulene. J. Amer. Chem. Soc. **75**, 3865 (1953).

137. MILLS, J. A.: Correlations between Monocyclic and Polycyclic Unsaturated Compounds from Molecular Rotation Differences. J. Chem. Soc. (London) **1952**, 4976.

138. MOFFETT, R. H. and D. ROGERS: The Molecular Configuration of Longifolene Hydrochloride. Chem. and Ind. **1953**, 916.

139. MONDON, A.: Zur Biogenese der Steroide. Angew. Chem. **65**, 333 (1953).

140. MUKHERJI, S. M. and N. K. BHATTACHARYYA: Terpenoids. I. Synthesis of the Gross Structure of Zingiberene. J. Amer. Chem. Soc. **75**, 4698 (1953).

141. NAFFA, P. and G. OURISSON: Chemical Approach to the Structure of Longifolene. Chem. and Ind. **1953**, 917.

142. NAVES, Y. R.: Études sur les matières végétales volatiles. CXVIII. Sur l'huile essentielle du Louro Inhamuy (*Nectandra elaiophora* BARB. RODR.) de l'Amazonie Brésilienne. Bull. soc. chim. France **1951**, 987.

143. — Études sur les matières végétales volatiles. LX. Présence d'epoxydihydrocaryophyllène (Caryophyllenoxyd de Treibs) dans l'extrait benzénique de bourgeons floraux du giroflier. Helv. Chim. Acta **31**, 378 (1948).

144. NAVES, Y. R. et G. MAZUYER: Les parfums naturels. Essences concrètes, résinoïdes, huiles et pommades, p. 292. Paris: Gauthier-Villars. 1939 [Chem. Abstr. **33**, 8362 (1939)].

145. NAYA, K.: The Chemical Constitution of Zingiberene. J. Japan. Chem. **5**, 648 (1951) [Chem. Abstr. **48**, 12024 (1954)].

146. NAYAK, U. R. and S. DEV: Structure of Longifolene. Chem. and Ind. **1954**, 989.

146a. NOVÁK, P. J., F. ŠORM and J. SICHER: On Terpenes. LXIII. Total Synthesis of Chamazulene. A Simple General Synthesis of 1,4,7-Trisubstituted Azulenes. Collect. Czech. Chem. Communs. **19**, 1264 (1954).

147. NOZOE, T.: Substitution Products of Tropolone and Allied Compounds. Nature (London) **167**, 1055 (1951).

148. NOZOE, T., T. MUKAI and I. MURATA: 1,3-Diazazulene. J. Amer. Chem. Soc. **76**, 3352 (1954).

149. NOZOE, T., S. SETO, S. MATSUMURA and T. TERASAWA: The Synthesis of 1-Aza-azulene and its Derivatives. Chem. and Ind. **1954**, 1357.

150. — — — — The Synthesis of 1-Aza-azulan-2-one and its Electrophilic Substitution. Chem. and Ind. **1954**, 1356.

151. OURISSON, G.: Molecular Rotations in the Series of Longifolene and β-Santalene. Chem. and Ind. **1953**, 918.

151a. PAUSON, P. L.: Tropones and Tropolones. Chem. Rev. **55**, 9 (1955).

151b. PELLETIER, S. W. and W. A. JACOBS: The Aconite Alkaloids. XXVII. The Structure of Atisine. J. Amer. Chem. Soc. **76**, 4496 (1954).

152. PENFOLD, A. R. and J. L. SIMONSEN: The Constitutions of Eremophilone, Hydroxyeremophilone, and Hydroxydihydroeremophilone. III. J. Chem. Soc. (London) **1939**, 87.

153. PFAU, A. ST. und PL. PLATTNER: Zur Kenntnis der flüchtigen Pflanzenstoffe. IV. Über die Konstitution der Azulene. Helv. Chim. Acta **19**, 858 (1936).

154. PLATTNER, PL. A.: Zur Kenntnis der Sesquiterpene. 50. Mitt. Konstitution und Farbe der Azulene. Helv. Chim. Acta **24**, 283 E (1941).

155. — Direkte Substitution im Azulen-Kern. Angew. Chem. **62**, 513 (1950).

156. PLATTNER, PL. A., A. FÜRST, A. ESCHENMOSER, W. KELLER, H. KLÄUI, ST. MEYER und M. ROSNER: Über Sesquiterpene und Azulene. 106. Mitt. Die Konstitution des Cedrens. Helv. Chim. Acta **36**, 1845 (1953).

157. PLATTNER, PL. A., A. FÜRST und J. HELLERBACH: Zur Kenntnis der Sesquiterpene. 80. Mitt. Über einige Abbauprodukte des Eudesmols. Helv. Chim. Acta **30**, 2158 (1947).

158. PLATTNER, PL. A., A. FÜRST, ST. MEYER und W. KELLER: Zur Kenntnis der Sesquiterpene und Azulene. 107. Mitt. Synthese einiger Abbauprodukte des Cedrens. Helv. Chim. Acta **37**, 266 (1954).

159. PLATTNER, PL. A., A. FÜRST, A. MÜLLER und A. R. SOMERVILLE: Zur Kenntnis der Sesquiterpene und Azulene. 96. Mitt. Über einige Azulen-carbonsäuren. Helv. Chim. Acta **34**, 971 (1951).

160. PLATTNER, PL. A. und E. HEILBRONNER: Über die Konstitution des Lactaroviolins. Experientia **1**, 233 (1945).

161. PLATTNER, PL. A., E. HEILBRONNER, R. W. SCHMID, R. SANDRIN and A. FÜRST: The Structure of Lactaroviolin. Chem. and Ind. **1954**, 1202.

162. PLATTNER, PL. A., E. HEILBRONNER und S. WEBER: Zur Kenntnis der Sesquiterpene und Azulene. 99. Mitt. Die basischen Eigenschaften der Azulene (Teil I). Helv. Chim. Acta **35**, 1036 (1952).

163. PLATTNER, PL. A. und H. KLÄUI: Zur Kenntnis der Sesquiterpene. 60. Mitt. Über den oxydativen Abbau der Nor-cedren-dicarbonsäure mit Salpetersäure. Helv. Chim. Acta **26**, 1553 (1943).

164. PLATTNER, PL. A., G. W. KUSSEROW und H. KLÄUI: Zur Kenntnis der Sesquiterpene. 55. Mitt. Über den stufenweisen Abbau der Nor-cedren-dicarbonsäure. Helv. Chim. Acta **25**, 1345 (1942).

165. POMMER, H.: Über den Stand der Forschung auf dem Gebiet der Azulene. Angew. Chem. **62**, 281 (1950).

166. — Ein neuer Weg zur Synthese von alkylierten Azulenen. Liebigs Ann. Chem. **579**, 47 (1953).

167. PRELOG, V.: Untersuchungen über asymmetrische Synthesen. I. Über den sterischen Verlauf der Reaktion von α-Ketosäure-estern optisch aktiver Alkohole mit Grignard'schen Verbindungen. Helv. Chim. Acta **36**, 308 (1953).

168. PRELOG, V. und H. L. MEIER: Untersuchungen über asymmetrische Synthesen. II. Über den sterischen Verlauf der Umsetzung von Phenylglyoxylsäureestern des Menthols, Neomenthols, Borneols und Isoborneols mit Methylmagnesiumjodid. Helv. Chim. Acta **36**, 320 (1953).

169. PRELOG, V. und K. SCHENKER: Zur Kenntnis des Kohlenstoffringes. 64. Mitt. Die Überführung von Derivaten des Cyclodecans in Naphtalin und Azulen. Einfaches Herstellungsverfahren für Azulen. Helv. Chim. Acta **36**, 1181 (1953).

169a. RAMAGE, G. R. and R. WHITEHEAD: The Caryophyllenes. X. Oxides from the Caryophyllenes. J. Chem. Soc. (London) **1954**, 4336.

169b. RAMAGE, G. R., R. WHITEHEAD and B. WILSON: The Caryophyllenes. XI. The Structure of Caryophyllene Nitrosite. J. Chem. Soc. (London) **1954**, 4341.

169c. RAO, A. S., K. B. DUTT, S. DEV and P. C. GUHA: Studies in Sesquiterpenes. XI. Sesquiterpenes of the Essential Oil of *Lansium annamalayanum*, BEDD. Stucture of α-Chigadmarene. J. Indian Chem. Soc. **29**, 620 (1952).

170. RINIKER, B., D. ARIGONI und O. JEGER: Über Steroide und Sexualhormone. 199. Mitt. Über die direkte konfigurative Verknüpfung der Steroide mit dem Citronellal, ein Beitrag zur Bestimmung der absoluten Konfiguration der Steroide. Helv. Chim. Acta 37, 546 (1954).

171. RINIKER, B., J. KALVODA, D. ARIGONI, A. FÜRST, O. JEGER, A. M. GOLD and R. B. WOODWARD: A Direct Stereochemical Correlation of a Sesquiterpene Alcohol with the Steroids. J. Amer. Chem. Soc. 76, 313 (1954).

172. ROBERTSON, J. M. and G. TODD: The Structure of β-Caryophyllene Alcohol Chloride and Bromide. An X-Ray Determination. Chem. and Ind. **1953**, 437.

173. ROY, J. K.: Synthesis of β-Cyperone. Chem. and Ind. **1954**, 1393.

174. RUZICKA, L.: Höhere Terpenverbindungen. VIII. Über die Konstitution des Nerolidols (Peruviol). Helv. Chim. Acta 6, 483 (1923).

175. — The Isoprene Rule and the Biogenesis of Terpenic Compounds. Experientia 9, 357 (1953).

176. RUZICKA, L. und FR. BALAS: Höhere Terpenverbindungen. XI. Zur Kenntnis der Dextro-primarsäure und über die Einteilung der Fichtenharzsäuren. Helv. Chim. Acta 6, 677 (1923).

177. RUZICKA, L. und E. BERNOLD: Zur Kenntnis der Diterpene. 48. Mitt. Über den Abbau der Agathen-disäure mit Kaliumpermanganat. Helv. Chim. Acta 24, 931 (1941).

178. RUZICKA, L., K. HUBER, PL. A. PLATTNER, S. S. DESHAPANDE und S. STUDER: Zur Kenntnis der Sesquiterpene. 43. Mitt. Zur Konstitution des Caryophyllengemisches. Abbau des Dihydro-caryophyllens. Helv. Chim. Acta 22, 716 (1939).

179. RUZICKA, L., D. R. KOOLHAAS und A. H. WIND: Polyterpene und Polyterpenoide. LX. Über den räumlichen Bau des Dekalinringes bei den Sesquiterpenen. Helv. Chim. Acta 14, 1171 (1931).

180. RUZICKA, L. und J. MEYER: Über Sesquiterpenverbindungen. I. Überführung des Cadinens in einen Naphthalinkohlenwasserstoff. Helv. Chim. Acta 4, 505 (1921).

181. RUZICKA, L., J. MEYER und M. MINGAZZINI: Höhere Terpenverbindungen. III. Über die Naphtalinkohlenwasserstoffe Cadalin und Eudalin, zwei aromatische Grundkörper der Sesquiterpenreihe. Helv. Chim. Acta 5, 345 (1922).

182. RUZICKA, L. und J. A. VAN MELSEN: Höhere Terpenverbindungen. XXXIX. Zur Kenntnis des Cedrens. Liebigs Ann. Chem. 471, 40 (1929).

183. RUZICKA, L., R. ZWICKY und O. JEGER: Zur Kenntnis der Diterpene. 57. Mitt. Die Konstitution der Agathendisäure. Helv. Chim. Acta 31, 2143 (1948).

184. RYDON, H. N.: The Synthesis of *cis*- and *trans-dl*-Norcaryophyllenic Acids and of Dehydronorcaryophyllenic Acid. J. Chem. Soc. (London) **1936**, 593.

185. SCHINZ, H. und C. F. SEIDEL: Zur Kenntnis des Lavendelöls. 1. Mitt. Über Lavandulol, einen neuen Monoterpenalkohol aus Lavendelöl. Helv. Chim. Acta 25, 1572 (1942).

186. SEBRELL, W. H. and R. S. HARRIS: The Vitamins. Vol. I. New York: Academic Press. 1954.

187. SEMMLER, F. W. und A. HOFFMANN: Zur Kenntnis der Bestandteile ätherischer Öle (Untersuchungen über das Sesquiterpen Cedren). Ber. dtsch. chem. Ges. 40, 3521 (1907).

188. SIMONSEN, J. L. and D. H. R. BARTON: The Terpenes. Vol. III. 2nd Ed. Cambridge: Univ. Press. 1952.

189. SOMASEKAR, A. and M. S. MUTHANNA: Synthesis of 5-Ethylazulene. Current Sci. (India) 21, 314 (1952) [Chem. Abstr. 48, 1323 (1954)].

190. SÖRENSEN, N. A. and F. HOUGEN: The Azulene from Oil of Elemi. Acta Chem. Scand. 2, 447 (1948).

191. ŠORM, F., V. BENEŠOVÁ und V. HEROUT: Über Terpene. LIV. Über die Struktur des Lactarazulens und des Lactaroviolins. Collect. Czech. Chem. Communs. **19**, 357 (1954).

192. ŠORM, F., V. BENEŠOVÁ, J. KRUPIČKA, V. ŠNEBERK, L. DOLEJŠ, V. HEROUT and J. SICHER: The Structure of Lactaroviolin. Chem. and Ind. **1954**, 1511.

193. ŠORM, F. and L. DOLEJŠ: On Terpenes. XXXIII. Proof of a Nine-membered Ring in Caryophyllene. Collect. Czech. Chem. Communs. **16**, 650 (1951).

193a. ŠORM, F., L. DOLEJŠ, O. KNESSL and J. PLÍVA: On Terpenes. XVI. On a Bicyclic Sesquiterpene and a New Azulene from the Oil of *Pogostemon Patchouli* P. Collect. Czech. Chem. Communs. **15**, 82 (1950).

194. ŠORM, F., L. DOLEJŠ and J. PLÍVA: On Terpenes. XX. A Note on the Constitution of β-Caryophyllene. Collect. Czech. Chem. Communs. **15**, 186 (1950).

195. ŠORM, F., J. GUT, J. HLAVNIČKA, J. KUČERA and L. ŠEDIVÝ: On Terpenes. XXV. The Total Synthesis of S-Guaiazulene. Collect. Czech. Chem. Communs. **16**, 168 (1951) [Chem. Abstr. **47**, 4313 (1953)].

196. ŠORM, F., J. GUT and J. KRUPIČKA: On Terpenes. XL. New Synthesis of 2,6-Dimethylazulene. Chem. Listy **46**, 240 (1952). [Chem. Abstr. **47**, 8702 (1953)].

197. ŠORM, F., V. HEROUT and K. TAKEDA: On Terpenes. LVI. On the Identity of Lindazulene and Chamazulene. Collect. Czech. Chem. Communs. **19**, 186 (1954) [Chem. Abstr. **48**, 12708 (1954)].

198. ŠORM, F., M. HOLUB, V. SÝKORA, J. MLEZIVA, M. STREIBL, J. PLÍVA, B. SCHNEIDER and V. HEROUT: On Terpenes. XLVI. Sesquiterpenic Hydrocarbons from Oil of Sweet Flag. Collect. Czech. Chem. Communs. **18**, 512 (1953).

199. ŠORM, F. and O. KNESSL: On Terpenes. IX. The Spectra of 2,6-Dimethylazulene and 4,6-Dimethylazulene in the Visible and Ultraviolet Region. Collect. Czech. Chem. Communs. **14**, 201 (1949) [Chem. Abstr. **44**, 5848 (1950)].

200. ŠORM, F., J. KUČERA and J. GUT: On Terpenes. XXVI. The Total Synthesis of Se-Guaiazulene. Collect. Czech. Chem. Communs. **16**, 184 (1951) [Chem. Abstr. **47**, 4314 (1953)].

201. ŠORM, F., J. MLEZIVA, Z. ARNOLD and J. PLÍVA: On Terpenes. XIII. On the Sesquiterpenes from the Essential Oil of Hops. Collect. Czech. Chem. Communs. **14**, 699 (1949).

202. ŠORM, F., J. NOVÁK and V. HEROUT: On Terpenes. LI. The Composition of Chamazulene. Preliminary Communication. Chem. Listy **47**, 1097 (1953); Collect. Czech. Chem. Comm. **18**, 527 (1953) [Chem. Abstr. **48**, 12707 (1954)].

203. ŠORM, F. and J. POKORNÝ: On Terpenes. XLVII. Synthesis of 1,7-Dimethylazulene. Chem. Listy **47**, 63 (1953).

204. ŠORM, F., M. STREIBL, V. JAROLÍM, L. NOVOTNÝ, L. DOLEJŠ and V. HEROUT: On Terpenes. LVIII. Total Synthesis of 1,1,4,8-Tetramethyl-cycloundecane (Humulane). Proof of the Eleven-membered Ring in Humulene. Collect. Czech. Chem. Communs. **19**, 570 (1954).

205. — — — — — — Synthesis of 1:1:4:8-Tetramethyl*cyclo*-undecane (Humulane). Proof of the Eleven-membered Ring in Humulene. Chem. and Ind. **1954**, 252.

206. ŠORM, F., M. STREIBL, J. PLÍVA and V. HEROUT: On Terpenes. XXXII. A Contribution to the Constitution of Humulene. Collect. Czech. Chem. Communs. **16**, 639 (1951).

207. ŠORM, F., M. SUCHÝ and V. HEROUT: On Terpenes. XXVIII. On the Constitution of the Aromatic Diterpene from Wormwood Oil. Collect. Czech. Chem. Communs. **16**, 278 (1951).

208. ŠORM, F., M. SUCHÝ, F. VONÁŠEK, J. PLÍVA and V. HEROUT: On Terpenes. XXVII. Sesquiterpenic and Diterpenic Components of Wormwood Oil (*Artemisia absinthium* L.). Collect. Czech. Chem. Communs. **16**, 268 (1951).

209. ŠORM, F., V. TOMÁŠEK and R. VRBA: On Terpenes. IX. On the Synthesis of 2,6-Dimethylazulene. Collect. Czech. Chem. Communs. 14, 345 (1949).

210. ŠORM, F., K. VEREŠ and V. HEROUT: On Terpenes. XXXVI. The Constitution of Calamenene. Collect. Czech. Chem. Communs. 18, 106 (1953).

211. ŠORM, F., M. VRANÝ and V. HEROUT: On Terpenes. XLII. Composition of the Oil of *Populus balsamifera*. Chem. Listy 46, 364 (1952).

212. ŠORM, F., M. ZAORAL and V. HEROUT: On Terpenes. XXXVIII. On the Constitution of Natural Bisabolol and Bisabolol Monoxide from Matricaria Oil. Collect. Czech. Chem. Communs. 18, 116 (1953).

213. STAFFORD, W. H. and D. H. REID: The Fine Structure of Azulene. Chem. and Ind. 1954, 277.

214. STAHL, E.: Über das Cham-Azulen und dessen Vorstufen, II. Mitt. Cham-Azulencarbonsäure aus Kamille. Chem. Ber. 87, 505 (1954).

214a. — Über das Cham-Azulen und dessen Vorstufen, III. Mitt. Zur Konstitution der Cham-Azulencarbonsäure. Ber. dtsch. chem. Ges. 87, 1626 (1954).

215. STOLL, M. et A. COMMARMONT: Odeur et constitution. II. α- et β-bicyclofarnésal et β-bicyclofarnésol. Helv. Chim. Acta 32, 1836 (1949).

216. STORK, G. and R. BRESLOW: The Structure of Cedrene. J. Amer. Chem. Soc. 75, 3291 (1953).

217. SÝKORA, V., J. ČERNÝ, V. HEROUT and F. ŠORM: On Terpenes. LV. Synthesis of Elemane (1-Methyl-1-ethyl-2,4-di*iso*propylcyclohexane). Collect. Czech. Chem. Communs. 19, 566 (1954).

218. SÝKORA, V., V. HEROUT, J. PLÍVA and F. ŠORM: On Terpenes. L. Contribution to the Constitution of Elemol. Coll. Czech. Chem. Comm. 19, 124 (1954).

219. TAKEDA, K., T. KUBOTA and W. NAGATA: Synthesis of 1,4-Dimethyl-7-ethylnaphthalene. Pharm. Bull. Japan 1, 241 (1953).

220. TAKEDA, K. and W. NAGATA: Components of the Root of *Lindera strychnifolia*. V. Azulenes Isolated from Linderene by Zinc-dust Distillation. Pharm. Bull. Japan 1, 164 (1953) [Chem. Abstr. 48, 7716 (1954)].

221. TAKEDA, K., F. ŠORM and V. HEROUT: The Identity of Lindazulene and Chamazulene. J. pharmac. Soc. Japan 74, 700 (1954) [Chem. Abstr. 48, 10718 (1954)].

222. TREIBS, W.: Über den oxydativen Abbau des Cedrens (III. Mitt. über Cedren und Cedrol). Ber. dtsch. chem. Ges. 76, 160 (1943).

223. — Über das Caryophyllenoxyd, seine Darstellung durch Autoxydation des Caryophyllens und sein Vorkommen in Pflanzenölen. Chem. Ber. 80, 56 (1947).

224. — Über bi- und polycyclische Azulene. 3. Mitt. Der Patchoulialkohol, ein tricyclischer Azulen-bildner. Liebigs Ann. Chem. 564, 141 (1949).

225. — Über das Calameon. Chem. Ber. 82, 530 (1949).

226. — Über bi- und polycyclische Azulene. 5. Mitt. Über den α-Kessylalkohol, ein natürliches azulenbildendes Oxy-sesquiterpenoxyd. Liebigs Ann. Chem. 570, 165 (1950).

227. — Über bi- und polycyclische Azulene 10. Mitt. Über ein tetracyclisches Azulen aus Tetrahydro-fluoranthen. Liebigs Ann. Chem. 574, 60 (1951).

228. — Über bi- und polycyclische Azulene. 11. Mitt. Die Dehydrierung von Hydroazulenen zu Azulenen mit Halogen. Liebigs Ann. Chem. 576, 110 (1952).

228a. — Über bi- und polycyclische Azulene. 12. Mitt. Das Germacrol, ein azulenbildendes Sesquiterpen-oxyd aus Germaniumöl. Liebigs Ann. Chem. 576, 116 (1952).

229. — Über bi- und polycyclische Azulene. 14. Das 1,2-4,5-Dibenzazulen, ein „vollaromatisches" tetracyclisches Azulen. Liebigs Ann. Chem. 577, 201 (1952).

229a. — Synthesen und Reaktionen von Azulenen. Angew. Chem. 67, 76 (1955).

230. TREIBS, W. und H. M. BARCHET: Über bi- und polycyclische Azulene. 4. Mitt. Das Aromadendren, sein chemischer Bau und seine Überführung in 5 Azulene. Liebigs Ann. Chem. **566**, 89 (1950).

231. TREIBS, W., H. M. BARCHET, G. BACH und W. KIRCHHOF: Über bi- und polycyclische Azulene. 9. Mitt. Versuche zur Darstellung bi- und tricyclischer Azaazulene. Liebigs Ann. Chem. **574**, 54 (1951).

232. TREIBS, W. und H. FROITZHEIM: Über bi- und polycyclische Azulene. 2. Mitt. Synthese des 1,8-Trimethylen-azulens; Versuch der Synthese eines Diazulenyls. Liebigs Ann. Chem. **564**, 43 (1949).

232a. TREIBS, W., W. KIRCHHOF und W. ZIEGENBEIN: Fortschritte der Azulenchemie seit 1950. Fortschr. chem. Forsch. **3**, 334 (1955).

232b. TREIBS, W. und R. KLIMKE: Über bi- und polyciclische Azulene. 19. Mitt. Die Synthese von 4,7-Dimethylazulen durch Dehydrierung von Butan. Liebigs Ann. Chem. **586**, 212 (1954).

233. TREIBS, W. und A. STEIN: Über bi- und polycyclische Azulene. 7. Mitt. Synthese einiger Derivate des 1,2-Benzazulens. Liebigs Ann. Chem. **572**, 165 (1951).

234. TREIBS, W., R. STEINERT und W. KIRCHHOF: Über bi- und polycyclische Azulene. 16. Darstellung von tri- und tetracyclischen Aza-azulenen durch Dehydrierung von Heptindolen mit Chloranil. Liebigs Ann. Chem. **581**, 54 (1953).

235. TREIBS, W., B. ULRICI und A. STEIN: Über bi- und polycyclische Azulene. 8. Mitt. Über die Buchner-Reaktion des Hydrindens und die Grignard-Umsetzung des Grundazulen-carbonsäureesters. Liebigs Ann. Chem. **573**, 93 (1951).

236. TSCHESCHE, R.: Neuere Vorstellungen auf dem Gebiete der Biosynthese der Steroide und verwandter Naturstoffe. Fortschr. chem. org. Naturstoffe **12**, 131 (1955).

236a. UKITA, T.: Structure of Kessoglycol. J. Pharmac. Soc. Japan **64**, 285 (1944), **65**, 458 (1945).

237. VESTERBERG, A.: Reten aus Abietinsäure. Ber. dtsch. chem. Ges. **36**, 4200 (1903).

237a. WENKERT, E.: Structural and Biogenetic Relationships in the Diterpene Series. Chem. and Ind. **1955**, 282.

237b. WETZEL, H.: Zur Kenntnis der Sesquiterpene im Pfefferminzöl. Dipl.-Arbeit, Univ. Leipzig. 1951.

237c. WIESNER, K., R. ARMSTRONG, M. F. BARTLETT and J. A. EDWARDS: Garrya Alkaloids. III. The Skeletas Structure of Garrya Alkaloids. J. Amer. Chem. Soc. **76**, 6068 (1954).

238. WILLSTAEDT, H.: Über die Farbstoffe des echten Reizkers (*Lactarius deliciosus* L.). 1. Mitt. Ber. dtsch. chem. Ges. **68**, 333 (1935).

239. WOODWARD, R. B. and P. YATES: The Stereochemistry of the Santonic Acids. Chem. and Ind. **1954**, 1391.

240. ZEISS, H. H. and M. ARAKAWA: The Structure of Longifolene. The Longifolic Acids. J. Amer. Chem. Soc. **76**, 1653 (1954).

(Received, January 19, 1955.)

Tetracyclic Triterpenes.

By E. R. H. Jones and T. G. Halsall, Manchester.

Contents.

I. Introduction.

Substances belonging to this group of organic compounds are widely distributed in Nature, being found in plants, both higher and lower fungi and, up to the present time, in one animal source, sheep wool-fat. For a longtime no real differentiation was possible between the tetra- and pentacyclic triterpenes and the sterols. Then the two latter became distinguishable by their selenium-dehydrogenation products, *i. e.*, picene and naphthalene derivatives from the pentacyclic triterpenes and DIELS' hydrocarbon from the sterols. It is now apparent that those compounds yielding predominantly 1 : 2 : 8-trimethylphenanthrene on dehydrogenation represent yet another class, and this property is regarded as typical of the tetracyclic triterpenes. The group contains both C_{30} and C_{31} compounds and, although the latter fall outside RUZICKA's strict definition of triterpenes (*142*), it seems desirable to permit this deviation. Members with thirty-two carbon atoms may well be discovered in due course.

In a most valuable account of the triterpenes written in 1949, JEGER (*113*) was able to summarize all that was known about the tetracyclic group in a very small compass. Most of the work discussed in the present article has been published during the last five years and in that time the structures of some twenty compounds have been elucidated. Of outstanding importance is the revelation of close structural relationships to the steroids, and the presence of C_8 and C_9 side-chains, skeletally identical with those of cholesterol and ergosterol. The compounds of the lanosterol group can be regarded as trimethyl-steroids, and lanosterol itself is actually 4 : 4 : 14-trimethylzymosterol. (At one time the names lanosterol, cryptosterol and agnosterol seemed most inappropriate for C_{30} compounds.) The discovery of this relationship with the steroids has inevitably stimulated much discussion as to the possibility of common or similar biogenetic routes and significant results in connection with these suggestions will undoubtedly accrue in the next few years.

It is convenient at this stage to divide the known compounds into groups based on lanosterol (A) and euphol (B), although further sub-

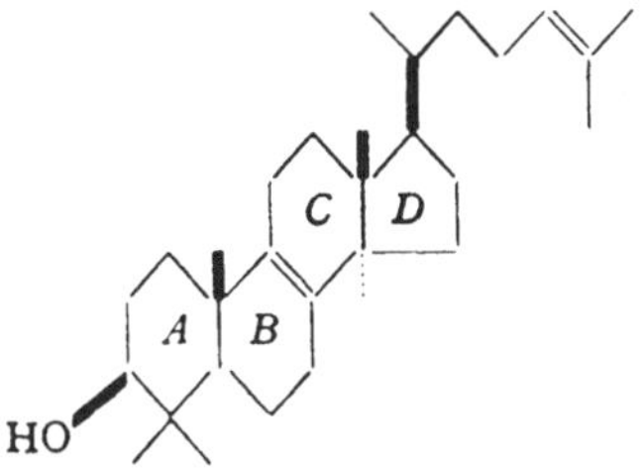

C D A B HO

Type A. Lanosterol.

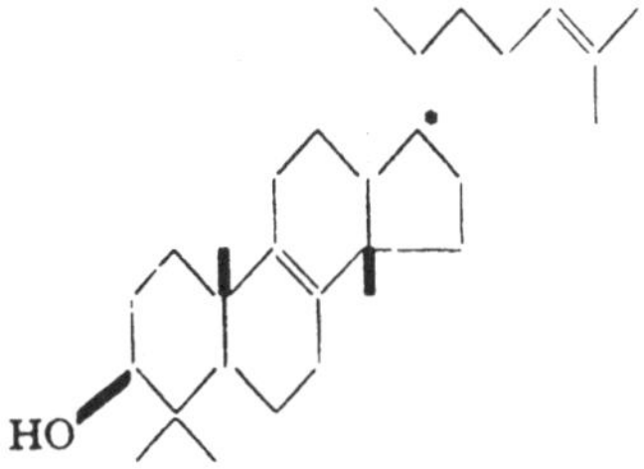

HO

Type B. Euphol.

* The α-configuration at $C_{(17)}$ is as yet not proved.

Table 1. Tetracyclic Triterpenes.

Substance	Molecular formula	C(3)-substituent	Other nuclear substituents	Nuclear unsaturation Δ	Side-chain R 
A. *With Lanostane Ring System.*					
Lanosterol	$C_{30}H_{50}O$	βOH	—	8	
Agnosterol	$C_{30}H_{48}O$	βOH	—	7 : 9(11)	,,
Dihydrolanosterol	$C_{30}H_{52}O$	βOH	—	8	,,
Dihydroagnosterol	$C_{30}H_{50}O$	βOH	—	7 : 9(11)	,,
Pinicolic Acid A	$C_{30}H_{46}O_3$	=O	—	8	 HOOC
Trametenolic Acid B	$C_{30}H_{48}O_3$	βOH	—	8	,,
Eburicoic Acid	$C_{31}H_{50}O_3$	βOH	—	8	 HOOC
Dehydroeburicoic Acid	$C_{31}H_{48}O_3$	βOH	—	7 : 9(11)	,,
Polyporenic Acid C	$C_{31}H_{46}O_4$	=O	16αOH	7 : 9(11)	,,
Tumulosic Acid	$C_{31}H_{48}O_4$	βOH	16αOH	8	,,
Dehydrotumulosic Acid	$C_{31}H_{46}O_4$	βOH	16αOH	7 : 9(11)	,,
Polyporenic Acid A	$C_{31}H_{48}O_4$	αOH	12αOH	8	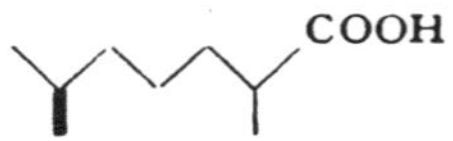 COOH
21-Hydroxylanostadienone	$C_{30}H_{48}O_2$	=O	—	8	 HO·H_2C
*cyclo*Artenol	$C_{30}H_{50}O$	βOH	—	[cyclo-9 : 19]	
*cyclo*Laudenol	$C_{31}H_{52}O$	βOH	—	[cyclo-9 : 19]	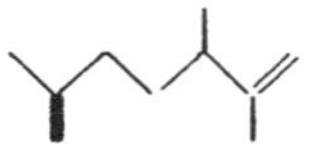

Substance	Molecular formula	C(3)-substituent	Other nuclear substituents	Nuclear unsaturation Δ	Side-chain R (M)
B. *With Euphane Ring system.*					
Euphol	$C_{30}H_{50}O$	βOH	—	8	
Butyrospermol	$C_{30}H_{50}O$	βOH	9βH (?)	7(?)	..
Tirucallol	$C_{30}H_{50}O$	βOH	—	8	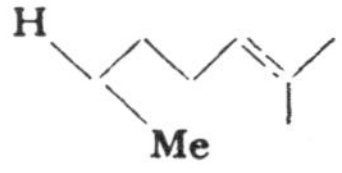 H, Me
Elemolic Acid	$C_{30}H_{48}O_2$	αOH	—	8	H, CO_2H
Elemonic Acid	$C_{30}H_{46}O_2$	=O	—	8	..
Dehydroelemolic Acid	$C_{30}H_{46}O_2$	αOH	—	7 : 9(11)	..
Dehydroelemonic Acid	$C_{30}H_{44}O_2$	=O	—	7 : 9(11)	..
Euphorbol	$C_{31}H_{52}O$	βOH	—	8	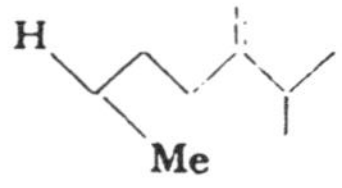 H, Me

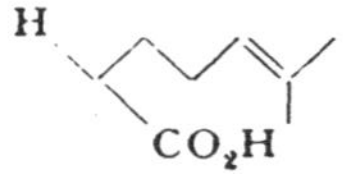

* There is still some uncertainty about the configuration at $C_{(17)}$ in this group.

division may well prove necessary later. In the accompanying *Table 1* the structures of all the members discussed in this article are indicated relevant to the basic lanosterol and euphol cyclic systems. It will be noted that, whereas the stereochemistry of the former corresponds with that of the sterols, in euphol the configurations at 13, 14, and possibly 17 are reversed.

For lanosterol and its immediate relatives the nomenclature based on the name lanostane for the parent hydrocarbon was proposed by RUZICKA, DENSS and JEGER (*143*) and has been adopted generally. When the connection with the steroids became clear it was obviously desirable to employ steroid numbering (p. 51) and stereochemical and other conventions. In the majority of papers prior to 1952 other numbering was employed (commencing from the *gem*-dimethyl group) and the formulae, as with the pentacyclic triterpenes, were inverted.

The general diagnostic utility of certain degradative procedures is evident in most of the structural studies; these are illustrated in detail in connection with lanosterol. Dehydrogenation, so wasteful of material, is not invariably employed. The characteristic retro-pinacolinic rearrangement brought about on phosphorus pentachloride dehydration of the alcohols provides evidence concerning the architecture of ring *A* and the ubiquitous oxygenation in the 3-position. Progressive dehydrogenation and oxidation have furnished evidence concerning rings *B* and *C* and the inert ethylenic linkage. Methods worked out in the steroid field have been put to good service in side-chain degradation, ultimately revealing the nature of ring *D*. Modern techniques and concepts such as infrared and ultraviolet spectroscopy, molecular rotation difference methods, and conformational and X-ray analysis have assisted notably in the phenomenally rapid development of the subject. Finally, as in the steroid and pentacyclic triterpene fields, by elimination and transformation of functional groups it has proved possible to establish precise structural relationships with the parent compounds, lanosterol and euphol.

II. Lanostane Type Group.

1. Lanosterol Group.

Occurrence and Isolation.

The unsaponifiable portion of wool-fat contains a considerable proportion of cholesterol accompanied by what was at first thought to be another pure alcohol, the so-called 'isocholesterol'. Following investigations by several workers it gradually became evident that 'isocholesterol' was not homogeneous and in 1930 WINDAUS and TSCHESCHE (*172*) resolved it into agnosterol (8%) and another component (92%) believed to be pure, which was named lanosterol (cf. *65*). Real clarification of our knowledge of the composition of the unsaponifiable portion of wool-fat came in 1944 when RUZICKA and his co-workers (*149*) separated four components from 'isocholesterol'. The fatty alcohols were removed by chromatography or boiling with petrol, and repeated treatment with boiling methanol extracted all of the cholesterol. The residual material, after acetylation, was subjected to detailed fractional crystallisation (chromatography was also employed on the less soluble portion) and, after hydrolysis, lanosterol (I), dihydrolanosterol (II), agnosterol (III), and dihydroagnosterol (γ-lanosterol) (IV) were obtained pure and fully characterised (*Table 2*). Approximate values for the composition of the cholesterol-free material are: lanosterol + dihydrolanosterol (50%), dihydroagnosterol (20%) and agnosterol (8%). It is possible that there are other minor components yet to be isolated.

Alternative procedures involving oxidation, and fractionation of the mixture of ketones by chromatography and by crystallisation, followed by regeneration of the alcohols by reduction, were described later (*143*, *144*). Separation of these substances is rendered extremely difficult by the close similarity of the constants in some cases and the absence of melting point depressions with different compounds possessing similar melting points.

Table 2. Constants of Lanosterol and Some Related Compounds.

	m. p.	$[\alpha]_D$	
Lanosterol (I) (Cryptosterol, lanosta-8 : 24-dien-3β-ol)	138–139° corr.	+ 61–62°	(*143*)
Dihydrolanosterol (II) (lanost-8-en-3β-ol) ..	145–146° corr.	+ 61°	(*143*)
	146–147°	+ 63°	(*51*)
Agnosterol (III) (lanosta-7 : 9 : 24-trien-3β-ol)	164,5–165,5° corr.	+ 66°	(*144*)
Dihydroagnosterol (IV) (γ-Lanosterol, lanosta-7 : 9-dien-3β-ol)	158–159°	+ 67,5°	(*51*)
	157–158° corr.	+ 67,5°	(*144*)

It can be seen from the structures of these four components that they have identical carbon skeletons and stereochemistry, the differences between them residing in the number and positions of ethylenic linkages. Although mixtures of at least two of these compounds were frequently employed in the earlier studies, the differences between the components were often eliminated by the processes employed, *e. g.*, hydrogenation, oxidation, dehydrogenation, and most of the results are therefore quite valid. The term 'isocholesterol' when used today connotes the mixture of lanosterol and its dihydro-derivative obtained by hydrolysis of the more soluble portion of the mixed acetates. It has been employed as starting material in many investigations and its use will not usually be commented upon in what follows.

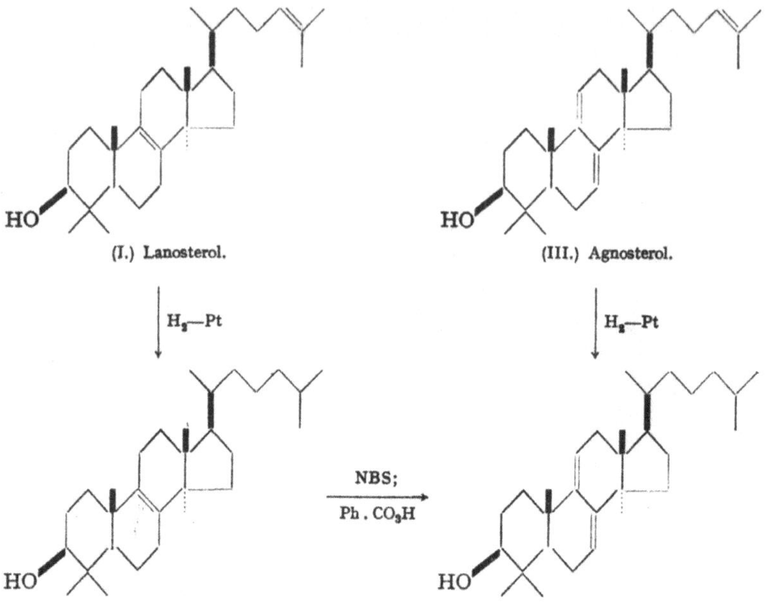

HO

(I.) Lanosterol.

HO

(III.) Agnosterol.

H_2—Pt

H_2—Pt

HO

(II.) Dihydrolanosterol.

NBS; Ph. CO_3H

HO

(IV.) Dihydroagnosterol.

Wieland, Pasedach and Ballauf (*171*) noted the similarity between lanosterol and cryptosterol, a constituent of the yeast sterol mixture, which could be obtained pure without too much difficulty, but large differences in melting points of two of their derivatives appeared to preclude the possibility of their identity with one another. On the other hand, between all the representatives of the dihydro-series there was the closest possible agreement. After comparing many new derivatives Ruzicka, Denss and Jeger (*143*) concluded that dihydrolanosterol and dihydrocryptosterol were identical. They realised that previous comparisons between the parent alcohols had been made, not with pure lanosterol derivatives, but with materials prepared from an approximately 1 : 1-mixture of lanosterol and its dihydro-compound and they showed that carefully purified lanosterol and cryptosterol were, in fact, one and the same compound.

Of considerable interest is the recent isolation of lanosterol from the latex of *Euphorbia balsamifera* Ait. (*83, 84*) in which the pentacyclic triterpene, germanicol, is also present.

Nomenclature. Although lanostane (V) is identical with 4 : 4 : 14α-trimethylcholestane, the nomenclature based on lanostane is much to be preferred. Steroid numbering is employed as also in reference to the four rings as *A*, *B*, *C* and *D*. The numbers 28 and 29 are reserved for carbon atoms attached to $C_{(24)}$.

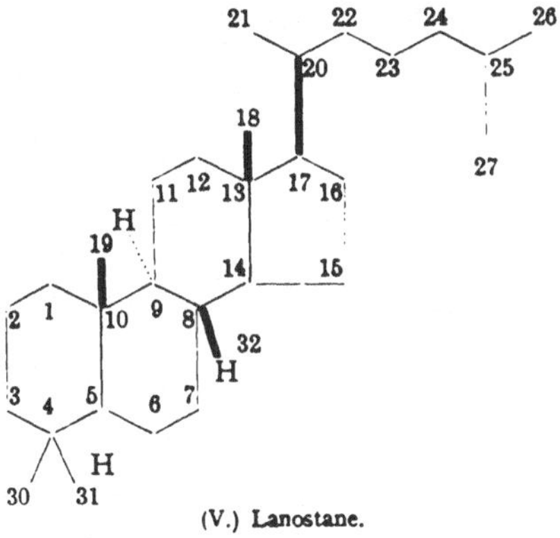

(V.) Lanostane.

The stereochemistry of the lanostane molecule is also indicated in (V), employing steroid conventions. In general throughout this article the configuration of the 5α-hydrogen atom will not be indicated; where asymmetry exists at $C_{(8)}$ and/or $C_{(9)}$ the configurations of the hydrogen atoms will either be indicated or else may be assumed to be 8 β and 9 α, *i. e.*, giving a stable ***trans-anti-trans*** arrangement. Because of possible and actual differences at $C_{(20)}$, the hydrogen atom is indicated in the formulae in Section 2 and subsequently.

Interrelationship of Lanosterol, Agnosterol and their Dihydro-derivatives. Both lanosterol (I) and agnosterol (III) and their derivatives are smoothly converted into dihydro-compounds (II, IV) by hydrogenation in the presence of platinum under the usual conditions. Quite a variety of reagents (p. 53) bring about the transformation of dihydrolanosterol (II) into dihydroagnosterol (IV) and both (III) and (IV), containing the heteroannular 7 : 9-diene system, exhibit characteristic absorption in the ultraviolet region of the spectrum (p. 53).

Structure of Lanosterol.

The separation of four pure compounds from saponified wool-fat (*149*) and the establishment of their inter-relationship made possible the major attack on the structural problem. Its complete solution, including the conversion of cholesterol into dihydrolanosterol, is one of the outstanding achievements of contemporary organic chemistry. Although many related studies undoubtedly assisted in the final clarification, the major chemical investigations during the 1949–1954 period were made by RUZICKA and JEGER (Zurich), BARTON (London) and MCGHIE (London) and their many collaborators but earlier, major contributions, in addition to those from the RUZICKA school, had been made by DORÉE and by WIELAND.

Lanosterol has the composition $C_{30}H_{50}O$ and since it contains two ethylenic linkages it must have a tetracyclic structure. The presence of a secondary hydroxyl group was indicated by the formation of a ketone on oxidation with chromic acid under mild conditions. Reduction of the ketone with aluminium *iso*propoxide and *iso*propanol regenerated lanosterol proving that the hydroxyl group is in the more stable configuration (*143*). Conversion of the ketone (in the dihydro-series) into its hydroxymethylene derivative and oxidation of this with alkaline hydrogen peroxide gave a dicarboxylic acid. The acid on heating yielded a *nor*-ketone, and it was inferred from the WIELAND-BLANC rule that the hydroxyl group was attached to a terminal ring containing at least six carbon atoms (*149*). Further detailed investigation of the environment of the hydroxyl group was rendered unnecessary when the product of the reaction of dihydrolanosterol (lanost-8-en-3β-ol) (VI) with phosphorus pentachloride (*60*) was studied. RUZICKA, MONTAVON and JEGER (*148*) were able to demonstrate that dehydration involving a retro-pinacolinic rearrangement had occurred, since the product (VII) yielded acetone and a C_{27}-ketone after lead tetraacetate fission of the stereoisomeric glycols (VIII) produced with osmium tetroxide (cf. *63*, *64*).

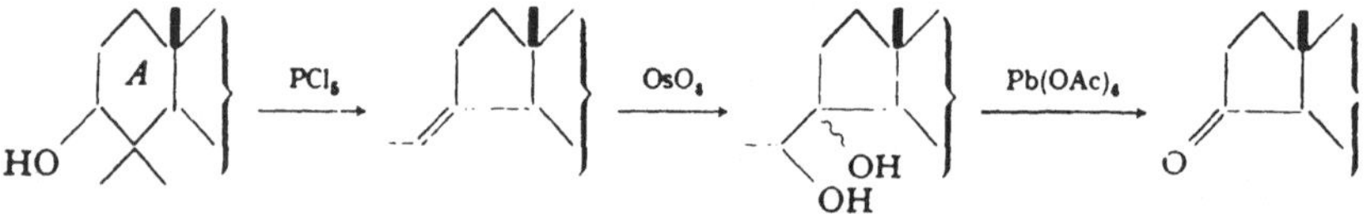

A PCl$_5$ OsO$_4$ Pb(OAc)$_4$ HO OH OH O

(VI.) Dihydrolanosterol. (VII.) (VIII.)

This dehydration reaction paralleled the behaviour of compounds of the α- and β-amyrin and lupeol series of pentacyclic triterpenes (cf. references cited in *148*) and it could be concluded very reasonably that the oxygen-containing *A*-ring of lanosterol was identical both structurally and stereochemically with that present in these compounds. The course of this reaction is readily explicable in terms of the conformation of the molecule (*12*), the coplanarity of the β-oxygen atom and $C_{(3)}$, $C_{(4)}$ and $C_{(5)}$ providing the most favourable arrangement for the occurrence of a synchronous skeletal rearrangement:

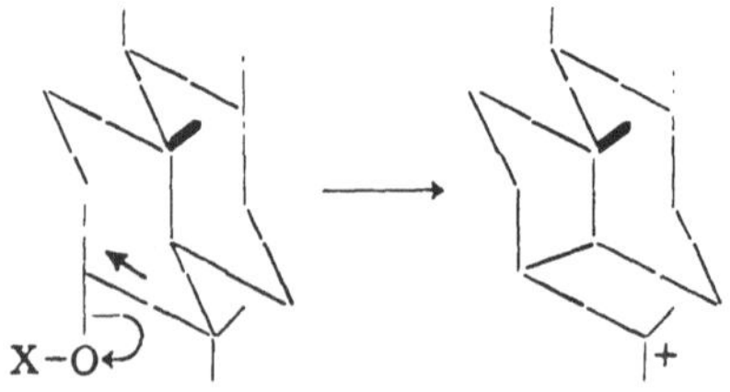

X–O +

Location of the Inert Double Bond. The environment of the less reactive double bond of lanosterol was largely deduced by studying the products arising from a variety of oxidation reactions; light absorption methods were extensively and decisively employed. The tetrasubstituted nature of this ethylenic linkage was deduced from infrared absorption measurements by ROTH and JEGER (*140*), and later confirmed by an ultraviolet examination of lanostene (*14a*). Simple dehydrogenation of dihydrolanosteryl acetate (3 β-acetoxylanost-8-ene) (IX) to the conjugated diene acetate (X) (*Chart 1*) is brought about by several reagents, *e. g.*, selenium dioxide in boiling acetic acid (*22*, *39*), heating with N-bromosuccinimide in carbon tetrachloride (*62*) or with perbenzoic acid in chloroform (*32*). This substance absorbs strongly in the 2400 Å. region, where three rather characteristic maxima are observed. DORÉE (*22*) recognised that in this and in other respects the hydrolysed dehydrogenation product resembled the so-called 'γ-lanosterol', the partial hydrogenation product of agnosterol and this suggested identity was subsequently established (*144*).

The ultraviolet light absorption of this heteroannular diene system (7 : 8–9 : 11) with its three maxima at ca. 2370, 2440, and 2510 Å., is now recognized as highly characteristic, being observed in all compounds containing such systems in the lanostane and steroid series (*58*). Furthermore the dehydrogenation process is invariably accompanied by a positive change of rotation (Δ $[M]_D$ ranging from + 14 to + 136). Similar conjugated dienes of the euphol, euphorbol, tirucallol and elemolic acid series exhibit maxima displaced to distinctly shorter wavelengths (ca. 2320, 2390, and 2470 Å.) and here dehydrogenation produces a very large negative change in rotation (Δ $[M]_D$ ranging from — 400 to — 700).

Oxidation of the diene-acetate (X) with monoperphthalic acid under mild conditions yielded the 7 : 8-epoxide (presumably α) (XII)* whilst with hydrogen peroxide in acetic acid the isomeric $\beta\gamma$-unsaturated ketone (XIV) was produced (*Chart 1*). Both compounds were isomerised by boron trifluoride in benzene into the $\alpha\beta$-unsaturated ketone (XI) (*136*), with light absorption (maximum 2550 Å.) compatible with such a structure (*149*). The reactivity of the carbonyl group in (XI) (oxime formation) is significant in regard to its allocation to the 7- rather than to the 11-position. (The final proof that it is a 7-ketone is discussed later.) The $\alpha\beta$-unsaturated ketone (XI) is formed directly by oxidation of the diene (X) with chromic (*39*) or perbenzoic acid (*32*), and it is also produced on oxidation of dihydrolanosteryl acetate (IX, p. 54) with chromic acid (*149*, cf. *32*, *40*, *41*) or with ozonised oxygen (*149*).

Further stepwise oxidation was achieved by mono-perphthalic acid oxidation of the enol acetate (XIII) prepared from (XI) or (XIV), followed

* In contrast, from ergosterol-D and related compounds of the cholestane and androstane series, the 9 : 11-epoxide is formed (cf. *108*).

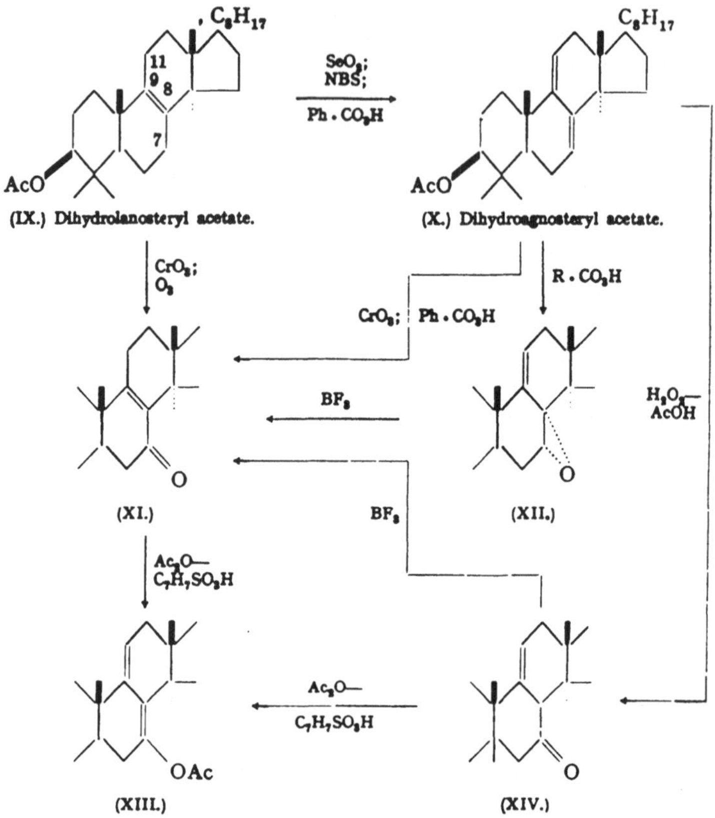

(IX.) Dihydrolanosteryl acetate. (X.) Dihydroagnosteryl acetate.

(XI.) (XII.)

(XIII.) (XIV.)

Chart 1. Oxidation of Dihydrolanosteryl Acetate.

by isomerisation of the epoxide (XV) to the γ-hydroxy-$\alpha\beta$-unsaturated ketone (XVI) (maximum 2520 Å.) with 1% acetic acid in warm methanol (*136*).

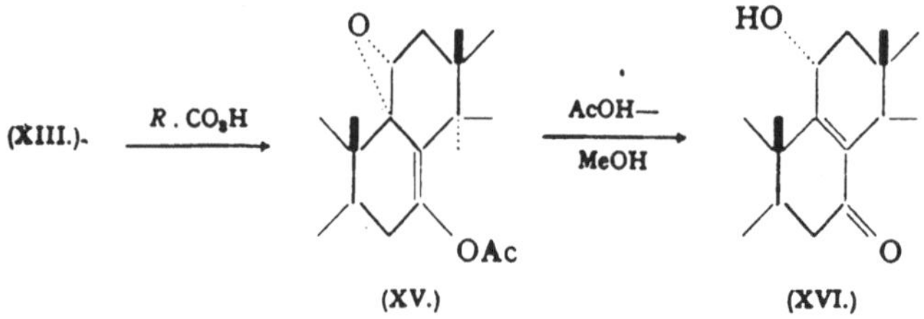

(XV.) (XVI.)

The most characteristic and most frequently encountered oxidation product of dihydrolanosterol derivatives is the yellow ene-dione (XVII). The acetate was originally obtained (*149*) by chromic acid oxidation in acetic acid at 40° (cf. *39*, *40*) and subsequently when using hydrogen peroxide in boiling acetic acid (*169*). It is also formed from the diene-acetate (X) with chromic acid (*41*, *149*), and similarly from both the

$\alpha\beta$-unsaturated ketone (XI) (*41*) and the enol acetate (XIII) (*136*). The presence of the —CO—C=C—CO— system was recognised on the basis of its light absorption (maximum 2720 Å.) (*149*), typical of the ene-dione system in a completely transoid arrangement as in the yellow 1 : 5-diketo-Δ^9-octalin (cf. *38*). Additional evidence was forthcoming later from infrared absorption studies (*169*).

Of the two carbonyl groups in (XVII) that in the 7-position is much more reactive, forming an oxime and an ethylene dithioketal (*136*) and being selectively reduced by the WOLFF-KISHNER method to the Δ^8-11-ketone (*129*, *130*). Both carbonyl groups can be reduced with lithium aluminium hydride (*40*). Treatment with zinc and acids and catalytic

Lanostenyl acetate (IX.); also (X.) and (XI.) —CrO_3→ (XVII.) —Zn—H^+; H_2—Pt→ (XVIII.)

hydrogenation with platinum in acetic acid leaves the keto-groups unaffected but the intervening double bond is reduced by these reagents, the resulting saturated diketone (XVIII) showing only low intensity ultraviolet absorption (at 2980 Å.) (*61*, *62*).

The formation of the ene-dione (XVII) suggested that the unreactive double bond of lanosterol was flanked by two methylene groups (*149*) and from the infrared absorption of the acetoxy-lanostanedione (XVIII) with a strong band at 1697 cm.$^{-1}$ (*169*), it could be inferred that the carbonyl groups were both in six-membered rings. More extensive oxidation experiments threw still more light on the environment of the inert ethylenic linkage.

Oxidation of the unsaturated diketone (XVII) with selenium dioxide in boiling acetic acid (*169*, cf. *39*) effected further dehydrogenation to the doubly-unsaturated diketone (XIX). This showed maximal ultraviolet absorption at the same wavelength (2750 Å.) as the starting material, but of appreciably higher intensity, consistent with the assigned structure. The infrared absorption spectrum contained a band at 812 cm.$^{-1}$, absent from that of the precursor, indicating the presence of a trisubstituted double bond. Further oxidation of (XIX) with selenium dioxide in boiling acetic acid (or in dioxan at 180°) yielded the diene-triketone (XX) (maximum 2850 Å.) (*14a*, *169*, cf. *62*). The α-diketone grouping failed to give a ferric chloride colour but its presence was demonstrated by oxidation with alkaline hydrogen peroxide (*14a*, *39*) which readily

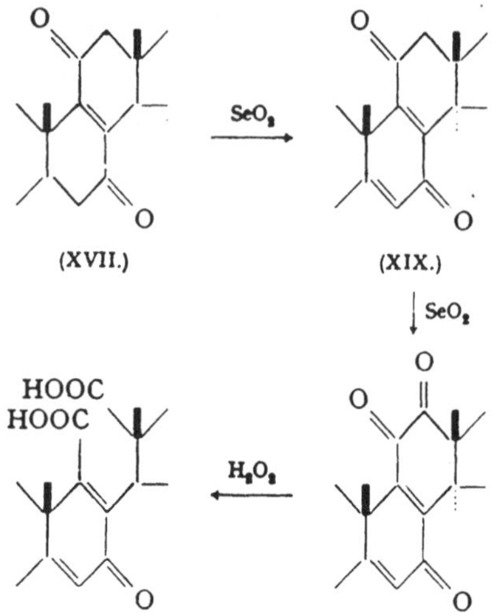

SeO$_2$

(XVII.) (XIX.)

SeO$_2$

HOOC
HOOC

H_2O_2

(XXI.) (XX.)

gave the dicarboxylic acid (XXI), the residual chromophoric system exhibiting maximal absorption at 2510 Å.

Confirmation of the environment of the inert double bond as so far revealed came from the production of the triply unsaturated ketone (XXII) by selenium dioxide oxidation of either lanostenyl acetate or the $\alpha\beta$-unsaturated ketone (XI). Its light absorption (maxima 2560 and 3270 Å.) is now known to be consistent with the structure given, as also is its conversion by chromic acid oxidation into the diene-triketone (XX) (*32, 39*). The real significance of this latter oxidation product (XX) and hence of all the intermediates, was emphasized by its conversion into the saturated diketone (XVIII) on CLEMMENSEN reduction (*61*), *i. e.*, the

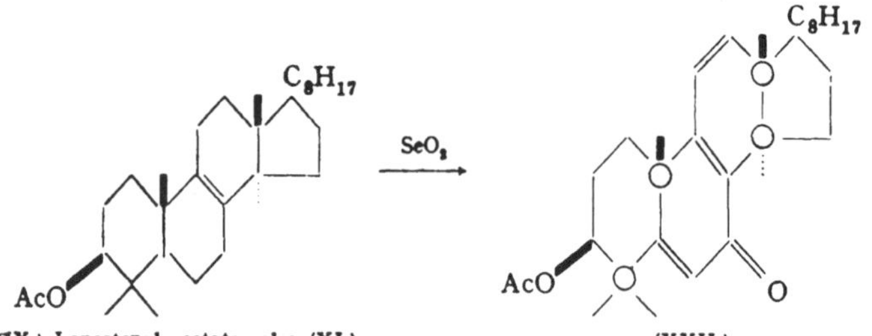

C_8H_{17} SeO$_2$ AcO

(IX.) Lanostenyl acetate; also (XI.) (XXII.)

occurrence of skeletal rearrangements seemed unlikely (*169*). Further evidence on this same point came from reduction of the $\alpha\beta$-unsaturated ketone (XI) with either sodium and propanol or lithium aluminium hydride, followed by acetic anhydride treatment, which gave the diene-acetate (X) (*39*).

The non-enolisation of the α-diketone system of (XX) and the absence of indications of the formation of more complex oxidation products made it probable that all the carbon atoms contiguous to the triene-ketone and diene-triketone systems in (XXII) and (XX) were quaternary (see XXII), and suggested that the inert double bond was between rings *B* and *C* rather than between rings *C* and *D*. [The recognition of the trisubstituted nature of the double bond of *iso*lanostenol (see p. 66) also pointed to the same conclusion.] Final proof was forthcoming from some incisive experiments by BARTON and his colleagues (*14a*). Treatment of the alcohol

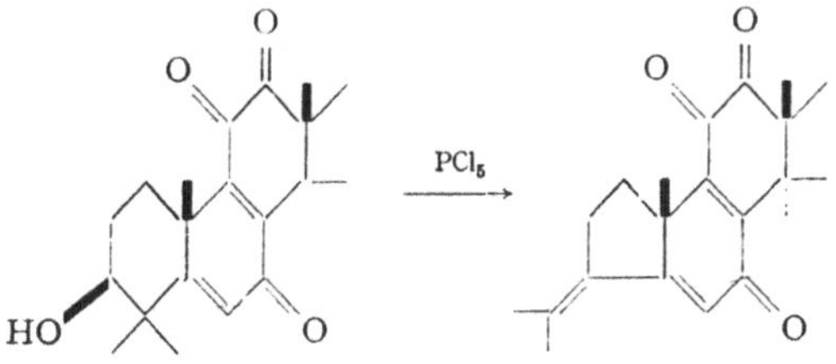

(XX.) $\xrightarrow{PCl_5}$ (XXIII.)

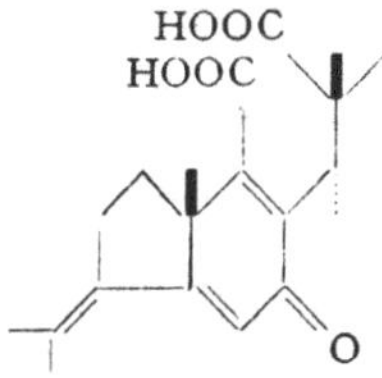

(XXIV.)

(XX) with phosphorus pentachloride, conditions under which the retro-pinacolinic dehydration occurs (cf. p. 52), yielded a triene-trione (XXIII) in which it could be concluded from the shift in the absorption maximum (from 2850 Å. to 2650 and 3630 Å.) that the additional double bond had become conjugated with the original chromophore. Moreover, since alkaline hydrogen peroxide oxidation now gave a dicarboxylic acid (XXIV) which clearly still contained an extended conjugated system, C=C—C=C—CO—C=C—COOH, (maxima, 2310 and 3210 Å.), the α-diketone grouping must reside in ring *C*, *i. e.*, farthest away from the site of the newly-introduced double bond. These conclusions were substantiated by the conversion of the saturated diketone (XXV) into the unconjugated ene-dione (XXVI) which with hydrochloric in boiling acetic acid gave the αβ-unsaturated isomer (XXVII). These latter experiments were simultaneously and independently performed by McGHIE and his colleagues (*41a*). Both of these unsaturated diketones, *i. e.*, (XXVI) and (XXVII) yielded *mono*-dinitrophenylhydrazones, but that from the latter contained the double bond in conjugation with the hydrazone residue (*i. e.*, —C=C—C=N—). Since the earlier-described $\alpha\beta$-unsaturated ketone (XI, p. 54) contains a reactive carbonyl group, this must

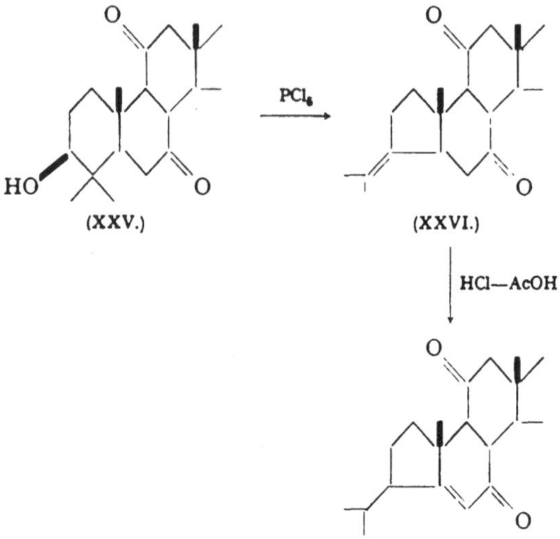

(XXV.) —PCl_5→ (XXVI.) —HCl—AcOH→ (XXVII.)

correspond to the carbonyl group in (XXV) which is nearer (in the 7-position) to the hydroxyl group in ring *A*. It was pointed out (*169*) that the reactivity of carbonyl groups in the 7-position, and their inertness in the 11-position, is precisely paralleled in the steroid series, as also are the properties of the hydroxyl group (analogous to steroid 11β-hydroxyl) resulting from lithium aluminium hydride reduction of the less-reactive carbonyl group.

It has been mentioned earlier that the main product of selenium dehydrogenation of the tetracyclic triterpenes is 1 : 2 : 8-trimethylphenanthrene. In the case of lanosterol this was isolated along with

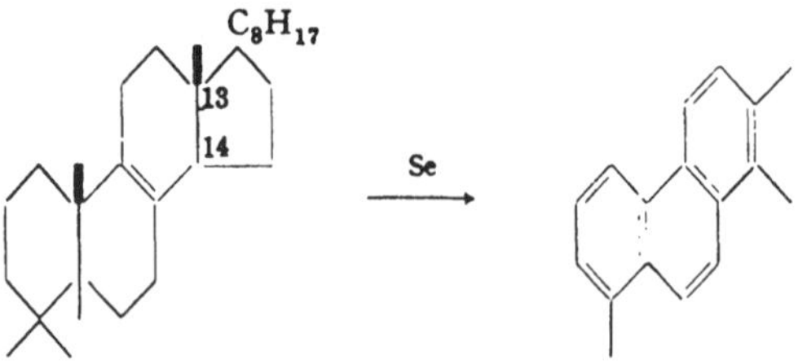

(XXVIII.) Lanostene. —Se→ (XXIX.) 1 : 2 : 8-Trimethylphenanthrene.

another hydrocarbon, $C_{22}H_{20}$ (?), the structure of which has yet to be elucidated (*149, 153*). The formation of the trimethylphenanthrene could be interpreted as arising from a methy group migration, instead of an elimination, in ring *A*, such a process being held responsible for the production of certain dehydrogenation products from pentacyclic triterpenes. This ambiguity was removed when it was shown (*14a, 167*) that (XXIX) was formed in superior yield from lanostene (XXVIII), in which the possibility of methyl group migration in ring *A* is largely eliminated. BARTON, FAWCETT and THOMAS (*14a*) therefore suggested that angular methyl groups were located at *both* $C_{(13)}$ and $C_{(14)}$ in lanosterol.

The Side-chain and Ring D. The presence of an *iso*propylidene group in the side-chain of lanosterol was demonstrated at an early stage in the investigations (*59, 149 170*), acetone and a small amount of the acetate of a C_{27}-acid being obtained on ozonolysis of the acetate of lanosterol. [An improved method of obtaining the C_{27}-acid, by controlled ozonolysis of 'isocholesterol', has been described (*54*).] In 1951, following a clue that a volatile fragment was produced on vigorous oxidation of dihydrolanosterol derivatives (*59, 132*), BARTON, MCGHIE and their colleagues (*10*) isolated 6-methylheptan-2-one from the products of chromic acid oxidation of lanostenyl acetate. At the same time JEGER, RUZICKA and their

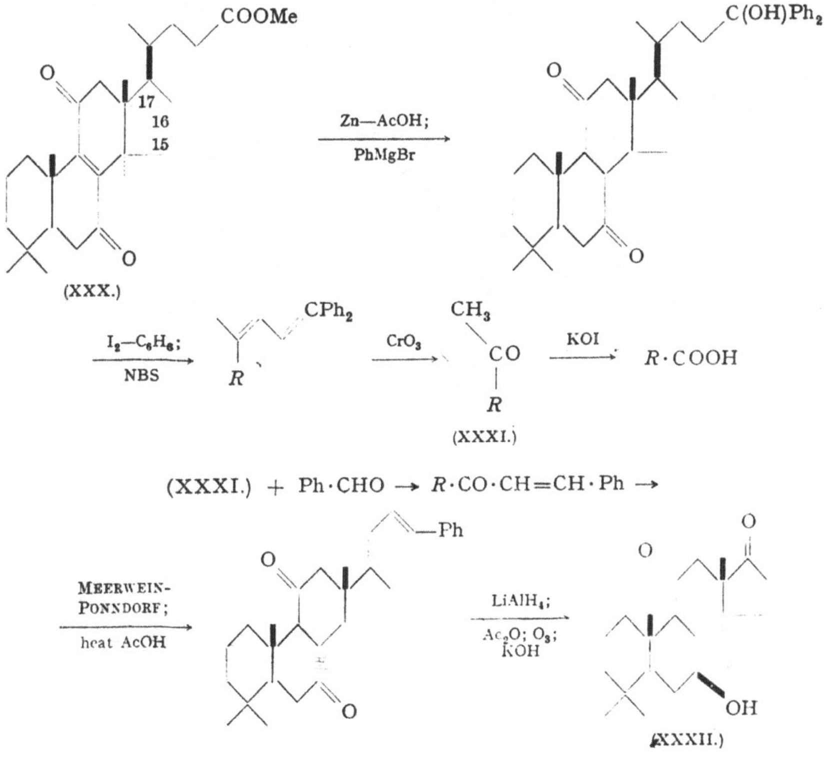

COOMe
O
17
16
15
O
(XXX.)

Zn—AcOH;
PhMgBr

C(OH)Ph$_2$
O
O

I_2—C_6H_6;
NBS

CPh$_2$
R

CrO_3

CH$_3$
CO
R
(XXXI.)

KOI

R·COOH

(XXXI.) + Ph·CHO → R·CO·CH=CH·Ph →

MEERWEIN-PONNDORF;
heat AcOH

—Ph
O
O

LiAlH$_4$;
Ac$_2$O; O$_3$;
KOH

O
O
OH
(XXXII.)

Chart 2. Degradation of the Side-chain of Lanosterol.

associates (*167*) carried out a stepwise degradation which also proved the *iso*-octenyl nature of the lanosterol side-chain. Starting from inhomogeneous 'isocholesterol' the oxygen function was removed by conversion to the ketone followed by WOLFF-KISHNER reduction. Chromic acid oxidation and purification of the acidic product *via* the methyl ester then gave the pure diketo-ester (XXX). The ester was converted with zinc and acetic acid to the saturated diketo-ester (*Chart 2*) and the side-chain was degraded by the WETTSTEIN-MIESCHER method (*135*) as indicated. In this way five carbon atoms, additional to the three forming the *iso*propylidene group could be accounted for (cf. *130*). Subsequently VOSER, GÜNTHARD, JEGER and RUZICKA (*163, 165, 166*) converted the C_{23}-triketone (XXXI) *via* the benzylidene derivative into the C_{20}-diketo-alcohol (XXXII). The latter had clearly defined carbonyl bands in its infrared spectrum at 1701 and 1745 cm.$^{-1}$. The former band arises from the 11-keto-group (in a 6-membered ring) and the latter, attributable to a carbonyl group in a five-membered ring, provided the first proof that ring D of lanosterol was five-membered. The *iso*octenyl side-chain in lanosterol was obviously attached to this five-membered ring at positions 15, 16 or 17 and it was pointed out that, although attachment at $C_{(17)}$ gave a structure analogous to that of zymosterol, the 'isoprene-rule' was only obeyed by the other two alternatives.

BARTON and his colleagues (*9*) prepared the triacetate (XXXIII) by reduction of the saturated diketone (XXV) with sodium and propanol (equilibrating conditions giving the di-equatorial diol), followed by acetylation. With the central rings thus protected against oxidative attack, three products of side-chain degradation, an acid, a lactone and a ketone (XXXIV) were isolated after chromic acid oxidation. Infrared data confirmed that this ketone had the carbonyl group in a five-membered ring. In addition evidence was obtained in favour of the environment

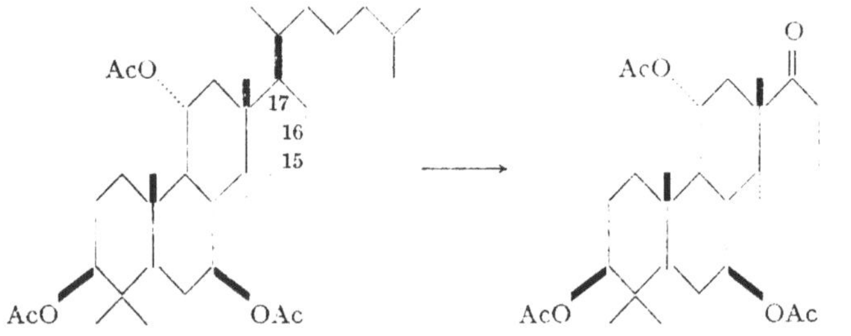

AcO, 17, 16, 15, AcO, OAc

(XXXIII.)

AcO, O, AcO, OAc

(XXXIV.)

—$CH_2 \cdot CO$—(15 or 17) as against —$CH_2 \cdot CO \cdot CH_2$—(16) for the carbonyl group by quantitative bromination experiments, and also by measurement of the intensity of the 1410 cm.$^{-1}$ band due to methylene groups adjacent to carbonyl groups. Preference for the location of the side-chain at $C_{(16)}$ was expressed on the grounds of the much lower reactivity of the carbonyl group towards 2 : 4-dinitrophenylhydrazine as compared with, *e. g.*, the acetate of androstanol-17-one, and also on the basis of conformity with the 'isoprene rule'.

Final chemical proof of the position of the side-chain came from the elegant experiments of VOSER, MIJOVIĆ, HEUSSER, JEGER and RUZICKA (*168*) (*Chart 3*). The starting material was the diketo-acetate (XXXV) prepared (*166*) by side-chain degradation of a C_{27}-ester (cf. XXX) from which the reactive 7-keto-group had been removed by RANEY-nickel reduction of the ethylene-dithioketal. Oxidation of (XXXV) with perbenzoic acid in the presence of sulphuric acid gave the 17-acetoxy-compound (XXXVI) which with selenium dioxide produced the hydroxy-α-diketone (XXXVII), as had already been demonstrated in model experiments (*162*). Fission of the α-diketo-system with alkaline hydrogen peroxide and of the resulting α-hydroxy-acid with lead tetraacetate gave, after hydrolysis, the dihydroxy-keto-acid (XXXVIII). Chromic acid oxidation of the latter gave the triketone (XXXIX) (*Chart 3*), elimination

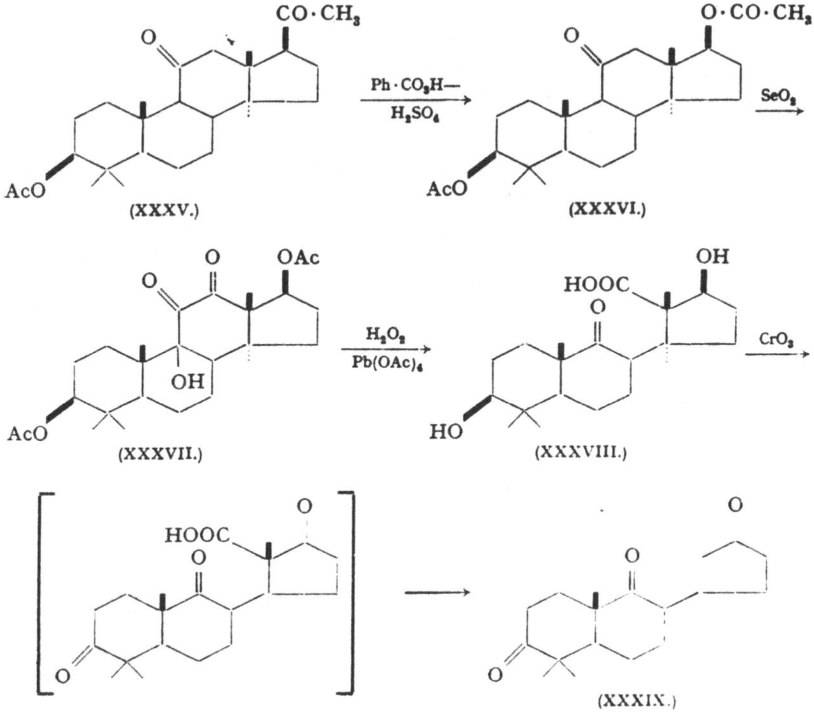

CO·CH₃ O AcO (XXXV.) $\xrightarrow[H_2SO_4]{Ph\cdot CO_3H—}$ O·CO·CH₃ O AcO (XXXVI.) $\xrightarrow{SeO_2}$

O O OAc OH AcO (XXXVII.) $\xrightarrow[Pb(OAc)_4]{H_2O_2}$ OH HOOC O HO (XXXVIII.) $\xrightarrow{CrO_3}$

[O HOOC O O] ⟶ O O O (XXXIX.)

Chart 3.

of carbon dioxide having occurred spontaneously from the intermediate β-keto-acid. It had already been shown that the related diketo-acid

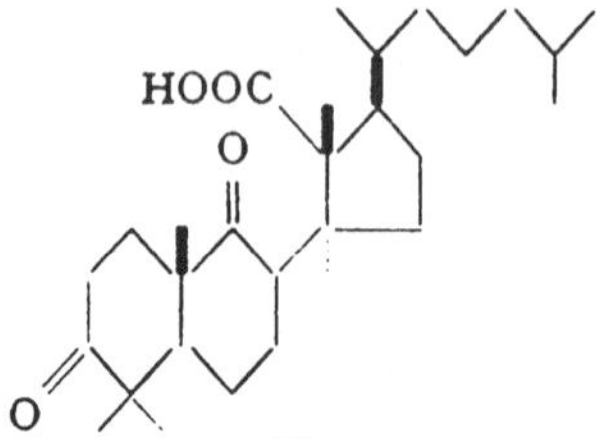

(XL.)

(XL) did not decarboxylate on heating with acetic acid (*162*), hence the keto-group in ring *D* must provide the necessary activation and be at $C_{(17)}$.

Shortly prior to the appearance of this convincing chemical proof Curtis, Fridrichsons and Mathieson (*52*) published a preliminary account of an X-ray diffraction analysis of the iodoacetate of lanostenol (*72*). This revealed the complete structure including the attachment of the side-chain at $C_{(17)}$, even though the chemical evidence available up to that time was ambiguous.

Stereochemistry of Lanosterol.

The X-ray studies referred to above also furnished full stereochemical information (cf. XLI), revealing the complete identity with the stereochemistry of the steroids (*72*). At the same time chemical evidence was presented by both Barnes, Barton, Fawcett and Thomas (*11*) and Voser, Mijović, Heusser, Jeger and Ruzicka (*168*) leading to identical conclusions.

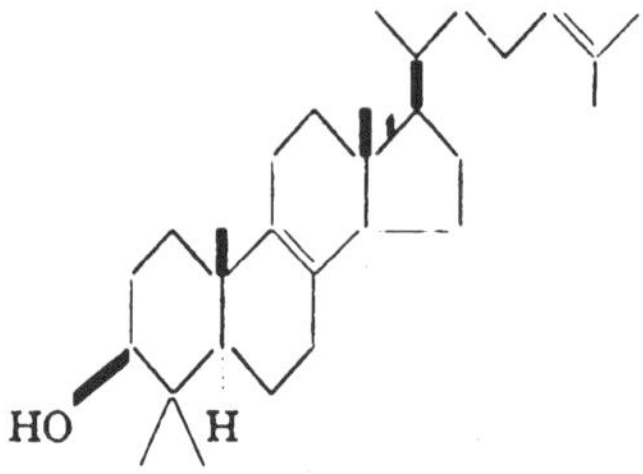

(XLI.) Lanosterol.

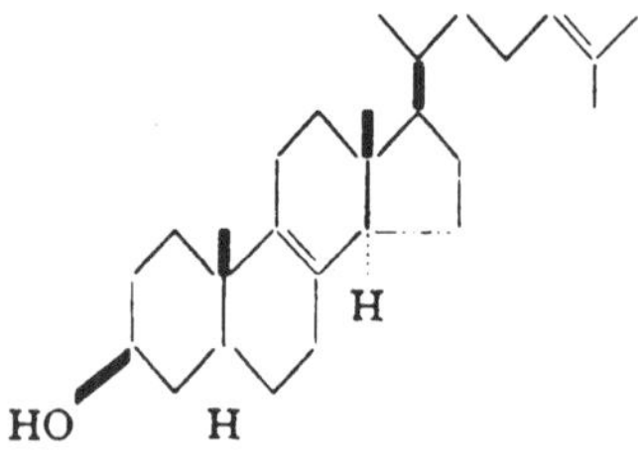

(XLI a.) Zymosterol.

Rings A and B. Molecular rotation evidence (*118*) strongly suggested that rings *A* and *B* are *trans*-fused, as in all the pentacyclic triterpenes, and that the stereochemistry corresponds to that of the steroids, and not to the enantiomeric form, *i. e.* (XLII). (It is now established that the conventional steroid formula represents the absolute configuration.) Further confirmation comes from the molecular rotation data for the

A-nor-ketones (XLIII) and probably from the identical behaviour of lanostenol and pentacyclic triterpenes in the phosphorus pentachloride rearrangement-dehydration reaction (p. 52).

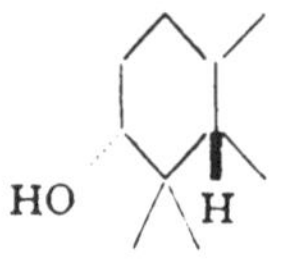

HO H (XLII.) O H (XLIII.)

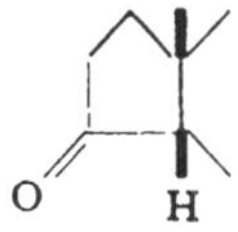

Subsequently, complete chemical proof of the identity of rings *A* and *B* of lanosterol and of the cyclic diterpenes and pentacyclic triterpenes was forthcoming (*124*). The triketone (XLIV) derived from lanostene was oxidised to the dicarboxylic acid (XLV) and, after reduction, the dimethyl ester was pyrolysed. The volatile ketonic fraction (GIRARD separation) on CLEMMENSEN reduction gave the acid (XLVI) also derived

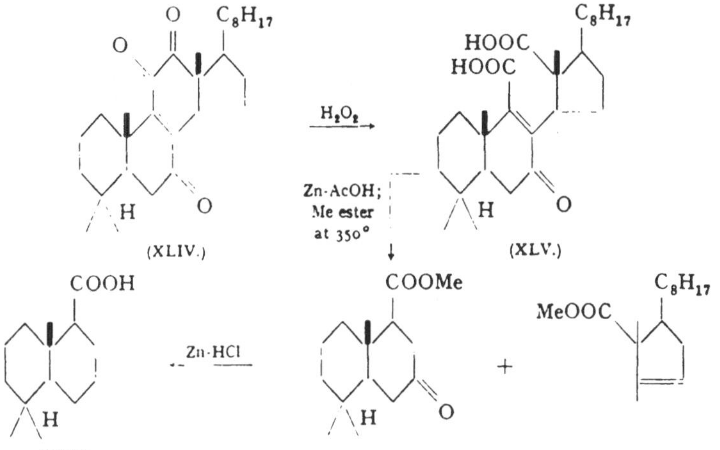

O O C_8H_{17} H O (XLIV.) H_2O_2 → HOOC HOOC C_8H_{17} H O (XLV.) Zn-AcOH; Me ester at 350° ↓ COOMe H O + MeOOC C_8H_{17} ← Zn-HCl COOH H (XLVI.)

from the diterpene, manool, by a procedure not involving the possibility of stereochemical change in rings *A* and *B*. Manool had previously been correlated with the bicyclic diterpene abietic acid, and the triterpenes ambrein and oleanolic acid.

The β(equatorial)-configuration of the 3-hydroxyl group follows from its stability towards epimerisation (*171*) and its regeneration from the corresponding ketones by sodium and alcohol reduction (*131*, *132*).

Rings C and D. Molecular rotation differences between the 17-keto-triacetate (XLVII) and the triacetate (XLVIII), the 17-carbonyl group having been removed by WOLFF-KISHNER reduction, followed by reacetylation, favour the stereochemistry indicated rather than that in which the 14-methyl group has the β-configuration (*i. e.*, a *C–D cis*-fusion) (*11*). No direct chemical evidence is available to indicate the inter-relationship

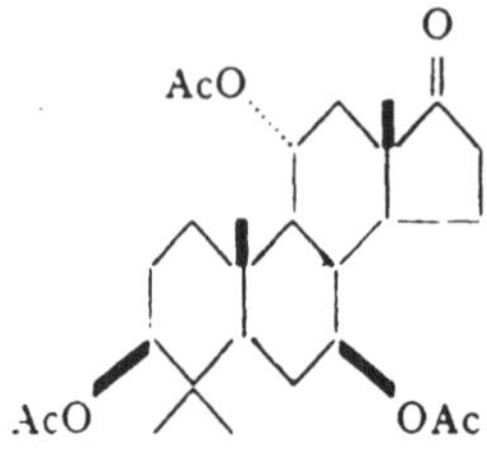

(XLVII.) 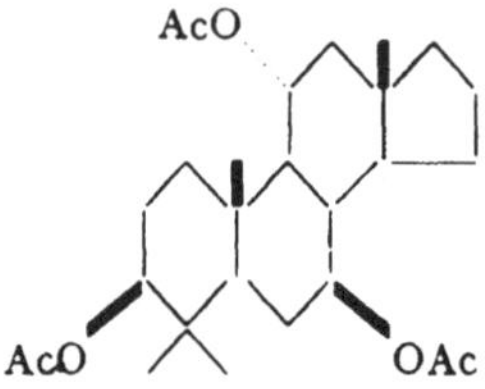 (XLVIII.)

of the $C_{(10)}$ and $C_{(13)}$ methyl groups although Barton (*14a*) suggested that the steroid-like properties of 11-oxygenated lanostane derivatives were best explained if both methyl groups were on the same side of the molecule. The point is of course clearly illustrated by the X-ray work.

Side-chain Configuration. Molecular rotation differences between steroids with the 17β-*iso*octyl side-chain and the 17β-acetyl side-chain are in good agreement with those observed in the lanosterol group and quite different from those expected for a 17α-side-chain (*11*). Again, steroids with a 17β-acetyl side-chain show large positive rotations as compared with their isomers with a 17β-acetoxy side-chain. This is also the case with the lanosterol derivatives, whereas steroid 17α-compounds exhibit large negative differences (*168*).

The configuration at the remaining centre of asymmetry ($C_{(20)}$) can be deduced to be identical with that in the steroid series by comparison of the molecular rotation differences between compounds containing —CHMe · CH_2 · CH_2 · COOMe and —CHMe · CH_2 · CH=CPh_2 side-chains (*11*). Looking from $C_{(17)}$ towards $C_{(20)}$ the configuration can be represented as (IL) or it can be projected as in (L).

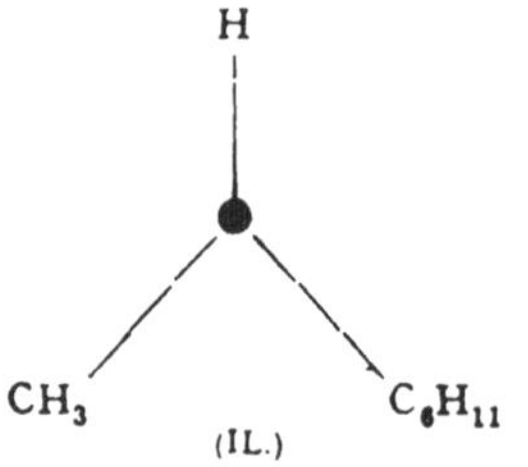

(IL.) 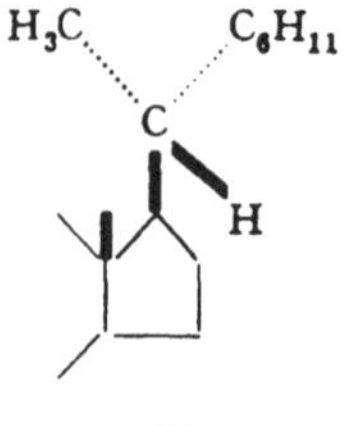 (L.)

Stereochemistry at $C_{(8)}$ and $C_{(9)}$ in Lanostane Derivatives. The inert double bond of lanosterol can only be reduced when carbonyl groups are present in the 7- and/or 11-position (*61, 62*), and lanostane and its derivatives are obtained by deoxygenation by Wolff-Kishner, Clemmensen and related procedures (*169*). Both catalytic hydrogenation and zinc-acid reduction (cf. *8*) probably give primarily the $8\alpha : 9\alpha$-dihydro-compound (LII) which isomerises under the experimental conditions to the more stable $8\beta : 9\alpha$-structure (LIV) possessing the *trans-anti-trans*-arrangement of rings *A*, *B* and *C*; the saturated diketone (LIV) is stable to fairly vigorous alkaline treatment (*11*). It is also relevant that selenium dioxide oxidation of (LIV) gives almost entirely the isomeric ene-dione (LIII) (*63*, *64*) and only a trace of the parent (LI). BARTON (*8*) has made the

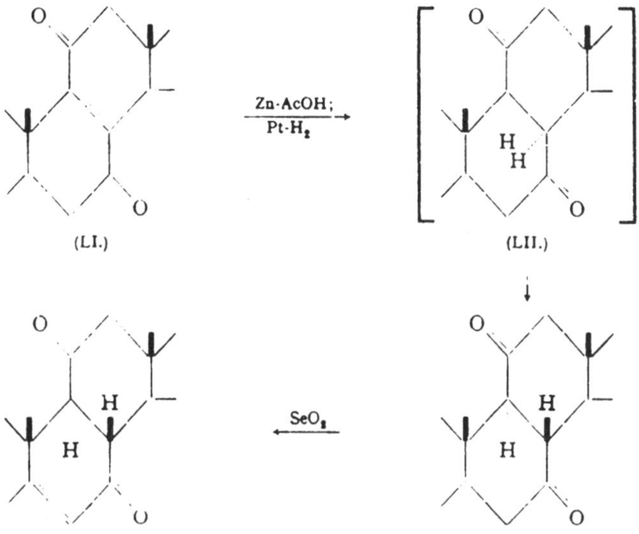

Zn-AcOH; Pt-H_2

(LI.) (LII.)

SeO_2

(LIII.) (LIV.)

generalisation that ready conversion of saturated 1 : 4-diketones into ene-1 : 4-diones requires the eliminated hydrogen atoms to bear a *cis*-relationship to one another.

The fact that the 14α-methyl-11-keto-progesterone (LV), prepared from lanosterol by removal of the ring *A* methyl groups and other obvious steps (*164*), exhibits activity in the highly structurally-specific CORNER-

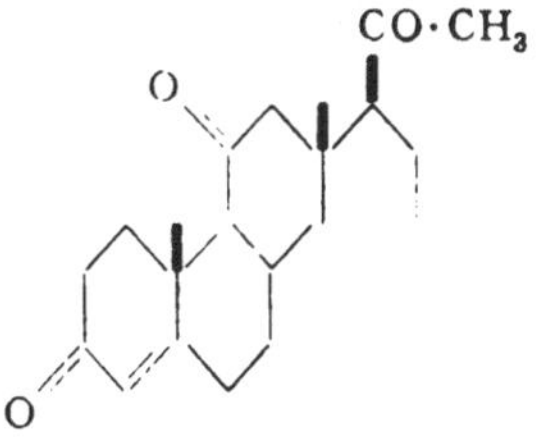

$CO \cdot CH_3$

(LV.)

ALLEN test equivalent to that of 11-ketoprogesterone, furnished still more proof of the stereochemical identity of cholestanol and lanostanol.

Lanost-7-enol (iso*Lanostenol*). Treatment of lanostenyl acetate (LVI) with hydrogen chloride in chloroform brings about double bond migration producing a mixture of starting material and an isomeric substance, now

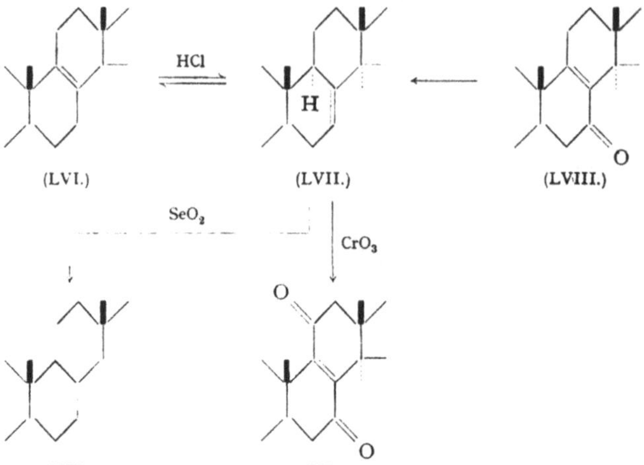

HCl

(LVI.) (LVII.) (LVIII.)

SeO_2 CrO_3

(LIX.) (LX.)

known to be lanost-7-enyl acetate (LVII) (*132*, cf. *41a*, *170*). The ethylenic linkage in (LVII) resists hydrogenation, but dehydrogenation with selenium dioxide or with N-bromosuccinimide furnished lanosta-7 : 9-dienyl acetate (dihydroagnosteryl acetate) (LIX), while chromic acid oxidation gave the well-known ene-dione (LX), both of which are produced from lanostenyl acetate (LVI) by similar procedures (*41a*, *62*). Infrared examination indicated that the ethylenic linkage was trisubstituted and its formation from the $\alpha\beta$-unsaturated 7-ketone (LVIII) on Wolff-Kishner reduction, the elimination of the carbonyl group being accompanied by the well-established double bond shift, provided final proof of its structure (*14a*). As is evident later this is of some significance in connection with similar studies with euphol and butyrospermol.

Lanosteryl acetate with hydrogen chloride yields a mono-addition product and this with alkali gives '*iso*lanosterol' (*64a*). Although its structure has not been proved, analogous studies suggest that it is probably to be represented by (LXI), *i. e.* lanosta-7 : 25-dien-3β-ol (*57*).

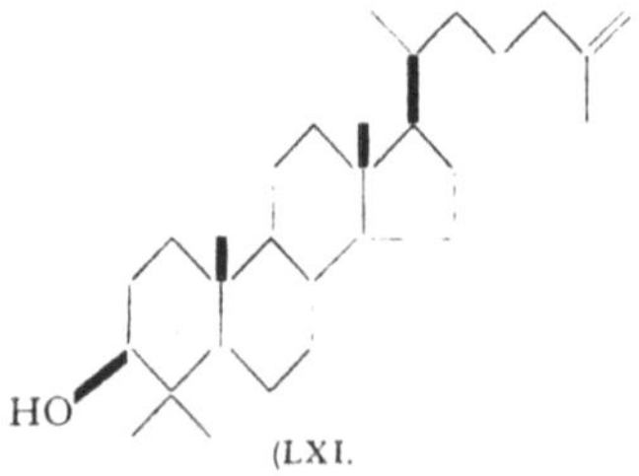

HO

(LXI.)

Synthesis of Lanostenol.

Barely a year after the complete structure of lanosterol had been elucidated a preliminary joint communication from WOODWARD and

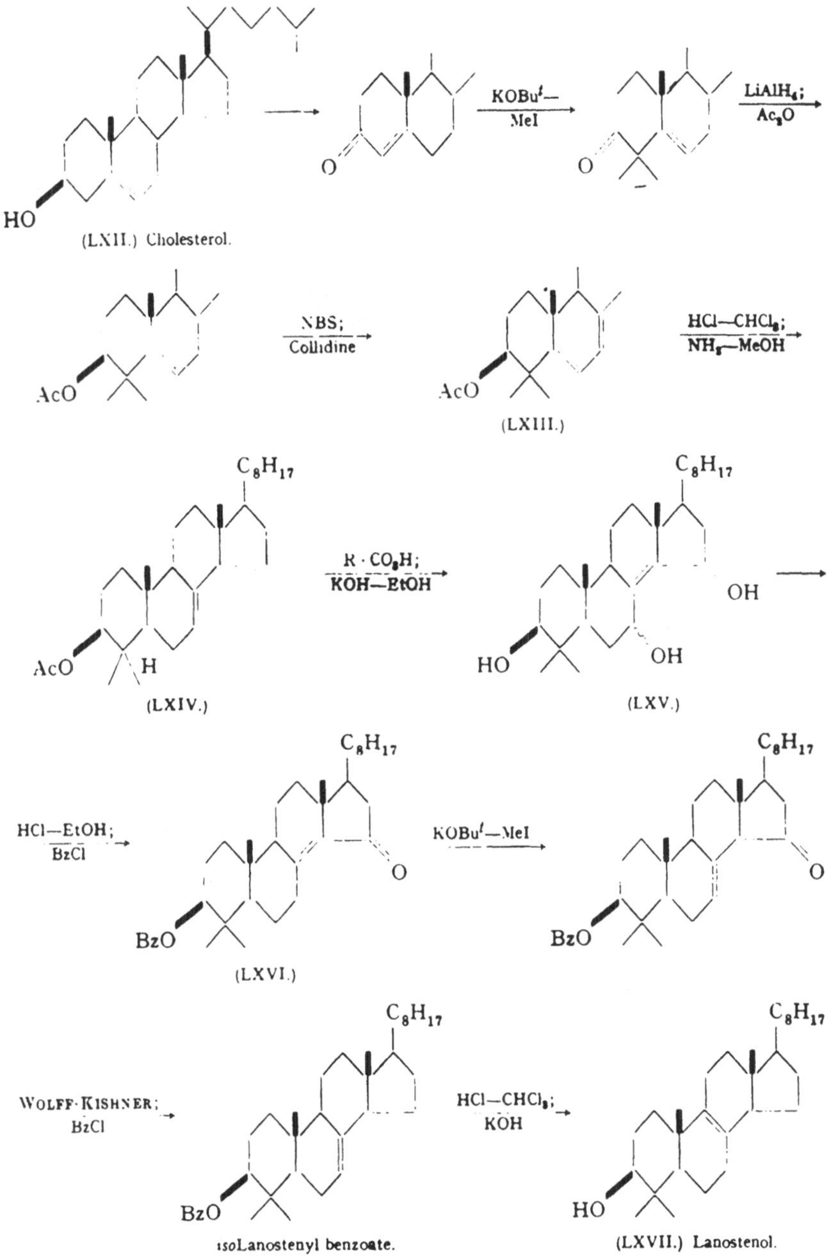
HO
(LXII.) Cholesterol.
O
KOBu^t—MeI
O
LiAlH$_4$; Ac$_2$O
AcO
NBS; Collidine
AcO
(LXIII.)
HCl—CHCl$_3$; NH$_3$—MeOH
C_8H_{17}
AcO
H
(LXIV.)
R · CO$_3$H; KOH—EtOH
C_8H_{17}
OH
HO
OH
(LXV.)
HCl—EtOH; BzCl
C_8H_{17}
O
BzO
(LXVI.)
KOBu^t—MeI
C_8H_{17}
O
BzO
WOLFF-KISHNER; BzCl
C_8H_{17}
BzO
isoLanostenyl benzoate.
HCl—CHCl$_3$; KOH
C_8H_{17}
HO
(LXVII.) Lanostenol.

Chart 4. Conversion of Cholesterol into Lanostenol.

BARTON and their respective collaborators (*173*) announced the conversion of cholesterol (LXII) into lanostenol (LXVII) (*Chart 4*). Cholesterol having already been prepared synthetically, this work constitutes the first total synthesis of a tetracyclic triterpene, since lanostenol accompanies lanosterol in wool-fat.

Although there are alternative sites at which methylation might be expected to occur, dimethylation of cholest-4-enone (or the 5-isomer) proceeded remarkably readily in the 4-position. Introduction of active centres into rings *C* and *D* was ingeniously achieved by double bond migrations, commencing with the production of a ring *B* diene (LXIII). This was then isomerised to the 7 : 14-diene (LXIV) which was oxygenated and converted into the required 15-ketone (LXVI) *via* the diol (LXV). Introduction of the angular methyl group required the use of a very large excess of strong base but then proceeded in good yield. WOLFF-KISHNER reduction under special conditions gave lanost-7-enol (*iso*lanostenol), and lanostenol (LXVII) itself was then obtained from the mixture produced by the action of hydrogen chloride on the benzoate.

In the final product the only centres where stereochemical dubiety exists are $C_{(5)}$ and $C_{(14)}$, elsewhere the situation is identical with that in cholesterol or else asymmetry has been eliminated. The reactions determining the configurations at these two centres would be expected to be largely stereo-specific. Isomerisation of the 5 : 7-diene (LXIII) into the 7 : 14-diene (LXIV) should lead to the more stable $C_{(5)}$ isomer, *i. e.*, *trans*-fusion of rings *A* and *B*; there are ample analogies in steroid chemistry. Introduction of the $C_{(14)}$ methyl group must occur from the rear in view of the considerable steric interference to the frontwise approach of a reagent.

2. Eburicoic Acid.

(3β-Hydroxyeburico-8 : 24(28)-dien-21-oic acid) $C_{31}H_{50}O_3$ (LXVIII).

In 1940 KARIYONE and KURONO isolated a monohydroxy-carboxylic acid from the wood-rotting fungus *Fomes officinalis* FR. (*117*). It was named eburicoic acid and analytical data suggested that it had the formula $C_{30}H_{48}O_3$.

About ten years later, during a study of the metabolic products of the higher fungi, GASCOIGNE, HOLKER, RALPH and ROBERTSON (*75*) examined decayed *Eucalyptus regnans* wood and isolated acetyleburicoic acid. The organism responsible for the decay was identified as *Polyporus anthracophilus* COOKE, from which LAHEY and STRASSER (*125*) independently isolated acetyleburicoic acid. Subsequently this fungus was grown on a synthetic medium when eburicoic acid and not the acetate was produced. Eburicoic acid has also been isolated from *Polyporus eucalyptorum* FR. (*76*), *Polyporus sulfureus* (BULL.) FR. (*76*), *Lentinus*

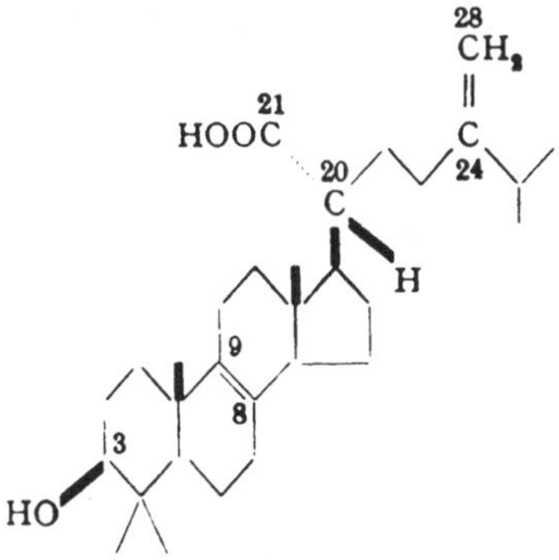

(LXVIII.) Eburicoic acid.

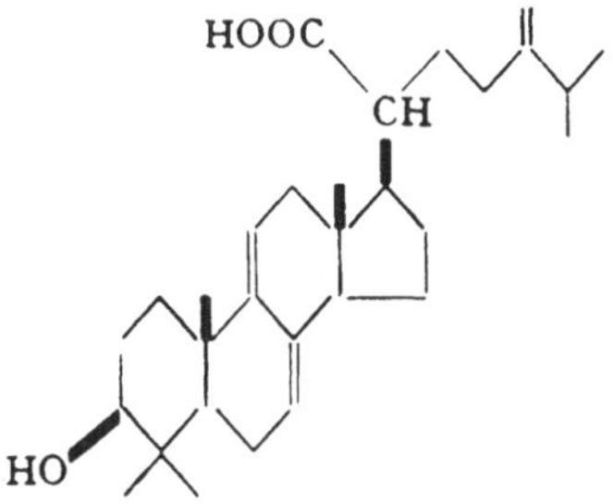

(LXIX.) Dehydroeburicoic acid.

dactyloides CLELAND (*76*), *Poria cocos* (SCHW.) WOLF (*48*), and *Polyporus hispidus* (BULL.) FR. (*48*). From the latter three fungi, as well as from *Fomes officinalis* FR., dehydroeburicoic acid (LXIX) was also isolated (*76*, *77*).

LAHEY and STRASSER (*125*) believed that eburicoic acid had the formula $C_{30}H_{48}O_3$. They showed that its hydroxyl group is secondary and that it has two double bonds, one of which can be hydrogenated. These results indicate that the acid is tetracyclic.

The more reactive double bond is present in a vinylidene group ($\rangle C = CH_2$), ozonolysis affording a satisfactory yield of formaldehyde (*125*). The less reactive double bond behaves analogously to the nuclear

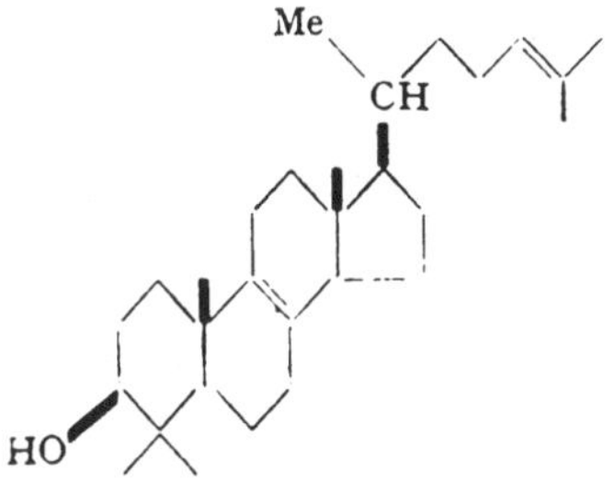

(LXX.) Lanosterol.

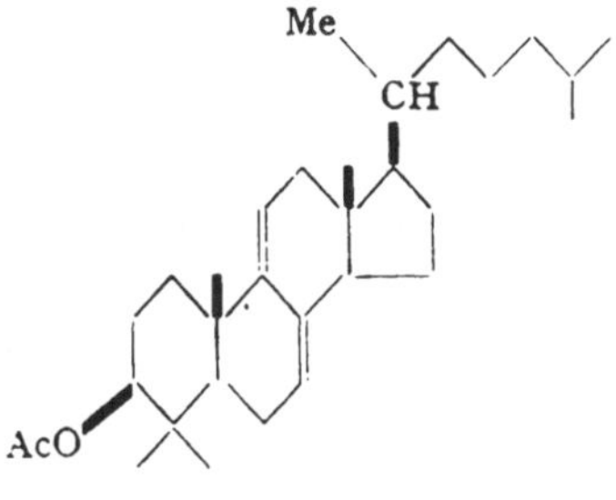

(LXXI.) Dihydroagnosteryl acetate.

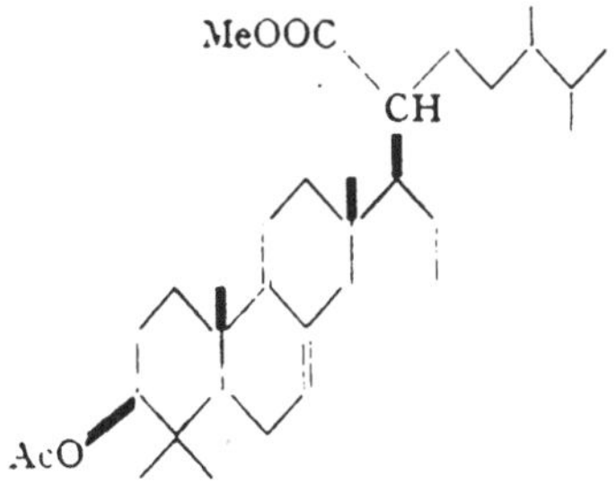

(LXXII.) Methyl acetyldehydro-dihydroeburicoate.

double bond of lanosterol (LXX) in a series of chromic acid and selenium dioxide oxidations (*77*, *110*). Other evidence suggesting a relationship between eburicoic acid and lanosterol is provided by a study of the molecular rotation changes occurring in comparable reactions with eburicoic acid and lanosterol derivatives (*77*). A comparison (*58*) of the ultraviolet absorption spectra of the appropriate agnosterol (*e. g.* LXXI) and dehydroeburicoic-acid (*e. g.* LXXII) derivatives, all of which exhibit characteristic maxima at 2360, 2430, and 2510 Å., supports this conclusion.

In an attempt to convert eburicoic acid into a known triterpene alcohol LAHEY and STRASSER (*125*) transformed the carboxyl group into a methyl group. Methyl acetyleburicoate (LXXIII) was reduced with

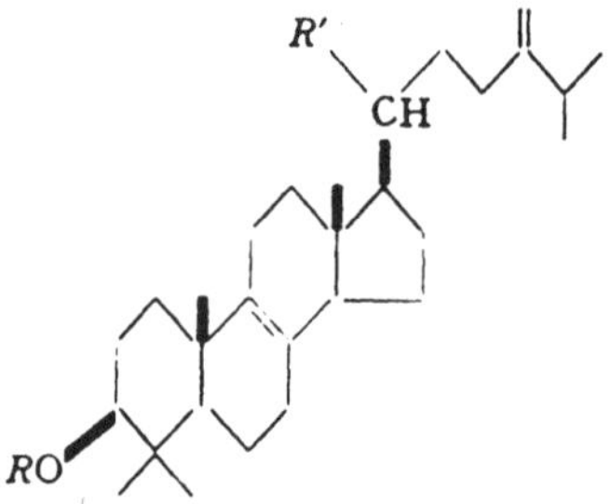

R' CH A RO

(LXXVII.)

(LXXIII.) Methyl acetyleburicoate. $R = \text{Ac}$; $R' = \text{COOMe}$.
(LXXIV.) $R = \text{H}$; $R' = \text{CH}_2\text{OH}$.
(LXXV.) $R = \text{Ac}$; $R' = \text{CHO}$.
(LXXVI.) Eburicol. $R = \text{H}$; $R' = \text{CH}_3$.

lithium aluminium hydride to the diol (LXXIV) which was then acetylated. The diacetate was hydrolysed partially to a monoacetate, oxidation of which gave the aldehyde (LXXV). WOLFF-KISHNER reduction then gave eburicol (LXXVI), not identical with any known compound.

Eburicoic acid has a typical triterpene-type ring *A* with a 3β-hydroxyl group. Treatment of methyl dihydroeburicoate with phosphorus pentachloride gave the characteristic retro-pinacolinic dehydration product (partial structure, LXXVII) ozonolysis of which produced acetone (*76*).

The first evidence revealing the nature of the side-chain of eburicoic acid was obtained when it was shown that ozonolysis of the vinylidene group gave a ketone which did not react with potassium hypoiodite at 80° (*111*, *125*). Treatment of methyl acetyleburicoate with hydrogen chloride in acetic acid gave an adduct which was converted by the action of acetic anhydride into two isomeric compounds, (LXXVIII) and (LXXIX), differing only in the position of a reactive double bond (*111*). Hydrogenation of both compounds gave the same dihydro-derivative, methyl 3β-acetoxyeburic-7-en-21-oate (LXXX) (*111*), different from the acetate of methyl dihydroeburicoate. Under the influence of acid the nuclear double bond moves into the 7:8-position, as when dihydrolanosteryl acetate is similarly treated. Ozonolysis of (LXXVIII) gave

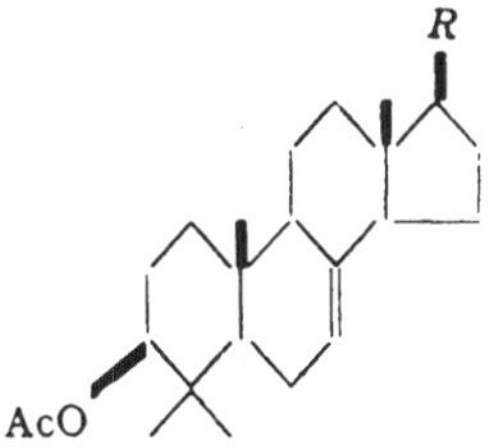

(LXXVIII.) $R = CH(COOMe)\cdot CH_2\cdot CH{=}C(Me)\cdot CHMe_2$.

(LXXIX.) $R = CH(COOMe)\cdot CH_2\cdot CH_2\cdot C({=}CH_2)\cdot CHMe_2$.

(LXXX.) $R = CH(COOMe)\cdot CH_2\cdot CH_2\cdot CH(Me)\cdot CHMe_2$.

methyl *iso*propyl ketone and an aldehydic product, and the side-chain of eburicoic acid must contain the grouping $Me_2CH\cdot C(:CH_2)\cdot CH_2$—.

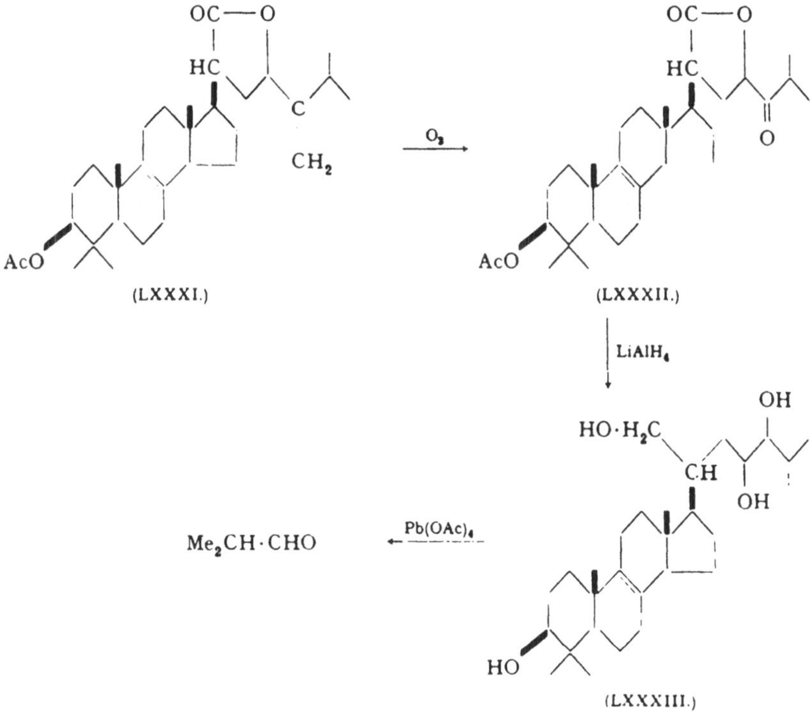

(LXXXI.) $\xrightarrow{O_3}$ (LXXXII.) $\xrightarrow{LiAlH_4}$ (LXXXIII.) $\xrightarrow{Pb(OAc)_4}$ $Me_2CH\cdot CHO$

Oxidation of acetyleburicoic acid with an equivalent of selenium dioxide gave a lactone (LXXXI), the infrared spectrum of which indicated that it was a γ-lactone and that it still contained a vinylidene group. Ozonolysis of (LXXXI) gave formaldehyde and a keto-lactone (LXXXII), reduced with lithium aluminium hydride to a tetrol (LXXXIII). Fission of the tetrol with lead tetraacetate gave *iso*butyraldehyde. These results enable the formulation of the side-chain of eburicoic acid to be extended to $\rangle C(COOH) \cdot CH_2 \cdot CH_2 \cdot C(: CH_2) \cdot CHMe_2$ with reasonable certainty (*111*).

The carbon skeleton of this side-chain bears the same relationship to that of lanostane as the carbon skeleton of the ergostane side-chain does to that of cholestane. Assuming a lanostane ring system in eburicoic acid and an ergostane type side-chain it was possible to visualise structure (LXVIII, p. 69) for the acid (*111*). Evidence in favour of the C_{31} formula was provided by a molecular weight determination using the X-ray method (77).

Final proof of structure (LXVIII) for eburicoic acid was provided by the conversion of the acid into lanost-8-ene (LXXXVIII) (75, *111*). Eburicoic acid was oxidised by the OPPENAUER method and the resulting ketone (LXXXIV) reduced to the deoxy-acid (LXXXV). Ozonolysis of the methyl ester of this acid gave the nor-ketone (XXXVI) and WOLFF-KISHNER reduction followed by esterification of the product gave methyl lanost-8-en-21-oate (LXXXVII). The methoxy-carbonyl group of this ester was converted to a methyl group by reduction with lithium aluminium hydride, oxidation, and then reduction of the aldehyde to (LXXXVIII) by the WOLFF-KISHNER method. This sequence of reactions

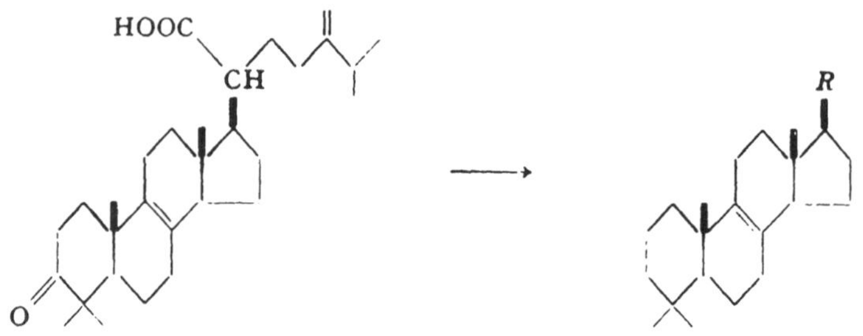

HOOC CH O R

(LXXXIV.) Eburiconic acid.

(LXVIII.) Eburicoic acid.

(LXXXV.) $R = CH(COOH) \cdot (CH_2)_2 \cdot C(=CH_2) \cdot CHMe_2$.

(LXXXVI.) $R = CH(COOH) \cdot (CH_2)_2 \cdot C(=O) \cdot CHMe_2$.

(LXXXVII.) $R = CH(COOMe) \cdot (CH_2)_3 \cdot CHMe_2$.

(LXXXVIII.) $R = CH(Me) \cdot (CH_2)_3 \cdot CHMe_2$; Lanost-8-ene.

together with the degradative evidence already discussed proves the structure of eburicoic acid as (LXVIII) apart from the configuration at $C_{(20)}$. The carboxyl group must, however, be in the lanostane configuration since HOLKER et al. (*111*) have shown that no inversion of the acid or its ester occurs at $C_{(20)}$ with alkali, and BOWERS et al. (*37*) have shown that the lanostane configuration is the more stable under these conditions.

The parent hydrocarbon of eburicoic acid is called eburicane (LXXXIX). So far, however, it has only been prepared from *cyclo*-laudenol (p. 101).

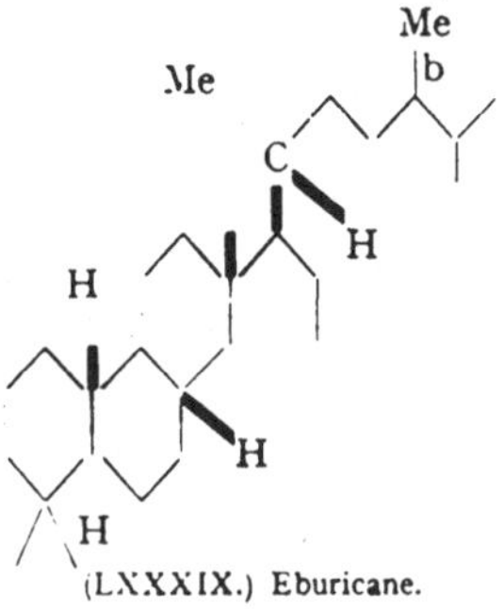

(LXXXIX.) Eburicane.

3. Polyporenic Acid A.

(3α : 12α-Dihydroxyeburico-8 : 24(28)-dien-26-oic acid) $C_{31}H_{50}O_4$ (XC, p. 74).

This acid is a constituent of *Polyporus betulinus* FR., a large white shelf-fungus parasitic on the birch tree (*Betula alba*). It was isolated in 1939 by CROSS, ELIOT, HEILBRON and JONES (*49*) who characterised it as a tetracyclic dihydroxy-carboxylic acid with two double bonds. Its molecular weight, as determined by titration, was ca. 485 corresponding to the formula $C_{31}H_{50}O_4$. Of the two double bonds only one could be hydrogenated at ordinary temperature and pressure, and it was concluded from ozonolysis experiments that the reactive double bond was present as a vinylidene group ($CH_2{=}C\diagup$) (*50*).

Unknown to CROSS et al. a chemical study of the constituents of *Polyporus betulinus* FR. had been carried out just previously by FRÈREJACQUE (*71*) who isolated a new compound which he designated ungulinic acid. Subsequently LOCQUIN, LOCQUIN and PRÉVOT (*127*) repeated the isolation of polyporenic acid A and confirmed its identity with ungulinic acid.

After a lapse of several years the investigation of polyporenic acid A was taken up again by CURTIS, HEILBRON, JONES and WOODS (*53*). The equivalent weight was redetermined and again corresponded to the formula $C_{31}H_{50}O_4$. Other analytical data, however, did not distinguish this formula from the formula $C_{30}H_{48}O_4$. A final decision in favour of the C_{31} formula only became possible with the complete elucidation of the structure. The two hydroxyl groups were further characterised and shown to be secondary; their different reactivities were reflected in the properties of the keto-groups formed on oxidation. One keto-group reacted with the usual carbonyl reagents such as hydroxylamine whereas the other was inert even under vigorous conditions. The absence of high intensity absorption at 2200–2600 Å. in any of the carbonyl derivatives of polyporenic acid A precluded the presence of $\alpha\beta$-unsaturation and showed that the hydroxyl groups are separated from the double bonds by at least one saturated carbon atom. The absence of selective absorption in the parent acid showed that the double bonds were not conjugated either with each other or with the carboxyl group.

Polyporenic acid A when melted loses one mole of carbon dioxide to give a decarboxy-compound (XCI) in good yield (*53*). The infrared spectrum of the decarboxy-compound shows a band at 815 cm.$^{-1}$ characteristic of a trisubstituted double bond, but nothing indicative of a

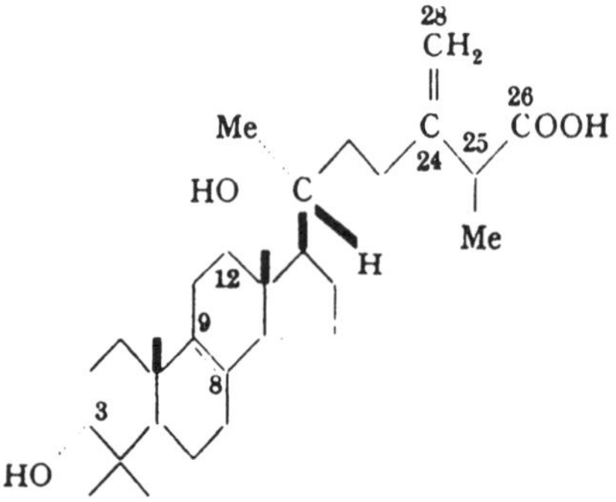

(XC.) Polyporenic acid A.

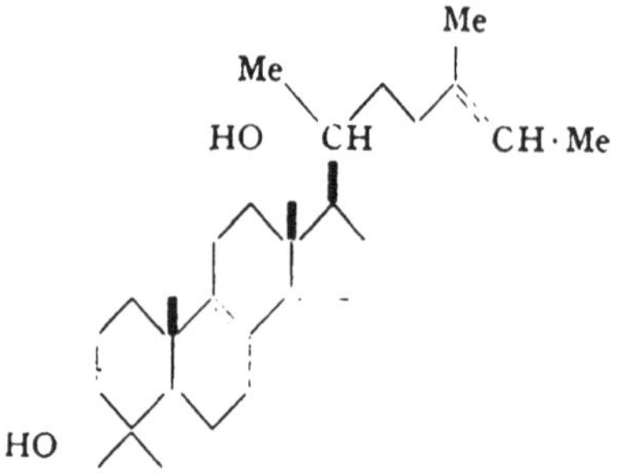

(XCI.)

vinylidene group. Hydrogenation of polyporenic acid A and its derivatives gives two series of dihydro-derivatives (stereoisomers). The mixed dihydro-acids do not decarboxylate. These results indicate that polyporenic acid A is a $\beta\gamma$-unsaturated acid, *i. e.*, the reactive double bond is in the $\beta\gamma$-position to the carboxyl group. Such acids readily decarboxylate when heated, the elimination of carbon dioxide being accompanied by a shift of the double bond to the former $\alpha\beta$-position (cf. *3*, *14*).

Evidence concerning the environment of the unreactive double bond was obtained by HALSALL, JONES and LEMIN (*98*) by oxidation of derivatives of the acid, when products with characteristic chromophores, similar to those of the oxidation products obtained from lanostenyl acetate, were formed (cf. p. 54). Oxidation of methyl polyporenate A diacetate (XCII) with chromic acid in acetic acid introduced two keto-groups and gave a yellow diketone (XCIII), the ultraviolet spectrum of which had a maximum at 2730 Å. characteristic of the system (XCIV). The analytical data concerning the diketone indicated that the reactive double bond was still present and this was confirmed by the presence of a band at 900 cm.$^{-1}$ in its infrared spectrum. When the diketone (XCV), formed by the oxi-

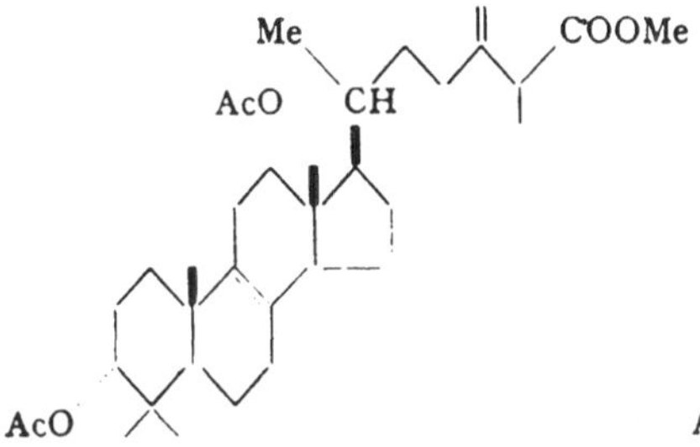

(XCII.) Methyl polyporenate A diacetate.

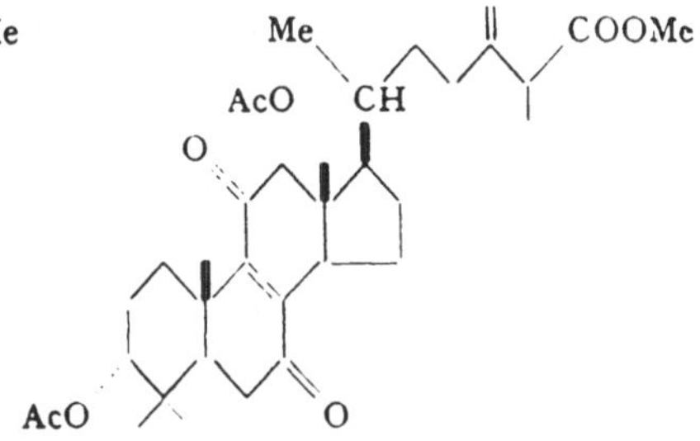

(XCIII.)

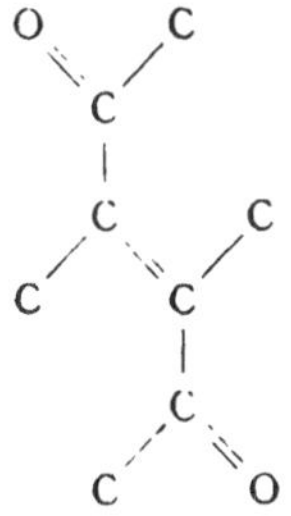

(XCIV.)

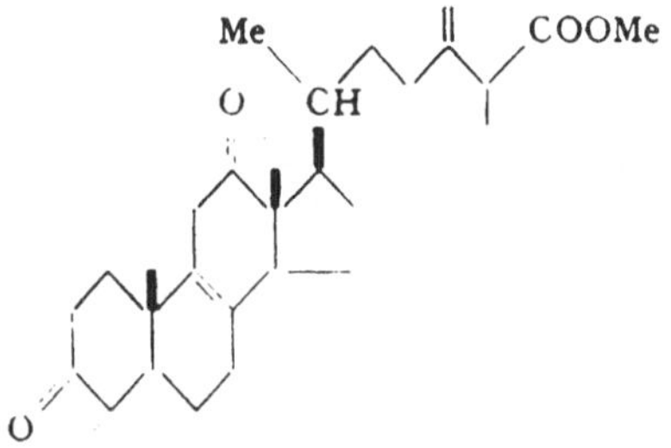

(XCV.)

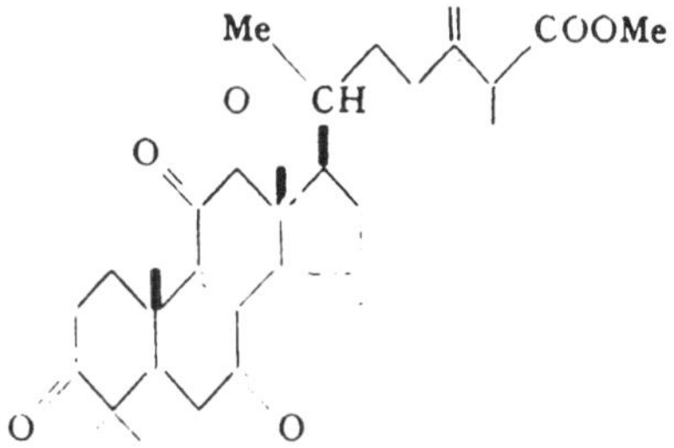

(XCVI.)

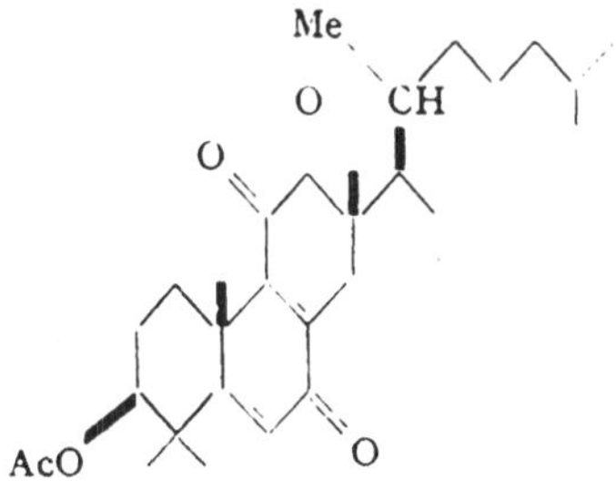

(XCVII.)

dation of the hydroxyl groups of methyl polyporenate A, was oxidised further, a tetraketone (XCVI) with an ultraviolet absorption maximum at 2830 Å. was obtained. Absorption at about this wavelength has been observed with 7 : 11 : 12-triketolanosta-5 : 8-dien-3β-yl acetate (XCVII) (*62*). It appeared probable, therefore, that one of the hydroxyl groups of polyporenic acid A, and hence the keto-group resulting from it, was in the α-position to one of the methylene groups undergoing oxidation in the formation of the tetraketone, *i. e.*, in the β-position to the unreactive double bond. The presence of the α-diketone grouping in the tetraketone (XCVI) was confirmed by oxidation with alkaline hydrogen peroxide whereupon the monomethyl ester of a tricarboxylic acid was formed without loss of carbon atoms, thus proving the cyclic location of the α-diketone group. These results enabled partial structure (XCIV) to be extended to (XCVIII).

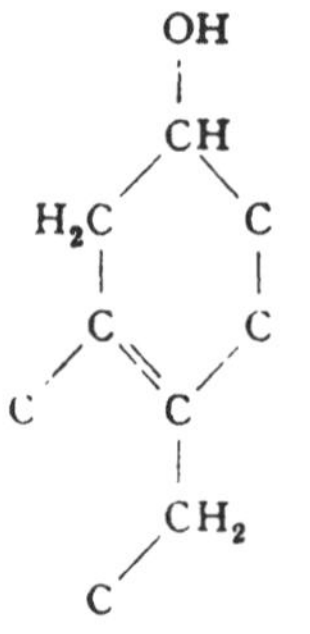

(XCVIII.)

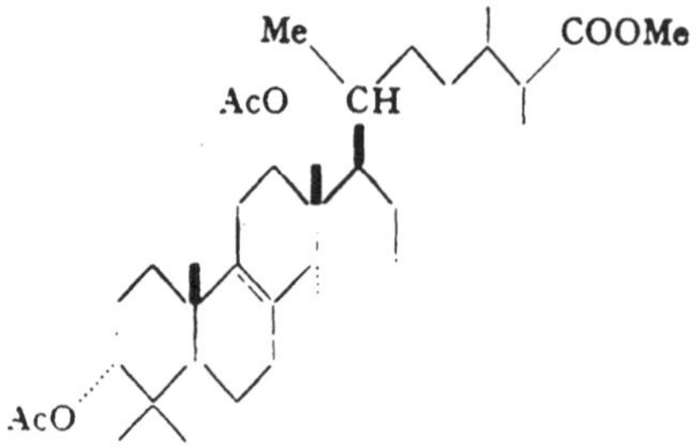

(IC.)

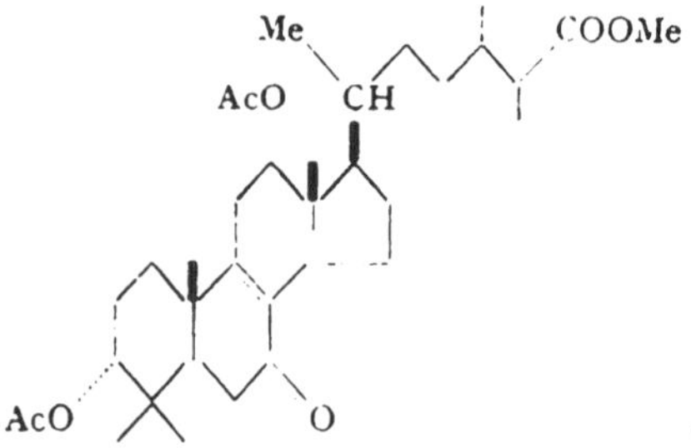

(C.)

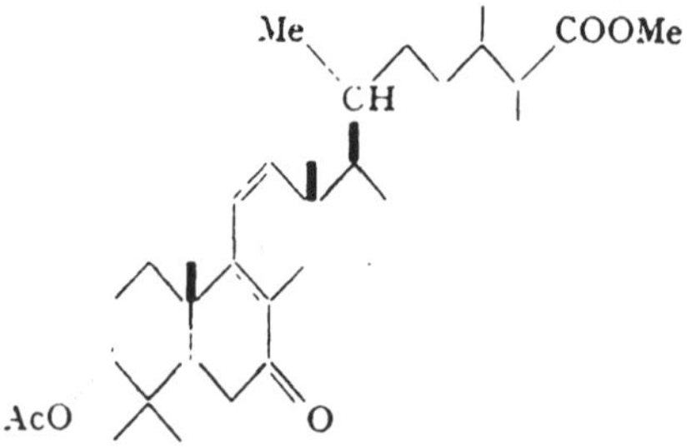

(CI.)

When one (IC) of the two dihydro-derivatives (dihydro-II series) of methyl polyporenate A diacetate was oxidised with chromic acid in acetic acid under mild conditions, one keto-group only was introduced. Treatment of the resulting ketone (C) with very mild alkali gave a conjugated dienone (CI) with the loss of the elements of acetic acid. The absorption maximum at 3180 Å. of the dienone is indicative of the conjugation of the carbonyl group with two double bonds in the *same* ring.

This result shows that the introduction of the first keto-group occurs at the α-methylene group on the opposite side of the double bond from the acetoxy group, leading to a δ-acetoxy-$\alpha\beta$-unsaturated ketone. Such a compound, a vinylogue of a β-acetoxy-ketone, would be expected readily to give a dienone on treatment with alkali.

The very close similarity of these oxidation reactions to those observed in the lanosterol series, and of the partial structure (XCVIII) to rings *B* and *C* of lanosterol (CII), together with the formation of 1 : 2 : 8-trimethylphenanthrene on selenium dehydrogenation of polyporenic acid A, led to the suggestion that the grouping (CIII) is present in the acid. The

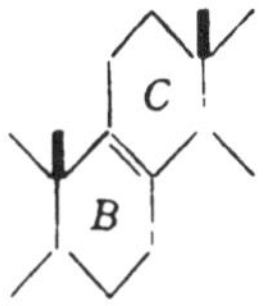

(CII.) 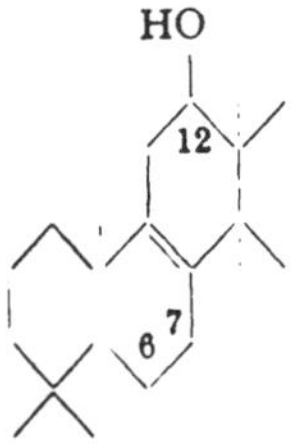 (CIII.)

hydroxyl group in the position β to the double bond is placed at $C_{(12)}$ rather than at $C_{(6)}$ because introduction of the first keto-group adjacent to the double bond should occur at $C_{(7)}$ rather than at $C_{(11)}$ by analogy with the reactions in the lanosterol series.

Halsall, Hodges and Jones (*96*) carried out a number of reductions with lithium aluminium hydride and sodium borohydride of the keto-groups formed by oxidation of the hydroxyl groups of polyporenic acid A. The unreactive keto-group at $C_{(12)}$ is reduced only by lithium aluminium hydride; the resulting hydroxyl group has the same configuration as the original $C_{(12)}$ hydroxyl group. Since reduction of hindered ketones under these conditions usually gives rise to hydroxyl groups in the axial conformation (*12*) it follows that the $C_{(12)}$ hydroxyl substituent in polyporenic acid A has the α-configuration. Acetylation of a 12α-hydroxyl group should produce an increase in molecular rotation (*119*) as compared with a decrease for a 12β-hydroxyl group. The value of ΔOAc actually varies from + 288° to + 322° depending upon the initial polyporenic acid A derivative studied (*53*).

Reduction of the more reactive keto-group with both lithium aluminium hydride and sodium borohydride gives rise to a hydroxyl group with a configuration different from that originally present. Since unhindered keto-groups give equatorial hydroxyl groups on reduction as described above the original hydroxyl group must be axial. In all the known penta- and tetra-cyclic triterpenes oxygen is found attached to $C_{(3)}$. It appeared likely therefore that the second hydroxyl group of polyporenic acid A was a 3α substituent, the α-configuration corresponding to the axial conformation. This was confirmed by oxidising methyl polyporenate A to the diketone (CIV) and reducing this with sodium

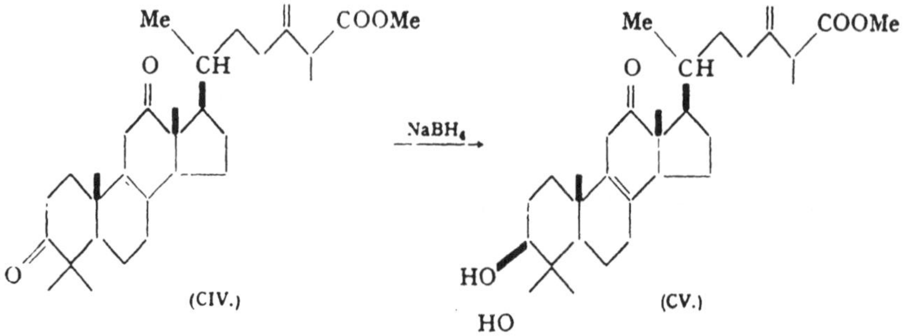

Me COOMe O CH NaBH$_4$ Me COOMe O CH

O (CIV.) HO (CV.)

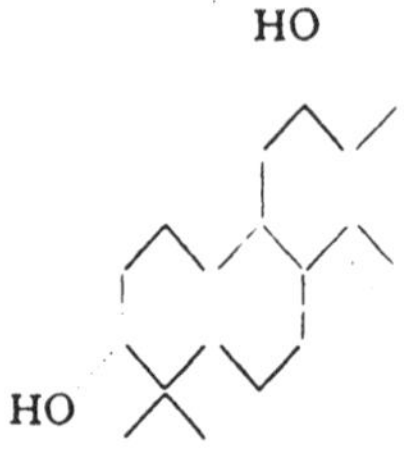

HO

HO (CVI.)

borohydride to the hydroxy-keto ester (CV), in which the hydroxyl group is epimeric to that at $C_{(3)}$ in polyporenic acid A. Treatment of (CV) with phosphorus pentachloride resulted in the retro-pinacolinic dehydration characteristic of triterpenes with a 3β-hydroxyl group. Partial structure (CIII) now becomes (CVI). Acetylation of the 3α-hydroxyl group produces the expected (*119*) negative shift in rotation [ΔOAc = about -200 (*53*)].

As already indicated polyporenic acid A is a $\beta\gamma$-unsaturated acid and is readily decarboxylated. By a series of degradations starting from

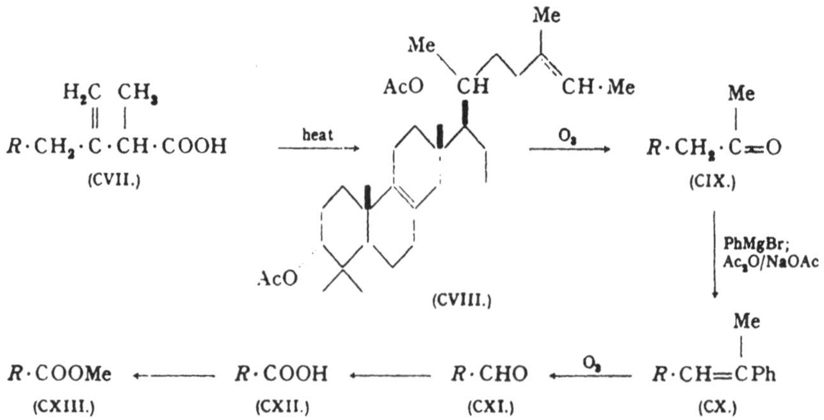

Me

Me

H_2C CH_3 AcO CH CH·Me Me

$R\cdot CH_2\cdot C\cdot CH\cdot COOH$ heat O_3 $R\cdot CH_2\cdot C{=}O$

(CVII.) (CIX.)

PhMgBr; Ac_2O/NaOAc

AcO (CVIII.)

Me

$R\cdot COOMe$ ← $R\cdot COOH$ ← $R\cdot CHO$ ←O_3— $R\cdot CH{=}CPh$

(CXIII.) (CXII.) (CXI.) (CX.)

the diacetate of the decarboxy-compound (CVIII) the grouping (CVII) was shown to be present. The diacetate on ozonolysis gave acetaldehyde and the methyl ketone (CIX). The latter was treated with phenylmagnesium bromide, followed by acetic anhydride and sodium acetate, and so yielded the styryl derivative (CX). This had an ultraviolet absorption maximum at 2460 Å. typical of an α-substituted styrene (*109*). On ozonolysis it gave acetophenone and an aldehyde (CXI) which was oxidised directly to an acid (CXII) characterised as its methyl ester (CXIII). Analysis of the ester indicated that five carbon atoms have been removed during the degradation of (CVII) to (CXII).

The six-carbon grouping (cf. CVII) present in the side-chain of polyporenic acid A was already known to be present in eburicoic acid (p. 68) and polyporenic acid C (p. 81). On the assumption that the biogenesis of polyporenic acid A follows the same general pattern as that of eburicoic acid and polyporenic acid C, combination of the partial structures (CVI) and (CVII) led to structure (XC, p. 74) for polyporenic acid A.

Proof of the correctness of this structure was obtained by the conversion of polyporenic acid A into derivatives of lanosterol by two different

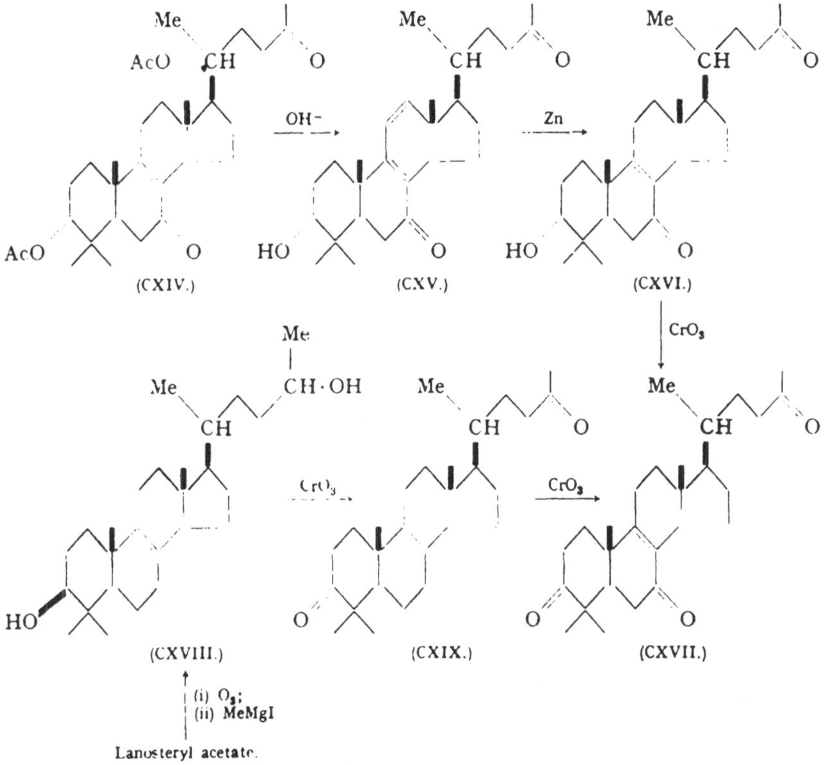

Chart 5. Relationship between Lanosterol and Polyporenic Acid A (i).

methods. HALSALL and HODGES (*95*) took the methyl ketone (CIX) formed by ozonolysis of the diacetate (CVIII) of decarboxylated polyporenic acid A and oxidised it for a short time with chromic acid. This introduced a second keto-group at $C_{(7)}$ giving (CXIV), converted to the dienone (CXV) by treatment with methanolic potassium hydroxide (*Chart 5*). Reduction of the dienone with zinc dust in acetic acid gave the $\alpha\beta$-unsaturated ketone (CXVI) which was oxidised to the triketone (CXVII). This was also obtained starting from "isocholesteryl" acetate (p. 49) which was ozonised, the crude product treated directly with methylmagnesium iodide and the resulting diol (CXVIII) oxidised with chromic acid in acetone to the corresponding diketone (CXIX). After purification this was oxidised with chromic acid in acetic acid to give the triketone (CXVII), already obtained from polyporenic acid A.

JEGER et al. (*141*) oxidised the diacetate (CVIII) of decarboxylated polyporenic acid A with chromic acid to the diacetoxy-triketone (CXX).

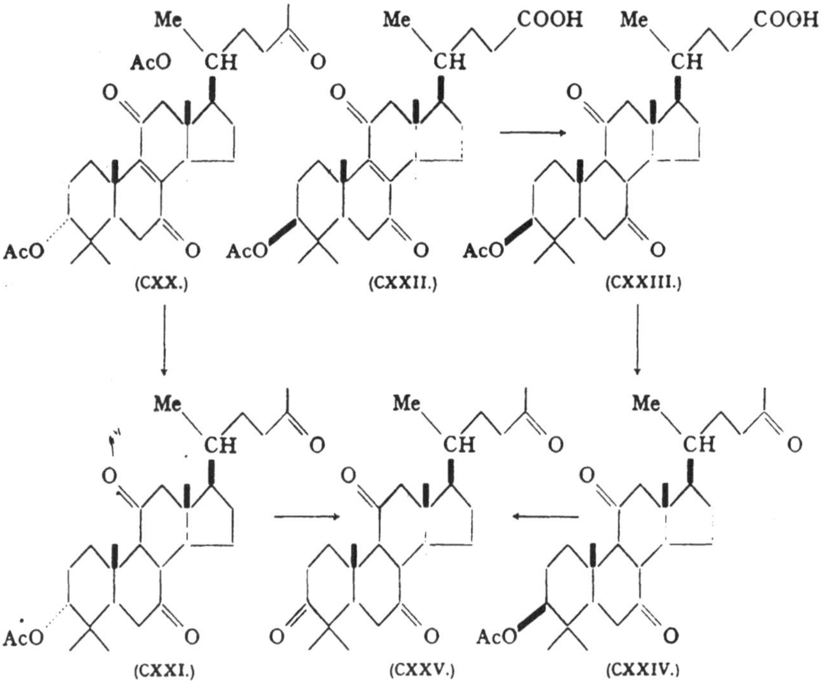

Me CH AcO O O AcO O (CXX.)
Me COOH CH O AcO O (CXXII.)
Me COOH CH O AcO O (CXXIII.)
Me CH O O AcO O (CXXI.)
Me CH O O O O (CXXV.)
Me CH O O AcO O (CXXIV.)

Chart 6. Relationship between Lanosterol and Polyporenic Acid A (ii).

This was reduced with zinc dust in acetic acid to the acetoxy-triketone (CXXI) (*Chart 6*). Vigorous oxidation of "isocholesteryl" acetate gave the trisnor-acid (CXXII) from which (CXXIII) was obtained by zinc dust reduction (*164*). The sodium salt of this acid was converted by oxalyl chloride to the acid chloride, treatment of which with dimethylcadmium gave the acetoxy-triketone (CXXIV). This compound differed from (CXXI) obtained from polyporenic acid A, having a 3β- instead of a 3α-acetoxy group. Hydrolysis of both acetates followed by oxidation of the resulting hydroxy-triketones led to the same tetraketone (CXXV).

These results finally establish structure (XC, p. 74) for polyporenic acid A, the only feature which has not yet been elucidated being the stereochemistry of $C_{(25)}$.

When methyl polyporenate A is heated with 5% methanolic potassium hydroxide, only part of it is hydrolysed. The remainder is isomerised to an $\alpha\beta$-unsaturated ester, methyl *iso*polyporenate (cf. CXXVI) which is hydrolysed much more slowly. When "polyporenic acid A", prepared by hydrolysis of methyl polyporenate A, is decarboxylated some ether-insoluble acid is obtained besides the expected decarboxy-compound. This acid has an ultraviolet absorption spectrum characteristic of an $\alpha\beta$-unsaturated acid, but its methyl ester (methyl ψ-polyporenate A) is not identical with methyl *iso*polyporenate A. However, the diacetates of both esters give on ozonolysis the same methyl ketone (CIX, p. 78) as is obtained from the diacetate of the decarboxy-compound (CVIII) (95). *iso*-Polyporenic acid A and ψ-polyporenic acid A must therefore be geometrical isomers. The formation of the ψ-acid is best explained by assuming that the hydrolysis product from methyl polyporenate A contains, in addition to acid A, some *iso*polyporenic acid A formed by hydrolysis of methyl *iso*polyporenate A, and that the *iso*-acid is thermally isomerised to the ψ-acid. The isomerisation of methyl polyporenate A to methyl *iso*-polyporenate A and the thermal isomerisation of the *iso*-acid to the ψ-acid are best explained if *iso*polyporenic acid is (CXXVI) ($3\alpha:12\alpha$-dihydroxyeburico-8 : 24-dien-26-oic acid) with the two biggest substituents about the double bond *cis* to one another, and if ψ-polyporenic acid is (CXXVII) ($3\alpha:12\alpha$-dihydroxyeburico-8 : 24-dien-27-oic acid) (95).

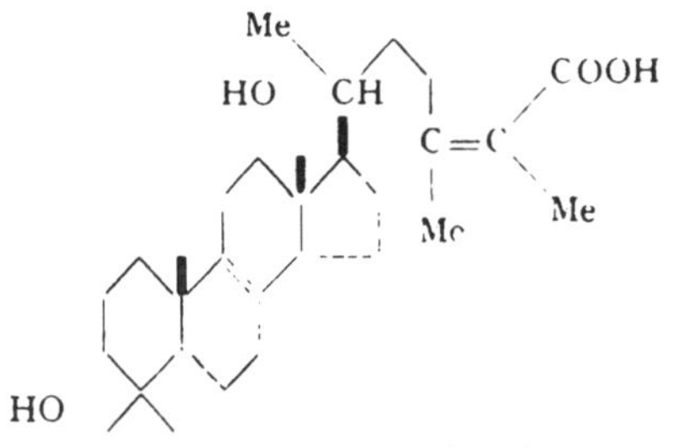

(CXXVI.) *iso*-Polyporenic acid.

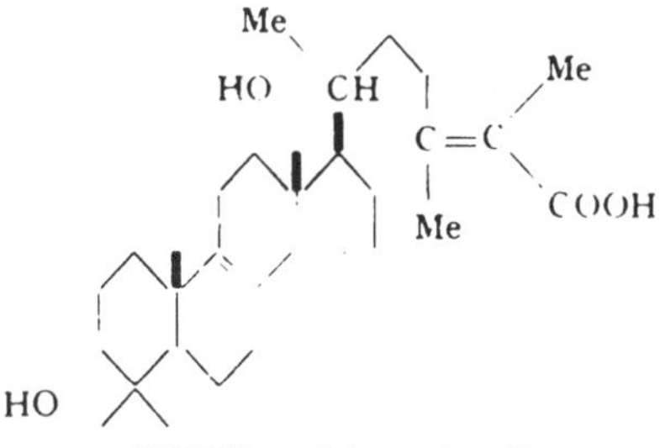

(CXXVII.) ψ-Polyporenic acid.

4. Polyporenic Acid C.

(16α-Hydroxy-3-oxoeburico-7 : 9(11) : 24(28)-trien-21-oic acid)
$C_{31}H_{46}O_4$ (CXXVIII, p. 82).

This acid was first isolated in 1939, along with polyporenic acids A and B, from *Polyporus betulinus* Fr. by Cross, Eliot, Heilbron and Jones (49). At that time very little was learned about the acid, the only pure derivative prepared being the methyl ester, analysis of which suggested a formula $C_{30}H_{46}O_4$ or $C_{30}H_{48}O_4$ for the acid.

In 1952 Birkinshaw, Morgan and Findlay (*33*) reported the isolation of a monocarboxylic acid containing four oxygen atoms from *Polyporus benzoinus* (Wahl) Fr. which appeared to be polyporenic acid C. The infrared spectra of the methyl esters of polyporenic acid C and of the *benzoinus* acid were identical (*36*). Birkinshaw et al., however, suggested that the *benzoinus* acid was a dihydroxy-monocarboxylic acid possessing a vinylidene group, although, as described below, polyporenic acid contains only one hydroxyl group, the fourth oxygen atom being present in a carbonyl group.

The chemistry of polyporenic acid C was greatly advanced when it was found to contain a conjugated diene system so that purification of the acid could be followed by spectrographic examination (*36*). The further discovery that it was a keto-acid led to a satisfactory method for its isolation based on a Girard separation.

The great diagnostic value of ultraviolet and infrared spectroscopic examination was strikingly illustrated with methyl polyporenate C since it enabled Bowers, Halsall, Jones and Lemin (*36*) to draw conclusions about all the reactive centres of the acid. The ultraviolet absorption spectrum showed maximal light absorption at 2360, 2430 and 2510 Å. These values, indicative of a heteroannular diene system, are very similar to those of methyl dehydroeburicoate (CXXIX) and of dehydrolanosterol (agnosterol) (CXXX) but different from those of dehydroeuphenyl acetate (CXXXI)

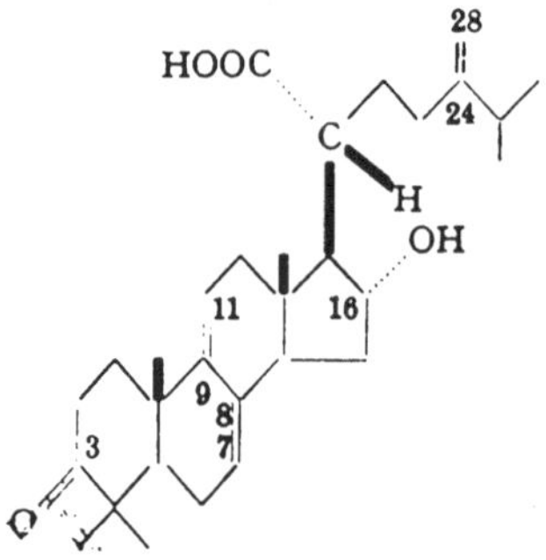

(CXXVIII.) Polyporenic acid C.

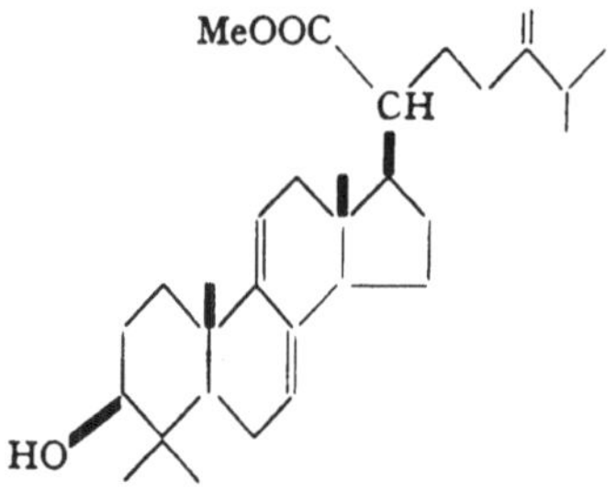

(CXXIX.) Methyl dehydroeburicoate.

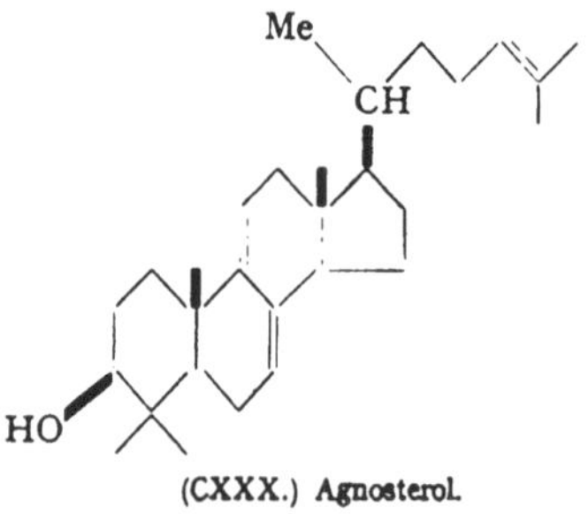

(CXXX.) Agnosterol.

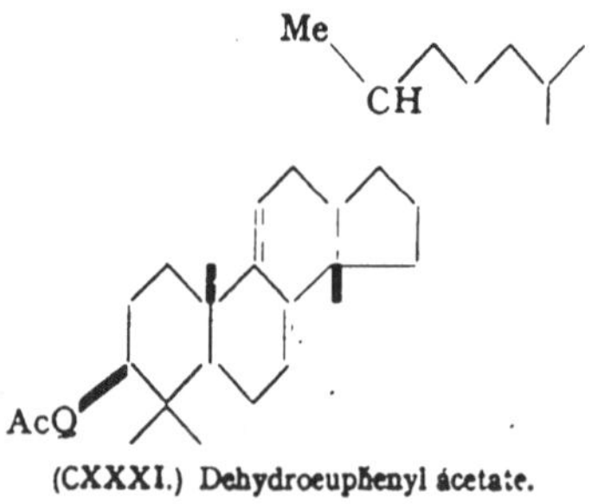

(CXXXI.) Dehydroeuphenyl acetate.

(cf. p. 104). This suggested that the lanostane type ring system might be present in polyporenic acid C. An inflexion in the spectrum at 2760–2820 Å. ($\varepsilon = 57$) indicated the presence of a keto-group which was confirmed by the preparation of a 2:4-dinitrophenylhydrazone. The infrared spectrum had bands at 1712 cm.$^{-1}$ indicative of a keto-group in a six-membered ring or on an aliphatic side chain, at 3461 and 1735 cm.$^{-1}$ indicative of hydroxyl and ester groups respectively, and at 891 and 1639 cm.$^{-1}$ indicative of a vinylidene group. The latter group, the presence of which was confirmed by ozonolysis, was not part of the conjugated diene system. Hydrogenation of methyl polyporenate C gave a dihydro-derivative which had the same ultraviolet absorption spectrum as the parent ester but no infrared absorption corresponding to a vinylidene group. Hence methyl polyporenate C is a hydroxy-keto-acid with three double bonds. Analytical data did not enable a clear-cut decision to be made between $C_{31}H_{46}O_4$ and $C_{32}H_{48}O_4$ for the methyl ester but both possibilities indicated a tetracyclic structure.

Oxidation of methyl polyporenate C gave a diketo-ester (CXXXII), the ultraviolet spectrum of which showed that the diene system was not conjugated with a carbonyl group. The original hydroxyl group could not therefore be adjacent to the diene system. It must, in fact, be attached to a five-membered ring since the infrared spectrum of the diketo-ester

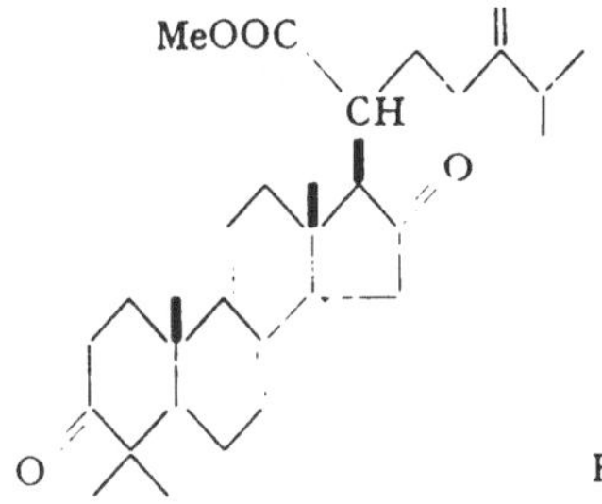

MeOOC CH O O

(CXXXII.)

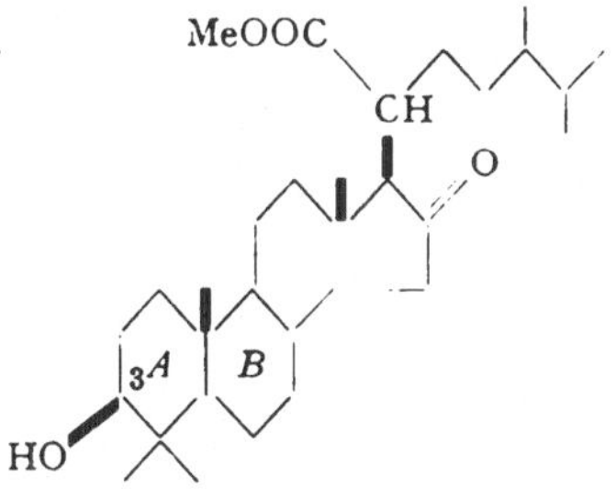

MeOOC CH O HO 3 A B

(CXXXIII.)

A B

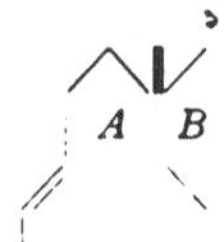

(CXXXIV.)

(CXXXII) has an additional band at 1743 cm.$^{-1}$, characteristic of a keto-group in a five-membered ring (cf. *116*). Reduction of the diketo-ester with sodium borohydride gave a keto-alcohol, effecting preferential reduction of the original keto-group, since the product now showed no infrared absorption band corresponding to a carbonyl group in a six-membered ring. The corresponding keto-alcohol (CXXXIII) derived from methyl dihydropolyporenate C was treated with phosphorus pentachloride and underwent the typical retro-pinacolinic dehydration, the product (CXXXIV) giving acetone on ozonolysis, indicating that the keto-group of polyporenic acid C was at $C_{(3)}$.

Polyporenic acid C, in contrast to polyporenic acid A, is not decarboxylated on melting, making a $\beta\gamma$-unsaturated acid formula improbable. The unconjugated double bond on the other hand migrates under acidic conditions; treatment of a chloroformic solution of methyl polyporenate C (CXXXV) with hydrogen chloride gave an *iso*-compound, methyl *iso*-polyporenate C (CXXXVI), ozonolysis of which gave acetone and a ketonic non-volatile product. These results are consistent with the structures (CXXXVII) and (CXXXVIII) and with the presence in polyporenic acid C of a side-chain skeleton terminating in (CXXXIX).

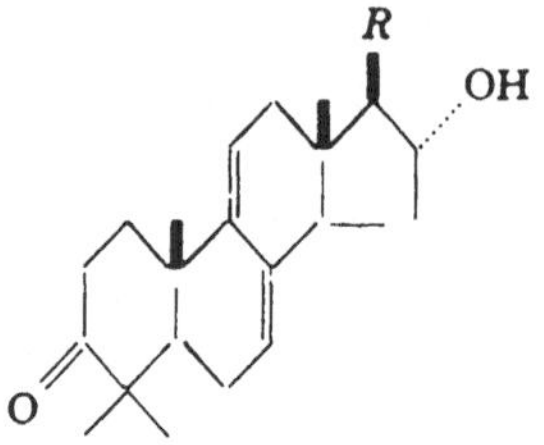

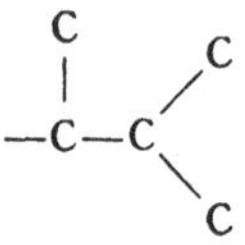

(CXXXIX.)

(CXXXV.) $R = MeOOC \cdot CH(-) \cdot (CH_2)_2 \cdot C(=CH_2) \cdot CHMe_2$, Methyl polyporenate C.

(CXXXVI.) $R = MeOOC \cdot CH(-) \cdot (CH_2)_2 \cdot C(CH_3){=}CMe_2$, Methyl *iso*polyporenate C.

(CXXXVII.) $-CH(R')-C(CH_3){=}CH_2 \longrightarrow -C(R){=}C(CH_3)-CH_3 \longrightarrow -C(R){=}O + Me_2CO \quad R' \neq H.$

(CXXXVIII.) $-C(={H_2C})-CH(CH_3)-CH_3 \longrightarrow -C(H_3C){=}C(CH_3)-CH_3 \longrightarrow -C(CH_3){=}O + Me_2CO.$

Treatment of methyl dihydropolyporenate C with phosphorus pentachloride brought about replacement of the hydroxyl group by chlorine rather than its elimination. Such a replacement would be unlikely to occur if either of the carbon atoms adjacent to the $>CH \cdot OH$ group were fully substituted, in which circumstance elimination would be expected as the alcohol would be of the *neo*pentyl type. If the lanostane system is present in polyporenic acid C, as the evidence already discussed indicates, then the only carbon atom in the five-membered ring *D* to which the hydroxyl group could be attached is $C_{(16)}$ (CXL). This conclusion is supported by the molecular rotation difference between the acetate of the keto-alcohol (CXXXIII) described above and the corresponding deoxo-compound. The value ($\Delta M_D = -498$) is characteristic of the grouping (CXLI) (*118*). Furthermore the hydroxyl group is at a position γ to the carboxyl group since methyl dihydropolyporenate C is converted by treatment with thionyl chloride in *iso*octane to a γ-lactone, the infrared spectrum of which has a characteristic band at 1766 cm.$^{-1}$. (The structure of this lactone is discussed later.)

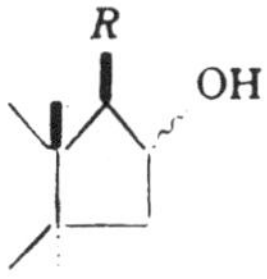

(CXL.) 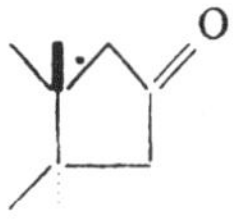 (CXLI.)

The evidence discussed above suggested that polyporenic acid C had a C_{31} formula and that its carbon skeleton (CXLII) bore the same relation to that of lanosterol (CXLIII) as does that of ergostanol (CXLIV) to that of cholestanol (CXLV). With such a carbon skeleton, (CXLVI) must represent dihydropolyporenic acid C (*36*). That this is correct follows from the conversion of methyl dihydropolyporenate C into a derivative of dehydro-dihydroeburicoic acid (CXLVII). The keto-alcohol (CXXXIII, p. 83) described above was reduced by the WOLFF-KISHNER method. The product was methylated and acetylated and proved to be identical with methyl acetyldehydrodihydro-eburicoate (CXLVIII). The position of the vinylidene group in polyporenic acid C as in [$\cdot C(: CH_2) \cdot CHMe_2$] was

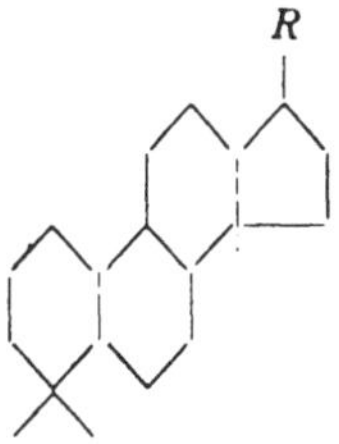

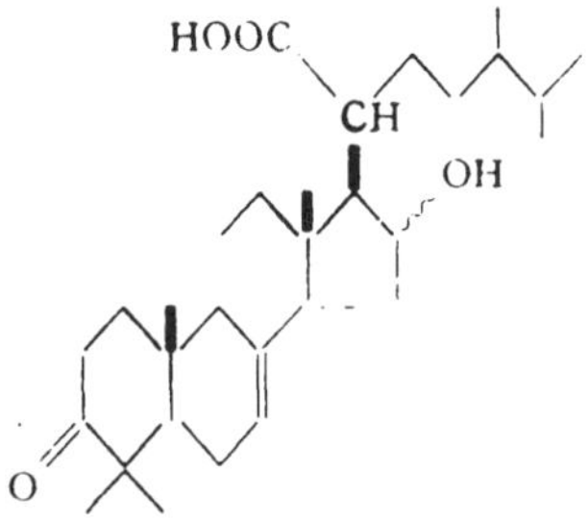

(CXLII.)
$R = CHMe \cdot (CH_2)_2 \cdot CHMe \cdot CHMe_2$.
(CXLIII.)
$R = CHMe \cdot (CH_2)_3 \cdot CHMe_2$

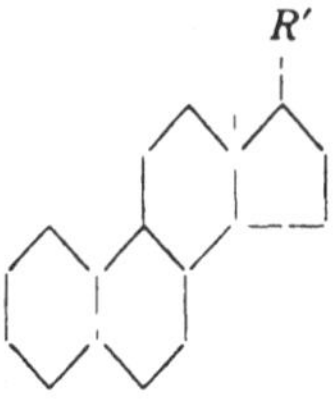

(CXLIV.)
$R' = CHMe \cdot (CH_2)_2 \cdot CHMe \cdot CHMe_2$.
(CXLV.)
$R' = CHMe \cdot (CH_2)_3 \cdot CHMe_2$.

(CXLVI.)

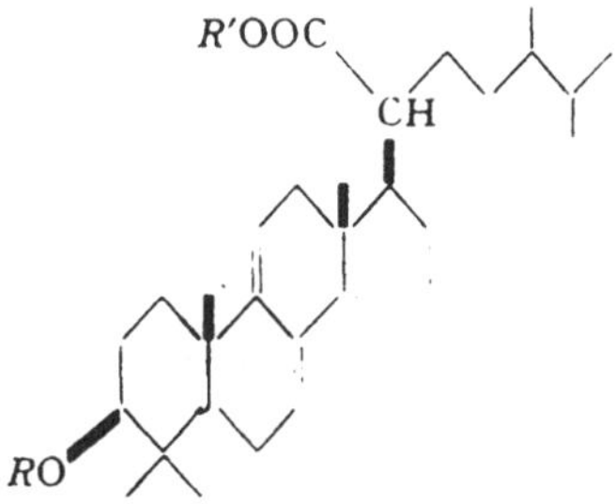

(CXLVII.) $R = H$; $R' = H$.
(CXLVIII.) $R = Ac$; $R' = Me$.

established by the similar conversion of methyl polyporenate C into methyl acetyleburicoate.

These transformations, along with the other evidence discussed above, establish all the features of structure (CXXVIII, p. 82) for polyporenic acid C except for the configurations of the $C_{(16)}$ hydroxyl group and of the substituents at $C_{(20)}$. The $C_{(16)}$ hydroxyl group must have the α-configuration since acetylation is accompanied by a large negative shift in rotation (*37*). Acetylation of 16 α-hydroxysteroids results in a large negative shift in molecular rotation while the reverse effect is observed with 16 β-hydroxysteroids on acetylation (*14, 119*).

The series of reactions leading to eburicoic acid derivatives described above provide no evidence about the initial stereochemistry at $C_{(20)}$ in polyporenic acid C because of the possibility of inversion under alkaline conditions. WOLFF-KISHNER reduction of the carbonyl group of methyl dihydropolyporenate C (CIL) followed by methylation of the product gave, rather unexpectedly, two hydroxy-esters [methyl 16α-hydroxy-eburico-7 : 9(11)-dien-21-oate (CL) and methyl 16α-hydroxy-20-*iso*-eburico-7 : 9(11)-dien-21-oate (CLI)]. These have been shown to differ only at $C_{(20)}$ (*37*). Both esters on treatment with thionyl chloride in a hydrocarbon solvent gave the *same* lactone (CLII) the infrared spectrum

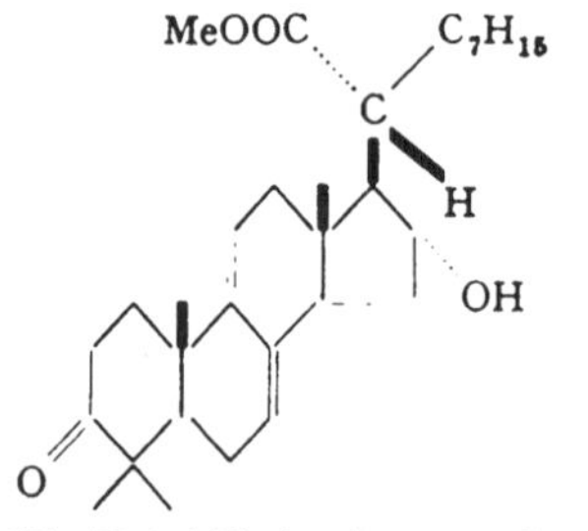

(CIL.) Methyl dihydropolyporenate C.

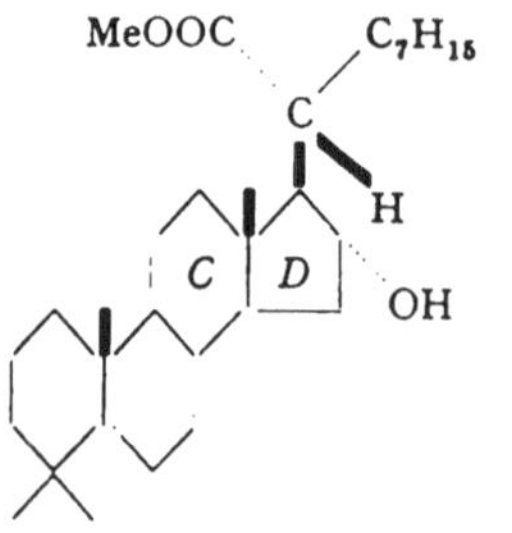

(CL.)

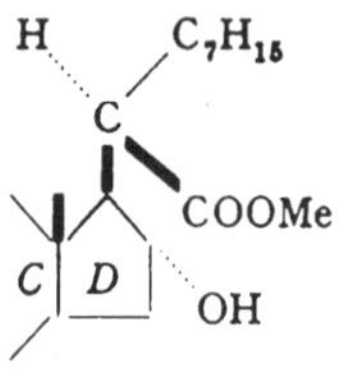

(CLI.)

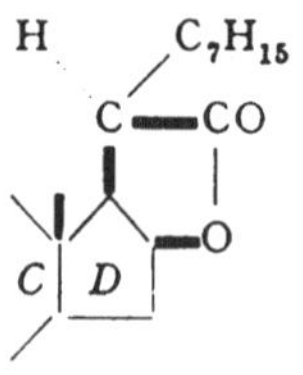

(CLII.)

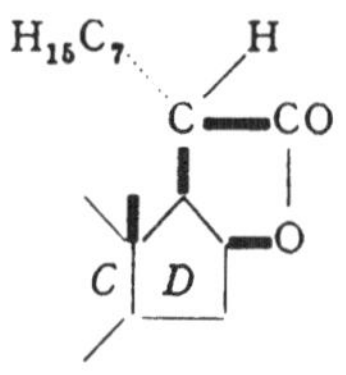

(CLIII.)

of which has a band at 1771 cm.$^{-1}$ characteristic of a saturated γ-lactone. Reduction of the lactone with lithium aluminium hydride gave a diol which was also obtained by similar reduction of the 16-keto-derivative of one of the two hydroxy-esters. The two possible structures for the lactone are (CLII) and (CLIII). Inspection of models indicates that the

20-*iso*eburicane structure (CLII) should be much more stable than the 20-(normal)-eburicane structure (CLIII), since in the latter there is very considerable compression between the 18-methyl and the 22-methylene groups. Structure (CLII) was therefore put forward for the lactone. and the hydroxy-ester which can be converted into the diol formed by reduction of the lactone is (CLI). The formation of the lactone from the ester with the 20-(normal)-eburicane structure involves epimerisation at $C_{(20)}$, probably through the intermediates (CLIV), (CLV), and (CLVI).

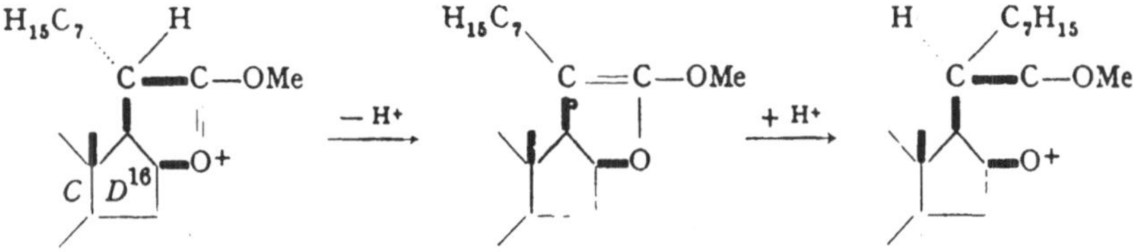

(CLIV.) (CLV.) (CLVI.)

WOLFF-KISHNER reductions have been carried out with eburicoic acid derivatives, *i. e.*, compounds devoid of the 16-oxygen atom, without the formation of isomers. The explanation suggested for this apparent discrepancy is that in the absence of a 16-hydroxyl group the normal (eburicane) configuration at $C_{(20)}$ is more stable than the *iso*-configuration. When a hydroxyl group is present, a 20-*iso*-acid (or ester *e. g.* CLI) may increase its stability by hydrogen-bond formation between the carboxyl (or methoxycarbonyl) and the hydroxyl groups (cf. the formation of the 20-*iso*-lactone). The 20-normal acid (or ester, *e. g.*, CL) is prevented from doing so, for hydrogen bonding can only occur in that conformation which leads to compression between the 22-methylene and 18-methyl groups. It may be noted that the infrared spectrum of the 20-*iso*-ester (CLI) has bands indicative of hydrogen bonding, but no such bands are found in the spectrum of the 20-normal-ester (CL).

These results emphasise that the isolation of deoxyeburicoic acid derivatives from WOLFF-KISHNER reductions does not necessarily prove that the carboxyl group originally has the more stable 20-normal (eburicane) configuration unless it can be shown by other means that the configuration at $C_{(20)}$ has not changed. In the conversion of polyporenic acid C into an eburicoic acid derivative such evidence is not available. However, evidence has been obtained which proves that in polyporenic acid C the configuration at $C_{(20)}$ is of the 20-normal series.

If polyporenic acid C were of the 20-*iso*-series, hydrogen bonding would occur in methyl polyporenate C between the methoxycarbonyl and the 16-hydroxyl groups as in the case of (CLI). The infrared spectrum of methyl polyporenate C, however, indicates that no such hydrogen bonding occurs.

When methyl dihydropolyporenate C (CIL) was treated with thionyl chloride in *iso*octane a lactone was obtained which, by analogy with the lactone (CLII) formed from methyl dihydro-3-deoxo-polyporenate C (CL) must be formulated as (CLVII). Reduction of the lactone with lithium aluminium hydride gave a triol which must be 3β : 16β : 21-trihydroxy-20-*iso*eburico-7 : 9(11)-diene (CLVIII). This triol was *not* identical with either of the triols (CLIX) and (CLX), obtained by the lithium aluminium

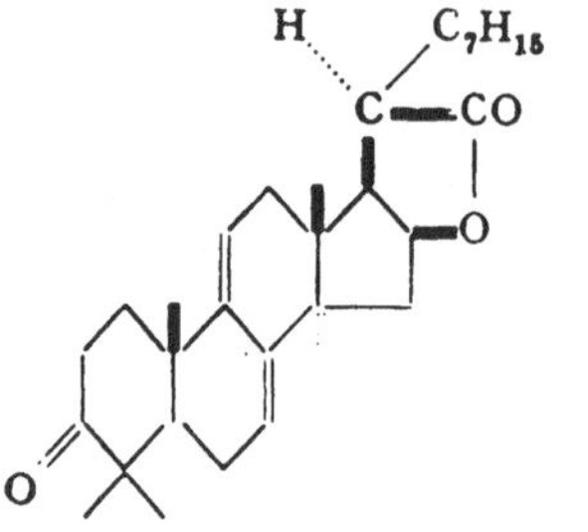

(CLVII.) 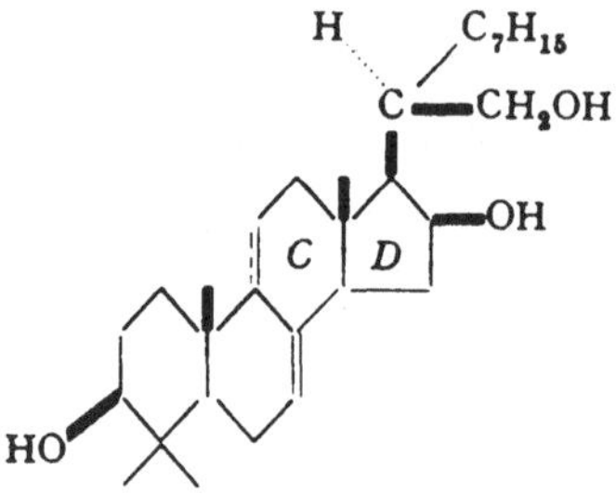 (CLVIII.)

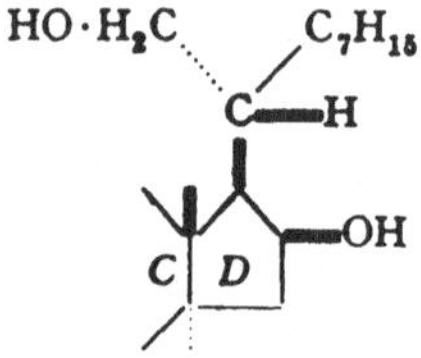

(CLIX.) 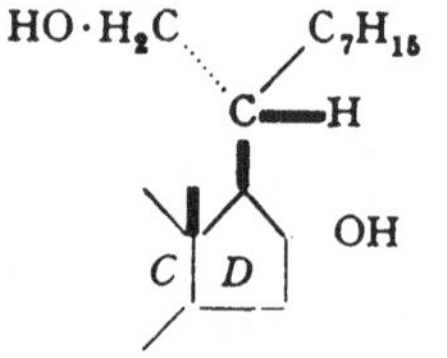 (CLX.)

hydride reduction of the 3β-hydroxy-16-oxo-derivative (CXXXIII, p. 83) of methyl dihydropolyporenate C. One of these triols must have both 3- and 16-hydroxyl groups in the β-configuration and hence must differ at $C_{(20)}$ from the triol from the lactone. Since the latter has the 20-*iso*-configuration the triols from (CXXXIII, p. 83), and hence polyporenic acid C (CXXVIII, p. 128) must have the 20-normal configuration and belong to the eburicoic acid series.

5. Tumulosic Acid.

(3β : 16α-Dihydroxyeburico-8 : 24(28)-dien-21-oic acid) $C_{31}H_{50}O_4$ (CLXI).

Tumulosic acid together with the corresponding dehydro-acid (CLXII) has been isolated by ROBERTSON and his co-workers (*48*) from *Polyporus tumulosus* COOKE, *P. australiensis* WAKEFIELD and *Poria cocos* WOLF. Polyporenic acid B, originally isolated by CROSS, ELIOT, HEILBRON and JONES (*49*) from *Polyporus betulinus* FR., has been shown by GUIDER et al. (*89*) to be a mixture of tumulosic acid and the dehydro-acid (CLXII).

The mixture of tumulosic and dehydrotumulosic acids was very difficult to resolve, but by exhaustive chromatography of the mixture

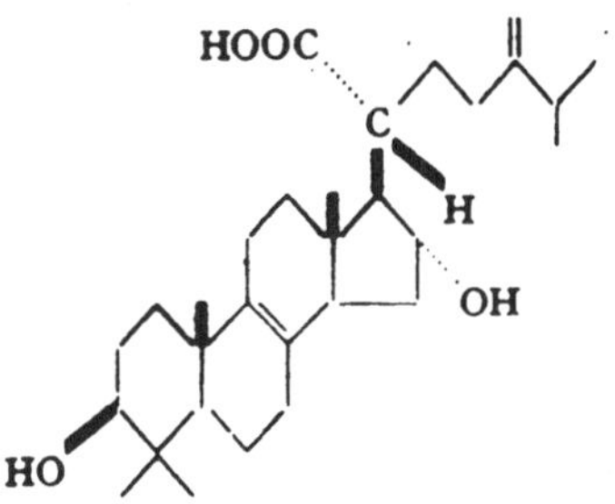

(CLXI.) Tumulosic acid.

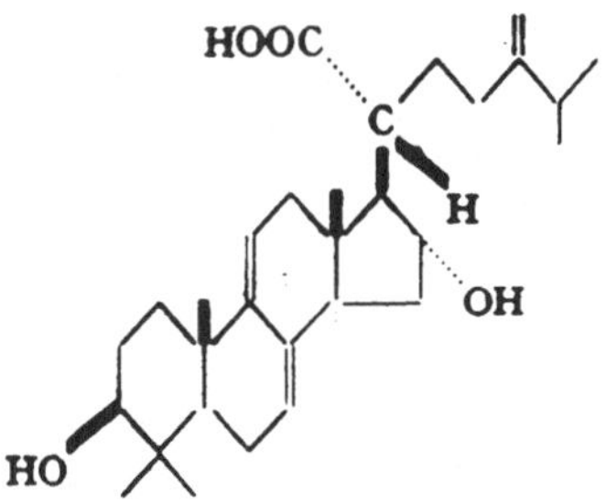

(CLXII.) Dehydrotumulosic acid.

of the acetylated acids a small amount of pure material was obtained and hydrolysed to tumulosic acid (*48*). This was shown to contain two hydroxyl groups, forming a diacetate, and two double bonds. One of these occurs in a vinylidene group, since oxidation with lead tetraacetate of the diol formed from methyl diacetyl-tumulosate with osmium tetroxide gave formaldehyde. It is this double bond which could be hydrogenated since methyl diacetyldihydro-tumulosate is not oxidised with osmium tetroxide and its infrared spectrum does not show a band at 89: cm.$^{-1}$ as does that of the unhydrogenated ester.

Dehydrotumulosic acid itself could not be separated from admixture with tumulosic acid. The mixture, however, showed ultraviolet absorption characteristic of the mixture of eburicoic and dehydroeburicoic acids

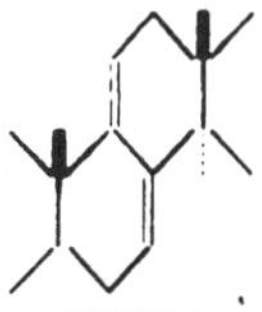

(CLXIII.)

(CLXIV.)

(p. 69), and indicative of the presence of the grouping (CLXIII) in the dehydro-acid and of probably (CLXIV) in tumulosic acid. The dehydro-derivative was prepared from the methyl ester diacetate of dihydro-

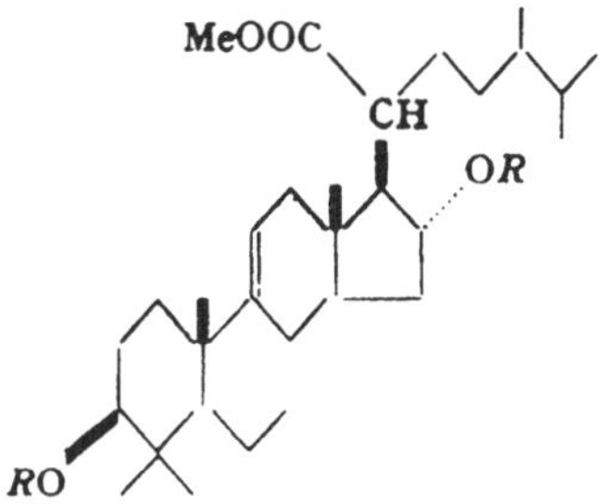

(CLXV.) *R* = Ac.
(CLXVI.) *R* = H

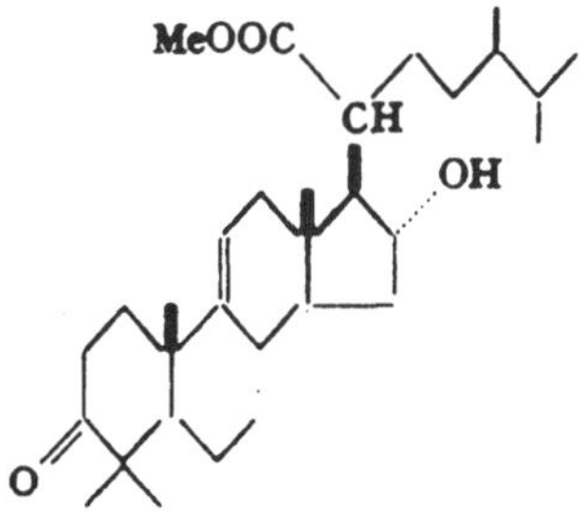

(CLXVII.) Methyl dihydropolyporenate C.

tumulosic acid by selenium dioxide oxidation, and proved to be identical with the diacetate (CLXV) of the sodium borohydride reduction product of methyl dihydropolyporenate C (CLXVII) Hydrolysis of the dehydro-derivative (CLXV) gave the corresponding ester (CLXVI) which was in turn identical with the dihydroxyester obtained from methyl dihydropolyporenate C by sodium borohydride reduction.

These results prove that dihydrotumulosic acid is $3\beta:16\alpha$-dihydroxy-eburic-8-en-21-oic acid and leave only the position of the vinylidene group to be established. Treatment of methyl diacetyltumulosate with chloroformic hydrogen chloride resulted in migration of the inert double bond from the 8 : 9- to the 7 : 8-position and addition of hydrogen chloride to the vinylidene group. By the action of acetic anhydride on the adduct (CLXVIII) hydrogen chloride was eliminated and a mixture of two isomeric compounds (CLXIX) and (CLXX), differing only in the position of the reactive double bond, was obtained. Ozonolysis of the mixture gave formaldehyde and methyl *iso*propyl ketone as the sole volatile products. This result is only compatible with the vinylidene group involving $C_{(24)}$ as in (CLXI) which must therefore be tumulosic acid.

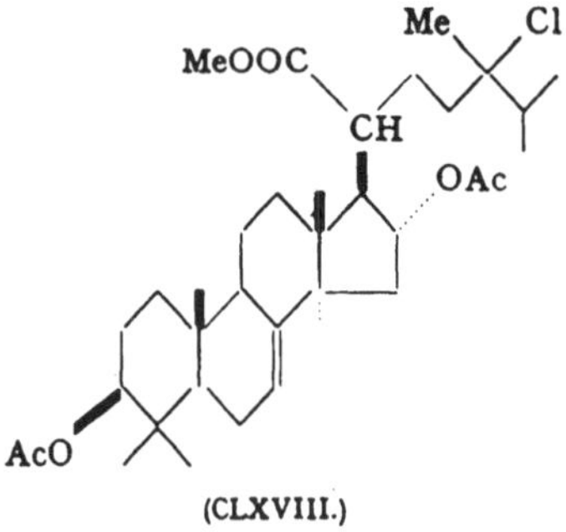

MeOOC CH Me Cl OAc AcO

(CLXVIII.)

R OAc AcO

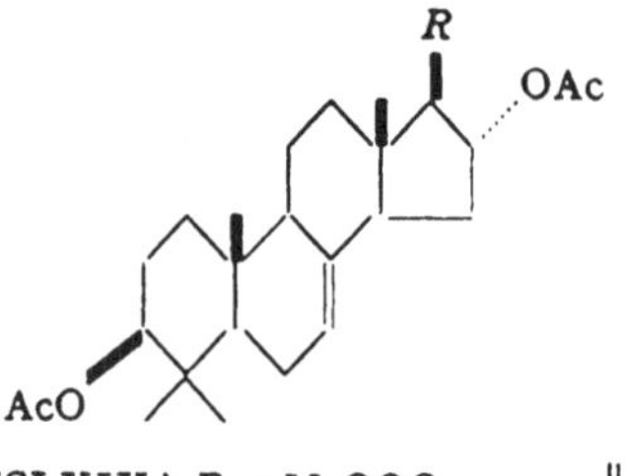

(CLXIX.) R = MeOOC

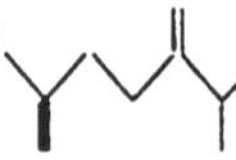

(CLXX.) R = MeOOC

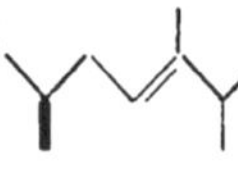

Polyporenic acid B was isolated as its methyl ester by Guider, Halsall, Hodges and Jones (*89*) from *Polyporus betulinus* Fr. and was shown by ultraviolet spectroscopic examination to be a mixture containing about 40% of a compound possessing the typical 7 : 9(11)-diene chromophore found in polyporenic acid C and dehydroeburicoic acid. The ester is, in fact, a mixture of methyl tumulosate (60%) and methyl dehydrotumulosate (40%). This was shown by hydrogenating the mixed methyl esters and converting the resulting dihydro-esters completely into the 7 : 9(11)-diene derivative (CLXXI) by oxidation with monoperphthalic acid and treatment of the product in boiling ethanol with concentrated hydrochloric acid. The diene-derivative (CLXXI) proved to be identical with the sodium borohydride reduction product of methyl dihydropoly-

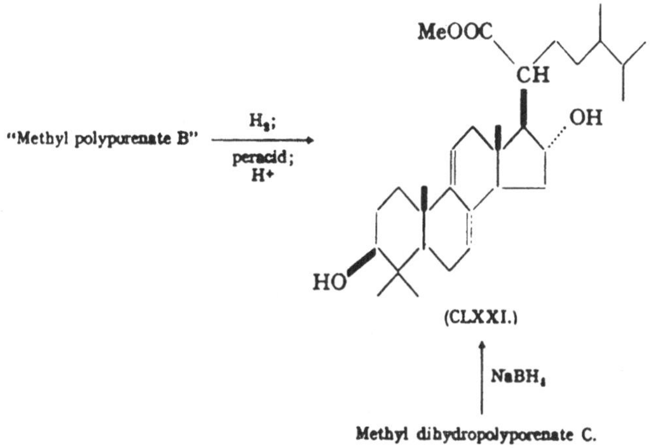

"Methyl polyporenate B" —H_2; peracid; H^+→

(CLXXI.)

↑ $NaBH_4$

Methyl dihydropolyporenate C.

porenate C. This interconversion proves the structure of the components of polyporenic acid B apart from the position of the reducible double bond. To locate this the diacetates of the mixed methyl esters were

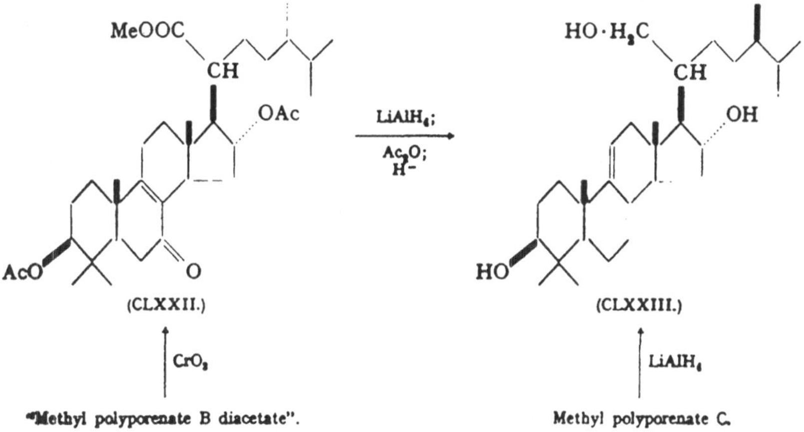

(CLXXII.) —$LiAlH_4$; Ac_2O; H^-→ (CLXXIII.)

↑ CrO_3

"Methyl polyporenate B diacetate".

↑ $LiAlH_4$

Methyl polyporenate C.

oxidised with chromic acid to give the 7-keto-derivative (CLXXII). This was reduced with lithium aluminium hydride and the product heated with acetic anhydride and then hydrolysed with methanolic potassium hydroxide to give $3\beta : 16\alpha : 21$-trihydroxyeburico-$7 : 9(11) : 24(28)$-triene (CLXXIII), identical with the substance prepared by the reduction of methyl polyporenate C with lithium aluminium hydride.

Besides "polyporenic acid B" there has also been isolated from *Polyporus betulinus* Fr. a mixture of the 3β-monoacetates (CLXXIV) and (CLXXV) of the two components of "polyporenic acid B" (*89*).

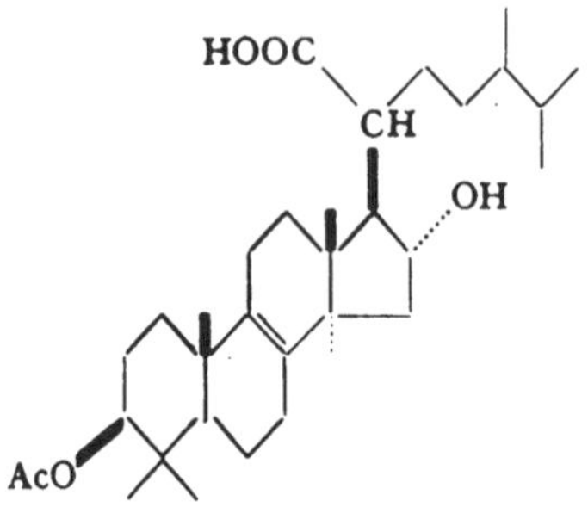

(CLXXIV.) 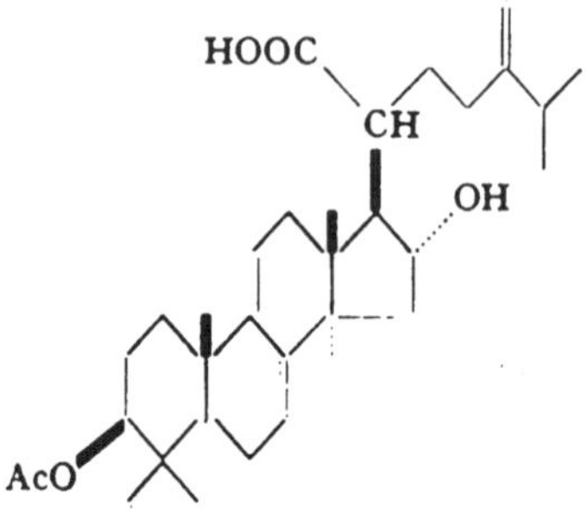 (CLXXV.)

6. Pinicolic Acid A.

(3-Oxolanosta-8 : 24-dien-21-oic acid) $C_{30}H_{46}O_3$ (CLXXVI).

Pinicolic acid A is the first triterpene fungal acid which has been shown to have a lanostane rather than an eburicane skeleton. It was isolated by Guider, Halsall and Jones (*90*) from the pine-rotting fungus, *Polyporus pinicola* Fr. Fröschl, Hartmann and Zellner had previously investigated the alcohol-soluble constituents of this fungus and had isolated two other acids, termed α- and β-pinicolic acids, m. p. 198–205.5°, $[\alpha]_D + 35.7°$, and m. p. 265–271°, $[\alpha]_D + 23.4°$, respectively (*73*, *102*). A common formula $C_{19}H_{30}O_2$ was suggested for these acids, but Cross et al. (*49*) indicated that the analytical data were consistent with a formula $C_{30}H_{50}O_3$, which suggests the possibility of these acids being trimethyl-steroids.

Pinicolic acid A was isolated as its methyl ester which analysed as either $C_{31}H_{48}O_3$ or $C_{32}H_{50}O_3$. Typical samples showed low-intensity light absorption at 2360, 2430, and 2520 Å., revealing the presence of a small amount (ca. 2–3%) of an impurity containing a conjugated diene system of the polyporenic acid C (CXXVIII, p. 82) and agnosterol (III, p. 50) type (*58*). In view of the co-existence in nature of a number of tetracyclic triterpenes, *e. g.*, lanosterol, eburicoic acid, and elemadienolic acid, with their dehydro-derivatives it was inferred that the diene impurity in methyl pinicolate A was the corresponding dehydro-ester and that pinicolic acid A had a ring system similar to that found in eburicoic acid (LXVIII, p. 69) and lanosterol (I, p. 50). The ultraviolet spectrum of methyl pinicolate A also had a broad maximum at 2740 Å. indicative of a carbonyl group, the presence of which was confirmed by the formation of a 2 : 4-dinitrophenylhydrazone. The infrared spectrum of the ester had a band at 1709 cm.$^{-1}$ corresponding to a keto-group in a six-membered ring. Hydrogenation of methyl pinicolate A gave a dihydro-derivative (CLXXVII). Pinicolic acid A is therefore a monoketo-acid with at least one double bond. The formula $C_{31}H_{48}O_3$ or $C_{32}H_{50}O_3$ for the methyl ester corresponds to a tetracyclic structure with two double bonds or a pentacyclic structure with one double bond.

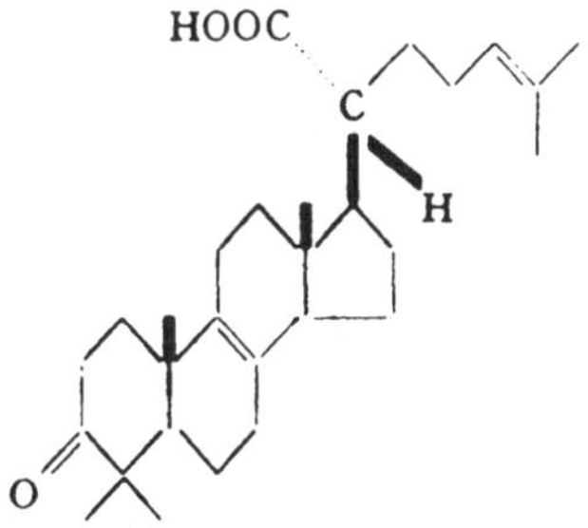

(CLXXVI.) Pinicolic acid A.

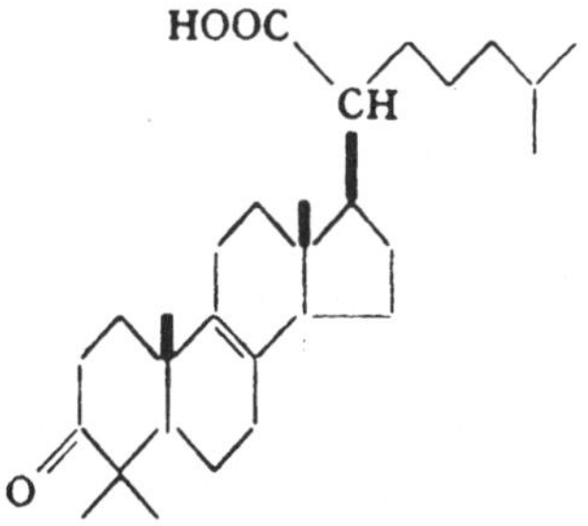

(CLXXVII.) Methyl dihydropinicolate A.

WOLFF-KISHNER red.; CH_2N_2

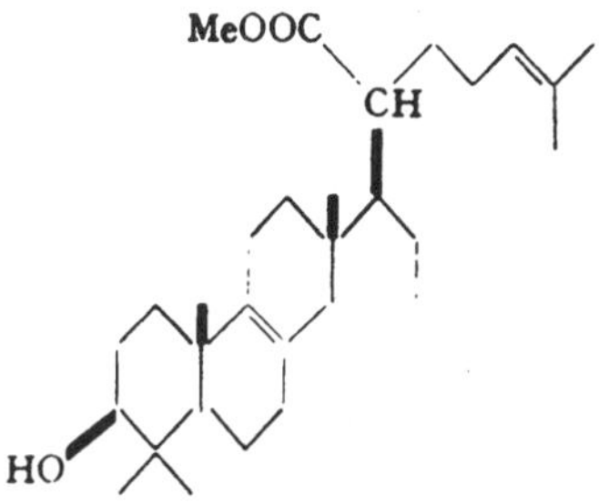

(CLXXVIII.)

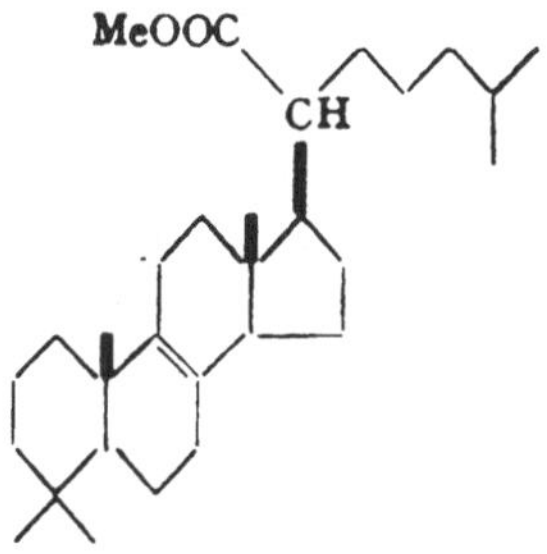

(CLXXXI.)

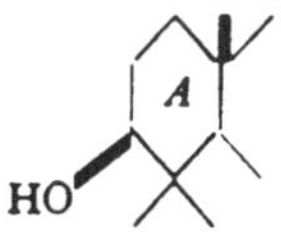

(CLXXIX.)

(CLXXX.)

Ozonolysis of methyl pinicolate A gave acetone and the non-volatile fragment was acidic. This conforms with a side-chain terminating in the grouping $—CH{=}CMe_2$ as in lanosterol rather than in the grouping $—C(:CH_2)—CHMe_2$ found in eburicoic acid. The ester was not easily hydrolysed and the free acid melted without decarboxylation. These properties are similar to those shown by methyl polyporenate C and the free acid, suggesting corresponding locations of the carboxyl group.

Reduction of methyl pinicolate A with sodium borohydride gave the hydroxy-ester (CLXXVIII) which was hydrogenated to a dihydro-derivative. This underwent the typical retro-pinacolinic dehydration on treatment with phosphorus pentachloride, indicating that the hydroxy-ester possesses the grouping (CLXXIX) and pinicolic acid A a typical triterpene ring A but containing a keto-group as in (CLXXX).

On the basis of the above evidence Guider, Halsall and Jones considered (CLXXVI) a possible structure for pinicolic acid A. Accordingly methyl dihydropinicolate A was reduced by the Wolff-Kishner method and the product methylated to give the expected ester of a C_{30} compound, methyl 28-noreburic-8-en-21-oate (CLXXXI) (*111*) (p. 93). Methyl pinicolate A and its derivatives do not undergo isomerisation with alkali, indicating that the carboxyl group of pinicolic acid A and its derivatives must be in the more stable lanostane configuration (cf. *37*). Since the positions of the carbonyl group and of the reducible double bond are known from the reactions described earlier, pinicolic acid A must be formulated as (CLXXVI).

Two acids, both probably $C_{30}H_{48}O_3$, and a neutral compound, $C_{30}H_{46}O_2$, have been isolated by Schmid and Czerny (*151 a*) from *Polyporus pinicola* Fr., and their properties have been described.

Recently Guider, Halsall and Jones (*91*) have isolated from the neutral portion of an extract of *Polyporus pinicola* Fr. a mixture (m. p. 114–117°, $[\alpha]_D + 68.5°$), as yet unresolved, of 3-oxo-21-hydroxylanosta-8:24-diene (CLXXXII) (75%), the primary alcohol corresponding to pinicolic acid A and the corresponding dehydro-derivative (25%). Reduction of this mixture with lithium aluminium hydride gives the corresponding mixed diols, m. p. 194–196°, $[\alpha]_D + 57°$. Pure 3β:21-dihydroxylanosta-8:24-diene, prepared from methyl pinicolate A, has m. p. 189–192°, $[\alpha]_D + 57°$.

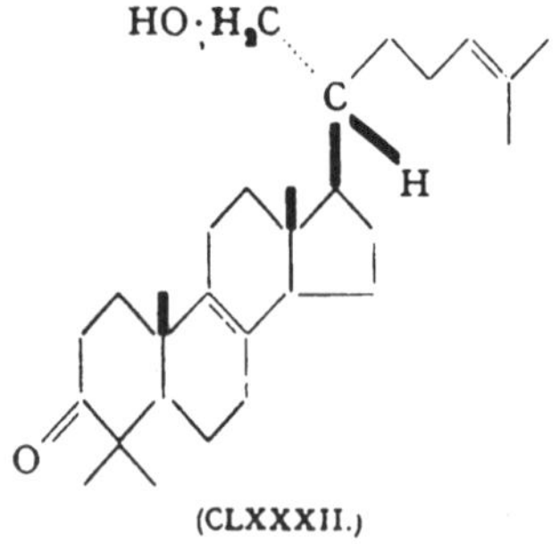

HO·H₂C — C — H — O

(CLXXXII.)

Evidence has been obtained recently by the authors that acids containing more than three oxygen atoms including polyporenic acid C occur in *Polyporus pinicola* Fr., while Shibamoto, Minami and Tajima (*155*) have described the isolation of two $C_{30}H_{46}O_4$ acids (these may be polyporenic acids B and C).

7. Trametenolic Acid A.

($C_{30}H_{48}O_3 \pm CH_2$.)

This secondary hydroxy-acid, the structure of which has not yet been elucidated, was isolated by GRUBER and PROSKE (*86*) from the wood-rotting fungus *Trametes odorata* (WULF.) FR. Equivalent weight determinations suggest that it has the C_{30} formulation. From the products of the dehydrogenation of trametenolic acid A with selenium GRUBER and PROSKE (*88*) isolated 1 : 2 : 5 : 6-tetramethylnaphthalene and 1 : 2 : 7 : 8-tetramethylphenanthrene. These results indicate that trametenolic acid A may be a tetracyclic triterpene with two double bonds although the dehydrogenation products are unusual. However, it has not yet been related to any known triterpene although it has been converted to the parent hydrocarbon by two routes as illustrated (*87*) in *Chart 7*.

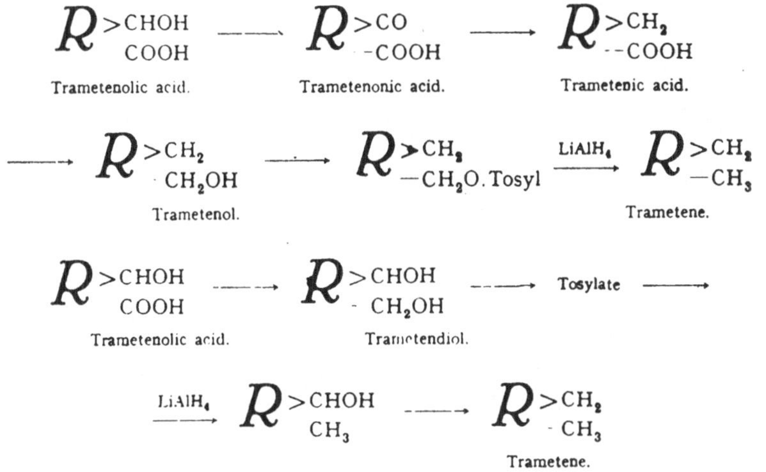

R>CHOH, COOH (Trametenolic acid.) ⟶ R>CO, -COOH (Trametenonic acid.) ⟶ R>CH_2, --COOH (Trametenic acid.)

⟶ R>CH_2, CH_2OH (Trametenol.) ⟶ R>CH_2, —CH_2O.Tosyl $\xrightarrow{LiAlH_4}$ R>CH_2, —CH_3 (Trametene.)

R>CHOH, COOH (Trametenolic acid.) ⟶ R>CHOH, - CH_2OH (Trametendiol.) ⟶ Tosylate ⟶

$\xrightarrow{LiAlH_4}$ R>CHOH, CH_3 ⟶ R>CH_2, - CH_3 (Trametene.)

Chart 7. Reactions of Trametenolic Acid.

In an investigation of the constituents of *Trametes odorata* at present being carried out at the University of Manchester 3β-hydroxy-lanosta-

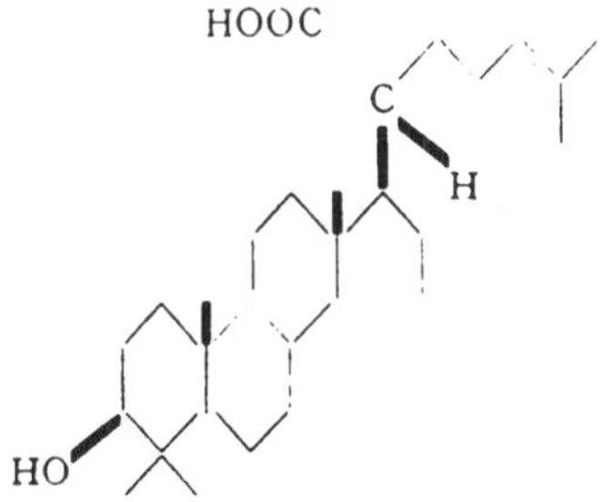

HOOC, C, H, HO

(CLXXXIII.)

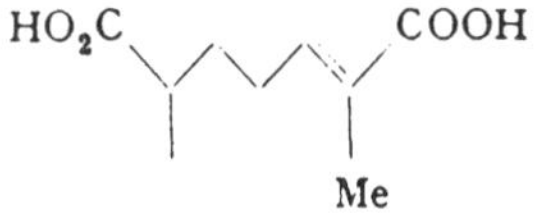

HO_2C, COOH, Me

(CLXXXIV.)

8 : 24-dien-21-oic acid (trametenolic acid B) (CLXXXIII) has been isolated from the fungus (*97*). Its methyl ester is identical with the sodium borohydride reduction product of methyl pinicolate A. A number of other acids, but not trametenolic acid A, have also been isolated and are being examined. One appears to be a dihydroxy-dicarboxylic acid $C_{30}H_{46}O_6$ with one of the carboxyl groups in conjugation with a trisubstituted double bond. Its side-chain may be (CLXXXIV) (*100*).

8. *cyclo*Artenol.

(Handianol; 9 : 19-*cyclo*lanost-24-en-3β-ol) $C_{30}H_{50}O$ (CLXXXV).

The name *cyclo*artenol was first given to a secondary alcohol which was isolated by BARTON (*13*) along with the corresponding ketone, *cyclo*-artenone, and butyrospermol from the fruit of *Artocarpus integrifolia*. The ketone had been originally isolated about fifteen years earlier by NATH and his collaborators (see *13* for a list of references) who had incorrectly concluded that it was an αβ-unsaturated steroidal ketone. *cyclo*Artenol was shown by GONZALEZ and his co-workers (*79*) to be identical with an alcohol called handianol which had been first isolated from the latex of *Euphorbia handiensis* BUCHARD (*81*, *82*) and which was later obtained from *E. obtusifolia* POIR (*78*) and *E. aphyla* BROUSS (*85*). *cyclo*-Artenol has also been isolated from *E. balsamifera* (*42*) and from the non-saponifiable fraction from the seed fat of *Strychnos nux-vomica* L (*26*, *27*).

BARTON (*13*) showed that *cyclo*artenol was a secondary alcohol containing one double bond present in an *iso*propylidene group. These facts, together with the formula $C_{30}H_{50}O$, indicated that *cyclo*artenol was pentacyclic. One of the rings was easily split by acid with the formation of a second double bond. This suggested the presence of a *cyclo*propane ring (*13*).

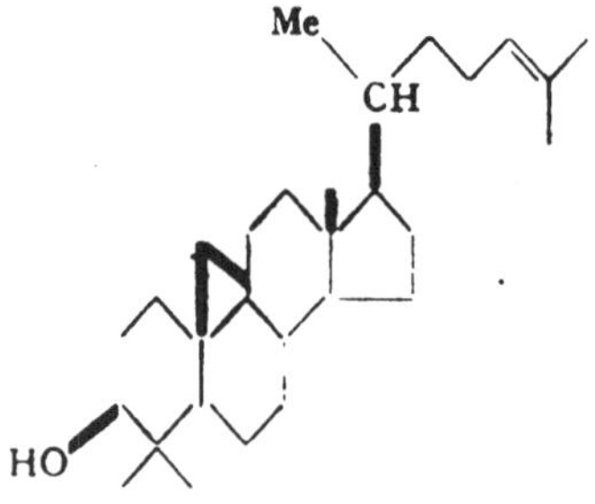

Me
CH
HO

(CLXXXV.) *cyclo*Artenol.

Treatment of *cyclo*artanyl acetate (dihydro-*cyclo*artenyl acetate) (CLXXXVI) with hydrogen chloride gave a mixture of lanost-9(11)-enyl acetate (CLXXXVII) (ca. 60%) together with smaller amounts of an equilibrium mixture of lanost-7-enyl and lanost-8-enyl acetates

(CLXXXVIII and CLXXXIX), from which it followed that *cyclo*artanol was a *cyclo*lanostan-3β-ol. Since lanost-9(11)-enyl acetate could not be isomerised under these conditions and since the equilibrium mixture of

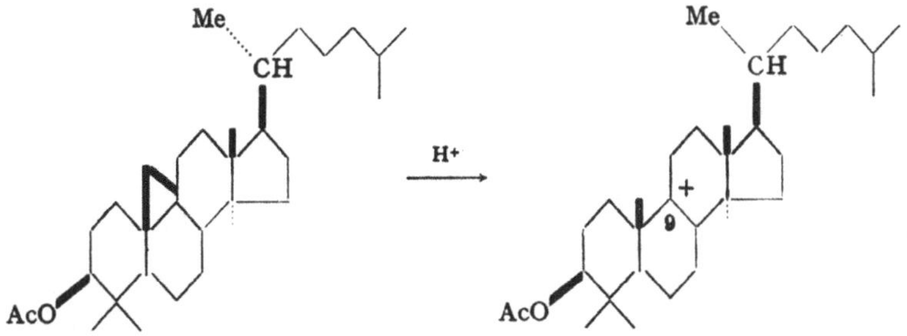

(CLXXXVI.) Dihydro-*cyclo*artenyl acetate.

(CXC.)

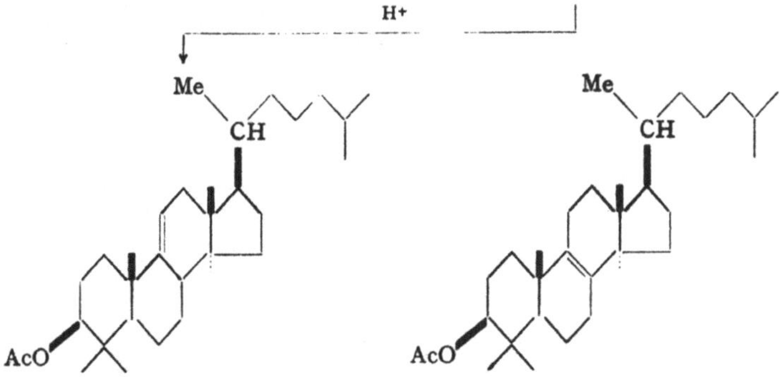

(CLXXXVII.) Lanost-9(11)-enyl acetate.

(CLXXXIX.) Lanost-8-enyl acetate.

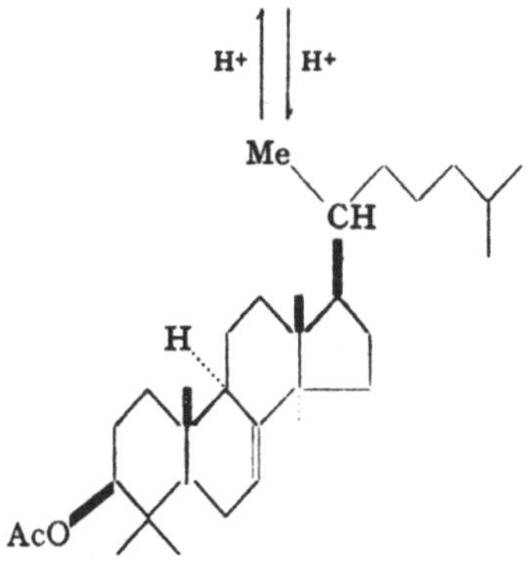

(CLXXXVIII.) Lanost-7-enyl acetate.

the lanost-7- and -8-enyl acetates was not converted to lanost-9(11)-enyl acetate, BENTLEY, HENRY, IRVINE and SPRING (*26*, *27*) concluded that the ion (CXC) was an intermediate in the isomerisation of *cyclo*artanyl acetate and that $C_{(9)}$ was part of the *cyclo*propane ring.

A limitation of the number of possible structures of *cyclo*artenol follows from COLE's observation that *cyclo*artenol and its derivatives have a band in their infrared spectra at about 3045 cm.$^{-1}$ characteristic of an unsubstituted methylene group in a *cyclo*propane ring (*46, 47*). The ultraviolet absorption spectrum of *cyclo*artenone shows that the $C_{(3)}$

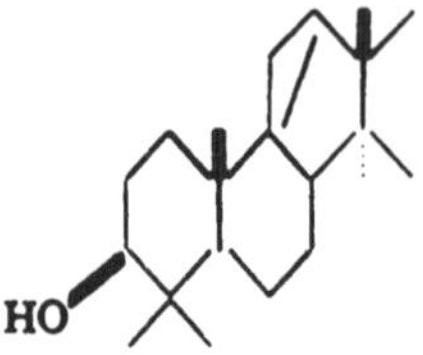

(CXCI.) 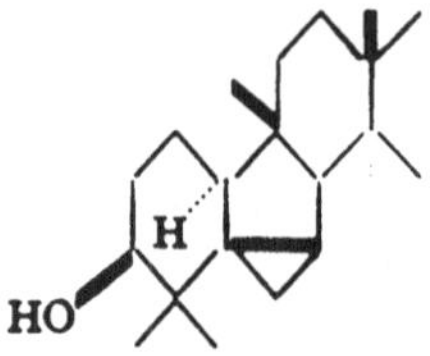 (CXCII.)

carbonyl group is not adjacent to the *cyclo*propane ring. With these restrictions, and allowing for the possibility of methyl migration during the acidic isomerisation, structures (CLXXXV), (CXCI), and (CXCII) have to be considered for *cyclo*artenol (*112*).

By treatment of *cyclo*artanyl (dihydro-*cyclo*artenyl) acetate (CLXXXIV) with deuterium chloride IRVINE, HENRY and SPRING (*112*) obtained a lanost-9(11)-enyl acetate (CXCIII) containing a deuterium atom attached to one of the carbon atoms of the original *cyclo*propane ring. Oxidation of this deuterated acetate gave a 12-keto-derivative (CXCIV) and, since this still contained deuterium, structure (CXCI) is excluded.

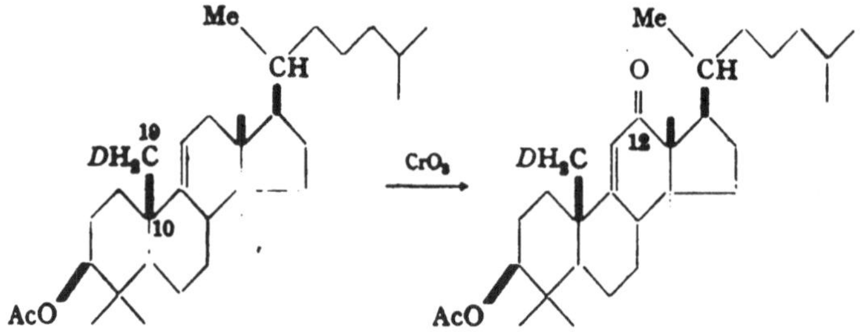

(CXCIII.) (CXCIV.)

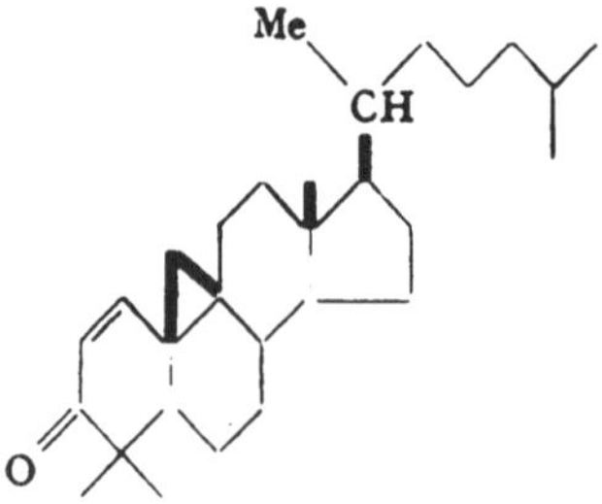

(CXCV.)

Of the two remaining possibilities (CLXXXV) and (CXCII) a decision in favour of (CLXXXV, p. 96) has been reached by both BARTON, WARN-

HOFF and PAGE (*17*, *18*) and IRVINE, HENRY and SPRING (*107*, *112*). The former workers concluded from the infrared spectrum of the deuterated lanostenes obtained by the action of deuterium chloride on *cyclo*artane that the deuterated carbon atom is in a deuterated methyl group (not one of the *gem.*-methyl groups). The deuterium is therefore attached to $C_{(19)}$ (cf. CXCIII) which is part of the *cyclo*propane ring. SPRING and his collaborators prepared an $\alpha\beta$-unsaturated ketone from *cyclo*artanone by bromination with N-bromosuccinimide followed by dehydrobromination with collidine. The ultraviolet absorption spectrum of the $\alpha\beta$-unsaturated ketone (CXCV) showed that the double bond was conjugated with the *cyclo*propane ring which must therefore include $C_{(10)}$. Structure (CLXXXV) follows for *cyclo*artenol.

9. *cyclo*Laudenol.

(9:19-*cyclo*Eburic-25-en-3β-ol) $C_{31}H_{52}O$ (CXCVI).

*cyclo*Laudenol occurs in the alkaloid-free fraction of opium from which it was isolated by BENTLEY, HENRY, IRVINE and SPRING (*25*). They showed that it is a secondary alcohol with only one double bond, present in a vinylidene group. A close similarity with *cyclo*artenol (CLXXXV) (p. 96) was suggested by the observation that the dihydro-derivatives of both compounds, although showing no light absorption in the ethylenic region, give a positive reaction with tetranitromethane. Further, on oxidation to the corresponding ketones, the molecular rotation change in both cases is strongly negative (cf. *13*).

The presence of a *cyclo*propane ring in *cyclo*laudenol was confirmed by treatment of the dihydro-derivative, *cyclo*laudanyl acetate (CXCVII), with hydrogen chloride when a mixture was obtained which showed ethylenic absorption between 2000–2200 Å. From this mixture laud-9(11)-enyl acetate (CXCVIII) was isolated. The infrared spectra of *cyclo*laudenol and its derivatives have bands at 3040 cm.$^{-1}$, typical of a methylene group in a *cyclo*propane ring (*46*, *47*). On oxidation with chromic acid laudenyl acetate behaved in exactly the same way as artenyl

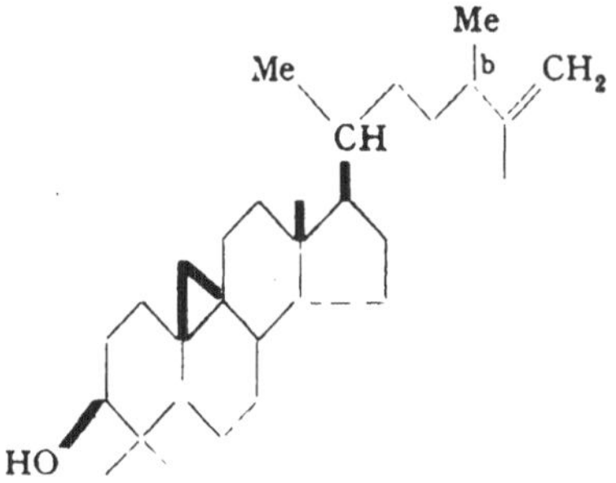

Me
Me
b
CH_2
CH
HO

(CXCVI.) *cyclo*Laudenol.

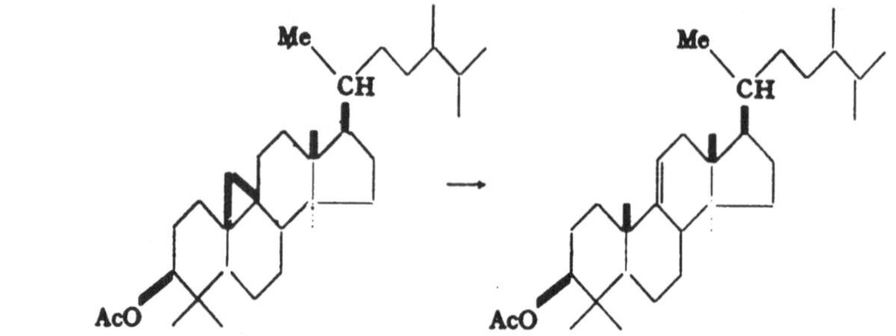

(CXCVII.) → (CXCVIII.)

(lanost-9(11)-enyl) acetate and it was concluded that *cyclo*laudenol and *cyclo*artenol had the same nuclear structure.

*cyclo*Laudenol contains a vinylidene group, its acetate giving formaldehyde and the *nor*ketone (CIC) on ozonolysis. The dihydro-derivatives of *cyclo*artenyl and *cyclo*laudenyl acetates proved not to be identical

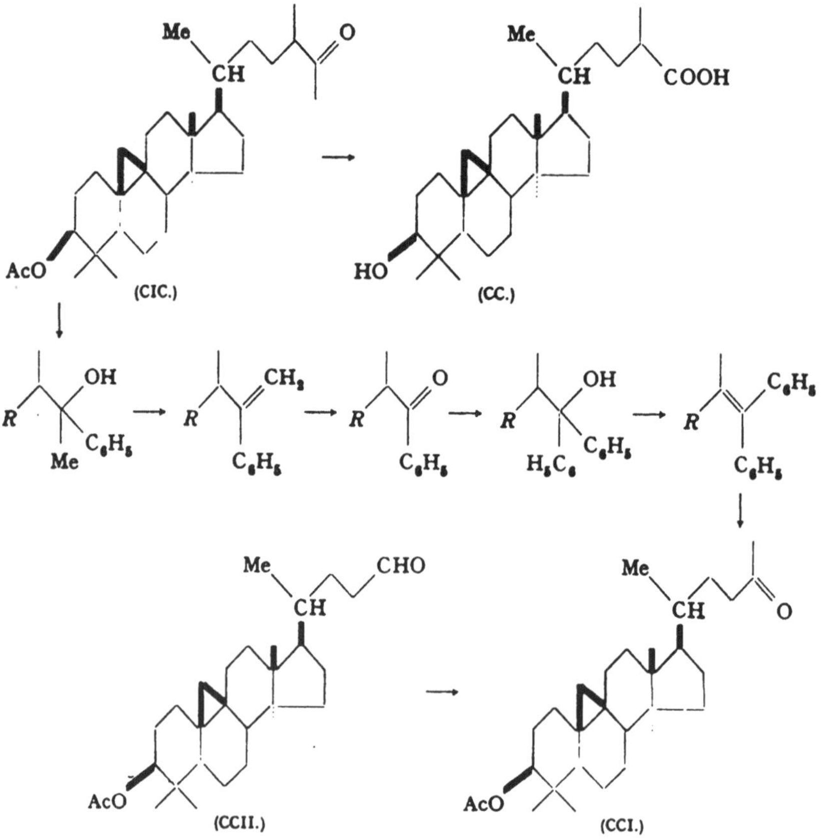

(CIC.) → (CC.)

(CCII.) ← (CCI.)

Chart 8. Relationship between *cyclo*Laudenol and *cyclo*Artenol.

which strongly suggested that they did not differ simply in the position of the double bond in the side-chain. HENRY, IRVINE and SPRING (*106*) have shown that the *cyclo*laudenol side-chain has nine carbon atoms, differing from *cyclo*artenol in containing the group —CHMe · C(Me) : CH_2 instead of an *iso*butenyl group, —CH : CMe_2. The *nor*ketone (CIC) described above was oxidised by potassium hypobromite to give the *bisnor*-acid (CC) characterised as its methyl ester. The structure (CIC) was confirmed by the degradation of the *nor*ketone by the BARBIER-WIELAND technique to a *trisnor*ketone (CCI). This was also obtained from *cyclo*-artenyl acetate by careful ozonolysis to the *trisnor*aldehyde (CCII), converted to (CCI) by treatment with diazomethane (*Chart 8*).

The stereochemistry at $C_{(24)}$ is the same as that in ergostane and eburicane (*106*, cf. *90*), the $C_{(24)}$ methyl group having the "b" configuration (CXCVI, p. 99). The molecular rotations of *cyclo*laudenol derivatives are more negative than those of the corresponding *cyclo*artanyl derivatives; ergostane derivatives differ in the same way from derivatives of cholestane. Laudane which has been prepared from *cyclo*laudenol, is thus identical with the hitherto-unprepared parent hydrocarbon of eburicoic acid, eburicane (LXXXIX, p. 73). *cyclo*Laudenol is therefore 9 : 19-*cyclo*-eburic-25-en-3β-ol.

III. Euphane Type Group.

1. Euphol.

(Eupha-8 : 24-dien-3β-ol) $C_{30}H_{50}O$ (CCIII, p. 102).

Euphol was first isolated in a state of purity by NEWBOLD and SPRING (*138*) from "euphorbium", a commercial resin (latex) from an unidentified *Euphorbia* species. It has since been found to occur in *E. resinifera* (*67*, *69*, *121*), *E. electa* (*114*), *E. canariensis* (*80*), *E. handiensis* BURCHARD (*82*), *E. ingens* MAY (*28*), and *E. tirucalli* (*128*).

NEWBOLD and SPRING (*138*) showed that euphol was a tetracyclic alcohol of formula $C_{30}H_{50}O$ and that it had two double bonds, one of which could be hydrogenated. This double bond was later found by JEGER and KRÜSI (*114*) to be present in an *iso*propylidene group. WARREN and his co-workers (*128*) confirmed this and further showed that the reducible double bond was not situated between two tertiary carbon atoms. These authors suggested that euphol might belong to the elemi-acid group (p. 116) since on selenium dehydrogenation 1 : 2 : 8-trimethylphenanthrene was the sole product isolated. This parallels the behaviour of elemolic acid (*150*) and lanosterol (*149*). Another similarity was revealed by the oxidation of dihydroeuphyl acetate (CCIV) with chromic acid to a ketone (CCV) and a diketone (CCVI) (*128*, *93*, *158*, *159*, *160*). These are analogous to compounds obtained by the oxidation of dihydrolanosteryl acetate

(*149*) (p. 54) and the acetate of methyl elemolate (*150*). However, euphol was not identical with any known compound related to either lanosterol

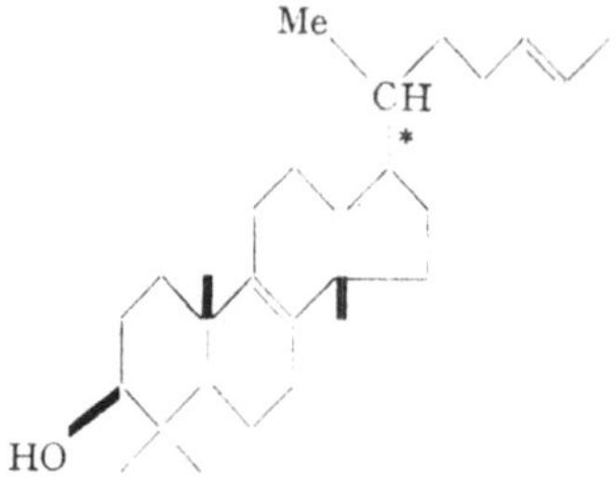

(CCIII.) Euphol.

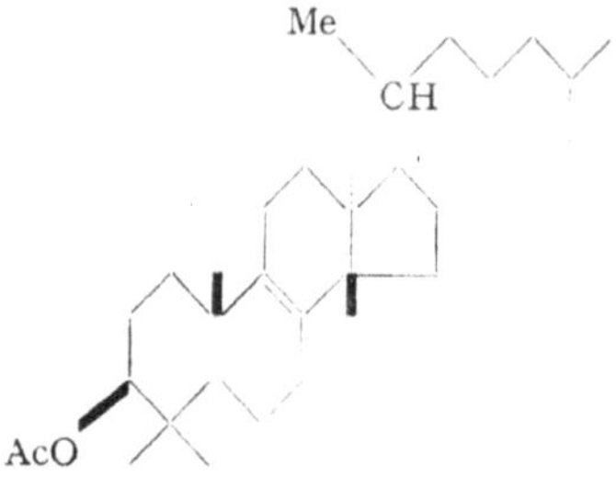

(CCIV.) Dihydroeuphyl acetate.

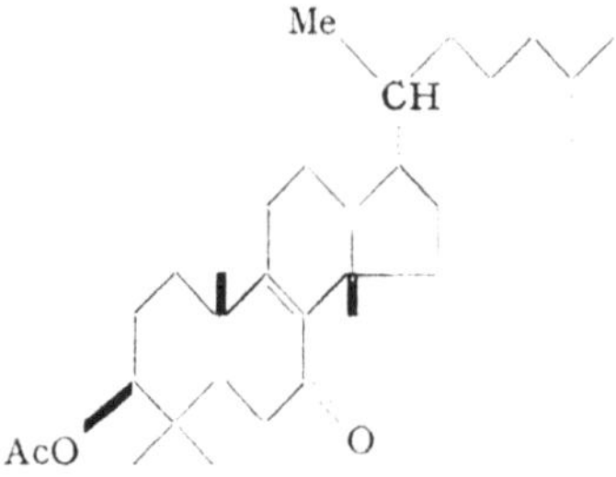

(CCV.)

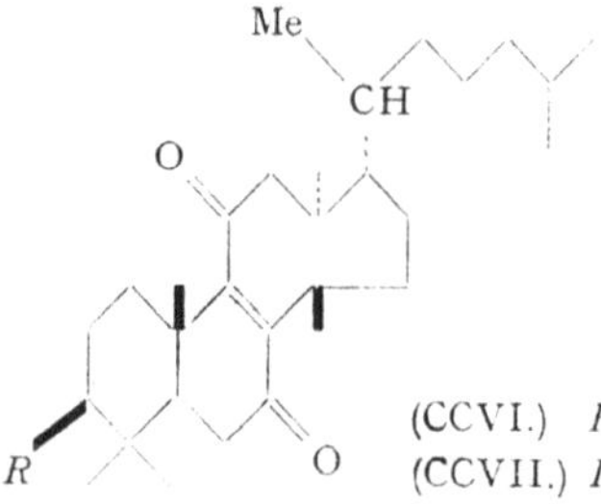

(CCVI.) R = OAc.
(CCVII.) R = H.

or elemolic acid, nor was *epi*euphol (eupha-8 : 24-dien-3α-ol) (*24*). Oxidation of euphene with chromic acid gave similar results, 7 : 11-diketoeuph-8-ene (CCVII) being obtained (*140*).

The nature of the side-chain of euphol was revealed by a series of degradations carried out by various workers. DUPONT, DULOU and VILKAS (*66*), as well as WARREN and his co-workers (*23, 122*), found that fission of the *iso*propylidene group with potassium permanganate gave a carboxylic acid, proving the presence of the $—CH{=}CMe_2$ grouping. CHRISTEN, JEGER and RUZICKA showed that this partial structure could be extended to $Me \cdot \underset{|}{CH} \cdot [CH_2]_2 \cdot CH{=}CMe_2$ (*44*). Euphene (CCVIII) was oxidised with chromic acid to give a diketo*trisnor*-acid which was isolated as its methyl ester (CCIX). The ene-dione grouping of (CCIX) was reduced with zinc dust in acetic acid and the resulting diketo-ester (CCX) converted through the intermediate products (CCXI), (CCXII), and (CCXIII) to the triketone (CCXIV) which gave iodoform on treatment with potassium hypoiodite.

The elucidation of the nature of the side-chain of euphol indicated the close relationship of euphol (CCIII) and lanosterol (I, p. 50). Further

* There is as yet no certainty about the α-configuration of the side-chain (cf. *1, 16*).

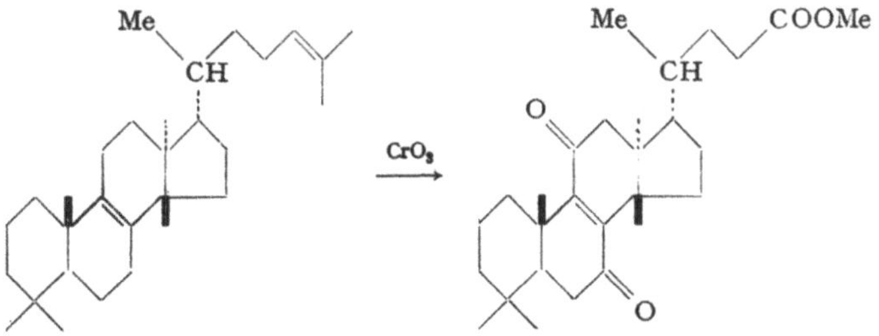

Me CH COOMe CH O CrO$_3$ O

(CCVIII.) Euphene. (CCIX.)

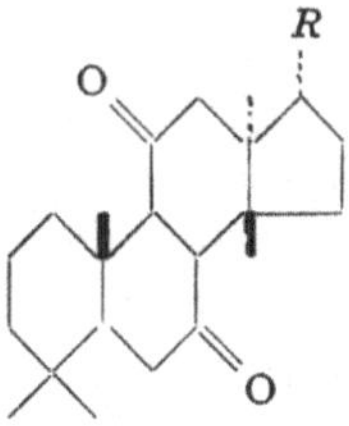

(CCX.) $R = Me\cdot CH\cdot[CH_2]_2\cdot COOMe$.

(CCXI.) $R = Me\cdot CH\cdot[CH_2]_2\cdot C(OH)(C_6H_5)_2$.

(CCXII.) $R = Me\cdot CH\cdot CH_2\cdot CH{=}C(C_6H_5)_2$.

(CCXIII.) $R = Me\cdot C{=}CH\cdot CH{=}C(C_6H_5)_2$.

(CCXIV.) $R = Me\cdot C{=}O$.

evidence pointing in the same direction was obtained when it was shown (*43*) by the characteristic phosphorus pentachloride reaction that the hydroxyl group of euphol is situated in a typical triterpene ring A (CCXV). In addition, oxidation of diketo-euphenyl acetate (CCVI) with selenium dioxide gave "triketo-euphadienyl acetate" (CCXVI) which showed maximal ultraviolet absorption at 2850 Å. ($\varepsilon = 10\,000$), indicative

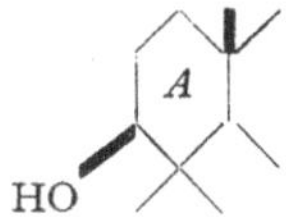

(CCXV.) 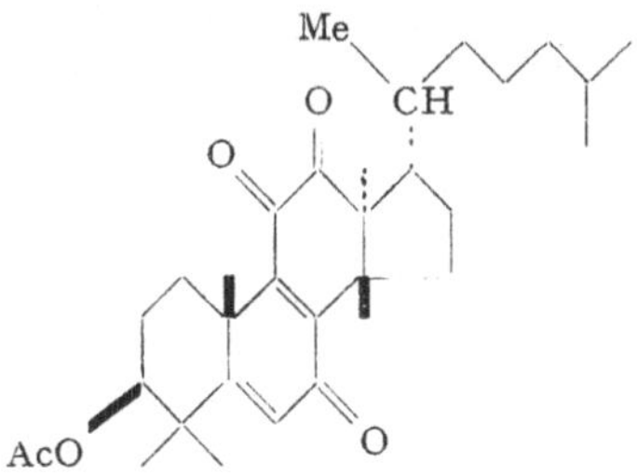 (CCXVI.) 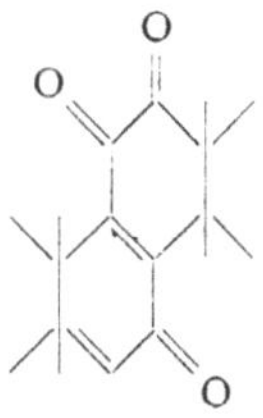 (CCXVII.)

of the grouping (CCXVII) (*43*). This selenium dioxide oxidation was studied in more detail by Christen et al. (*43*) who also isolated the intermediate dione-diene (CCXVIII). They further converted (CCXVI) to a dicarboxylic acid (CCXIX) by fission of the α-diketone grouping with alkaline hydrogen peroxide. These reactions were all similar to those already carried out in the lanosterol series (p. 55).

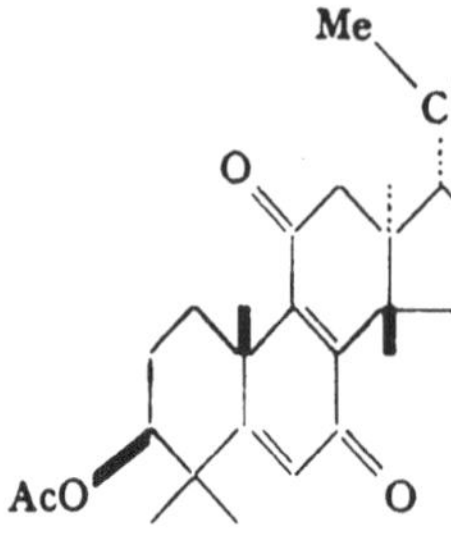

(CCXVIII.) 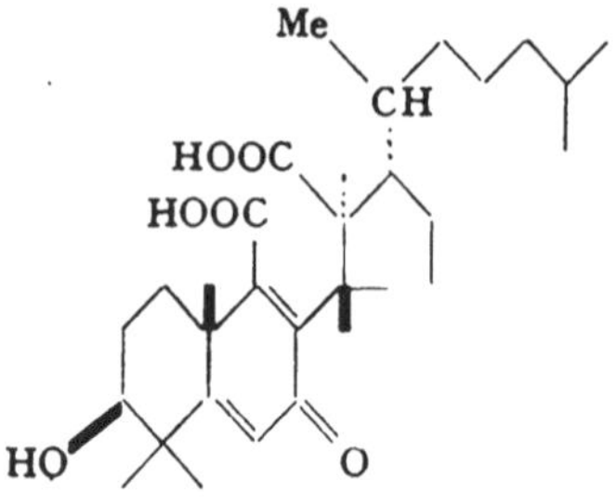 (CCXIX.)

Although the above evidence indicated a close relationship of euphol to lanosterol the nature of the difference between them was, for a time, obscure. Gradually evidence accumulated which indicated that the inert double bond of euphol has somewhat different properties from that of lanosterol and that ring *C* of euphol must be fused to the terminal ring in a manner different from that of lanosterol. For instance, derivatives of lanosterol and eburicoic acid (p. 68) are much more readily converted by selenium dioxide into dehydro-derivatives than is dihydroeuphyl acetate (*160*, cf. *77*). Further, the epoxides of the inert double bonds of lanosteryl and dihydrolanosteryl acetates (*31*) and acetyl-dihydroeburicoic acid (*125*) are most unstable, giving the dehydro-compounds very easily, whereas the epoxides of dihydroeuphyl, dihydrotirucallyl, and dihydro-euphorbyl acetates are only converted to the corresponding dehydro-compounds such as eupha-7 : 9(11)-dienyl acetate (CCXX) on treatment with an acidic dehydrating agent (*5*). These conjugated dienes (cf. CCXX) of the euphol group show maximal ultraviolet absorption at shorter wavelengths (2320, ca. 2395, and ca. 2470 Å.) than those (*e. g.* CCXXI) of the lanosterol group (cf. p. 53). They also have very large negative rotations (*58*). Most important of all, euphenyl acetate undergoes an

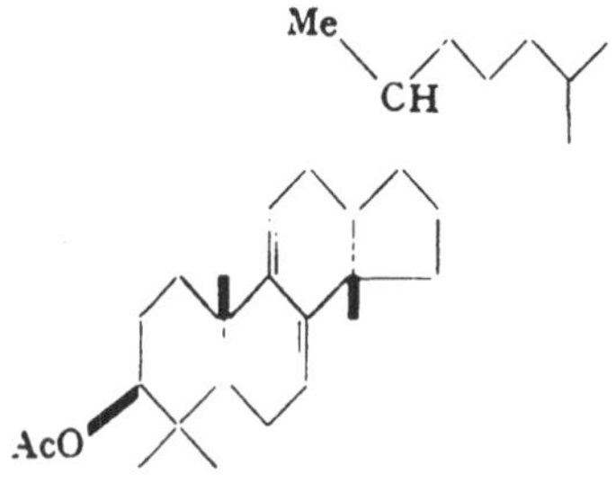

(CCXX.) 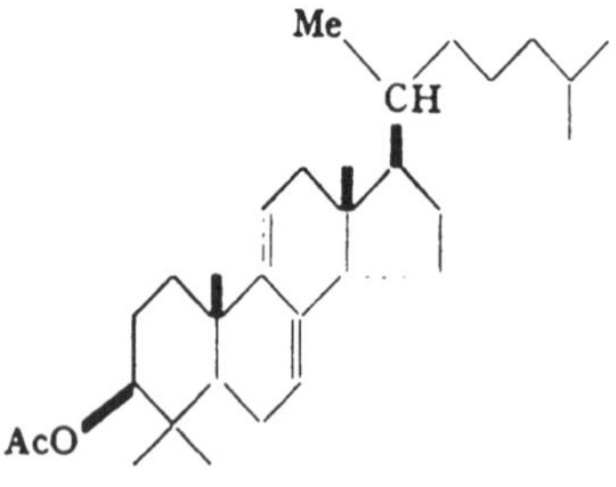 (CCXXI.)

acid-catalysed rearrangement to give *iso*euphenyl acetate (*43*, *58*, *160*), the double bond of which is differently situated from the trisubstituted double bond of lanost-7-enyl acetate (CCXXII), the corresponding re-arrangement product from lanost-8-enyl acetate (dihydro-lanosteryl acetate, p. 66). Both ultraviolet light absorption (*58*) and chemical data (*43, 158*) indicated that the double bond of *iso*euphenyl acetate was tetrasubstituted and exocyclic to one ring. Treatment of *iso*euphenyl acetate with osmium tetroxide produced an α-glycol which with lead tetraacetate gave a diketone, also obtained by ozonolysis of *iso*euphenyl acetate (*43*). The diketone formed a dioxime and did not react with sodium hypoiodite solution. JEGER et al. (*43*) explained these reactions on the basis of the partial structures (CCXXIII), (CCXXIV), and (CCXXV) for euphol, *iso*euphenol, and the diketone, and suggested (CCXXVI) as a complete structure for euphol (cf. *142*). This structure differs from that of lanosterol in having a hydrogen atom at $C_{(14)}$. It does not, however,

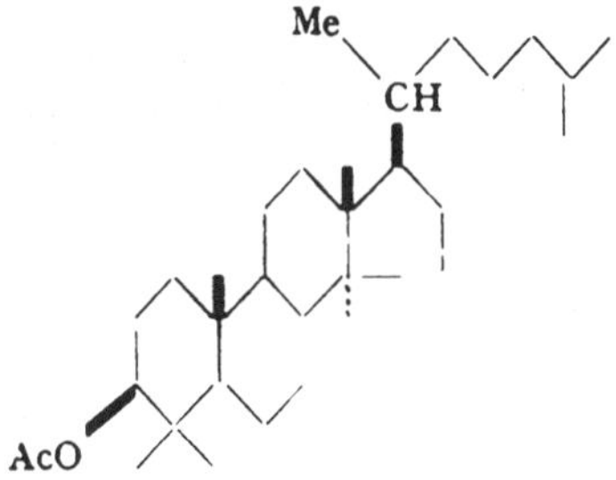

Me CH AcO

(CCXXII.)

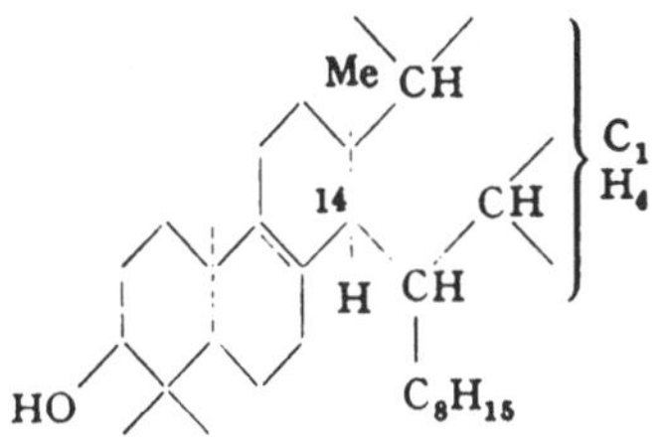

Me CH 14 CH H CH C_8H_{15} HO C_1 H_4

(CCXXIII.)

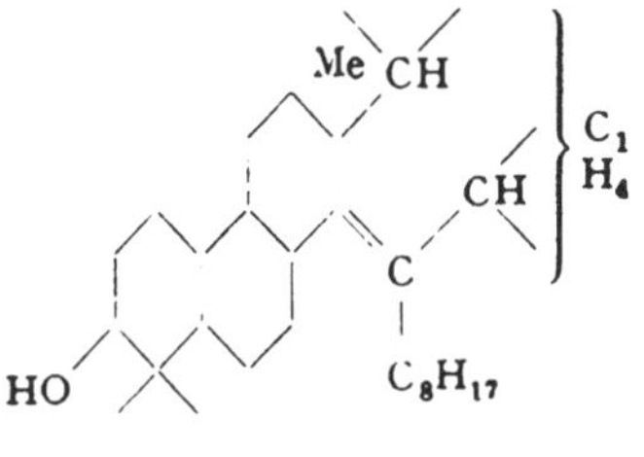

Me CH CH C C_8H_{17} HO C_1 H_4

(CCXXIV.)

Me CH O AcO O CH C C_8H_{17} C_1 H_4

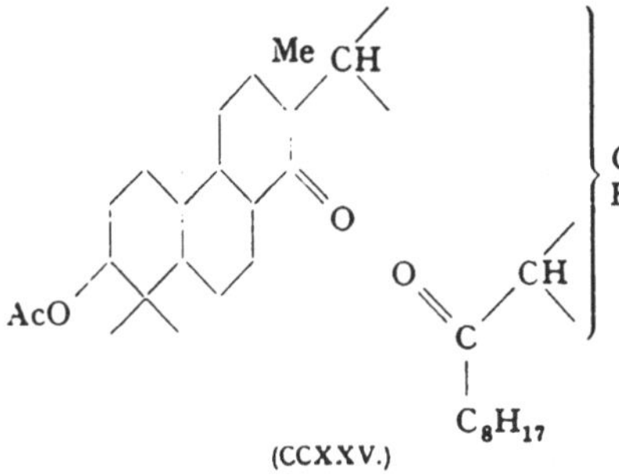

(CCXXV.)

satisfactorily explain the formation of 1 : 2 : 8-trimethylphenanthrene on dehydrogenation. Later RUZICKA (*142*) suggested that structure (CCXXVII) should be considered for euphol, but did not exclude a structure stereoisomeric with lanosterol, *i. e.*, having a *cis* *C/D* ring junction.

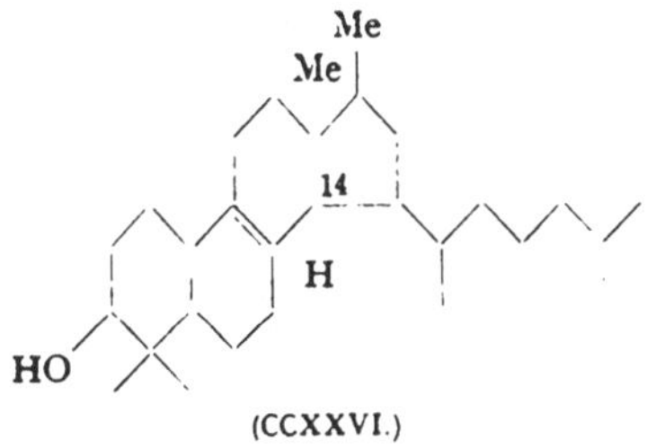

Me Me 14 H HO

(CCXXVI.)

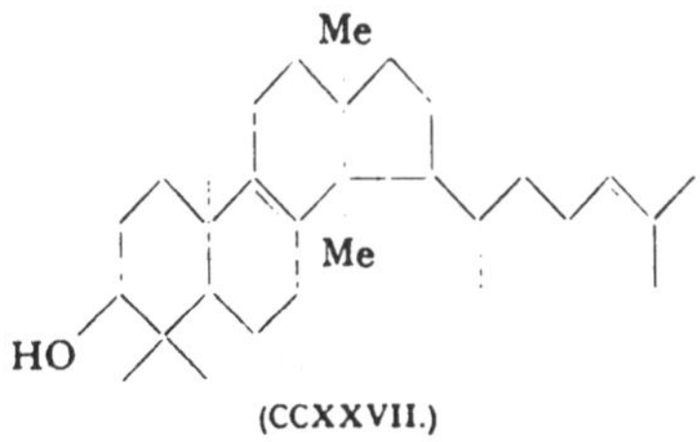

Me Me HO

(CCXXVII.)

In the light of some new evidence Barton, McGhie, Pradhan and Knight (*16*) have recently discussed the chemistry of euphol and have formulated it as (CCXXVIII) leaving open the question of configuration at $C_{(17)}$. By infrared measurements they have shown that euphene and lanostene have the same number of methyl groups and that, after allowing for the rings *A*, *B*, and *C* of euphol and its side-chain, only three carbon atoms are unaccounted for. Ring *D* cannot therefore be more than five-membered. On dehydrogenation of *iso*euphadiene under conditions whereby euphadiene gave 1 : 2 : 8-trimethylphenanthrene, only 1 : 2 : 5-trimethylnaphthalene was obtained. These results indicate that methyl groups are attached to $C_{(13)}$ and $C_{(14)}$ of euphadiene and are in accord with the migration of a methyl group from $C_{(14)}$ to $C_{(8)}$. *iso*Euphadiene may then be formulated either as (CCXXIX) or as (CCXXXI) depending upon whether or not a $C_{(13)}$ methyl group moves to $C_{(14)}$. The diketone formed on fission of the double bond of *iso*euphenyl acetate (CCXXX) would then have partial structures (CCXXXII) or (CCXXXIII). Barton, McGhie and their co-workers argued that (CCXXXII) must be correct for the diketone since the latter took up 5 moles of bromine. This led to (CCXXIX) for *iso*euphadiene and suggested (CCXXVIII) for euphol. The configuration of the methyl groups at $C_{(13)}$ and $C_{(14)}$ follows from a consideration of the reason for the isomerisation of euphenol. The stereochemistry of (CCXXVIII) provides a conformational driving force which, following proton addition at $C_{(9)}$ from the α-face of the molecule, causes a concerted migration of the $C_{(13)}$ and $C_{(14)}$ methyl groups to give the favourable ***trans-anti-trans-anti*** arrangement of (CCXXIX). The stereochemistry of the *C*/*D* ring fusion of euphol (CCXXVIII) can account for

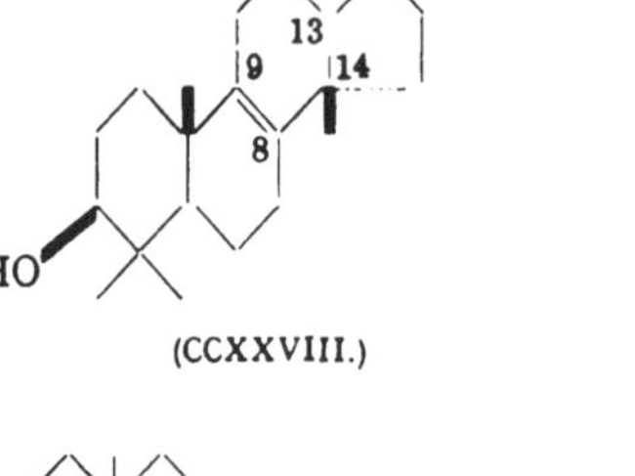

(CCXXVIII.)

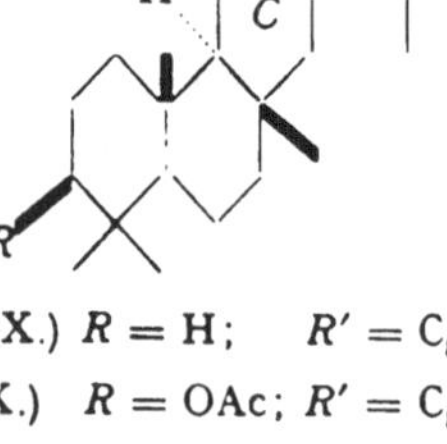

(CCXXIX.) $R = H$; $R' = C_8H_{15}$.
(CCXXX.) $R = OAc$; $R' = C_8H_{17}$.

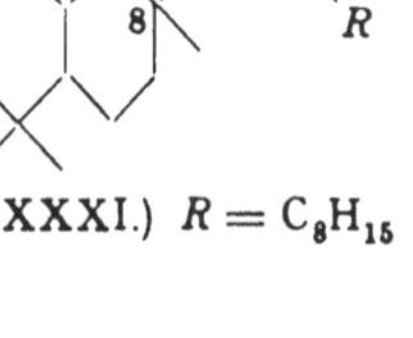

(CCXXXI.) $R = C_8H_{15}$

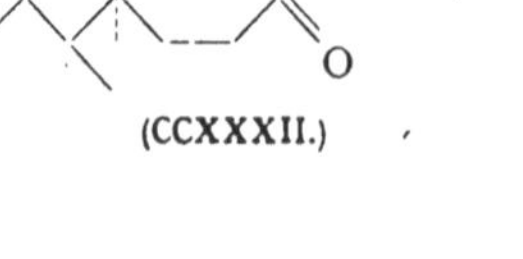

(CCXXXII.)

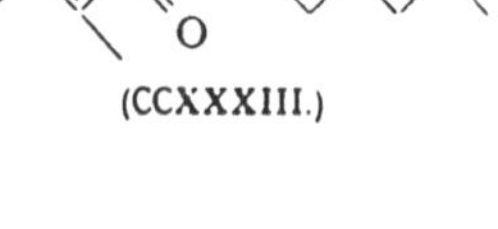

(CCXXXIII.)

'abnormal' hydrogenation of 7 : 11-diketoeuphenyl acetate (CCVI, p. 102) (*6, 120*) and for the zinc-dust reduction of the same compound to a 7 : 11-diketoeuphenyl acetate with the $C_{(8)}$ and $C_{(9)}$ hydrogen atoms *cis* to each other (*120*). BARTON, MCGHIE, et al. (*16*) from a comparison of rotation data in the lanosterol (*11*) and euphol (*44*) series concluded initially that the side-chain was attached to $C_{(17)}$ on the α-side and that the configuration at $C_{(20)}$ was opposite to that at $C_{(20)}$ in lanosterol.

The structure (CCXXVIII) for euphol has been supported by the work of ARIGONI, VITERBO, DÜNNENBERGER, JEGER and RUZICKA (*1*). They have shown that a methylene group in a six-membered ring must be situated adjacent to the double bond of *iso*euphenyl acetate, since oxidation of the latter with *tert.*-butyl chromate gave the keto-epoxide

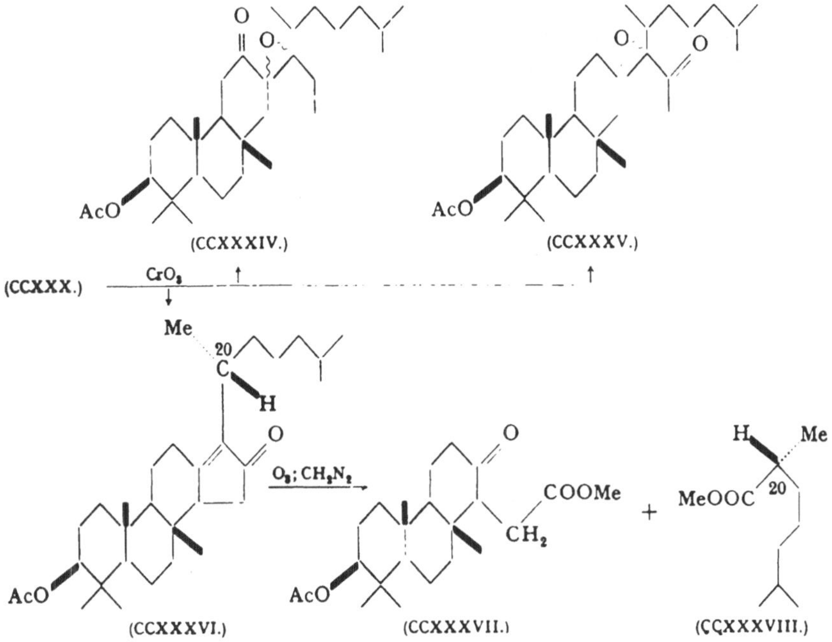

(CCXXXIV.) (CCXXXV.)

(CCXXX.) CrO_3

(CCXXXVI.) O_3; CH_2N_2 (CCXXXVII.) + (CCXXXVIII.)

(CCXXXIV) which had an infrared spectrum characteristic of a six-membered ring ketone. Also formed was the αβ-unsaturated ketone (CCXXXVI), the ultraviolet spectrum of which showed the hypsochromic effect typical of a *cyclo*pentenone. These two compounds, along with the other keto-epoxide (CCXXXV) [m. p. 179–181°, $[\alpha]_D + 14°$ (*c.*, 0,8 in $CHCl_3$)], have also been obtained by DAWSON and HALSALL (*55*) by oxidation of *iso*euphenyl acetate with chromic acid in acetic acid. Ozonolysis of the αβ-unsaturated ketone (CCXXXVI), followed by esterification of the acidic products with diazomethane gave the keto-ester (CCXXXVII) and optically pure *D*(—)-methyl 2 : 6-dimethylheptanoate (CCXXXVIII)

($[\alpha]_D - 17°$) of known configuration (*1*). From this it follows that the configuration at $C_{(20)}$ in euphol is the same as that at $C_{(20)}$ in lanosterol.

Earlier Ruzicka (*142*) had described the isolation of $L(+)$-methyl 2 : 6-dimethylheptanoate ($[\alpha]_D + 6°$) following a degradation of the diketone (CCXXXII). This was condensed with ethyl formate in the presence of sodium methoxide to yield a hydroxymethylene derivative which was split by alkaline hydrogen peroxide to give methyl 2 : 6-dimethylheptanoate with the dextro-rotation quoted above (*1*). The formation of the dextro-rotatory product is believed to be due to inversion occurring at $C_{(20)}$ in the preparation of the hydroxymethylene derivative. On treatment of the diketone (CCXXXII) with alkali it underwent a reverse Michael reaction to give (CCXXXIX) and (CCXL). These products account for 29 carbon atoms, the remaining carbon atom presumably being lost as formaldehyde.

The structure of euphol can now be regarded as being very probably (CCIII, p. 102) ($13\alpha : 14\beta : 17\alpha$-lanosterol) apart from uncertainty about the configuration at $C_{(17)}$. Evidence based on molecular rotation data is

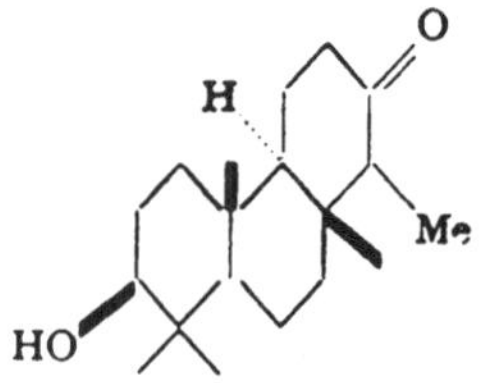

(CCXXXIX.) 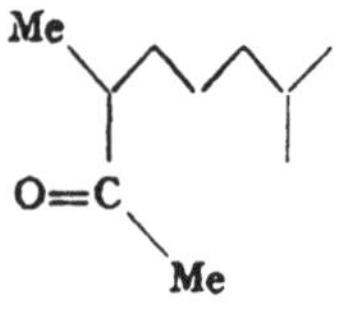 (CCXC.)

inconclusive (*16*). Arguments in favour of the 17α-configuration of the side-chain, involving the mechanism of the rearrangement of euphenol to *iso*euphenol (*1*), have not yet received experimental support.

2. Butyrospermol.

$C_{30}H_{50}O$ (CCXLI).

Just over twenty years ago Heilbron, Moffet and Spring (*105*) described the isolation of a tetracyclic diethenoid alcohol, basseol, from the non-saponifiable matter of shea-nut fat from the West African tree *Butyrospermum parkii*. Later this alcohol was reported to possess the remarkable property of isomerising with acidic reagents to the pentacyclic triterpene, β-amyrin (*28*, *29*). Such an isomerisation would be of very great importance and further investigation of it was obviously indicated. However, in recent examinations of the non-saponifiable matter of shea-nut fat the isolation of basseol has never been reported (*57*, *66*, *103*, *154*) except by Pinto Coelho (*139*). Instead, a new tetracyclic triterpene, butyrospermol, has been isolated (*103*, *154*). This triterpene has also been isolated along with *cyclo*artenol (p. 96) from the fruits of *Artocarpus integrifolia* (*13*). BAUER and MOLL (*19*) claim to have isolated basseol during an investigation of the non-saponifiable matter of shea-nut fat, but since no detailed examination of its properties was made it could well have been butyrospermol.

HEILBRON, JONES and ROBINS (*103*) showed that butyrospermol was a tetracyclic diethenoid secondary alcohol. Of the two double bonds, one is present in an *iso*propylidene group, and is easily hydrogenated. The other cannot be hydrogenated but it reacts with perbenzoic acid. These results were confirmed by SEITZ and JEGER (*154*) who also prepared the hydrocarbon, butyrospermene, derived from dihydrobutyrospermol. They concluded from its infrared spectrum that the less reactive double bond of butyrospermol was fully substituted. This conclusion differs from that recently reached by the writers (see below). In fact the infrared spectrum published by SEITZ and JEGER (*154*) has a small peak at ca. 830 cm.$^{-1}$ indicative of a trisubstituted double bond.

When butyrospermyl acetate was treated with hydrogen chloride in chloroform, addition occurred at the *iso*propylidene group and simultaneously isomerisation of the less reactive double bond took place (*57*).

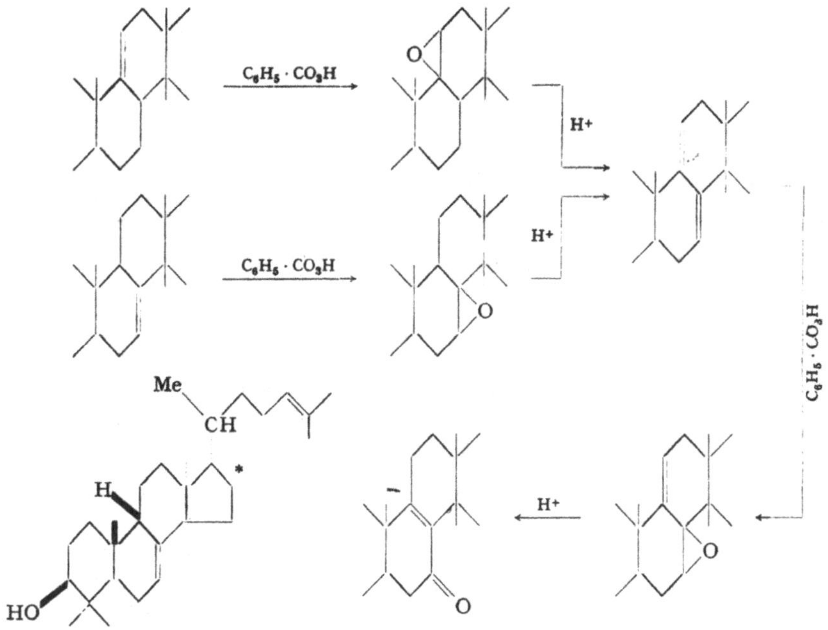

$C_6H_5 \cdot CO_3H$ $C_6H_5 \cdot CO_3H$ H+ H+ $C_6H_5 \cdot CO_3H$ H+

(CCXLI.) Butyrospermol.

* See Footnote p. 102.

Dehydrochlorination of the "hydrochloride" gave an isomer of butyrospermyl acetate (*iso*butyrospermyl acetate) containing a vinylidene group. Hydrogenation of the *iso*-acetate gave a dihydro-derivative also obtained directly from dihydrobutyrospermyl acetate by acidic isomerisation. The properties of the double bond in the isomeric dihydro-acetate were different from those of the double bond in *iso*euphenol. It appeared to be extremely unreactive towards per-acids while the ultraviolet light absorption in the region 2100–2250 Å. was less intense than with *iso*euphenol (57). Accepting the conclusions of Seitz and Jeger concerning the less reactive double bond of butyrospermol, Dawson, Halsall, Jones and Robins (57) suggested that the isomerisation of dihydrobutyrospermol was analogous to the isomerisation of lanost-8-enol to lanost-7-enol (cf. p. 66). However, recent detailed examination of the infrared spectra of derivatives of dihydrobutyrospermol and of dihydro*iso*butyrospermol

(CCXLII.) 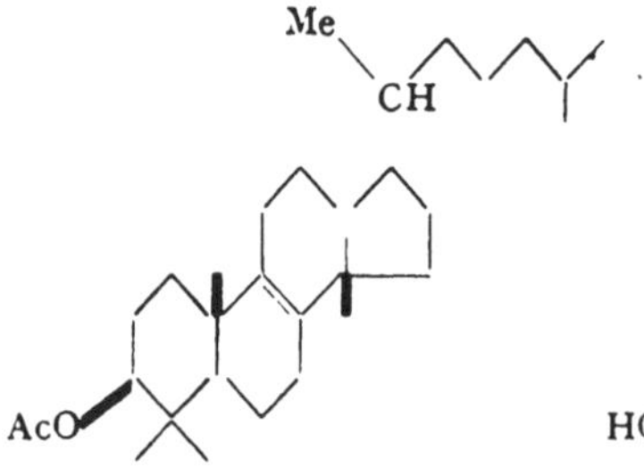 (CCXLIII.) 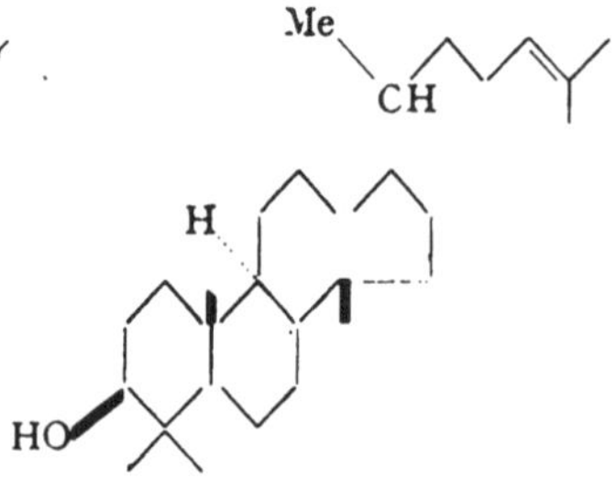 (CCXLIV.)

by Dawson, Halsall, Jones and Meakins (56) indicates that the less reactive double bond of butyrospermol is *trisubstituted* and that of *iso*-butyrospermol is *tetrasubstituted*.

Butyrospermol has a normal triterpene ring *A* (*12*), dihydrobutyrospermol undergoing the typical retro-pinacolinic dehydration to a hydrocarbon which gave acetone on ozonolysis (55). In an attempt to convert dihydrobutyrospermyl acetate into the corresponding conjugated diene, to determine from its ultraviolet spectrum whether it belonged to the euphol or lanosterol group, the acetate has been treated with perbenzoic acid in chloroform. The main product was 7-ketoeuph-8-en-3β-yl acetate (55). Butyrospermol therefore must have a euphol-type carbon skeleton and differ from euphol (CCIII, p. 102) in having a 7 : 8- or a 9 : 11-double bond. Either possibility would account for the formation of 7-ketoeuph-8-en-3β-yl acetate by the mechanism indicated.

Additional evidence in favour of butyrospermol being a double bond isomer of euphol has been obtained by further consideration of the isomer of dihydrobutyrospermyl acetate described above. This is, in fact, dihydroeuphyl acetate (CCXLIII). Since dihydroeuphyl acetate is itself isomerised by acid, its isolation might appear to be rather unexpected but the isomerisation of dihydrobutyrospermyl acetate was carried out for only two hours at 0° whereas dihydroeuphyl acetate was kept for four days at 20° in chloroformic hydrogen chloride.

So far no conclusive evidence has been obtained to indicate whether the less reactive double bond of butyrospermol is at the 7 : 8- or the 9 : 11-position. In the case of lanost-7-ene and lanost-9(11)-ene derivatives the double bond of the former, but not that of the latter can be isomerised to the 8 : 9-position (cf. p. 66). If the nuclear double bond of butyrospermol is between $C_{(7)}$ and $C_{(8)}$ there are two possible structures, (CCXLIV) with the 9α- and (CCXLI) with the 9β-configuration. The former is excluded since the dihydro-acetate corresponding to it, now prepared from euphol (*16*), is not identical with dihydrobutyrospermyl acetate. Structure (CCXLI) with its 9β-substituent provides an explanation of the unusual but similar molecular rotation differences observed when butyrospermol and *cyclo*artenol are oxidised to the corresponding ketones since *cyclo*artenol also has a 9β-substituent, in this case the $C_{(19)}$ methylene group (p. 96). BARTON commented on the similar rotation differences and suggested that they might be significant (*13*). In the light of this striking molecular rotation difference butyrospermol is tentatively formulated as (CCXLI) (p. 109).

When butyrospermol is isolated as its acetate from shea-nut fat, great difficulty is experienced in eliminating β-amyrin acetate, even with the aid of chromatographic techniques, and it is possible to obtain samples of seemingly nearly pure butyrospermyl acetate, m. p. ca. 140–143° (*i. e.* 3–4° low) ($[\alpha]_D$ ca. + 20°) still containing 15–20% of β-amyrin acetate. Recently Dr. G. D. MEAKINS (*134*) has examined samples of fat from many different varieties of shea-nuts but in no case has any evidence been obtained of the presence of "basseol". It now seems probable that the original "basseol acetate" was, in fact, butyrospermyl acetate contaminated with β-amyrin acetate. The constants reported for "basseol acetate" (m. p. 141°; $[\alpha]_D$ + 22.4°) agree well with those of the β-amyrin acetate-contaminated butyrospermyl acetate described above. The presence of the β-amyrin acetate could explain the isolation of some of this compound in acidic isomerisation reactions with "basseol acetate"

A sample of "dihydrobasseol acetate" has been reported to be identical with dihydroeuphyl acetate (*66*). An explanation of this observation is that catalytic hydrogenation of the *iso*propylidene group of butyrospermyl acetate can also, under certain conditions, bring about isomerisation of the less reactive double bond to the 8 : 9-position, giving dihydroeuphyl acetate (*99*).

3. Tirucallol.

(Tirucalla-8 : 24-dien-3β-ol; Elema-8 : 24-dien-3β-ol) $C_{30}H_{50}O$ (CCXLV).

Tirucallol was first isolated a few years ago by HAINES and WARREN (92) from *Euphorbia tirucalli* resin in which euphol and the pentacyclic triterpene, taraxasterol, also occur, and later from *E. triangularis* DESF. (7). HAINES and WARREN showed that tirucallol was a tetracyclic alcohol with the formula $C_{30}H_{50}O$, containing an *iso*propylidene group and an inert ethylenic linkage (92). The hydroxyl group was later shown to be secondary, whilst the inert double bond was found to be in a position similar to that in lanosterol or euphol, since prolonged chromic acid oxidation of dihydrotirucallyl acetate (CCXLVI) gave a typical ene-dione (CCXLVII) (93). Oxidation of tirucallyl acetate with potassium permanganate introduced two keto-groups adjacent to the inert double bond and split off the *iso*propylidene group with the formation of a carb-

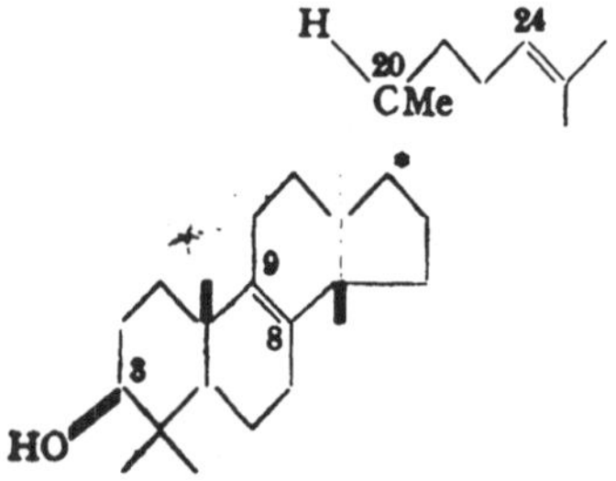

(CCXLV.) Tirucallol.

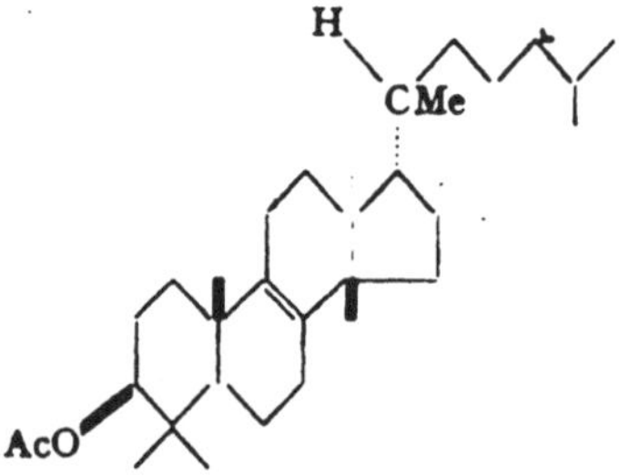

(CCXLVI.) Dihydrotirucallyl acetate.

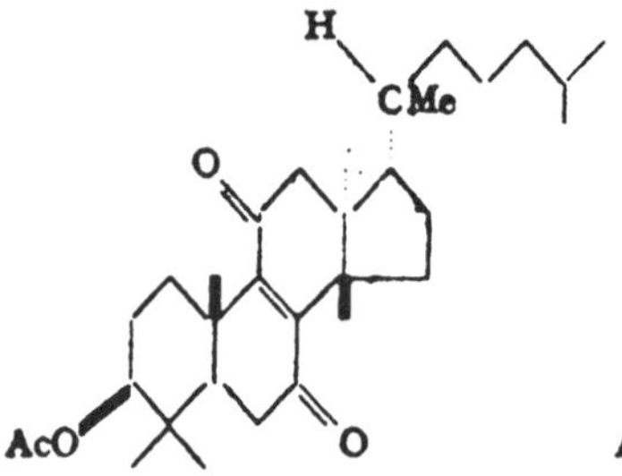

(CCXLVII.)

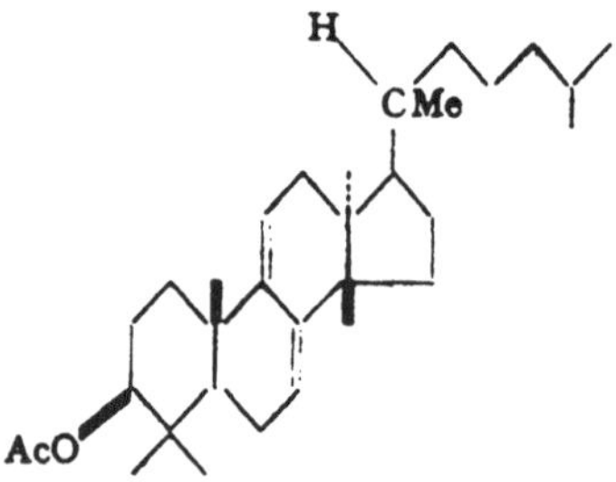

(CCXLVIII.)

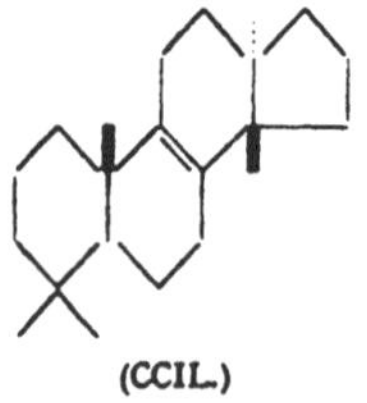

(CCIL.)

* The α-configuration at $C_{(17)}$ is not yet proved.

oxylic acid (*23*). The grouping —CH=CMe_2 is therefore present in tirucallol.

Dihydrotirucallyl acetate gives a stable epoxide with perbenzoic acid converted by sulphuric acid in acetic acid into the conjugated dehydro-compound (CCXLVIII) with a very large negative rotation (*5*). The ultraviolet light absorption of this compound is identical with that of eupha-7 : 9(11)-dienyl acetate and indicates, together with the large negative rotation, that the environment of the inert double bond of tirucallol is the same as that of euphol (CCIII, p. 102) (*58*) and that its basic ring system is (CCIL).

JEGER, RUZICKA, and their co-workers (cf. *1*, *1a*) have recently isomerised dihydrotirucallyl acetate (CCXLVI) to *iso*tirucallenyl acetate (CCL). This was oxidised to the $\alpha\beta$-unsaturated ketone (CCLI), ozonolysis of which, followed by esterification of the acidic products with diazomethane gave $L(+)$-methyl 2 : 6-dimethylheptanoate (CCLII) and the keto-ester (CCLIII) identical with the keto-ester (p. 107) obtained from *iso*euphenyl acetate. Tirucallol is therefore 20-*iso*euphol (CCXLV), assuming that the $C_{(17)}$ configuration is the same as in euphol, but it has now been suggested that tirucallol is 17-*iso*-20-*iso*euphol (*5a*).

In addition ARIGONI, WYLER and JEGER (*2*) have shown that dihydrotirucallol is identical with 3β-hydroxyelem-8-ene derived from elemonic acid with structure (CCLXII, p. 116). In systematically naming tirucallol it appears preferable to base the nomenclature on the basic hydrocarbon tirucallane rather than on elemane, since use of the latter may lead to confusion with the parent hydrocarbon of the sesquiterpene alcohol elemol.

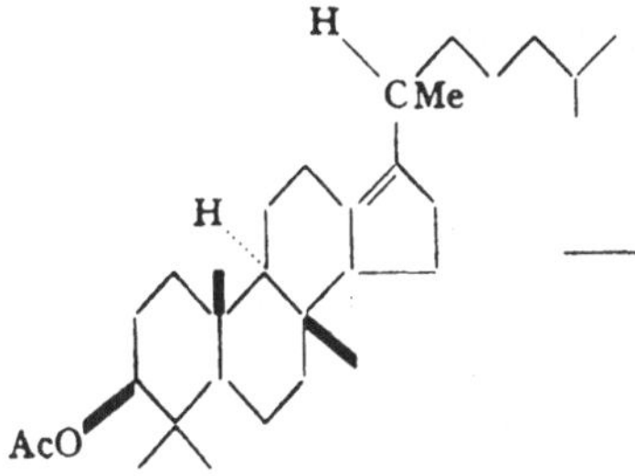

H CMe H AcO

(CCL.) *iso*Tirucallenyl acetate.

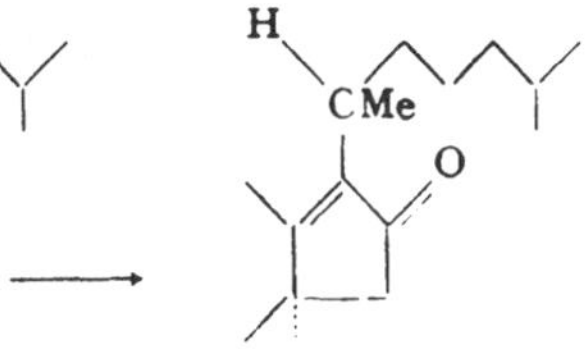

H CMe O

(CCLI.)

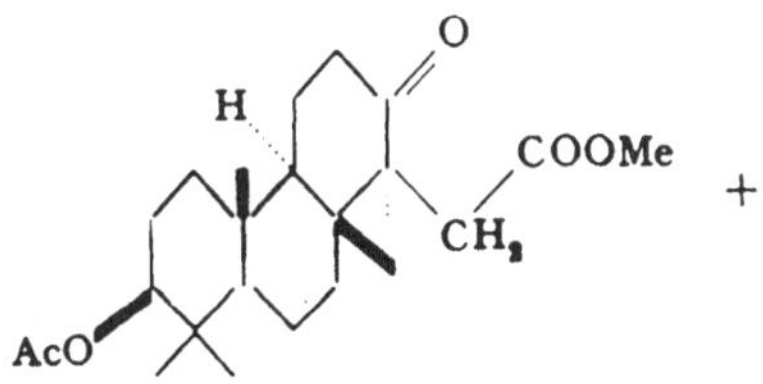

O H COOMe CH_2 AcO + H Me MeOOC

(CCLIII.) 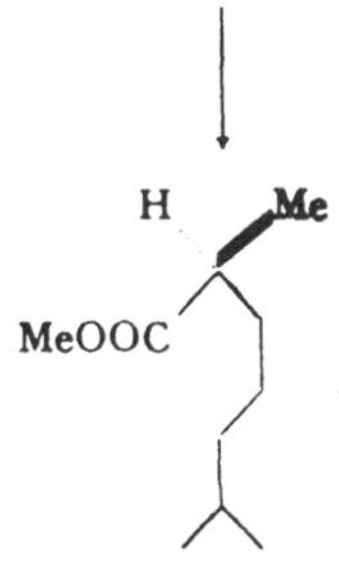 (CCLII.)

4. Euphorbol.

(α-Euphorbol, Euphorba-8 : 24(28)-dien-3β-ol) $C_{31}H_{52}O$ (CCLIV).

Euphorbol was first isolated by Bauer and Schenkel (*20*) from "Euphorbone", the light petroleum extract from the latex (resin) of *Euphorbia spp.*, and a little later from the same source by Schmid and Zacherl (*152*). Bauer and Schröder (*21*) found that ozonolysis of euphorbol gave formaldehyde indicating the presence of a vinylidene group $\rangle C{=}CH_2$. They also suggested that two alcohols isolated by Müller (*137*) from "euphorbone", to which the names vitorbol and novorbol were given, were, respectively, impure euphorbol and inhomogeneous. Euphorbol has recently been isolated from *Euphorbia resinifera* (*67*), *E. triangularis* (*7*), and *E. ingens* (*7*).

Euphorbol was later characterised by Newbold and Spring (*138*) as a tetracyclic triterpene alcohol with two double bonds, one of which could be hydrogenated, and the formula $C_{30}H_{50}O$ was suggested for it. As will be seen later the correct formula is $C_{31}H_{52}O$.

The presence of a vinylidene group was confirmed by Barbour, Warren and Wood (*7*). Ozonolysis of euphorbol gave formaldehyde

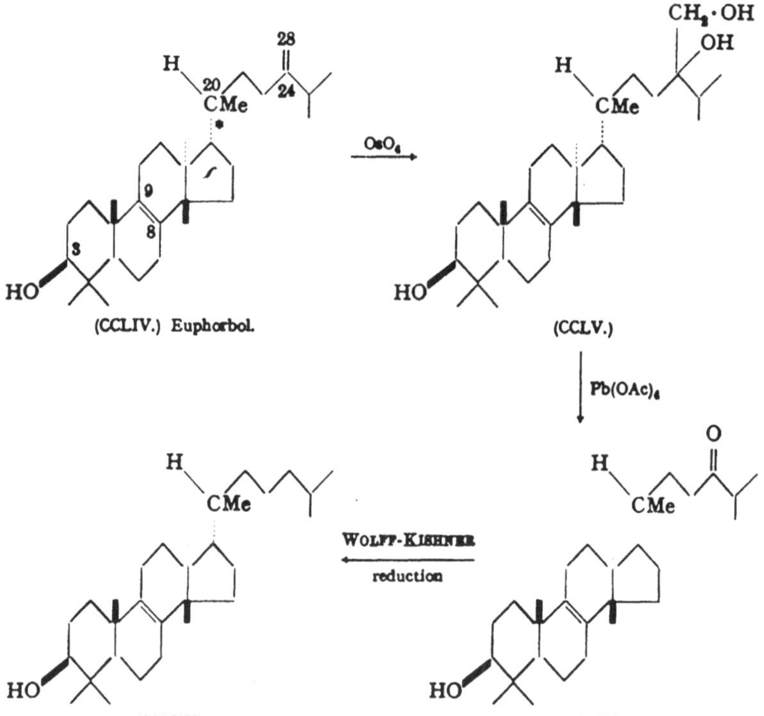

(CCLIV.) Euphorbol. $\xrightarrow{OsO_4}$ (CCLV.) $\xrightarrow{Pb(OAc)_4}$ (CCLVI.) $\xrightarrow{\text{Wolff-Kishner reduction}}$ (CCLVII.)

* The α-configuration at $C_{(17)}$ is not yet proved.

whilst treatment with osmic acid followed by hydrolysis of the product yielded euphorbenetriol (CCLV). Fission of the triol with lead tetra-acetate gave a *nor*-ketone (keto*nor*euphorbenol; CCLVI). These workers also showed that a chromic acid oxidation of euphorbenyl acetate yielded diketoeuphorbenyl acetate (CCLVIII) with the typical ene-dione chromophore. Further oxidation of diketoeuphorbenyl acetate with selenium dioxide gave the trione-diene (CCLIX) (*5*). Dihydroeuphorbyl (euphorbenyl) acetate gave a stable epoxide converted by sulphuric acid in acetic acid into the conjugated dehydro-compound (CCLX) with a large negative rotation (*5*). The ultraviolet light absorption of this compound is identical with that of eupha-7 : 9(11)-dienyl acetate and indicates, together with the large negative rotation, that the environment of the inert double bond of euphorbol is the same as that of euphol (CCIII, p. 102) and tirucallol (CCXLV) (*58*). The results described above were later confirmed by VOGEL et al. (*161*).

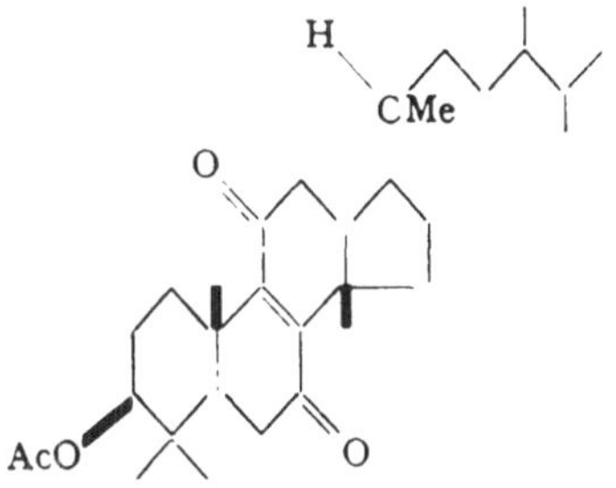

(CCLVIII.) $\xrightarrow{SeO_2}$ (CCLIX.)

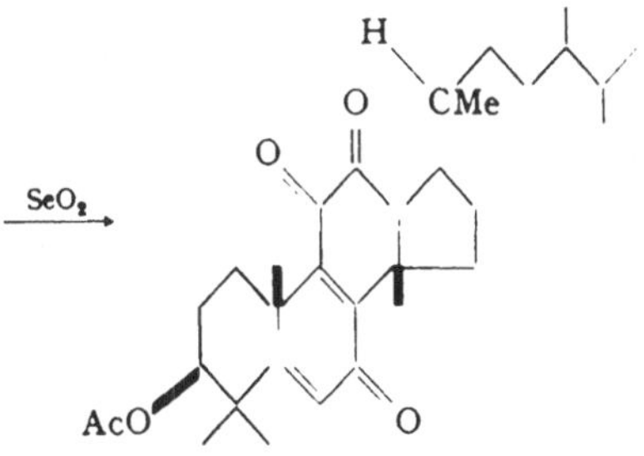

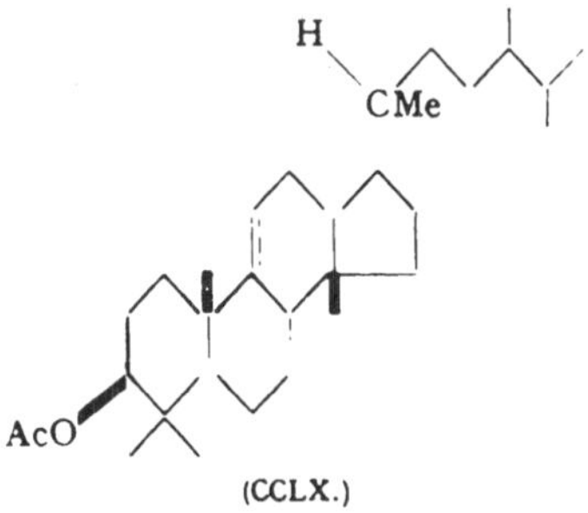

(CCLX.)

The C_{31} composition of euphorbol and the very probable structure (CCLIV) were recently established by ARIGONI, WYLER and JEGER (*2*). They showed that reduction of keto*nor*euphorbenol (CCLVI) by the WOLFF-KISHNER method yielded dihydrotirucallol, now known to be (CCLVII) (p. 114). The only unproven feature of (CCLIV) is the position of the methylene group. However, since all the other C_{31} tetracyclic triterpenes such as eburicoic acid (LXVIII; p. 69), the polyporenic acids and *cyclo*laudenol (CXCVI; p. 99) have the additional carbon atom attached to $C_{(24)}$, it is extremely likely that the methylene group of euphorbol is similarily located.

The molecular rotation difference between euphorbenyl acetate ($[M]_D \pm 0$) and tirucallenyl acetate ($[M]_D - 25$) is of approximately the same size and sign as the difference found between campestane and cholestane derivatives, but is of approximately the same size and opposite sign as the differences found between ergostane and cholestane derivatives and eburicane and lanostane derivatives (cf. *90*, *106*). The configuration at $C_{(24)}$ in euphorb-8-en-3β-yl acetate is therefore the same as that at $C_{(24)}$ in campestane and euphorb-8-en-3β-yl acetate is 13-*iso* : 14-*iso* : 17-*iso* (?) : 20-*iso* : 24-*iso*-eburic-8-en-3β-yl acetate.

5. The Acids of Manila Elemi Resin.

The two main acidic constituents of Manila elemi resin (ex *Canarium luzonicum*) are, a hydroxy-acid (3α-hydroxy-elema-8 : 24-dien-21-oic acid; $C_{30}H_{48}O_3$) (CCLXI), originally called α-elemolic acid, and the corresponding keto-acid (3-ketoelema-8 : 24-dien-21-oic acid; $C_{30}H_{46}O_3$) (CCLXII), originally called β-elemonic acid (*126*, *145*, *150*, *30*, *151*). In addition the dehydro-derivatives of these acids are also present. Before the discovery of the true relationship of elemolic and elemonic acids by RUZICKA,

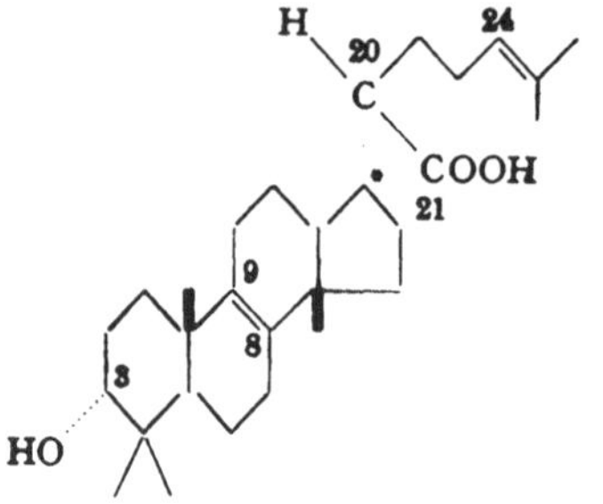

(CCLXI.) Elemolic acid.

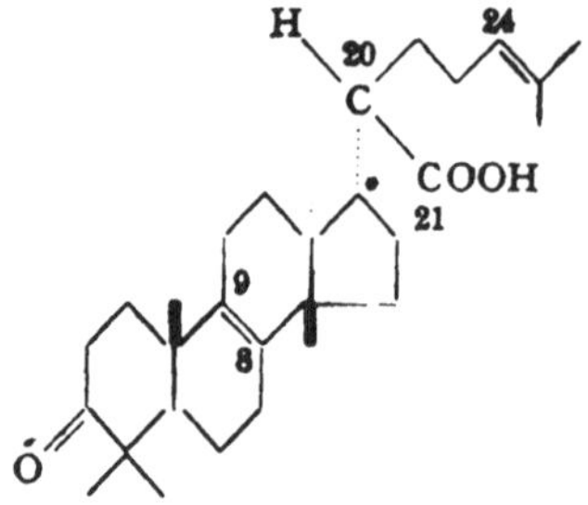

(CCLXII.) Elemonic acid.

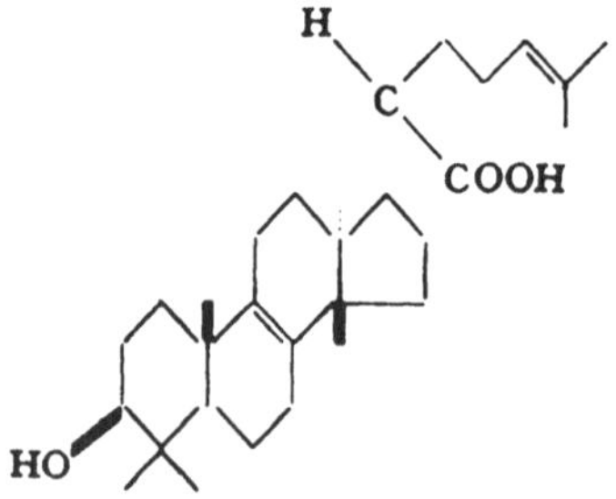

(CCLXIII.)

* The α-configuration at $C_{(17)}$ is not yet proved.

REY, SPILLMANN and BAUMGARTNER (*151*) much confusion existed in the literature of the elemi acids. The reasons were two-fold. Reduction of the keto-acid usually gives the more stable *epi*(3β)-hydroxy acid (*151*) (CCLXIII), originally called β-elemolic acid, whilst oxidation of the hydroxy-acid often gives rise to a mixture of the keto-acid and its dehydro-derivative (*101*). This mixture was originally thought to be a homogeneous compound and was called α-elemonic acid (*150, 151*). A review of the early literature concerning the elemi acids before their true relationship had been established is to be found in ELSEVIER's Encyclopaedia (*70a*), Vol. **14**, p. 598; Vol. **14S**, p. 1203 S. In the discussion of these acids the naturally occurring hydroxy- and keto-acids will be called elemolic acid and elemonic acid for convenience.

Elemolic acid contains two double bonds, only one of which can be hydrogenated. The acid is therefore tetracyclic (*126, 30, 150*). The more

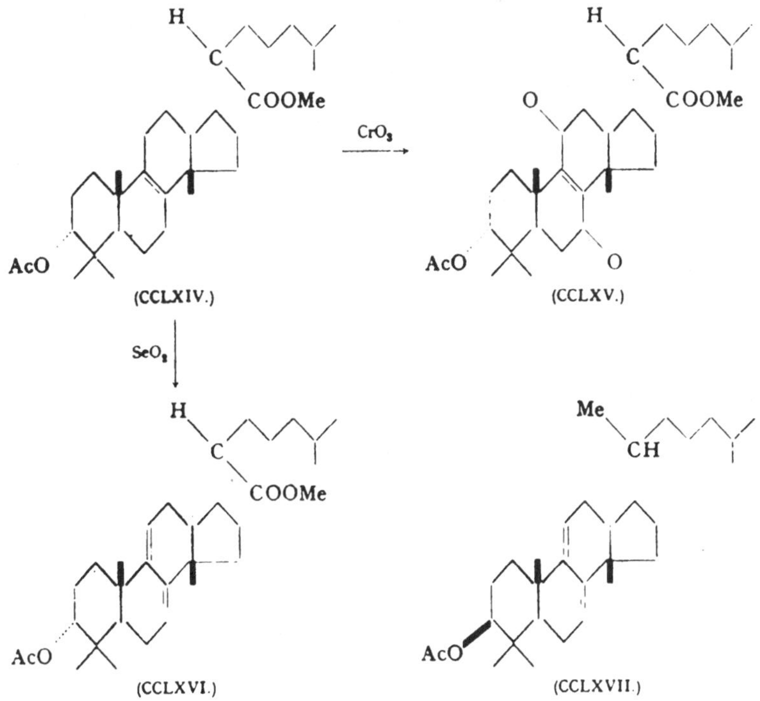

(CCLXIV.) —CrO_3→ (CCLXV.)

↓ SeO_2

(CCLXVI.) (CCLXVII.)

reactive double bond is present in an *iso*propylidene group (*146, 150*). The reactions of the less-reactive double bond and its ultraviolet light absorption indicate that it is situated in a similar position to that of the majority of the tetracyclic triterpenes and to that of the euphol group in particular (*101*). Oxidation of methyl dihydroelemolate acetate (CCLXIV) with chromic acid gives a compound (CCLXV) with the characteristic ene-dione chromophore (maximum 2720 Å.) (*101*, *150*), whilst oxidation with selenium dioxide gives the corresponding dehydro-compound (CCLXVI) (*30*, *101*, *150*), the ultraviolet absorption spectrum of which (maxima, 2320, 2390, and 2470 Å.) is characteristic of the conjugated dehydro-derivatives obtained from euphenyl, euphorbenyl, and tirucallenyl acetates such as eupha-7 : 9(11)-dienyl acetate (CCLXVII) (*58*, *101*). These results, together with the formation of 1 : 2 : 8-trimethylphenanthrene on the selenium dehydrogenation of elemolic acid (*150*) indicate that the elemi acids must have partial structure (CCLXVIII) identical, as far as the carbon skeleton is concerned, with rings *A*, *B* and *C* of euphol.

The hydroxyl group of elemolic acid is at the characteristic triterpene $C_{(3)}$ position, but it has the uncommon less-stable α (axial) configuration (cf. polyporenic acid A). Sodium borohydride reduction of methyl dihydroelemonate (CCLXIX) gives the epimer (CCLXX) of methyl dihydroelemolate. This *epi*-compound must have a 3β-hydroxyl group since it undergoes the characteristic retro-pinacolinic dehydration with phosphorus pentachloride to give a product which yields acetone on ozonolysis (*101*). It follows that elemolic acid has a 3α-hydroxyl group. The negative molecular rotation change occurring on acetylation supports this conclusion, being similar to that found on acetylation of the 3α-

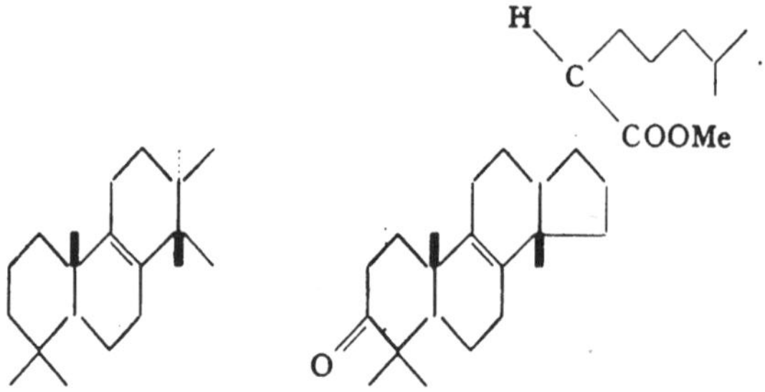

(CCLXVIII.)

(CCLXIX.)

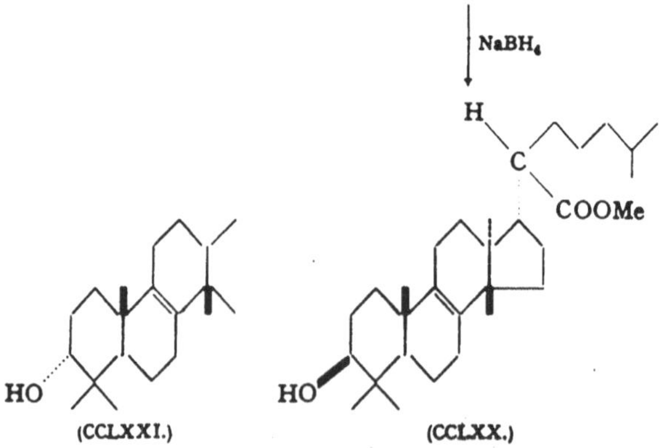

$NaBH_4$

(CCLXXI.)

(CCLXX.)

hydroxyl group of polyporenic acid A (*53*, *96*, cf. *119*) (cf. p. 77). Partial structure (CCLXVIII) can now be extended to (CCLXXI) (*123*, *101*).

The nature of the side-chain of the elemi acids and the position of the carboxyl group relative to the *iso*propylidene group have been determined by ARNOLD, KOLLER and JEGER (*4*), MAZUR, KOLLER, JEGER and RUZICKA (*133*) and HALSALL, MEAKINS and SWAYNE (*101*). The Swiss workers showed that the side-chain must be either (CCLXXII) or (CCLXXIII) by carrying out the sequence of reactions indicated (*Chart 9*), starting initially with acetyl-dihydroelemolic acid (CCLXXIV). The position of the carboxyl group is not fixed uniquely since it could be attached originally at either carbon atom of the double bond of (CCLXXV). The English workers established that (CCLXXII) was correct for the side-chain by studying the products from the action of phosphoric oxide on acetylelemolic acid. This reaction was first carried out by RUZICKA, HIESTAND, BAUMGARTNER and JEGER (*147*) who obtained an $\alpha\beta$-unsaturated ketone. In the later work it was

$HOOC\cdot CH\cdot (CH_2)_2\cdot CH{=}CMe_2$ (CCLXXII.)

$CH_2\cdot CH(COOH)\cdot CH_2\cdot CH{=}CMe_2$ (CCLXXIII.)

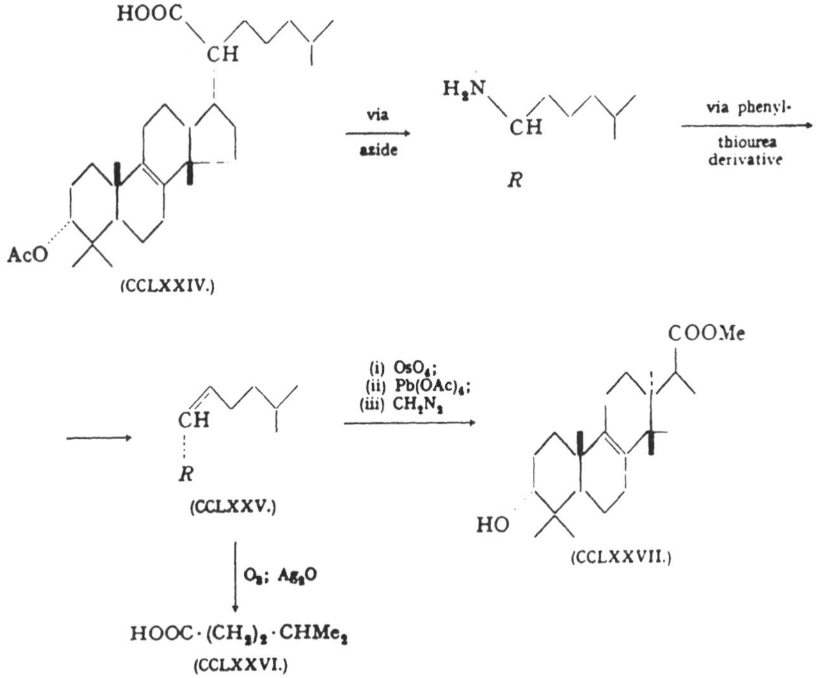

HOOC, CH, AcO (CCLXXIV.) — via azide → H_2N, CH, *R* — via phenylthiourea derivative →

CH, *R* (CCLXXV.) — (i) OsO_4; (ii) $Pb(OAc)_4$; (iii) CH_2N_2 → COOMe, HO (CCLXXVII.)

(CCLXXV.) — O_3; Ag_2O → $HOOC\cdot (CH_2)_2\cdot CHMe_2$ (CCLXXVI.)

Chart 9. Degradation of the Side-chain of Elemolic Acid.

shown that two *cyclo*pentenone derivatives are formed by the sequence of reactions in *Chart 10* and that the carboxyl group is situated in the γ-position to the double bond.

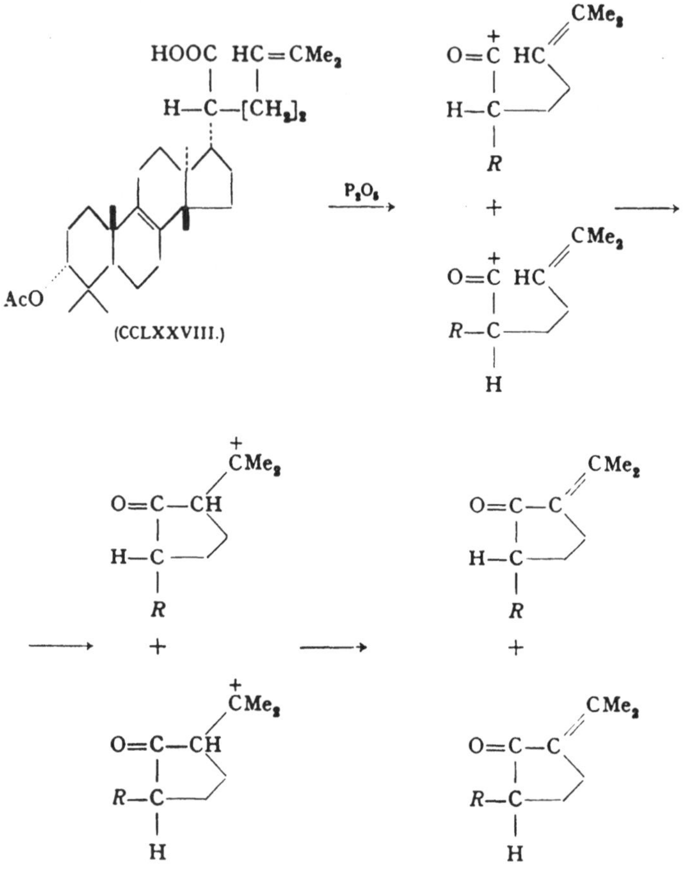

(CCLXXVIII.)

(CCLXXIX.)

Chart 10.

The results described above, in the light of the new structure for euphol, suggest that elemolic acid is (CCLXXX). This has now been rendered highly probable by Arigoni, Wyler and Jeger (*2*) who have shown that a compound (CCLXXXII), first obtained by Ruzicka and Häusermann (*145*) from elemonic acid (CCLXXXI) by the route indicated *(Chart 11)* is, in fact, dihydrotirucallol (p. 121). This correlation not only shows that (CCLXXX) is correct, but also clarifies the stereochemistry of the elemi acids and enables the structure (CCLXI) to be formulated for elemolic acid (*1a*).

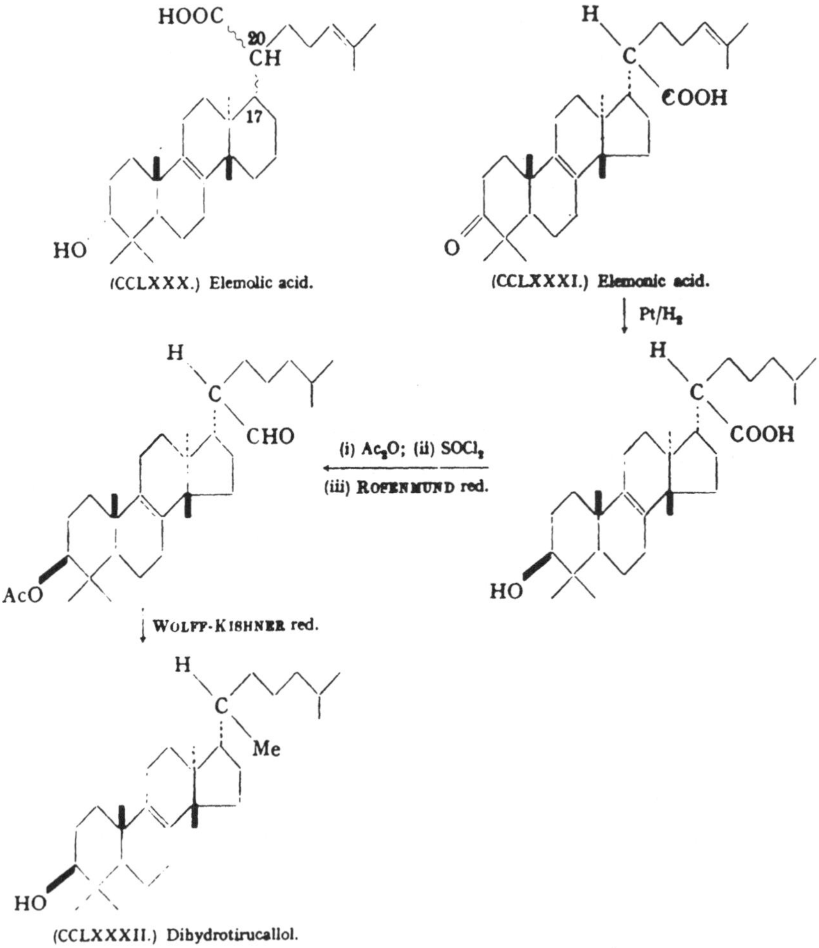

HOOC CH 20 17 HO

(CCLXXX.) Elemolic acid.

H C COOH O

(CCLXXXI.) Elemonic acid.

Pt/H_2

H C COOH HO

(i) Ac_2O; (ii) $SOCl_2$

(iii) ROSENMUND red.

H C CHO AcO

WOLFF-KISHNER red.

H C Me HO

(CCLXXXII.) Dihydrotirucallol.

Chart 11. Relationship between Elemolic Acid and Tirucallol.

IV. Compounds of Unknown Structure.

Three compounds of unknown structure have been isolated from *Euphorbia* species. These are aphylol from *E. aphyla* BROUSS (*85*), obtusifoliol from *E. obtusifolia* POIR (*78*), and resiniferol from *E. resinifera* (*68*). Resiniferol may be identical with β-balalbaresinol (*45*, *157*) and balatol (*156*). All these compounds are alcohols with the probable formula $C_{30}H_{50}O$ ($\pm CH_2$) and they may be tetracyclic triterpenes.

It is possible that compounds of the tritisterol group including the tritisterols, orysterols, and theosterol, isolated from wheat germ, rice germ, and cocoa germ oil, respectively, and of the α-sitosterol group are tetracyclic triterpenes.

An admirable summary of the somewhat confused literature concerning these latter compounds is to be found in Elsevier's Encyclopaedia (*70a*), Vol. 14 S, pp. 1301 S and 1307 S.

References.

1. Arigoni, D., R. Viterbo, M. Dünnenberger, O. Jeger und L. Ruzicka: Konstitution und Konfiguration von Euphol und iso-Euphenol. Helv. Chim. Acta 37, 2306 (1954).

1a. Arigoni, D., O. Jeger und L. Ruzicka: Über die Konstitution und Konfiguration von Tirucallol, Euphorbol und Elemadienolsäure. Helv. Chim. Acta 38, 222 (1955).

2. Arigoni, D., H. Wyler und O. Jeger: Über die gegenseitigen Beziehungen bei Elemadienolsäure, Tirucalladienol und Euphorbadienol. Helv. Chim. Acta 37, 1553 (1954).

3. Arnold, R. T., O. C. Elmer and R. M. Dodson: Thermal Decarboxylation of Unsaturated Acids. J. Amer. Chem. Soc. 72, 4359 (1950).

4. Arnold, R. T., E. Koller und O. Jeger: Über Abbaureaktionen an der Carboxyl-Gruppe der Elemadienolsäure. Helv. Chim. Acta 34, 555 (1951).

5. Barbour, J. B., R. N. E. Bennett and F. L. Warren: The *Euphorbia* Resins. VIII. Epoxides from, and Oxidations at the Inert Double Bond of, Euphol, Euphorbol, and Tirucallol. J. Chem. Soc. (London) 1951, 2540.

5a. Barbour, J. B., W. A. Lourens, F. L. Warren and K. H. Watling: The Structure of Tirucallol, Euphorbol and the Elemi Acids. Chem. and Ind. 1955, 226.

6. Barbour, J. B. and F. L. Warren: The *Euphorbia* Resins. The Structure of Euphol and Euphorbol. Chem. and Ind. 1952, 295.

7. Barbour, J. B., F. L. Warren and D. A. Wood: The *Euphorbia* Resins. VII. The Characterisation of the Groups in Euphorbol. J. Chem. Soc. (London) 1951, 2537.

8. Barnes, C. S. and D. H. R. Barton: Triterpenoids. XI. Some Stereospecific Reactions in the Lanostadienol (Lanosterol) Series. J. Chem. Soc. (London) 1953, 1419.

9. Barnes, C. S., D. H. R. Barton, A. R. H. Cole, J. S. Fawcett and B. R. Thomas: Triterpenoids. IX. The Constitution of Lanostadienol (Lanosterol). J. Chem. Soc. (London) 1953, 571.

10. Barnes, C. S., D. H. R. Barton, J. S. Fawcett, S. K. Knight, J. F. McGhie, M. K. Pradhan and B. R. Thomas: Nature of the Lanosterol Side Chain. Chem. and Ind. 1951, 1067.

11. Barnes, C. S., D. H. R. Barton, J. S. Fawcett and B. R. Thomas: Triterpenoids. X. The Stereochemistry of Lanostadienol (Lanosterol). J. Chem. Soc. (London) 1953, 576.

12. Barton, D. H. R.: The Conformation of the Steroid Nucleus. Experientia 6, 316 (1950); Stereochemistry of *cyclo*Hexane Derivatives. J. Chem. Soc. (London) 1953, 1027.

13. — Triterpenoids. III. *cyclo*Artenone, a Triterpenoid Ketone. J. Chem. Soc. (London) 1951, 1444.

14. Barton, D. H. R. and C. J. W. Brooks: Triterpenoids. I. Morolic Acid, a New Triterpenoid Sapogenin. J. Chem. Soc. (London) 1951, 257.

14a. Barton, D. H. R., J. S. Fawcett and B. R. Thomas: Triterpenoids. IV. Some Observations on the Constitution of Lanostadienol (Lanosterol). J. Chem. Soc. (London) 1951, 3147.

15. Barton, D. H. R. and N. J. Holness: Triterpenoids. V. Some Relative Configurations in Rings C, D, and E of the β-Amyrin and Lupeol Group of Triterpenoids. J. Chem. Soc. (London) 1952, 83.
16. Barton, D. H. R., J. F. McGhie, M. K. Pradhan and (in part) S. A. Knight: The Constitution and Stereochemistry of Euphol. Chem. and Ind. 1954, 1325.
17. Barton, D. H. R., J. E. Page and E. W. Warnhoff: Triterpenoids. XVIII. The Constitutions of Phyllanthol and *cyclo*Artenol. J. Chem. Soc. (London) 1954, 2715.
18. Barton, D. H. R., E. W. Warnhoff and J. E. Page: The Constitutions of Phyllanthol and *cyclo*Artenol. Chem. and Ind. 1954, 220.
19. Bauer, K. H. und H. Moll: Zur Kenntnis der unverseifbaren Bestandteile des Sheafettes. Fette und Seifen 46, 560 (1939).
20. Bauer, K. H. und P. Schenkel: Zur Kenntnis des Euphorbiumharzes. Arch. Pharmaz. 266, 633 (1928).
21. Bauer, K. H. und G. Schröder: Über Euphorbol. 2. Mitt. über Euphorbiumharz. (5. Mitt. über die Chemie der Harzbestandteile.) Arch. Pharmaz. 269, 209 (1931).
22. Bellamy, L. J. and C. Dorée: Lanosterol. III. The Action of Selenium Dioxide and of Perbenzoic Acid on Lanosterol. J. Chem. Soc. (London) 1941, 176.
23. Bennett, R. N. E., Hs. K. Krüsi and F. L. Warren: The *Euphorbia* Resins. VI. The Oxidative Degradation of the Side Chain of Euphol and Tirucallol. J. Chem. Soc. (London) 1951, 2534.
24. Bennett, R. N. E. and F. L. Warren: The *Euphorbia* Resins. III. The Epimerisation and Dehydration of Euphol. J. Chem. Soc. (London) 1950, 697.
25. Bentley, H. R., J. A. Henry, D. S. Irvine, D. Mukerji and F. S. Spring: *cyclo*Laudenol, a Triterpenoid Alcohol from Opium. J. Chem. Soc. (London) 1955, 596.
26. Bentley, H. R., J. A. Henry, D. S. Irvine and F. S. Spring: The Structure of *cyclo*Artenol. Chem. and Ind. 1953, 217.
27. — — — — Triterpene Resinols and Related Acids. XXVIII. The Non-saponifiable Fraction from *Strychnos nux-vomica* Seed Fat: The Structure of *cyclo*Artenol. J. Chem. Soc. (London) 1953, 3673.
28. Beynon, J. H., I. M. Heilbron and F. S. Spring: The Characterisation of Basseol, a Tetracyclic Triterpene Alcohol, and its Isomerisation to β-Amyrenol. J. Chem. Soc. (London) 1937, 989.
29. — — — Structure of the Triterpenes. Nature (London) 142, 434 (1938).
30. Bilham, P. and G. A. R. Kon: Sapogenins. XVI. The Acids of Elemi Resin. J. Chem. Soc. (London) 1942, 544.
31. Birchenough, M. T. and J. F. McGhie: Lanosterol. VIII. The Action of Perbenzoic Acid on Derivatives of Lanosterol. J. Chem. Soc. (London) 1949, 2038.
32. — — Lanosterol. IX. Ketodihydrolanosterol. J. Chem. Soc. (London) 1950, 1249.
33. Birkinshaw, J. H., E. N. Morgan and W. P. K. Findlay: Biochemistry of the Wood-rotting Fungi. 7. Metabolic Products of *Polyporus Benzoinus* (Wahl). Fr. Biochemic. J. 50, 509 (1952).
34. Bladon, P., H. B. Henbest and G. W. Wood: Olefinic Light Absorption Measurements in the 2050—2250 Å Region. Chem. and Ind. 1951, 866.
35. — — — Studies in the Sterol Group. LV. Ultra-violet Absorption Spectra of Ethylenic Centres. J. Chem. Soc. (London) 1952, 2737.
36. Bowers, A., T. G. Halsall, E. R. H. Jones and (in part) A. J. Lemin: The Chemistry of the Triterpenes and Related Compounds. XVIII. Elucidation of the Structure of Polyporenic Acid C. J. Chem. Soc. (London) 1953, 2548.

37. BOWERS, A., T. G. HALSALL and (in part) G. C. SAYER: The Chemistry of the Triterpenes and Related Compounds. XXV. Some Stereochemical Problems concerning Polyporenic Acid C. J. Chem. Soc. (London) **1954**, 3070.
38. CAMPBELL, W. P. and G. C. HARRIS: Oxidation Products of $\Delta^{9,10}$-Octalin. J. Amer. Chem. Soc. **63**, 2721 (1941).
39. CAVALLA, J. F. and J. F. McGHIE: Lanosterol. X. Ketodihydrolanosterol (continued). J. Chem. Soc. (London) **1951**, 744.
40. — — Lanosterol. XI. Reduction of Ketones in the Lanosterol Series. J. Chem. Soc. (London) **1951**, 834.
41. CAVALLA, J. F., J. F. McGHIE, E. C. PICKERING and R. A. REES: Lanosterol. XII. The Constitution of Some Oxidation Products from Lanostadienol and Lanosterol. J. Chem. Soc. (London) **1951**, 2474.
41a. CAVALLA, J. F., J. F. McGHIE and M. K. PRADHAN: Lanosterol. XIII. Further Experiments on the Constitution of Lanostadienol. J. Chem. Soc. (London) **1951**, 3142.
42. CHAPON, S. et S. DAVID: Sur un constituant secondaire du latex d'*Euphorbia balsamifera*. Bull. soc. chim. France **1952**, 456.
43. CHRISTEN, K., M. DÜNNENBERGER, C. B. ROTH, H. HEUSSER und O. JEGER: Über weitere Beiträge zur Konstitution des Euphadienols und seine nahe Verwandtschaft mit dem Lanostadienol. Helv. Chim. Acta **35**, 1756 (1952).
44. CHRISTEN, K., O. JEGER und L. RUZICKA: Über die Konstitution der ungesättigten Seitenkette des Euphols. Helv. Chim. Acta **34**, 1675 (1951).
45. COHEN, N. H.: Über Phytosterine aus Balata. Arch. Pharmaz. **246**, 510 (1908).
46. COLE, A. R. H.: The Structure of *cyclo*Artenol. Chem. and Ind. **1953**, 946.
47. — Infra-Red Spectra of Natural Products. III. *cyclo*Artenol and Phyllanthol. J. Chem. Soc. (London) **1954**, 3810.
48. CORT, L. A., R. M. GASCOIGNE, J. S. E. HOLKER, B. J. RALPH, A. ROBERTSON and J. J. H. SIMES: The Chemistry of Fungi. XXIII. Tumulosic Acid. J. Chem. Soc. (London) **1954**, 3713.
49. CROSS, L. C., C. G. ELIOT, I. M. HEILBRON and E. R. H. JONES: Constituents of the Higher Fungi. I. The Triterpene Acids of *Polyporus betulinus* FR. J. Chem. Soc. (London) **1940**, 632.
50. CROSS, L. C. and E. R. H. JONES: Constituents of the Higher Fungi. II. The Unsaturated System of Polyporenic Acid A. J. Chem. Soc. (London) **1940**, 1491.
51. CURTIS, R. G.: The Chloro- and Iodo-acetates of Dihydrolanosterol and Dihydroagnosterol. J. Chem. Soc. (London) **1950**, 1017.
52. CURTIS, R. G., J. FRIDRICHSONS and A. McL. MATHIESON: Structure of Lanosterol. Nature (London) **170**, 321 (1952).
53. CURTIS, R. G., SIR IAN HEILBRON, E. R. H. JONES and G. F. WOODS: The Chemistry of the Triterpenes. XIII. The Further Characterisation of Polyporenic Acid A. J. Chem. Soc. (London) **1953**, 457.
54. CURTIS, R. G. and H. SILBERMAN: The Degradation of the Side Chain of Lanostadienol. J. Chem. Soc. (London) **1952**, 1187.
55. DAWSON, M. C., T. G. HALSALL and E. R. H. JONES: unpublished. [DAWSON, M. C.: The Chemistry of Some Tetracyclic Triterpenes. Ph. D. Thesis. Victoria Univ., Manchester. 1953.]
56. DAWSON, M. C., T. G. HALSALL, E. R. H. JONES and G. D. MEAKINS: unpublished.
57. DAWSON, M. C., T. G. HALSALL, E. R. H. JONES and P. A. ROBINS: The Chemistry of the Triterpenes. XVI. The Action of Hydrogen Chloride on Butyrospermol. J. Chem. Soc. (London) **1953**, 586.

58. DAWSON, M. C., T. G. HALSALL and R. E. H. SWAYNE: The Chemistry of the Triterpenes. XVII. Some Aspects of the Chemistry of the Tetracyclic Triterpenes. J. Chem. Soc. (London) **1953**, 590.

59. DORÉE, C. and D. C. GARRATT: *iso*Cholesterol. I. Lanosterol and a New Method for its Preparation. J. Soc. Chem. Ind. **52**, 141 (1933); II. The Chemistry of Lanosterol. J. Soc. Chem. Ind. **52**, 355 (1933).

60. DORÉE, C., J. F. MCGHIE and F. KURZER: Lanosterol. IV. Hydrocarbons formed by the Action of Dehydrating Agents. J. Chem. Soc. (London) **1947**, 1467.

61. — — — Lanosterol. V. Hydrogenation of the Inert Double Bond in Lanosterol Derivatives. J. Chem. Soc. (London) **1948**, 988.

62. — — — Lanosterol. VI. Further Dehydrogenation and Oxidation Reactions. J. Chem. Soc. (London) **1949**, 570.

63. — — — Position of the Hydroxyl Group in Lanosterol. Nature (London) **163**, 140 (1949).

64. — — — Lanosterol. VII. The Position of the Hydroxyl Group in the Lanosterol Molecule. J. Chem. Soc. (London) **1949**, Suppl. No. 1, S 167.

64a. DORÉE, C. and V. A. PETROW: Lanosterol. I. J. Chem. Soc. (London) **1936**, 1562.

65. DRUMMOND, J. C. and L. C. BAKER: The Composition of Wool Fat. J. Soc. Chem. Ind. **48**, 232 (1929).

66. DUPONT, G., R. DULOU et M. VILKAS: Contribution à l'étude des résines d'Euphorbiacées. IV. Sur l'α-euphol. Bull. soc. chim. France [5] **16**, 809 (1949).

67. DUPONT, G., M. JULIA et W. R. WRAGG: Contribution à l'étude des résines d'Euphorbiacées. VII. Séparation par chromatographie de quelques-uns des constituants du latex de l'*Euphorbia resinifera*. Indentification de l'un d'eux avec l'α-euphorbol. Bull. soc. chim. France [5] **18**, 643 (1951).

68. — — — Contribution à l'étude des résines d'Euphorbiacées. VIII. L'identification du tarascérol, de la β-amyrine et d'un nouvel alcool triterpénique, le résiniférol, comme constituants peu abondants du latex de l'*Euphorbia resinifera*. Bull. soc. chim. France [5] **20**, 852 (1953).

69. DUPONT, G., W. KOPACZEWSKI et BRODSKI: Contribution à l'étude des résines d'Euphorbiacées. II. Latex de l'*Euphorbia resinifera*. Bull. soc. chim. France [5] **14**, 1068 (1947).

70. Editorial Report on Nomenclature, 1951: Appendix C. Steroid Nomenclature. J. chem. Soc. (London) **1951**, 3526.

70a. ELSEVIER's Encyclopaedia of Organic Chemistry, Vol. **14**, New York-Amsterdam. 1940; Vol. **14 S**, Amsterdam, Houston, London, New York. 1951–52.

71. FRÈREJACQUE, M.: Note sur l'acide ungulinique, acide cristallisé isolé de *Polyporus (Ungulina) betulinus* FR. Rev. mycol. (Paris) **3**, 95 (1938).

72. FRIDRICHSONS, J. and A. McL. MATHIESON: Triterpenoids. The Crystal Structure of Lanostenyl Iodoacetate. J. Chem. Soc. (London) **1953**, 2159.

73. FRÖSCHL, N. und J. ZELLNER: Zur Kenntnis der Pilzharze. Monatsh. Chem. **53—54**, 146 (1929).

74. FUKUSHIMA, D. K. and T. F. GALLAGHER: The Action of Alcoholic Potassium Hydroxide on Δ^{16}-20-Ketosteroids. J. Amer. Chem. Soc. **73**, 196 (1951).

75. GASCOIGNE, R. M., J. S. E. HOLKER, B. J. RALPH and A. ROBERTSON: Occurrence of Eburicoic Acid. Nature (London) **166**, 652 (1950).

76. — — — — The Chemistry of Fungi. XVI. Eburicoic Acid. J. Chem. Soc. (London) **1951**, 2346.

77. GASCOIGNE, R. M., A. ROBERTSON and J. J. H. SIMES: The Chemistry of Fungi. XVII. Dehydroeburicoic Acid. J. Chem. Soc. (London) **1953**, 1830.

78. González, A. G. and J. L. Bretón: Aportación al estudio del latex de las "Euphorbias Canarias". V. Latex de la *Euphorbia Obtusifolia* Poir. Anales soc. españ. fís. quím. **47** B, 365 (1951).
79. — — Aportación al estudio del latex de las "Euphorbias Canarias". IX. Identidad de los Triterpenos Handianol y *ciclo*Artenol. Anales soc. españ. fís quím. **49** B, 237 (1953).
79a. González, A. G., J. L. Bretón and C. Bretón: The Constitution of Handianol. Chem. and Ind. **1955**, 416.
80. González, A. G. and A. Calero: Aportación al estudio del latex de las "Euphorbias Canarias". I. Latex de la *Euphorbia Canariensis*. Anales soc. españ. fís. quím. **45** B, 269 (1949).
81. — — Aportación al estudio del latex de las "Euphorbias Canarias". III. Sobre el Eufol y el Handianol. Anales soc. españ. fís. quím. **46** B, 175 (1950).
82. González, A. G., A. Calero and R. Calero: Aportación al estudio del latex de las "Euphorbias Canarias". II. Latex de la *Euphorbia Handiensis* Burchard. Anales soc. españ. fís. quím. **45** B, 1441 (1949).
83. González, A. G. and M. L. G. Mora: Aportación al estudio del latex de las "Euphorbias Canarias". VI. Obtención del Germanicol y Lanosterol del latex de la *Euphorbia balsamifera* Ait. Anales soc. españ. fís. quím. **48** B, 475 (1952).
84. González, A. G. and A. H. Toste: Aportación al estudio del latex de las "Euphorbias Canarias". VII. Sobre el Lanosterol procedente de la *Euphorbia balsamifera* Ait. Anales soc. españ. fís. quím. **48** B, 487 (1952).
85. — — Aportación al estudio del latex de las "Euphorbias Canarias". XII. Latex de la *Euphorbia aphyla* Brouss. Anales soc. españ. fís. quím. **50** B, 597 (1954).
86. Gruber, W. und G. Proske: Über Trametenolsäure. [Aus *Trametes odorata* (Wulf.) Fr.] Monatsh. Chem. **81**, 837 (1950).
87. — — Über Trametenolsäure. II. Mitt. Die Überführung in den Grundkohlenwasserstoff. Monatsh. Chem. **81**, 1024 (1950).
88. — — Über Trametenolsäure. III. Mitt. Selendehydrierung. Monatsh. Chem. **82**, 255 (1951).
89. Guider, (Miss) J. M., T. G. Halsall, R. Hodges and E. R. H. Jones: The Chemistry of the Triterpenes and Related Compounds. XXVI. The Nature of Polyporenic Acid B. J. Chem. Soc. (London) **1954**, 3234.
90. Guider, (Miss) J. M., T. G. Halsall and E. R. H. Jones: The Chemistry of the Triterpenes and Related Compounds. XXVII. Pinicolic Acid A. J. Chem. Soc. (London) **1954**, 4471.
91. — — — unpublished. [Guider, J. M.: Studies in the Triterpene Field. Ph. D. Thesis, Victoria Univ., Manchester. 1954.]
92. Haines, D. W. and F. L. Warren: The *Euphorbia* Resins. II. The Isolation of Taraxasterol and a New Triterpene, Tirucallol, from *E. tirucalli*. J. Chem. Soc. (London) **1949**, 2554.
93. — — The *Euphorbia* Resins. IV. A Comparative Study of Euphol and Tirucallol. J. Chem. Soc. (London) **1950**, 1562.
94. Halsall, T. G.: The Use of Ultra-Violet Absorption Spectra for the Investigation of Double Bonds in Triterpenes. Chem. and Ind. **1951**, 867.
95. Halsall, T. G. and R. Hodges: The Chemistry of the Triterpenes and Related Compounds. XXIV. The Conversion of Polyporenic Acid A into a Lanosterol Derivative. J. Chem. Soc. (London) **1954**, 2385.
96. Halsall, T. G., R. Hodges and E. R. H. Jones: The Chemistry of the Triterpenes and Related Compounds. XIX. Further Evidence concerning the Structure of Polyporenic Acid A. J. Chem. Soc. (London) **1953**, 3019.

97. HALSALL, T. G., R. HODGES, E. R. H. JONES and G. C. SAYER: unpublished.

98. HALSALL, T. G., E. R. H. JONES and A. J. LEMIN: The Chemistry of the Triterpenes. XV. The Environment of the Unreactive Double Bond of Polyporenic Acid A. J. Chem. Soc. (London) **1953**, 468.

99. HALSALL, T. G., E. R. H. JONES and P. C. PHILIPS: unpublished.

100. HALSALL, T. G., E. R. H. JONES and G. C. SAYER: unpublished.

101. HALSALL, T. G., G. D. MEAKINS and R. E. H. SWAYNE: The Chemistry of the Triterpenes and Related Compounds. XXI. Some Aspects of the Chemistry of the Elemi Acids. J. Chem. Soc. (London) **1953**, 4139.

102. HARTMANN, E. und J. ZELLNER: Zur Chemie der höheren Pilze. XIX. Über *Polyporus pinicola* FR. Monatsh. Chem. **50**, 193 (1928).

103. HEILBRON, SIR IAN, E. R. H. JONES and P. A. ROBINS: The Non-saponifiable Matter of Shea Nut Fat. IV. A New Tetracyclic Diethenoid Alcohol, Butyrospermol. J. Chem. Soc. (London) **1949**, 444.

104. HEILBRON, I. M., E. D. KAMM and W. M. OWENS: The Unsaponifiable Matter from the Oils of Elasmobranch Fish. I. A Contribution to the Constitution of Squalene (Spinacene). J. Chem. Soc. (London) **129**, 1630 (1926).

105. HEILBRON, I. M., G. L. MOFFET and F. S. SPRING: The Non-saponifiable Matter of Shea Nut Fat. J. Chem. Soc. (London) **1934**, 1583.

106. HENRY, J. A., D. S. IRVINE and F. S. SPRING: The Constitution of *cyclo*-Laudenol. J. Chem. Soc. (London) **1955** (in print).

107. HENRY, J. A. and F. S. SPRING: The Structure of *cyclo*Artenol. Chem. and Ind. **1954**, 189.

108. HEUSSER, H., K. HEUSLER, K. EICHENBERGER, C. G. HONEGGER und O. JEGER: Ein neuer Weg zur Synthese von 11-Keto-Steroiden. II. Versuche in der Androstan- und der Cholestan-Reihe. Helv. Chim. Acta **35**, 295 (1952).

109. HIRSCHBERG, Y.: The Influence of Substituents on the Ultraviolet Absorption Spectrum of Styrene. J. Amer. Chem. Soc. **71**, 3241 (1949).

110. HOLKER, J. S. E., A. D. G. POWELL, A. ROBERTSON, J. J. H. SIMES and R. S. WRIGHT: The Chemistry of Fungi. XVIII. The Cyclic System of Eburicoic Acid. J. Chem. Soc. (London) **1953**, 2414.

111. HOLKER, J. S. E., A. D. G. POWELL, A. ROBERTSON, J. J. H. SIMES, R. S. WRIGHT and (in part) R. M. GASCOIGNE: The Chemistry of Fungi. XIX. The Structure of Eburicoic Acid. J. Chem. Soc. (London) **1953**, 2422.

112. IRVINE, D. S., J. A. HENRY and F. S. SPRING: Triterpenoids. XXXIV. The Constitution of *cyclo*Artenol. J. Chem. Soc. (London) **1955**, 1316.

113. JEGER, O.: Über die Konstitution der Triterpene. Fortschr. Chem. organ. Naturstoffe **7**, 1 (1950).

114. JEGER, O. und Hs. K. KRÜSI: Nachweis einer Isopropyliden-Gruppe im Euphol. Helv. Chim. Acta **30**, 2045 (1947).

115. JONES, E. R. H. and G. F. WOODS: The Chemistry of the Triterpenes. XIV. Further Evidence concerning the Unsaturated Centres of Polyporenic Acid A. J. Chem. Soc. (London) **1953**, 464.

116. JONES, R. N., V. Z. WILLIAMS, M. J. WHALEN and K. DOBRINER: The Characterization of Carbonyl and Other Functional Groups in Steroids by Infrared Spectrometry. J. Amer. Chem. Soc. **70**, 2024 (1948).

117. KARIYONE, T. and G. KURONO: Constituents of *Fomes officinalis* (FR.) J. Pharmac. Soc. Japan **60**, 110, 318 (1940).

118. KLYNE, W.: The Molecular Rotations of Polycyclic Compounds. I. General Principles and the Correlation of the Triterpenoids with the Steroids. J. Chem. Soc. (London) **1952**. 2916.

119. KLYNE, W. and W. M. STOKES: The Molecular Rotations of Polycyclic Compounds. III. Polycyclic Alcohols and their Derivatives. J. Chem. Soc. (London) **1954**, 1979.

120. KNIGHT, S. A. and J. F. MCGHIE: Studies in the Euphadienol Group. Chem. and Ind. **1953**, 920; Experiments on the Side-Chain of Euphadienol. Chem. and Ind. **1954**, 24.

121. KOPACZEWSKI, W. et G. DUPONT: Contribution à l'étude des résines d'Euphorbiacées. I. Extraction et historique. Bull. soc. chim. France [5] **14**, 909 (1947).

122. KRÜSI, Hs. K.: The *Euphorbia* Resins. V. Oxidative Degradation in the Side Chain of Euphol. J. Chem. Soc. (London) **1950**, 2864.

123. KYBURZ, E., M. V. MIJOVIĆ, W. VOSER, H. HEUSSER, O. JEGER und L. RUZICKA: Versuche zur Verknüpfung des tetracyclischen Lanostadienols mit bi- und tricyclischen Diterpenen und pentacyclischen Triterpenen. Helv. Chim. Acta **35**, 2073 (1952).

124. KYBURZ, E., B. RINIKER, H. R. SCHENK, H. HEUSSER und O. JEGER: Über die konfigurative Verknüpfung des Lanostadienols mit cyclischen Diterpenen und Triterpenen. Helv. Chim. Acta **36**, 1891 (1953).

125. LAHEY, F. N. and P. H. A. STRASSER: Eburicoic Acid. J. Chem. Soc. (London) **1951**, 873.

126. LIEB, H. und M. MLADENOVIĆ: Über die Elemisäure aus Manila-Elemiharz. Monatsh. Chem. **58**, 59 (1931).

127. LOCQUIN, M., J. LOCQUIN et A. R. PRÉVOT: Recherches sur l'acide ungulinique (acide polyporénique A), antibiotique produit par *Ungulina betulina*. Rev. mycol. (Paris) **13**, 3 (1948).

128. MCDONALD, A. D., F. L. WARREN and J. M. WILLIAMS: The *Euphorbia* Resins. I. Euphol. J. Chem. Soc. (London) **1949**, Suppl. No. 1, S 155.

129. MCGHIE, J. F., M. K. PRADHAN and (in part) J. F. CAVALLA: Lanosterol. XIV. Further Experiments on the Reduction of Ketones in the Lanosterol Series. J. Chem. Soc. (London) **1952**, 3176.

130. MCGHIE, J. F., M. K. PRADHAN and (in part) J. F. CAVALLA and S. A. KNIGHT: The Side Chain of Lanosterol; and the Conversion of Acetoxylanostanedione to Acetoxylanostane. Chem. and Ind. **1951**, 1165.

131. MARKER, R. E. and E. L. WITTLE: Sterols. XXI. Lanosterol and Agnosterol. J. Amer. Chem. Soc. **59**, 2289 (1937).

132. MARKER, R. E., E. L. WITTLE and L. M. MIXON: Sterols. XVI. Lanosterol and Agnosterol. J. Amer. Chem. Soc. **59**, 1368 (1937).

133. MAZUR, Y., E. KOLLER, O. JEGER und L. RUZICKA: Über die ungesättigte Seitenkette der Elemadienolsäure. Helv. Chim. Acta **35**, 181 (1952).

134. MEAKINS, G. D.: private communication.

135. MEYSTRE, CH., H. FREY, A. WETTSTEIN und K. MIESCHER: Ein einfacher Abbau der Gallensäuren-Seitenkette zur Methylketonstufe. Helv. Chim. Acta **27**, 1815 (1944).

136. MIJOVIĆ, M. V., W. VOSER, H. HEUSSER und O. JEGER: Über weitere Umwandlungen in den Ringen B und C des Lanostenols. Helv. Chim. Acta **35**, 964 (1952).

137. MÜLLER, J. A.: Zur Kenntnis des Euphorbons aus Euphorbiumharz. J. prakt. Chem. [2] **121**, 97 (1929).

138. NEWBOLD, G. T. and F. S. SPRING: The Isolation of Euphol and α-Euphorbol from Euphorbium. J. Chem. Soc. (London) **1944**, 249.

139. PINTO COELHO, F.: The Determination of the Structure of the Tetracyclic Triterpene Alcohol, Basseol. II. The Dehydrogenation of Basseol and of Bassenol-Triterpenes of the β-Amyrenol Group. Rev. faculd. cienc. Univ. Coimbra **18**, 71 (1949).

140. Roth, M. und O. Jeger: Vergleich des Euphols mit Kryptosterin. Helv. Chim. Acta **32**, 1620 (1949).

141. Roth, M., G. Saucy, R. Anliker, O. Jeger und H. Heusser: Die Konstitution der Polyporensäure A. Helv. Chim. Acta **36**, 1908 (1953).

142. Ruzicka, L.: The Isoprene Rule and the Biogenesis of Terpenic Compounds. Experientia **9**, 357 (1953).

143. Ruzicka L., R. Denss und O. Jeger: Beweis der Identität von Lanosterin und Kryptosterin. Helv. Chim. Acta **28**, 759 (1945).

144. — — — Nachweis der Identität von Dihydro-agnosterin und γ-Lanosterin, und über die Lage der hydrierbaren Doppelbindung im Agnosterin. Helv. Chim. Acta **29**, 204 (1946).

145. Ruzicka, L. und H. Häusermann: Über die β-Elemonsäure. Helv. Chim. Acta **25**, 439 (1942).

146. Ruzicka, L., H. Häusermann und Ed. Rey: Über die Elemonsäure. Helv. Chim. Acta **25**, 1403 (1942).

147. Ruzicka, L., A. Hiestand, H. Baumgartner und O. Jeger: Überführung der tetracyclischen Elemadienolsäure in einen pentacyclischen Kohlenwasserstoff $C_{30}H_{50}$. Helv. Chim. Acta **30**, 2119 (1947).

148. Ruzicka, L., M. Montavon und O. Jeger: Über den Bau des hydroxylhaltigen Ringes des Lanosterins. Helv. Chim. Acta **31**, 818 (1948).

149. Ruzicka, L., Ed. Rey und A. C. Muhr: Über verschiedene Umwandlungsprodukte des Lanosterins. Helv. Chim. Acta **27**, 472 (1944).

150. Ruzicka, L., Ed. Rey und M. Spillmann: Über die α-Elemolsäure. Helv. Chim. Acta **25**, 1375 (1942).

151. Ruzicka, L., Ed. Rey, M. Spillmann und H. Baumgartner: Über die Beziehungen zwischen der α-Elemolsäure und der sogenannten β-Elemonsäure. Helv. Chim. Acta **26**, 1638 (1943).

151a. Schmid, L. und H. Czerny: Chemische Untersuchung des *Polyporus pinicola* Fries. II. Mitt. Monatsh. Chem. **85**, 1307 (1954).

152. Schmid, L. und M. K. Zacherl: Über das Euphorbiumharz. Monatsh. Chem. **57**, 177 (1931).

153. Schulze, H.: Über die Abgrenzung der Sterine gegenüber anderen Alkoholen der Polyterpenreihe und über den Bau des Lanosterins und Onocerins. Z. physiol. Chem. (Hoppe-Seyler) **238**, 35 (1936).

154. Seitz, K. und O. Jeger: Über die Isolierung eines unbekannten tetracyclischen Alkohols $C_{30}H_{50}O$ aus "shea nut"-Öl. Helv. Chim. Acta **32**, 1626 (1949).

155. Shibamoto, T., K. Minami and T. Tajima: Constituents of Fruit-body of *Fomes pinicola*. J. Japan. Forest Soc. **35**, 56 (1953).

156. Tanaka, Y., T. Kuwata and T. Suzuki: Balata Resin. I. Crystalline Components of Surinum Sheet Balata Resin. J. Soc. Chem. Ind. Japan **38**, Suppl. binding 504 (1935) [Chem. Abstr. **30**, 1260 (1936).]

157. Tschirch, A. und E. Schereschewski: Über Balata. Arch. Pharmaz. **243**, 358 (1905).

158. Vilkas, M.: Contribution à l'étude des résines d'Euphorbiacées. VI. Sur la double liaison inerte de l'α-euphol. Bull. soc. chim. France [5] **17**, 582 (1950).

159. — Contribution à l'étude des triterpènes tétracycliques. Sur la structure de l'α-euphol. Ann. Chim. [12] **6**, 325 (1951).

160. Vilkas, M., G. Dupont et R. Dulou: Contribution à l'étude des résines d'Euphorbiacées. V. Sur l'α-euphol (suite). Bull. soc. chim. France [5] **16**, 813 (1949).

161. Vogel, C., O. Jeger und L. Ruzicka: Über Euphorbadienol. Helv. Chim. Acta **35**, 510 (1952).

162. VOSER, W., Hs. H. GÜNTHARD, H. HEUSSER, O. JEGER und L. RUZICKA: Ein neuer Weg zur Öffnung des Ringes C beim Lanostadienol. Helv. Chim. Acta **35**, 2065 (1952).

163. VOSER, W., Hs. H. GÜNTHARD, O. JEGER und L. RUZICKA: Über die Größe des Ringes D im Lanostadienol. Helv. Chim. Acta **35**, 66 (1952).

164. VOSER, W., H. HEUSSER, O. JEGER und L. RUZICKA: 14-Methyl-11-keto-progesteron aus Lanosterin. Helv. Chim. Acta **36**, 299 (1953).

165. VOSER, W., O. JEGER und L. RUZICKA: Über die Herstellung von Trisnor-acetoxy-lanostandionsäuremethylester aus „Isocholesterin"-acetat. Helv. Chim. Acta **35**, 497 (1952).

166. — — — Über die Herstellung eines weiteren Ring-D-Ketons aus Lanostadienol. Helv. Chim. Acta **35**, 503 (1952).

167. VOSER, W., M. V. MIJOVIĆ, O. JEGER und L. RUZICKA: Über die Vervollständigung der Teilformel des Lanostadienols. Helv. Chim. Acta **34**, 1585 (1951).

168. VOSER, W., M. V. MIJOVIĆ, H. HEUSSER, O. JEGER und L. RUZICKA: Über die Konstitution des Lanostadienols (Lanosterins) und seine Zugehörigkeit zu den Steroiden. Helv. Chim. Acta **35**, 2414 (1952).

169. VOSER, W., M. MONTAVON, Hs. H. GÜNTHARD, O. JEGER und L. RUZICKA: Zur Konstitution des Lanostadienols. Helv. Chim. Acta **33**, 1893 (1950).

170. WIELAND, H. und W. BENEND: Die Beziehungen zwischen Lanosterin und Kryptosterin. Über die Nebensterine der Hefe. X. Z. physiol. Chem. (Hoppe-Seyler) **274**, 215 (1942).

171. WIELAND, H., H. PASEDACH und A. BALLAUF: Über die Nebensterine der Hefe. IV. Kryptosterin. Liebigs Ann. Chem. **529**, 68 (1937).

172. WINDAUS, A. und R. TSCHESCHE: Über das sogenannte „Isocholesterin" des Wollfettes. Z. physiol. Chem. (Hoppe-Seyler) **190**, 51 (1930).

173. WOODWARD, R. B., A. A. PATCHETT, D. H. R. BARTON, D. A. J. IVES and R. B. KELLY: The Synthesis of Lanostenol. J. Amer. Chem. Soc. **76**, 2852 (1954).

174. ZELLNER, J.: Zur Chemie der höheren Pilze. X. Mitt. Über *Armillaria mellea* VAHL., *Lactarius piperatus* L., *Pholiota squarrosa* MULL und *Polyporus betulinus* FR. Monatsh. Chem. **34**, 321 (1913).

(Received, February 8, 1955.)

Neuere Vorstellungen auf dem Gebiete der Biosynthese der Steroide und verwandter Naturstoffe.

Von R. TSCHESCHE, Hamburg.

Inhaltsübersicht.

I. Einleitung.

Nachdem in den Dreißigerjahren die Konstitution der Sterole* und Gallensäuren und anschließend bald auch die der anderen Steroide (Sexualhormone, Nebennierenrinden-Hormone, Spirostanole, Cardenolide, Bufadienolide und Steranalkaloide) in den wesentlichen Grundzügen

* In der deutschen Literatur wird im allgemeinen die Bezeichnung Ster*in* gegenüber Ster*ol* bevorzugt. Um die Alkoholnatur dieser Verbindungen hervorzuheben, ist in diesem Aufsatz die Bezeichnung Sterol verwendet worden.

geklärt war, hat es nicht an Versuchen gefehlt, Einblick in den Weg der Biogenese dieser Stoffklasse zu gewinnen. Solange die Technik des biologischen Experiments mit markierten Verbindungen (Isotope) wenig entwickelt war, bestand keine Aussicht, die gemachten Hypothesen durch Versuche entscheidend zu kontrollieren. So gründeten sich diese vor allem auf Ähnlichkeiten gewisser Naturstoffe im chemischen Aufbau mit den Steroiden und auch auf das gemeinsame Vorkommen in biologischem Material. Nunmehr ist es möglich geworden, diese Vorstellungen mit den Ergebnissen der Tracer-Technik zu prüfen und auf ihren Wert hinsichtlich der Klärung des vorliegenden Problems zu untersuchen. Trotzdem sind wir heute noch weit davon entfernt, ein klares Bild von der Biogenese dieser Stoffklasse zu besitzen. Es dürfte nützlich sein, das gesamte Tatsachenmaterial einmal kritisch zu sichten, um einen Überblick über den derzeitigen Stand unseres Wissens zu geben. Es kann dabei nicht ausbleiben, daß diesem Artikel gewisse Subjektivität in der Bewertung der Tatsachen zukommt, wie es stets der Fall ist, wenn auf einem Gebiet abschließende Erkenntnisse ausstehen. Weiter können einzelne Erörterungen stark hypothetische Züge zeigen, was nach Ansicht des Verfassers aber nicht unbedingt ein Fehler zu sein braucht, da damit Anregungen zu neuen experimentellen Ansätzen auf diesem weiten Gebiet vermittelt werden mögen.

II. Steroidbiogenese mit markierten Verbindungen.

1. Essigsäure als Baustein.

Die ersten Versuche zur Klärung der Frage, welche Bausteine zum Aufbau der Steroide bei der Biosynthese verwendet werden, wurden von Smedley MacLean und Hoffert (*90*) gemacht. Sie zeigten, daß Essigsäure durch nicht wachsende Hefe sowohl in Sterole wie in Fettsäuren eingebaut wird. Dieser Befund wurde von Sonderhoff und Thomas (*92*) bestätigt, die Hefe in einer Nährlösung mit Natriumacetat wachsen ließen, dessen Wasserstoff teilweise durch Deuterium ersetzt worden war. Es zeigte sich, daß die gebildeten Sterole einen hohen Gehalt an Deuterium aufwiesen. Analoge Untersuchungen wurden später von Bloch und Rittenberg (*17*—*19*) an Tieren durchgeführt und festgestellt, daß das gebildete Cholesterol ebenfalls stark deuteriumhaltig war. Diese Ergebnisse wurden in neuerer Zeit ergänzt und bestätigt durch die Anwendung von markierter Essigsäure, wobei entweder die Methyl- oder die Carboxylgruppe mit ^{13}C oder ^{14}C bzw. mit beiden versehen worden war (*13*, *16*—*19*). Diese Versuche lassen wenig Zweifel, daß eine Zwei-Kohlenstoff-Einheit zum Aufbau der Sterole sowohl in der Hefe wie im Tierkörper herangezogen wird. Außer Essigsäure (*6*, *23*) werden auch Äthanol (*36*), Aceton (*77*, *109*), Acetaldehyd und in geringerem

Ausmaß Brenztraubensäure hierbei verwertet, wobei an Stelle des Tierversuches auch Gewebepräparate als Testobjekt sich geeignet erwiesen. Ferner zeigten die Versuche, daß sowohl das Methyl wie das Carboxyl der Essigsäure verwendet wird und ihre C-Atome in den Sterolen wieder erscheinen (*73*).

Eine besondere Stütze findet die Anschauung, daß Essigsäure ein Baustein der Biosynthese ist, durch die Arbeit von OTTKE, SIMMONDS und TATUM (*70, 71*) an einer Mutante von *Neurospora crassa*. Diese Mutante erfordert Essigsäure zum Wachstum in der Nährlösung und kann nicht Glucose zu dieser Säure abbauen. Ließ man diesen Stamm auf einem Medium mit Glucose und $^{14}CH_3{}^{13}COOH$ wachsen, so enthielt das gebildete Ergosterol fast dieselbe Isotopenverteilung wie das angewandte Acetat. Auch Isovaleriansäure kann von diesem Organismus zur Steroidsynthese ausgenutzt werden, wobei die Essigsäurestufe wahrscheinlich nicht durchlaufen wird (*40, 63*).

Eine weitere Stütze für die Essigsäurehypothese als Baustein der Steroide stellt die Feststellung dar, daß für die Verwertung in Schnitten überlebender Gewebe, insbesondere der Leber (*16*), Pantothensäure notwendig ist. Offenbar geht die Essigsäure als Acetyl-Coenzym A in die Biosynthese der Sterole ein. Hefe, die an Pantothensäure und damit an Coenzym A-Mangel leidet, ist nach KLEIN (*55—57*) nicht der Ergosterolsynthese fähig. Bei den Versuchen mit Leberschnitten sind aerobe Bedingungen notwendig (*63*), ferner wurde eine Förderung des Aufbaus durch 0,0001 molare Mangansulfatlösung beobachtet (*63*). Auch Leberhomogenate und sogar zellfreie Extrakte können noch die Cholesterolsynthese durchführen (*26, 27, 78, 79*). Es ergibt sich aus diesen Befunden die enge Verwandtschaft von Fettsäure- und Sterolaufbau, die schon frühzeitig festgestellt worden war (*80, 106*). Andererseits verläuft aber die Cholesterolsynthese sicher unabhängig vom Aufbau der Fettsäuren, da die Bildung der letzteren im diabetischen Organismus bis gegen Null absinken kann, während die Synthese des Cholesterols unverändert oder eher vermehrt abläuft. Ein weiterer Unterschied besteht darin, daß das α-Kohlenstoffatom von Brenztraubensäure $CH_3—^{14}CO—COOH$ merkwürdigerweise kaum im Cholesterolmolekül erscheint, während es in Fettsäuren sehr wohl eingebaut wird.

Auch Glucose kann vermutlich über Essigsäure an der Cholesterolsynthese teilhaben (*76*). Außer der Leber besitzen auch andere Organe, wie Niere, Nebenniere, Testis, Dünndarm, Haut, Aorta und das Gehirn neugeborener Tiere, die Fähigkeit zur Cholesterolsynthese, während das Gehirn erwachsener Tiere nicht mehr dazu in der Lage ist (*93, 94*). Das Blutcholesterol stammt vermutlich fast vollständig aus der Leber, da hepatektomierte Hunde 4 Stunden nach der Verabreichung von ^{14}C-Acetat im Serumcholesterol keine Radioaktivität aufweisen (*42*). Die Cholesterolbildung in Gewebeschnitten verläuft recht schnell, was erstaunlich erscheint, da zwischen Essigsäure und dem Endprodukt der Synthese auf jeden Fall eine größere Anzahl von Stufen durchlaufen werden muß.

Die Tracer-Technik mit $^{14}CH_3{}^{13}COOH$ gestattet auch noch eine Aussage zu machen, wieviele C-Atome des Cholesterols aus dem Methyl und wieviele aus dem Carboxyl der Essigsäure herrühren. Das gefundene Verhältnis ist 15 aus der Methyl- und 12 aus der Carboxylgruppe, im Kern liegt es bei 10 : 9, in der Seitenkette bei 5 : 3. Durch systematischen Abbau des markierten Cholesterols und Prüfung jedes abgetrennten C-Atoms auf seine Radioaktivität gelang es mehreren Arbeitskreisen (*14, 33, 34, 37, 65, 107, 108*) bisher 20 von den 27 C-Atomen hinsichtlich ihrer Herkunft festzulegen. Das nachstehende Formelbild gibt diese Befunde wieder:

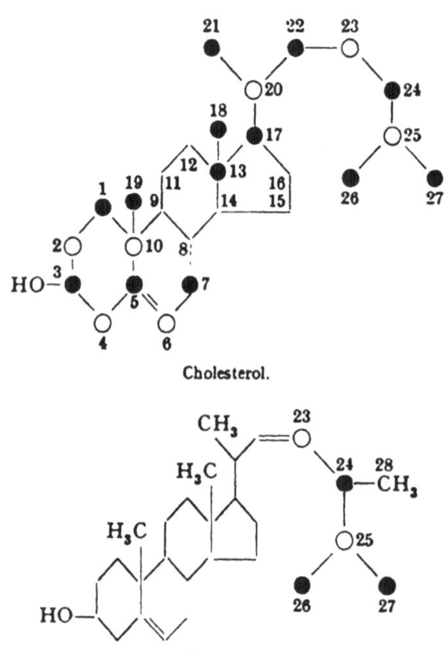

Cholesterol.

Ergosterol.

● Stammt aus Methyl-Kohlenstoff der Essigsäure.
○ Stammt aus Carboxyl-Kohlenstoff der Essigsäure.

Es fehlt also nur noch die Zuordnung der C-Atome 8, 9, 11, 12, 14, 15 und 16 des Cholesterols, von denen sich zwei aus der Methyl- und fünf aus der Carboxylgruppe herleiten müssen.

Entsprechende Versuche sind am Ergosterol der Hefe von Hanahan und Wakil (*50, 51*) durchgeführt worden. Ergosterol hat in der Seitenkette eine zusätzliche Methylgruppe $C_{(28)}$ an $C_{(24)}$ und der bisher durchgeführte systematische Abbau in der Seitenkette ergibt an diesem Sterol ein entsprechendes Bild wie beim Cholesterol. Ausgehend von $CH_3{}^{14}COOH$ ließ sich in dem durch Hefe gebildeten ^{14}C-Ergosterol bei $C_{(24)}$, $C_{(26)}$ und $C_{(27)}$ keine Radioaktivität feststellen, während $C_{(23)}$ und $C_{(25)}$ radioaktiv waren. Diese C-Atome müssen daher aus dem Carboxyl der Essigsäure stammen. Die nachfolgende *Tabelle 1* gibt die gefundenen Werte wieder.

Tabelle 1. Verteilung des Carboxyls-^{14}C aus CH_3-$^{14}COOH$ in der Seitenkette von ^{14}C-Ergosterol.

	Gef. Ausschläge/min. mmol ^{14}C	Ber. Ausschlage/min. mmol ^{14}C
$C_{(23)}$	4200	4200
$C_{(25)}$	4080	4080
$C_{(24)}$	0	0
$C_{(26)}$, $C_{(27)}$	0	0
$C_{(28)}$	150	0

Auffällig ist die gefundene geringe Radioaktivität von $C_{(28)}$, worauf noch zurückzukommen sein wird.

In dem Bestreben, weiteren Einblick in den Weg der Essigsäure zum Sterolsystem zu erhalten, wurden von Bloch und Mitarb. (*13*) eine Reihe einfacher mit ^{14}C markierter Verbindungen hinsichtlich ihrer Verwertbarkeit bei der Cholesterolsynthese geprüft. Wie sich aus der nachfolgenden *Tabelle 2* ergibt, scheinen Buttersäure und teilweise auch Isovaleriansäure besser als Essigsäure ausnutzbar zu sein. Diese Befunde an Leberschnitten wurden durch entsprechende Versuche am Tier ergänzt. Auch hier erwiesen sich Buttersäure, Isobuttersäure und Isovaleriansäure als gut für die Biosynthese des Cholesterols verwertbar (*44*, *58*).

Tabelle 2. Ausnutzung von ^{14}C-Verbindungen für die Cholesterolsynthese (*13*).

Verfutterte Verbindung	Relativer Gehalt an ^{14}C im Lebercholesterol
2-^{14}C-Essigsäure	100
1-^{14}C-Essigsäure	80
1,3-^{14}C-Aceton	66
2-^{14}C-Brenztraubensäure	16
1-^{14}C-Buttersäure	135
3-^{14}C-Buttersäure	180
1-^{14}C-Isovaleriansäure	60
4,5-^{14}C-Isovaleriansäure	550

Zabin und Bloch (*109*—*111*) haben aus ihren Versuchen den Schluß gezogen, daß Buttersäure vermutlich erst nach β-Oxydation zu Acetacetat in die Biosynthese eingeht, das gleiche gilt auch für Capronsäure und Caprylsäure (*23*). Wird ein Butyrat mit doppelter Markierung verwendet, im Carboxyl ^{13}C, im β-C-Atom ^{14}C, so entsteht mit Leberschnitten ein Acetacetat mit entsprechender Verteilung der radioaktiven C-Atome. Wird ein solches Butyrat an Ratten verfüttert, so entsteht ein Cholesterol, dessen Isotopenverteilung anzeigt, daß das Acetacetat ohne vorherige Spaltung verwertet worden ist. Diese Befunde stehen in guter Übereinstimmung mit Ergebnissen von CURRAN (*35*) und von BRADY und GURIN (*24*), die zeigen, daß Rattenleberschnitte Acetacetat-^{14}C in Gegenwart von großen Mengen nicht markierter Essigsäure zu stark radioaktivem Cholesterol umsetzen. Wurde jedoch die Acetessigsäure zunächst gespalten und so mit viel nicht isotopem Acetat verdünnt, so resultiert ein Cholesterol mit niedriger Aktivität (*25*).

Acetacetat muß daher ein direkter Vorläufer des Cholesterols sein. Aus Tabelle 2 ergibt sich, daß Isovaleriansäure besser als Essigsäure ausgenützt wird. Jedoch scheint der Isopropylteil des Moleküls vor allem verwertet zu werden, so daß die Vermutung besteht, daß die Isovaleriansäure zunächst durch β-Oxydation zu einer C_3-Einheit abgebaut wird, die nicht Aceton oder Brenztraubensäure ist (*32, 75*). Diese geht dann möglicherweise mit CO_2 wieder in Acetessigsäure über. Auch Isobuttersäure mag vielleicht erst decarboxyliert werden, um durch weiteren Abbau über eine C_2-Einheit in die Sterolsynthese einzugehen.

2. Die Rolle des Squalens.

Auf der Suche nach höher molekularen Zwischenstufen der Biogenese wurde von LANGDON und BLOCH (*59—61*) das Squalen (I) (siehe S. 138) in den Kreis der Untersuchungen mit markierten Molekülen einbezogen. Schon 1926 hatte CHANNON (*30*) aus dem gemeinsamen Vorkommen von Squalen und Cholesterol in Fischleberölen den Schluß gezogen, daß eine genetische Beziehung zwischen diesen Stoffen bestehen müßte. Aus dieser Annahme heraus verfütterte er Squalen an Ratten und fand in der Leber der Tiere eine doppelt so große Menge von Cholesterol als normalerweise beobachtet wird. Ferner zeigten ANDRÉ und CANAL (vgl. *63*), daß im Leberöl junger Haie das Squalen gegenüber dem Cholesterol weit überwiegt, während sich das Verhältnis im Alter umkehrt. Von diesem Kohlenwasserstoff ist bekannt, daß er sich in der menschlichen Haut (*66*) und im Sebum (*91*) findet, er wurde ferner von LANGDON und BLOCH (*59—61*) in der Rattenleber, von SCHWENK, ALEXANDER, STOUDT und FISH (*85*) in der Schweineleber aufgefunden, TÄUFEL und Mitarbeiter wiesen Squalen in der Hefe (*95*) und THORBJARNARSON und DRUMMOND (*96*) in einigen Pflanzenölen nach.

Von BONNER und ARREGUIN (*5, 20*) stammt eine Hypothese der Biogenese von Isoprenpolymeren, z. B. des Kautschuks, die auch für

Squalen verwendbar ist. Die genannten Autoren nehmen folgendes Schema an:

1. $2\,CH_3\cdot COOH \rightarrow CH_3\cdot CO\cdot CH_2\cdot COOH$
2. $CH_3\cdot CO\cdot CH_2\cdot COOH \rightarrow CH_3\cdot CO\cdot CH_3 + CO_2$
3. $CH_3\cdot CO\cdot CH_3 + CH_3\cdot COOH \rightarrow \begin{matrix} CH_3 \diagdown \\ C{=}CH\cdot COOH \\ CH_3 \diagup \end{matrix}$
4. $\begin{matrix} CH_3 \diagdown \\ C{=}CH\cdot COOH \\ CH_3 \diagup \end{matrix} \rightarrow \begin{matrix} CH_3 \diagdown \\ C\cdot CH{=}CH_2 \\ CH_2 /\!\!/ \end{matrix} \rightarrow \left[\begin{matrix} CH_3 \diagdown \\ C{=}CH\cdot CH_2{-} \\ {-}CH_2 \diagup \end{matrix}\right]_x$

Dieses würde mit der durch die Isotopentechnik festgestellten Vorläuferposition des Acetacetats gegenüber der Essigsäure in guter Übereinstimmung sein und steht auch mit der Herkunftsanalyse der Sterol-Kohlenstoffatome in der Seitenkette in Einklang, die ebenfalls Isoprenstruktur besitzt. Da nach SCHOPFER und Mitarb. (*45*, *84*) auch die Carotinsynthese bei *Mycophyces blakesleeanus* über Essigsäure abläuft, wird der Ort der Verzweigung von Fettsäure- und Carotinoidbiogenese ersichtlich (*4*).

Es dürfte in diesem Zusammenhang wichtig sein darauf hinzuweisen, daß β,β-Dimethylacrylsäure und β-Methyl-β-oxyglutarsäure aus Rattenleberhomogenaten, aus intakten Tieren und aus den Blättern australischer Eucalyptusarten isoliert worden sind (*85*). Die β-Methyl-β-oxyglutarsäure läßt sich aus dem Schema von BONNER und ARREGUIN ableiten, wenn bei der Kondensation von Acetessigsäure mit Essigsäure weder Decarboxylierung noch Wasserabspaltung erfolgt.

LANGDON und BLOCH (*59*—*61*) verfütterten ^{14}C-Acetat (entweder im Methyl oder Carboxyl markiert) gleichzeitig mit Squalen an Ratten und isolierten ein radioaktives Squalen aus der Leber dieser Tiere. Wurde dieses Material wieder an Mäuse verfüttert, so konnte radioaktives Cholesterol aus der Leber isoliert werden. Unter Zugrundelegung der Größe der stattgehabten Resorption des Squalens im Magen-Darm-Trakt, der Verteilung des gebildeten Cholesterols im Organismus und unter gleichzeitiger Inrechnungstellung der Halblebenszeit des Cholesterols bei diesen Tieren, ließ sich zeigen, daß Squalen 10—20mal wirksamer als Vorstufe des Cholesterols ist als Essigsäure. Die niedrige Radioaktivität der ebenfalls isolierten Fettsäuren dokumentiert, daß vor der Umwandlung in Sterol kein Abbau zu Zweikohlenstoff-Einheiten erfolgt ist. Diese Untersuchungen sagen aber nichts aus darüber, ob wirklich Squalen als Ganzes oder nur Teile des Moleküls zur Steroidbiogenese verwertet werden.

Eine gewisse Klärung der Rolle des Squalens bei der Biogenese der Steroide bringt eine neue Arbeit von SCHWENK und Mitarb. (*85*). Die

Autoren arbeiteten mit nicht wachsender Hefe, *Saccharomyces cerevisiae*, die mit ^{14}C-markiertem Natriumacetat inkubiert wurde. Die Aufarbeitung mit Hilfe von Digitonin ergab eine Sterolfraktion, aus der durch Umkristallisieren und über das Acetat reines radioaktives Ergosterol (II) isoliert wurde. Weiter gelang es, nach Entfernung des Ergosterols aus den Mutterlaugen durch Kondensation mit Maleinsäureanhydrid, radioaktives Zymosterol (III) zu gewinnen. Schließlich ergab der nicht mit Digitonin fällbare Anteil nach Chromatographie an Aluminiumoxyd eine Pentan-, Benzol-, Äther- und Methanoleluat-Fraktion. Aus dem Pentaneluat ließ sich radioaktives Squalen erhalten. Squalen und die anderen Fraktionen wurden an Ratten verfüttert und aus dem Körper dieser Tiere, nur mit der Benzolfraktion sowie aus dem Squalen selbst, radioaktives Cholesterol-^{14}C isoliert; die Methanolfraktion jedoch ergab kein Sterol, die Ätherfraktion nur wenig.

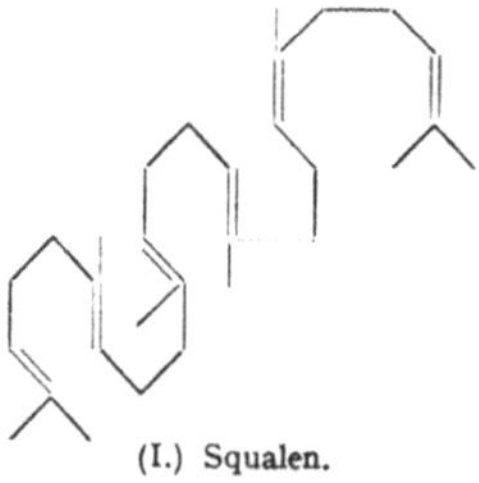

(I.) Squalen.

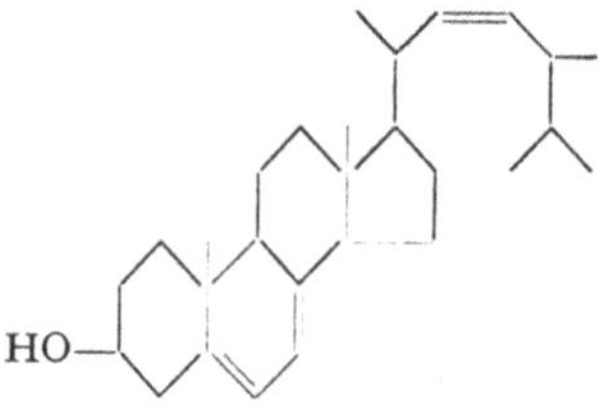

HO—

(II.) Ergosterol.

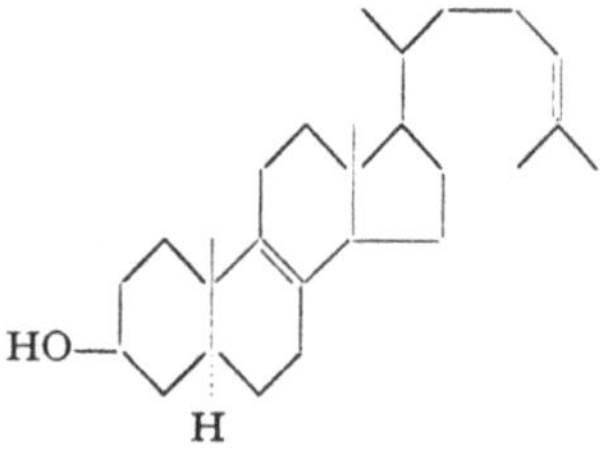

HO—

H

(III.) Zymosterol.

Der Vergleich der Radioaktivität der einzelnen isolierten Substanzen und Fraktionen zeigte, daß das erhaltene Squalen aktiver als das isolierte Ergosterol war, während am Zymosterol eine doppelt so große spezifische Aktivität wie beim Ergosterol gefunden wurde. Schwenk und Mitarb. nehmen an, daß die Benzolfraktion Zwischenprodukte der Biogenese der Steroide enthält, die in der Hefe in Ergosterol bzw. in Zymosterol übergehen, im tierischen Organismus jedoch in Cholesterol. Möglicherweise ist Zymosterol auf Grund seiner höheren Aktivität Vorstufe des Ergosterols. Bemerkenswert ist die geringere Aktivität des Squalens gegenüber Zymosterol. Dies läßt vielleicht den Schluß zu, daß Squalen nicht im Hauptweg der Sterolsynthese liegt, sondern eine Abzweigung darstellt, daß es aber infolge der möglichen Umkehrbarkeit biochemischer Reaktionen wieder in die Sterolsynthese eingeschleust werden kann. In der Benzolfraktion von SCHWENK dürfte vermutlich das wahre Intermediärprodukt zu suchen sein. In diesem Zusammenhang mag noch darauf hingewiesen werden, daß SCHWENK und WERTHESSEN (*87*) nach der Perfusion von Schweinelebern oder anderen Organen mit Carboxyl-markierter Essigsäure ein Roh-Cholesterol erhalten, dessen Aktivität bedeutend größer ist als die von reinem Cholesterol, das nach Reinigung über das Dibromid daraus gewonnen wird (*86*).

Folgendes Schema der Steroidbiogenese läßt sich aus diesen Befunden mit Hilfe der Tracer-Technik ableiten:

Essigsäure → Acetessigsäure → β-Oxy-β-methylglutarsäure → β,β-Dimethylacrylsäure → X → X → X —(Hefe)→ Zymosterol → Ergosterol

$X \to X$ ↓ ↑ Squalen; X —(Tier)→ Cholesterol

Die Menge des synthetisierten Cholesterols wird beim Tier durch die Zusammensetzung der Nahrung beeinflußt. So wird der Übergang von Essigsäure in Cholesterol in Leber, Haut, Nebenniere und Dünndarm unterdrückt (*43*, *98*), wenn die Menge des Cholesterols in der Diät wenigstens 3% ist, wie Versuche mit radioaktivem Acetat erweisen. Auch die Höhe der Blutcholesterolwerte beeinflußt die Synthese dieses Stoffes in der Leber (*97*). Ferner sollen Squalen, Cholestenon, Δ^7-Cholestenol, 7-Dehydrocholesterol und Dehydroisoandrosteron die Cholesterolbildung vermindern oder unterdrücken (*62*), nicht aber Lanosterol, Koprosterol, Ergosterol, Farnesol und Isomere des Squalens. Besonders der Einfluß des Squalens in diesen Versuchen erscheint bemerkenswert und unterstützt die Annahme, daß nahe Beziehungen zwischen der Biogenese dieses Stoffes und der des Cholesterols bestehen müssen.

III. Unsere Kenntnisse der biologischen Beziehungen der Steroide untereinander.

1. Ein C_{27}-Sterol als Vorstufe der Pflanzensterole mit 28 und 29 C-Atomen.

Ehe auf die Theorien der Biogenese des Grundsystems der Steroide näher eingegangen wird, empfiehlt es sich, einige Fragen zu untersuchen, die zunächst mehr am Rande des Problems zu liegen scheinen. Da drängt sich die Überlegung auf, *ob die Biosynthese im Tier- und Pflanzenreich im wesentlichen gleichlaufend erfolgt oder ob grundlegende Unterschiede zu erwarten sind.* Während das typische Sterol der Tiere, das Cholesterol, 27 C-Atome enthält, zeigen die meisten *Pflanzensterole* 28 oder 29 C-Atome im Molekül (*38*, *63*). Das zusätzliche C-Atom z. B. im Ergosterol (II, S. 138) mit 28 C-Atomen findet sich als Methylgruppe an $C_{(24)}$ in der Seitenkette, im Stigmasterol (IV) mit 29 Kohlenstoffatomen die beiden weiteren C-Atome am gleichen C-Atom als Äthylgruppe vor. Das einzige Pflanzensterol, das 27 C-Atome enthält, ist das Mycosterol *Zymosterol* (III). Aus den Versuchen von SCHWENK (*85*) an Hefe mit ^{14}C-Acetat ergibt sich, daß das gebildete Zymosterol eine höhere Radioaktivität besitzt als das gleichzeitig entstandene Ergosterol, so daß vermutet wird, daß es vielleicht eine Vorstufe des Ergosterols ist. Die Untersuchungen von HANAHAN und WAKIL (*51*) zeigen, daß das zusätzliche $C_{(28)}$-Atom der Seitenkette im Ergosterol eine geringfügige Radioaktivität besaß. Beide Ergebnisse würden sich gut vereinen lassen, wenn man annimmt, daß das $C_{(28)}$-Atom nachträglich in ein Sterol mit 27 C-Atomen eingebaut worden ist. Die beobachtete geringe Radioaktivität von $C_{(28)}$ kann vielleicht so gedeutet werden, daß dieses C-Atom nicht unmittelbar aus der Essigsäure stammt, sondern zum Teil durch einen nebenher laufenden Stoffwechselvorgang der Hefe geliefert wird, der diese Säure zunächst zu einer C_1-Einheit abbaut. Für das spätere $C_{(28)}$ wird vornehmlich das CH_3 der Essigsäure herangezogen, zu einem geringen Ausmaß kann aber auch das Carboxyl ^{14}C beitragen.

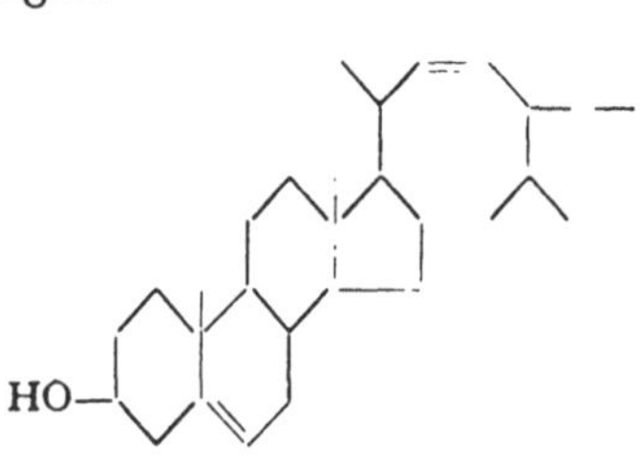

(IV.) Stigmasterol. 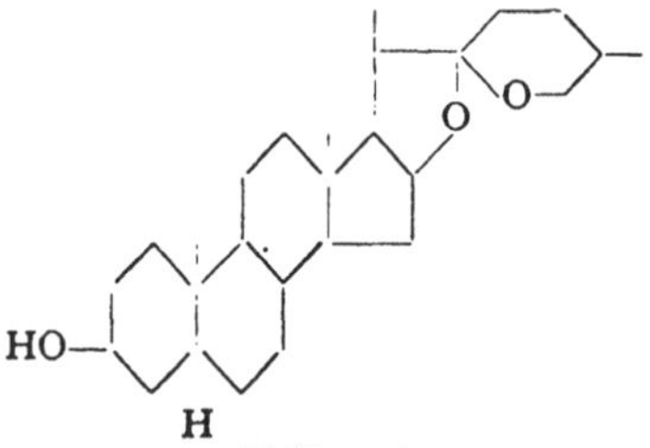 (V.) Tigogenin.

Sieht man sich nun im Pflanzenreich nach ***weiteren Steroiden mit 27 C-Atomen*** um, so stößt man auf die Aglykone der neutralen Saponine, die *Spirostanole* (*38, 63*). Von diesem Typ, von dem als einfachster Vertreter das Tigogenin (V) wiedergegeben sei, kennt man heute schon zirka 30 verschiedene natürlich vorkommende Derivate, die sich außer in der Zahl und Stellung von HO-Gruppen, noch durch Vorliegen einer Ketogruppe an $C_{(12)}$ und von Doppelbindungen im Kern unterscheiden. Auch die sterischen Verhältnisse an $C_{(5)}$ und $C_{(25)}$ können Unterschiede aufweisen. Von TSCHESCHE und KORTE (*99*) ist die Ansicht vertreten worden, daß den Spirostanolen ein Sterol mit 27 C-Atomen vorausgeht und die typische Anordnung der beiden Sauerstoffringe in der Seitenkette durch Anlagerung von 2 Mol. Wasser an ein $\Delta^{22,25}$-Dien und anschließenden Ringschluß zustande gekommen ist (VIa). Hierbei mag vorher ein Austausch der Oxydationsstufen zwischen der Ketogruppe an $C_{(16)}$ und der Hydroxylgruppe an $C_{(22)}$ stattgefunden haben (VIb).

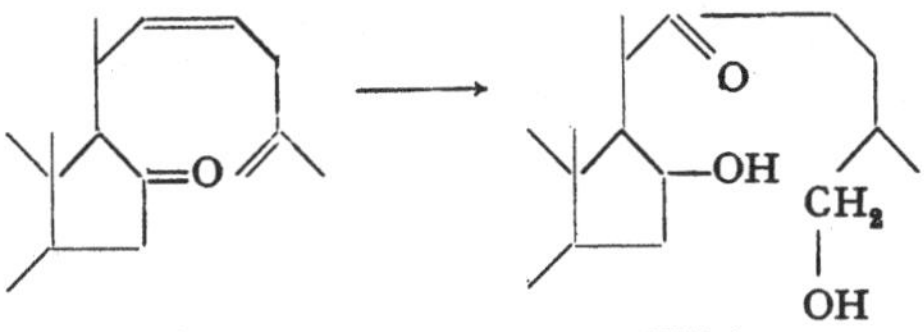

NH$_2$ OH —OH CH$_2$ OH O —OH CH$_2$ OH =O

(VI a.) (VI b.)

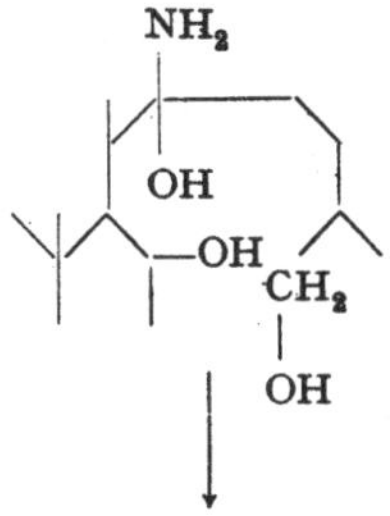

(VI.) Mögliches Intermediärprodukt.

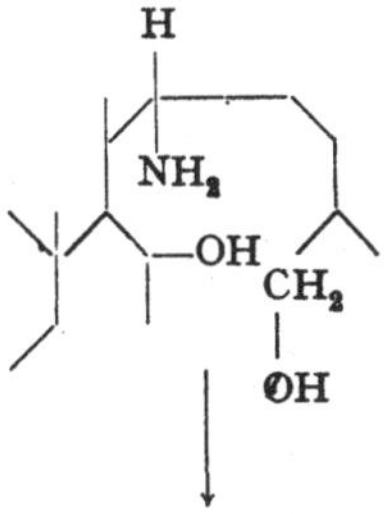

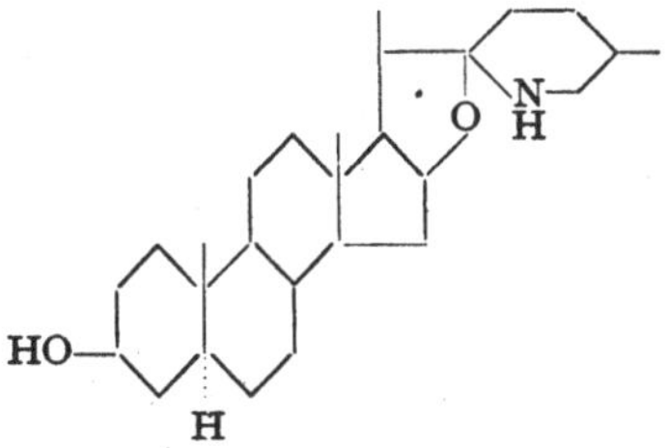

H NH$_2$ —OH CH$_2$ OH

N O H HO— H

(VII.) Tomatidin.

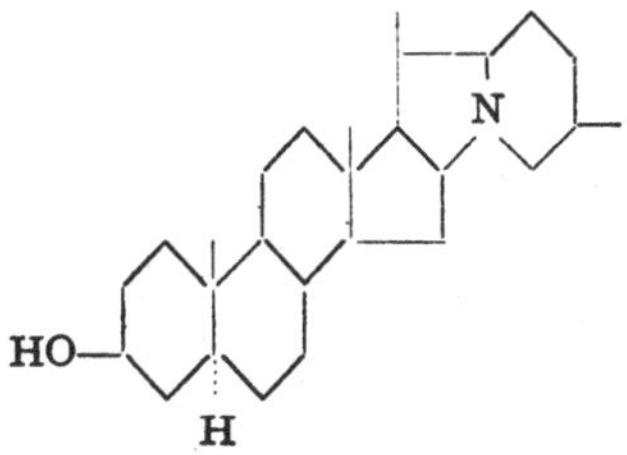

N HO— H

(VIII.) Demissidin.

H N HO HO—

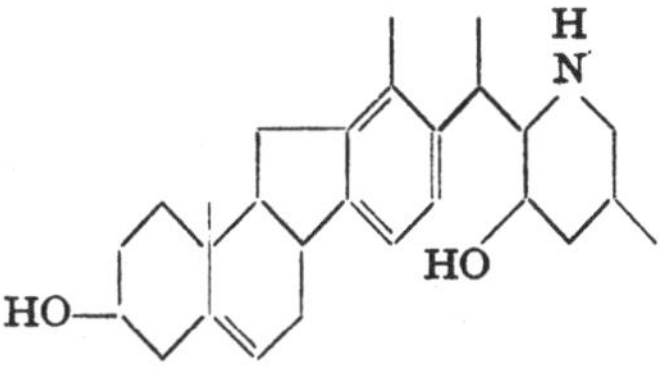

(IX.) Veratramin.

Ein Sterol mit 27 C-Atomen von Typ (VI) dürfte auch den Geninen der Steranalkaloide zugrunde liegen, soweit sie 27 C-Atome enthalten. Von ihnen sind schon über 20 bekannt. Das *Tomatidin* (VII) z. B. könnte aus (VI b) nach Anlagerung von Ammoniak an die Ketogruppe $C_{(22)}$ unter anschließender zweifacher Wasserabspaltung entstehen, *Demissidin* z. B. aus dem Ketimin an $C_{(22)}$ nach Reduktion zum Amin und anschließende Dehydratisierung. *Veratramin* (IX) und *Veracevin* (X) dürften sich durch Ringverengung von Ring *C* und Ringerweiterung von *D* bilden, wobei nachher an (IX) unter Wasserabspaltung die Aromatisierung von Ring *D* eintritt. Der angenommene Vorgang ist auf rein chemischem Wege am *Rockogenin* (XI) durchgeführt worden (*63*). Erhitzt man das $C_{(12)}$-Methansulfoderivat mit Alkalialkoholat in *t*-Butanol, so tritt die genannte Umlagerung außerordentlich leicht ein (XII), vorausgesetzt, daß wie im Rockogenin eine 12 (β)-Oxy-Verbindung vorliegt. Von solchen Verbindungen aber sind unter den Spirostanolen mehrere bekannt. In bezug auf die erforderlichen Ringschlüsse zu N-haltigen Ringen, wie sie im

Veratramin und Veracevin vorliegen, bestehen keine Schwierigkeiten, da sich die Folge der Kohlenstoffatome in der Seitenkette in ihnen wie im Cholesterol unverändert wiederfindet.

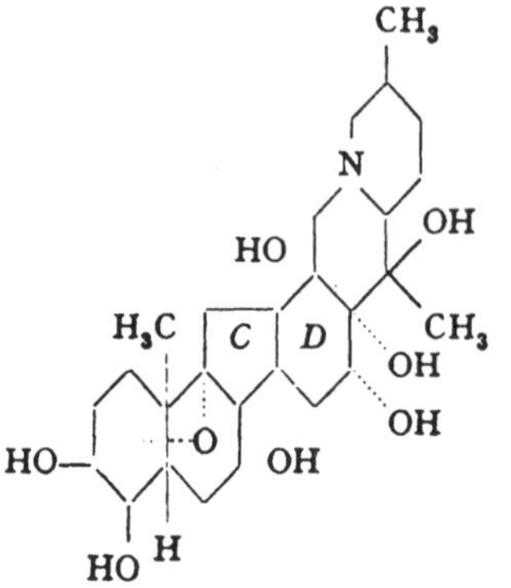

(X.) Veracevin.

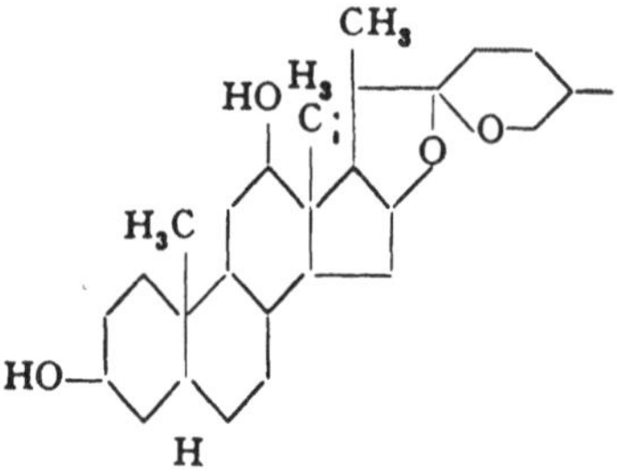

(XI.) Rockogenin.

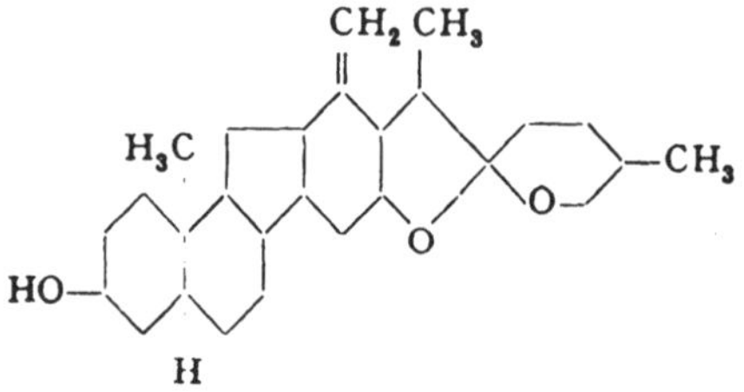

(XII.)* Umlagerungsprodukt.

Die große Zahl von C_{27}-Steroid-derivaten aus dem Pflanzenreich mit stets gleicher Kohlenstoffanordnung wie im Cholesterol macht ebenfalls die Vermutung plausibel, daß die zusätzlichen C-Atome 28, bzw. 28 und 29, in den pflanzlichen Sterolen erst nachträglich in ein schon fertig gebildetes C_{27}-Sterol eingeführt worden sind. TSCHESCHE und KORTE (99) nehmen an, daß die durch die benachbarten Doppelbindungen Δ^{22} und Δ^{25} aktivierte CH_2-Gruppe $C_{(24)}$ [in (VI a)] in der Pflanze mit ein oder zwei Kohlenstoffeinheiten unter Wasserverlust reagiert. Es würde sich aus (VI a) ein System wie in (XIII a) und (XIII b) ausbilden, aus dem sich durch teilweise oder vollständige Hydrierung sämtliche bekannten Pflanzensterole ableiten lassen.

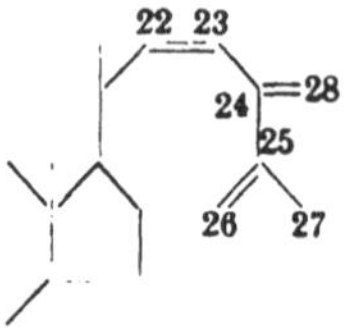

(XIII a.)

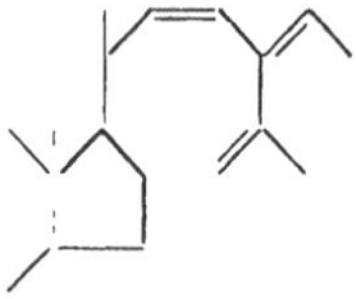

(XIII b.)

Zymosterol enthält die Doppelbindung in der Seitenkette in Δ^{24} an entsprechender Stelle wie Squalen. Zur Bildung eines $\Delta^{22,25}$-Diens wäre eine Verschiebung dieser Doppelbindung in die endständige Position und eine Dehydrierung an $C_{(22)}$, $C_{(23)}$ notwendig — ein Vorgang, der biologisch nicht unmöglich erscheint. Auch aus dieser Überlegung heraus könnte Zymosterol sehr gut eine Vorstufe des Ergosterols sein, so wie SCHWENK und Mitarb. (*85*) vermuten.

Durch teilweise oder vollständige Hydrierung der drei Doppelbindungen in (XIIIa) und (XIIIb) läßt sich nun leicht die Bildung aller in der Natur aufgefundener Sterole mit 28 bzw. 29 C-Atomen erklären (*63*) (*Tabelle 3*). Dabei scheint die Absättigung der endständigen Doppelbindung Δ^{25} erwartungsgemäß am leichtesten zu erfolgen.

Tabelle 3. C_{28}- und C_{29}-Sterole.

C_{28}-*Reihe:* vollständige Hydrierung der Seitenkette		C_{29}-*Reihe:* vollständige Hydrierung der Seitenkette	
Campesterol	24 α	Clionasterol	24 α
Fungisterol	24 β	β-Sitosterol	24 β
Haliclonasterol	24 α	γ-Sitosterol	24 α
Δ^{22}-Verbindungen		*Δ^{22}-Verbindungen*	
Brassicasterol	24 β	Chondrillasterol	24 α
Chalinasterol	24 α	Poriferasterol	24 α
14-Dehydro-ergosterol	24 β	Stigmasterol	24 β
Ergosterol	24 β	α-Spinasterol	24 β
Neospongosterol	24 α		
Stellasterol	24 α		
$\Delta^{24:28}$-Dien-Verbindungen		*$\Delta^{24:28}$-Dien-Verbindungen*	
Episterol		Fucosterol	
Faecosterol			

Je nach der Anlagerung des Wasserstoffs bei der biochemischen Hydrierung entstehen die Sterole der α- und β-Reihe an $C_{(24)}$, da hierbei ein neues Asymmetriezentrum gebildet wird. Für das Hefesterol *Ascosterol* mit einer Doppelbindung Δ^{23} bleibt die 1,4-Addition des Wasserstoffs an das konjugierte Doppelbindungssystem der Seitenkette oder eine nachträgliche Verschiebung einer bei der Absättigung verbliebenen Doppelbindung in diese Position als Erklärung. Ascosterol wäre übrigens das einzige Sterol, das eine solche zusätzliche Annahme erfordern würde. Auch ist die Δ^{23}-Stellung der Doppelbindung im Ascosterol, einem Nebensterol der Hefe, nicht völlig gesichert. Daß das Cholesterol auf der C_{27}-Stufe verharrt, mag vielleicht an der fehlenden Dehydrierung an $C_{(22)}$—$C_{(23)}$ im tierischen Organismus liegen oder aber, daß die endständige Doppelbindung der Seitenkette nicht verlagert werden kann.

2. Gibt es Unterschiede der Biosynthese von Steroiden bei Tieren und Pflanzen?

Daß keine grundsätzlichen Unterschiede der Steroidsynthese im pflanzlichen und tierischen Organismus bestehen können, ergibt sich ferner aus der Auffindung des Oestrons (XIV) und Oestriols (XV) im Tier- und Pflanzenreich. Die beiden *Hormone* wurden außer aus tierischem Material von BUTENANDT auch aus Palmöl, von SKARZYNSKI aus Weidenkätzchen isoliert (vgl. *38*). Ferner finden sich Vertreter der *Bufadienolid-Reihe* an Suberylarginin gebunden in Kröten als Giftstoffe der Parotisdrüsen, bei Liliaceen und Ranunculaceen als Glykoside vor; *Bufotoxin* (XVI) und *Scillaren A* (XVII) seien als Beispiele angeführt. Daß die Krötengifte durch Aufnahme von Bufadienolid-glykosiden mit der Nahrung in diese Tiere gelangen, ist nicht anzunehmen.

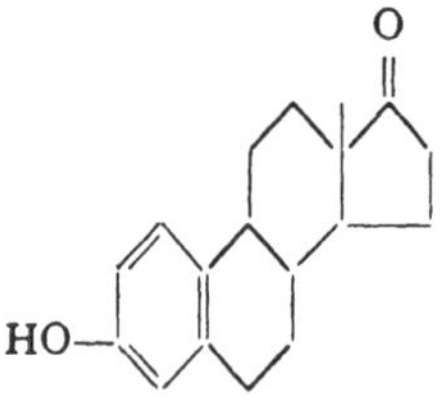

(XIV.) Oestron.

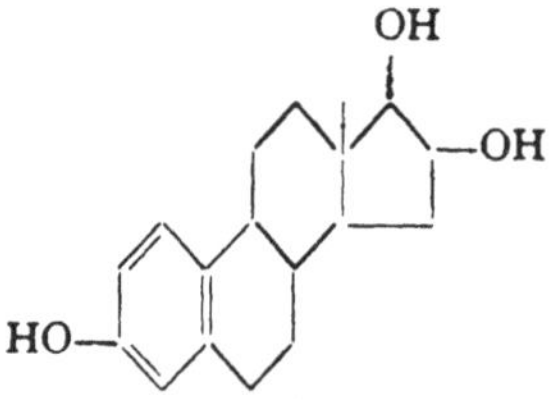

(XV.) Oestriol.

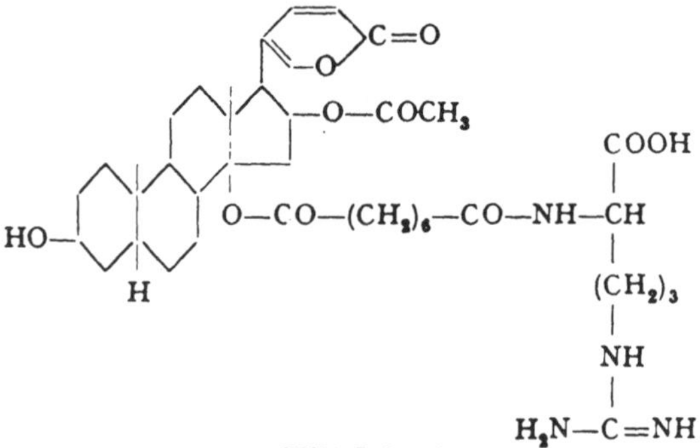

(XVI.) Bufotoxin.

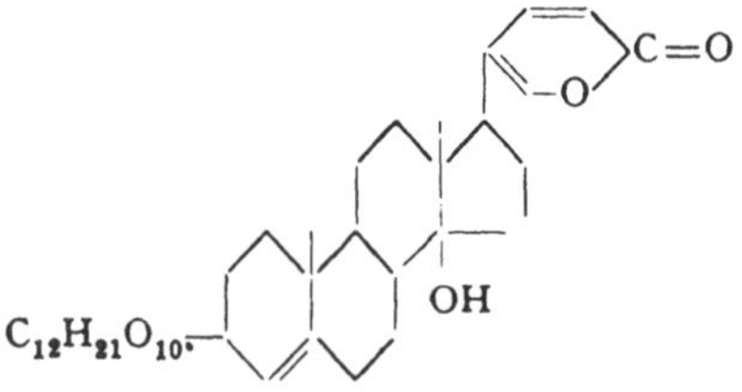

(XVII.) Scillaren A.

Es muß daher mit einer gleichartigen Bildung der genannten Steroide bei Tier und Pflanze gerechnet werden; die gefundenen Unterschiede dürften auf Prozessen beruhen, die dem eigentlichen Syntheseweg überlagert oder von ihm abgezweigt sind. Solche können in der späteren Einfügung von C-Atomen in die Seitenkette — wie bei den pflanzlichen Sterolen — beruhen, oder in Hydrierungen, Dehydrierungen oder in der Einführung von Sauerstoffunktionen an unterschiedlichen Stellen in das Grundsystem bestehen. Mit den vielfältigen Variationen der Steroide in der Seitenkette wird sich der nächste Abschnitt befassen.

3. Ist ein C_{27}-Sterol obligates Zwischenprodukt bei der Biogenese aller Steroide?

Die gute Ausnutzung des Squalens bei der Biogenese des Cholesterols und Ergosterols hat zunächst zu der Anschauung geführt, daß dieser Kohlenwasserstoff als Ganzes cyclisiert wird und dabei drei überzählige C-Atome, die er gegenüber einem C_{27}-Sterol besitzt, nachträglich entfernt werden (siehe S. 153). Das würde bedeuten, daß alle Steroide, die weniger als 27 C-Atome im Molekül enthalten, sich durch weiteren Abbau der Seitenkette von dieser C_{27}-Verbindung ableiten müssen. Im Tierreich handelt es sich dabei einmal um die *Gallensäuren* mit 24 C-Atomen, z. B. die Cholsäure (XVIII), das *Progesteron* (XIX) und die *Nebennierenrindenhormone*, z. B. Desoxycorticosteron (XX) mit 21 C-Atomen, *Testosteron* (XXI) mit 19 und *Oestron* (XIV) und *Oestriol* (XV) mit 18 C-Atomen. Die *Bufadienolide* (XVI) mit ebenfalls 24 C-Atomen wurden bereits erwähnt.

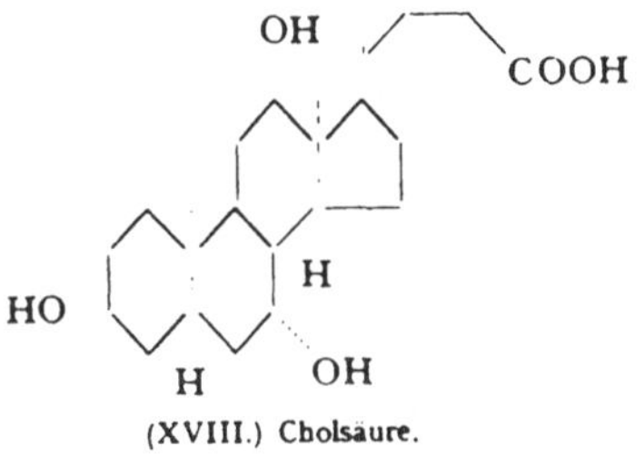

OH COOH HO H H OH

(XVIII.) Cholsäure.

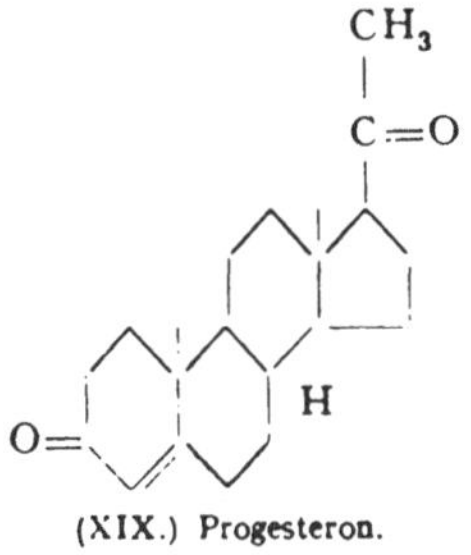

CH_3 C=O O= H

(XIX.) Progesteron.

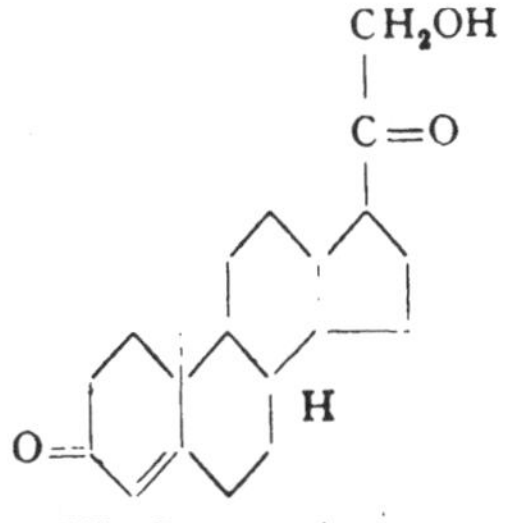

CH_2OH C=O O= H

(XX.) Desoxycorticosteron.

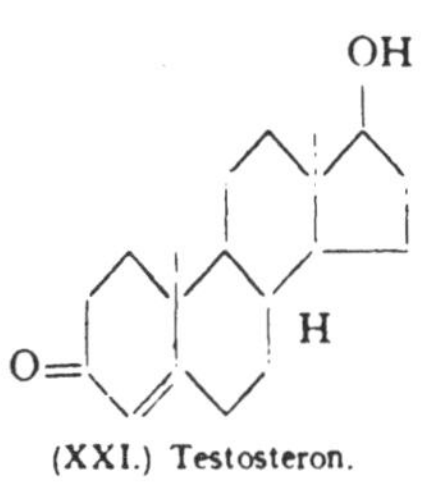

OH O= H

(XXI.) Testosteron.

Im Pflanzenreich stehen diesen gegenüber außer den Bufadienoliden, dem Oestron und Oestriol, die *Cardenolide*, z. B. Digitoxigenin (XXII) mit 23 C-Atomen, die *Holarrhena-alkaloide* mit 21 C-Atomen (nach Abzug der CH_3-Gruppen am N), z. B. *Conessin* (XXIII) und *Diginigenin*

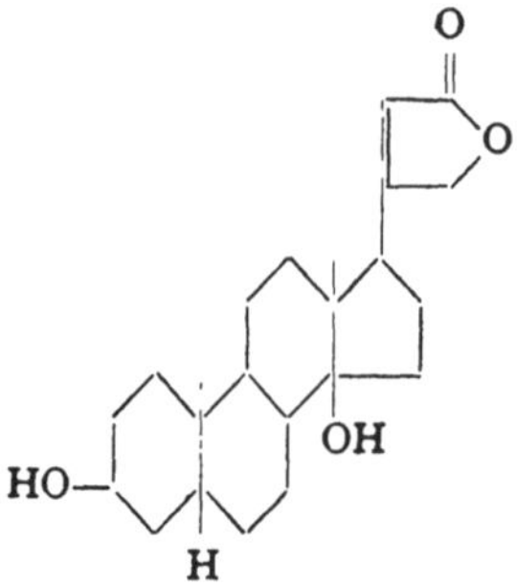

(XXII.) Digitoxigenin.

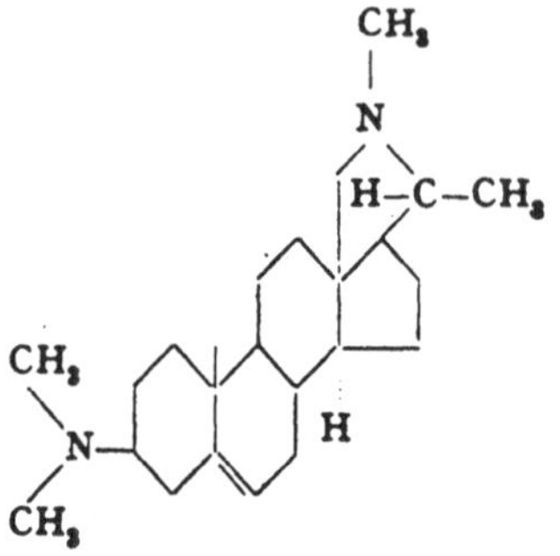

(XXIII.) Conessin.

(21 C-Atome), dessen endgültige Konstitutionsermittlung noch aussteht. Diginigenin enthält aber sicher das Kohlenstoffgerüst des Pregnans oder Allopregnans (*63*).

Gibt es experimentelle Befunde, die die Anschauung rechtfertigen, daß alle diese kohlenstoffärmeren Steroide über ein C_{27}-Steroid als obligate Zwischenstufe entstehen? Untersuchungen zur Klärung dieser Frage sind bisher nur an tierischen Steroiden durchgeführt worden, da hier wesentlich einfacher als an Pflanzen Ergebnisse mit Hilfe der Isotopen-Technik erzielbar sind.

Gallensäuren. 1943 zeigten Bloch, Berg und Rittenberg (*15*), daß deuteriertes Cholesterol beim Hund in Cholsäure übergeht. Der Austausch von Wasserstoff gegen Deuterium wurde mittels Platin-Katalysator vorgenommen (*1*). Spätere Untersuchungen von Fukushima und Gallagher (*39*) stellten fest, daß sich zirka 40% des in dieser Weise eingeführten Isotopes an $C_{(6)}$ befinden, 3% an $C_{(3)}$, während sich 50% auf die C-Atome 24, 25, 26 und 27 verteilen. Sie berechneten, daß beim Übergang von Cholesterol in Cholsäure im Tierkörper wenigstens 87% dieser Säure aus dem Cholesterol stammen müssen. In Ergänzung dieser Versuche fanden Chaikoff, Dauben und Mitarb. (*29*) mit an $C_{(4)}$ und $C_{(26)}$ markiertem Cholesterol, daß erwartungsgemäß ein großer Teil des eingeführten Isotopes als CO_2 in der Atemluft erscheint, jedoch nicht, wenn die Markierung allein auf $C_{(4)}$ beschränkt ist. Später ließ sich zeigen (*88*), daß die Radioaktivität von an $C_{(4)}$ markiertem Cholesterol über die Galle und die Faeces eliminiert wird. Bergström (*11*) stellte entsprechende Versuche an Ratten an und fand, daß Cholesterol-4-^{14}C in Cholsäure übergeht, die den Hauptteil der Aktivität enthält. Praktisch kein Isotop erschien in dem ausgeatmeten CO_2. Zu entsprechenden Ergebnissen kamen Byers und Biggs (*28*) ebenfalls mit ^{14}C markiertem Cholesterol. *Es kann daher kein Zweifel sein, daß dieses Sterol im tierischen Organismus unter Verkürzung der Seitenkette um drei C-Atome abgebaut werden kann.* Die dabei beobachtete hohe Ausbeute läßt diesen Weg als einzigen oder weitaus bevorzugten der Bildung der Gallensäuren im tierischen Organismus erscheinen (*89*).

Es gibt einige Naturstoffe, die die Rolle von Zwischenstufen dieser Verkürzung der Seitenkette am Cholesterol übernehmen können. Es handelt sich dabei in bezug auf den Aufbau des interessierenden Teils des Moleküls um *Cerebrosterol* (XXIV), *Ranol* (XXV), *Gigifisch-gallensäure* (XXVI), *Scymnol* (XXVII) (*63*). Auch *Cholegenin* (XXVIII) (*2, 67*) ist in diesem Zusammenhang interessant, da es ebenfalls die Einführung einer OH-Gruppe in eine CH_3-Gruppe zeigt.

```
 CH3               H  CH3
 |                 |  |
R—CH—CH2—CH2—C—CH
                   |  |
                   OH CH3
```

(XXIV.) Cerebrosterol.

```
 CH3               H  CH3
 |                 |  |
R—CH—CH2—CH2—C—C—OH
                   |  |
                   OH CH3
```

(XXV.) Ranol (?).

```
 CH3               H  COOH
 |                 |  |
R—CH—CH2—CH2—C—CH
                   |  |
                   OH CH3
```

(XXVI.) Gigifisch-gallensäure.

```
 CH3               H      CH2OH
 |                 |      |
R—CH—CH2—CH2—C——C
                   \   /|
                    O  CH3
```

(XXVII.) Scymnol.

```
      CH3
      |     /‾‾‾\
      |   O\    >—CH2OH
      |  O   O—/
  \ / \/
   |   |
   |___|
  /
```

(XXVIII.) Cholegenin.

Progesteron. Dieses Hormon wird vom weiblichen Organismus in Form von Pregnandiol im Harn ausgeschieden und Bloch (*12*) hat die Frage geprüft, ob die Verabreichung von deuteriertem Cholesterol zum Auftreten von deuteriertem Pregnandiol führt. Dies ist in der Tat der Fall und unter Berücksichtigung der Konzentration des Deuteriums im Ausgangs- und Endprodukt, des Verlustes von D bei der Verkürzung der Seitenkette und unter der Annahme, daß vor allem das Blutcholesterol als Vorstufe in Frage kommt, wurde eine 65—70proz. Bildung aus letzterem errechnet. Unter Zugrundelegung der Art und Größe des Deuterium-Austausches nach Fukushima und Gallagher (*39*) am Platinkatalysator ließ sich dieser Wert auf 68% präzisieren.

Corticosteroide. Während die Herkunft der Gallensäuren und weniger vollständig auch die des Progesterons aus Cholesterol gesichert erscheint, ist die der Nebennierenrinden-hormone weniger eindeutig geklärt. Der augenblickliche Stand unserer Kenntnisse dürfte durch die Feststellung wiederzugeben sein, daß zwar ein Übergang von Cholesterol in diese Hormone möglich ist (*103, 113*), der bevorzugte Weg aber der Aufbau direkt aus Essigsäure darstellt (*47*), bei dem die Cholesterolstufe nicht durchlaufen wird. Lieberman und Teich (*64*) haben ein mögliches Schema der Biosynthese der Nebennierenrinden-hormone aufgestellt, das nachfolgend wiedergegeben sei (*Tabelle 4*).

Tabelle 4. Schema der Biosynthese der Nebennierenrinden-hormone (*64*).

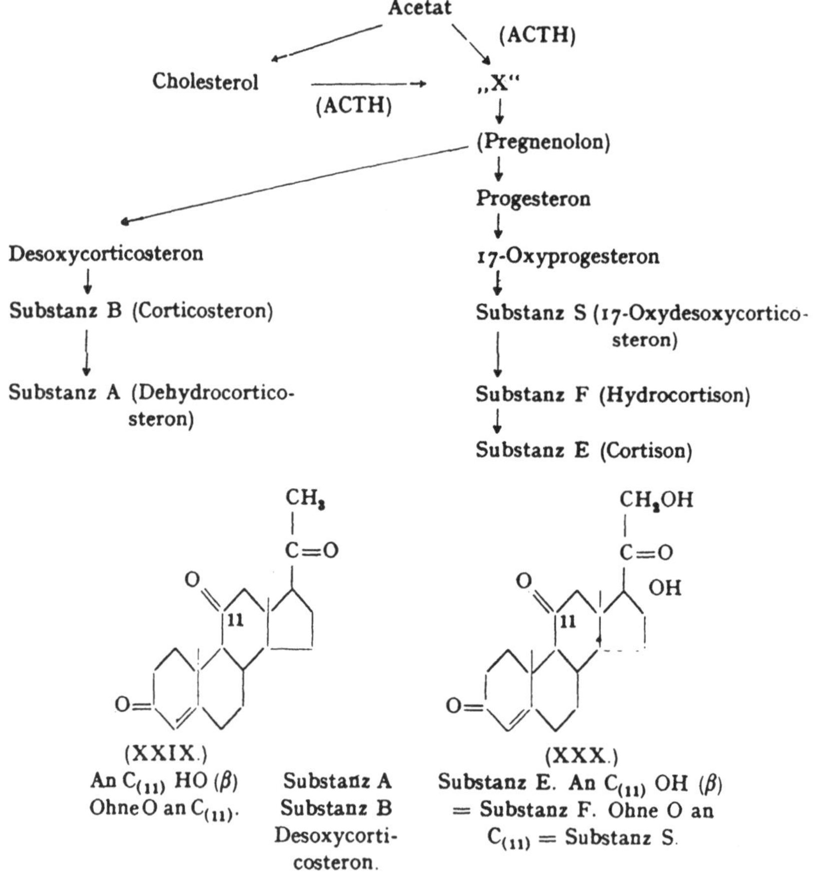

(XXIX.)
An $C_{(11)}$ HO (β) Substanz A
Ohne O an $C_{(11)}$. Substanz B
Desoxycorticosteron.

(XXX.)
Substanz E. An $C_{(11)}$ OH (β) = Substanz F. Ohne O an $C_{(11)}$ = Substanz S.

Für die Bildung von Corticosteroiden (XXIX und XXX) aus Cholesterol sprechen vor allem Perfusionsversuche der Nebenniere mit markiertem Cholesterol. So fanden Zaffaroni, Hechter und Pincus (*112, 113*), daß die Perfusion von Rindernebennieren mit Cholesterol-3-^{14}C zu radioaktivem Corticosteron und Cortison führte. Auch ist bekannt, daß unter dem Einfluß von Adrenocorticotropen-Hormon (ACTH) der Gehalt der Nebennierenrinde an Cholesterol abnimmt, dafür die Menge an Corticosteroiden zunimmt (*74*). ZAFFARONI und Mitarb. fanden jedoch, daß, wenn sie die Nebennieren mit ^{14}C-Acetat durchströmten, Corticosteron und 17-Oxycorticosteron einen sechsfach höheren Isotopengehalt aufwiesen als das freie Cholesterol der Drüse. Letzteres war im übrigen aktiver als das Blutcholesterol oder das veresterte Cholesterol der Nebennieren. Wurde jedoch ^{14}C-Cholesterol für die Durchströmungsversuche verwendet, war die Aktivität der Hormone nur 40% derjenigen des freien Cholesterols in der Drüse, noch geringer war die Aktivität von Blut- und gebundenem Cholesterol.

UNGAR und DORFMAN (*102*) haben versucht, das Ausmaß der Einfügung von ^{14}C aus markiertem Acetat und aus markiertem Cholesterol in die Ketosteroide des Harns beim Menschen zu bestimmen. Es wurden zwei weibliche Patienten mit Nebennierentumor herangezogen, da bei derartiger Krankheit die Ausscheidung von C_{19}- und C_{21}-Steroiden im Harn erhöht ist. Während bei der Gabe von ^{14}C-Acetat die Radioaktivität des ausgeschiedenen Dehydro-isoandrosterons, Aetiocholanolons, Dehydroepiandrosterons, 3(β)-Chlor-Δ^4-androsten-17-ons, Δ^5-Androsten-3(β), 17(β)-diols, 3(β)-Chlor-Δ^5-androsten-17(β)-ols, Δ^5-Pregnen-3(β), 20(α)-diols und zweier weiterer nicht identifizierter Verbindungen hoch war, ergab die Verabreichung von ^{14}C-Cholesterol nur eine sehr geringe Aktivität der Harnsteroide. Da bekannt ist, daß von zugeführtem Cholesterol zirka 20% vom Menschen resorbiert werden, zeigt dieser Versuch, daß die Umwandlung von Cholesterol in C_{19}- und C_{21}-Verbindungen beim ganzen Organismus wenig bedeutend ist. Der Aufbau der Nebennierenrinden-hormone erscheint in diesem Falle einen direkten Weg und nicht den über Cholesterol einzuschlagen.

Nebennieren enthalten ein kompliziertes Fermentsystem, bestehend aus einer 6-β-Hydroxylase, einer 11-β-Hydroxylase, einer 17-α-Hydroxylase und einer 21-Hydroxylase. Ferner findet sich ein Ferment, das 3(β)-Oxy-Δ^5-steroide in 3-Keto-Δ^4-Derivate verwandelt (KAHNT und WETTSTEIN (*54 a*); SAMUELS und Mitarb. (*74 a, 82 a, 82 b*); HAYANO und DORFMAN (*51 a, 51 b, 51 c*).

Testosteron. Die Biosynthese dieses Hormons aus Acetat durch Hodengewebe ist sowohl für Tiere wie für den Menschen demonstriert worden (*22, 83*). BRADY (*22*) beobachtete die Bildung von radioaktivem Testosteron aus ^{14}C-Acetat bei der Inkubation mit Schweine-, Kaninchen- und menschlichem Hodengewebe. Ebenso ergab die Perfusion des Testes mit ^{14}C-Acetat und Gonadotropin nach SAVARD und Mitarb. (*83*) markiertes Testosteron und Δ^4-Androsten-3,17-dion. Aber BRADY (*22*) fand, daß das bei der Inkubation gebildete Cholesterol weniger radioaktiv war als das Testosteron. Führte er die Versuche in Gegenwart von Chorion-gonadotropin durch, vergrößerte sich der Einbau des Acetats in Testosteron auf das Zehnfache, während die Aktivität des Cholesterols unbeeinflußt blieb. Auch hier erscheint also der Hauptweg der Biosynthese nicht über Cholesterol zu laufen. Neuerdings fanden Gallagher und Mitarb. (*41*), daß ^{14}C-Cholesterol beim Menschen in Androsteron und Aetiocholanon übergeht.

Für die Verbindung „X“ (S. 148), die sowohl vom Cholesterol wie vom Acetat her erreicht werden soll, wird von Hirschmann und Hirschmann (*54*) angenommen, daß es sich dabei um eine C_{19}-Verbindung handelt. Diese soll anschließend in eine C_{21}-Verbindung übergeführt werden. Folgende Feststellungen sind für diese Annahme beigebracht worden.

In vielen Fällen von Nebennierentumor wird als überwiegendes Harnsteroid Dehydro-isoandrosteron (XXXI), eine C_{19}-Verbindung ausgeschieden; die anderen Harnsteroide treten weit zurück. Bei gesunden Personen jedoch ist die Menge des Dehydro-isoandrosterons sehr klein. Stoffwechselstudien mit ^{14}C-Acetat an Tumorkranken mit Nebennierencarcinom ergaben fast dieselbe spezifische Aktivität von Serum-Cholesterol wie Harn-Dehydro-isoandrosteron. Wurden weiter Nebennieren mit nicht radioaktivem Dehydro-isoandrosteron und ^{14}C-Acetat durchströmt, so waren die gebildeten C_{21}-Corticosteroide radioaktiv (*40*). Die anschließende Aboxydation der Seitenkette führte zu einem 90proz. Verlust der Aktivität; hierdurch wird dargetan, daß die markierte Essigsäure zur Angliederung einer C_2-Einheit an $C_{(17)}$ verwendet wurde. Von einem Kranken mit Adrenocarcinom wurden im Harn $3(\beta),17(\alpha)$-Dioxy-Δ^5-pregnen-20-on (XXXII) und $3(\beta),21$-Dioxy-Δ^5-pregnen-20-on ausgeschieden (XXXIII). Ihre bemerkenswerte Ähnlichkeit im Aufbau in den Ringen *A* und *B* mit Cholesterol ist auffallend. Alle diese Befunde stellen aber keine sichere Grundlage dar, um anzunehmen, daß „X“ wirklich eine C_{19}-Einheit darstellt; auf eine andere Möglichkeit wird später hingewiesen (S. 160).

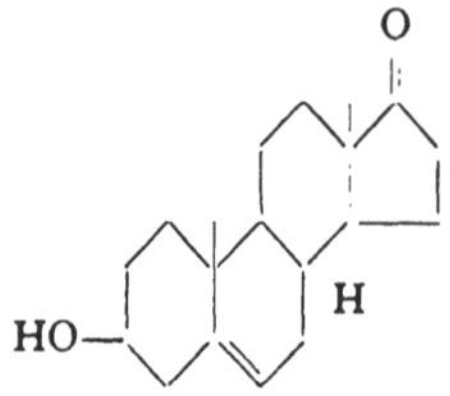

(XXXI.) Dehydro-isoandrosteron. (XXXII.) 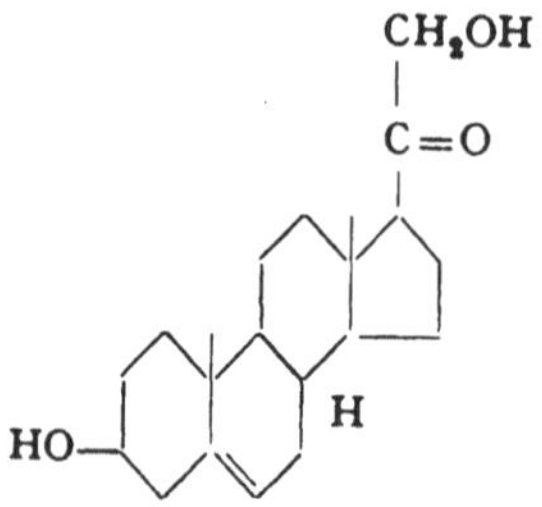 (XXXIII.)

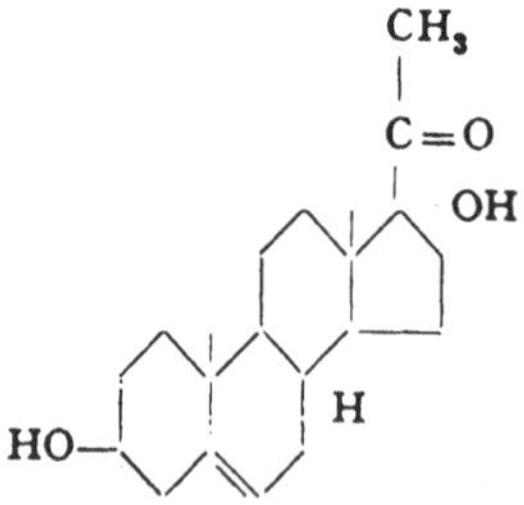

Oestrogene Hormone (XIV und XV, S. 144). Noch größere chemische Umwandlungen würde die Biosynthese des Oestrons und Oestriols erfordern, wenn für diese Hormone Cholesterol die Muttersubstanz wäre. Um diese Frage zu prüfen, haben Heard und O'Donnell (*51d*)

Cholesterol-4-^{14}C einer tragenden Stute injiziert, da gerade unter den gewählten Umständen die Ausscheidung von oestrogenen Hormonen im Harn hoch ist. Das Ergebnis war, daß innerhalb von 5 Tagen nach der Injektion im isolierten Oestron keinerlei Radioaktivität nachweisbar wurde, während die neutrale Keton- und Nichtketon-Fraktion beträchtliche Aktivität aufwiesen. Damit ist dargetan, daß Oestron sicher nicht aus dem Cholesterol entsteht.

4. Die C_{30}- und C_{31}-Steroide.

Die letzten Jahre haben die Auffindung weiterer Steroide als Naturstoffe gebracht, die sich durch einen noch höheren Kohlenstoffgehalt (30 und 31 C-Atome) von den pflanzlichen Sterolen unterscheiden. Die zusätzlichen C-Atome sind jedoch nicht oder nicht allein in der Seitenkette zu finden, sondern als Methylgruppen im Kern untergebracht. Eine Untersuchung über die Möglichkeiten der Steroidbiogenese kann diese Stoffe nicht ohne eingehende Würdigung lassen.

Die Arbeiten des Züricher Arbeitskreises um RUZICKA und JEGER (*82, 104, 105*) sowie von BARTON (*8*) und von HALSALL u. a. (*21, 46, 48, 49*) in England und von WOODWARD in den Vereinigten Staaten haben zur Klärung der Konstitution mehrerer dieser Verbindungen geführt.* Nur zwei dieser Stoffe sind tierischen Ursprungs, nämlich *Lanosterol* (XXXIV) und *Agnosterol* (XXXV) aus dem Wollfett der Schafe, während die meisten anderen aus höheren Pflanzen oder Pilzen isoliert worden sind. Hierzu rechnen: *Polyporensäure A (XXXVI) und C (XXXVII)* aus *Polyporus betulinus* FR. und *Eburiconsäure (XXXVIII)* und *Dehydroeburiconsäure* aus verschiedenen Polyporusarten sowie aus *Lentinus dactyloides* CLEL. Aus den Früchten von *Strychnos nux-vomica* sowie Früchten und Latex von *Artocarpus integrifolia* und *Euphorbia balsamifera* stammt das *Cyclocartenol* (XXXIX) (*7, 9, 10*).

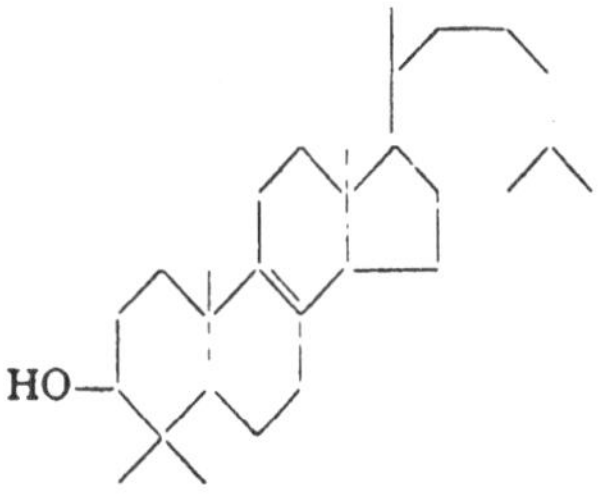

HO—

(XXXIV.) Lanosterol.

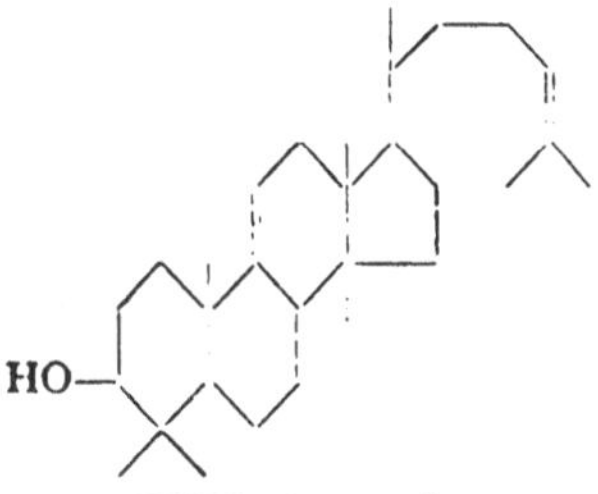

HO—

(XXXV.) Agnosterol.

Von weiteren Verbindungen dieses Typs sind zu nennen (*3 b*): *Euphol, Euphorbol* und *Tirucallol* aus Euphorbiaceen, *Butyrospermol* aus „shea nut"-Öl und *α-Elemi-*

* Für eine ausführlichere Besprechung dieses Gebietes s. die voranstehende Abhandlung von E. R. H. JONES und T. G. HALSALL (S. 44).

säure aus Manila-Elemi-Harz. Für diese Verbindungen wird neuerdings gegenüber Lanosterol eine an $C_{(13)}$, $C_{(14)}$ und $C_{(17)}$ konfigurativ gegensätzliche Konstitution angenommen. Euphorbol enthält an $C_{(24)}$ ebenfalls ein zusätzliches C-Atom (*3, 3a*).

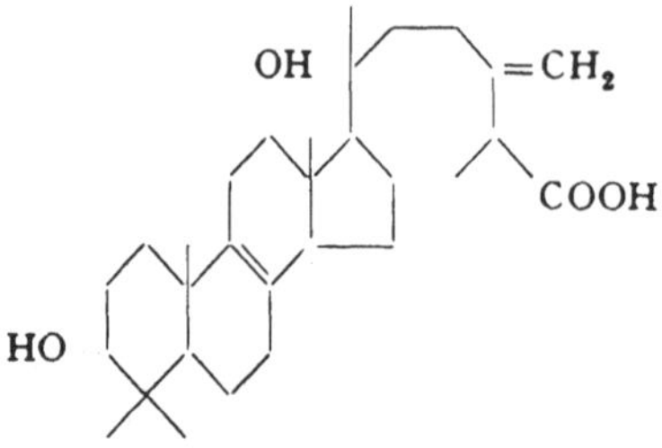

(XXXVI.) Polyporensäure A.

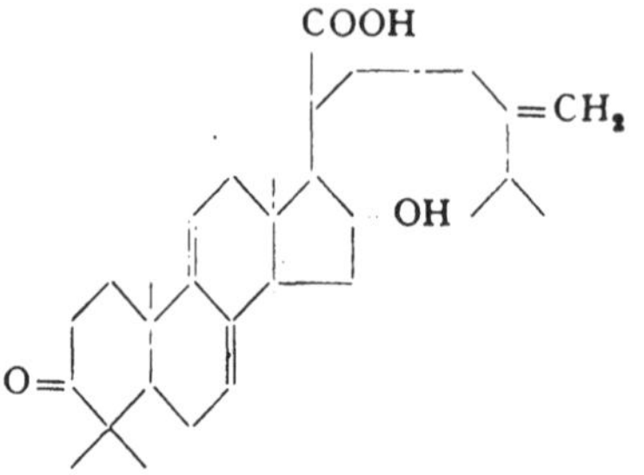

(XXXVII.) Polyporensäure C.

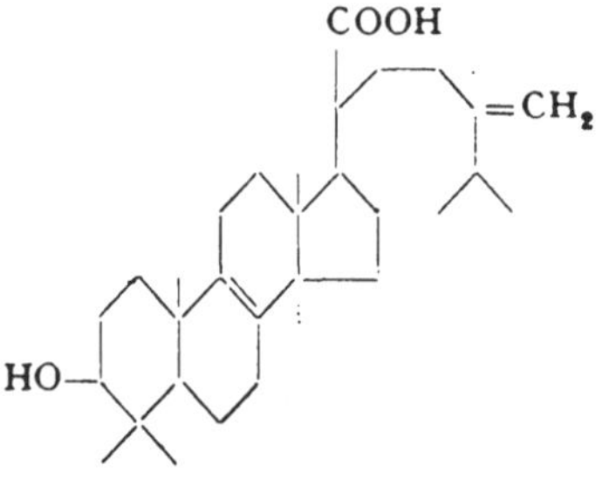

(XXXVIII.) Eburiconsäure.

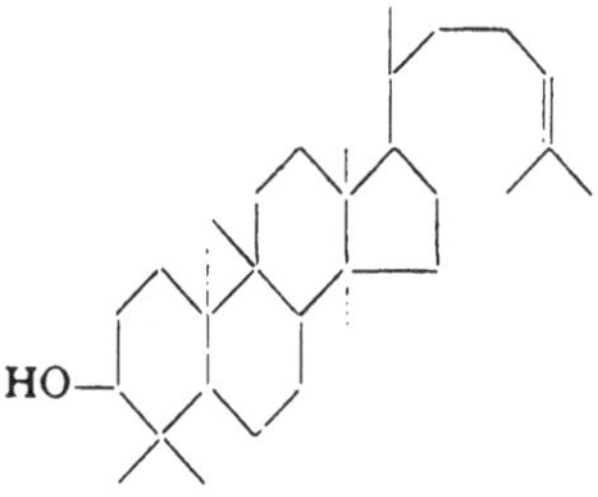

(XXXIX.) Cycloartenol.

Bei der Betrachtung der Formeln (XXXIV)—(XXXIX) fällt auf, daß die Seitenkette des Lanosterols, Agnosterols und Cycloartenols derjenigen des Zymosterols (III, S. 138) entspricht. Wenn der Vorgang zutrifft, der für die Einführung des einen oder der beiden zusätzlichen C-Atome an $C_{(24)}$ in die pflanzlichen Sterole erörtert wurde (S. 139), wäre das Auftreten solcher C-Atome in der Seitenkette auch für die C_{30}-Verbindungen vom Lanosteroltyp zu erwarten. Die zusätzlichen Methylgruppen im Kern sollten diesen Vorgang nicht wesentlich hindern. Es ist bemerkenswert, daß in der Polyporensäure A und C und in der Eburiconsäure solche Verbindungen vorliegen, die in der Seitenkette den C_{28}-Pilzsterinen Episterol und Faecosterol entsprechen. Der Unterschied besteht allein in der Oxydation einer Methylgruppe zu Carboxyl, solche Umwandlungen sind aber von der Gigifisch-gallensäure (XXVI, S. 147) und von der Koprostan-27-säure aus Krokodilgalle biologisch nicht unbekannt.

Daß zwischen den üblichen Steroiden und den C_{30}- bzw. C_{31}-Verbindungen vom Lanosteroltyp enge biologische Beziehungen bestehen müssen, ergibt sich ferner aus der sterischen Identität (*105*). Es muß daher erwartet werden, daß sich auch die Biosynthese dieser Verbindungen nicht grundsätzlich von derjenigen der übrigen Steroide unterscheidet.

IV. Die Hypothesen der Steroidbiogenese.

Schon bald nach der Ermittlung der chemischen Konstitution der Steroide sind Hypothesen entwickelt worden, um den biologischen Aufbau dieses komplizierten Ringsystems zu deuten. Eine der ersten Vorstellungen, die geäußert wurden, war die von WINDAUS (vgl. *63*), daß möglicherweise Ölsäure (XL) und Zibeton (XLI) Zwischenstufen sein könnten. Diese Annahme ist heute nicht mehr haltbar, nachdem verzweigte Säuren und Verbindungen vom Typ des Squalens als unmittelbare Vorstufen erkannt worden sind.

1. Squalen-Hypothese.

Die Vermutung, daß die Steroide durch Cyclisierung von Squalen entstehen könnten, wurde zuerst 1934 von ROBINSON (*81*) und HEILBRON (*52, 53*) auf Grund der Befunde von CHANNON (*30*) und von ANDRÉ und CANAL (vgl. *63*) diskutiert. ROBINSON nimmt eine Cyclisierung wie in (XLII) an, während neuerdings von WOODWARD und BLOCH (*107*) eine andere Art der Ringbildung (XLIII) bevorzugt wird. Wenn man diese beiden genannten Annahmen vergleicht, so zeigt sich, daß nur die letztere allein eine einigermaßen befriedigende Übereinstimmung mit der Herkunftsanalyse der C-Atome des Cholesterols aufweist. Allerdings ist es auch hierbei nötig, eine Wanderung von Methylgruppen anzunehmen.

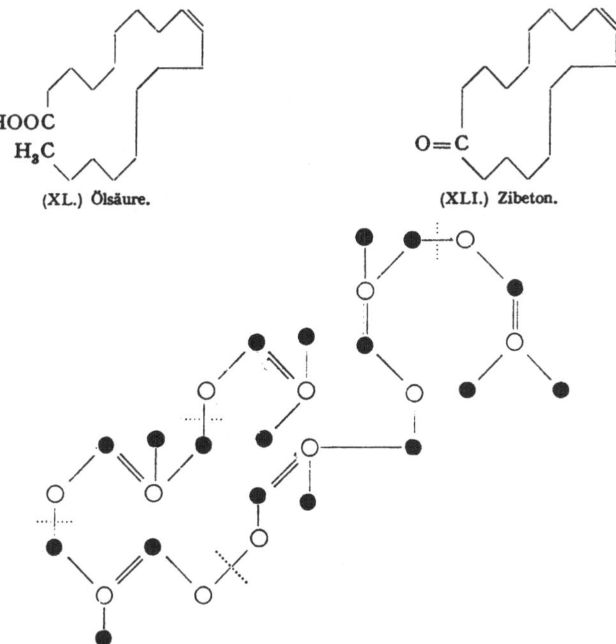

(XL.) Ölsäure. (XLI.) Zibeton.

(XLII.) Cyclisierungshypothese von ROBINSON.

Für die C_{30}- und C_{31}-Steroide vom Lanosterol-Typ ist eine Verschiebung dieser Gruppen von $C_{(8)}$ nach $C_{(14)}$ und von $C_{(14)}$ nach $C_{(13)}$ notwendig, oder aber die Verlagerung allein der CH_3-Gruppe an $C_{(8)}$ nach $C_{(13)}$, während die an $C_{(14)}$ an ihrem Ort bleiben könnte. Mit dieser Wanderung wäre die Feststellung von Woodward und Bloch (*107*) in Einklang, daß die Methylgruppe $C_{(18)}$ und das C-Atom 13 beide gleichermaßen aus dem CH_3 der Essigsäure herrühren. Demgegenüber würde die Cyclisierungshypothese von Robinson auch noch in bezug auf die C-Atome 7 und 13 mit der Herkunftsanalyse in Konflikt geraten. Zweifellos stellt die Hypothese von Woodward und Bloch (*107*) eine für die Biogenese der C_{30}- und C_{31}-Steroide plausible Erklärung dar, die eine gute Grundlage für weitere Versuche darstellt.

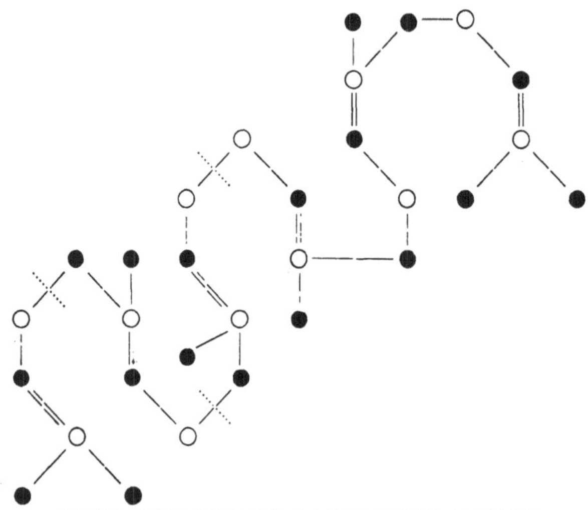

(XLIII.) Cyclisierungshypothese von Woodward und Bloch.

Schwierigkeiten treten jedoch sofort auf, wenn man versucht, die Vorstellungen von Woodward und Bloch (*107*) auch auf die üblichen Steroide ohne geminale Methylgruppen an $C_{(4)}$ und ohne die zusätzliche Methylgruppe an $C_{(14)}$ zu übertragen. Es ergibt sich dann die Notwendigkeit, ihre Herausnahme aus einem C_{30}-Primärprodukt durch Oxydation und nachfolgende CO_2-Abspaltung anzunehmen. Wenn auch die biologische Oxydation von Methyl zu Carboxyl keine grundsätzlichen Schwierigkeiten bereitet, so ist doch ihre dreimalige Wiederholung, wobei noch bei den C_{21}- und C_{19}-Steroiden die teilweise oder vollständige Entfernung der Seitenkette käme, nicht sehr wahrscheinlich. Dazu kommt, daß keinerlei Zwischenprodukte bekannt sind, bei denen eine oder zwei dieser fraglichen Methylgruppen fehlen oder aber auch nur die Einführung einer Hydroxyl- oder Aldehydgruppe in sie bekannt wäre. Betrachtet man die Verhältnisse bei den chemisch sehr ähnlich gebauten Triterpenen, so zeigt sich, daß sie alle der Isoprenregel gehorchen und die Zahl der Kohlenstoffatome durch 5 teilbar ist. Bei ihnen finden sich zwar die Umwandlungen derartiger Methyle zu CH_2OH, $HC{=}O$ oder COOH-Gruppen häufig, eine Abspaltung von Kohlenstoffatomen erfolgt jedoch nicht. Es scheint daher notwendig, nach einem Weg zu suchen, der einerseits der Bedeutung des Squalens für die Biogenese der Steroide ihr Recht werden läßt, aber das Auftreten der zusätzlichen Kohlenstoffatome an $C_{(4)}$ und $C_{(14)}$ von vornherein vermeidet.

2. Isosqualen-Hypothese.

Um die Schwierigkeiten zu umgehen, die die Annahme der Wanderung eines Methyls von $C_{(8)}$ nach $C_{(13)}$ bzw. von $C_{(14)}$ nach $C_{(13)}$ macht, hat

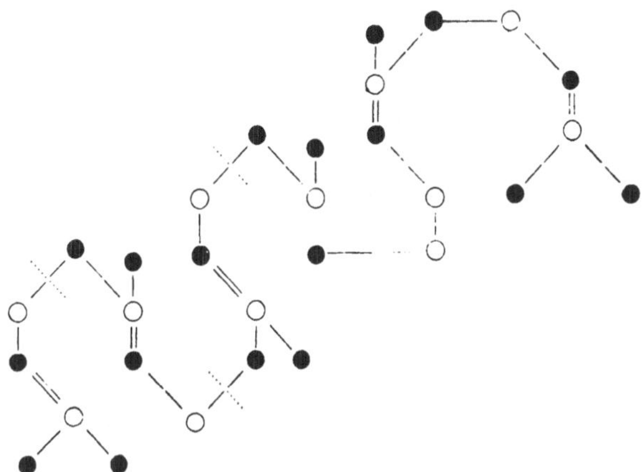

(XLIV.) Isosqualen-Hypothese von MONDON.

MONDON (*69*) die Hypothese aufgestellt, daß nicht Squalen, sondern Isosqualen die Vorstufe der Steroide wäre. Hierbei wird die Nahtstelle des Moleküls aus symmetrischen Hälften zwischen $C_{(11)}$ und $C_{(12)}$ nach $C_{(15)}$ und $C_{(16)}$ verlegt (XLIV). Damit kommt MONDON aber mit der Herkunftsanalyse der C-Atome im Cholesterol in Konflikt, nach der das C-Atom 13 aus dem Methyl der Essigsäure stammt. Eine Entscheidung für oder gegen diese Vorstellung würde ferner die Bestimmung der Herkunft der C-Atome 12 und 15 bringen. Im übrigen treten die Schwierigkeiten der Entfernung dreier C-Atome an $C_{(4)}$ und $C_{(14)}$ für die Bildung der üblichen Steroide bei der Hypothese von MONDON in gleicher Weise wie bei der von WOODWARD und BLOCH (*107*) auf.

3. Hypothese von MIESCHER.

Die Vorstellungen von MIESCHER (*68*) (1950) gehen von der Bedeutung der Acetessigsäure für die Biogenese der Steroide aus. Er zerlegt das

Molekül des Cholesterols in 5 Acetessigsäure-, 1 Oxalessigsäure- und 4 Formaldehyd-Bausteine, wie in (XLV). Das unbekannte Primärprodukt soll dann unter Wasserabspaltung in B übergehen (XLVI). In der Formulierung A sind die C-Atome der C_4-Moleküle durch schwarze Punkte, die des

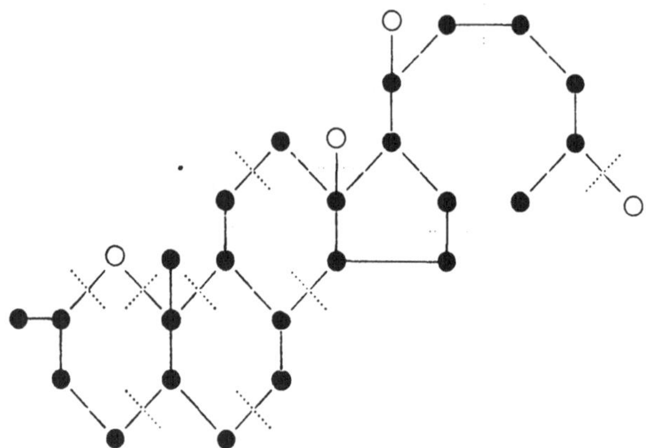

(XLV.) Vorstufe von Miescher. Formel A. Aufbau aus Acetessigsäure.

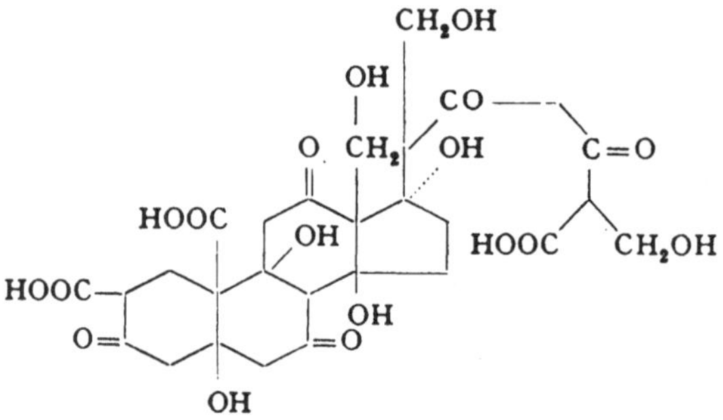

(XLVI.) Formel B von Miescher.

Formaldehyds durch Kreise wiedergegeben. Der größte Nachteil der Hypothese von Miescher ist der, daß sie keine wesentlichen Ansatzpunkte liefert, mit welchen Zwischenprodukten man bei der Biogenese der Steroide rechnen kann.

4. Hypothese von Tschesche und Korte.

Diese Autoren gehen von der Konstitutionsbestimmung des Picrotoxinins (XLVII) aus *Anaminta cocculus* (Menispermacee) durch Conroy (*31*) aus. Man sieht sofort, daß diese Verbindung in der Anordnung der beiden Ringe mit derjenigen von Ring *C* und *D* der Steroide übereinstimmt. Die Ringe sind *cis*-verknüpft wie bei den Cardenoliden und Bufadienoliden, bei denen sich ebenfalls eine Hydroxyl-gruppe am

C-Atom 14 wie im Picrotoxinin befindet. Die Verteilung von Hydroxylen entspricht im übrigen der, wie sie sich auch bei Cardenoliden und Bufadienoliden wiederfindet, allerdings nicht bei einem einzigen Vertreter. Schon CONROY hat darauf aufmerksam gemacht, daß durch Angliederung von weiteren fünf C-Atomen an das Picrotoxinin das Vierringskelett der

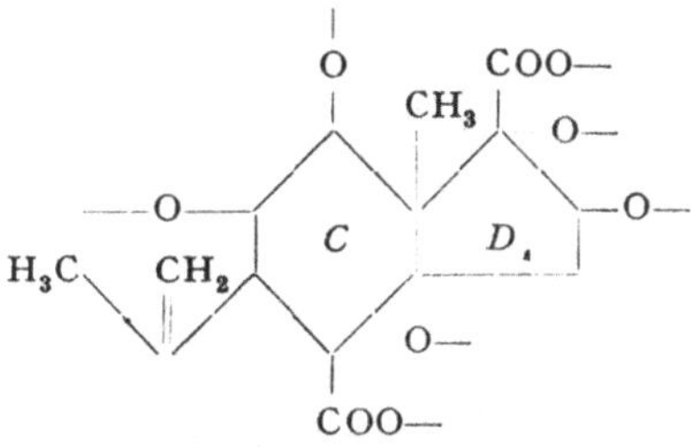

(XLVII.) Picrotoxinin.

Steroide vervollständigt werden kann. Anderseits aber hat er offengelassen, ob nicht diese Verbindung ein Abbauprodukt der Steroide sein könnte. Diese letzte Annahme erscheint aber wenig wahrscheinlich. Einmal spricht dagegen die Isopropylengruppe des Picrotoxinins, die sich in vielen Naturstoffen wiederfindet, so z. B. auch im Bitterstoff Gentiopikrin (XLVIII). Es ist schwer zu verstehen, wie sie sich beim Abbau von Steroiden bilden soll. Weiter aber ist auf Grund unserer Kenntnisse des Stoffwechselgeschehens assimilierender Pflanzen ein so weitgehender Abbau von Steroiden nicht zu vermuten; im allgemeinen werden unerwünschte Produkte als solche oder wenig verändert in Pflanzenteilen abgelagert, die

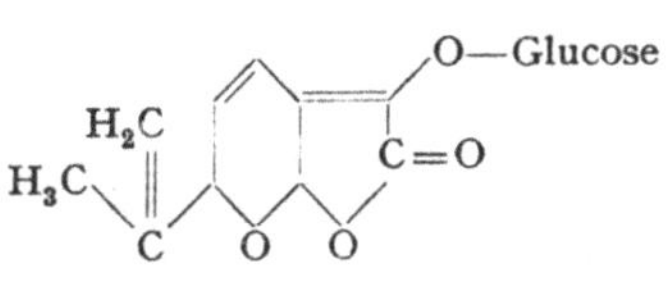

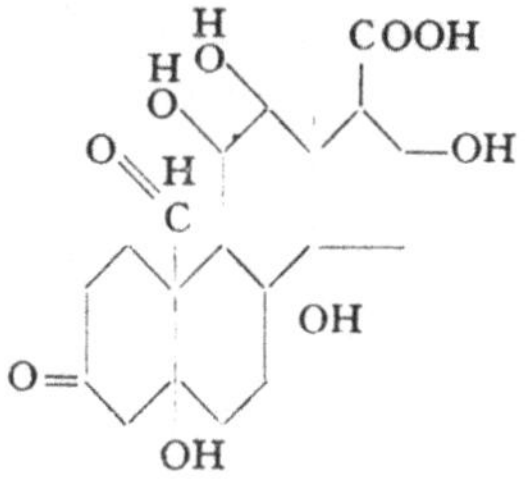

nicht mehr intensiv am allgemeinen Umsatz teilhaben. TSCHESCHE und KORTE (99) haben daher angenommen, daß sich Picrotoxinin vom Aufbauweg der Steroide abzweigt und daß möglicherweise ein Intermediärprodukt wie (XLIX) Zwischenstufe der Biosynthese sein könnte. Für eine Angliederung der fünf fehlenden C-Atome an das Picrotoxinin in einer die Forderungen der Herkunftsanalyse erfüllenden Weise bestehen sicherlich gedanklich keine Schwierigkeiten. Vielleicht steht die Di-isopren-Kohlenstoffkette des Gentiopikrins einer Vorstufe des Picrotoxinins nahe.

Gegen eine solche Vorstellung ist besonders von MONDON (*69*) eingewendet worden, daß Cholesterol als ein Primärprodukt der Biosynthese anzusehen ist, eine Behauptung, die nicht in jedem Falle zutreffend ist, wie dargelegt wurde. Auch ist zu bedenken, daß sich an Stelle der Hydroxyle im Intermediärkörper auch Doppelbindungen finden können, aus denen die hydroxylierten Verbindungen erst durch Wasser-Anlagerung entstehen. Es ist schwer etwas darüber auszusagen, ob die ungesättigte oder hydroxylierte Stufe die primäre ist. So könnten die Doppelbindungen Δ^4 oder Δ^5 sowie Δ^{11} und Δ^{14} durchaus an Stelle der Hydroxyle stehen. Eine Doppelbindung Δ^5 findet sich in vielen Steroiden wieder und eine solche Δ^{14} ist z. B. im *14-Dehydro-ergosterol* aus *Aspergillus niger* von BARTON und BRUUN (vgl. *63*) festgestellt worden. Die Anlagerung von Wasserstoff an diese Doppelbindung wird voraussichtlich zu der bei den Steroiden üblichen *trans*-Verknüpfung der Ringe *C* und *D* führen, jedoch könnte eine enzymatische Wasseranlagerung sehr wohl zu einer *cis*-Anordnung der Ringe wie in Cardenoliden und Bufadienoliden mit Hydroxyl an $C_{(14)}$ führen.

Es erscheint wichtig, in diesem Zusammenhang darauf hinzuweisen, daß neuerdings von TSCHESCHE und Mitarb. (*100, 101*) im *Neriantogenin*, *Smalogenin* und *Xysmalogenin* an $C_{(14)}$ hydroxylfreie Cardenolide als Aglykone von natürlich vorkommenden Glykosiden beobachtet worden sind. Auf welcher Stufe der Biogenese die Dehydrierung und Wasseranlagerung erfolgt, ist vorerst nicht zu entscheiden. Entsprechende Verhältnisse sind an $C_{(5)}$ zu erwarten. Verbindungen mit OH an $C_{(5)}$ sind z. B. *Periplogenin*, *Strophanthidin*, *Telocinobufagin* und *Hellebrigenin*, Verbindungen mit Doppelbindung Δ^4, z. B. das Aglykon des Acofriosids, *Scillarenin*, mit Doppelbindung Δ^5, z. B. *Diosgenin* und sehr viele Sterole (*63*).

5. Neue Vorstellungen.

Während für die C_{30}- und C_{31}-Steroide die Cyclisierungshypothese von WOODWARD und BLOCH (*107*) eine befriedigende Erklärung besitzt, ist noch eine Variation dieses Prozesses zu suchen, welche die nachherige Herausnahme von drei C-Atomen aus diesen ausschließt. Wie schon ausgeführt wurde (S. 139), ist Squalen vermutlich nicht die direkte Vorstufe des Cholesterols und der C_{21}-Steroide, sondern zweigt sich von einer noch unbekannten Verbindung ab. Macht man nun die Annahme, daß dieser Vorstufe in der „linken“ Hälfte des Moleküls die C-Atome zweier Isopren-Reste des Squalens fehlen, so läßt sich Cholesterol erhalten, wenn man die verkürzte Isoprenkette in folgender Weise durch sieben C-Atome ergänzt (L). Hierbei ist eine etwas andere Faltung der Kette notwendig.

Es ist ersichtlich, daß man bei der Einfügung der fehlenden sieben C-Atome nicht mit den bisherigen Befunden der Herkunftsanalyse in Konflikt zu geraten braucht, wenn man sie in der wiedergegebenen Weise vornimmt. Über den Mechanismus dieser Reaktion ist es zur Zeit nicht möglich, Aussagen zu machen.

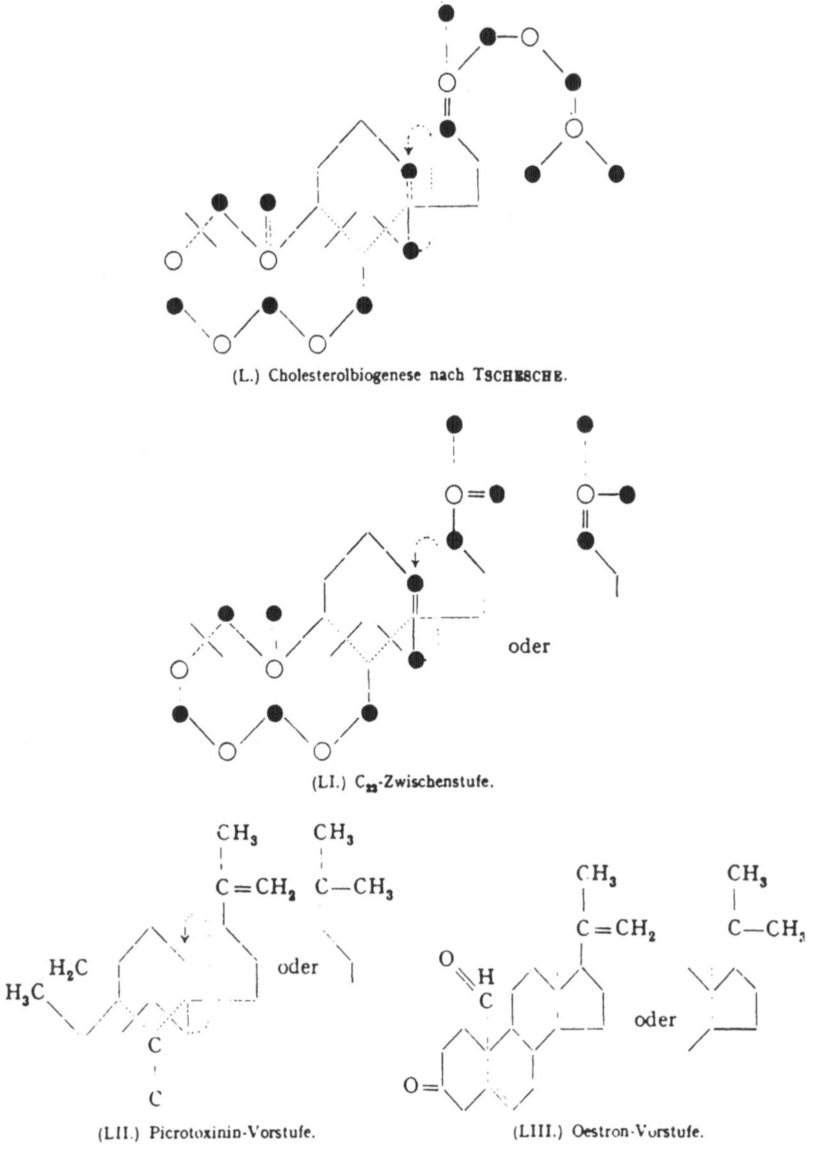

(L.) Cholesterolbiogenese nach TSCHESCHE.

(LI.) C_{23}-Zwischenstufe.

(LII.) Picrotoxinin-Vorstufe.

(LIII.) Oestron-Vorstufe.

Entfernt man auch noch aus der Seitenkette die C-Atome einer Isopren-Einheit, so würde bei entsprechendem Verlauf des Ringschlusses wie in (L) eine C_{22}-Verbindung entstehen, die gut die gesuchte Zwischenstufe „X" der Biogenese der C_{21}-Steroide sein könnte (LI). In ihr würde an $C_{(17)}$ eine Isopropylidengruppe auftauchen, von der durch Oxydation ein Übergang in Verbindungen der C_{21}-Steroidreihe ohne weiteres vorstellbar wäre. Picrotoxinin könnte aus der C_{15}-Verbindung vom Polypren-Typ durch Einschiebung von zwei C-Atomen wie in (LII) gebildet werden. Weiterhin müßte ein Abbau der Isopropylidengruppe am $C_{(17)}$ der Steroide zu Carbonyl erfolgen; hierfür liegen ebenfalls keine grundsätzlichen Schwierigkeiten vor.

Damit fallen nunmehr die großen Schwierigkeiten fort, die bisher für eine Erklärung der Biogenese der aromatischen Steroide vom Oestrontyp bestanden. Aus der Vorstufe (LI) könnte durch direkte Oxydation oder Wasseranlagerung und Oxydation eine Zwischenstufe (LIII) mit HC=O an $C_{(10)}$ gebildet werden, von der vermutet werden darf, daß sie leicht unter Abspaltung von Formaldehyd oder unter Dehydrierung und Abtrennung von Formyl den aromatischen Ring des Oestrons entstehen lassen könnte. Eine Aldehydgruppe ist in vielen Cardenoliden an $C_{(10)}$ nachgewiesen worden, so z. B. im Strophanthidin, Corotoxigenin, Adonitoxigenin, Bovogenin A und Hellebrigenin (*63*). Jedoch wurde niemals bisher eine Carboxylgruppe an dieser Stelle in Naturstoffen beobachtet. Auch eine CH_2OH-Gruppe an $C_{(10)}$ wird leicht als Formaldehyd abgetrennt, wenn Hydroxyle an $C_{(1)}$, $C_{(3)}$ und $C_{(5)}$ im Ring *A*, wie z. B. im Ouabagenin, enthalten sind. FLOREY und EHRENSTEIN (*38a*) sowie TAMM (*94a*) haben in letzter Zeit die Aromatisierung des Ringes *A* an diesem Cardenolid eingehender studiert; schon früher war von JACOBS dieser Vorgang unter drastischeren Bedingungen beobachtet worden (*38,63*). Das Oestron würde damit keine Sonderstellung wie bisher in der Biogenese einnehmen.

Für die Angliederung des α,β-ungesättigten Butenolidringes an $C_{(17)}$ in den Cardenoliden und des zweifach ungesättigten Lactonsechsringes in den Bufadienoliden ist von TSCHESCHE und KORTE (*99*) nachfolgende Möglichkeit diskutiert worden: Für die Cardenolide wird angenommen, daß Acetessigsäure über γ-Oxy-acetessigsäure mit einem Derivat der 16-Keto-aetiocholansäure unter Wasserabspaltung kondensiert wird (LIVa—d). Für die Bufadienolide wird Oxalacetessigsäure herangezogen, die in analoger Weise über die γ-Oxysäure reagieren soll (LVa—e). Für eine solche Möglichkeit spricht einmal das Vorkommen von 16-Oxyverbindungen unter den Herzgift-aglykonen, so im Gitoxigenin, Oleandrigenin, Adonitoxigenin und Bufotalin, ferner die von STOLL (*94b*) vorgeschlagene Scillirosidin-Struktur (LVI), in dem die Ketogruppe in α-Stellung zum Carboxyl des Lactonringes noch als acetyliertes Enol erhalten geblieben ist. Auch muß darauf hingewiesen werden, daß der Abbau von der C_{21}-Stufe, z. B. vom Desoxycorticosteron zu 3-Keto-Δ^4-aetiocholensäure (LVII) in Rattenleber von PICHA und Mitarb. (*72*) festgestellt worden ist und auch von Bodenbakterien Cholesterol zu der gleichen Säure abgebaut wird.

Eine andere Möglichkeit wäre die Verkürzung der Seitenkette bei einem C_{27}-Sterol auf 23 C-Atome in den Cardenoliden und auf 24 C-Atome in den Bufadienoliden. Durch Hydroxylierung der Methylgruppe $C_{(21)}$, Lactonringschluß mit der Carboxylgruppe $C_{(23)}$, bzw. $C_{(24)}$ und Dehydrierung im Lactonring könnten diese Steroide ebenfalls gebildet werden. Eine Entscheidung zwischen den beiden Möglichkeiten ist zur Zeit nicht zu treffen.

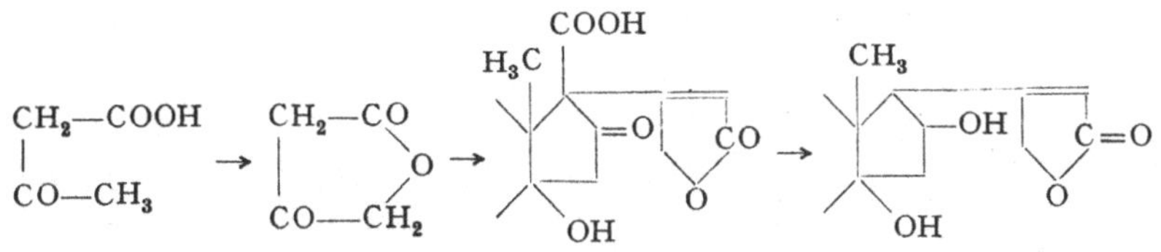

(LIV a.) (LIV b.) (LIV c.) (LIV d.)

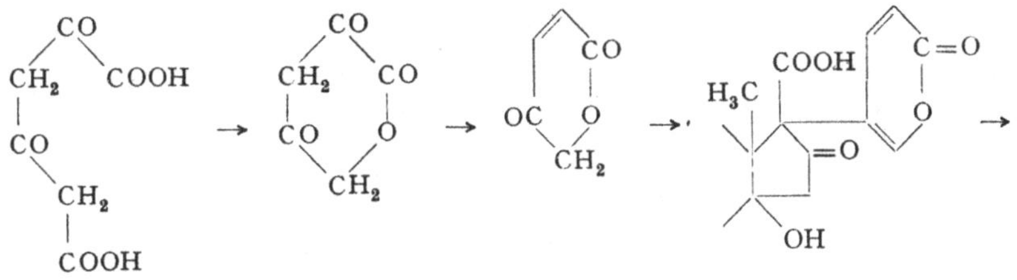

(LV a.) (LV b.) (LV c.) (LV d.)

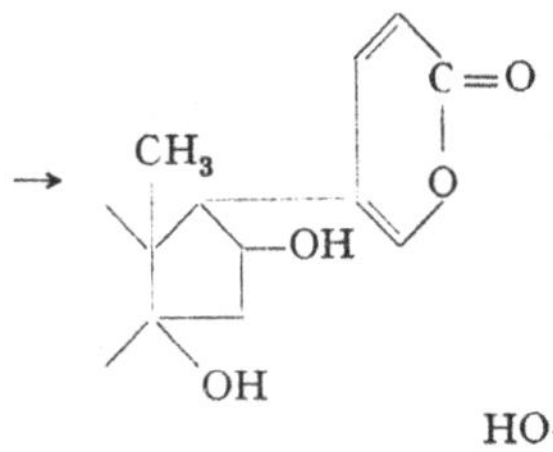

(LV e.) 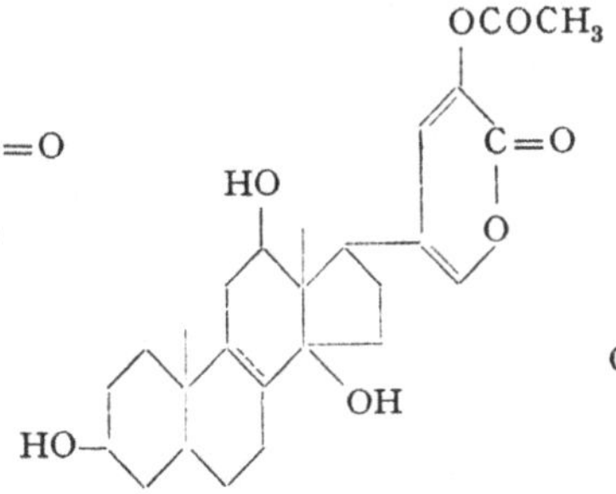 (LVI.) 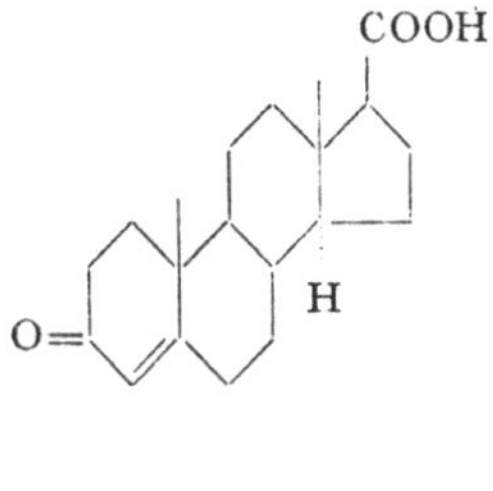 (LVII.)

Ein noch wenig geklärtes Problem der Biogenese ist die Frage, in welcher Weise die in den natürlich vorkommenden Steroiden sich findenden Hydroxyl- bzw. Oxogruppen in das Ringsystem gelangen. Eine Sauerstoffunktion an $C_{(3)}$ findet sich in allen Verbindungen wieder und ihr Auftreten muß daher eng mit der Synthese verknüpft sein. Eine mögliche Annahme ist die enzymatisch gelenkte Anlagerung von Wasser an eine von $C_{(3)}$ ausgehende Doppelbindung, wofür die Hypothese von WOODWARD und BLOCH für die C_{30}- und C_{31}-Steroide theoretisch einen

Anhaltspunkt bietet. Für C_{27}-Steroide würde sich ebenfalls eine solche ergeben, wenn die Angliederung der sieben C-Atome, die den „unteren" Teil von Ring *A* und *B* bilden, durch eine Dienkondensation erfolgen würde (LVIII und LIX).

(LVIII.) (LIX.)

Für die Gallensäuren ist die Einführung weiterer Hydroxylgruppen in das Ringsystem an $C_{(6)}$, $C_{(7)}$ und $C_{(12)}$ vom Cholesterol her gesichert (*89*). Über den Mechanismus dieser Oxydation ist nichts Näheres bekannt, aber man darf vielleicht vermuten, daß auch hier der erste Schritt eine Dehydrierung ist, die von einer Wasser-Anlagerung gefolgt wird. Ein entsprechender Vorgang könnte für die Nebennierenrinden-Hormone mit Sauerstoffunktion an $C_{(11)}$ gelten; die Einführung von Hydroxyl an dieser Stelle ist für die Nebenniere nachgewiesen (*40*). Auch Pilze und andere Mikroorganismen können diese Reaktion durchführen. Vermutlich dürfte die Oxydation bei den teilweise noch stärker im Ringsystem mit Sauerstoff beladenen Steroiden des Pflanzenreiches auf ähnliche Weise erreicht werden. Daneben ist auch die direkte Einführung von Hydroxyl in Betracht zu ziehen.

V. Schlußbetrachtung.

Die soeben geäußerten Vorstellungen tragen im wesentlichen hypothetischen Charakter, aber sie scheinen eine Möglichkeit zu sein, um die bestehenden Schwierigkeiten der Biogenese in einer alle Steroide umfassenden Hypothese zu überwinden. Soweit ich feststellen konnte, bestehen zur Zeit keinerlei Befunde, die nicht mit den vorgetragenen neuen Vorstellungen in Einklang gebracht werden können. Es bleibt abzuwarten, welchen Einfluß neue experimentelle Ergebnisse auf sie haben werden. Selbst wenn sich zeigen sollte, daß sie auf die Dauer nicht ohne wesentliche Änderungen aufrechtzuerhalten ist, wäre ihr Zweck erfüllt, wenn sie Anregungen zu wertvollen neuen Ansätzen auf diesem Gebiet vermittelt hätte.

Literaturverzeichnis.

1. Anker, H. S. and K. Bloch: The Action of Platinum on Cholesterol in Acetic Acid Solution. J. Amer. Chem. Soc. **66**, 1752 (1944).

2. Antia, N. J., Y. Mazur, R. R. Wilson and F. S. Spring: Steroids. XI. Isolation of Cholegenin and *iso*Cholegenin from Ox-bile. J. Chem. Soc. (London) **1954**, 1218.

3. Arigoni, D., O. Jeger und L. Ruzicka: Zur Kenntnis der Triterpene. 183. Mitt. Über die Konstitution und Konfiguration von Tirucallol, Euphorbol und Elemadienolsäure. Helv. Chim. Acta **38**, 222 (1955).

3a. ARIGONI, D., R. VITERBO, M. DÜNNENBERGER, O. JEGER und L. RUZICKA: Zur Kenntnis der Triterpene. 182. Mitt. Konstitution und Konfiguration von Euphol und iso-Euphenol. Helv. Chim. Acta **37**, 2306 (1954).

3b. ARIGONI, D., H. WYLER und O. JEGER: Zur Kenntnis der Triterpene. 179. Mitt. Über die gegenseitigen Beziehungen bei Elemadienolsäure, Tirucalladienol und Euphorbadienol. Helv. Chim. Acta **37**, 1553 (1954).

4. ARNAKI, M. und Z. STARY: Untersuchungen über die Biosynthese der Carotinoide. Biochem. Z. **323**, 376 (1952).

5. ARREGUIN, B., J. BONNER and B. J. WOOD: Studies on the Mechanism of Rubber Formation in the Guayule. III. Experiments with Isotopic Carbon. Arch. Biochem. Biophys. **31**, 234 (1951).

6. BAKER, C. G. and D. M. GREENBERG: Studies with Radioactive Carbon-labeled Acetate on Cholesterol Metabolism in Rats Fed *p*-Dimethylaminoazobenzene. Cancer Res. **9**, 701 (1949).

7. BARTON, D. H. R.: Triterpenoids. III. *cyclo*Artenone, a Triterpenoid Ketone. J. Chem. Soc. (London) **1951**, 1444.

8. BARTON, D. H. R., D. A. J. IVES and B. R. THOMAS: Triterpenoids. XVII. The Transformation of Lanostadienol (Lanosterol) into 14-Methylcholestan-3(β)-ol. J. Chem. Soc. (London) **1954**, 903.

9. BARTON, D. H. R., J. E. PAGE and E. W. WARNHOFF: Triterpenoids. XVIII. The Constitutions of Phyllanthol and *cyclo*Artenol. J. Chem. Soc. (London) **1954**, 2715.

10. BENTLEY, H. R., J. A. HENRY, D. S. IRVINE and F. S. SPRING: Triterpene Resinols and Related Acids. XXVIII. The Non-saponifiable Fraction from *Strychnos nux-vomica* Seed Fat: The Structure of *cyclo*Artenol. J. Chem. Soc. (London) **1953**, 3673.

11. BERGSTRÖM, S.: The Formation of Bile Acids from Cholesterol in the Rat. Kongl. Fysiogr. Sällsk. Lund, Förhandl. **22**, No. 16, p. 1 (1952).

12. BLOCH, K.: The Biological Conversion of Cholesterol to Pregnanediol. J. Biol. Chem. **157**, 661 (1945).

13. — The Biological Synthesis of Cholesterol. Recent Progr. Hormone Res. **6**, 111 (1951).

14. — Über die Herkunft des Kohlenstoff-Atoms 7 in Cholesterin. Ein Beitrag zur Kenntnis der Biosynthese der Steroide. Helv. Chim. Acta **36**, 1611 (1953).

15. BLOCH, K., B. N. BERG and D. RITTENBERG: The Biological Conversion of Cholesterol to Cholic Acid. J. Biol. Chem. **149**, 511 (1943).

16. BLOCH, K., E. BOREK and D. RITTENBERG: Synthesis of Cholesterol in Surviving Liver. J. Biol. Chem. **162**, 441 (1946).

17. BLOCH, K. and D. RITTENBERG: On the Utilization of Acetic Acid for Cholesterol Formation. J. Biol. Chem. **145**, 625 (1942).

18. — — The Preparation of Deuterio Cholesterol. J. Biol. Chem. **149**, 505 (1943).

19. — — An Estimation of Acetic Acid Formation in the Rat. J. Biol. Chem. **159**, 45 (1945).

20. BONNER, J. and B. ARREGUIN: The Biochemistry of Rubber Formation in the Guayule. I. Rubber Formation in Seedlings. Arch. Biochemistry **21**, 109 (1949).

21. BOWERS, A., T. G. HALSALL and (in part) G. C. SAYER: The Chemistry of the Triterpenes and Related Compounds. XXV. Some Stereochemical Problems Concerning Polyporenic Acid C. J. Chem. Soc. (London) **1954**, 3070.

22. BRADY, R. O.: Biosynthesis of Radioactive Testosterone *in vitro*. J. Biol. Chem. **193**, 145 (1951).

23. BRADY, R. O. and S. GURIN: The Biosynthesis of Radioactive Fatty Acids and Cholesterol. J. Biol. Chem. **186**, 461 (1950).

24. — — The Synthesis of Radioactive Cholesterol and Fatty Acids *in vitro*. J. Biol. Chem. **189**, 371 (1951).

25. BUCHANAN, J. M., W. SAKAMI and S. GURIN: A Study of the Mechanism of Fatty Acid Oxidation with Isotopic Acetoacetate. J. Biol. Chem. **169**, 411 (1947).

26. BUCHER, N. L. R.: The Formation of Radioactive Cholesterol and Fatty Acids from C^{14}-Labeled Acetate by Rat Liver Homogenates. J. Amer. Chem. Soc. **75**, 498 (1953).

27. BUCHER, N. L. R., N. H. MCGOVERN, R. KINGSTON and M. H. KENNEDY: Factors Affecting Biosynthesis of Cholesterol from Acetate by Rat Liver Homogenates. Federat. Proc. (Amer. Soc. exp. Biol.) **12**, 184 (1953).

28. BYERS, S. O. and M. W. BIGGS: Cholic Acid and Cholesterol: Studies Concerning Possible Intraconversion. Arch. Biochem. Biophys. **39**, 301 (1952).

29. CHAIKOFF, I. L., M. D. SIPERSTEIN, W. G. DAUBEN, H. L. BRADLOW, J. F. EASTHAM, G. M. TOMKINS, J. R. MEIER, R. W. CHEN, S. HOTTA and P. A. SRERE: C^{14}-Cholesterol. II. Oxidation of Carbons 4 and 26 to Carbon Dioxide by the Intact Rat. J. Biol. Chem. **194**, 413 (1952).

30. CHANNON, H. J.: The Biological Significance of the Unsaponifiable Matter of Oils. I. Experiments with the Unsaturated Hydrocarbon, Squalene (Spinacene). Biochemic J. **20**, 400 (1926).

31. CONROY, H.: Picrotoxin. II. The Skeleton of Picrotoxin. The Total Synthesis of *dl*-Picrotoxadiene. J. Amer. Chem. Soc. **74**, 3046 (1952).

32. COON, M. J.: The Metabolic Fate of the Isopropyl Group of Leucine. J. Biol. Chem. **187**, 71 (1950).

33. CORNFORTH, J. W., G. D. HUNTER and G. POPJÁK: Distribution of Acetate Carbon in the Ring-structure of Cholesterol. Biochemic. J. **53**, XXXIV (1953).

34. — — — Studies of Cholesterol Biosynthesis. I. A New Chemical Degradation of Cholesterol. II. Distribution of Acetate Carbon in the Ring Structure. Biochemic. J. **54**, 590, 597 (1953).

35. CURRAN, G. L.: Utilization of Acetoacetic Acid in Cholesterol Synthesis by Surviving Rat Liver. J. Biol. Chem. **191**, 775 (1951).

36. CURRAN, G. L. and D. RITTENBERG: The Role of Ethyl Alcohol in the Biological Synthesis of Cholesterol. J. Biol. Chem. **190**, 17 (1951).

37. DAUBEN, W. G., S. ABRAHAM, S. HOTTA, I. L. CHAIKOFF, H. L. BRADLOW and A. H. SOLOWAY: On the Incorporation of Acetate into Cholesterol. J. Amer. Chem. Soc. **75**, 3038 (1953).

38. FIESER, L. F. and M. FIESER: Natural Products Related to Phenanthrene, 3rd ed. Amer. Chem. Soc. Monograph Series, No. 70. New York: Relnhold Publ. 1949.

38a. FLOREY, K. and M EHRENSTEIN: Investigations on Steroids. XXII. Studies on Ouabagenin. I. J. Organ. Chem. (USA) **19**, 1174 (1954).

39. FUKUSHIMA, D. K. and T. F. GALLAGHER: Isotopic Distribution in Cholesterol after Platinum-catalyzed Hydrogen-Deuterium Exchange. J. Biol. Chem. **198**, 861 (1952).

40. FUKUSHIMA, D. K. and R. S. ROSENFELD: Sterol and Steroid Metabolism. Chem. Pathways Metabolism **1**, 349 (1954).

41. GALLAGHER, T. F., H. L. BRADLOW, D. K. FUKUSHIMA, C. T. BEER, T. H. KRITCHEVSKY, M. STOKEM, M. L. EIDINOFF, L. HELLMAN and K. DOBRINER: Studies of the Metabolites of Isotopic Steroid Hormones in Man. Recent Progr. Hormone Res. **9**, 411 (1954).

42. GOULD, R. G., D. J. CAMPBELL, C. B. TAYLOR, F. B. KELLY, Jr., I. WARNER and C. B. DAVIS, Jr.: Origin of Plasma Cholesterol using Carbon[14]. Federat Proc. (Amer. Soc. exp. Biol.) **10**, 191 (1951).

43. GOULD, R. G., C. B. TAYLOR, J. S. HAGERMAN, I. WARNER and D. J. CAMPBELL: Cholesterol Metabolism. I. Effect of Dietary Cholesterol on the Synthesis of Cholesterol in Dog Tissue *in vitro*. J. Biol. Chem. **201**, 519 (1953).

44. GRAY, I., P. ADAMS and H. HAUPTMANN: The Utilization of the Branched Chain of Isobutyric Acid Studied with ^{14}C. Experientia **6**, 430 (1950).

45. GROB, E. C., G. G. PORETTI, A. v. MURALT et W. H. SCHOPFER: Recherches sur la biosynthèse des caroténoïdes chez un microorganisme. Production de caroténoïdes marqués par *Phycomyces blakesleeanus*. Experientia **7**, 218 (1951).

46. GUIDER, J. M., T. G. HALSALL, R. HODGES and E. R. H. JONES: The Chemistry of the Triterpenes and Related Compounds. XXVI. The Nature of Polyporenic Acid B. J. Chem. Soc. (London) **1954**, 3234.

47. HAINES, W. J., E. D. NIELSON, N. A. DRAKE and O. R. WOODS: Biosynthesis of 17-α-Hydroxycorticosterone from Acetate. Arch. Biochem. Biophys. **32**, 218 (1951).

48. HALSALL, T. G. and R. HODGES: The Chemistry of the Triterpenes and Related Compounds. XXIV. The Conversion of Polyporenic Acid A into a Lanosterol Derivative. J. Chem. Soc. (London) **1954**, 2385.

49. HALSALL, T. G., R. HODGES and E. R. H. JONES: The Chemistry of the Triterpenes and Related Compounds. XIX. Further Evidence concerning the Structure of Polyporenic Acid A. J. Chem. Soc. (London) **1953**, 3019.

50. HANAHAN, D. J. and S. J. AL-WAKIL: The Biosynthesis of Ergosterol from Isotopic Acetate. Arch. Biochem. Biophys. **37**, 167 (1952).

51. HANAHAN, D. J. and S. J. WAKIL: The Origin of Some of the Carbon Atoms of the Side Chain of C[14]-Ergosterol. J. Amer. Chem. Soc. **75**, 273 (1953).

51a. HAYANO, M. and R. I. DORFMAN: The Enzymatic C-11β-Hydroxylation of Steroids. J. Biol. Chem. **201**, 175 (1953).

51b. — — The Action of Adrenal Homogenates on Progesterone, 17-Hydroxyprogesterone, and 21-Desoxycortisone. Arch. Biochem. Biophys. **36**, 237 (1952).

51c. HAYANO, M., R. I. DORFMAN and E. Y. YAMADA: The Conversion of Desoxycorticosterone to Glycogenic Material by Adrenal Homogenates. J. Biol. Chem. **193**, 175 (1951).

51d. HEARD, R. D. H. and V. J. O'DONNELL: Biogenesis of the Estrogens: The Failure of Cholesterol-4-C[14] to give Rise to Estrone in the Pregnant Mare. Endocrinology **54**, 209 (1954).

52. HEILBRON, I. M., T. P. HILDITCH and E. D. KAMM: The Unsaponifiable Matter from the Oils of Elasmobranch Fish. II. The Hydrogenation of Squalene in the Presence of Nickel. J. Chem. Soc. (London) **1926**, 3131.

53. HEILBRON, I. M., E. D. KAMM and W. M. OWENS: The Unsaponifiable Matter from the Oils of Elasmobranch Fish. I. A Contribution to the Study of the Constitution of Squalene (Spinacene). J. Chem. Soc. (London) **1926**, 1630.

54. HIRSCHMANN, H. and F. B. HIRSCHMANN: Steroid Excretion in a Case of Adrenocortical Carcinoma. III. The Isolation of Δ^5-Pregnenediol-3(β),17(β)-one-20 and of 17a-Methyl-Δ^5-D-homoandrostenediol-3(β),17a(α)-one-17. J. Biol. Chem. **167**, 7 (1947).

54a. KAHNT, F. W. und A. WETTSTEIN: Die 11-Oxydation von Desoxy-corticosteron und Reichstein's Substanz S mit Hilfe tierischer Organhomogenate. Bildung von Corticosteron und 17-Oxy-corticosteron. Helv. Chim. Acta **34**, 1790 (1951).

55. KLEIN, H. P.: Relation of Co-enzyme A to Steroid and Total Lipid Synthesis in Yeast. Federat. Proc. (Amer. Soc. exp. Biol.) **10**, 209 (1951).

56. KLEIN, H. P. and F. LIPMANN: The Relationship of Co-enzyme A to Lipide Synthesis. I. Experiments with Yeast. J. Biol. Chem. **203**, 95 (1953).

57. — — The Relationship of Co-enzyme A to Lipide Synthesis. II. Experiments with Rat Liver. J. Biol. Chem. **203**, 101 (1953).

58. KRITCHEVSKY, D. and I. GRAY: Biosynthesis of Cholesterol from Isobutyrate. Experientia **7**, 183 (1951).

59. LANGDON, R. G. and K. BLOCH: Biosynthesis of Squalene and Cholesterol. J. Amer. Chem. Soc. **74**, 1869 (1952).

60. — — The Biosynthesis of Squalene. J. Biol. Chem. **200**, 129 (1953).

61. — — The Utilization of Squalene in the Biosynthesis of Cholesterol. J. Biol. Chem. **200**, 135 (1953).

62. — — The Effect of some Dietary Additions on the Synthesis of Cholesterol from Acetate *in vitro*. J. Biol. Chem. **202**, 77 (1953).

63. LETTRÉ, H., H. H. INHOFFEN und R. TSCHESCHE: Über Sterine, Gallensäuren und verwandte Naturstoffe. Bd. I. Stuttgart: F. Enke. 1954.

64. LIEBERMAN, S. and S. TEICH: Recent Trends in the Biochemistry of the Steroid Hormones. Pharmacol. Rev. **5**, 285 (1953).

65. LITTLE, H. N. and K. BLOCH: Studies on the Utilization of Acetic Acid for the Biological Synthesis of Cholesterol. J. Biol. Chem. **183**, 33 (1950).

66. MACKENNA, R. M. B., V. R. WHEATLEY and A. WORMALL: The Composition of the Surface Skin Fat (Sebum) from the Human Forearm. J. Invest. Dermatol. **15**, 33 (1950).

67. MAZUR, Y. and F. S. SPRING: Steroids. XII. The Structures of Cholegenin and *iso*Cholegenin. J. Chem. Soc. (London) **1954**, 1223.

68. MIESCHER, K. und P. WIELAND: Über Steroide. 100. Mitt. Zur Biosynthese der Steroide. Helv. Chim. Acta **33**, 1847 (1950).

69. MONDON, A.: Zur Biogenese der Steroide. Angew. Chem. **65**, 333 (1953); **66**, 32 (1954).

70. OTTKE, R. C., S. SIMMONDS and E. L. TATUM: Deuteroacetate in the Biosynthesis of Ergosterol by Neurospora. J. Biol. Chem. **186**, 581 (1950).

71. OTTKE, R. C., E. L. TATUM, I. ZABIN and K. BLOCH: Isotopic Acetate and Isovalerate in the Synthesis of Ergosterol by Neurospora. J. Biol. Chem. **189**, 429 (1951).

72. PICHA, G. M., F. J. SAUNDERS and D. M. GREEN: An Oxydative Metabolite of Desoxycorticosterone. Science (Washington) **115**, 704 (1952).

73. PIHL, A., K. BLOCH and H. S. ANKER: The Rates of Synthesis of Fatty Acids and Cholesterol in the Adult Rat Studied with the Aid of Labeled Acetic Acid. J. Biol. Chem. **183**, 441 (1950).

74. PINCUS, G., O. HECHTER and A. ZAFFARONI: The Effect of ACTH upon Steroidogenesis by the Isolated Perfused Adrenal Gland. Proc. 2nd Clinical ACTH Conf. Research **1**, 40 (1951).

74a. PLAGER, J. E. and L. T. SAMUELS: Synthesis of C^{14}-17-Hydroxy-11-desoxycorticosterone and 17-Hydroxycorticosterone by Fractionated Extracts of Adrenal Homogenates. Arch. Biochem. Biophys. **42**, 477 (1953).

75. PLAUT, G. W. E. and H. A. LARDY: Enzymatic Incorporation of C^{14}-Bicarbonate into Acetoacetate in the Presence of Various Substrates. J. Biol. Chem. **192**, 435 (1951).

76. POPJÁK, G.: Lipid Synthesis from Small Molecules. Biochemic. J. **51**, XIV (1952).

77. PRICE, T. D. and D. RITTENBERG: The Metabolism of Acetone. I. Gross Aspects of Catabolism and Excretion. J. Biol. Chem. **185**, 449 (1950).

78. RABINOVITZ, M. and D. M. GREENBERG: Incorporation of Radioacetate into the Cholesterol of Fetal Rat Liver Homogenates. Arch. Biochem. Biophys. **40**, 472 (1952).

79. RABINOWITZ, J. L. and S. GURIN: The Biosynthesis of Radioactive Cholesterol by Particle-free Extracts of Rat Liver. Biochim. Biophys. Acta **10**, 345 (1953).

80. RITTENBERG, D. and K. BLOCH: The Utilization of Acetic Acid for the Synthesis of Fatty Acids. J. Biol. Chem. **160**, 417 (1945).

81. ROBINSON, R.: Structure of Cholesterol. J. Soc. Chem. Ind. **53**, 1062 (1934).

82. ROTH, M., G. SAUCY, R. ANLIKER, O. JEGER und H. HEUSSER: Über Steroide und Sexualhormone. 195. Mitt. Die Konstitution der Polyporensäure A. Helv. Chim. Acta **36**, 1908 (1953).

82a. SAMUELS, L. T., M. L. HELMREICH, M. B. LASATER and H. REICH: An Enzyme in Endocrine Tissues which Oxydizes Δ^5-3-Hydroxy Steroids to $\alpha\beta$-Unsaturated Ketones. Science (Washington) **113**, 490 (1951).

82b. SAMUELS, L. T. and C. D. WEST: The Intermediary Metabolism of the Non-benzoid Steroid Hormones Vitamins and Horm. **10**, 251 (1952).

83. SAVARD, K., R. I. DORFMAN and E. POUTASSE: Biogenesis of Androgens in Human Testis. J. Clin. Endocrinol. and Metabolism **12**, 935 (1952).

84. SCHOPFER, W. H. et E. C. GROB: Sur la biosynthèse du β-carotène par *Phycomyces* cultivé sur un milieu contenant de l'acétate de sodium comme unique source de carbone. Experientia **8**, 140 (1952).

85. SCHWENK, E., G. J. ALEXANDER, T. H. STOUDT and C. A. FISH: Studies on the Biosynthesis of Cholesterol. VII. Formation of Cholesterol Precursors by Yeast. Arch. Biochem. Biophys. **55**, 274 (1955).

86. SCHWENK, E. and N. T. WERTHESSEN: Studies on the Biosynthesis of Cholesterol. III. Purification on C^{14}-Cholesterol from Perfusions of Livers and Other Organs. Arch. Biochem. Biophys. **40**, 334 (1952).

87. — — Studies on the Biosynthesis of Cholesterol. IV. Higher Counting Substances Accompanying C^{14}-Cholesterol in the Intact Rat. Arch. Biochem. Biophys. **42**, 91 (1953).

88. SIPERSTEIN, M. D. and I. L. CHAIKOFF: C^{14}-Cholesterol. III. Excretion of Carbons 4 and 26 in Feces, Urine, and Bile. J. Biol. Chem. **198**, 93 (1952).

89. SIPERSTEIN, M. D., F. M. HAROLD, I. L. CHAIKOFF and W. G. DAUBEN: C^{14}-Cholesterol. VI. Biliary End-products of Cholesterol Metabolism. J. Biol. Chem. **210**, 181 (1954).

90. SMEDLEY MACLEAN, I. and D. HOFFERT: The Carbohydrate and Fat Metabolism of Yeast. III. The Nature of the Intermediate Stages. Biochemic. J. **20**, 343 (1926).

91. SOBEL, H.: Squalene in Sebum and Sebumlike Materials. J. Invest. Dermatol. **13**, 333 (1949).

92. SONDERHOFF, R. und H. THOMAS: Die enzymatische Dehydrierung der Trideutero-Essigsäure. Liebigs Ann. Chem. **530**, 195 (1937).

93. SRERE, P. A., I. L. CHAIKOFF and W. G. DAUBEN: The *in vitro* Synthesis of Cholesterol from Acetate by Surviving Adrenal Cortical Tissue. J. Biol. Chem. **176**, 829 (1948).

94. SRERE, P. A., I. L. CHAIKOFF, S. S. TREITMAN and L. S. BURSTEIN: The Extrahepatic Synthesis of Cholesterol. J. Biol. Chem. **187**, 629 (1950).

94a. TAMM, CH.: Glykoside und Aglykone. 141. Mitt. Die Konstitution des Ouabagenin-monoacetonids. Helv. Chim. Acta. **38**, 147 (1955).

94b. STOLL, A. und J. RENZ: Der enzymatische Abbau des Scillirosids zum Scillirosidin. Helv. Chim. Acta **33**, 268 (1950).

95. TÄUFEL, K., H. THALER und H. SCHREYEGG: Squalen als Bestandteil des Hefefettes. Fette und Seifen **43**, 26 (1936).

96. THORBJARNARSON, T. and J. C. DRUMMOND: Occurrence of an Unsaturated Hydrocarbon in Olive Oil. Analyst **60**, 23 (1935).

97. TOMKINS, G. M., H. SHEPPARD and I. L. CHAIKOFF: Cholesterol Synthesis by Liver. III. Its Regulation by Ingested Cholesterol. J. Biol. Chem. **201**, 137 (1953).

98. — — — Cholesterol Synthesis by Liver. IV. Suppression by Steroid Administration. J. Biol. Chem. **203**, 781 (1953).

99. TSCHESCHE, R. und F. KORTE: Zum biochemischen Syntheseweg der Steroide. Angew. Chem. **64**, 633 (1952); **65**, 81 (1953); **66**, 32 (1954).

100. TSCHESCHE, R., M.-E. RÜHSEN und G. SNATZKE: Zur Konstitution des Urezigenins und Xysmalogenins. Chem. Ber. (im Druck).

101. TSCHESCHE, R. und G. SNATZKE: Über Adynerin und Neriantin. Chem. Ber. **88**, 511 (1955).

102. UNGAR, F. and R. I. DORFMAN: Incorporation of C^{14} in the Urinary Steroids *in vivo*. J. Biol. Chem. **205**, 125 (1953).

103. VESTLING, C. S. and G. F. LATA: Steroid Changes in Incubating Adrenal Homogenates. Science (Washington) **113**, 582 (1951).

104. VOSER, W., Hs. H. GÜNTHARD, H. HEUSSER, O. JEGER und L. RUZICKA: Zur Kenntnis der Triterpene. 175. Mitt. Ein neuer Weg zur Öffnung des Ringes C beim Lanostadienol. Helv. Chim. Acta **35**, 2065 (1952).

105. VOSER, W., M. V. MIJOVIĆ, H. HEUSSER, O. JEGER und L. RUZICKA: Über Steroide und Sexualhormone. 186. Mitt. Über die Konstitution des Lanostadienols (Lanosterins) und seine Zugehörigkeit zu den Steroiden. Helv. Chim. Acta **35**, 2414 (1952).

106. WEINHOUSE, S.: The Structure of "Active Acetyl" and a Theory of Fatty Acid Catabolism. Arch. Biochem. Biophys. **37**, 239 (1952).

107. WOODWARD, R. B. and K. BLOCH: The Cyclization of Squalene in Cholesterol Synthesis. J. Amer. Chem. Soc. **75**, 2023 (1953).

108. WÜERSCH, J., R. L. HUANG and K. BLOCH: The Orgin of the Isooctyl Side Chain of Cholesterol. J. Biol. Chem. **195**, 439 (1952).

109. ZABIN, I. and K. BLOCH: The Utilization of Isovaleric Acid for the Synthesis of Cholesterol. J. Biol. Chem. **185**, 131 (1950).

110. — — Studies on the Utilization of Isovaleric Acid in Cholesterol Synthesis. J. Biol. Chem. **192**, 267 (1951).

111. — — The Utilization of Butyric Acid for the Synthesis of Cholesterol and Fatty Acids. J. Biol. Chem. **192**, 261 (1951).

112. ZAFFARONI, A., O. HECHTER and G. PINCUS: Adrenal Conversion of C^{14}-Labeled Cholesterol and Acetate to Adrenal Cortical Hormones. J. Amer. Chem. Soc. **73**, 1390 (1951).

113. — — — Conversion of C^{14}-Labeled Acetate and Cholesterol to Adrenocortical Hormones by Perfused Adrenal Glands. Federat. Proc. (Amer. Soc. exp. Biol.) **10**, 150 (1951).

(Eingelaufen am 28. Dezember 1954.)

Some Biochemical Aspects of Fungal Carotenoids.

By **F. T. Haxo**, La Jolla, California.

Contents.

I. Introduction.

The term "*carotenoid*" as used in this paper, refers to the colored members of the C_{40}-polyene series.

Although many species of fungi are apparently devoid of carotenoids, there are conspicuous instances in nature of yellow, orange, or red coloration attributable to the presence of such pigments, notably, in the red yeasts (*Rhodotorula*), the red bread mold (*Neurospora*), and the chantarelle mushrooms (*Cantharellus*). The occurrence of carotenoids in a wide variety of fungi is known from the early surveys of Zopf (*116*), Kohl (*67*) and van Wisselingh (*106*). In a limited number of fungi, examined with the modern techniques of chromatography and spectroscopic analysis, the kinds and amounts of carotenoids present have been established. Such studies have revealed widely different bio-synthetic capacities for C_{40}-polyenes. The water molds *Phycomyces* and *Allomyces* produce only carotenoids of the hydrocarbon type (carotenes), and predominantly one carotene each, *i. e.*, β-carotene and γ-carotene, respectively. Other fungi contain a variety of both carotenes and xanthophylls (oxygen-containing carotenoids), *e. g.*, *Neurospora crassa* elaborates at least six xanthophylls and nine polyene hydrocarbons.

All of the stable all-*trans* carotenes found in higher plants have also been observed among the fungi. Poly-*cis* carotenoids such as pro-γ-carotene and prolycopene, which occur in flowers and fruits (*110*), have not been reported present in any cryptogam. Whereas most higher plants contain a complex mixture of xanthophylls, including carotenoid epoxides, xanthophyll formation in fungi is less diversified, and such epoxides are apparently absent. Production of unique carotenoid acids, *e. g.*, torularhodin, is characteristic of some fungi, and formation of neutral xanthophylls such as torulene, spirilloxanthin (rhodoviolascin) and canthaxanthin, is a capacity apparently shared only with certain bacteria.

Among the algae, surveys of carotenoid distribution have been useful in ascertaining phylogenetic relationships, the major taxonomic groups displaying characteristic differences, especially in xanthophylls (*96*). Patterns of carotenoid composition have not been explored sufficiently among the fungi to appraise fully the usefulness of this feature as a supplementary taxonomic tool. Among the asporogenous yeasts, which lack distinguishing morphological features, the genus *Rhodotorula* has long been separated from related genera such as *Cryptococcus*, primarily because of its production of red to yellow pigments of carotenoid nature. However, NAKAYAMA et al. (*78*) have noted a considerable variability in the kinds and amounts of carotenoids in various species of *Rhodotorula* and *Cryptococcus*, and have concluded that "a separation of the above genera on the basis of manifest carotenoid is vague and arbitrary". CANTINO and HYATT (*13*) have recently used carotenoid content and oxidative enzyme activity to substantiate phylogenetic relationships, presumed mainly on morphological grounds, between two groups of aquatic fungi, the *Blastocladiales* and the *Chytridiales*. Whereas wild type *Blastocladiella* lacked pigment and possessed α-ketoglutarate oxidase activity, mutant strains of *Blastocladiella*, like wild type *Rhizophlyctis*, a representative of the *Chytridiales*, contained almost exclusively γ-carotene and showed no α-ketoglutarate oxidase effect.

The conspicuous concentration of carotenoids in specific structures of some fungi, *e. g.*, light-sensitive sporangiophores of *Pilobolus* and *Phyco-*

myces and male gametangia of *Allomyces*, has long attracted interest. However, attempts to establish a function for carotenoids in phototropic responses or in the sexuality of fungi are as yet inconclusive.

Of greater significance is the use made of carotenoid-rich fungi for studies on the mechanism of C_{40}-polyene formation in plants, a field in which marked advances have been made in recent years. Fungi have proven advantageous for such studies because their growth is rapid and the conditions of culture and nutrition may be controlled. Furthermore, carotenoid formation is not complicated by the presence of chlorophyllous pigments as is true of algae and higher plants. There is, however, no reason *a priori* for assuming that the basic pathways of carotenoid formation differ in such widely separated groups as fungi and green plants.

It is the purpose of the present review to survey the occurrence of carotenoid pigments among the fungi and to indicate wherein studies of fungal carotenoids have contributed to our understanding of the biochemistry and physiology of the C_{40}-polyenes. Various aspects of this field have been most recently reviewed by GOODWIN (*32*, *33*, *35*), KARRER and JUCKER (*59*), and MACKINNEY (*72*).

II. Occurrence of Carotenoids in Fungi.

Investigations of the carotenoid composition of cryptogams have frequently been handicapped by the availability of a limited amount of starting material. Complete removal of carotenoids from some fungi can be effected only by exhaustive extraction, and in some cases, application of specially designed techniques. For such reasons it may be assumed that in many instances referred to in *Table 1* (p. 172), only the major components have been identified. In few cases has the pigment in question been isolated in the pure crystalline form. However, even with only a few micrograms of chromatographically purified pigment, it is possible to establish identity with reasonable certainty. Determination of partition behavior and spectral characteristics usually point to a tentative identification, and this can be confirmed by a mixed chromatogram with an authentic sample. A tool which may find use in the detection of some minor polyene components in fungi is growth in the presence of diphenylamine, which inhibits formation of xanthophylls and the more unsaturated carotenes, while enhancing production of partially hydrogenated carotenes and the colorless C_{40}-polyenes [GOODWIN (*34*); TURIAN (*99*)]. In extracts from diphenylamine-treated cells of *Mycobacterium phlei*, it was possible to detect phytofluene and a flavacin-like carotene, neither of which accumulated in sufficient quantity in untreated bacterial cells for easy detection [TURIAN and HAXO (*101*)].

Table 1. Distribution of
(Instances of tentative identification

Source / C_{40}-Polyenes	α-carotene	β-carotene	γ-carotene	δ-carotene	lycopene	rhodopurpurin
Myxomycetes						
Lycogola epidendron		+				
Phycomycetes						
Allomyces arbuscula			+			
A. javanicus		+	+			
A. macrogyna			+			
A. moniliformis			+			
Blastocladiella sp.			+			
Rhizophlyctis (Karlingia) rosea			+			
Mucor hiemalis	?	+				
Phycomyces blakesleeanus	+	+	+		+	
P. nitens	+	+	+		+	
Ascomycetes						
Aleuria aurantia		+	+			
Neurospora crassa		+	+	?	+	?
N. sitophila		+	+		+	?
Polystigma rubrum						
Basidiomycetes						
Anthurus aseroformis					+	
Cantharellus cibarius	+	+	?	?	+	
C. cinnabarinus		+				
C. infundibiliformis					+	
C. lutescens					+	
Coleosporium senecionis	+	+	+			
Dacromyces stillatus	+	+	?			
D. ellisii		+	?		?	
Gymnosporangium juniperi-virginianae		+	+			
Puccinia coronifera	+	+	+			
Tremella mesenterica		+				
Fungi Imperfecti						
Cryptococcus sps.		+	+		+	
Rhodotorula aurea		+	+		+	
R. flava		+	+		+	
R. glutinis		+	+			
R. peneus		+	+		+	
R. rubra	+	+	+			
R. sanniei		+	+		+	
Sporobolomyces roseus						
S. salmonicolor						

C_{40}-Polyenes in Fungi.

are indicated by a question mark.)

	neurosporene	ζ-carotene	θ-carotene	phytofluene	phytoene	lycoxanthin	rubixanthin	canthaxanthin	cryptoxanthin	zeaxanthin	spirilloxanthin-rhodoviolascin	torulene	torularhodin	unnamed acidic carotenoids	References
											+	+			69
															22
		?		+											22, 102
															22
															22
															13
															13
															18, 45
	+	+		+	?										6, 34, 60, 88
	+	+		+	?										38
							?								69
	+		+	+	+	+	?				+			+	51, 55, 94, 108
	+		+	+	+	+	?				+	?		+	1, 19, 54
						+								+	69
															21
															105
				+	?			+							52
	+														105
	+														105
															69
		+		+	?				+	+		+			36
															48
															95, 107
														+	69
															69
															78
															78
															78
												+	+		78
															78
	+	?		+								+	+	+	10, 23, 61, 68
												+	+		27
												+	+		69
												+	+		69

1. Carotenes with Formula $C_{40}H_{56}$.

The distribution of carotenoids in the various groups of fungi is summarized in Table 1. As noted above, all of the common carotenes have been detected in fungi.

α-Carotene has been reported in a relatively few fungi and always as a minor component of a mixture. Early reports (*14*) that α-carotene was the major pigment of *Phycomyces blakesleeanus* have not been confirmed by later workers (*30, 90*). *β-Carotene* and *γ-carotene* are the carotenoids most frequently encountered among the fungi. The former was identified by SCHOPFER (*88, 93*) as the major constituent of *Phycomyces*. On a standard asparagine-glucose medium, *Phycomyces* produces 140 mg. of β-carotene/100 g. dry mold, the minor polyenes constituting less than 5% of the total (*34, 41*). Crystalline β-carotene has also been prepared from *Cantharellus cibarius* (*105*) and from *Rhodotorula rubra* (*69*).

γ-Carotene, first noted by LEDERER (*69*) in the ascomycete *Aleuria aurantia* and by WILLSTAEDT (*105*) in *Cantharellus cibarius*, has since been observed in a variety of fungi, both as the predominant carotenoid and as a minor component of a mixture. EMERSON and FOX (*22*) have reported the high localized accumulation of this carotene in the male gametangia of several species of the aquatic phycomycete *Allomyces*. In *Allomyces javanicus* the concentration of γ-carotene was recently estimated at 2.6 mg./100 g. dried gametangia, but the starting material included, besides male gametangia, non-pigmented female gametangia and probably some mycelia (*102*). *Rhizophlyctis rosea*, another aquatic phycomycete, has recently been reported to contain almost exclusively γ-carotene (*13*). A strain of the same fungus, examined in the writer's laboratory, however, contained in addition to γ-carotene rather appreciable amounts of an unidentified neutral xanthophyll which could be recovered from the mycelium only under special conditions of extraction (*81*). In *Neurospora sitophila* γ-carotene is present in concentrations as high as 13.7 mg./100 g. dry mold and constitutes about 40% of the total pigment (*54*).

The presence of *δ-carotene*, a rare carotenoid first detected in the fruit hulls of a higher plant, has never been unequivocably demonstrated in fungi. Reported occurrences are in *Cantharellus cibarius* (*105*) and in *Neurospora crassa* (*51*).

Lycopene occurs as the major pigment of the phalloid fungus *Anthurus aseroformis* (*21*) and in rather considerable amounts in *Cantharellus lutescens* and *C. infundibiliformis* (*105*). Crystalline lycopene was isolated from *Neurospora crassa*, where it occurs in concentrations up to 10 mg./100 g. (*51*). This pigment is a minor carotenoid constituent in red yeasts (*27*), *Phycomyces* (*6, 34*), and *Allomyces* (*102*).

Rhodopurpurin, $C_{40}H_{56}$ (or $C_{40}H_{58}$), first isolated by KARRER and SOLMSSEN (*63*) from rhodovibrio bacteria, and little known otherwise, is possibly present in *Neurospora crassa* (*51*).

2. Carotenoid Hydrocarbons more Saturated than $C_{40}H_{56}$.

The occurrence in fungi of carotenes and colorless C_{40}-polyene hydrocarbons more saturated than common carotenes, is now well established. *Neurosporene*, first described from *Neurospora crassa*, where it is found in concentrations up to 6 mg./100 g. dry mold, has also been observed in several other fungi and in higher plants. The positions of the absorption maxima indicate that neurosporene contains nine conjugated double bonds, and the shape of the absorption curve may suggest an open-chain structure for the molecule, as in lycopene. Elementary analysis of this pigment gave best agreement with $C_{40}H_{60}$, and the tentative formula $C_{40}H_{58} \pm 2$ H was provisionally assigned (*51*). A carotene hydrocarbon with very similar properties has recently been isolated from tomatoes by TROMBLY and PORTER (*98*), and termed "tetrahydro-lycopene". The identity of the latter pigment with neurosporene, as suggested by these workers, has been confirmed by the writer on the basis of mixed chromatograms. Similarly, pigment *A* of red yeasts has been identified as neurosporene (*51*). GOODWIN (*32, 34*) found neurosporene to be a minor component of *Phycomyces blakesleeanus* and *P. nitens*, and considers that the major pigment of *Cantharellus lutescens* and *C. infundibiliformis*, tentatively identified by WILLSTAEDT as β-apo-8-carotenal, may well be neurosporene.

Pigments resembling *ζ-carotene*, $C_{40}H_{64}$, have been observed in *Rhodotorula rubra* (*10*), *Neurospora crassa* (*53, 94*), *Phycomyces* (*34*), *Dacromyces stillatus* (*36*), and *Allomyces javanicus* (*102*). ζ-Carotene, best known from tomato preparations, has absorption maxima in hexane at 425, 400 and 378 mμ and possesses an open-chain structure with seven conjugated double bonds (*79*). GOODWIN and OSMAN (*43*) have recently reported that the anomalous ζ-carotene of *Phycomyces* is a mixture of "ζ-carotene" and an unidentified carotene having spectral maxima at 452, 421 and 400 mμ in hexane. The ζ-carotene-like fraction of *Neurospora* is now known to be a mixture of two yellow pigments which can be separated on a magnesia column (*55, 102*). One carotene has an absorption spectrum similar to flavacin with maxima in hexane at 454, 427 and 406 mμ; however, the other is a new carotene, tentatively named *Θ-carotene*, which resembles ζ-carotene closely but is chromatographically and spectrally different (max. at 421, 397 and 375.5 mμ). In view of the similarity of Θ-carotene to ζ-carotene, the occurrence of the latter pigment in fungi remains uncertain.

Phytofluene, $C_{40}H_{64} \pm 2$ H, a colorless, fluorescent polyene containing the carotenoid skeleton and five conjugated double bonds (*114, 115, 66a, 81a*), has been detected in all carotenoid-rich fungi wherein a careful search has been made. This compound is distributed widely among higher plants and has attracted attention as a possible precursor of the colored C_{40}-polyenes (*34, 83, 115*). Among the fungi, phytofluene was first detected by BONNER et al. (*10*) in red yeast (*Rhodotorula rubra*), and has since been reported present in the major groups of fungi and in bacteria.

Occurrence of a "*phytoene*-like fraction" in *Phycomyces blakesleeanus* has been reported by GOODWIN (*34*), but the presence of impurities precluded unequivocal identification. A similar fraction was observed in the basidiomycete *Dacromyces* (*36*) and more recently in *Cantharellus cinnabarinus* (*55*). *Neurospora* has since been found to contain phytoene as its most abundant polyene (*55, 101*). Employing fractional-flow chromatography on magnesia columns, HAXO and TURIAN (*55*) isolated purified phytoene from *Neurospora* which had an absorption curve identical with that of tomato phytoene. Wild type strains of *N. crassa* contain up to 70 mg./100 g. dry mold, whereas an albino strain, devoid of carotenoids, contained 200 mg. Non-pigmented cultures of *Ascobolus furfuraceous* and *Nectria cinnabarinus* are reported as lacking phytoene (*34*). PORTER and LINCOLN (*83*) consider phytoene to be $C_{40}H_{72}$ and to have an open-chain structure similar to lycopene, with only three conjugated double bonds.

3. Neutral Xanthophylls.

Three monohydroxy carotenoids have been reported as present in fungi. A pigment resembling *lycoxanthin* (3-hydroxy-lycopene) was found by LEDERER (*69*) in *Polystigma rubrum*. A minor pigment of *Neurospora crassa*, tentatively identified as lycoxanthin or rhodopin, appears on re-examination to be most probably lycoxanthin (*53*). A mono-hydroxy-γ-carotene also occurs in *Neurospora*. The latter has been shown not to be gazaniaxanthin, making identification as *rubixanthin*, (3-hydroxy-γ-carotene) most probable (*54*). A recent study by GOODWIN (*36*) of the carotenoids of *Dacromyces stillatus* has provided the first reported occurrence among fungi of two typical leaf xanthophylls, viz. *cryptoxanthin* (3-hydroxy-β-carotene) and *zeaxanthin* (3,3'dihydroxy-β-carotene).

A new carotenoid, *canthaxanthin* has been isolated in crystalline form (m. p. 218°) from the edible mushroom *Cantharellus cinnabarinus*, where it is accompanied by smaller amounts of β-carotene, phytofluene, phytoene, and several unidentified carotenoids, probably all new xanthophylls (*52*). The recent investigation of SAPERSTEIN and STARR (*86*) has established the occurrence of canthaxanthin in the bacterial phytopathogen

Corynebacterium michiganense and has provided further information concerning its chemical nature. Canthaxanthin lacks provitamin A activity and hence does not contain an unsubstituted β-ionone ring. Comparison with other carotenoids exhibiting a single absorption maximum in the visible spectrum suggested that canthaxanthin is a carotenoid having a carbonyl group cross-conjugated in the system of double bonds. The formula $C_{40}H_{54}O_2$ ($\pm$ 2 H) and a structure were tentativelyproposed.

Carotenoids originally indentified as either *rhodoviolascin* or *spirilloxanthin* have been reported in the slime mold *Lycogola epidendron* (*69*) and *Neurospora crassa* (*51*). The spirilloxanthin content of *Neurospora* cultures as a whole approaches 9 mg./100 g. dry weight, and is especially high, viz. 159 mg., in the asexual spores (*108*). Rapid biosynthesis of spirilloxanthin is mainly responsible for the first pink color that develops, upon exposure to light, of colorless dark-grown mycelium. Spirilloxanthin was first isolated by VAN NIEL and SMITH (*80*) from the purple bacterium *Rhodospirillum rubrum*, whereas rhodoviolascin was obtained from mass cultures of *Rhodovibrio* by KARRER and SOLMSSEN (*63*). More recent work has shown that the observed features of the chemical composition of rhodoviolascin, $C_{40}H_{54}(OCH_3)_2$, also hold true for spirilloxanthin, and that the two carotenoids are undoubtedly identical (*82*).

Torulene, so named because it was first isolated from *Rhodotorula* (*Torula*) *rubra*, is of fairly common occurrence among the fungi. Reported occurrences are among various species of red yeasts, *Sporobolomyces* sp., *Lycogola* (*69*) and recently *Dacromyces stillatus* (*36*). FROMAGEOT and TCHANG (*27*) reported the concentrations of torulene in *Rhodotorula sanniei* as 14.6 mg./100 g. dry weight. The molecular structure of torulene has not been established with certainty. LEDERER (*69*) suggested a relationship of torulene to rhodoviolascin and assigned the provisional formula, 3,3'dimethoxy-γ-carotene. It is of interest that torulene, perhaps originating from red yeasts, has been reported as present in marine sediments from the ocean floor (*25*).

4. Acidic Xanthophylls.

Torularrodin, probably $C_{37}H_{48}O_2$, is a monocarboxylic carotenoid acid, so far found only in fungi. First noted by LEDERER (*68*) as a major constituent of *Rhodotorula rubra*, torularhodin has since been isolated in the pure form and its chemical constitution studied by KARRER and RUTSCHMANN (*61*, *62*). Torularhodin displays a remarkably long wavelength absorption spectrum, containing twelve conjugated double bonds in the molecule. It exhibits weak vitamin A activity and is therefore presumed to have an unsubstituted β-ionone ring. The structural formula (I) has been provisionally assigned.

(I.) Torularhodin.

Torularhodin has been most frequently encountered among the red yeasts, although it is not an invariable component (*10, 23, 78*). In *Rhodotorula sanniei*, concentrations as high as 290 mg./100 g. dry weight have been found (*27*). LEDERER (*69*) reports the presence in the rust fungus *Puccinia coronifera* of an acid pigment resembling torularhodin.

Acidic carotenoids of undetermined nature have been detected in several fungi. One such pigment, observed by LEDERER (*69*) in extracts of the ascomycete *Polystigma rubrum*, showed maxima in petroleum ether at 516 and 485 mμ. Most strains of *Neurospora crassa* contain appreciable amounts of an acidic carotenoid fraction which includes at least three components, the predominant one resembling somewhat the *Polystigma* pigment (*51*). A yellow mutant of *Neurospora*, differing from the wild type by a single gene change, has been encountered which is incapable of synthesizing acidic carotenoids (*53*).

III. Carotenoid Formation in Fungi.

What we know about carotenoid formation in plants has been learned from studies on the intact organism. Use has not been made of cell-free preparations and may prove not feasible. Carotenoid accumulation in fungi is usually a slow process and is generally most marked after active growth has ceased (*30, 71*). In fungi as well as in several other microorganisms, a great deal is known concerning the effects of various environmental and nutritional factors on carotenogenesis (cf. *32, 33*). Some studies along these lines, together with inhibitor studies and analyses of mutant strains, have provided valuable clues as to the interrelationships of the various C_{40}-polyenes. It will be of help in discussing these investigations to consider first what the current theories are.

1. Interrelationships of the C_{40}-Polyenes.

In 1934 ZECHMEISTER (*109*) suggested that the final chemical step in the formation of polyene pigments might involve the dehydrogenation of a colorless precursor. The discovery in more recent years of phytofluene, a widely distributed C_{40}-polyene with a higher degree of hydrogenation than the carotenes, led to the further suggestion that this colorless hydrocarbon might constitute an intermediate product in the formation of carotenoid pigments. At this time the idea was also entertained that phytofluene and the polyene pigments might have a common precursor (*115*). The "phytofluene theory" of ZECHMEISTER has since been expanded by PORTER and LINCOLN (*83*) to include other hydrogenated carotenes and polyenes now known to occur in plants. The first C_{40}-polyene formed would be tetrahydro-phytoene, $C_{40}H_{76}$, which would give rise successively to phytoene, phytofluene, ζ-carotene, neurosporene and lycopene by individual steps involving the removal of four hydrogens. The *in vitro* conversion of colorless polyenes to carotenoid pigments has been effected by ZECHMEISTER and KOE (*113*).

In fungi, wherein tetrahydro-phytoene has not been detected, the interrelationships of the C_{40}-polyenes may be indicated as follows:

Scheme A.

Colorless C_{40}-polyenes are precursors to pigments.

phytoene $C_{40}H_{72}$ —(dehydrogenation / chromogenesis)→ lycopene $C_{40}H_{56}$ —(+O)→ neutral xanthophylls —(+O)→ acidic xanthophylls

precursor → phytoene

An alternative possibility, namely that the C_{40}-polyenes are derived independently from a common precursor, with little or no interconversion between them, has recently been stressed by GOODWIN (*34*) and MACKINNEY (*58, 72*). This "parallel mechanism" may be summarized as follows:

Scheme B.

Colorless polyenes and carotenoids arise independently.

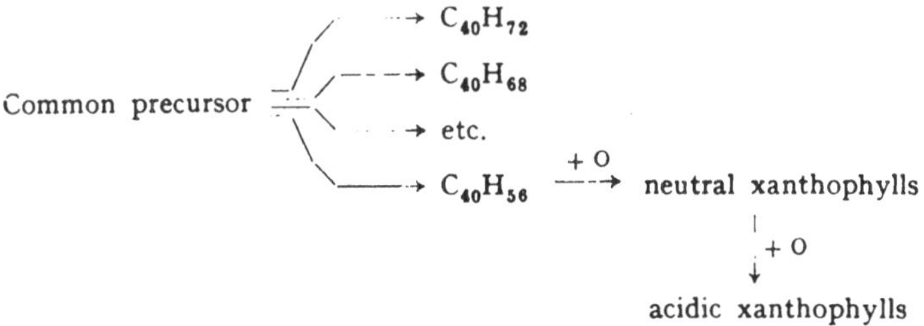

A third possibility (Scheme *C*) is that the PORTER-LINCOLN series operates in reverse, *i. e.*, that colorless polyenes are formed by hydrogenation of carotenes (*39*).

a) *Mutant Studies.*

Following the pioneering studies of BEADLE and TATUM (*4*, *5*), on *Neurospora*, conspicuous success has attended the use of mutant strains in investigations of biosynthetic processes in plants. It is now well established that biochemical reactions are under genetic control and that gene mutation may result in the blocking of a specific step in a chain of biosynthetic reactions [cf. also BEADLE (*4a*) and p. 466 of the present Volume]. Whereas failure to synthesize a given vitamin or amino acid precludes growth on unsupplemented media, failure to synthesize carotenoids in fungi has not been associated with deficiencies in either growth or any other physiological function. Additional difficulties are presented by the fact that carotenoids are relatively complex molecules and water-insoluble, and it has not been feasible to feed mutant strains suspected products of the blocked reaction or to carry out cross-feeding experiments with different mutants.

BONNER et al. (*10*) investigated the polyene composition of seven ultraviolet-induced mutants of *Rhodotorula rubra* varying in color from colorless (III and V), pale orange (I), brownish orange (II), to red (IV, VI, VII). With the exception of the colorless mutants which synthesized either no polyenes (V) or only barely detectable amounts of torulene (III), the polyene composition of the mutants was altered mainly in a quantitative way, compared to the original strain. In the red mutants, synthesis of all components was enhanced, that of γ-carotene, β-carotene, neurosporene (pigment *A*) and phytofluene by ca. 200–400%, and that of torulene, almost 25%. The pale colored mutants (I and II) showed one common feature, a great decrease in torulene. In addition, mutant II showed a four-fold increase in γ-carotene and contained a pale yellow pigment *B* (ζ-carotene?) not detected in the other strains. Making the assumption (admittedly not proven) that phytofluene is an intermediate in the synthesis of carotenoid pigments, the position of the genetic blocks were assigned as follows:

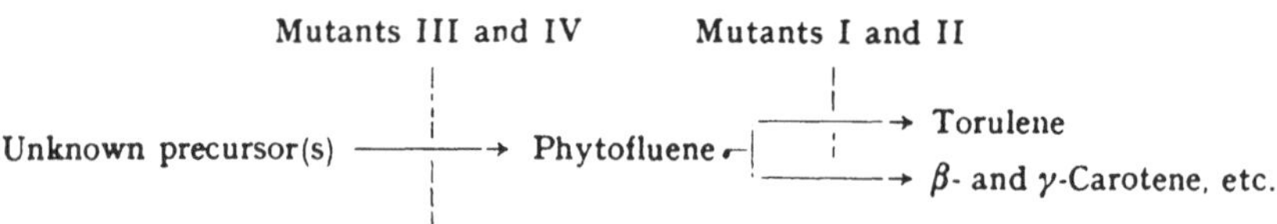

It was noted that after a few years in laboratory culture, the mutants showed a marked decrease in phytofluene content, and that in at least one case this was associated with increased pigment production (*97*).

The carotenoid composition of a number of morphological variants of *Phycomyces blakesleeanus* was examined with a view toward elucidating the mechanism of β-carotene formation (*38*). This study proved to be of little value, since the mutants contained the same polyenes as did the parent strain and in about the same proportion.

Of greater interest in this connection are studies of color mutants of the red bread mold, the carotenoid composition of which is now well known (*51*). HUNGATE (*56*) found that certain albino mutants of *Neurospora crassa*, which differ from the wild type by a single gene change, never produced visible amounts of carotenoids when cultured separately but did so when reared together as heterocaryons. In the heterocaryotic condition, a common plant body (mycelium) is formed in which the haploid nuclei from both parents occur side by side in the same cytoplasm but do not fuse. This indicates that the mutated gene of each parent albino affects a different step in the chain of reactions involved in carotenoid synthesis. Growing each albino on an extract of the other, however, failed to produce pigment as would have been expected if different precursors were being accumulated by the two mutant strains. Extracts of various carotenoid-containing plant tissues and mixtures of vitamins and amino acids were without effect on pigmentation.

SHENG and SHENG (*94*) have characterized effects on carotenoid synthesis in *Neurospora* of the "C" and the "I" genes, termed the major color factor and the intensifier factor, respectively. In crosses of wild type *Neurospora crassa* (# 1 A) and an albino strain (# 15300a), four discernable color types could be obtained: the parental types, salmon (CI) and white (ci) respectively, and in addition, pink (cI) and peach (Ci). The CI genotype, possessing the wild type allele of both genes, synthesizes the pigment mixture characteristic of wild type strains (see Table 1, p. 172). Introduction of the recessive allele of the intensifier gene (genotype Ci) depresses synthesis of all components to 1/10–1/28. In the genotype cI only β-carotene, γ- carotene and an acidic carotenoid were detected, and these at 1/90 or less the concentration found in the wild type strain. In the ci genotype, neither carotenoids nor phytofluene were found, and this was also true of the non-allelic albino mutant # 4637. Assuming phytofluene to be a carotenoid precursor, all three genes were believed to affect polyene synthesis prior to formation of phytofluene, although the exact sequence was not determined.

HAXO (*53*, *54*) has examined the polyene composition of eight mutant strains of *Neurospora*, including the two albino strains studied by the SHENGS (*94*). Three new categories of polyene deficiencies were observed: (a) specific pigment block, *e. g.*, complete absence of the acidic carotenoid fraction in a yellow mutant; (b) multiple pigment deficiencies characterized by elimination of xanthophylls and most carotenes, reduced synthesis of hydrogenated polyenes, and a normal phytoene content; (c) carotenoid-less mutants with a normal or enhanced phytoene content. In a fourth category, including the albino mutants 15300 and L-1, polyene synthesis was greatly reduced. Phytofluene and most of the wild type carotenoids could be detected in trace amounts only, and it is probable that the qualitative composition of these mutants has not been altered by the mutation.

Table 2. Types of Polyene Deficiencies Observed in Some Mutant Strains of *Neurospora crassa*.

[(+) signifies presence in reduced amounts.]

	Strain	Colorless polyenes	Hydrogenated carotenes	Common carotenes	Neutral xanthophylls	Acidic xanthophylls
	Wild-Type	+ +	+	+	+	+
I	"Yellow"	+ +	+	+	+	—
II	34508 L-2	+ (+)	(+)	—?	—	—
III	4637 9698 216	+ —	—	—	—	—
IV	15 300 L-1	—? (+)	—?	(+)	(+)	(+)

The patterns of polyene deficiencies observed, as summarized in *Table 2*, can equally well be explained by the two opposing theories of carotenoid synthesis mentioned, *i. e.*, that the carotenoids are formed sequentially from the colorless polyenes or that the C_{40}-polyenes originate through parallel lines of synthesis from a common precursor. If the first hypothesis (Scheme *A*, p. 179) is accepted, then the genetic blocks can be assigned somewhat as follows (*Chart 1*).

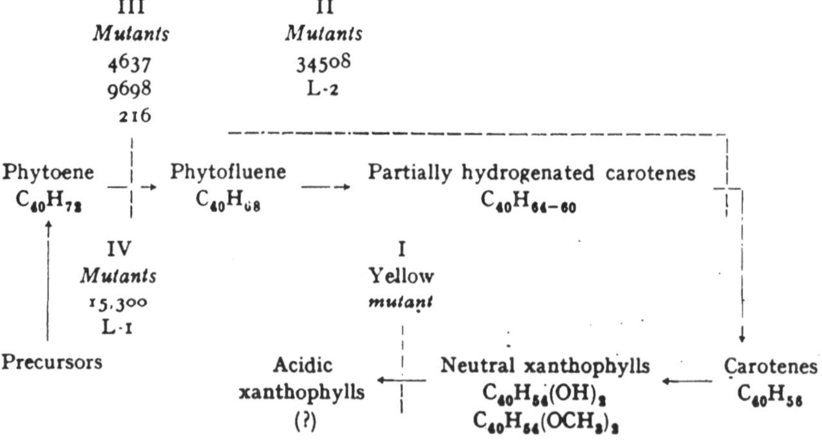

Chart 1. Hypothetical Scheme of Carotenoid Biosynthesis in *Neurospora*.

Accordingly, in group IV (*Chart 1*), the general availability of some common precursor has been reduced although not eliminated. In group III the block would occur between phytoene and phytofluene, a supposition that is further supported by the accumulation of phytoene in some mutants. In group II the block is not specific in that the formation of phytofluene and partially hydrogenated carotenes is also reduced. Location of block I is arbitrary. Alternatively, the acidic and neutral xanthophylls could arise from carotenes through a branched system, or they could arise independently from a more elementary precursor.

The occurrence of carotenoid-less mutants with a normal phytoene content (group III) would seem clearly in any case to eliminate the possibility that phytoene is formed in *Neurospora* by hydrogenation of carotenes (Scheme *C*, cf. p. 179). Similarly, acid carotenoids could not give rise to neutral xanthophylls or to the carotenes.

b) Inhibitor Studies.

A valuable tool for the study of carotenogenesis in micro-organisms was introduced by TURIAN (*99*) who found that diphenylamine, previously known through the work of KHARASCH et al. (*66*) to inhibit chromogenesis in various bacteria, had a specific effect on carotenoid formation in *Mycobacterium phlei*. In the presence of diphenylamine, growth was affected only slightly, but production of keto-enol polyenes such as chrysoflein was markedly reduced. It was suggested that diphenylamine acted by inhibiting dehydrogenation of phytofluene to carotenoid pigments. The actual accumulation of colorless polyenes concommittant to inhibition was first demonstrated for *Phycomyces* (*34*) and has since been observed in *Neurospora* and *Mycobacterium* (*101*). In *Phycomyces* diphenylamine almost completely inhibits production of α-, β-, and γ-carotene and lycopene, while stimulating production of the more saturated polyenes, phytofluene, neurosporene, ζ-carotene and, possibly, phytoene. In *Neurospora* either diphenylamine inhibition or growth in darkness alters the polyene composition in much the same way (*101*,*55*). Biosynthesis of xanthophylls and the more unsaturated carotenes is greatly reduced, whereas synthesis of phytofluene and phytoene is enhanced.

Although these findings would appear to confirm the "phytofluene precursor hypothesis" (Scheme *A*, p. 179), it is equally plausible that accumulation of the colorless polyenes results from shunting of a common precursor normally converted to the individual carotenoid pigments. GOODWIN et al. (*40*) have concluded that separate routes of synthesis for β-carotene and the phytofluene series (Scheme *B*, p. 179) is more likely in *Phycomyces*, from the finding that the synthesis of phytofluene and of β-carotene are not invariably linked together. In normal cultures, phytofluene is synthesized during the early stages of growth and remains at a constant level after growth has ceased, the period during which β-carotene synthesis is most marked. In diphenylamine cultures to which adenosine monophosphate or riboflavin was added, there was a reversal of β-carotene inhibition, but phytofluene still accumulated, as in cultures containing only diphenylamine. Furthermore, it was not possible to demonstrate conversion of phytofluene into β-carotene in mycelial mats washed free of diphenylamine.

The mechanism of action of diphenylamine in inhibiting carotenogenesis has not been established. An anti-oxidant effect has been proposed (*99*), but substances with oxidation-reduction potentials similar to diphenylamine were found to be inactive in *Phycomyces* (*40*). Ferrous sulfate counteracts the effects of diphenylamine in *Mycobacterium* but not in *Phycomyces* (*40*) or in *Neurospora* (*55*). Diphenylamine inhibition in *Phycomyces* is partially overcome by β-ionone (*74*, *75*). Goodwin et al. (*40*) consider that diphenylamine acts by blocking the transfer of high-energy phosphate to adenosine monophosphate and that β-carotene synthesis involves such a reaction. Riboflavin is regarded as allowing the organism to by-pass the blocked reaction.

Other inhibitors may provide a selective means of disentangling synthetic steps in carotenoid formation. In *Phycomyces* there apparently exist two synthetic routes, one of which is insensitive to streptomycin (*35*, *37*). Chloramphenicol and isonicotinic acid hydrazide also inhibit carotenoid formation in this fungus, but the qualitative effects on polyene composition have not yet been reported. Methyl-heptenone appears to alter the pattern of polyene formation in *Phycomyces* in somewhat the same way as does diphenylamine (*73*).

c) Studies of Light-Activated Carotenoid Synthesis.

Carotenoid formation in some fungi apparently involves one or more light-dependent reactions. Lederer (*69*) found that light was required for full pigmentation in *Rhodotorula rubra*, but this does not appear to apply to the strain studied by Bonner et al. (*10*). In *Phycomyces* Schopfer (*89*, *92*) reported a specific requirement for blue light; no carotene was formed in red light or in darkness. At variance with this report are the recent findings in two laboratories that *Phycomyces* synthesizes appreciable β-carotene when grown in darkness and that exposure to light at most doubles the concentration of β-carotene (*17*, *30*). Garton et al. (*30*) were unable to detect any wavelength specificity for the light effect, but a more critical testing of this point should be made because the red cellophane filter employed would appear to transmit some blue light, and this may be adequate to allow full pigment development.

In *Neurospora*, the effect of light on carotenogenesis is quite striking, as was first noted by Went (*104*). Cultures grown in darkness are essentially colorless but upon exposure to white light become light pink-colored within an hour. The color change is most marked in the mycelial mat; the asexual spores (conidia) produced in darkness already contain visible amounts of pigment. Mycelia exposed to white light for only 8 seconds show a visible yellowing after 12 hours further incubation in darkness; and ultimate full pigment development equivalent to that obtained with continuous illumination may be effected by exposure of a few minutes duration (*50*, *108*). The quantity of pigment produced is roughly proportional to the light dosage, clearly establishing the participation of a photochemical reaction (*50*).

Enzymatic reactions have also been implicated. Carotenoid formation in *Neurospora* following light activation shows a marked temperature dependence (*50*) and requires the presence of oxygen (*19*, *108*). The recent work of ZALOKAR (*108*) indicates that oxygen is required both for the light activation process and for subsequent dark synthesis of the pigment. The photochemical reaction could involve transformation of a precursor or activation of an enzyme involved in carotenoid synthesis. The requirement of oxygen for the light-independent reaction suggests that oxygen may serve as a final hydrogen acceptor for a dehydrogenase system involved in pigment formation. This speculation would be in line with the hypothesis that carotenoids are derived from more saturated compounds (Scheme *A*, p. 179). A requirement for oxygen during carotenogenesis has also been shown for *Rhodotorula* (*76*, *78*), but this may not be true of all fungi. GOODWIN and WILLMER (*44*) have suggested that *Phycomyces* forms carotene by fermentative assimilation.

In *Neurospora*, the effective wavelengths for light activation fall in the spectral region, 365–510 mμ (cf. *19*, *51*). Red light alone is completely inactive. It has been suggested that carotenoid pigments, perhaps present only in trace amounts, act as sensitizers in the photochemical reaction leading to their further formation (*50*). This does not exclude the possibility that other yellow pigments are the effective light absorbers for this reaction.

Does light-activated carotenoid formation involve *dé novo* synthesis of C_{40}-polyenes from simple building units, *e. g.*, C_5-fragments, or conversion of colorless polyenes accumulated in darkness into pigment? That the latter mechanism may operate in *Neurospora* was first suggested by the finding that the ratio, pigment/phytofluene, increases greatly during illumination of dark-grown cultures (*94*, *112*). The absolute amount of phytofluene, a minor polyene component in this fungus, does not change appreciably and the new pigment formed cannot be accounted for in this way.

Phytoene, the major polyene component of *Neurospora*, accumulates in cultures grown in darkness and shows a marked decrease following illumination (*55, 101, 108*). ZALOKAR (*108*) found the decrease in phytoene to be most marked after 3 hours illumination of mature, dark-grown mycelia and to coincide with an equivalent rise in the acidic carotenoid fraction, suggesting that phytoene might be a specific precursor for the carotenoid acids. The antiparallel nature of the curves suggested that phytoene might be a specific precursor for the carotenoid acids. The independent formation rates observed for the other polyenes do not fit into the concept of sequential synthesis of all carotenoids *via* the PORTER-LINCOLN series, and ZALOKAR suggests "parallel synthesis" of these components as more likely. Nevertheless, by recalculating ZALOKAR's data or polyene changes during the first 3 hours of illumination, employing the recently established extinction coefficient of phytoene (*98*) and assuming a more probable value for the carotenoid acids, all carotenoid formation can reasonably be accounted for in terms of phytoene losses. It is noteworthy that in *Neurospora* as in *Phycomyces* β-carotene formation appears to proceed independently of other carotenoids.

d) Temperature Studies.

In some fungi, *e. g., Rhodotorula sanniei* (*28*), *R. rubra* (*78*) and *Phycomyces* (*26*), temperature variations appear to have little effect on the kinds and relative amounts of carotenoids produced. However, in *R. glutinis* (*78*) growth at low temperatures considerably depresses formation of torulene and torularhodin while enhancing production of β- and γ-carotene, indicating that in some fungi the temperature may provide a means of blocking specific steps involved in carotenogenesis.

2. Precursors of the C_{40}-Polyenes.

Schematically, the carotenoids may be considered as being formed by condensation of eight isoprene units, $CH_2{=}C(CH_3){-}CH{=}CH_2$. Thus, the problem of carotenoid biosynthesis is analogous to that of other isoprenoids such as the terpenes or rubber, all of which contain the isoprene carbon skeleton as the basic repeating unit. Isoprene itself, however, has never been detected in plants and has no effect either on carotenogenesis in fungi (*28*), or on rubber formation in higher plants (*8*).

Recent developments in the field of rubber biosynthesis provide a rational framework for considering the more uncertain pathways of carotenoid synthesis. BONNER and co-workers (*2, 8, 9*) have shown that the guayule plant can make all of its rubber from the simple two carbon fragment acetate. Plants fed C^{14}-labeled acetate produce rubber in which all carbon atoms are marked. Synthesis apparently proceeds through the intermediate formation of the branched-chain 5-carbon compound β-methyl-crotonic acid, $(CH_3)C{=}CH{-}COOH$. The latter compound, like glycerol, acetate, acetoacetate and acetone, specifically stimulates rubber formation and appears to be the basic repeating unit. Analogy with some known enzyme systems in plants suggests that formation of β-methyl-crotonic acid from acetate first involves conversion of the acetate into an active 2-carbon fragment by combination with co-enzyme A to give acetyl co-enzyme A, condensation of two molecules of which would yield, after hydrolysis, acetoacetate. The branched-chain characteristic of the isoprene skeleton is introduced by joining to acetoacetate another acetyl-CoA molecule, to form a 6-carbon intermediate, from which carbon dioxide and water are removed to leave β-methyl-crotonyl Co-A (*9*). The latter, like acetyl Co-A, could condense, "head to tail", to form longer chains. Subsequent reduction and dehydration could then provide the conjugated double bond system characteristic of rubber and carotenoids (*70*).

The most extensive investigations into the nature of carotenoid precursors in plants have been conducted on *Phycomyces*, which, as SCHOPFER (*88, 93*) first pointed out, elaborates copious amounts of one predominant carotenoid, β-carotene. A related fungus, *Mucor*, presents similar advantages.

In *Phycomyces*, pigmentation and growth are favored especially by cultivation on media containing glucose as the carbon source. SCHOPFER and GROB (*90, 91*) have shown that the processes of growth and carotenogenesis in *Phycomyces* can be separated by substituting lactate for glucose as the only source of assimilable carbon. Under this condition, growth is considerably reduced, while carotenoid formation is apparently suppressed completely. Of particular significance is the further observation that a preferential enhancement of β-carotene formation is afforded by supplements of acetate to the basal lactate medium. Pyruvate, butyrate, palmitic, tartaric, and citric acids were less effective than acetate, whereas glycerol, acetone, and β-methyl-crotonic ester were individually ineffective. *Phycomyces* also synthesizes β-carotene on a medium containing acetate as the only carbon source, and on this basis, acetate has been suggested as the fundamental carotenoid precursor.

Support for this suggestion has come from the finding of GROB et al. (*47*) that both carbon atoms of acetate enter into the carbon skeleton of β-carotene. If *Phycomyces* or *Mucor* is grown on media containing C^{14}-labeled acetate as the only source of organic carbon, about 70% of the carbon atoms of carotene can be accounted for on the basis of the acetate provided. The 30% carbon deficit has not as yet been explained, but fixation of atmospheric CO_2 is one possibility (*47*), although probably a remote one. The recent experiments of GROB and BÜTLER (*45*) suggest the manner in which acetate may enter into the β-carotene molecule. By degradation of radioactive β-carotene synthesized by *Mucor* from labeled acetate, it was found that the four methyl side-groups of the main chain located between ionone end-groups, as well as the side methyl group (positions 5 and 5') on each ionone ring, are derived from the methyl groups of acetate. The carbon atoms immediately adjacent to the above-mentioned methyl groups stem from the carboxylic groups of acetate. Although the derivation of the remaining carbon atoms of β-carotene is yet to be determined, the above results are consistent with the mechanism proposed for utilization of acetate in rubber formation, in particular the formation of a branched C_5 repeating unit via a head-to-tail condensation of an active acetyl group with acetoacetate. In this connection it is noteworthy that pantothenic acid and related compounds, which are constituents of Co-enzyme A, have recently been shown to stimulate production of β-carotene in *Mucor* (*46*). It has also been suggested (cf. *35*) that acetate may be incorporated into the C_5 repeating unit via the tricarbocylic acid cycle. This would help to explain the failure to observe preferential utilization of acetate for carotenoid synthesis in the presence of glucose, in that glycolysis would more readily lead to formation of "active acetate", *i. e.* acetyl coenzyme A.

GOODWIN and his colleagues (*35, 41, 42*) have devoted particular attention to the nature of the C_5 repeating unit in *Phycomyces*. Their approach has been to culture *Phycomyces* on media containing sufficient glucose to permit moderate growth but production of only small amounts of carotenoids. Suspected carotenoid precursors were tested either during the growth period or in mature mycelial pads during a subsequent incubation period.

Leucine and β-methyl-crotonaldehyde showed a marked stimulation of carotene synthesis, whereas valine, α-hydroxy-isocaproic acid and

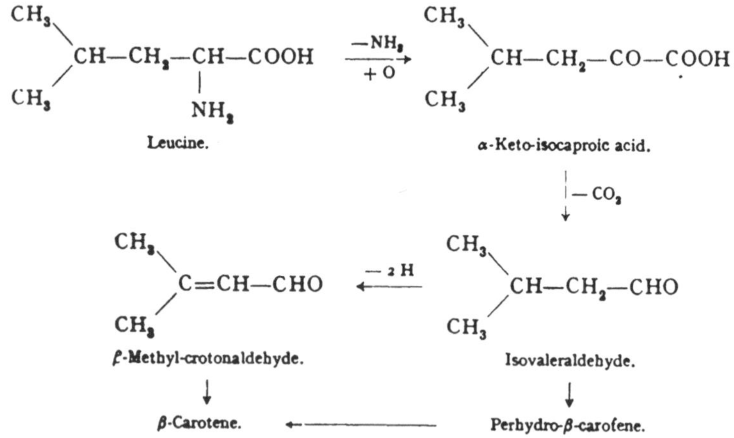

$$(CH_3)_2CH{-}CH_2{-}CH(NH_2){-}COOH \xrightarrow[+O]{-NH_3} (CH_3)_2CH{-}CH_2{-}CO{-}COOH$$

Leucine. → α-Keto-isocaproic acid.

$$\downarrow -CO_2$$

$$(CH_3)_2C{=}CH{-}CHO \xleftarrow{-2H} (CH_3)_2CH{-}CH_2{-}CHO$$

β-Methyl-crotonaldehyde. ← Isovaleraldehyde.

↓ ↓

β-Carotene. ← Perhydro-β-carotene.

Chart 2. Postulated Scheme for the Biosynthesis of β-Carotene from Leucine in *Phycomyces*.

α-keto-isocaproic acid were only slightly stimulatory. Isovaleraldehyde and β-methyl-crotonic acid were without significant effect, as were combinations of C_2- and C_3-fragments which might conceivably condense to give a C_5 repeating unit. From these observations GOODWIN (*35, 41, 42*) has concluded that the repeating unit is most likely β-methyl-crotonaldehyde. Leucine could be considered to give rise to this aldehyde and β-carotene as indicated in *Chart 2*.

Alternative routes have been proposed for the synthesis of β-methyl-crotonaldehyde from valine, but this, like the scheme outlined above, will require verification from tracer experiments or studies on isolated enzyme systems.

Although the marked effectiveness of a branched-chain compound such as leucine in promoting carotenogenesis suggests a rather direct conversion into a β-methyl-crotonyl repeating unit, it is conceivable that such compounds are incorporated into the carotenoid molecule *after* preliminary breakdown to simpler units. The latter seems probable from the tentative observation of CHICHESTER et al. (*16*) that there is no selective incorporation into β-carotene of the $C_{(1)}$ or $C_{(2)}$ carbon atoms of leucine.

Nevertheless, no evidence is available for deciding definitely that the repeating unit for carotenoid synthesis in fungi is β-methyl-crotonaldehyde or that the C_5-unit is derived exclusively from acetate. In this connection, the failure of acetoacetate or mixtures of acetone and acetaldehyde to stimulate carotenogenesis in *Phycomyces* is puzzling. It should be noted that these compounds and other likely precursors are also apparently inactive in *Neurospora* (*108*). On the other hand, glycerol is a highly effective carbon source for carotenogenesis in red yeasts (*28, 57*).

That the biosynthesis of β-carotene may involve incorporation of preformed ionone groups has been suggested by MACKINNEY (*73–75*) from the finding that addition of β-ionone to the basal sugar medium greatly increased production of β-carotene in *Phycomyces*. A recent study (*17*) has shown, however, that β-carotene formed on a medium containing radioactive fructose in the presence of β-ionone has the same specific activity as that of control cultures grown in the absence of β-ionone, suggesting that the ionone effect on carotenogenesis is probably an indirect one.

IV. Physiological Role of Fungal Carotenoids.

Specific physiological or biochemical functions have at various times been attributed to the carotenoid pigments of plants. At present the most securely established role is that of fucoxanthin as an accessory light absorber in the photosynthesis of diatoms and brown algae (cf. *7*). Carotenoids of fungi have been implicated mainly through circumstantial evidence: (a) as antioxidants in respiration (*15, 71, 76*); (b) as protective light filters for photosensitive enzymes (*104*); (c) as playing a role in sexual and reproductive processes, and (d) as photosensitizers in tropistic responses. Only the latter two roles will be considered further here.

1. Carotenoids in Sexual and Reproductive Processes.

That carotenoids may be involved in sexuality and reproduction has been suggested by the observations that in some fungi there is either an accumulation of carotenoids in reproductive structures or a selective distribution between sexes. In *Neurospora tetrasperma* the two mating types can readily be distinguished by the color of the conidia, one producing golden-orange conidia and the other, conidial masses more pinkish in color (*20*). Whether this involves qualitative differences in carotenoid composition has not been ascertained.

The (+) strains of *Mucor hiemalis* (*18*) and of *Phycomyces blakesleeanus* (*88, 89*) were found in early studies to contain more carotene than the (—) strains. In a more recent study Goodwin and Griffiths (*38*) found in general that (—) strains of *Phycomyces blakesleeanus* produce about twice as much carotene as do the (+) strains. A (+) strain of the variety *gracilis*, however, behaved as (—) strain as far as carotene content was concerned, an observation which led these workers to doubt that carotene production had any direct function in the reproduction of *Phycomyces*.

In *Allomyces* the ability to synthesize carotenoids as well as colorless polyenes is restricted to the sexual (gametophytic) phase of the life cycle and specifically to the male gametangia. The asexual phase and the female gametangia are apparently devoid of C_{40}-polyenes, whereas the orange colored male gametangia contain predominantly γ-carotene, accompanied by small amounts of β-carotene, lycopene, and phytofluene (*22, 102*). Preferential synthesis of γ-carotene in the male gametangia of *Allomyces* appears to be the consequence of a more fundamental change accompanying sexual differentiation, rather than indicating a specific role of carotenoids in sexuality. In liquid cultures of *Allomyces* the male gametes are only very pale colored, and in diphenylamine-treated cultures, γ-carotene formation may be almost completely suppressed without apparently altering the reproductive capacity (*100, 102*). It is pertinent to recall that carotenoid accumulation in sexual structures is the exception rather than the rule among fungi. Furthermore, albino strains of the red bread mold *Neurospora* do not show a diminished capacity for sexual reproduction (*56*). Such observations do not, of course, eliminate the possibility that carotenoids may function in trace amounts in the sexual process.

It is interesting to note that in the rust fungus *Gymnosporangium juniperi-virginianae* the kinds and amounts of carotenoid pigments (γ-carotene and β-carotene in the ratio of 2 : 1) are essentially identical in two different stages of the life cycle upon different hosts (*107*).

2. Carotenoids as Photosensitizers.

Early evidence (*12*, *14*) for the participation of β-carotene as the primary light absorber in the phototropic responses of *Pilobolus* and *Phycomyces* was based mainly upon the similarity in the spectral sensitivity curve (action spectrum) for phototropic bending and the absorption curve of β-carotene, the major pigment present in the light-sensitive sporangiophores of these fungi (cf. *103*). Re-examination of phototropism in higher plant seedlings, where a similar correlation has been drawn, has directed attention also to the flavins, which have somewhat similar absorption spectra in the visible region. GALSTON (*29*) has demonstrated that riboflavin, which is present throughout the *Avena* coleoptile, exerts a strong sensitizing effect on the decomposition of the growth hormone indole-acetic acid in the light, and he has suggested that the bending response of the coleoptile base "probably involves riboflavin as the receptor pigment and auxin destruction as the mechanism". Riboflavin could, with equal justification, be evoked as the sensitizer for phototropic responses in fungi. Single-peaked action spectra have been reported for *Pilobolus* and *Phycomyces*, a finding more in accord with absorption by riboflavin than by β-carotene. It is noteworthy that sporangiophores of *Phycomyces* in which the β-carotene content has been greatly reduced by growth in the presence of diphenylamine, still show a strong phototropic response (*34*), as do those made "carotene-free" by growth on a medium containing lactate (*85*). Carotenoid-deficient albino strains of *Neurospora crassa* have been found to display the same tropistic response to blue light (*53*) that is characteristic of the perithecial necks of wild type strains (*3*).

Such observations, although suggestive, are not conclusive, since it may be questioned whether carotenoids are completely absent from the photosensitive organs. Further studies of albino mutants of *Neurospora* as well as those in which the capacity for riboflavin synthesis has been impaired might provide definite clues as to the identity of the sensitizing pigment in phototropic responses of fungi.

V. Conclusion.

The capacity for carotenoid synthesis found in some fungi parallels in quantity and in diversity that of higher plants. Inasmuch as some unique carotenoids have been encountered among the relatively few fungi investigated by modern methods, this group of lower plants seems to present a fertile field for the descriptive biochemist.

Fungi, like bacteria, have provided favorable experimental material for studies on the biosynthesis of carotenoid pigments, although agreement has yet to be reached on the pathways involved. At the present time it seems most probable that the carotenoid molecule is elaborated by sequential condensation of acetate-derived β-methyl-crotonyl units, at least to the level of a C_{20}-compound, following which dimerization of two similar C_{20}-units could occur. Whether the unsaturated character of the common carotenes is acquired at the C_{40} level or at an earlier stage is as yet an unresolved question. That some carotenoids, *e. g.*, β-carotene, may be formed independently of other C_{40}-polyenes, seems highly probable.

The functional significance of carotenoids in fungi is likewise uncertain and it will remain for future studies to decide whether these pigments may possess specific physiological roles.

References.

1. AKAKI, M. and R. ISHII: *Monilia sitophila*. IV. The Pigment of *Monilia sitophila*. J. Ferment. Technol. (Japan) **28**, 63 (1950) [Chem. Abstr. **47**, 2824 (1953)].
2. ARREGUIN, B., J. BONNER and B. J. WOOD: Studies on the Mechanism of Rubber Formation in the Guayule. III. Experiments with Isotopic Carbon. Arch. Biochemistry **31**, 234 (1951).
3. BACKUS, M. P.: Phototropic Response of Perithecial Necks in *Neurospora*. Mycologia **29**, 383 (1937).
4. BEADLE, G. W.: Biochemical Genetics. Chem. Reviews **37**, 15 (1945).

4a. — Some Recent Developments in Chemical Genetics. Fortschr. Chem. organ. Naturstoffe **5**, 300 (1948). Cf. also p. 466 of the present Volume.

5. BEADLE G. W. and E. L. TATUM: Genetic Control of Biochemical Reactions in *Neurospora*. Proc. Nat. Acad. Sci. (USA) **27**, 499 (1941).
6. BERNHARD, K. und H. ALBRECHT: Untersuchung der Lipidsynthese des Schimmelpilzes *Phycomyces blakesleeanus* mit Hilfe der Isotopentechnik. Helv. Chim. Acta **31**, 2214 (1948).
7. BLINKS, L. R.: The Photosynthetic Function of Pigments other than Chlorophyll. Annu. Rev. Plant Physiol. **5**, 93 (1954).
8. BONNER, J. and B. ARREGUIN: The Biochemistry of Rubber Formation in the Guayule. I. Rubber Formation in Seedlings. Arch. Biochem. **21**, 109 (1949).
9. BONNER, J., M. W. PARKER and J. C. MONTERMOSO: Biosynthesis of Rubber. Science (Washington) **120**, 549 (1954).
10. BONNER, J., A. SANDOVAL, Y. W. TANG and L. ZECHMEISTER: Changes in Polyene Synthesis Induced by Mutation in a Red Yeast (*Rhodotorula rubra*). Arch. Biochemistry **10**, 113 (1946).
11. BRAUNER, L.: Tropisms and Nastic Movements. Annu. Rev. Plant Physiol. **5**, 163 (1954).
12. BÜNNING, E.: Phototropismus und Carotinoide. III. Weitere Untersuchungen an Pilzen und höheren Pflanzen. Planta **27**, 583 (1937).
13. CANTINO, E. C. and M. T. HYATT: Carotenoids and Oxidative Enzymes in the Aquatic Phycomycetes *Blastocladiella* and *Rhizophlyctis*. Amer. J. Bot. **40**, 688 (1953).
14. CASTLE, E. S.: Photic Excitation and Phototropism in Single Plant Cells. Cold Spring Harbor Sympos. Quant. Biol. **3**, 224 (1935).

15. CHAMPEAU, M. F. et P. J. LUTERAAN: Sur quelques données histochimiques et physiologiques concernant des champignons levuriformes. Ann. Parasit. **21**, 345 (1946).

16. CHICHESTER, C. O., T. NAKAYAMA, G. MACKINNEY and T. W. GOODWIN: On Incorporation of Leucine-Carbon into Carotene by *Phycomyces*. Abstr. Pacific Slope Biochem. Conf., Berkeley (1954).

17. CHICHESTER, C. O., P. S. WONG and G. MACKINNEY: On the Biosynthesis of Carotenoids. Plant Physiol. **29**, 238 (1954).

18. CHODAT, F. et W. H. SCHOPFER: Carotène et séxualité. C. R. Soc. Phys. Hist. Nat. Genève **44**, 176 (1927).

19. DEVENTER, W. F. VAN: De Kleurstoffen van *Neurospora* (*Monilia*) *sitophila* SHEAR et DODGE. Thesis, Utrecht 1930.

20. DODGE, B. O.: Some Problems in the Genetics of the Fungi. Science (Washington) **90**, 379 (1939).

21. EGLE, K.: Über das Pigment von *Anthurus aseroeformis* MACALPINE. Planta **38**, 233 (1950).

22. EMERSON, R. and D. L. FOX: γ-Carotene in the Sexual Phase of the Aquatic Fungus *Allomyces*. Proc. Roy. Soc. (London) **128**, 275 (1940).

23. FINK, H. und E. ZENGER: Zur Biochemie der Farbstoffe der roten Hefe. Wschr. Brauerei **51**, 129 (1934).

24. FOX, D. L.: The Carotenoid Pigments of Cryptogams. Chron Bot. **7**, 196 (1942).

25. FOX, D. L., D. M. UPDEGRAFF and G. D. NOVELLI: Carotenoid Pigments in the Ocean Floor. Arch. Biochemistry **5**, 1 (1944).

26. FRIEND, J. and T. W. GOODWIN: Carotenogenesis. XII. Effect of Temperature and Thiamine Concentration on Carotenogenesis by *Phycomyces blakesleeanus*. Biochemic. J. **57**, 434 (1954).

27. FROMAGEOT, C. et J. L. TCHANG: Sur les pigments caroténoïdes de *Rhodotorula sanniei*. Arch. Mikrobiol. **9**, 424 (1938).

28. — — Sur la synthèse des pigments caroténoïdes par *Rhodotorula sanniei*. Arch. Mikrobiol. **9**, 434 (1938).

29. GALSTON, A. W.: Phototropism. Bot. Rev. **16**, 361 (1950).

30. GARTON, G. A., T. W. GOODWIN and W. LIJINSKY: Studies in Carotenogenesis. I. General Conditions Governing β-Carotene Synthesis by the Fungus *Phycomyces blakesleeanus* BURGEFF. Biochemic. J. **48**, 154 (1951).

31. GOODWIN, T. W.: Carotenoids and Reproduction. Biol. Rev. **25**, 391 (1950).

32. — Fungal Carotenoids. Bot. Rev. **18**, 291 (1952).

33. — The Comparative Biochemistry of the Carotenoids. London: Chapman and Hall 1952.

34. — Studies in Carotenogenesis. III. Identification of the Minor Polyene Components of the Fungus *Phycomyces blakesleeanus* and a Study of Their Synthesis Under Various Cultural Conditions. Biochemic. J. **50**, 550 (1952).

35. — The Biogenesis of Carotenoids. J. Sci. Food Agric. **4**, 209 (1953).

36. — Studies in Carotenogenesis. VIII. The Carotenoids Present in the Basidiomycete *Dacromyces stillatus*. Biochemic. J. **53**, 538 (1953).

37. GOODWIN, T. W. and L. A. GRIFFITHS: Carotene Synthesis by Some Naturally Occurring Mutants of *Phycomyces blakesleeanus* and by *Phycomyces nitens*: Inhibition of Carotenogenesis by Streptomycin. Biochemic. J. **51**, XXXIII (1952).

38. — — Studies in Carotenogenesis. V. Carotene Production by Mutants of *Phycomyces blakesleeanus* and by *Phycomyces nitens*. Biochemic. J. **52**, 499 (1952).

39. GOODWIN, T. W. and M. JAMIKORN: Biosynthesis of Carotenes in Ripening Tomatoes. Nature (London) **170**, 104 (1952).

40. GOODWIN, T. W., M. JAMIKORN and J. S. WILLMER: Studies in Carotenogenesis. VII. Further Observations Concerning the Action of Diphenylamine in Inhibiting the Synthesis of β-Carotene in *Phycomyces blakesleeanus*. Biochemic. J. 53, 531 (1953).

41. GOODWIN, T. W. and W. LIJINSKY: Studies in Carotenogenesis. II. Carotene Production by *Phycomyces blakesleeanus*: the Effect of Different Amino-acids When Used in Media Containing Low Concentrations of Glucose. Biochemic. J. 50, 268 (1951).

42. GOODWIN, T. W., W. LIJINSKY and J. S. WILLMER: Studies in Carotenogenesis. VI. The Effect of Some Possible Carotene Precursors on Growth, Lipogenesis and Carotenogenesis in the Fungus *Phycomyces blakesleeanus*. Biochemic. J. 53, 208 (1953).

43. GOODWIN, T. W. and H. G. OSMAN: ζ-Carotene. Arch. Biochem. Biophys. 47, 215 (1953).

44. GOODWIN, T. W. and J. S. WILLMER: Studies in Carotenogenesis. IV. Nitrogen Metabolism and Carotene Synthesis in *Phycomyces blakesleeanus*. Biochemic. J. 51, 213 (1952).

45. GROB, E. C. und R. BÜTLER: Über die Biosynthese des β-Carotins bei *Mucor hiemalis* WEHMER. Die Beteiligung der Essigsäure am Aufbau der Carotinmolekel, insbesondere in den Jonongruppierungen, untersucht mit Hilfe von ^{14}C-markierter Essigsäure. Helv. Chim. Acta 37, 1908 (1954).

46. GROB, E. C., V. GRUNDBACHER und W. H. SCHOPFER: Der Einfluß der Pantothensäure, des Pantethins und des phosphorylierten Pantethins auf die Carotinbildung bei *Mucor hiemalis*. Experientia 10, 378 (1954).

47. GROB, E. C., G. G. PORETTI, A. v. MURALT et W. H. SCHOPFER: Recherches sur la biosynthèse des caroténoïdes chez un microorganisme. Production de caroténoïdes marqués par *Phycomyces blakesleeanus*. Experientia 7, 218 (1951).

48. HANNA, C. and T. J. BULAT: Pigment Study of *Dacrymyces ellisii*. Mycologia 45, 143 (1953).

49. HASKINS, R. H. and W. H. WESTON: Studies in the Lower Chytridiales. I. Factors Affecting Pigmentation, Growth, and Metabolism of a Strain of Karlingia (*Rhizophlyctis rosea*). Amer. J. Bot. 37, 739 (1950).

50. HAXO, F.: The Carotenoid Pigments of *Neurospora*. Thesis. Stanford University. 1947.

51. — Studies on the Carotenoid Pigments of *Neurospora*. I. Composition of the Pigment. Arch. Biochemistry 20, 400 (1949).

52. — Carotenoids of the Mushroom *Cantharellus cinnabarinus*. Bot. Gaz. 112, 228 (1950).

53. — Carotenoid Formation by Mutant Strains of *Neurospora crassa*. Biol. Bull. 103, 286 (1952).

54. — (unpublished).

55. HAXO, F. and G. TURIAN: Additional Polyene Components of *Neurospora crassa* (unpublished).

56. HUNGATE, M. V.: A Genetic Study of Albino Mutants of *Neurospora crassa*. Thesis, Stanford University. 1945.

57. ISHII, R.: The Carotene Production by Microorganisms. VI. The Influence of Culture Conditions on the Carotene Production of *Torula shibitana*. J. Ferment. Technol. (Japan) 30, 390 (1951) [Chem. Abstr. 48, 3449 (1954)].

58. JENKINS, J. A. and G. MACKINNEY: Inheritance of Carotenoid Differences in the Tomato Hybrid Yellow X Tangerine. Genetics 38, 107 (1953).

59. KARRER, P. and E. JUCKER: Carotenoids. New York: Elsevier. 1950.

60. KARRER, P. und E. KRAUSE-VOITH: Einige weitere Beobachtungen bezüglich der Verbreitung der Carotinoide, insbesondere Carotinoid-epoxyde. Helv. Chim. Acta **31**, 802 (1947).

61. KARRER, P. und J. RUTSCHMANN: Ein Carotinfarbstoff von neuartigem Charakter aus roter Hefe (*Torula rubra*). Helv. Chim. Acta **26**, 2109 (1943).

62. — — Torularhodin. Helv. Chim. Acta **29**, 355 (1946).

63. KARRER, P. und U. SOLMSSEN: Die Carotinoide der Purpurbakterien. I. Helv. Chim. Acta **18**, 1306 (1935).

64. — — Die Carotinoide der Purpurbakterien. II. Über Rhodoviolascin. Helv. Chim. Acta **19**, 3 (1936).

65. KARRER, P., U. SOLMSSEN und H. KOENIG: Carotinoide aus Purpurbakterien. IV. Helv. Chim. Acta **21**, 454 (1938).

66. KHARASCH, M. S., E. A. CONWAY and W. BLOOM: Some Chemical Factors Influencing Growth and Pigmentation of Certain Microörganisms. J. Bacteriol. **32**, 533 (1936).

66 a. KOE, B. K. and L. ZECHMEISTER: Preparation and Spectral Characteristics of all-*trans* and a *cis* Phytofluene. Arch. Bioch. Biophys. **46**, 100 (1953).

67. KOHL, F. G.: Untersuchungen über das Carotin und seine physiologische Bedeutung in der Pflanze. Leipzig: Bornträger. 1902.

68. LEDERER, E.: Sur les caroténoïdes d'une levure rouge (*Torula rubra*). C. R. hebd. Séances Acad. Sci. **197**, 1694 (1933).

69. — Sur les caroténoïdes des cryptogames. Bull. soc. chim. biol. (Paris) **20**, 611 (1938).

70. LIPMANN, F.: Development of the Acetylation Problem, a Personal Account. Science (Washington) **120**, 855 (1954).

71. LUTERAAN, P. J. et J. CHOAY: Données expérimentales sur l'origine, la formation, et le rôle des pigments caroténoïdes chez les *Rhodotorula*. Ann. Parasitol. **22**, 89 (1947).

72. MACKINNEY, G.: Carotenoids. Annu. Rev. Biochem. **21**, 473 (1952).

73. MACKINNEY, G., C. O. CHICHESTER and P. S. WONG: Carotenoids in *Phycomyces*. J. Amer. Chem. Soc. **75**, 5428 (1953).

74. MACKINNEY. G., T. NAKAYAMA, C. D. BUSS and C. O. CHICHESTER: Carotenoid Production in Phycomyces. J. Amer. Chem. Soc. **74**, 3456 (1952).

75. — — — — Biosynthesis of Carotene in *Phycomyces*. J. Amer. Chem. Soc. **75**, 236 (1953).

76. MÉRY, J.: Action de quelques facteurs sur la croissance et la pigmentation des Rhodotorulacées. Ann. Parasitol. **24**, 180 (1949).

77. MRAK, E. M., H. J. PHAFF and G. MACKINNEY: A Simple Test for Carotenoid Pigments in Yeasts. J. Bacteriol. **57**, 409 (1949).

78. NAKAYAMA, T., G. MACKINNEY and H. J. PHAFF: Carotenoids in Asporogenous Yeasts. J. Microbiol. Serol. (Antonie van Leeuwenhoek) **20**, 217 (1954).

79. NASH, H. A., F. W. QUACKENBUSH and J. W. PORTER: Studies on the Structure of ζ-Carotene. J. Amer. Chem. Soc. **70**, 3513 (1948).

80. NIEL, C. B. VAN and J. H. C. SMITH: Studies on the Pigments of the Purple Bacteria. Arch. Mikrobiol. **6**, 219 (1935).

81. NORRIS, P. M. and F. HAXO: (unpublished).

81 a. PETRACEK, F. J. and L. ZECHMEISTER: Stereoisomeric Phytofluenes. J. Amer. Chem. Soc. **74**, 184 (1952).

82. POLGÁR, A., C. B. VAN NIEL and L. ZECHMEISTER: Studies on the Pigments of the Purple Bacteria. II. A Spectroscopic and Stereochemical Investigation of Spirilloxanthin. Arch. Biochemistry **5**, 243 (1944).

83. PORTER, J. W. and R. E. LINCOLN: *Lycopersicon* Selections Containing a High Content of Carotene and Colorless Polyenes. II. The Mechanism of Carotene Biosynthesis. Arch. Biochemistry **27**, 390 (1950).

84. RABOURN, W. J., F. W. QUACKENBUSH and J. W. PORTER: Isolation and Properties of Phytoene. Arch. Biochem. Biophys. **48**, 267 (1954).

85. REINERT, J.: Über die Bedeutung von Carotin und Riboflavin für die Lichtreizaufnahme bei Pflanzen. Naturwiss. **39**, 47 (1952).

86. SAPERSTEIN, S. and M. P. STARR: The Ketonic Carotenoid Canthaxanthin Isolated from a Colour Mutant of *Corynebacterium michiganense*. Biochemic. J. **57**, 273 (1954).

87. SAPERSTEIN, S., M. P. STARR and J. A. FILFUS: Alterations in Carotenoid Synthesis Accompanying Mutation in *Corynebacterium michiganense*. J. Gen. Microbiol. **10**, 85 (1954).

88. SCHOPFER, W. H.: Étude et identification d'un caroténoïde de Champignon. C. R. Séances Soc. Biol. **118**, 3 (1935).

89. — Plants and Vitamins. Waltham, Mass.: Chronica Botanica. 1943.

90. SCHOPFER, W. H. et E. C. GROB: Recherches sur la biosynthèse des caroténoïdes chez un Microorganisme. Experientia **6**, 419 (1950).

91. — — Sur la biosynthèse du β-carotène par *Phycomyces* cultivé sur un milieu contenant de l'acétate de sodium comme unique source de carbone. Experientia **8**, 140 (1952).

92. SCHOPFER, W. H. et A. JUNG: Recherches sur l'activité vitaminique A du thalle d'une Mucorinée. C. R. Séances Soc. Biol. **120**, 1093 (1935).

93. SCHOPFER, W. H. et V. KOCHER: Sur la cristallisation du carotène de *Phycomyces*. Actes Soc. Helv. Sci. Nat. p. 320 (1936).

94. SHENG, T. C. and G. SHENG: Genetic and Non-genetic Factors in Pigmentation of *Neurospora crassa*. Genetics **37**, 264 (1952).

95. SMITS, B. L. and W. J. PETERSON: Carotenoids of Telial Galls of *Gymnosporangium juniperi-virginianae* LK. Science (Washington) **96**, 210 (1942).

96. STRAIN, H. H.: The Pigments of Algae. Manual of Phycology. Waltham, Mass.: Chronica Botanica. 1951.

97. TANG, Y. W., J. BONNER and L. ZECHMEISTER: Some Further Experiments on Red Yeast Polyenes. Arch. Biochemistry **21**, 455 (1949).

98. TROMBLY, H. H. and J. W. PORTER: Additional Carotenes and a Colorless Polyene of Lycopersicon Species and Strains. Arch. Biochem. Biophys. **43**, 443 (1953).

99. TURIAN, G.: Recherches sur la biosynthèse des caroténoïdes chez un Bacille paratuberculeux. 3. Inhibition de la pigmentation par la diphénylamine. Helv. Chim. Acta **33**, 1988 (1950).

100. — Caroténoïdes et différentiation sexuelle chez *Allomyces*. Experientia **8**, 302 (1952).

101. TURIAN, G. and F. HAXO: Further Use of Diphenylamine for the Study of Carotenoid Biosynthesis in *Mycobacterium phlei*. J. Bacteriol. **63**, 690 (1952).

102. — — Minor Polyene Components in the Sexual Phase of *Allomyces javanicus*. Bot. Gaz. **115**, 254 (1954).

103. WALD, G.: The Photoreceptor Function of the Carotenoids and Vitamin A. Vitamins and Horm. **1**, 195 (1943).

104. WENT, F. A.: Über den Einfluß des Lichtes auf die Entstehung des Carotins und auf die Zersetzung der Enzyme. Rec. trav. bot. Néerl. **1904**, 1.

105. WILLSTAEDT, H.: Pilzfarbstoffe. III. Über die Carotinoide einiger Canthareilus-Arten. Svensk Kem. Tidskr. **49**, 318 (1937).

106. WISSELINGH, C. VAN: Über die Nachweisung und das Vorkommen von Carotinoiden in der Pflanze. Flora **107**, 371 (1915).

107. WOLF, F. T. and F. A. WOLF: The Carotenoid Pigments of the Cedar Rust Fungus (unpublished).

108. ZALOKAR, M.: Studies on Biosynthesis of Carotenoids in *Neurospora crassa*. Arch. Biochem. Biophys. **50**, 71 (1954).

109. ZECHMEISTER, L.: Carotinoide. Berlin: J. Springer. 1934.

110. — *cis-trans* Isomerization and Stereochemistry of Carotenoids and Diphenylpolyenes. Chem. Rev. **34**, 267 (1944).

111. ZECHMEISTER, L. und L. v. CHOLNOKY: Lycoxanthin und Lycophyll, zwei natürliche Derivate des Lycopins. Ber. dtsch. chem. Ges. **69**, 422 (1936).

112. ZECHMEISTER, L. and F. HAXO: Phytofluene in *Neurospora*. Arch. Biochemistry **11**, 539 (1946).

113. ZECHMEISTER, L. and B. K. KOE: Stepwise Dehydrogenation of the Colorless Polyenes Phytoene and Phytofluene with N-Bromosuccinimide to Carotenoid Pigments. J. Amer. Chem. Soc. **76**, 2923 (1954).

114. ZECHMEISTER, L. and A. POLGÁR: On the Occurrence of a Fluorescing Polyene with a Characteristic Spectrum. Science (Washington) **100**, 317 (1944).

115. ZECHMEISTER, L. and A. SANDOVAL: Phytofluene. J. Amer. Chem. Soc. **68**, 197 (1946).

116. ZOPF, W.: Die Pilze. Breslau: Trewendt. 1890.

(*Received, January 19, 1955.*)

The Pyrrolizidine Alkaloids.

By F. L. WARREN, Pietermaritzburg, Natal, South Africa.

Contents.

I. Introduction: Origin, Occurrence and Nature of Pyrrolizidine Alkaloids.

Considerable advances have been made in the study of the pyrrolizidine alkaloids during the few years which have elapsed since the last admirable reviews on this subject were made by HENRY (*87*) in 1949, and by LEONARD (*111*) in 1950, and even since ADAMS (*1*) in 1953 gave his excellent summary of the then significant recent findings. The present work attempts to correlate the wealth of chemical evidence which has accumulated on the structures of the alkaloids, as well as their acid and basic fission products, many of which have now been completely elucidated Furthermore, precise methods have revealed previously supposed individuals to be mixtures or identical with known substances; and where this evidence is seemingly conclusive the correction has been made and new lists compiled.

These alkaloids were first detected in *Senecio*, one of the largest genera of the family Compositae, and have since been found in Leguminosae and Boraginaceae, as shown in *Table 1*. No identification of these bases has yet been made amongst Monocotyledons. The three families in which these alkaloids have been found show little relationship with each other. Within the families, however, the plants which have been found to contain alkaloids are more closely related. The genera of Leguminosae and Compositae containing this type of alkaloid belong to one tribe within each of these families, and the alkaloid-containing members of Boraginaceae are restricted to two tribes of that family.

Table 1. List of Plant Genera Containing Pyrrolizidine Alkaloids.

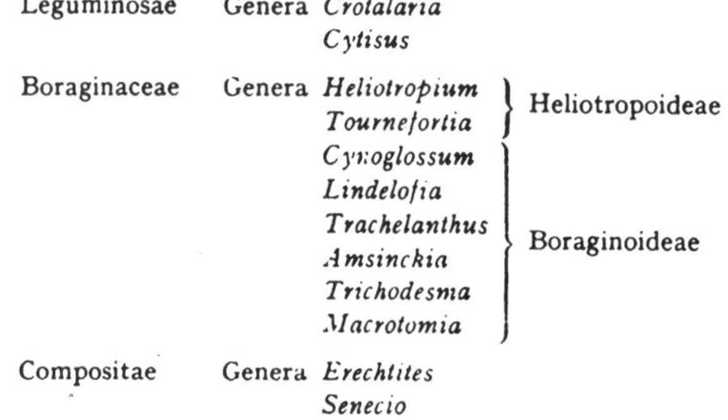

Family		Genus	Subfamily
Leguminosae	Genera	*Crotalaria*	
		Cytisus	
Boraginaceae	Genera	*Heliotropium*	Heliotropoideae
		Tournefortia	
		Cynoglossum	Boraginoideae
		Lindelofia	
		Trachelanthus	
		Amsinckia	
		Trichodesma	
		Macrotomia	
Compositae	Genera	*Erechtites*	
		Senecio	

The *Senecio* alkaloids assumed from the very beginning a special significance in that they were the cause of liver cirrhosis in cattle, and the finding of the same alkaloids in certain *Crotalaria* species had agricultural implications in that several *Crotalaria* species are grown specifically for cattle feed (*39*).

Poisoning of cattle, horses and sheep by *Senecio* species has been found in many parts of the globe where it has been described under various local names. Adami (cf. *32*) in 1902 was the first to describe the pathology of the Pictou disease of Nova Scotia as an extreme condition of cirrhosis of the liver, whilst in the same year Gilruth (cf. *32*) associated Winton disease of New Zealand with poisoning by *Senecio* plants, although ragwort (*S. jacobaea*) had been suspected in Canada as far back as 1882 (*32*). The year following Gilruth's findings Chance attributed Molteno's disease in cattle grazing in the South African veld to *S. burchellii*, whilst planned feeding experiments conducted by Pethick (cf. *32*) established conclusively that Pictou disease was due to *S. jacobaea*. In 1911 these diverse concepts were carefully co-ordinated by Cushny (*56*, see also *58*, *67*) who, using pure senecifoline nitrate prepared by Watt (*192*) from *S. latifolius*, established that Pictou, Winton and Molteno diseases were more or less chronic poisoning by *Senecio* alkaloids. Since then *Senecio* poisoning has been suspected in the upper plains and in the foothills of Kurdistan where it is known as 'Gulilkah Zar' (little yellow flower) (*91*); and it occurs in Norway as 'Sirasyke' disease due also to *S. jacobaea* (*58*). The alkaloid content was shown (*33, 91*) to be highest in the young *Senecios* and before flowering; and it is on the succulent small plant that the cattle graze when rains are late and the spring grasses are delayed. Bread poisoning in South Africa has also been attributed to contamination of wheat by seeds of *Senecio* species (*33, 194*).

Nature of the Pyrrolizidine Alkaloids.

The pyrrolizidine alkaloids comprise a group of basic substances which have a methylpyrrolizidine nucleus with one or more hydroxyl groupings esterified by organic acids. The alkaloids found in the different plants which contain this type of nitrogenous base are set out in *Table 2*.

Table 2. The Alkaloid Content of the Plant Species which have been Investigated.

Compositae	Alkaloid	Reference
Cacalia hastata = *Senecio sagittatus* SCH. BIP.	Hastacine	(*94*)
Erechtites hieracifolia (L) RAFIN.	Senecionine Seneciphylline Alkaloid $C_{20}H_{17(19)}O_6N$	(*55*)
E. quadridentata D. C.	Senecionine Seneciphylline Retrorsine	(*55*)
Senecio adnatus D. C.	Platyphylline	(*189*)
S. ampullaceus HOOK.	Seneciphylline Retrorsine Senecionine	(*6, 15, 191*)
S. aquaticus HILL	Jacodine	(*40*)
S. aureus L.	Senecionine	(*89, 37, 121*)
S. brachypodus D. C.	Rosmarinine	(*159*)
S. brasiliensis D. C.	Brasilinecine Alkaloid, $C_{18}H_{25}O_5N$	(*70, 150*)
S. campestris D. C. var. *maritimus*	Campestrine	(*40*)
S. carthamoides GREENE	Seneciphylline Senecionine	(*6, 15*)
S. cineraria D. C.	Senecionine Jacobine Jacodine	(*5, 37*)
S. douglasii D. C.	Retrorsine Seneciphylline Riddelliine	(*6, 15*)

Compositae	Alkaloid	Reference
S. eremophilus RICHARDS	Seneciphylline	(*6, 15*)
	Riddelliine	
	Senecionine	
	Retrorsine	
S. erucifolius L.	Alkaloid, $C_{18}H_{27}O_{6}N$	(*40*)
S. fuchsii C. C. GMEL.	Fuchsisenecionine	(*148*)
	Alkaloid, $C_{9}H_{15}O_{2}N$	
S. glabellus D. C.	Senecionine	(*25*)
S. glaberrimus D. C.	Retrorsine	(*40*)
S. graminifolius JACQ.	Retrorsine	(*182*)
	Graminifoline	
S. hygrophilus DYER and SM.	Platyphylline	(*159*)
	Rosmarinine	
	Alkaloid, $C_{18}H_{27}O_{6}N$	
S. ilicifolius THUNB.	Retrorsine	(*181, 55*)
	Senecionine	
	Seneciphylline	
S. integerrimus NUTT.	Integerrimine	(*121*)
	Senecionine	
S. isatideus D. C.	Isatidine	(*40, 179*)
	Retrorsine	
S. jacobaea L.	Jacobine	(*118, 40, 119, 37, 42*)
	Seneciphylline	
	Jaconine	
	Jacozine	
	Jacoline	
S. kaempferi D. C.		
= *Ligularia tussilaginea*	Senecioic acid	(*166, 34*)
	(= Tiglic acid)	
S. kirkii HOOK.	Senkirkine	(*43*)
S. longilobus BENTH.	Seneciphylline	(*6, 15, 191*)
	Retrorsine	
	Riddelliine	
S. mikanioides (WALP) OTTO	Mikanoidine	(*120, 37*)
	Senecioic acid	
S. othonnae BIEB.	Otosenine	(*195*)
S. paludosus L.	Jacobine	(*40*)
	Jacodine	
S. palustris HOOK.	Alkaloid, $C_{18}H_{27}O_{5}N$	(*42*)
*S. parksii**	Retrorsine	(*6*)
	Riddelliine	
	Alkaloid	
S. pauciligulatus DYER and SM.	Rosmarinine	(*159*)

* This has not been raised to the rank of species, and is best designated *S. riddellii* T. and G., var. *Parksii* CORY described in Rhodore **45**. 164 (1943). (Private communication from Dr. LLOYD H. SHINNERS and from Sir EDWARD SALISBURY.) (Added in proof.)

Compositae	Alkaloid	Reference
S. platyphyllus D. C.	Platyphylline Seneciphylline and other alkaloids	(*153, 102, 101*)
S. pseudo-arnica LESS.	Senecionine	(*121*)
S. pterophorus D. C.	Retrorsine Senecionine Seneciphylline	(*177, 55*)
S. renardi WINKL.	Seneciphylline Renardine Otosenine	(*59*)
S. retrorsus BENTH.	Isatidine Retrorsine	(*119, 38*)
S. riddellii TORR. and GRAY	Riddelliine	(*10*)
S. rosmarinifolius L.	Rosmarinine	(*181*)
S. ruderalis HARV.	Retrorsine	(*108*)
S. ruwenzoriensis S. MOORE	Ruwenine Ruzorine	(*162*)
S. sagittatus (see *Cacalia hastata* SCH. BIP.)		
S. sarracenius L.	Sarricine Alkaloid, $C_{13}H_{21}O_3N$ Alkaloid, $C_8H_{13}ON$ Alkaloid, C_5H_9ON	(*61, 40*)
S. sceleratus SCHWEIKERDT	Isatidine Retrorsine Sceleratine	(*187*)
S. solidago RYDB.	Alkaloid	(*120*)
S. spartioides T. and G.	Seneciphylline Spartioidine Senecionine	(*121, 36*)
S. squalidus L.	Senecionine Integerrimine	(*105*)
S. stenocephallus MAXIM	Seneciphylline	(*101*)
S. sylvaticus L.	Silvasenecine	(*148*)
S. venosus HARV.	Retrorsine	(*40*)
S. viscosus L.	Senecionine and another alkaloid	(*36*)
S. vulgaris L.	Senecine Senecionine Seneciphylline Retrorsine Riddelliine (?)	(*75, 100, 36, 120*)

Leguminosae	Alkaloid	Reference
Crotalaria dura WOOD and EVANS	Dicrotaline	(*123*)
C. globifera E. MEY.	Dicrotaline	(*123*)
C. grantiana HARV.	Grantianine	(*3*)

Leguminosae	Alkaloid	Reference
C. incana L.	Integerrimine	(*25*)
C. retusa L.	Monocrotaline	(*76, 18*)
C. spectabilis ROTH.	Monocrotaline	(*149, 18*)
C. striata SCHRANK	Alkaloid (?)	(*123*
C. usaramoensis E. G. BAKER	Usaramoensine	(*25*)
Cytisus laburnum = *Laburnum laburnum* DÖRFLER	Laburnine	(*72*)

Boraginaceae	Alkaloid	Reference
Cynoglossum viridiflorum WILLD.	Viridiflorine	(*136*)
Heliotropium europaeum L.	Heliotrine Supinine Lasiocarpine Lasiocarpic ester of Heliotridine Heliotric ester of Supinidine	(*54, 53*)
H. lasiocarpum FISCH. and MEY. = *H. eichwaldi* STEUD.	Heliotrine Lasiocarpine	(*125, 126, 129*)
H. supinum L.	Supinine	(*141, 53*)
Lindelofia anchusoides = *Paracaryum heliocarpum* KERN.	Lindelofine Lindelofamine	(*106*)
Macrotomia echoides BOISS.	Makrotomine	(*144*)
Tournefortia sibirica L.	Turneforcine	(*139*)
Trachelanthus korolkovi LIPSKY	Trachelanthamine Trachelanthine	(*137*)
Trichodesma incanum BUNGE	Trichodesmine	(*145, 129*)

In addition to the list in Table 2, the following species of *Senecio* have also been shown to contain alkaloids: *S. brachychaetus* D. C., *S. candolleanus* WALL., *S. grandidentatus* LEDEB. (= *S. erucifolius* L.), *S. jacquinianus* REICHB. (= *S. nemorensis* L.), *S. massagetovii* (sp. *nova*), *S. orientalis* WILLD., *S. pedunculosus* TRAUTV., *S. platyphylloides* SOMM. and LEVIER, *S. thyrsophorus* KOCH (= *S. racemosus* D. C.) (*153*).

Table 3 gives a list of plants which are reported to be toxic and which have not yet been studied chemically. The alkaloid content of these species would be of interest.

The *Senecio* alkaloids are C_{18}-compounds which on hydrolysis give C_8-bases and C_{10}-acids for which MANSKE (*122*) suggested the names *necines* and *necic* acids respectively.

$$\underset{\text{Retrorsine.}}{C_{18}H_{25}O_6N} + H_2O \rightarrow \underset{\text{Retronecine.}}{C_8H_{13}O_2N} + \underset{\text{Retronecic acid.}}{C_{10}H_{16}O_6}.$$

A less descriptive nomenclature has been applied to the *Crotalaria* alkaloids (*149*), which have been called monocrotaline and dicrotaline

Table 3. Species which have been Studied for Toxicity only.

Plant species	Effect on animals	Reference
Senecio angustifolius WILLD.	Produces staggers in horses	(*168*)
S. burchellii D. C.	Responsible for bread poisoning. Negative (?)	(*181, 168*)
S. bupleuroides D. C.	Transient fever but no positive poisoning	(*168*)
S. fremonti TORR. and GRAY.	Produces 'Dunsiekte' in horses (?)	(*168*)
S. pubigerus L.	Causes liver cirrhosis in mules	(*168*)
Crotalaria burkeana BENTH.	Toxic	(*168*)
C. juncea L.	Causes death in sheep	(*149*)
C. sagittalis L.	Causes Missouri bottom disease in horses	(*149*)
Amsinckia intermedia FISCH. and MEY.	Seeds produce cirrhosis in swine, horses and cattle	(*124*)

In addition, the following species have been reported as having no effect on animals: *Senecio albanensis* D. C. (*168*), *S. glutinosus* THUNB. (*168, 172*), *S. laevigatus* THUNB. (*168*), *S. othonnaeflorus* D. C. (*168*), *S. pinnulatus* THUNB. (*168*), *Crotalaria allenii* VERDOERN (*168*), *and C. distans* BENTH. (*168*).

as the first and second alkaloids isolated from this genus. These two alkaloids hydrolyse to retronecine and two new acids termed monocrotalic and dicrotalic acid, respectively. However, when the alkaloid is of the C_{18}-type as found in *Senecio* species, MANSKE's terminology is now usually employed. A list of the known pyrrolizidine alkaloids and their hydrolysis products is shown in *Table 4*, p. 206.

Several alkaloids which were originally considered to be individuals but now found to be mixtures or identical with previously known alkaloids have been placed in a separate list in *Table* 5, p. 209.

Isolation.

The extraction and isolation of the pyrrolizidine alkaloids has proved at times a difficult task. The difficulty of separation by crystallisation has been emphasised by several workers in this field, and as may be readily realised from Table 5, mixtures of senecionine and seneciphylline have given misleading results. The recent model experiments of CULVENOR, DRUMMOND and PRICE (*54*) could well find general application. These authors separated successfully the complex mixture of ten alkaloids in *Heliotropium europaeum* by chromatography on kieselguhr, buffered at p_H 8, with different eluting solvents, whilst paper chromatograms were used to detect mixtures and characterise individuals. Furthermore, the occurrence of *N*-oxides in the plant was demonstrated by KOEKE-

Table 4. The Pyrrolizidine Alkaloids and their Hydrolysis Products.
[Solvents used for spec. rotation measurements: (c) chloroform; (e) ethanol; (m) methanol; and (w) water.]

Alkaloid	Melting point	$[\alpha]_D$	Base	Acid	Reference
Brasilinecine	171° (d)	— 68.2° (c)	Retronecine	Acid, m. p. 143°	(*70*)
Base-C $C_{16}H_{27}O_4N$	67–68°	— 12.1° (e)	Supinidine	Heliotric acid.	(*53*)
Campestrine, $C_{13}H_{19}O_3N$	93°				(*40*)
Dicrotaline, $C_{14}H_{19}O_5N$	170° (d)		Retronecine	Dicrotalic acid	(*123*)
Fuchsisenecionine $C_{12}H_{21}O_3N$	225–227° (HCl salt)				(*148*)
Base-G, $C_{16}H_{27}O_6N$	gum	+ 10.9° (e)	Heliotridine	Lasiocarpic acid	(*53*)
Base-G *N*-oxide	171°	+ 25.3° (e)			(*53*)
Graminifoline, $C_{18}H_{23}O_5N$	236°				(*182*)
Grantianine, $C_{18}H_{23}O_7N$	205° (d)	+ 50.6° (c)	Retronecine	Grantianic acid	(*3*)
Hastacine, $C_{18}H_{27}O_5N$	171°	— 72.3°	Hastanecine	Hastanecic acid	(*94*)
Heliotrine, $C_{16}H_{27}O_5N$	126°	+ 63.8° (c) + 17.6° (e)	Heliotridine	Heliotrinic acid	(*125, 54, 53*)
Heliotrine *N*-oxide $C_{16}H_{27}O_6N$	171–172°	+ 26.6° (e)			(*54, 53*)
Integerrimine, $C_{18}H_{25}O_5N$	172.5°	+ 4.3° (c)	Retronecine	Integerrinecic acid	(*121, 105*)
Isatidine, $C_{18}H_{25}O_7N$	145°	— 8° (w)	Isatinecine	Retronecic or Isatinecic acid	(*40, 179, 48*)
Jacobine, $C_{18}H_{25}O_6N$	226°	— 46.3° (c)	Retronecine	Jaconecic acid	(*37, 42, 41*)
Jacoline, $C_{18}H_{27}O_7N$	221°	+ 48° (c)			(*42*)
Jaconine, $C_{20}H_{32}O_7NCl$	147°	+ 52.5° (c)	Retronecine	Jaconecic acid	(*37, 42, 41*)
Jacozine	228°	— 140°			(*42*)
Laburnine, $C_8H_{15}ON$	b. p.$_{0.01}$ 80–90°	+ 15.45° (e)			(*72*)
Lasiocarpine, $C_{21}H_{33}O_7N$	95.5°	— 4° (e)	Heliotridine	Lasiocarpic acid and Angelic acid	(*125, 146, 54, 53*)
Lasiocarpine *N*-oxide $C_{21}H_{33}O_8N \cdot H_2O$	133–135°				(*54, 53*)

Alkaloid	Melting point	$[\alpha]_D$	Base	Acid	Reference
Lindelofamine, $C_{20}H_{33}O_5N$ ..	88°		*D-iso*Retronecanol	Trachelanthic acid + + Tiglic acid	(*106*)
Lindelofine, $C_{15}H_{27}O_4N$	107°	+ 50° (e)	*D-iso*Retronecanol	Trachelanthic acid	(*106*)
Makrotomine, $C_{15}H_{27}O_5N$ ...	97°	— 6.9°	Trachelanthamidine		(*144*)
Mikanoidine, $C_{21}H_{29}O_6N$ ...	Amorphous		Mikanecine	Mikanecic acid	(*120*)
Monocrotaline, $C_{16}H_{23}O_6N$ ..	198°	— 54.7° (c)	Retronecine	Monocrotalic or Monocrotic acid	(*149, 18*)
Otosenine, $C_{19}H_{27}O_7N$	221°	+ 20.8° (c)	Otonecine	Jaconecic acid	(*195*)
Platyphylline, $C_{18}H_{27}O_5N$ ..	129°	— 56° (c) — 59° (e)	Platynecine	Senecic or Platynecic acid	(*153, 60, 152*)
Platyphylline *N*-oxide ..... $C_{18}H_{27}O_6N$	195°	— 59° (w)			(*92*)
Renardine	193°	— 2.23° (c)	Unidentified necine	Acid, m. p. 149°, giving lactone identical senecic acid lactone	(*59*)
Retrorsine, $C_{18}H_{25}O_6N$	217°	18° (e)	Retronecine	Isatinecic or Retronecic acid	(*119, 38, 48*)
Riddelliine, $C_{18}H_{23}O_6N$	198° (d)	— 109.5° (c)	Retronecine	Riddellic acid	(*121, 10*)
Rosmarinine, $C_{18}H_{27}O_6N$ ...	209°	— 120° (c) — 94° (e)	Rosmarinecine	Senecic acid	(*181, 159*)
Ruwenine, $C_{18}H_{27}O_6N$	175.5–179°				(*162*)
Ruzorine, $C_{18}H_{27}O_8N$	161–163°				(*162*)
Sarricine, $C_{18}H_{27}O_5N$	51–52°	— 129.7°	Platynecine		(*61*)
Sarricine *N*-oxide .... $C_{18}H_{27}O_6N$	123–124°	— 81.6°			(*61*)
Sceleratine, $C_{18}H_{27}O_7N$	178°	+ 54° (e)	Retronecine	Sceleranecic acid	(*188, 187*)
Senecifolidine, $C_{18}H_{25}O_7N$ ..	212°	— 13.9° (e)			(*192, 38*)
Senecifoline, $C_{18}H_{27}O_8N$	195°	+ 28.8° (e)	Retronecine	Senecifolic acid	(*192*)
Senecionine, $C_{18}H_{25}O_5N$	245°	— 56° (c)	Retronecine	Senecic acid	(*75, 36*)

Alkaloid	Melting point	$[\alpha]_D$	Base	Acid	Reference
Seneciphylline, $C_{18}H_{25}O_5N$ ..	216–218°	— 139° (c)	Retronecine	Seneciphyllic acid or *iso*-Seneciphyllic acid	(*101, 98, 153*)
Senkirkine, $C_{18}H_{25}O_6N$	198°	— 6.2° (c)	$C_8H_{13}O_3N$ (?)	Senecic acid lactone	(*43*)
Silvasenecine hydrochloride $C_{12}H_{22}O_4NCl$					(*148*)
Spartioidine, $C_{18}H_{23}O_5N$ ...	178°				(*121*)
Squalidine, $C_{18}H_{23}O_5N$	217–218°	— 54.6° (c)	Retronecine	Squalinecic acid	(*36*)
Supinine, $C_{15}H_{25}O_4N$	147.5°	— 23.8° (e)	Supinidine	(+)-Trachelanthic acid	(*141, 54*)
Trachelanthamine .... $C_{15}H_{27}O_4N$	93°	— 18.1° (w)	Trachelanthamidine	Trachelanthic acid	(*135, 137, 138*)
Trachelanthine, $C_{15}H_{27}O_5N$..	167°	— 22.5° (w)	Trachelanthidine	Trachelanthic acid	(*135, 137, 138*)
Trichodesmine, $C_{15}H_{27}O_6N$ ..	161°	+ 38° (e)	Trichodesmidine — Retronecine (*101*)	$C_7H_{12}O_3$ and lactic acid	(*145, 129*)
Turneforcine, $C_{13}H_{21}O_3N$ ...	oil		Turneforcidine	Angelic acid	(*139*)
Usaramoensine, $C_{18}H_{25}O_5N$..	221° (d)	— 25° (c)	Retronecine	Usaramoensinecic acid	(*25*)
Viridiflorine, $C_{15}H_{27}O_4N$	103.5°		Trachelanthamidine	Viridifloric acid	(*136*)

MOER and WARREN (*92*) in several *Senecio* species and their prediction that these *N*-oxides are wider spread in other genera has been substantiated by CULVENOR (*54*). These *N*-oxides are not always readily extracted, and their reduction to the tertiary base prior to extraction has led to much improved yields.

II. The Basic Hydrolysis Products.

The known bases obtained from the hydrolysis of pyrrolizidine alkaloids are listed in *Table 6*, p. 210.

The conversion of retronecine, platynecine and heliotridine to heliotridane and the early realisation that these bases had a common structure was of considerable significance. The determination of the structure of heliotridane itself by degradation and synthesis opened up the whole field of study of the bases. One interesting base, trachelanthamine, however, gave on removal of the hydroxyl group a new nitrogen compound, pseudoheliotridane, which was stereoisomeric with heliotridane. The structures and stereochemistry of all but four of the basic hydrolysis products have now been elucidated.

Table 4a. Properties of the Unnamed Pyrrolizidine Alkaloids.

Source	Formula	Melting point	$[\alpha]_D$	Reference
E. hieracifolia (L.) RAFIN ...	$C_{20}H_{17(19)}O_6N$	237°		(*55*)
H. europaeum L.	$C_{12}H_{17(19)}O_4N$	158–159°		(*53*, *54*)
S. brasiliensis D. C.	$C_{18}H_{25}O_6N$	232–244°	66.8°	(*150*)
S. erucifolius L.	$C_{18}H_{27}O_5N$	222°		(*40*)
S. hygrophilus DYER and SM. .	$C_{18}H_{27}O_6N$	175–176°	−62.4° (m)	(*189*)
S. palustris HOOK.	$C_{18}H_{27}O_5N$	169°		(*42*)

Table 5. Alkaloids Originally Considered to be Individuals and now Found to be Mixtures or Identical with Previously Known Forms.

Original name	Reference	Present identification	Reference
Aureine	(*120*)	Senecionine	(*111*)
Carthamoidine	(*81*)	Seneciphylline	(*6*, *15*)
		Senecionine	
Douglasiine	(*111*)	Seneciphylline	(*6*, *15*)
		Senecionine	(*191*)
		Riddelliine	
		Retrorsine	
Eremophiline	(*111*)	Senecionine	(*6*, *191*, *15*)
		Seneciphylline	
		Retrorsine	
Hieracifoline.................	(*122*)	Senecionine	(*122*, *55*)
		Seneciphylline	
α-Longilobine.................	(*4*)	Seneciphylline	(*15*)
β-Longilobine.................	(*4*)	Retrorsine	(*191*)
Pterophine	(*182*)	Senecionine	(*55*)
		Seneciphylline	
Jacodine	(*37*)	Seneciphylline	(*42*)

Retronecine, Platynecine and Heliotridine as Derivatives of Heliotridane.

Retronecine, $C_8H_{13}O_2N$, first isolated by WATT (*192*) as senecifolinine from senecifoline, was obtained and named by MANSKE (*119*) from retrorsine. BARGER, SESHARDI, WATT and YABUTA (*38*) showed this base to be a tertiary amine containing one double bond and two hydroxyl groups. These authors drew attention to the interesting catalytic reduction of retronecine and its derivatives, and this became of great importance in the structural studies which were to follow. Retronecine on catalytic reduction gave retronecanol, $C_8H_{15}ON$, whilst diacetylretronecine hydrogenated more readily to give acetic acid and acetylretronecanol. The alkaloid retrorsine, which hydrolysed to retronecine and retronecic acid, gave on catalytic reduction and hydrolysis, retronecanol and retronecic acid.

Retronecine $C_8H_{11}N(OH)_2$	⟶	Retronecanol $C_8H_{14}N(OH)$
↓		
Diacetylretronecine $C_8H_{11}N(O \cdot CO \cdot CH_3)_2$	⟶	Acetylretronecanol + Acetic acid $C_8H_{14}N(O \cdot CO \cdot CH_3)$

Table 6. The Bases from the Hydrolysis of Pyrrolizidine Alkaloids*. [Solvents used for spec. rotation measurements: (c) chloroform; (e) ethanol; (m) methanol; and (w) water.]

Base	$[\alpha]_D$	Melting points			Reference
		Base	Picrate	Hydro-chloride	
Hastanecine, $C_8H_{15}O_2N$..	— 9.1°	113–114°			(*94*)
Heliotridine, $C_8H_{13}O_2N$..	+ 31° (m)	116–118°			(*125*)
Isatinecine, $C_8H_{13}O_3N$...	+ 22.4° (w)	212–215°	147°		(*182*)
Mikanecine, $C_8H_{15}O_2N$...		oil	186°		(*111*)
Otonecine, $C_9H_{15}O_3N$....				146–148°	(*195*)
Platynecine, $C_8H_{15}O_2N$ ..	— 56.8° (c)	148–148.5°	184°		(*152*)
Retronecine (Senecifolinine), $C_8H_{13}O_2N$ ..	+ 50° (e)	121–122°	142°	164–165°	(*38, 3, 192*)
Rosmarinecine $C_8H_{15}O_3N$	— 118.5° (m)	171–172°	175°		(*182, 159*)
Supinidine, $C_8H_{13}ON$....	— 9.45° (e)	158–159°	142–143°		(*141*)
Trachelanthidine .... $C_8H_{15}O_2N$				107–108°	(*137, 138*)
Trachelanthamidine .... $C_8H_{15}ON$	— 12.9°	b. p. 114–115°	174°	110–112°	(*138*)
Turneforcidine $C_8H_{15}O_2N$	— 10.5°	118.5–120°			(*139*)

Only a short period elapsed before MEN'SHIKOV (*127*) described similar reactions for heliotrine. This alkaloid which hydrolysed to heliotrinic acid and heliotridine, $C_8H_{13}O_2N$ (*125*), gave on hydrogenolysis heliotrinic acid and oxyheliotridane, $C_8H_{15}ON$. Furthermore, dibenzoyl-heliotridine on hydrogenolysis gave benzoic acid and benzoyl-oxyheliotridane.

Two years previously MEN'SHIKOV (*126*) had treated heliotridine with thionyl chloride to obtain a dichloro-compound which was catalytically reduced to chloroheliotridane, which, treated with sodium ethoxide to give heliotridene and then reduced catalytically, gave heliotridane. This was also formed from oxyheliotridane by dehydration with sulphuric acid to heliotridene which was then hydrogenated catalytically (*127*).

KONOVALOVA and OREKHOV (*100*) similarly converted retronecanol to heliotridane and so established that retronecine was an isomer of heliotridine, and hence that alkaloids of the *Senecio* species were closely related to those of Boraginaceae.

* Senecifolinine (*192*) and trichodesmidine (*145*) are identical with retronecine (*38, 101*).

The previous year these authors (*99*) had also converted platynecine with thionyl chloride to dichloroplatynecine, which was reduced with sodium and alcohol to a mixture that gave heliotridane with platinum oxide and hydrogen (*Chart 1.*).

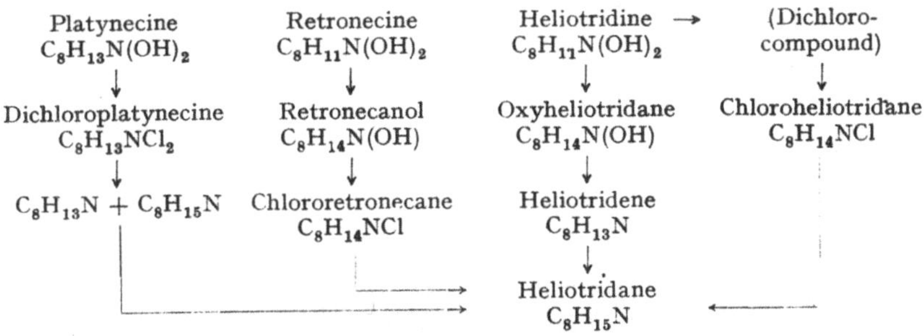

Chart 1. The Degradation of Heliotridine, Retronecine and Platynecine to Heliotridane.

Heliotridane obtained by these different routes was always the same, as shown by the preparation of derivatives; but the rotation varied according to the method, heliotridane from oxyheliotridane had $[\alpha]_D$-99.5° whilst that from the chloro-compound had $[\alpha]_D$-10.5°. This difference was undoubtedly due to partial racemisation. To establish that rearrangement had not occurred on dehydration of retronecanol, ADAMS and ROGERS (*19*) repeated the conversion by less drastic conditions through chlororetronecane.

The Degradation of Heliotridane.

Heliotridane is a tertiary base not containing an *N*-methyl group and MEN'SHIKOV (*125*) concluded that the nitrogen atom was common to two rings. Exhaustive methylation and reduction of the product gave *L*-dihydro-des-*N*-methylheliotridane which was dehydrogenated to a substituted pyrrole, which in turn rehydrogenated to the *DL*-substituted pyrrolidine yielding a picrate, m. p. 116°. The four possibilities for heliotridane are shown in *Table 7*, together with the seven products expected from the Hofmann reaction. 8-Methylpyrrolizidine is not included in this list as it would give a 2 : 2-disubstituted pyrrolidine which would not dehydrogenate without the loss of a group. Six of these reaction products were synthesised by MEN'SHIKOV and found to be different from *DL*-dihydrodes-*N*-methylheliotridane which he concluded must be the seventh form, namely, 1 : 3-dimethyl-2-*n*-propyl-pyrrolidine (II), so that heliotridane was almost certainly 1-methylpyrrolizidine (I). He proved the structure by carrying out a second Hofmann degradation on *DL*-dihydro-des-*N*-methylheliotridane and reduced the product to *DL*-tetrahydro-des-

N-dimethylheliotridane identical with 4-dimethylamino-3-methylheptane (IV) obtained from 1-methyl-2-*sec*butylpyrrolidine (III).

(I.) 1-Methyl-pyrrolizidine. (II.) 1:3-Dimethyl-2-*n*-propyl-pyrrolidine. or (III.) 1-Methyl-2-*sec*butyl-pyrrolidine.

$CH_3 \cdot CH_2 \cdot CH_2 \cdot CHN(CH_3)_2 \cdot CH(CH_3) \cdot CH_2 \cdot CH_3$

(IV.) 4-Dimethylamino-3-methylheptane.

Table 7. Substituted 1-Methylpyrrolidines Obtained from the Hofmann Reaction.

Parent substance	Substituted 1-methylpyrrolidine
Perhydropyrrocholine	2-*n*-Butyl- (*147*)
1-Methylpyrrolizidine	2-*sec*Butyl- (*131*)
	3-Methyl-2-propyl-
2-Methylpyrrolizidine	2-*iso*Butyl- (*147*) and
	2-Methyl-4-*n*-propyl- (*130*)
3-Methylpyrrolizidine	2-*n*-Butyl- (*147*) and
	2-Methyl-5-*n*-propyl- (*129*)

In confirmation of this concept MEN'SHIKOV (*147*, *130*) synthesised 2-methylpyrrolizidine and established that the Hofmann degradation proceeded as envisaged, the unsubstituted pyrrolidine ring opening to give 1 : 4-dimethyl-2-*n*-propyl-pyrrolidine. Furthermore *DL*-1-methylpyrrolizidine was synthesised in small yield to give a picrate, m. p. 236°, which was the same as that of the picrate of *L*-heliotridane, with which is gave no depression.

ADAMS and ROGERS (*19*) criticised some of the evidence presented by MEN'SHIKOV (*147*, *130*) although not his final conclusions. Firstly, the identity of *DL*-1-methylpyrrolizidine with *L*-heliotridane based on the similarity of their picrates seemed unjustified, since *DL*- and optically active forms frequently have different melting points. Furthermore similar substances may exhibit no melting point depression when mixed. Secondly, the conclusion was not justified that the Hofmann degradation of heliotridane (I) to give (IV) proceeded by way of (II), which had not been synthesised, simply because the picrate of (III) was not identical with *DL*-dihydro-des-*N*-methylheliotridane. Compound (III) can exist in two diastereoisomeric modifications and the other form of (III) could be identical with *DL*-dihydro-des-*N*-methylheliotridane. These authors synthesised the two forms of 1 : 3-dimethyl-2-*n*-propyl-pyrrolidine (II) which gave picrates, m. p. 126° and 116°, the latter being identical with *DL*-dihydro-des-*N*-methylheliotridane picrate, m. p. 116°. The identity of heliotridane with 1-methylpyrrolizidine and the course of the Hofmann degradation was thus fully established.

The Structure of Retronecine and Platynecine.

ADAMS and his collaborators in a series of studies established conclusively the disposition of the hydroxyl groups and the unsaturated centre in retronecine. The reduction experiments of ADAMS and ROGERS (*20*) were of special significance. Retronecine and its mono-esters were known to be reduced catalytically to retronecanol (*38*). A similar reduction of monocrotaline with ADAMS' platinum or RANEY's nickel was shown to proceed by the removal of the ester group to give first desoxyretronecine and then retronecanol. The reduction of retronecine itself with RANEY's nickel gave platynecine and so established that the two hydroxyl groups in these two necines occupied the same position, a possibility which had been envisaged by OREKHOV and TIDEBEL (*153*).

Retronecine $C_8H_{11}N(OH)_2$ $\xrightarrow[H_2]{Ni}$ Platynecine $C_8H_{13}N(OH)_2$

↑ OH^-

Monocrotaline $\xrightarrow[H_2]{Ni\ or\ PtO_2}$ Desoxyretronecine $C_8H_{12}N(OH)$ → Retronecanol $C_8H_{14}N(OH)$

The wealth of evidence which led to the assignment of structures is accumulated in *Chart 2* (p. 215) and is summarised below.

The primary hydroxyl group. One hydroxyl was necessarily attached to the 1-methyl group because retronecine (IV) and platynecine (V) contained no C-methyl group whilst the reduced products like retronecanol (VII) possessed this grouping. Furthermore KONOVALOVA and OREKHOV (*99*) had prepared monobenzoyl platynecine (IX) and converted it to a chloro-compound (X) which was successfully reduced (*9*) to benzoyl *iso*retronecanol (XI). This was hydrolysed to (XII) and oxidised to pyrrolizidine-1-carboxylic acid (XIII).

The secondary hydroxyl group. The presence of a secondary hydroxyl was established by ADAMS and HAMLIN (*9*) who successfully oxidised retronecanol (VII) to retronecanone (VIII). This secondary hydroxyl group was limited to the $C_{(6)}$ or $C_{(7)}$ from the dehydration of platynecine to anhydroplatynecine. The 7-position, however, followed from the elegant experiments of ADAMS, CARMACK and MAHAN (*2*). Cyanogen bromide and retronecanol (VII) gave an addition product which, after rearrangement to (XIV) and boiling with alkali, yielded a neutral cyanamide ether (XV) which could be hydrolysed to a secondary amine. The ease of formation of an ether ring indicated a six-membered ring as in compound (XV), and hence the secondary hydroxyl in the 7-position. If the secondary hydroxyl were at $C_{(6)}$, a seven-membered ether ring would have to be formed. The opening of the non-methylated pyrrolizidine ring in the von Braun reaction would give unstable ethers. Confirmation of this $C_{(7)}$-hydroxyl came from the synthesis by ADAMS and LEONARD (*13*) of retronecanone (VIII), which is described in a later section.

The double bond. The double bond was associated with the labile hydroxyl groups by ADAMS and ROGERS (*20*) who pointed out that when the labile hydroxyl of retronecine was esterified with an organic acid, or replaced by chlorine, the organic acid or the chlorine atom was removed by reduction. Furthermore, one hydroxyl group in retronecine could be esterified more readily than the other (*119*) and this was the labile hydroxyl group. In addition retronecine was shown by ADAMS et al. (*2*) to be less basic than platynecine so that it was not a tertiary vinyl amine (*16*). Final proof of the position of the double bond was given by ADAMS and MAHAN (*17*). Desoxyretronecine (VI) was treated with thionyl chloride and the chloro-compound (XVI) reduced with chromous chloride to *iso*-heliotridene (XVII). Ozonolysis of (XVII) yielded the keto-acid (XVIII) which gave the iodoform-reaction characteristic of a methyl ketone. Reduction of the keto-acid gave a hydroxyacid (XIX), which with acetic anhydride gave the lactone (XX) and with diazomethane yielded a betain.

The Structure of Heliotridine.

Heliotridine, $C_8H_{13}O_2N$, was first shown to be the stereoisomer of retronecine by MEN'SHIKOV and KURZOKOV (*142*) as late as 1950 by the oxidation of oxyheliotridane to retronecanone (VIII, p. 215) obtained previously by ADAMS and HAMLIN (*9*) from the oxidation of retronecanol (VII). The difference between oxyheliotridane and retronecanol was accordingly associated with the configuration of the $C_{(7)}$-hydroxyl group, as is shown in *Chart 5*. Heliotridine was also reduced catalytically (*142*) in the presence of nickel to give dihydroxyheliotridane, a reaction which parallels the conversion of retronecine (IV) to platynecine (V). The monobenzoyl derivative of dihydroxyheliotridane (IX) reacted with thionyl chloride, and the arising chloro-compound (X) was reduced and hydrolysed to *iso*retronecanol. Finally, heliotrine with thionyl chloride and reduction of the reaction product with chromium chloride and zinc amalgam resulted in the formation of supinidine.

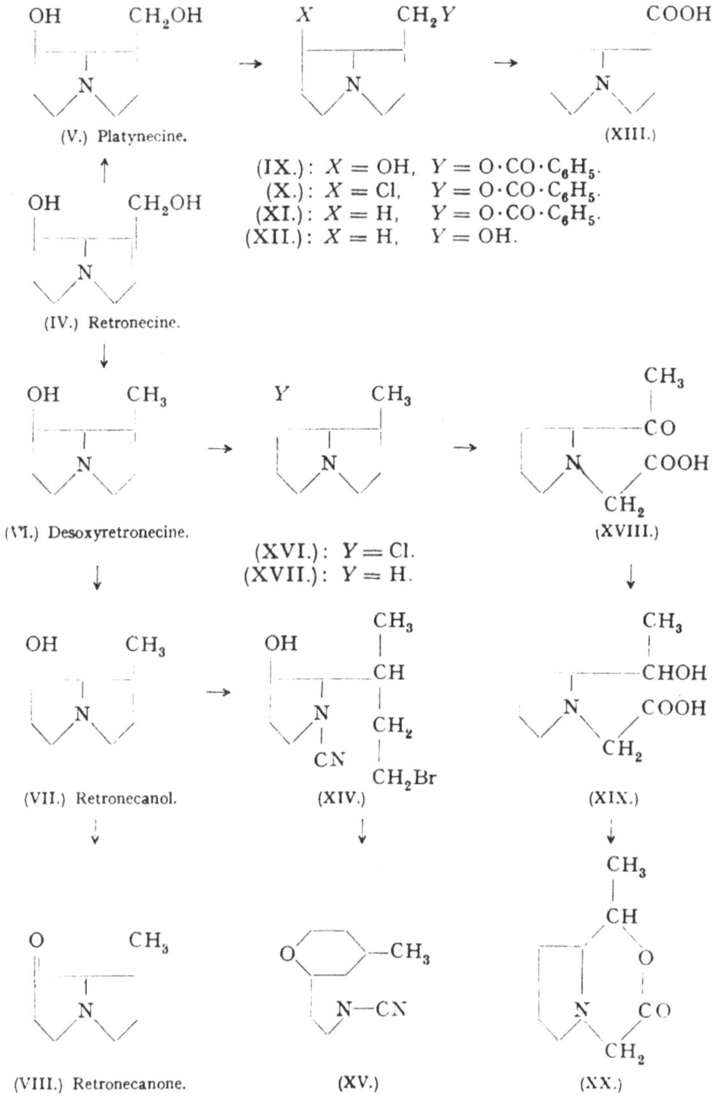

(V.) Platynecine.

(XIII.)

(IX.): $X = OH$, $Y = O \cdot CO \cdot C_6H_5$.
(X.): $X = Cl$, $Y = O \cdot CO \cdot C_6H_5$.
(XI.): $X = H$, $Y = O \cdot CO \cdot C_6H_5$.
(XII.): $X = H$, $Y = OH$.

(IV.) Retronecine.

(VI.) Desoxyretronecine.

(XVI.): $Y = Cl$.
(XVII.): $Y = H$.

(XVIII.)

(VII.) Retronecanol.

(XIV.)

(XIX.)

(VIII.) Retronecanone.

(XV.)

(XX.)

Chart 2. Reactions which Establish the Positions of the Hydroxyl Groups and the Ethylenic Bond in Retronecine and Platynecine.

Monohydroxy-methylpyrrolizidines: Trachelanthamidine and Supinidine.

Trachelanthamidine (XXI), $C_8H_{15}ON$, was shown by MEN'SHIKOV (*134*) to contain a primary hydroxyl group by oxidation to an amino-acid (XXII) which was decarboxylated to pyrrolizidine (XXIII), identical with that prepared by PRELOG (*156*). However, when the hydroxyl group was replaced by chlorine, and the chlorocompound reduced, a compound isomeric with heliotridane was obtained. The Hofmann degradation of this compound did not follow that of heliotridane. The difference was attributed (*134*) to diastereoisomerism, and the new base was termed pseudo-heliotridane (XXIV). This reasonable assumption was later established by the synthesis of Leonard and Felley (*114*).

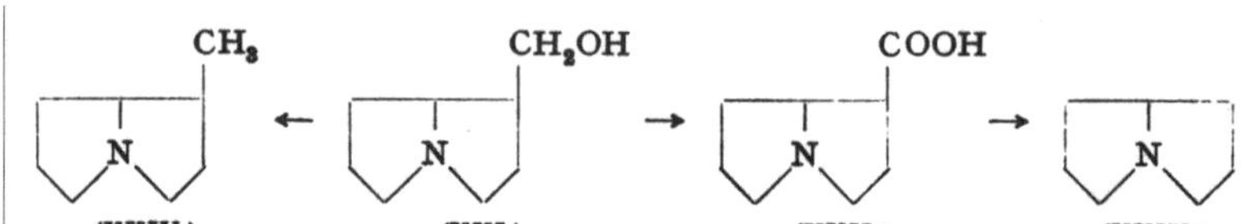

CH_3 ← CH_2OH → COOH →

(XXIV.) Pseudo-heliotridane. (XXI.) Trachelanthamidine. (XXII.) (XXIII.) Pyrrolizidine.

Supinidine, $C_8H_{13}ON$, is the necine base from supinine described by Men'shikov and Gurevich (*141*) who reduced it to *iso*retronecanol. The structure is accordingly (XXV).

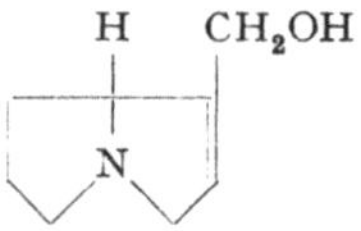

H CH_2OH

(XXV.) Supinidine.

N-Oxides: Trachelanthidine and Isatinecine.

Trachelanthidine, $C_8H_{15}O_2N$, was isolated as the basic product from the hydrolysis of trachelanthine, which Men'shikov and Borodina (*137, 138*) showed was trachelanthamine *N*-oxide. Trachelanthine gave trachelanthamine with sulphur dioxide and was formed from trachelanthamine by hydrogen peroxide. Accordingly, trachelanthidine is trachelanthamidine *N*-oxide (XXVI).

Trachelanthine $C_{15}H_{27}O_5N$ → Trachelanthidine $C_8H_{15}O_2N$ + Trachelanthic acid

H_2O_2 ⇅ SO_2

Trachelanthamine $C_{15}H_{27}O_4N$ → Trachelanthamidine $C_8H_{15}ON$ + Trachelanthic acid

Isatinecine, $C_8H_{13}O_3N$, was originally considered different from the other necines, since de Waal (*182, 183*) found that hydrogenation over platinum gave tetrahydroisatinecine, $C_8H_{17}O_3N$, which was reported to have no tertiary nitrogen. This same substance was also formed from the hydrogenation of isatidine to hexahydro-desoxyisatidine and hydrolysis. Leisegang and Warren (*109*) found that isatidine was readily reduced by zinc dust and acetic acid to retronecine from which it was obtained by the action of hydrogen peroxide. Isatinecine was accordingly designated

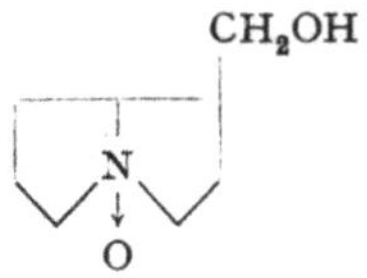

(XXVI.) Trachelanthamidine *N*-oxide. 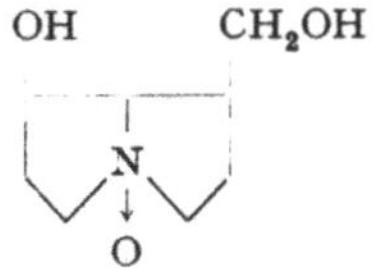 (XXVII.) Retronecine *N*-oxide.

retronecine *N*-oxide (XXVII), and confirmation (*109*) was afforded by the direct formation of platynecine with RANEY's nickel and complete reduction to retronecanol.

It is difficult to account for DE WAAL's tetrahydroisatinecine $C_8H_{17}O_3N$, m. p. 174.5°, $[\alpha]_D$ — 88°, unless it be identical with retronecanol hydrochloride, $C_8H_{15}ON \cdot HCl$, m. p. 184–185°, for which the analyses would fit. Such a formulation would be in accord with its formation from hexahydro-desoxyisatidine, the structure of which follows from that of isatidine.

Syntheses of Pyrrolizidine Bases.

MEN'SHIKOV (*147*, *131*) synthesised 1-methyl- (XXVIII) and 2-methylpyrrolizidine (XXIX) from 2-*sec*.- (XXX) and 2-*iso*butylpyrrolidine (XXXI), respectively, by treatment with sodium hypobromite and then concentrated sulphuric acid. The yields were small but sufficient to confirm the identity of heliotridane and 1-methylpyrrolizidine.

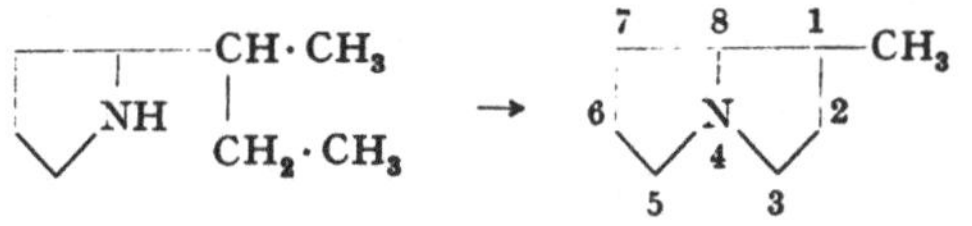

(XXX.) 2-*sec*Butylpyrrolidine. (XXVIII.) 1-Methyl-pyrrolizidine.

(XXXI.) 2-*iso*-Butylpyrrolidine. (XXIX.) 2-Methyl-pyrrolizidine.

CLEMO and his co-workers (*51*, *50*) effected ring closure by the Dieckmann reaction and the resulting ketone was made to react with a Grignard reagent to bring about further substitution. Diethyl pyrrolidine-1 : 2-diacetate (XXXII) gave pyrrolizid-2-one (XXXIII) which with methyl magnesium iodide and hydrolysis gave the tertiary alcohol (XXXIV) which in turn was reduced to 2-methylpyrrolizidine (XXIX).

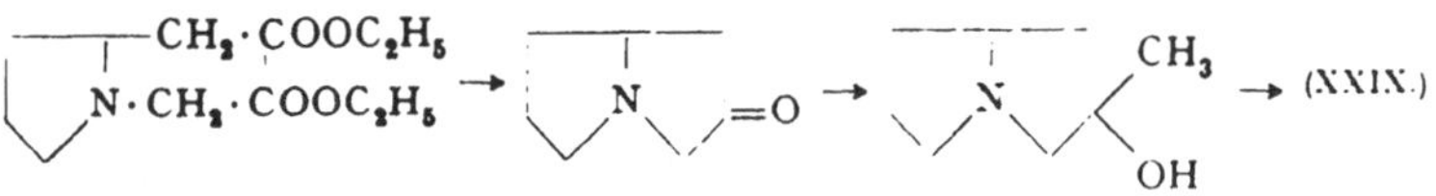

(XXXII.) Diethyl pyrrolidine-1 : 2-diacetate. (XXXIII.) Pyrrolizid-2-one. (XXXIV.) → (XXIX.)

PRELOG and HEIMBACH (*156*) carried through the synthesis of pyrrolizidine by the malonic ester condensation followed by ring closure by PRELOG's twofold intramolecular alkylation (*155*). Diethyl 4-ethoxybutane-1 : 1-dicarboxylate was condensed as the sodium salt with 3-ethoxypropyl bromide to give diethyl 1 : 7-diethoxyheptane-4 : 4-dicarboxylate. This ester was hydrolysed and decarboxylated to the corresponding monocarboxylic acid which in sulphuric acid with sodium azide was converted to 1 : 7-diethoxy-4-amino-heptane. Fission of the ether linkages with hydrobromic acid and treatment of the resulting 1 : 7-dibromo-4-amino-heptane with alkali yielded pyrrolizidine:

$$C_2H_5O \cdot CH_2 \cdot CH_2 \cdot CH_2 \cdot CH(COOR)_2 \rightarrow (C_2H_5O \cdot CH_2 \cdot CH_2 \cdot CH_2)_2C(COOR)_2$$

$$\downarrow$$

$$(C_2H_5O \cdot CH_2 \cdot CH_2 \cdot CH_2)_2CH \cdot NH_2 \leftarrow (C_2H_5O \cdot CH_2 \cdot CH_2 \cdot CH_2)_2CH \cdot COOH$$

$$\downarrow$$

$$(Br \cdot CH_2 \cdot CH_2 \cdot CH_2)_2CH \cdot NH_2 \rightarrow \text{(XXIII, p. 216)}$$

PRELOG and ZALÁN (*157*) then modified the synthesis to prepare 1-methylpyrrolizidine by way of the Grignard reaction. 4-Phenoxy-2-cyano-butane and 3-ethoxypropanyl magnesium bromide reacted and the product hydrolysed to 1-phenoxy-7-ethoxy-3-methylheptane-4-one, the oxime of which was reduced to the amine. The same procedure for fission of the ether and ring-formation as described above gave 1-methylpyrrolizidine.

$$NC \cdot CH(CH_3) \cdot [CH_2]_2 \cdot OC_6H_5 \rightarrow C_2H_5O \cdot [CH_2]_3 \cdot CO \cdot CH(CH_3) \cdot [CH_2]_2 \cdot OC_6H_5$$

$$\downarrow$$

$$C_2H_5O \cdot [CH_2]_3 \cdot CH(NH_2) \cdot CH(CH_3) \cdot [CH_2]_2 \cdot OC_6H_5$$

$$\downarrow$$

$$Br \cdot [CH_2]_3 \cdot CH(NH_2) \cdot CH(CH_3) \cdot [CH_2]_2 \cdot Br \rightarrow \text{(XXVIII)}.$$

GALINOVSKY (*73*) contributed a further synthesis by effecting ring closure by amide formation:

$$CH_2 \cdot CH_2 \cdot COOC_2H_5 \quad NH$$

→ N, O → N

The elegant synthesis of 7-keto-1-methylpyrrolizidine (XLI) by ADAMS and LEONARD (*13*) established conclusively the structure of retronecanone (*Chart 3*). 1-*m*-Nitrobenzoyl-4-methylpiperidine (XXXV) was oxidised to the amino-acid (XXXVI) which on bromination gave the bromo-amino-acid (XXXVII), the constitution of which follows from the researches of FISCHER and ZEMPLÉN (69). The acid (XXXVII) was cyclised with alkali to the pyrrolidine-carboxylic acid (XXXVIII) which by hydrolysis and esterification yielded the ester (XXXIX). This ester

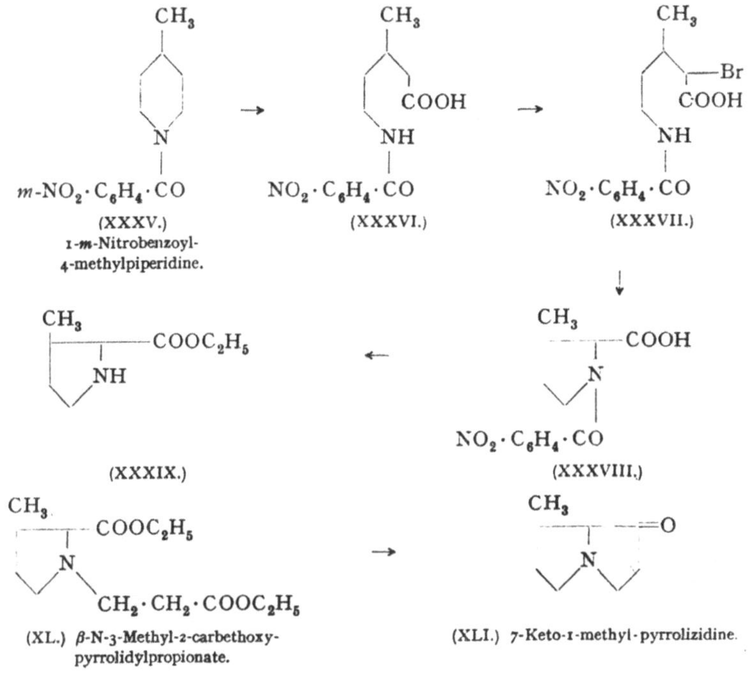

$m\text{-}NO_2\cdot C_6H_4\cdot CO$ (XXXV.) 1-*m*-Nitrobenzoyl-4-methylpiperidine. → $NO_2\cdot C_6H_4\cdot CO$ (XXXVI.) → $NO_2\cdot C_6H_4\cdot CO$ (XXXVII.)

(XXXIX.) ← $NO_2\cdot C_6H_4\cdot CO$ (XXXVIII.)

$CH_2\cdot CH_2\cdot COOC_2H_5$

(XL.) β-N-3-Methyl-2-carbethoxy-pyrrolidylpropionate. → (XLI.) 7-Keto-1-methyl-pyrrolizidine.

Chart 3. Synthesis of Retronecanone.

added quantitatively to ethyl acrylate to form β-*N*-3-methyl-2-carbethoxy-pyrrolidylpropionate (XL) which by a Dieckmann cyclisation and hydrolysis gave 7-keto-1-methylpyrrolizidine (XLI). Only single derivatives were obtained from the oily product although theoretically two racemic forms were possible. The *L*-menthylhydrazone of (XLI) was different from that of retronecanone and hydrolysed back to optically inactive (XLI). The synthesis was repeated using optically active 1-β-methyl-δ-*m*-nitrobenzoylamino-valeric acid (XXXVI) to give an optically active 7-keto-1-methylpyrrolizidine (XLI) from which was isolated a single individual oxime identical with retronecanone oxime.

The Illinois school added a further synthesis which had wide applicability and gave good yields. LEONARD, HRUDA and LONG (*115*) as well as LEONARD and BECK (*112*) effected a Michael condensation of a primary nitroalkane (XLII) with two moles of ethyl acrylate to yield ethyl γ-alkyl-γ-nitropimelate (XLIV; $R' = H$). Hydrogenation at 2 atm. pressure with platinum oxide as catalyst brought about reduction of the nitro group with consequent amide formation (XLV; $R' = H$) and further reduction in the presence of copper chromite at 250° and 300° atmospheres gave 8-alkylpyrrolizidine (XLVI; $R' = H$). Alkyl substituted pyrrolizidines resulted from the usė of substituted acrylic esters (*116*). LEONARD and FELLEY (*113*) condensed nitromethane with ethyl crotonate and then with ethyl acrylate to give ethyl β-methyl-γ-nitropimelate (XLIV; $R = H$, $R' = CH_3$) which by reductive ring closure gave 1-methylpyrrolizidine (XLVI; $R = H$, $R' = CH_3$).

$$R \cdot CH_2 \cdot NO_2 \rightarrow R \cdot CH(NO_2) \cdot CHR' \cdot CH_2 \cdot COOC_2H_5 \rightarrow$$

(XLII.) (XLIII.)

$$\rightarrow C_2H_5OOC \cdot CH_2 \cdot CH_2 \cdot CR(NO_2) \cdot CHR' \cdot CH_2 \cdot COOC_2H_5 \rightarrow$$

(XLIV.) Ethyl γ-alkyl-γ-nitropimelate ($R' = H$).

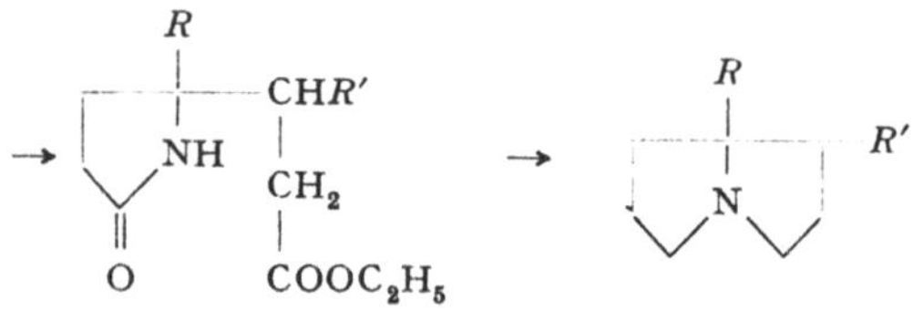

(XLV.) (XLVI.) 8-Alkyl-pyrrolizidine ($R' = H$).

This synthesis has been extended by LEONARD and SHOEMAKER (*116*, *117*) for the preparation of 2-methyl-, 2:5-dimethyl-, 8-hydroxymethyl- and 8-chloromethyl-pyrrolizidines. The 8-chloromethyl-derivatives were of interest in that they remained unchanged under conditions which cause rearrangement of acyclic (*90*) and monocyclic (*71*) β-chloramines.

Stereochemistry and Syntheses of Heliotridane, Pseudo-heliotridane and their Derivatives.

The 1-methylpyrrolizidine (XLVI, $R = H$, $R' = CH_3$) was identical with that synthesised by MEN'SHIKOV (*147*) and PRELOG (*156*) and designated as "*DL*-heliotridane". This synthesis should give two stereoisomers and LEONARD and FELLEY (*114*), employing chromatography and distillation, isolated two picrates, m. p. 234–236° and 243–244°, corresponding to the picrates of *DL*-pseudoheliotridane and *DL*-heliotridane, respectively. The *DL*-pseudoheliotridane was successfully resolved to give *D*-pseudoheliotridane, $[\alpha]_D + 6.9°$ (homogeneous) which compared favourably with *L*-heliotridane, $[\alpha]_D - 8.25°$ (homogeneous) obtained by MEN'SHIKOV and BORODINA (*138*). The identity was also confirmed by the preparation of derivatives. The slightly different figures given by various authors have been collected by LEONARD and FELLEY (*114*) who have also recorded the infrared spectra for pseudoheliotridane and heliotridane. For clarity the data presented in *Table 8* have been selected

Table 8. Physical Constants of the Diastereoisomeric 1-Methylpyrrolizidines and their Derivatives.

	B. p.	n_D^{20}	$[\alpha]_D$ (homogeneous)	Melting points: picrate	picrolonate	methiodide
L-Pseudoheliotridane	165–171°	1.464	— 8.25°	232°	162–163°	275°
L-Heliotridane	159–160°	1.462	— 92°, — 68° — 51°	243–244°	153–154°	240–250°

and only where large discrepancies exist has more than one figure been included.

The last synthesis gave predominantly *DL*-pseudoheliotridane and this was seemingly true also for PRELOG's synthesis (*156*). A substantial increase in the yield of *DL*-heliotridane from 5% to 13% was effected by employing a modification of the synthesis using ethyl ethylidenemalonate instead of ethyl crotonate.

The synthesis undoubtedly gave only 1-methylpyrrolizidine, and the successful separation of the two diastereoisomers proved conclusively that *L*-pseudoheliotridane was indeed the diastereoisomer of *L*-heliotridane as MEN'SHIKOV (*138, 134*) had previously considered probable.

The synthesis was extended to the preparation of *DL*-hydroxymethylpyrrolizidine (XLVI; $R = H$, $R' = CH_2OH$, p. 220) by effecting the Michael reaction first with ethyl acetoxycrotonate and then with ethyl acrylate. The product (XLIV; $R = H$, $R' = —CH_2OOC \cdot CH_3$; p. 220) on gentle reductive cyclisation gave a mixture from which *DL*-hydroxymethylpyrrolizidine (XLVI; $R = H$, $R' = CH_2OH$) could be separated as a picrate having a melting point similar to that reported for *L*-trachelanthamidine.

LEONARD and FELLEY (*114*) pointed out that the ready dehydration of platynecine to anhydro-platynecine (XLIX) demands that the hydroxymethyl group at $C_{(1)}$, and therefore the methyl group in heliotridane (LI), must be *trans* to the $C_{(8)}$-hydrogen. It follows that the hydroxymethyl group in trachelanthamidine (LIII) and the methyl group in pseudoheliotridane (LII) must be *cis* to the $C_{(8)}$-hydrogen.

No definitive configuration could be assigned to the hydroxyl group at $C_{(7)}$ in platynecine because, with the reagents employed in the dehydration, inversion could not be discounted. DRY, KOEKEMOER and WARREN (*64*) have recently shown, however, that toluene-*p*-sulphonyl chloride and platynecine reacted readily to give anhydro-platynecine. These authors interpreted the reaction as taking place first by the esterification of the primary hydroxyl group of platynecine to give the toluene-*p*-sulphonyl ester (XLVII), and then, under the basic conditions and the close proximity of the hydrogen of the sec. hydroxyl group, the splitting of toluene-*p*-sulphonic acid to leave anhydro-platynecine (XLIX). The hydroxyl group at $C_{(7)}$ was accordingly placed also *trans* to the $C_{(8)}$-hydrogen. These reactions are shown in *Chart 4*, and permit further definition of the stereochemistry of platynecine (XLVII) and retronecine (L) and their derivatives*.

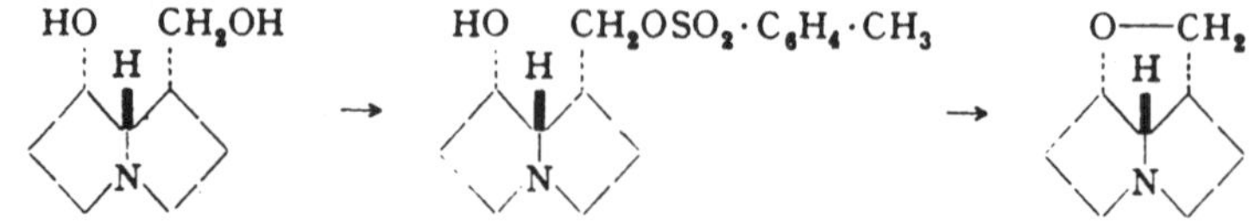

HO CH₂OH → HO CH₂OSO₂·C₆H₄·CH₃ → O—CH₂

(XLVII.) Platynecine. (XLVIII.) (XLIX.) Anhydro-platynecine.

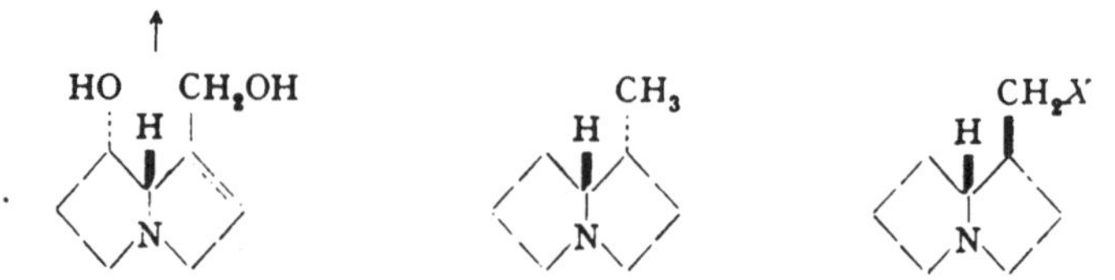

(L.) Retronecine. (LI.) Heliotridane. (LII, X = H) Pseudo-heliotridane. (LIII, X = OH) Trachelanthamidine.

Chart 4. Reactions Showing the Stereochemistry of Heliotridane and its Derivatives**.

Trihydroxy-methylpyrrolizidine: Rosmarinecine.

Rosmarinecine, $C_8H_{15}O_3N$, was first obtained as an oil by DE WAAL (*182*) from the hydrolysis of rosmarinine. RICHARDSON and WARREN (*159*) obtained it crystalline and showed that it was a tertiary amine which was not reduced by sodium amalgam or catalytically. The existence of a methylpyrrolizidine nucleus for the known necine bases indicated that rosmarinecine could be a trihydroxy-pyrrolizidine $C_8H_{12}N(OH)_3$. The failure to reduce precluded any hydroxyl being on a carbon atom vicinal to the nitrogen. In addition rosmarinecine was not oxidised by periodic acid so that no glycol grouping existed. This left two possibilities, 2 : 7-

* ADAMS and VAN DUUREN (*28 a*) also arrived at the same conclusion by the preparation of platynecine sulphite and deduced the steric configurations of the other pyrrolizidine bases in agreement with *Chart 5*, p. 224. (Added in proof.)

** The N–$C_{(8)}$ bond is assumed to be in the plane of the paper, and the two rings are inclined towards each other and in a direction opposite to that of the $C_{(8)}$-hydrogen atom. The hydrogen atoms which may be attached to the other carbon atoms are not shown. Lines in heavy type are intended to indicate a configuration opposite to that of the dotted lines:

H C N H C N

(or 2:6)-dihydroxy-1-hydroxymethyl-pyrrolizidine. The authors favoured structure (LIV) because it represented rosmarinecine as the hydroxyl derivative of platynecine with which it was associated in some *Senecio* species, and dehydration of rosmarinecine could account for retronecine (L).

Proof of this structure has recently been afforded by DRY, KOEKEMOER and WARREN (*64*) who prepared the triacetate (LV), and by the action of thionyl chloride a trichloro-compound (LVI). It is of interest that only a dibenzoyl derivative (LVII) could be prepared, the structure of which follows from the observation that the two hydroxyl groups of platynecine can be benzoylated (*151*). Concentrated sulphuric acid gave anhydro-rosmarinecine (LVIII) which yielded a monacetate (LIX) and

X CH_2X → O — CH_2 ; N, —Y

(LIV; $X = Y =$ OH)
(LV; $X = Y = OOC \cdot CH_3$)
(LVI; $X = Y =$ Cl)
(LVII; $X = O \cdot CO \cdot C_6H_5$, $Y =$ OH)

(LVIII; $Y =$ OH)
(LIX; $Y = O \cdot CO \cdot CH_3$)
(LX; $Y =$ Cl)
(XLIX; $Y =$ H)

which with thionyl chloride gave chloroanhydro-platynecine (LX); the latter was reduced catalytically to anhydro-platynecine (XLIX).

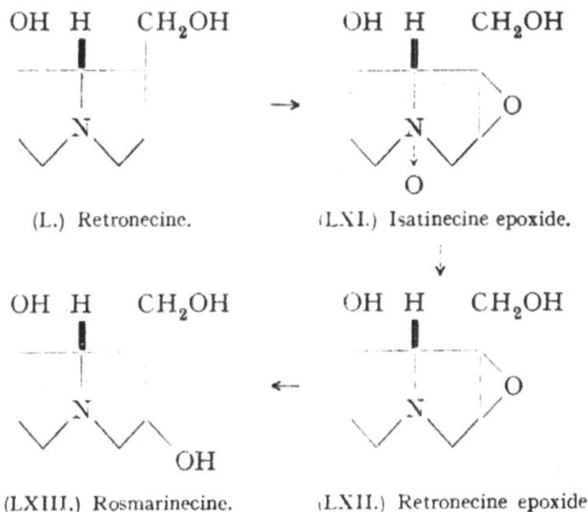

(L.) Retronecine. (LXI.) Isatinecine epoxide.

(LXIII.) Rosmarinecine. (LXII.) Retronecine epoxide.

The orientation of the groups at the 1-, 7-, and 8-positions follows from those of platynecine. The configuration of the 2-hydroxyl group has been established by DRY, KOEKEMOER and WARREN (*64*) by synthesis. Perbenzoic acid on retronecine (L) readily gave retronecine *N*-oxide (XXVII, p. 217) (isatinecine) which reacted slowly to give isatinecine epoxide (LXI). The epoxide group is placed *cis* to the $C_{(8)}$-hydrogen as the outer fold of the pyrrolizidine structure will be less structurally hindered. Reduction by zinc dust in acid solution or with hydrogen and Adams' catalyst gave retronecine epoxide (LXII), which was further reduced catalytically in the presence of Raney's nickel. Although the epoxide ring could be opened theoretically in two ways, the bulk of the product, which contained no glycol grouping, was rosmarinecine. This places the 2-hydroxyl group *cis* to the $C_{(8)}$-hydrogen and the configuration of rosmarinecine is completely defined (LXIII).

Bases of Unknown Structure: Mikanecine, Hastanecine, Turneforcidine and Otonecine.

Mikanecine, $C_8H_{15}O_2N$, is the amorphous base obtained from the hydrolysis of mikanoidine by Manske (*120*) who deduced the formula from that of the crystalline picrate, m. p. 186°, and suggested that it was dihydroretronecine. Leonard (*111*) elaborates this concept by drawing attention to the possible identity of the picrate with that of platynecine (dihydroretronecine) picrate, m. p. 184°. — *Hastanecine*, $C_8H_{15}O_2N$, which Konovalov and Men'shikov (*94*) isolated from the alkaline hydrolysis of hastacine, has not been further studied. — *Turneforcidine*, $C_8H_{15}O_2N$, derived

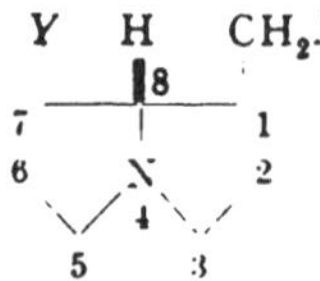

	X	Y
Retronecine:	X = OH	Y = OH.
Supinidine:	X = OH	Y = H.
Desoxyretronecine:	X = H	Y = OH.

Y H CH_2X

N

Heliotridine X = OH, Y = OH.

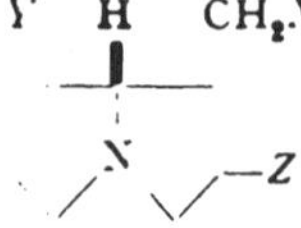

	X	Y	Z
Heliotridane:	H	H	H.
Retronecanol:	H	OH	H.
*L-iso*Retronecanol:	OH	H	H.
Platynecine:	OH	OH	H.
Rosmarinecine:	OH	OH	OH.

OH H CH_3

N

Oxyheliotridane.

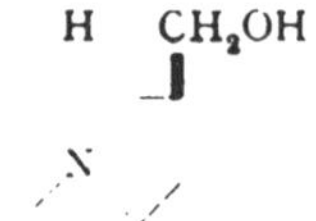

*D-iso*Retronecanol

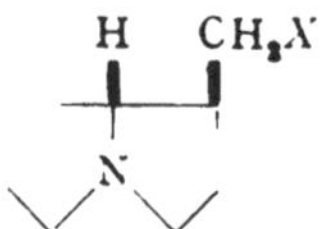

Pseudoheliotridane: X = H.
L-Trachelanthamidine: X = OH.
(*D*-Trachelanthamidine ⩵ Laburnine.)

Chart 5. Stereochemistry of the Hydroxy-methylpyrrolizidines.

by MEN'SHIKOV, DENISOVA and MASSAGETOV (*139*) by the alkaline hydrolysis of turneforcine, has a melting point and rotation similar to that of hastanecine; but since these two necines were obtained in the same laboratory their identity cannot here be implied. — *Otonecine* hydrochloride, $C_9H_{15}O_3N \cdot HCl$, from the acid hydrolysis of otosenine, was shown by ZHDANOVICH and MEN'SHIKOV (*195*) to contain an *N*-methyl group. It was reduced catalytically to dihydro-desoxyotonecine, $C_9H_{17}O_2N$, which contained one active hydrogen and a carbonyl group. A pyrrolidine nucleus has been suggested (*195*), and this necine is an exception to the general necine structure.

Summary of the Structures of the Bases.

The different structures of the bases set out in *Chart 5* follow from the configuration for platynecine (*64*) (*Chart 4*, p. 222). The structure, however, for pseudoheliotridane and hence also for trachelanthamidine may be either as shown or the mirror image of this form. In considering these structures it is interesting to note that heliotridine and retronecine do not behave similarly on reduction. It was established by ADAMS and LEONARD (*13*) that retronecine and retronecine esters gave on reduction one isomer only, as for example retronecanol or platynecine. The reduction of heliotrine gives oxyheliotridane which CULVENOR, DRUMMOND and PRICE (*54*) conclude is a mixture of two possible isomers resulting from the hydrogenation of the double bond.

III. The Acid Hydrolysis Products.

The acids associated with the pyrrolizidine alkaloids are set out in *Table 9*, p. 226. These acids are obtained from the alkaloids by hydrolysis or hydrogenolysis, except that sceleratinic acid and senecioic acid have only been found free in the plant and not in combination with the derivatives of pyrrolizidine. It was early observed that the 'necic' acids from the *Senecio* alkaloids were C_{10}-acids and BARGER (*36*) envisaged an organic unity when he predicted that these acids were terpene acids. The difficulties of separation have resulted in mixtures being characterised as pure substances; and ease of isomerisation has caused some known acids to be rediscovered under different names. This confusion has been resolved and the corrections are shown in *Table 10*, p. 227.

C_{10}-Acids.

The first indication that the necic acids were capable of isomerism came from the observation that isatidine was retrorsine *N*-oxide (*48*), and yet these two alkaloids, isatidine and retrorsine, were reported as yielding different acids on hydrolysis, namely isatinecic and retronecic acids, respectively. The experimental conditions, however, were not identical, and in fact DE WAAL (*180*) reported that the hydrolysis of isatidine with alcoholic potassium hydroxide gave a new isatinecic acid, dewalinic

Table 9. The Acids from the Hydrolysis of Pyrrolizidine Alkaloids.

[Solvents used for spec. rotation measurements: (c) chloroform; (e) ethanol; (m) methanol; (w) water.]

Acid	Formula	Melting point	$[\alpha]_D$	References: Isolation	References: Structure	References: Synthesis
Dicrotalic acid	$C_6H_{10}O_5$	109°	0°	(123)	(24)	(24)
Grantianic acid	$C_{10}H_{14}O_7$	—	—	(3)	—	—
Hastanecic acid	$C_{10}H_{16}O_5$	149°	+ 4.6°	(94)	—	—
Heliotrinic acid	$C_8H_{16}O_4$	94.5°	— 12° (w)	(125, 54)	(133, 26)	—
Integerrinecic acid	$C_{10}H_{16}O_5$	151°	+ 15.9° (e)	(121)	(105)	—
Isatinecic acid	$C_{10}H_{16}O_6$	148.5°	+ 88.2° (w)	(40)	(49, 182)	—
Jaconecic acid	$C_{10}H_{16}O_6$	182°	+ 31.7°	(37)	(41)	—
Lasiocarpic acid	$C_8H_{16}O_5$	97°	+ 10.6° (e)	(146)	(54, 63)	—
Mikanecic acid	$C_{12}H_{16}O_5$	240°	—	(120)	—	—
Monocrotalic acid	$C_8H_{12}O_5$	182°	— 5.3° (w)	(18)	(7)	(29)
Monocrotic acid	$C_7H_{12}O_3$	b. p. 146°/18 mm	0°	(18)	(22)	(22)
Platynecic acid	$C_{10}H_{16}O_5$	133–135°	—	(60)	(60)	—
Retronecic acid	$C_{10}H_{16}O_6$	181°	— 11.4° (w)	(38)	(49)	—
Riddellic acid	$C_{10}H_{14}O_6$	103°	2.6° (e)	(10)	(27)	—
Sceleranecic acid	$C_{10}H_{14}O_5$	156°	— 9.3° (w)	(187)	(188)	—
Sceleratinic acid	$C_{10}H_{13}O_4Cl$	208°	—	(186)	(188)	—
Senecic acid (Senecionic)	$C_{10}H_{16}O_5$	152°	+ 11.8° (e)	(36)	(36, 104)	—
Senecifolic acid	$C_{10}H_{16}O_6$	199°	+ 28.4° (e)	(192)	—	—
Seneciphyllic acid	$C_{10}H_{14}O_5$	145°	0°	(98)	(8)	—
*iso*Seneciphyllic acid	$C_{10}H_{14}O_5$	105–108°	— 8.6° (e)	(98)	(8, 105)	—
Senecioic acid (αβ-dimethylacrylic)	$C_5H_8O_2$	69–70°	—	(34)	(34)	—
Squalinecic acid	$C_{10}H_{14}O_4$	129°	—	(36)	—	—
Trachelanthic acid	$C_7H_{14}O_4$	93–95°	+ 1.3° (w) + 2.3° (e)	(137, 135)	(138, 53)	(28, 65)
Acid from trichodesmine	$C_7H_{12}O_3$	—	—	(129)	—	—
Usaramoensinecic acid	$C_{10}H_{16}O_5$	170°	+ 6.66° (e)	(25)	(25)	—
Viridifloric acid	$C_7H_{14}O_4$	121°	0°	(136)	(136)	(28, 65)

Table 10. List of Acids which were Formerly Characterised as New Individuals and are now Known to be Mixtures or Identical with Previously Known Forms.

Original name	Reference	Present characterisation	Reference
Carthamoidinecic acid	(*111*)	Mixture of acids from senecionine and seneciphylline	(6, *15*)
Dewalinic acid	(*179*, *180*)	Retronecic acid	(*48*)
Hieracinecic acid	(*122*)	Mixture of acids from senecionine and seneciphylline	(*55*)
α-Longinecic acid	(*121*)	*iso*Seneciphyllic acid	(*15*)
β-Longinecic acid	(*6*)	Isatinecic acid	(*191*)
Pterophinecic acid	(*182*)	Mixture of acids from senecionine and seneciphylline	(*55*)

acid, m. p. 181°, whereas BLACKIE (*40*) had obtained isatinecic acid by hydrolysis with barium hydroxide. The identity of dewalinic acid with retronecic acid was pointed out by CHRISTIE, KROPMAN, LEISEGANG and WARREN (*48*) and later confirmed by DE WAAL (*185*).

Isatinecic and Retronecic Acids.

Retronecic acid had been shown by BARGER (*38*) to be a monoethylenic dicarboxylic acid which gave a monolactone. The difference between isatinecic and retronecic acids was attributed by CHRISTIE et al. (*48*) to geometrical isomerism since both these acids were αβ-unsaturated acids which on reduction gave the same dihydro-acid.

CHRISTIE, KROPMAN, NOVELLIE and WARREN (*49*) established that isatinecic and retronecic acids were *cis-* and *trans-1 : 2-dihydroxy-3-methylhept-5-ene-2 : 5-dicarboxylic acid* (LXIV), respectively. The glycol moiety containing a primary hydroxyl group was established by the rapid absorption of two equivalents of oxygen and the formation of formaldehyde with periodic acid and with lead tetraacetate. The slower consumption of a third equivalent of oxygen with these reagents indicated an α-hydroxy-acid grouping, which was confirmed by the quantitative production of carbon monoxide on treatment with concentrated sulphuric acid and of carbon dioxide on treatment with lead tetraacetate.

The two acids gave with lead tetraacetate in aqueous solution *cis-* and *trans-*α-methyl-γ-ethylideneglutaric acid (LXV) both of which yielded the *trans*-imide (*Chart 6*). The *trans*-acid was hydrogenated to α-methyl-γ-ethylglutaric acid, (LXVI) which in turn was obtained directly by treatment of both dihydroisatinecic (LXVII) and dihydroretronecic (LXVII) acids with lead tetraacetate. Ozonolysis of isatinecic and retronecic acids gave acetaldehyde and an acid (LXVIII) which by treatment with lead tetraacetate gave methylsuccinic acid (LXIX).

Structure (LXX) would also be capable of giving the degradation products (LXVI) and (LXIX); but this formula was excluded in that oxidation of isatinecic acid with osmium tetroxide gave a compound with equivalent weight corresponding to dihydroxy-dihydroisatinecic acid, $C_{10}H_{18}O_8$ (LXXI).

(LXVI.) $HOOC \cdot CH(C_2H_5) \cdot CH_2 \cdot CH(CH_3) \cdot COOH + CH_2O + CO_2$

α-Methyl-γ-ethyl-glutaric acid.

↑ $Pb(OOC \cdot CH_3)_4$

(LXVII.) $HOOC \cdot CH(C_2H_5) \cdot CH_2 \cdot CH(CH_3) \cdot C(OH)(COOH) \cdot CH_2OH$

Dihydroisatinecic and dihydroretronecic acids.

↑

(LXXI.) $CH_3 \cdot CH(OH) \cdot C(OH)(COOH) \cdot$
$\cdot CH_2 \cdot CH(CH_3) \cdot C(OH)(COOH) \cdot CH_2OH$

↑ OsO_4

(LXIV.) $CH_3 \cdot CH : C(COOH) \cdot CH_2 \cdot CH(CH_3) \cdot C(OH)(COOH) \cdot CH_2OH$

↓ $Pb(OOC \cdot CH_3)_4$ ↓ O_3

(LXV.) $CH_3 \cdot CH : C(COOH) \cdot CH_2 \cdot CH(CH_3) \cdot COOH + CO_2$

α-Methyl-γ-ethylidene-glutaric acid.

(LXVIII.) $CH_3 \cdot CHO + HOOC \cdot CO \cdot CH_2 \cdot CH(CH_3) \cdot C(OH)(COOH) \cdot CH_2OH$

↓ $Pb(OOC \cdot CH_3)_4$

(LXIX.) $2\,CO_2 + CH_2O + HOOC \cdot CH_2 \cdot CH(CH_3) \cdot COOH$

Methylsuccinic acid.

(LXX.) $HOOC \cdot CH(CH_3) \cdot CH_2 \cdot C(: CH \cdot CH_3) \cdot C(OH)(COOH) \cdot CH_2OH$

Chart 6. Reactions Revealing the Structures of Retronecic and Isatinecic Acids.

The dihydroxy-compound from structure (LXX) would yield a γ-hydroxy-acid which would be expected to give a lactone. Furthermore, formula (LXIV) is in agreement with spectroscopic data for both isatinecic and retronecic acids (*48*).

Senecic, Integerrinecic and Usaramoensinecic Acids.

Senecic acid lactone, $C_{10}H_{14}O_4$, was first obtained pure under the name of platynecic acid by OREKHOV and TIDEBEL (*153*). BARGER and BLACKIE (*36*) prepared this lactone from the hydrolysis of senecionine and called it 'senecic acid', an unsaturated lactonic acid (readily reduced to 'dihydrosenecic' acid), containing three C-methyl groups. DE WAAL (*181*) showed that it was identical with platynecic acid.

RICHARDSON and WARREN (*159*) obtained senecic acid as a hydroxy-dicarboxylic acid, $C_{10}H_{16}O_5$, as well as its lactone, and since both showed considerable stability it was reasonable to assume that the hydroxyl group was in the δ-position to one of the carboxyls.

KROPMAN and WARREN (*104*) established that *senecic acid* was *2-hydroxy-3-methylhept-5-ene-2:5-dicarboxylic acid* (LXXII) (*Chart 7*). Reduction of senecic acid gave dihydrosenecic acid (LXXIII) which was oxidised with lead tetraacetate to carbon dioxide and a keto-acid (LXXIV) which with sodium hypobromite gave methyl-ethyl-glutaric acid (LXXV). Ozonisation of senecic acid (LXXII) yielded acetaldehyde and an acid presumably (LXXVI) which was degraded by oxidation to (+)-methyl-succinic acid (LXXVII).

(LXXVII.) $HOOC \cdot CH_2 \cdot CH(CH_3) \cdot COOH \leftarrow HOOC \cdot CH_2 \cdot CH(CH_3) \cdot CO \cdot CH_3 + CO_2$
Methylsuccinic acid.

↑

$CH_3 \cdot CHO + HOOC \cdot CO \cdot CH_2 \cdot CH(CH_3) \cdot C(OH)(COOH) \cdot CH_3$ (LXXVI.)

↑

(LXXII.) $CH_3 \cdot CH : C(COOH) \cdot CH_2 \cdot CH(CH_3) \cdot C(OH)(COOH) \cdot CH_3$
Senecic acid.

↓

(LXXIII.) $CH_3 \cdot CH_2 \cdot CH(COOH) \cdot CH_2 \cdot CH(CH_3) \cdot C(OH)(COOH) \cdot CH_3$
Dihydrosenecic acid.

↓

$CH_3 \cdot CH_2 \cdot CH(COOH) \cdot CH_2 \cdot CH(CH_3) \cdot CO \cdot CH_3$ (LXXIV.)

↓

$HOOC \cdot CH(C_2H_5) \cdot CH_2 \cdot CH(CH_3) \cdot COOH$ (LXXV.)
α, α'-Methyl-ethyl-glutaric acid.

Chart 7. Reactions Showing the Structure of Senecic Acid.

The lactone from senecic acid was shown by KROPMAN and WARREN (*105*) to give *trans*-senecic acid, which was identified as integerrinecic acid obtained by MANSKE (*121*) by the hydrolysis of integerrimine. The geometrical isomerism of senecic and integerrinecic acids thus parallels

$$\begin{array}{c} H-C-CH_3 \\ \| \\ R-C-COOH \end{array} \qquad\qquad \begin{array}{c} CH_3-C-H \\ \| \\ R-C-COOH \end{array}$$

Senecic acid, m. p. 152°. Isatinecic acid, m. p. 148.5°.

Integerrinecic acid m. p. 151°. Retronecic acid m. p. 181°.

$R = CH_2X \cdot C(OH)(COOH) \cdot CH(CH_3) \cdot CH_2 \cdot$,
in which $X = H$ for senecic and integerrinecic acids
and $X = OH$ for isatinecic and retronecic acids.

that found for isatinecic (*cis*-hydroxysenecic) and retronecic (*trans*-hydroxysenecic) acids.

It is of interest here to note that the ultraviolet extinction curves for the *cis*- and *trans*-acids showed corresponding maxima (*Table 11*) which as KROPMAN and WARREN (*105*) indicated could be of significance in assigning a geometrical configuration when only one form of an $\alpha\beta$-unsaturated necic acid was available.

Table 11. Absorption Maxima in the Ultraviolet for *cis*- and *trans*-'Necic' Acids.

cis-acids	λ_{max}	ε_{max}	*trans*-acids	λ_{max}	ε_{max}
Senecic acid	215 mμ	4100	Integerrinecic acid	218 mμ	9333
Isatinecic acid	215	4500	Retronecic acid	218	9400

It is of significance that senecioic (2-methylprop-1-ene-1-carboxylic) acid (*34*) possesses the *iso*pentane unit; and KROPMAN and WARREN (*105*) indicated that it was not improbable that all the 'necic' acids possessed the same carbon skeleton.

A general formula for these *'necic' acids* was advanced (*105*) and represented a class of monoterpenes with a new 'irregular' isoprene chain (LXXVIII). Thus, the concept put forward in 1936 by BARGER and BLACKIE (*36*) that these C_{10}-acids were examples of terpene acids is fully justified.

$$\begin{array}{l} \text{C—C(COOH)—C—C} \\ \qquad\qquad\qquad\quad | \\ \qquad\qquad\qquad\quad \text{C—C(COOH)—C—C} \end{array} \qquad \text{(LXXVIII.)}$$

The basic carbon skeleton of the necic acids so far studied seemed established and the found variation in structure was limited to the number of hydroxyl groups and the geometrical configuration, although obviously isomerism was possible at the two asymmetric centres. Actual isomerism at $C_{(2)}$ of these necic acids was established by the elegant work of ADAMS and his co-workers (*25*).

Usaramoensinecic acid from the *Crotalaria* alkaloid usaramoensine was isomeric with senecic and integerrinecic acids and like senecic acid (*105*) gave integerrinecic acid lactone. Usaramoensinecic acid showed in the ultraviolet a maximum absorption at 215 mμ (ε_{max} 6000) which was similar to that of senecic acid for which the *cis*-configuration was assigned (*105*). Since senecic, integerrinecic and usaramoensinecic acids give the same lactone, namely integerrinecic acid lactone, ADAMS (*25*) concluded that the isomerism is not associated with the asymmetry at $C_{(3)}$ but more probably with that at $C_{(2)}$.

The orientation of the methyl group at $C_{(3)}$ is defined by the isolation by KROPMAN and WARREN (*104*) of (+)-methylsuccinic acid on oxidative degradation of senecic acid and must be the same in all acids. The orienta-

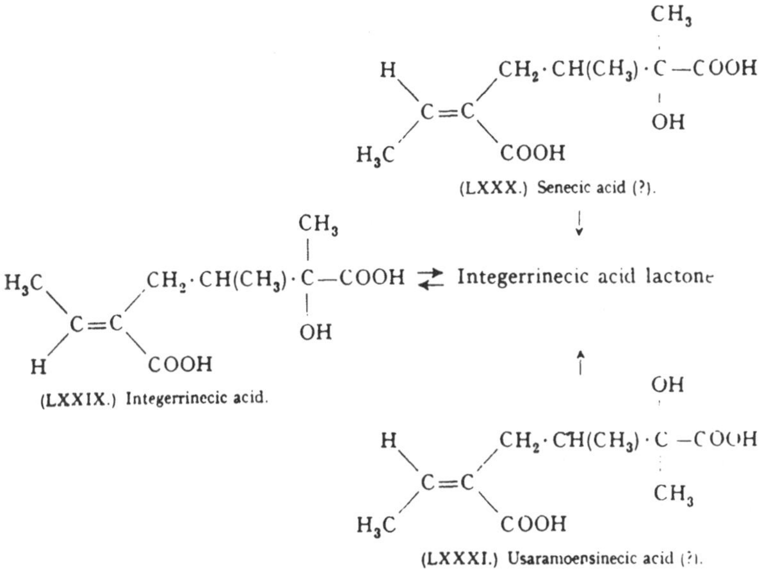

$$\begin{array}{c} \mathrm{H} \\ \mathrm{H_3C} \end{array} \!\!>\!\! C{=}C \!\!<\!\! \begin{array}{l} \mathrm{CH_2\cdot CH(CH_3)\cdot C(CH_3)(OH)-COOH} \\ \mathrm{COOH} \end{array}$$

(LXXX.) Senecic acid (?).

$$\begin{array}{c} \mathrm{H_3C} \\ \mathrm{H} \end{array} \!\!>\!\! C{=}C \!\!<\!\! \begin{array}{l} \mathrm{CH_2\cdot CH(CH_3)\cdot C(CH_3)(OH)-COOH} \\ \mathrm{COOH} \end{array} \rightleftarrows \text{Integerrinecic acid lactone}$$

(LXXIX.) Integerrinecic acid.

$$\begin{array}{c} \mathrm{H} \\ \mathrm{H_3C} \end{array} \!\!>\!\! C{=}C \!\!<\!\! \begin{array}{l} \mathrm{CH_2\cdot CH(CH_3)\cdot C(OH)(CH_3)-COOH} \\ \mathrm{COOH} \end{array}$$

(LXXXI.) Usaramoensinecic acid (?).

tion at $C_{(3)}$ is not shown in the formulae because the configuration at $C_{(3)}$ relative to that at $C_{(2)}$ is not known. If integerrinecic acid be defined as shown in formula (LXXIX) then senecic acid and usaramoensinecic acid may be represented by (LXXX) and (LXXXI) respectively or the other way round.

According to DANILOVA and KONOVALOVA (*60*) platyphylline gives on hydrolysis with alcoholic alkali senecic acid and a new acid, platynecic acid, $C_{10}H_{16}O_5$, m. p. 133–135°, which lactonises to integerrinecic acid lactone.

ADAMS and VAN DUUREN (*25*) speculate that this acid is the fourth stereoisomer and is identical with hieracinecic acid obtained by MANSKE (*122*) from hieracifoline, which, however, has since been shown by CULVENOR (*55*) to be a mixture.

Seneciphyllic and iso*Seneciphyllic Acids.*

A further structural modification within the same carbon skeleton was forthcoming from ADAMS, GOVANDACHARI, LOOKER and EDWARDS (*8*) who established the structure of seneciphyllic acid (α-longinecic acid) (*15*), m. p. 111–113°, which was originally obtained by the hydrolysis of seneciphylline as 2-hydroxy-3-methylhepta-3 : 5-diene-2 : 5-dicarboxylic acid (LXXXII). This acid was shown to be a dibasic α-hydroxy-acid which was reduced catalytically to a tetrahydro-acid (LXXXIII).

Lead tetraacetate on this reduced acid gave a keto-acid (LXXXIV) which, with sodium hypobromite yielded α-methyl-α'-ethyl-glutaric acid (LXXXV). Seneciphyllic acid itself with lead tetraacetate gave an optically inactive keto-acid (LXXXVI) and carbon dioxide.

$$CH_3 \cdot CH : C(COOH) \cdot CH : C(CH_3) \cdot CO \cdot CH_3 \quad (LXXXVI) + CO_2$$

$$\uparrow$$

$$CH_3 \cdot CH : C(COOH) \cdot CH : C(CH_3) \cdot C(OH)(COOH) \cdot CH_3$$

(LXXXII.) Seneciphyllic acid.

$$\downarrow$$

$$C_2H_5 \cdot CH(COOH) \cdot CH_2 \cdot CH(CH_3) \cdot C(OH)(COOH) \cdot CH_3 \quad (LXXXIII.)$$

$$\downarrow$$

$$C_2H_5 \cdot CH(COOH) \cdot CH_2 \cdot CH(CH_3) \cdot CO \cdot CH_3 \quad (LXXXIV.)$$

$$\downarrow$$

$$HOOC \cdot CH(C_2H_5) \cdot CH_2 \cdot CH(CH_3) \cdot COOH$$

(LXXXV.) α-Methyl-α'-ethyl-glutaric acid.

KONOVALOVA and DANILOVA (*98*) reported that the hydrolysis of seneciphylline with aqueous and alcoholic alkali gave *iso*seneciphyllic, m. p. 105–108°, and seneciphyllic acid, m. p. 144–145°, respectively. The two acids were reported as geometrical isomers, and the *iso*-acid the less stable. KROPMAN and WARREN (*105*) assigned to *iso*seneciphyllic and seneciphyllic acids the *cis*- and *trans*-configurations, respectively, on which basis ADAMS' structure (LXXXII) (*8*) for seneciphyllic acid may now be elaborated as follows.

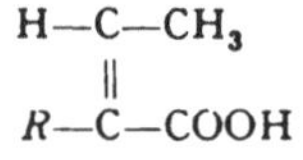

```
H—C—CH3                CH3—C—H
  ||                       ||
R—C—COOH               R—C—COOH
```

*iso*Seneciphyllic acid (*cis*). 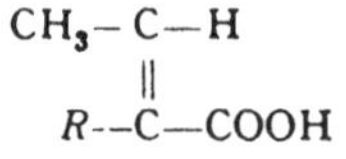 Seneciphyllic acid (*trans*).

$$R = CH : C(CH_3) \cdot C(OH)(COOH) \cdot CH_3$$

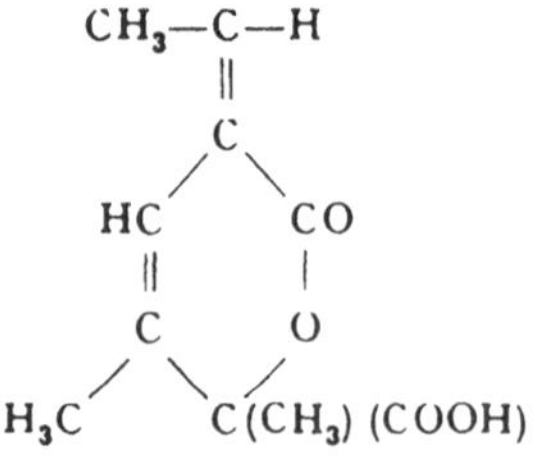

```
      CH3—C—H
          ||
          C
        /   \
      HC     CO
      ||     |
      C      O
    /   \   /
H3C      C(CH3)(COOH)
```

(LXXXVII.) Seneciphyllic acid lactone.

No reference has ever been made to the configuration at the double bond at $C_{(3)}$ of this acid. Undoubtedly the arrangement in seneciphyllic acid lactone (LXXXVII) is *cis* with respect to the carbon chain, and presumably seneciphyllic acid is similarly *cis*. Inspection of a model of the alkaloid seneciphylline, which is elaborated in a later section reveals also that only the *cis*-configuration at $C_{(3)}$ is possible, so the *iso*seneciphyllic acid is seemingly similarly oriented.

Riddellic Acid.

Using essentially these same reactions riddellic acid, the hydrolysis product of riddelliine, obtained by ADAMS, HAMLIN, JELINEK and PHILLIPS (*10*), was shown by ADAMS and VAN DUUREN (*27*) to be 1 : 2-dihydroxy-3-methylhepta-3 : 5-diene-2 : 5-dicarboxylic acid (LXXXVIII). The acid was degraded by lead tetraacetate to give formaldehyde, carbon dioxide and 2-methylhexa-2 : 4-diene-1 : 3-dicarboxylic acid (LXXXIX) which reduced to methyl-ethyl-glutaric acid (LXXXV). An interesting and additional proof of the 1 : 2-glycol grouping was the formation of a sulphite ester (XC) by the action of thionyl chloride on the bis-*p*-phenylphenacyl ester of riddellic acid. Furthermore, infrared and ultraviolet spectra revealed the same unsaturated chromophore as in seneciphyllic acid (*27*).

$$HOOC \cdot CH(C_2H_5) \cdot CH_2 \cdot CH(CH_3) \cdot COOH$$

(LXXXV.) α, α′-Methyl-ethyl-glutaric acid.

↑

$$CH_3 \cdot CH : C(COOH) \cdot CH : C(CH_3) \cdot COOH + CO_2 + CH_2O$$

(LXXXIX.) 2-Methylhexa-2 : 4-diene-1 : 3-dicarboxylic acid.

↑

$$CH_3 \cdot CH : C(COOH) \cdot CH : C(CH_3) \cdot C(OH)(COOH) \cdot CH_2OH$$

(LXXXVIII.) Riddellic acid.

↓

$$CH_3 \cdot CH : C(COOR) \cdot CH : C(CH_3) \cdot \underset{|}{C}(COOR) \cdot \underset{|}{CH_2}$$
$$O—SO—O$$

(XC.)

Sceleranecic and Sceleratinic Acids.

Sceleranecic acid dilactone, $C_{10}H_{14}O_5$, m. p. 156°, was obtained by DE WAAL and PRETORIUS (*187*) from sceleratine by alkaline hydrolysis and by catalytic hydrogenation, whilst DE WAAL and LOUW (*186*) found sceleratinic acid, $C_{10}H_{13}O_4Cl$, m. p. 208°, as a free acid in the extracts from *S. sceleratus*.

One lactone ring of sceleranecic acid dilactone was slowly opened with water whilst the other required hot alkali. Furthermore the dilactone contained two C-methyl groups. DE WAAL and CROUS (*184*) established that the remaining oxygen atom was present as a primary alcohol group in that it readily gave a monoacetate and a monobenzoate, and was oxidised by nitric acid to a carboxylic acid, $C_{10}H_{12}O_6$, m. p. 216°. Attempts to replace the hydroxyl group by chlorine using thionyl chloride gave a sulphite $(C_{10}H_{13}O_5)_2SO$, but thionyl chloride on the potassium sceleranecate gave sceleratinic acid and a new dicarboxylic acid, $C_8H_{14}O_2(COOH)_2$, m. p. 192°, which was also obtained direct from sceleratinic acid with alkali. All attempts to regenerate sceleranecic acid lactone by replacing the chlorine atom in sceleratinic acid failed. The stability of the lactone rings in the parent acids contrasted with that of the two free carboxylic acid groups in the acid, m. p. 192°, and DE WAAL, SERFONTEIN and GARBERS (*188*) envisaged the formation of a cyclic ether after inversion.

This new acid, m. p. 192°, which is shown to be a hydroxy-acid by conversion to a chloro-acid, $C_8H_{13}OCl(COOH)_2$, m. p. 131°, reacts with concentrated hydrochloric acid to give formaldehyde and an acid, $C_9H_{14}O_5$, m. p. 128°.

In a series of oxidation experiments a monolactonic, monocarboxylic acid, $C_8H_{12}O_4$, m. p. 100°, seemed to be the stable degradation product. This acid was obtained by potassium periodate or lead tetraacetate on both potassium sceleranecate and sceleratinate, and from sceleranecic

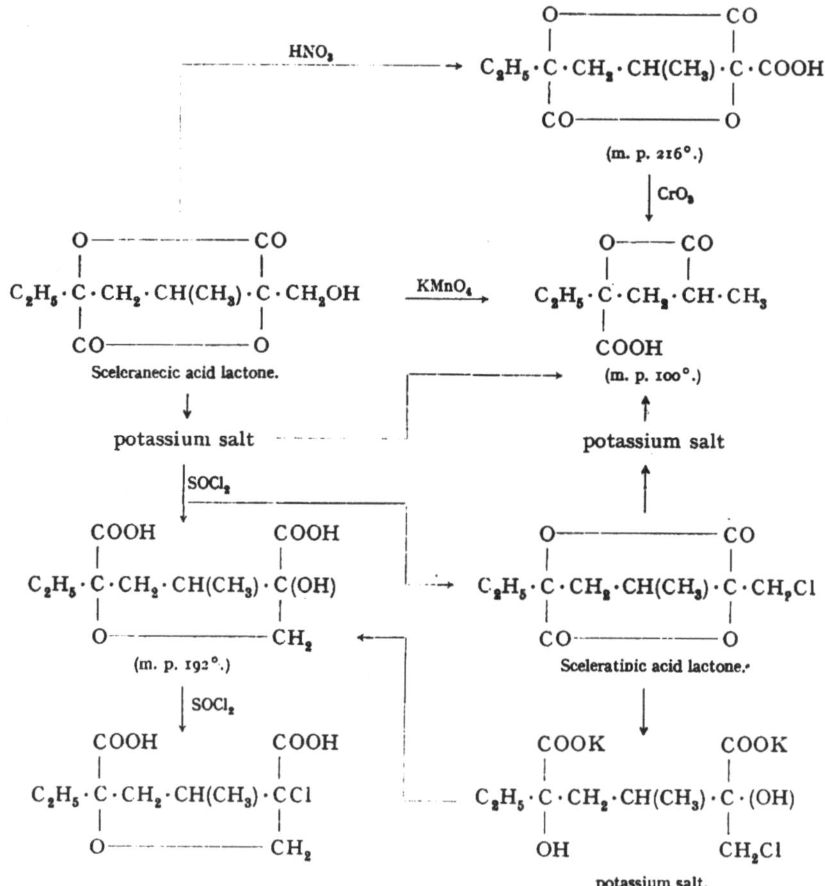

Chart 8. Reactions of Sceleranecic and Sceleratinic Acids.

acid lactone direct by alkaline potassium permanganate or by way of the nitric acid oxidation product with chromic acid.

DE WAAL (*188*) envisaged the same carbon skeleton as was shown for retronecic and isatinecic acids (*49*) and incidentally later found in other acids (*104, 105, 27, 8*). The several reactions are interpreted in *Chart 8.*

Jaconecic Acid and the Neutral Lactone from Jaconine.

Jaconecic acid, $C_{10}H_{16}O_6$, originally obtained by BARGER and BLACKIE (*37*) by alkaline hydrolysis of jaconine, was later obtained by ZHDANOVICH and MEN'SHIKOV (*195*) from otosenine and by BRADBURY (*41*) from jaconine. The hydrolysis of these alkaloids with hydrochloric acid, however, gave a neutral substance, $C_{10}H_{13}O_4NCl$ (*41, 195*).

Jaconecic acid has been shown by BRADBURY to be a saturated monohydroxy dibasic acid containing three carbon-methyl groups. It gives a ferric chloride reaction and liberates one mole of carbon dioxide with lead tetraacetate. It is not oxidised by periodic acid so that it does not contain a glycol group. The infrared spectrum indicates that the acid contains an epoxide ring. The presence of an epoxide ring permits an explanation of the ready conversion of jacobine to jaconine. Jacobine with one mole of hydrochloric acid gives jacobine hydrochloride, whilst with two moles of hydrochloric acid and crystallisation from alcohol it yields jaconine hydrochloride from which jaconine may be obtained. This conversion is explained as the addition of HCl to the epoxide ring, and the reverse conversion of jaconine to jacobine by silver oxide as the formation of an epoxide ring. The interconversion and hydrolyses of jacobine and jaconine are shown in *Chart 9.*

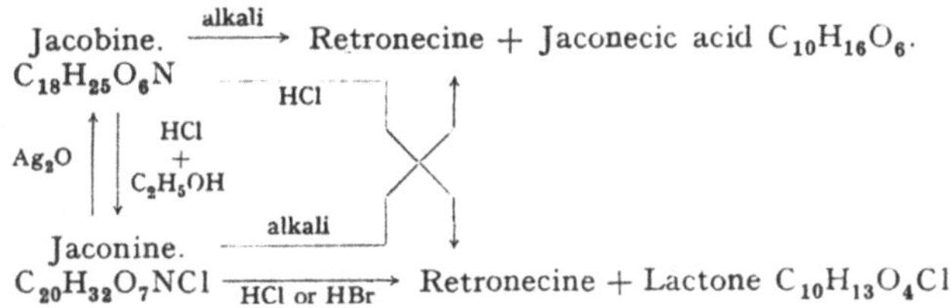

Chart 9. The Interconversion and Hydrolyses of Jacobine and Jaconine.

The neutral chloro-compound is a di-γ-lactone which gives jaconecic acid on treatment with alkali, but which could not be obtained from jaconecic acid by boiling with hydrochloric acid. The dilactone carbon skeleton which accommodates the above properties is (XCI). Assuming the applicability of the 'isoprene rule' to the necic acids as has been suggested by KROPMAN and WARREN (*105*), BRADBURY places the methyl groups to give tentative structures for the dilactone (XCII) and jaconecic acid (XCIII).

```
C——C——C——C——C          CH(CH3)·CCl(CH3)·C·CH(CH3)·CH2
 \    /  \    /         |              / \          |
  CO—O    CO—O          CO————————————O   CO————————O
    (XCI.)                            (XCII.)
```

```
CH(CH3)(COOH)·C(CH3)·C(COOH)·CH(CH3)·CH2OH
                  \  /
                   O
```

(XCIII.) Jaconecic acid (?).

C_{10}-Acids of Undetermined Structure.

Senecifolic acid, $C_{10}H_{16}O_6$, was prepared by Watt (*192*) in 1909 from the hydrolysis of senecifoline which has not again been obtained. — *Squalinecic acid*, $C_{10}H_{14}O_4$, was obtained from squalidine by Barger and Blackie (*36*) who indicated that the acid was not pure. It contained three C-methyl groups. The suggestion that it was a mixture of integerrinecic acid lactone and integerrinecic acid (*105*) has not received support from the isolation of integerrimine picrate (*25*) which is different from squalidine picrate. — *Mikanecic acid*, $C_{13}H_{16}O_5$, has not been studied since it was isolated by Manske (*120*) by the hydrolysis of mikanine. — *Grantianic acid*, $(C_8H_{12}O_3)(COOH)_2$, has not been obtained pure and its formula has been deduced from that of the alkaloids grantianine and tetrahydrograntianine. Adams, Carmack and Rogers (*3*) who studied this acid draw attention to the possible relation with monocrotalic acid $C_7H_{11}O_3(COOH)$ from another *Crotalaria* species. — *Hastanecic acid*, $C_{10}H_{16}O_5$, m. p. 148–149°, $[\alpha]_D + 4.6°$, was the acid separated by Konovalov and Men'shikov (*94*) after the hydrolysis of hastacine. It is a monohydroxy-dibasic acid, $C_8H_{13}(OH)(COOH)_2$, and resembles in this respect senecic acid, m. p. 151°, $[\alpha]_D + 11.8°$.

Monocrotalic and Dicrotalic Acids.

Monocrotalic acid, $C_8H_{12}O_5$, was obtained by Adams and Rogers (*18*) by the hydrogenolysis of monocrotaline, the alkaline hydrolysis of which gave retronecine, monocrotic acid and carbon dioxide. Monocrotalic acid contained a lactone group and on heating with alkali gave monocrotic acid.

$$C_{16}H_{23}O_6N + 2\,H_2 \rightarrow C_8H_{15}ON + C_8H_{12}O_5$$

$$C_{16}H_{23}O_6N + H_2O \xrightarrow{Ba(OH)_2} C_8H_{13}O_2N + C_7H_{12}O_3 + CO_2.$$

Monocrotic acid was shown by Adams, Rogers and Sprules (*22*) to be a monobasic acid containing a methyl-keto group which with sodium hypobromite gave *meso* and *racemic* dimethylsuccinic acid (XCIV). Monocrotic acid was therefore 3-methylpentane-4-one-2-carboxylic acid (XCV) in agreement with the dehydration and ready formation on heating of a neutral lactone, (XCVI), which gave a saturated lactone (XCVII) on reduction.

Adams and Long (*14*) obtained this same unsaturated lactone (XCVI) by a two-step procedure. Methyl monocrotalate was heated in a vacuum when it lost water to leave methyl anhydromonocrotalate, $C_7H_9O_2(COOCH_3)$,

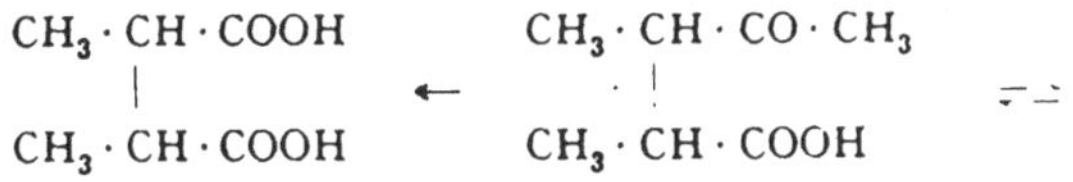

$$\begin{array}{c} CH_3\cdot CH\cdot COOH \\ | \\ CH_3\cdot CH\cdot COOH \end{array} \leftarrow \begin{array}{c} CH_3\cdot CH\cdot CO\cdot CH_3 \\ | \\ CH_3\cdot CH\cdot COOH \end{array} \rightleftharpoons$$

(XCIV.) Dimethyl-succinic acid. (XCV.) Monocrotic acid.

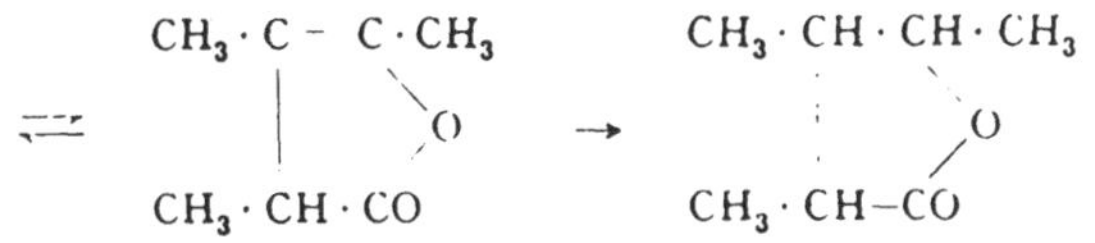

$$\rightleftharpoons \begin{array}{c} CH_3\cdot C - C\cdot CH_3 \\ |\quad\quad \searrow O \\ CH_3\cdot CH\cdot CO \end{array} \rightarrow \begin{array}{c} CH_3\cdot CH\cdot CH\cdot CH_3 \\ \quad\quad\quad O \\ CH_3\cdot CH-CO \end{array}$$

(XCVI.) (XCVII.)

which on hydrolysis gave the lactone (XCVI), but which on hydrogenation and hydrolysis was converted to dihydroanhydro-monocrotalate, $C_7H_{11}O_2$ (COOH).

After Adams and his co-workers (*21, 30*) had made a detailed study of the reactions, Adams and Govindachari (*7*) established the structure of monocrotalic acid by reduction with lithium aluminium hydride. Methyl monocrotalate with this reducing agent gave an almost quantitative yield of a neutral product, $C_8H_{18}O_4$. This reduction consumed exactly two mole equivalents when oxidised with sodium periodate to yield formaldehyde, acetic acid and 2-methylbutane-3-one-1-ol (XCVIII), which was isolated as its 2,4-dinitrophenylhydrazone and partly as the 2,4-dinitrophenylhydrazone of methyl *iso*propenyl ketone (XCIX). The neutral reduction product gave a dibenzoate (C) which with periodate gave methyl *iso*propenyl ketone (XCIX) and the benzoate of hydroxy-acetone (CI). These findings could be explained if monocrotalic acid had structure (CII), and the ester would then be expected to be reduced to the tetrahydroxy compound (CIII) which would yield a dibenzoate (C). Methyl anhydro- (CIV) and dihydroanhydro-monocrotalate (CV) were similarly reduced with lithium aluminium hydride to compounds presumably (CVI) or (CVII) respectively, each of which gave formaldehyde on glycol fission with periodate (*Chart 10*, p. 238).

Confirmation of these structures was advanced by Adams and Hauserman (*11*) by the synthesis of dihydroanhydro-monocrotalic acid. 3-Methyl-pentane-4-one-2-carboxylic acid (CVIII) with hydrocyanic acid gave the cyano-lactone (CIX) which hydrolysed to an oily acid. This acid was probably a mixture of racemates but gave a pure *p*-bromophenacyl ester showing an infrared spectrum almost identical with that of the corresponding ester of the dihydroanhydro-monocrotalic acid, $[\alpha]_D$-60°. Partial hydrolysis of the cyano-lactone with alkaline hydrogen peroxide gave

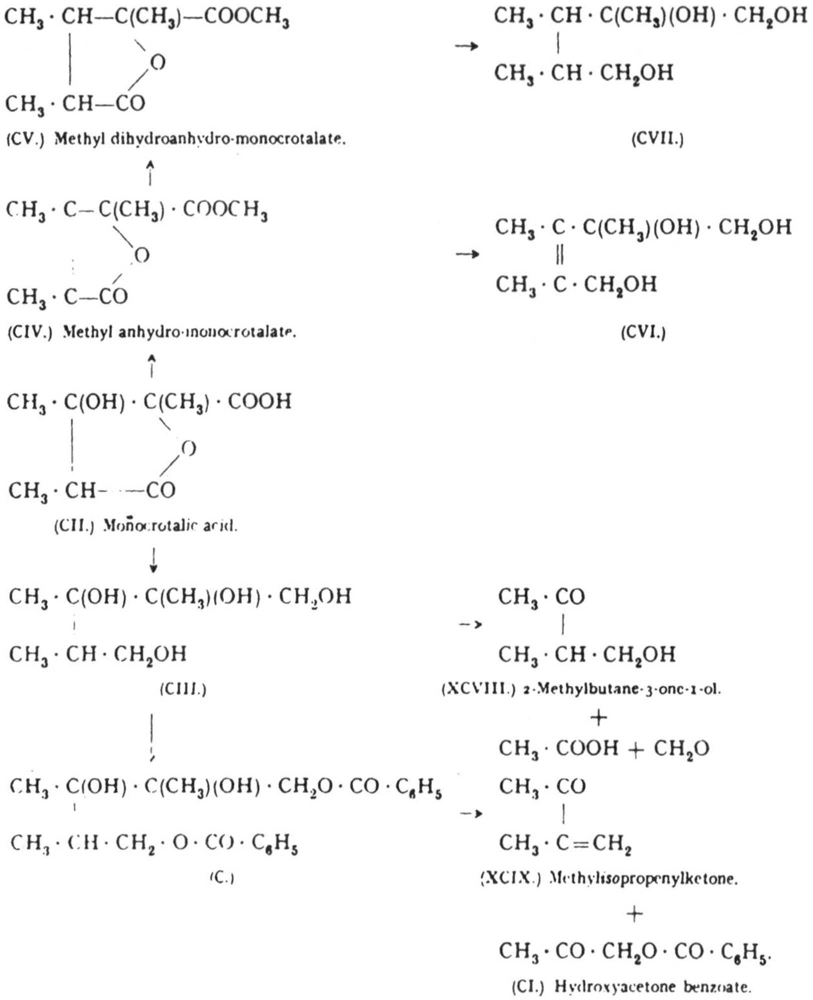

$$CH_3\cdot CH{-}C(CH_3){-}COOCH_3 \quad (\text{ring O}) \quad CH_3\cdot CH{-}CO \rightarrow CH_3\cdot CH\cdot C(CH_3)(OH)\cdot CH_2OH \,|\, CH_3\cdot CH\cdot CH_2OH$$

(CV.) Methyl dihydroanhydro-monocrotalate. (CVII.)

↑

$$CH_3\cdot C{-}C(CH_3)\cdot COOCH_3 \quad (\text{ring O}) \quad CH_3\cdot C{-}CO \rightarrow CH_3\cdot C\cdot C(CH_3)(OH)\cdot CH_2OH \,\|\, CH_3\cdot C\cdot CH_2OH$$

(CIV.) Methyl anhydro-monocrotalate. (CVI.)

↑

$$CH_3\cdot C(OH)\cdot C(CH_3)\cdot COOH \quad (\text{ring O}) \quad CH_3\cdot CH{-}CO$$

(CII.) Monocrotalic acid.

↓

$$CH_3\cdot C(OH)\cdot C(CH_3)(OH)\cdot CH_2OH \,|\, CH_3\cdot CH\cdot CH_2OH \rightarrow CH_3\cdot CO \,|\, CH_3\cdot CH\cdot CH_2OH$$

(CIII.) (XCVIII.) 2-Methylbutane-3-one-1-ol.

$$+\; CH_3\cdot COOH + CH_2O$$

↓

$$CH_3\cdot C(OH)\cdot C(CH_3)(OH)\cdot CH_2O\cdot CO\cdot C_6H_5 \,|\, CH_3\cdot CH\cdot CH_2\cdot O\cdot CO\cdot C_6H_5 \rightarrow CH_3\cdot CO \,|\, CH_3\cdot C{=}CH_2$$

(C.) (XCIX.) Methyl*iso*propenylketone.

$$+\; CH_3\cdot CO\cdot CH_2O\cdot CO\cdot C_6H_5.$$

(CI.) Hydroxyacetone benzoate.

Chart 10. Degradations Leading to the Structure of Monocrotalic Acid.

an impure amide (CX) the hydrolysis of which yielded an oily acid (CXI). This was resolved by way of the brucine salt to a solid acid, m. p. 117.6–119.3°, $[\alpha]_D$ − 60.0°, showing an infrared spectrum identical with the natural stereoisomer, $[\alpha]_D$ — 6 ·.0°.

The hydrolysis of methyl dihydroanhydro-monocrotalate gave mixtures. Nevertheless, it was possible to isolate by acid hydrolysis an acid, m. p. 132–134°, $[\alpha]_D$ + 5.6°, and by potassium cyanide treatment an isomeric acid, m. p. 117.6–119.5°, $[\alpha]_D$ — 60.0°, whilst sodium hydroxide gave a mixture of these two acids which were difficult to separate. A study of the isomerism using the pure isomeric acids revealed an interconversion to an equilibrium mixture, $[\alpha]_D$ -56° ($\pm 1^\circ$), containing predominantly the -60° isomer, provided the treatment was long enough. It was concluded that the methyl dihydroanhydro-crotalate (CV, p. 238)

```
                                          CH3·CH·C(CN)·CH3
CH3·CH·CO·CH3          HCN                  |        \
      |            ---------->              |         O
CH3·CH·COOH                                 |        /
                                          CH3·CH·CO
(CVIII.) 3-Methylpentan-4-one-2-carboxylic acid.      (CIX.)
                                                 | H2O2
                                                 | OH-
                                                 v
CH3·CH—C(CH3)·COOH                        CH3·CH·C(CONH2)·CH3
   |        \                                |        \
   |         O           <--------           |         O
   |        /                                |        /
CH3·CH—CO                                 CH3·CH·CO
     (CXI.)                                    (CX.)
```

had the same configuration as the isomeric acid, m. p. 132.4–134.4°. The mechanism of this interconversion was shown to be epimerisation of the carbon atom alpha to the lactone and explained as acid-catalysed enolisation and base-catalysed ionisation.

ADAMS, VAN DUUREN and BRAUN (*29*) synthesised monocrotalic acid by the *trans*-hydroxylation of *cis*-trimethyl-glutaconic acid (CXII) to give either (CXIII) or (CXIV).

```
                                       O                          O
                                     /   \                      /
CH3—C—COOH                     CH3—C—COOH                 CH3—C—COOH
    ||              ------>        |       CO    or                      CO
CH3—C—CH(CH3)·COOH             CH3—C—OH                   CH3—C—OH
                                    \   /                      \   /
                                  H—C—CH3                   CH3—C—H
(CXII.) cis-Trimethyl-glutaconic acid.   (CXIII.)              (CXIV.)
```

The catalytic reduction of methyl anhydro-monocrotalate (CIV) will proceed by *cis*-addition so that methyl dihydroanhydro-monocrotalate will have structures (CVa) or (CVb). If epimerisation proceeds at the α-carbon the equilibrium will favour the new epimeric form in which the methyl groups will be *trans* to one another. Finally, since the dehydration of methyl monocrotalate is effected thermally it is probable that *cis*-elimination of water occurs and monocrotalic acid has structure (CXIV).

Dicrotalic acid, obtained by MARAIS (*123*) by the hydrolysis of dicrotaline, was shown by ADAMS and VAN DUUREN (*24*) to be 2-hydroxy-2-methylpentane-1 : 3-dicarboxylic acid (CXV). It was readily acetylated to an acetyldicrotalic anhydride which lost acetic acid at 100° to give

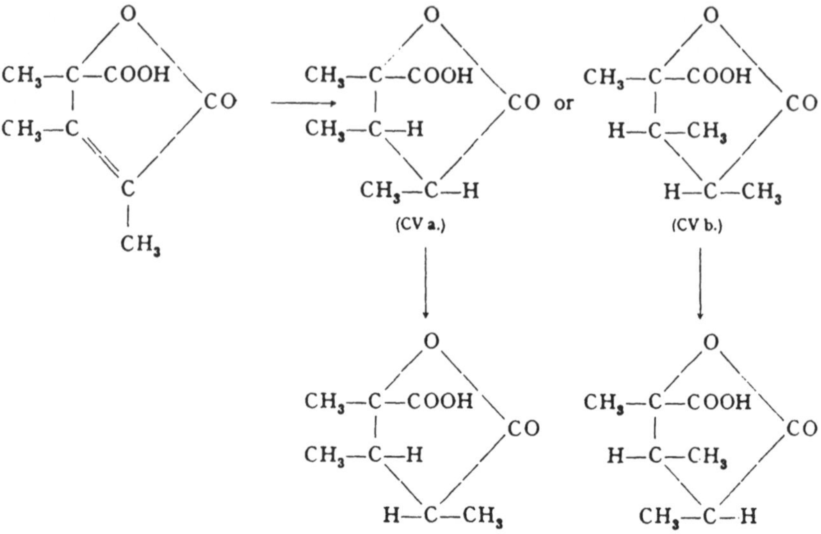

```
              O                             O                              O
             / \                           / \                            / \
CH3—C—COOH    \              CH3—C—COOH     \             CH3—C—COOH       \
    |          CO    ———→        |          CO    or          |            CO
CH3—C         /              CH3—C—H       /               H—C—CH3        /
     \\      /                    \       /                     \        /
       C                           CH3—C—H                        H—C—CH3
       |                            (CV a.)                        (CV b.)
      CH3                              |                              |
                                       ↓                              ↓
                                       O                              O
                                      / \                            / \
                             CH3—C—COOH  \                  CH3—C—COOH  \
                                 |        CO                    |        CO
                             CH3—C—H     /                   H—C—CH3    /
                                  \     /                         \    /
                                  H—C—CH3                       CH3—C—H
```

anhydro-dicrotalic anhydride, the acid from which was identical with *cis*-β-methylglutaconic acid (CXVI).

(CXV.) $HOOC \cdot CH_2 \cdot C(OH)\,(CH_3) \cdot CH_2 \cdot COOH$

(CXVI.) $HOOC \cdot CH{=}C(CH_3) \cdot CH_2 \cdot COOH$

Acids having Seven-Carbon Skeleton Structures.

Trachelanthic acid, obtained by MEN'SHIKOV and his colleagues from trachelanthine (*135, 137, 138*), lindelofine (*106*), lindelofamine (*106*) and supinine (*141*, cf. also *53*), and *viridifloric acid* from viridiflorine (*136*) have been shown to be diastereoisomers of 2 : 3-dihydroxy-4-methylpentane-3-carboxylic acid (CXVII) (*135, 136*). Both acids on oxidation with mercuric oxide gave 4-methylpentane-2 : 3-dione (CXVIII), whilst reduction with hydriodic acid yielded 4-methylpentane-3-carboxylic acid (CXIX).

$CH_3 \cdot CH(OH) \cdot C(OH)\,(COOH) \cdot CH(CH_3)_2 \rightarrow CH_3 \cdot CO \cdot CO \cdot CH(CH_3)_2$

(CXVII.) 2 : 3-Dihydroxy-4-methylpentane-3-carboxylic acid. (CXVIII.)

↓ ↑

$CH_3 \cdot CH_2 \cdot CH(COOH) \cdot CH(CH_3)_2$ (CXIX.) $CH_3 \cdot CH{=}C(COOH) \cdot CH(CH_3)_2$ (CXX.)

ADAMS and HERZ (*12*) synthesised one form of (CXVII), m. p. 119–121°, by *cis*-hydroxylation of 2-methylpent-3-ene-3-carboxylic acid (CXX), and identified their synthetic product as optically inactive viridifloric acid, m. p. 119–121° (*136*). ADAMS and VAN DUUREN (*28*) subsequently synthesised the diastereoisomer by *trans*-hydroxylation of (CXX) by pertungstic acid and resolved both acids. These authors concluded that viridifloric acid is probably the partially racemised optical isomer of the *trans*-hydroxylated acid, whilst trachelanthic acid is the partially racemised optical isomer of the *cis*-hydroxylated acid. This conclusion was prompted by the report (*141*) that supinine gave by alkaline hydrolysis *racemic* trachelanthic acid whilst hydrogenolysis yielded (+)-trachelanthic acid; and in support of this they described the conversion of the (+)-acid to the racemate by boiling alkali (*28*).

Dry and Warren (*65*) have also synthesised and resolved the two forms of (CXVII). These authors assign the *cis*-configuration (CXXI) with respect to the carbon-chain to the 2-methylpent-3-ene-3-carboxylic acid (CXX). Thus *cis*-addition with permanganate and *trans*-addition by way of the epoxide (CXXII) gives the (±)-*threo* (CXXIII) and (±)-*erythro* (CXXIV) forms of 3 : 4-dihydroxy-2-methylpentane-3-carboxylic acid respectively (*Chart 11*, p. 242). They conclude that trachelanthic acid is identical with the (+)-*threo*-acid (CXXIII) and viridifloric acid with either one of the optical forms of the *erythro*-acid (CXXIV).

Table 12. Comparison of Natural and Synthetic Trachelanthic Acid
[The melting points are uncorrected; (e) ethanol; (w) water.]

	Adams and van Duuren		Dry and Warren		Culvenor	
	m. p.	$[\alpha]_D$	m. p.	$[\alpha]_D$	m. p.	$[\alpha]_D$
Racemate	119°	—	119–121°			
(+)-Form	89°	2.9° (w)	92°	2.2° (e)	93–94°	2.3° (e)
Brucine salt	217–220°	—	225°	—	230°	

The two sets of results are in conflict (see *Table 12*). It is significant, however, that trachelanthic acid obtained by alkaline hydrolysis has the same melting point as the active form from hydrolysis and cannot be the racemate, as indicated earlier (*141*). Secondly, it is difficult to envisage simultaneous inversion of the two groups without which viridifloric acid must give some trachelanthic acid and vice versa; and no such inversion has been found. This racemisation by alkali has been refuted by Culvenor (*53*) who suggests that the lack of rotation reported for trachelanthic acid may be due to the low specific rotation coupled with a decrease in specific rotation in aqueous solution as the concentration is increased.

Heliotrinic acid was obtained first by Men'shikov (*125*) from heliotrine, later by Trautner and Neufeld (*175*) and Culvenor, Drummond and Price (*54*). Men'shikov (*133*) showed that the acid was a monohydroxy-monomethoxy acid which on oxidation with phosphoric acid and lead dioxide gave methoxy-4-methylpentane-3-one (CXXVI) which with potassium permanganate gave *iso*butyric acid. The ketone (CXXVI) with phenyl magnesium bromide and hydrolysis gave a tertiary alcohol (CXXVII) which by oxidation yielded phenyl *iso*propyl ketone (CXXVIII).

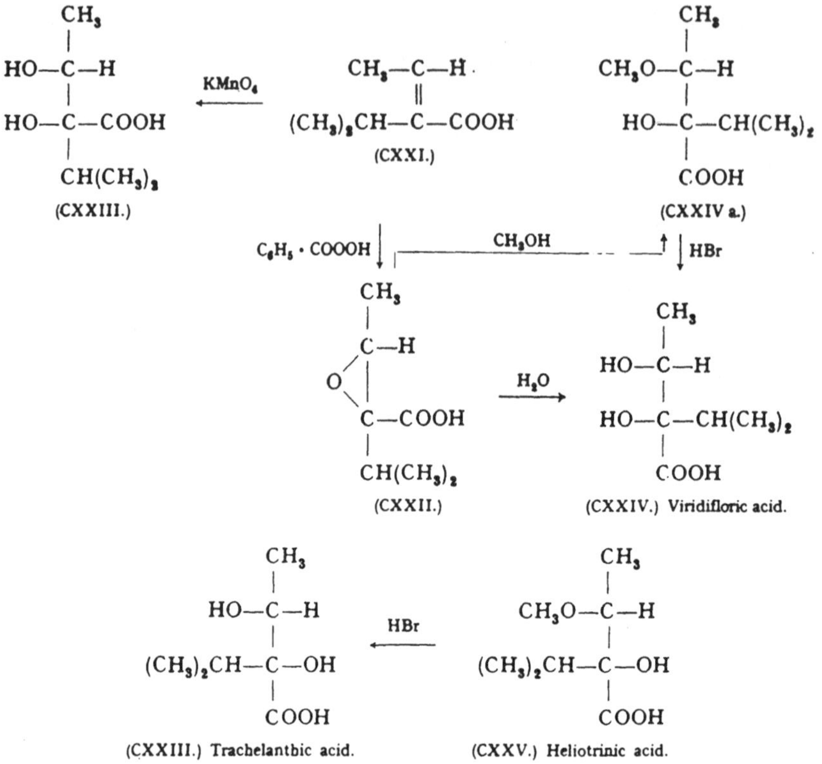

$$\begin{array}{c} CH_3 \\ | \\ HO{-}C{-}H \\ | \\ HO{-}C{-}COOH \\ | \\ CH(CH_3)_2 \end{array} \xleftarrow{KMnO_4} \begin{array}{c} CH_3{-}C{-}H \\ \| \\ (CH_3)_2CH{-}C{-}COOH \end{array} \qquad \begin{array}{c} CH_3 \\ | \\ CH_3O{-}C{-}H \\ | \\ HO{-}C{-}CH(CH_3)_2 \\ | \\ COOH \end{array}$$

(CXXIII.) (CXXI.) (CXXIV a.)

$C_6H_5 \cdot COOOH$ ↓ — CH_3OH → ↑ ↓ HBr

$$\begin{array}{c} CH_3 \\ | \\ C{-}H \\ O \,| \\ C{-}COOH \\ | \\ CH(CH_3)_2 \end{array} \xrightarrow{H_2O} \begin{array}{c} CH_3 \\ | \\ HO{-}C{-}H \\ | \\ HO{-}C{-}CH(CH_3)_2 \\ | \\ COOH \end{array}$$

(CXXII.) (CXXIV.) Viridifloric acid.

$$\begin{array}{c} CH_3 \\ | \\ HO{-}C{-}H \\ | \\ (CH_3)_2CH{-}C{-}OH \\ | \\ COOH \end{array} \xleftarrow{HBr} \begin{array}{c} CH_3 \\ | \\ CH_3O{-}C{-}H \\ | \\ (CH_3)_2CH{-}C{-}OH \\ | \\ COOH \end{array}$$

(CXXIII.) Trachelanthic acid. (CXXV.) Heliotrinic acid.

Chart 11. Stereochemistry of Trachelanthic, Viridifloric and Heliotrinic Acids.

Heliotrinic acid is accordingly 3-hydroxy-2-methoxy-4-methylpentane-3-carboxylic acid (CXXV). The *threo*-configuration, shown in Chart 11, follows from the work of Adams (*26*) on the demethylation by hydrobromic acid of heliotrinic acid to trachelanthic acid.

$$CH_3 \cdot CH(OCH_3) \cdot C(OH)\,(COOH) \cdot CH(CH_3)_2$$

(CXXV.) Heliotrinic acid.

$$\downarrow PbO_2$$

$$CH_3 \cdot CH(OCH_3) \cdot CO \cdot CH(CH_3)_2 \rightarrow (CH_3)_2CH \cdot COOH$$

(CXXVI.) Methoxy-4-methylpentane-3-one.

$$\downarrow C_6H_5MgBr \text{ and hydrolysis.}$$

$$CH_3 \cdot CH(OCH_3) \cdot C(OH)\,(C_6H_5) \cdot CH(CH_3)_2 \rightarrow C_6H_5 \cdot CO \cdot CH(CH_3)_2$$

(CXXVII.) (CXXVIII.)

The racemic *erythro* isomer (CXXIV a) has also been synthesised (*66*) by way of the epoxide (CXXII) and demethylated to the *erythro*-glycol acid (CXXIV).

Lasiocarpic (lăsiocarpinic) *acid* was the acid found by MEN'SHIKOV and ZHDANOVICH (*129*, *146*) from the hydrogenolysis of lasiocarpine. DRUMMOND (*63*) demonstrated that this acid contained one methoxyl and two hydroxyl groups and was oxidised with one mole of periodic acid to give acetone which would result from a glycol moiety accommodated in the grouping $(CH_3)_2 \cdot C(OH) \cdot C(OH)\langle$. The acid gave a ferric chloride reaction and evolved carbon dioxide with sulphuric acid so that it was an α-hydroxy acid. Kuhn-Roth oxidation which gave 1.57 moles of acetic acid indicated a methyl in addition to the *gem*-dimethyl group. Finally, lasiocarpic acid readily yielded acetone on heating with dilute alkali leaving an acid $C_5H_{10}O_4$. DRUMMOND concluded that lasiocarpic acid was 2 : 3-dihydroxy-4-methoxy-2-methylpentane-3-carboxylic acid (CXXIX). CULVENOR, DRUMMOND and PRICE (*54*) point out that the C_5-acid is presumably 2-hydroxy-3-methoxy-butyric acid (CXXX). The (±)-*erythro* form (CXXIV) of this acid (CXXX) has been synthesised by DRY and WARREN (*65a*) but the melting points of the acids and the *p*-phenylphenacyl esters are different.

$$CH_3 \cdot CH(OCH_3) \cdot C(OH)(COOH) \cdot C(OH)(CH_3)_2$$

(CXXIX.)

↓

$$(CH_3)_2CO + CH_3 \cdot CH(OCH_3) \cdot CHOH \cdot COOH$$

(CXXX.)

Another interesting *acid* within this group is that in makrotomine. MEN'SHIKOV and PETROVA (*144*) found that makrotomine on alkaline hydrolysis gave trachelanthamidine, acetic acid and tarry products. Makrotomine in sulphuric acid was oxidised with one mole of periodic acid to acetaldehyde, oxalic acid, acetone and trachelanthamidine (*144*), so that makrotomine is the ester of trachelanthamidine with 2 : 3 : 4-trihydroxy-2-methylpentane-3-carboxylic acid (CXXXI).

$$CH_3 \cdot CH(OH) \cdot C(OH)(COOH) \cdot C(OH)(CH_3)_2$$

(CXXXI.) 2 : 3 : 4-Trihydroxy-2-methylpentane-3-carboxylic acid.

↓

$$CH_3 \cdot CHO + (COOH)_2 + (CH_3)_2CO$$

The *acid moiety of trichodesmine* has also not been isolated. MEN'SHIKOV (*129*) showed that trichodesmine (CXXXII) on hydrolysis gave trichodesmidine (i. e. retronecine, cf. *101*) (CXXXIII), *DL*-lactic acid and methyl *iso*butyl ketone (CXXXIV), which it was concluded could come from the decarboxylation of either (CXXXV) or (CXXXVI), p. 244.

The acid (CXXXVI) is of interest in that it has the same carbon-skeleton as that of viridifloric, trachelanthic and heliotrinic acids and accordingly the writer considers this as the more probable precursor of the methyl *iso*butyl ketone.

$$C_{18}H_{27}O_6N \rightarrow C_8H_{13}O_3N + CH_3 \cdot CH(OH) \cdot COOH + CH_3 \cdot CO \cdot CH_2 \cdot CH(CH_3)_2 + (CO_2)$$

(CXXXII.) Trichodesmine. (CXXXIII.) Trichodesmidine. (CXXXIV.)

$$HOOC \cdot CH_2 \cdot CO \cdot CH_2 \cdot CH(CH_3)_2 \quad \text{(CXXXV.)} \qquad CH_3 \cdot CO \cdot CH(COOH) \cdot CH(CH_3)_2 \quad \text{(CXXXVI.)}$$

The close relationship between these different acids is readily seen in the following *summary* of the structures.

Trachelanthic acid Viridifloric acid	$CH_3 \cdot CH(OH) \cdot C(OH)\,(COOH) \cdot CH(CH_3)_2$
Heliotrinic acid	$CH_3 \cdot CH(OCH_3) \cdot C(OH)\,(COOH) \cdot CH(CH_3)_2$
Acid from makrotomine	$CH_3 \cdot CH(OH) \cdot C(OH)\,(COOH) \cdot C(OH)\,(CH_3)_2$
Lasiocarpic acid	$CH_3 \cdot CH(OCH_3) \cdot C(OH)\,(COOH) \cdot C(OH)\,(CH_3)_2$
Acid from trichodesmine	$CH_3 \cdot CO \cdot CH(COOH) \cdot CH(CH_3)_2$ (?)

IV. The Structures of the Alkaloids.

The structures of the alkaloids themselves may be considered where those of the base and acid or acids obtained by hydrolysis are known. Where the basic moiety is a monohydroxylated methylpyrrolizidine and the acid fraction is a monocarboxylic acid the structure of the total alkaloid follows logically.

When the base is an unsaturated dihydroxy-methylpyrrolizidine, retronecine or heliotridine, the orientation of the acids is shown by the mode of hydrogenolysis. Where two different acids are joined to the two hydroxyl groups (CXXXVII), the acid which is liberated on reduction is that linked to the allylic hydroxyl group. When a dibasic acid is attached to the two hydroxyls as in (CXXXVIII) the process of hydro-

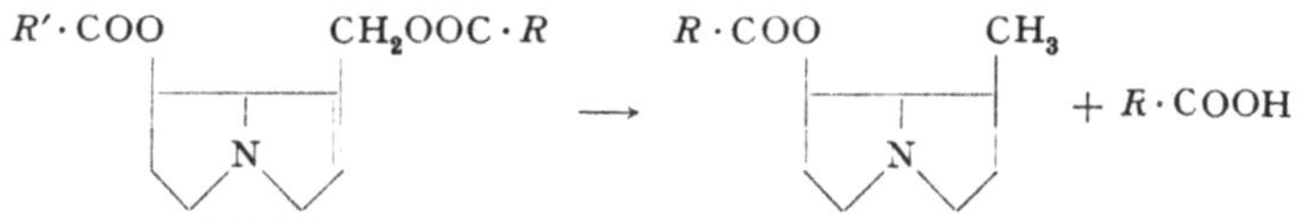

$R' \cdot COO$ $CH_2OOC \cdot R$ → $R \cdot COO$ CH_3 + $R \cdot COOH$

N N

(CXXXVII.)

CO—(R)—CO → CO—(R)—COOH

O CH_2O O CH_3

N N

(CXXXVIII.) (CXXXIX.)

genation does not in general effect complete fission and the problem resolves itself into the determination of the nature of the free carboxyl grouping. Nevertheless, caution must be exercised in the interpretation of the results because a compound (CXXXIX) may under certain conditions fission.

The similarity in structure found in the acids and bases is again further exemplified in the total alkaloids. The varied combinations of acid and base make possible the classification of the alkaloids as derivatives of either acid or base. The system adopted here is to consider the alkaloids as derivatives first of the monohydroxy- and then of the dihydroxy-methylpyrrolizidines. There is one trihydroxy-methylpyrrolizidine alkaloid, rosmarinine, which, however, is more conveniently included in the latter group.

Alkaloids from Monohydroxy-methylpyrrolizidines.

A. Alkaloids from Trachelanthamidine.

Trachelanthamine, $C_{15}H_{27}O_4N$, was isolated by MEN'SHIKOV and BORODINA (*137, 138*) from *Trachelanthus korolkovi* in which it occurred together with trachelanthine, $C_{15}H_{27}O_5N$, that was the first *N*-oxide detected as such in this group of alkaloids. Hydrolysis of trachelanthamine gave trachelanthamidine and trachelanthic acid. The structure assigned by MEN'SHIKOV (*135*) may now be given as (CXL).

CH_3 | HO—C—H | $(CH_3)_2CH$—C—OH | H CH_2OOC N

(CXL.) Trachelanthamine.

CH_3 | HO—C—H | HO—C—$CH(CH_3)_2$ | H CH_2OOC N

(CXLI.) Viridiflorine.

CH_3 | CHOH | C(OH)·C(OH)$(CH_3)_2$ | H CH_2OOC N

(CXLII.) Makrotomine.

Viridiflorine, $C_{15}H_{27}O_4N$, from *Cyanoglossum viridiflorum*, was shown by MEN'SHIKOV (*136*) to hydrolyse to trachelanthamidine and viridifloric acid. The structure is accordingly (CXLI), p. 245.

Makrotomine, $C_{15}H_{27}O_5N$, was studied by MEN'SHIKOV and PETROVA (*144*) who isolated by hydrolysis trachelanthamidine, acetic acid and tarry products. The structure, however, of the acid fraction was deduced from the oxidation of the alkaloid itself which may now be assigned structure (CXLII).

B. Alkaloids from Supinidine.

Supinine, $C_{15}H_{25}O_4N$, was isolated by MEN'SHIKOV and GUREVICH (*141*) who reported that it gave supinine and racemic trachelanthic acid on hydrolysis whilst on complete reduction it yielded slightly (+)-trachelanthic acid and *L*-heliotridane. This supposed racemisation has been refuted by CULVENOR (*53*) who identified supinine with 'Base F' which was isolated as one of the alkaloids from *Heliotropium europaeum* by CULVENOR, DRUMMOND and PRICE (*54*). By partial reduction with only one mole of hydrogen supinine gave *iso*heliotridene. The formula of supinine (CXLV) is in complete agreement with these reactions.

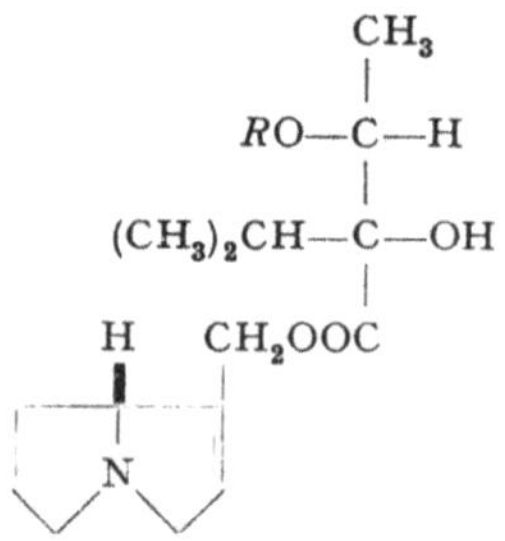

```
               CH3
                |
            RO—C—H
                |
   (CH3)2CH—C—OH
                |
     H    CH2OOC
     |     |
     N
```

(CXLV.) Supinine ('Base F'): $R = H$.

(CXLVI.) 'Base C': $R = CH_3$.

'Base C', $C_{16}H_{27}O_4N$, was found as its *N*-oxide as one of the minor alkaloids in *H. europaeum* by CULVENOR, DRUMMOND and PRICE (*54*). It was shown by CULVENOR (*53*) to be an ester of heliotrinic acid with supinidine and accordingly has formula (CXLVI).

C. Alkaloids from D-isoRetronecanol.

Lindelofine, $C_{15}H_{27}O_4N$, was isolated by LABENSKII and MEN'SHIKOV (*106*) who hydrolysed it to *D-iso*retronecanol and trachelanthic acid so that lindelofine has the structure (CXLIII).

Lindelofamine, $C_{20}H_{33}O_5N$, was found (*106*) associated with lindelofine, and hydrolysed to *D-iso*retronecanol, trachelanthic and tiglic acids. The structure of lindelofamine (CXLIV) is accordingly that of lindelofine with one of the hydroxyl group of its trachelanthate moiety esterified by tiglic acid.

CH₃ | *R*O—C—H | (CH₃)₂CH—C—OH | H CH₂OOC, N

(CXLV.) Supinine ('Base F'): *R* = H.

(CXLVI.) 'Base C': *R* = CH_3.

(CXLIV.) Lindelofamine: *R'* = H, and *R* = $CH_3 \cdot CH : C(CH_3)COO$ (or vice versa).

Alkaloids from Heliotridine: Heliotrine, 'Base G', Lasiocarpine.

Heliotrine, $C_{16}H_{27}O_5N$, was isolated from *Heliotropium lasiocarpum* by MEN'SHIKOV (*125*) and his extensive studies on this alkaloid (*126, 127, 128, 129, 132, 133*) pioneered our knowledge of the pyrrolizidine alkaloids. On hydrolysis this alkaloid gave heliotrinic acid and heliotridine. The attachment of the heliotrinic acid to the primary hydroxyl group of heliotridine as in (CXLVII) followed from the hydrogenolysis of the alkaloid to oxyheliotridane and heliotrinic acid.

'Base G', $C_{16}H_{27}O_6N$, was one of the major alkaloids from *H. europaeum* in which it occurred both as the free base and as its *N*-oxide (*54*). CULVENOR (*53*) isolated heliotridine by hydrolysis but failed to obtain a pure acid. He showed (*53*), however, that the reduction of the *N*-oxide in the presence of palladised charcoal gave hydroxyheliotridane and lasiocarpic acid so that 'Base G' is (CXLVIII).

CH₃ | CH₃O—C—H | (CH₃)₂CH—C—OH | OH H CH₂OOC, N

(CXLVII.) Heliotrine.

CH₃ | CH(OCH₃) | C(OH)·C(OH)(CH₃)₂ | OH H CH₂OOC, N

(CXLVIII.) 'Base G'.

Lasiocarpine, $C_{21}H_{33}O_7N$, was first isolated by MEN'SHIKOV (*125*) from *H. lasiocarpum* and found later by CULVENOR, DRUMMOND and

PRICE (*54*) in *H. europaeum*, in which it also occurred as its *N*-oxide. MEN'SHIKOV and ZHDANOVICH (*146*) showed that this alkaloid on alkaline hydrolysis gave heliotridine and angelic acid; but on catalytic reduction with platinum it yielded lasiocarpic acid and a base, $C_8H_{13}O_2N$, $[\alpha]_D + 3.8°$, characterised as a picrate, m. p. 157–159°. This base hydrolysed to butane-2-carboxylic acid and oxyheliotridane. Lasiocarpine may accordingly have the structure (CXLIX) which would give (CL) on reduction.

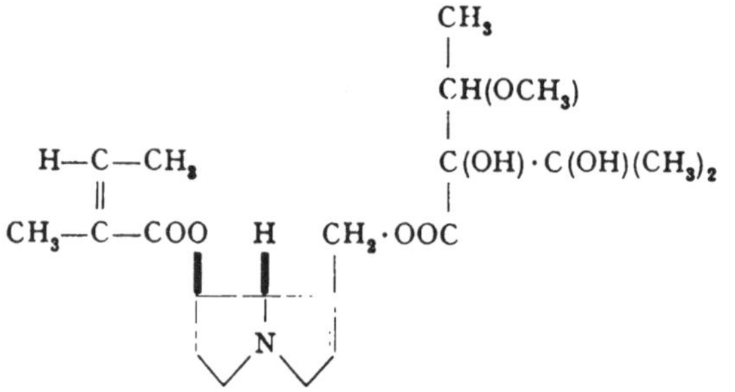

```
                                        CH3
                                        |
                                        CH(OCH3)
                                        |
      H—C—CH3                           C(OH)·C(OH)(CH3)2
        ||                              |
    CH3—C—COO    H    CH2·OOC
                          N
```

(CXLIX.) Lasiocarpine.

↓

```
              C2H5
              |
        CH3—CH—COO    H    CH3
                          N
```

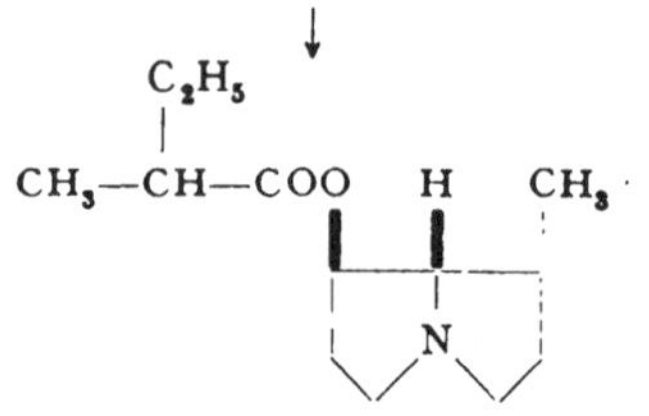

(CL.)

C_{18}-Alkaloids from Retronecine, Platynecine and Rosmarinecine.

Retrorsine, $C_{18}H_{25}O_6N$, was first isolated by BARGER, SESHADRI, WATT and YABUTA (*38*) who hydrolysed it to retronecine and retronecic acid and hydrogenated it to tetrahydro-retrorsine, which in turn gave retronecanol and retronecic acid on hydrolysis. LEISEGANG and WARREN (*110*) reduced retrorsine to hexahydro-retrorsine which was shown to have formula (CLII). This hexahydro-compound was treated with lead tetraacetate under anhydrous conditions to fission the glycol grouping. The addition of a little water then caused the evolution of one mole of carbon dioxide in accordance with the existence of a free keto-acid (CLIII), which would be oxidised to (CLIV), so that retrorsine is (CLI).

Isatidine, $C_{18}H_{25}O_7N$, which was originally isolated by BLACKIE (*40*) from *S. isatideus* was found in *S. retrorsus* by DE WAAL (*179*), who reduced it to hexahydro-desoxyisatidine. CHRISTIE, KROPMAN, NOVELLIE and WARREN (*49*) showed that isatidine was retrorsine *N*-oxide, so that DE WAAL's hexahydro-desoxycompound would be (CLII) (*Chart 12*).

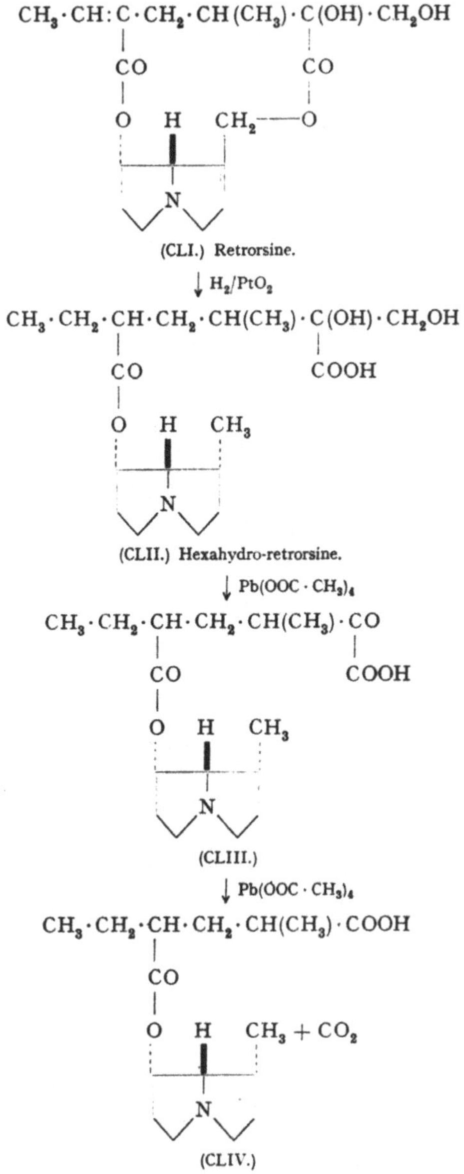

$$CH_3\cdot CH{:}C(CO\text{–}O\text{–})\cdot CH_2\cdot CH(CH_3)\cdot C(OH)(CO\text{–}O\text{–}CH_2\text{–})\cdot CH_2OH$$

(CLI.) Retrorsine.

↓ H_2/PtO_2

$$CH_3\cdot CH_2\cdot CH(CO\text{–}O\text{–})\cdot CH_2\cdot CH(CH_3)\cdot C(OH)(COOH)\cdot CH_2OH \quad (\text{with } CH_3 \text{ on the pyrrolizidine nucleus})$$

(CLII.) Hexahydro-retrorsine.

↓ $Pb(OOC\cdot CH_3)_4$

$$CH_3\cdot CH_2\cdot CH(CO\text{–}O\text{–})\cdot CH_2\cdot CH(CH_3)\cdot CO\cdot COOH \quad (\text{with } CH_3 \text{ on the pyrrolizidine nucleus})$$

(CLIII.)

↓ $Pb(OOC\cdot CH_3)_4$

$$CH_3\cdot CH_2\cdot CH(CO\text{–}O\text{–})\cdot CH_2\cdot CH(CH_3)\cdot COOH \quad (\text{with } CH_3 \text{ on the pyrrolizidine nucleus}) + CO_2$$

(CLIV.)

Chart 12. Reactions Showing the Orientation of the Ester Groups in Retrorsine.

Riddelliine, $C_{18}H_{23}O_6N$, was isolated and studied by ADAMS and his co-workers (*10, 27*) to whom we owe our complete knowledge of this base. The alkaloid hydrolysed to retronecine and riddellic acid. On catalytic reduction in the presence of palladium on strontium carbonate riddelliine absorbed one mole of hydrogen to give a crystalline product (CLVI) which showed in the infrared a band at 1707 cm.$^{-1}$ corresponding to a conjugated ester carbonyl.

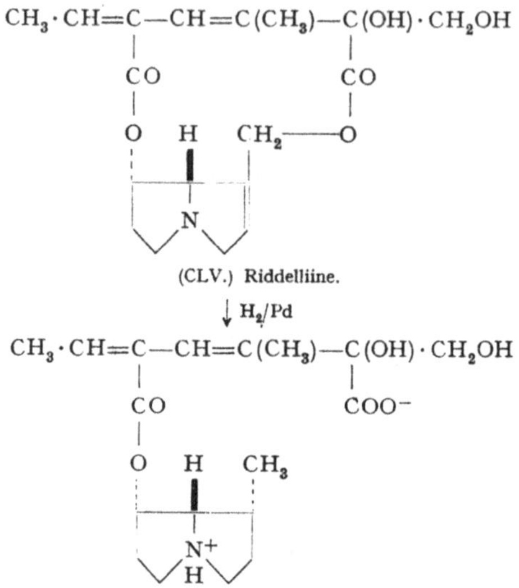

$$\begin{array}{l}CH_3\cdot CH{=}C{-}CH{=}C(CH_3){-}C(OH)\cdot CH_2OH\\ \quad\quad\;\; |\quad\quad\quad\quad\quad\quad\quad\;\; |\\ \quad\quad\; CO \quad\quad\quad\quad\quad\quad\; CO\\ \quad\quad\;\; |\quad\quad\quad\quad\quad\quad\quad\;\; |\\ \quad\quad\; O\quad H\quad CH_2{-}{-}{-}O\\ \quad\quad\quad\quad\quad N \end{array}$$

(CLV.) Riddelliine.

↓ H_2/Pd

$$\begin{array}{l}CH_3\cdot CH{=}C{-}CH{=}C(CH_3){-}C(OH)\cdot CH_2OH\\ \quad\quad\;\; |\quad\quad\quad\quad\quad\quad\quad\;\; |\\ \quad\quad\; CO \quad\quad\quad\quad\quad\quad\; COO^-\\ \quad\quad\;\; |\\ \quad\quad\; O\quad H\quad CH_3\\ \quad\quad\quad\quad\quad N^+H \end{array}$$

(CLVI.)

This is interpreted as indicative of the structure (CLVI) of the reduction product from riddelliine (CLV). Further proof was advanced by the treatment of the dihydro-compound with lead tetraacetate in aqueous solution whereby glycol fission and degradation of the arising keto-acid occurs in one operation, when 0.74 mole of carbon dioxide was evolved.

It is significant that in these alkaloids it is the carboxyl group which is likely to have the stronger ionisation constant that is esterified with the less hindered primary hydroxyl group. It is logical to assume that

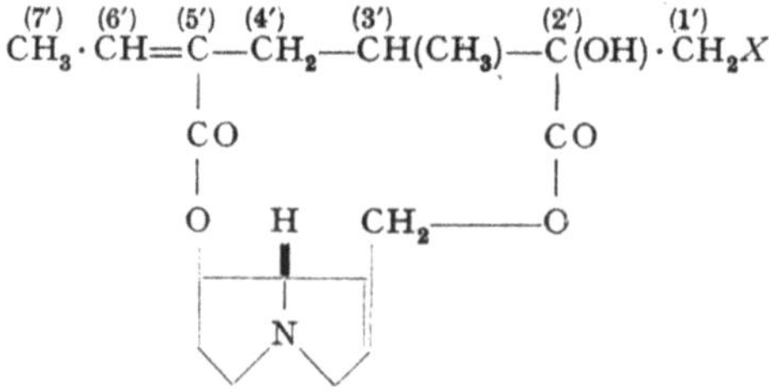

$$\begin{array}{l}\overset{(7')}{CH_3}\cdot\overset{(6')}{CH}{=}\overset{(5')}{C}{-}\overset{(4')}{CH_2}{-}\overset{(3')}{CH}(CH_3){-}\overset{(2')}{C}(OH)\cdot\overset{(1')}{CH_2}X\\ \quad\quad\quad\quad\;\; |\quad\quad\quad\quad\quad\quad\quad\quad\quad\; |\\ \quad\quad\quad\quad CO \quad\quad\quad\quad\quad\quad\quad\quad CO\\ \quad\quad\quad\quad\;\; |\quad\quad\quad\quad\quad\quad\quad\quad\quad\; |\\ \quad\quad\quad\quad O\quad H\quad CH_2{-}{-}{-}{-}O\\ \quad\quad\quad\quad\quad\quad\quad N \end{array}$$

		Ethylidene group:
(CLI, X = OH)	Retrorsine:	*cis*
(CLVII, X = H)	Senecionine:	*cis*
(CLVIII, X = H)	Integerrimine:	*trans*
(CLIX, X = H)	Usaramoensine:	*trans*

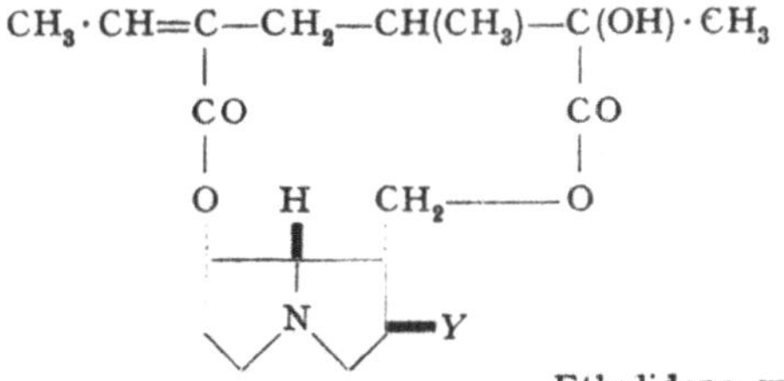

$CH_3 \cdot CH{=}C{-}CH_2{-}CH(CH_3){-}C(OH) \cdot CH_3$

CO CO

O H CH_2—O

N Y

Ethylidene group:

(CLX, $Y = H$)	Platyphylline:	*cis*
(CLXI, $Y = OH$)	Rosmarinine:	*cis*
(CLXII, $Y = p\text{-}CH_3 \cdot C_6H_4 \cdot SO_3{-}$):		—

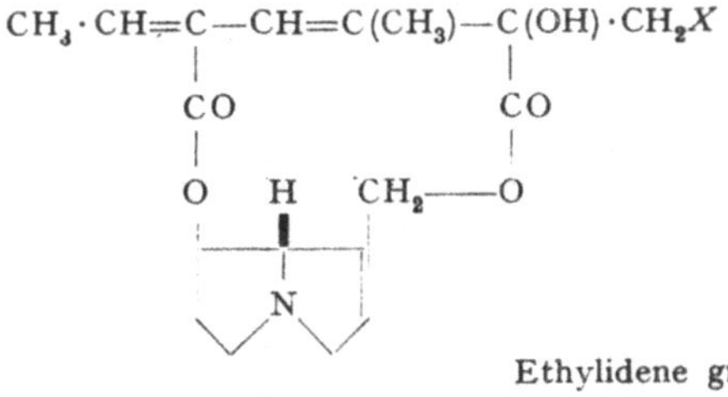

$CH_3 \cdot CH{=}C{-}CH{=}C(CH_3){-}C(OH) \cdot CH_2X$

CO CO

O H CH_2—O

N

Ethylidene group:

(CLV, $X = OH$)	Riddelliine:	—
(CLXIII, $X = H$)	Seneciphylline:	*cis*

the other C_{18}-pyrrolizidine alkaloids are similarly oriented (*110, 93*) so that it is possible now to give general formulae as above.

The assumption that the acids are similarly oriented and may be represented by these general formulae has recently received some confirmation from the dehydration of rosmarinine (CLXI) to senecionine (CLVII) by KOEKEMOER and WARREN (*93*). Rosmarinine readily gave rosmarinine toluene-*p*-sulphonate (CLXII), which by heating with dry pyridine lost toluene-*p*-sulphonic acid to yield senecionine (CLVII). This establishes that the geometrical configuration at $C_{(5)}$ and the optical modification at $C_{(2)}$ of the acid moiety in these two alkaloids are identical. It is of interest that the hydrolysis of rosmarinine toluene-*p*-sulphonate did not give rosmarinecine but an isomer which was probably 2-*epi*rosmarinecine which would be expected from the inversion during hydrolysis of the toluenesulphonyl group.

Senecionine, $C_{18}H_{25}O_5N$, was isolated by GRANDVAL and LAJOUX (*75*) as early as 1895 and was one of the first *Senecio* alkaloids prepared. BARGER and BLACKIE (*36*) showed that it hydrolysed to retronecine and senecic acid, and with the structures of these two hydrolysis products known, senecionine is (CLVII, p. 250) (*93*).

Integerrimine, $C_{18}H_{25}O_5N$, was found by MANSKE (*121*) as the minor alkaloid after the separation of senecionine from *S. integerrimus*, and

ADAMS and VAN DUUREN (*25*) isolated it as the sole alkaloid from *Crotalaria incana*. Hydrolysis (*121*) of integerrimine gave retronecine and integerrinecic acid, which was *trans*-senecic acid (*105*). This alkaloid (CLVIII) therefore differed from senecionine (CLVII) in the configuration of the ethylidene group (*105*) whilst there could be a difference in the arrangement of the groups around $C_{(2')}$ (*25*).

Usaramoensine, $C_{18}H_{25}O_5N$, was isolated by ADAMS and VAN DUUREN (*25*) and hydrolysed to retronecine and usaramoensinecic acid which is isomeric with senecic and integerrinecic acids and differs from senecic acid only in the configuration of the 2'-hydroxyl group. From a study of the ultraviolet spectra ADAMS and VAN DUUREN (*25*) concluded that usaramoensine has the acid in the *trans*-form, although LEISEGANG (*108*) was unable to deduce the configuration of the acid in other alkaloids from ultraviolet spectra. Usaramoensine may now be formulated as (CLIX, p. 250). ADAMS attributes the difference between usaramoensine and integerrimine to a different optical configuration at $C_{(2')}$.

Platyphylline, $C_{18}H_{27}O_5N$, was first isolated and studied by OREKHOV, KONOVALOVA and TIDEBEL (*153, 152, 95, 102, 96*). The hydrolysis of platyphylline was reported (*102, 95*) as giving platynecine and platynecic acid which DE WAAL (*189*) identified as senecic acid lactone. Later DANILOVA and KONOVALOVA (*60*) found that platyphylline on hydrolysis with alcoholic and aqueous alkali gave senecic acid, m. p. 151°, and platynecic acid, m. p. 133–135°. The formula (CLX) for platyphylline is advanced on the assumption of similarity of structure of these alkaloids. KONOVALOVA (*97*) made the interesting observation that the yield of platyphylline from *S. platyphyllus* was increased by treatment of the acid extract with zinc dust, and attributed this to the presence of genplatyphylline. This gen-compound is almost certainly platyphylline *N*-oxide that was prepared by KOEKEMOER and WARREN (*92*) and which they showed was the form in which the alkaloid was present in *S. hygrophilus*.

Rosmarinine, $C_{18}H_{27}O_6N$, was isolated by DE WAAL (*181, 182*) who hydrolysed it to rosmarinecine and senecic acid. KOEKEMOER and WARREN (*92*) prepared the *N*-oxide as a hygroscopic amorphous solid and showed that it was predominantly in this form in which the alkaloid occurred in several *Senecio* species. The formula (CLXI) which was advanced on the basis of similarity of structure has some justification, outlined above (*93*), in that rosmarinine (CLXI) has recently been converted to senecionine (CLVII).

Seneciphylline, $C_{18}H_{25}O_5N$, was isolated by OREKHOV and TIDEBEL (*153*). KONOVALOVA and DANILOVA (*98*) showed that it was catalytically hydrogenated to an ester of retronecanol which was hydrolysed with aqueous and alcoholic potassium hydroxide to *iso*seneciphyllic and seneciphyllic acids, respectively. These authors therefore concluded that it was a cyclic diester of seneciphyllic acid with retronecine, and this conclusion may be now formulated (CLXIII). Seneciphylline is widespread in the *Senecio* species and frequently occurs with senecionine and this mixture is difficult to separate.

cis-trans-Isomerization. In all these alkaloids there exists an ethylidene group as part of the acid moiety and thus geometrical isomerisation is possible. It is easy to understand that if the alkaloid contains the acid in the labile *cis*-form, the acid obtained on hydrolysis will either retain its *cis*-configuration or isomerise to its stable *trans*-form according to the conditions of hydrolysis. On the other hand, integerrimine with the acid in the stable *trans*-state gives the stable *trans*-acid (*121*).

Table 13. Geometrical Isomerisation on Hydrolysis of Some Pyrrolizidine Alkaloids.

Alkaloid	Acids obtained on hydrolysis		Reference
	cis-acid	*trans*-acid	
Retrorsine	Isatinecic acid	Retronecic acid	(*48*)
Isatidine	Isatinecic acid	Retronecic acid	(*48*)
Senecionine	Senecic acid	Integerrinecic acid	(*105*)
Seneciphylline	*iso*Seneciphyllic acid	Seneciphyllic acid	(*98*)
Integerrimine	—	Integerrinecic acid	(*121*)

Adams and Van Duuren (*25*) have shown that integerrimine and usaramoensine have in the infrared a strong band at 1665 cm.$^{-1}$, and senecionine a band at 1645 cm.$^{-1}$, from which it is concluded that the ethylidene group is *trans* in integerrimine and usaramoensine but *cis* in senecionine. It is not easy to appreciate thermodynamically why usaramoensine will by the ordinary process of hydrolysis give the less stable *cis*-usaramoensinecic acid, which is the geometrical configuration reported for the hydrolysed acid (*25*).

Grantianine, $C_{18}H_{23}O_7N$, isolated from *Crotalaria grantiana* by Adams, Carmack and Rogers (*3*), gave on hydrolysis retronecine and an acid which could not be obtained pure. Grantianine on hydrogenation over platinum oxide absorbed two moles of hydrogen to give tetrahydrograntianine (CLXIV), $C_{18}H_{27}O_7N$, m. p. 242–245°, and accordingly the alkaloid was formulated with two carboxyl groups attached to the necine. With the stereoconfiguration established for retronecine grantianine is (CLXIII) which on reduction will give (CLXIV). (A by-product of the reduction was a substance $C_{18}H_{27}O_6N$, which was isolated as its picrate, m. p. 156–157°.) The acid in grantianine is seemingly saturated, and Adams (*3*) drew attention to the possible relationship between monocrotalic acid, $C_7H_{11}O_3 \cdot COOH$, and grantianic acid, $C_8H_{12}O_3(COOH)_2$, which might be written $C_7H_{10}O_3(COOH)\,(CH_2 \cdot COOH)$.

```
CO·(C8H12O3)·CO            CO·(C8H12O3)·COOH
|            |             |
O         CH2·O            O         CH3

   (pyrrolizidine ring)      (pyrrolizidine ring)
```

(CLXIII.) Grantianine. (CLXIV.)

Sarracine, $C_{18}H_{27}O_5N$, was isolated from *Senecio sarracenius* by Konovalova, Massagetov and Garina (*61*) who found it occurred in the plant as the *N*-oxide which under certain conditions could be isolated. Sarracine was unsaturated, contained one hydroxyl group, and gave on hydrolysis platynecine. The necic acid was not isolated. Sarracine is seemingly isomeric with platyphylline.

Sceleratine, $C_{18}H_{27}O_7N$, was first isolated by de Waal and Pretorius (*187*) who showed that alkaline hydrolysis gave retronecine and sceleranecic acid dilactone whilst catalytic hydrogenation effected fission to retronecanol and the same dilactone. These authors concluded that sceleranecic acid was attached by one carboxylic group. In view of the finding of Adams and Govindachari (*7*), there is the possibility that both carboxylic groups are attached and complete fission occurs by a process of *trans*-esterification.

Senecifoline, $C_{18}H_{27}O_8N$, m. p. 194–195°, was obtained by Watt (*192*) as one of the alkaloids from *Senecio latifolius* D. C. This species has since been divided and the origin of the material remains unknown. On hydrolysis it gives senecifolinine, which is possibly identical with retronecine (*38*) and senecifolic acid.

Alkaloids from S. brasiliensis. Two seemingly different alkaloids have been isolated from *S. brasiliensis*. Novelli and de Varela (*150*) found an alkaloid, $C_{18}H_{25}O_5N$, m. p. 232–244°, $[\alpha]_D$ 66.8°, which they hydrolysed to retronecine and an acid, $C_{10}H_{16}O_5$, m. p. 140–141°, which was dextrorotatory. Fonseca (*70*) reported, however, an alkaloid unfortunately named brasilinecine, m. p. 169–171°, $[\alpha]_D$ — 68.2°, which hydrolysed to retronecine and an acid, m. p. 141–143°, after distillation. The alkaloid, m. p. 232–244°, has a melting point similar to that of senecionine, m. p. 234° (*36*), 245° (*55*), and brasilinecine gives a picrate, m. p. 191–192°, a methiodide, m. p. 235.5°, and aurichloride, m. p. 146–148°, whilst senecionine yields a picrate, m. p. 191° (*36*), a methiodide, m. p. 243° (*5*) and an aurichloride, m. p. 186°.

Monocrotaline.

Monocrotaline, $C_{16}H_{23}O_6N$, first isolated by Neal, Rusoff and Ahmann (*149*), was hydrolysed by Adams and Rogers (*18*) to retronecine and monocrotic acid, whilst catalytic hydrogenation resulted in the consumption of two moles or one mole of hydrogen leaving monocrotalic acid (CLXIX) and retronecanol (CLXX) or desoxyretronecine, respectively, according to the catalyst used. The complete fission of

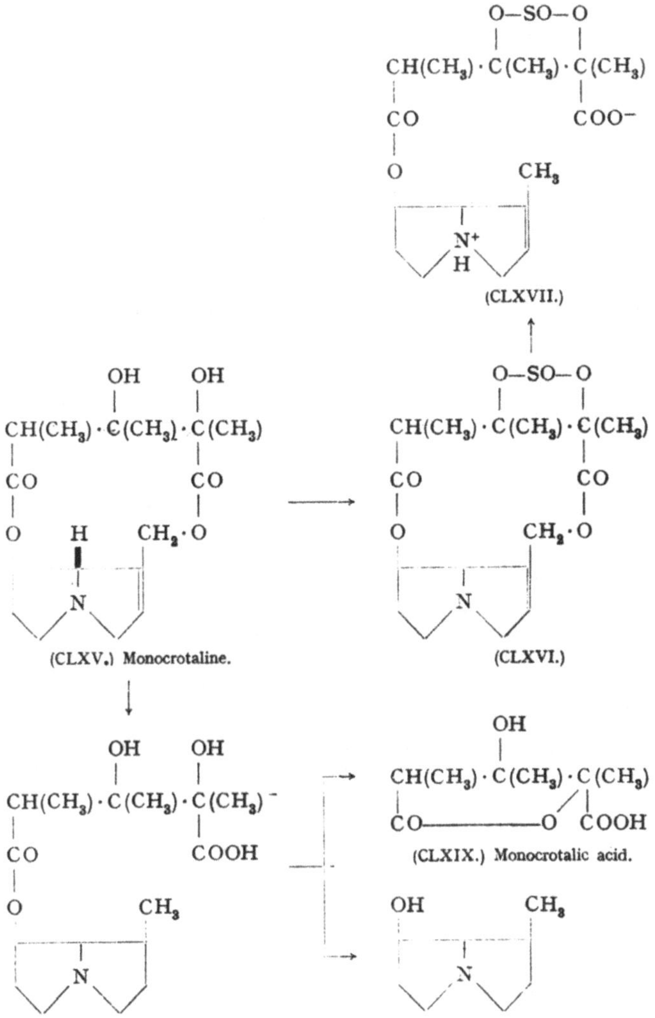

(CLXV.) Monocrotaline.

(CLXVI.)

(CLXVII.)

(CLXVIII.)

(CLXIX.) Monocrotalic acid.

(CLXX.) Retronecanol.

$-CH_2 \cdot O \cdot CO \cdot CO \cdot CH_3$

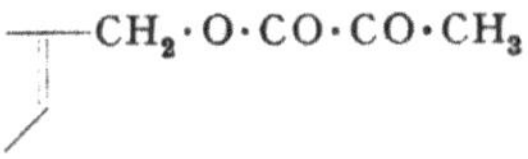

(CLXXI.)

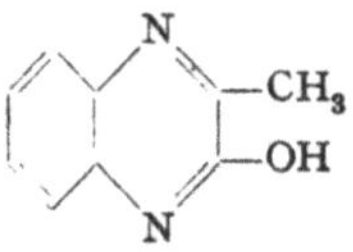

(CLXXII.) 2-Hydroxy-3-methylquinoxaline.

the acid moiety from the base on hydrogenolysis led to the assumption by ADAMS and GOVINDACHARI (*7*) that the alkaloid had a lactone ester structure. The infrared spectrum could best be explained, however, on the diester structure (CLXV) characteristic of the *Senecio* alkaloids from retronecine; and ADAMS, SHAFER and BRAUN (*23*) advanced convincing evidence to establish this.

Monocrotaline reacted with thionyl chloride to give a cyclic sulphite ester (CLXVI) (*Chart 13*) showing no bands in the infrared characteristic of a lactone-carbonyl or hydroxyl groups. The reduction of this sulphite ester (CLXVI) resulted in the formation of a zwitterion containing an ester linkage, and this reduction product may be formulated (CLXVII). The necic acid is accordingly joined to the pyrrolizidine nucleus by two ester links, one of which is easily cleaved by hydrogenolysis, whilst the other survives only when the α-hydroxyl group is blocked. Hydrogenolysis of monocrotaline proceeds by fission of the allylic ester group to give (CLXVIII) and then intramolecular *trans*-esterification with the formation of the lactone ring of monocrotalic acid. Additional proof of the presence of the glycol group in monocrotaline, and hence of the structure of the acid, was the oxidation with lead tetraacetate. The product with *o*-phenylenediamine gave 2-hydroxy-3-methylquinoxaline (CLXXII) expected from the pyruvic ester (CLXXI).

Dicrotaline.

Dicrotaline, $C_{14}H_{19}O_5N$, isolated originally by MARAIS (*123*), was hydrolysed to retronecine and dicrotalic acid, which is 2-hydroxy-2-methylpentane-1:3-dicarboxylic acid (*24*), the symmetrical nature of which permits the assignment of structure (CLXXIII).

$CH_2 \cdot C(CH_3)(OH) \cdot CH_2$ | CO | O ... CO | $CH_2 \cdot O$ ... H, N (pyrrolizidine nucleus)

(CLXXIII.) Dicrotaline.

The Alkaloids as *N*-Oxydes.

Several of these alkaloids have been isolated as their *N*-oxides whilst there is evidence that many more exist in the plant in this form (*92, 53, 54*). The presence of these amine oxides may be detected in a number of ways. POLONOVSKI and POLONOVSKI (*154*) have shown that acetic anhydride becomes coloured in the presence of *N*-oxides. Secondly,

CULVENOR, DRUMMOND and PRICE (*54*) have observed that the R_f value of *N*-oxides are altered only very slightly when butanol-ammonia is substituted for butanol-acetic acid as chromatographing solvent, whereas that of the tertiary amines is much higher in the butanol-ammonia, as shown in Table 14.

Table 14. R_f Values of Alkaloids and their *N*-oxides.

Alkaloid	Butanol-acetic acid	Butanol-ammonia
Heliotrine	0.42	0.85
Heliotrine *N*-oxide	0.52	0.52
Heliotridine	0.16	0.61
Heliotridine *N*-oxide	0.22	0.15
Lasiocarpine	0.59	0.90
Lasiocarpine *N*-oxide	0.69	0.70

Not all *N*-oxides are soluble in chloroform which is frequently used in extraction. They may, however, be detected by the method of KOEKEMOER and WARREN (*92*), namely, by estimating the quantity of alkaloid obtained before and after reduction of the crude alcoholic extract with zinc. Sulphur dioxide has been used successfully by MEN'SHIKOV and BORODINA (*137*, *138*) for the conversion of trachelanthine to trachelanthamine. Sulphur dioxide may react, however, with *N*-oxides to cause degradation (*107*, *44*), and in fact lasiocarpine was obtained in low yield from its *N*-oxide with this reagent (*54*).

V. Pharmacology.

(With M. E. VON KLEMPERER.)

Toxicological studies on the pyrrolizidine alkaloids over the last half century have shown that, whereas most pharmacologically active alkaloids confine their attack to the central nervous system, the *Senecio* and related alkaloids have a selective action on the liver and produce hepatic necrosis in various species of animals.

Using pure samples of senecionine and senecifolidine supplied by WATT (*192*, cf. *193*) for his feeding experiments, CUSHNY (*56*, *57*, *58*) arrived at the conclusion that death by *Senecio* poisoning could be brought about in three ways depending on the amount ingested. Large quantities affected the central nervous system and induced acute symptoms lasting only two to three hours, but produced no hepatic cirrhosis. In subacute poisoning death by respiratory failure followed intestinal disturbances lasting one to two days. Chronic poisoning, however, was associated with haemorrhage sometimes in the alimentary canal but always in the liver where it is responsible for the secondary effects and forms the starting point for cirrhosis which is now recognised as so characteristic of poisoning by pyrrolizidine alkaloids.

Pioneering work in South Africa was undertaken by STEYN and his co-workers (*167, 168, 169, 170, 171, 172*) whose toxicological studies extended over nearly twenty years. Similar feeding experiments to study palatability, toxicity and the extent of liver injury have been carried out by workers all over the world (*32, 33, 39, 67, 70, 77, 88, 124, 161, 174, 176–178, 183*).

The action of individual alkaloids on various laboratory animals has been extensively studied by CHEN, HARRIS, HENDERSON and their collaborators at Indianopolis (*45–47, 78–86, 160, 190*). They found that all the alkaloids, other than platyphylline, caused hepatic necrosis, generally with congestion and haemorrhage and in extreme cases with lesions. These lesions were also observed by DAVIDSON (*62*). It is of interest that, whereas necrosis was predominantly central, there was a resemblance between carthamoidine (*81*) and pterophine (*80*) where the effect was mainly periportal. Both these supposed alkaloids have since been shown to consist of a mixture of senecionine and seneciphylline (*6, 15, 55*). Monocrotaline and isatidine proved to be considerably less toxic than the other alkaloids (*68*). Isatidine is retrorsine *N*-oxide (*110*) and the question arises whether this lowering of toxicity always accompanies *N*-oxide formation, since it has been shown that the *N*-oxide form may predominate in several of the *Senecio* species (*92, 53, 54*). Retronecic acid monolactone, monocrotalic acid and retronecine have been tested in an attempt to determine whether liver injury is due to the whole alkaloid or the hydrolysis products. In each case the effect was the same as with platyphylline and no damage of the liver was observed. SCHOENTAL (*164*), however, is of the opinion that this problem has yet to be solved.

Working with a mixture of alkaloids from *S. jacobaea*, COOK, DUFFY and SCHOENTAL (*52*) produced liver tumours in rats which survived eight months administration of the poison. Using modified doses to ensure better survival, SCHOENTAL, HEAD and PEACOCK (*164*) induced profound liver damage in rats, which were fed over periods of more than one year on mixed alkaloids of *S. jacobaea* and pure samples of isatidine and retrorsine. The animals showed liver changes ranging from nodular hyperplasia and fibrosis to neoplasia, whilst metastases were found in one rat treated with isatidine.

These findings of the Glasgow school, which link the *Senecios* with cancer research (*163*) are a significant advance in this branch of study of the pyrrolizidine alkaloids. One implication is that the aetiological factor in the high incidence of primary liver carcinoma among African negroes may be the consumption of *Senecio* alkaloids as native herbal remedies.

Medicinal properties, including oestrogenic and antibiotic action have been claimed for these alkaloids; but SCHOENTAL and PULLINGER (*165*) have established that isatidine, retrorsine, monocrotaline, riddelliine and the alkaloids of *S. jacobaea* have no such properties. In fact, testosterone propionate increased the toxicity of monocrotaline (*158*).

On the other hand, considerable interest has been shown in platyphylline (*31*) which has been found to exert an atropine-like effect (*74*). CHEN, HARRIS and ROSE (*46*) observed that platyphylline produced either immediate death or complete recovery and no hepatic or renal damage resulted. BABSKIĬ (*35*) and KOVYREV (*103*) both showed that the drug produced a drop in blood pressure and inhibited the transmission of excitations in the parasympathetic nerves, but that the dose required was twenty-five to thirty times that of atropine. Human clinical experiments were carried out by GOLDENHERSHEL (*74*) using platyphylline hydrochloride or tartrate. He observed that therapeutic doses of the drug were well tolerated over long periods: it had no effect on the heart action, it avoided the disagreeable secondary effects of atropine and its antispasmodic action was superior. SYRNEVA (*173*) confirmed this last finding. He proved that the hypertensive action is due to depression of vasomotor centres in the *medulla oblongata* and noted a mydriatic effect similar to that of atropine. It is of interest to record here that KONOVALOV and MEN'SHIKOV (*94*) attribute pronounced spasmolytic properties to the alkaloid hastacine.

In the pseudo-heliotridane series some interesting synthetic substances have been prepared by MEN'SHIKOV et al. (*140*, *143*). The *p*-aminobenzoyl derivative of trachelanthamidine (CLXXIV) is comparable with cocaine, whereas the quinoline derivative (CLXXV) has antimalarial activity.

H $CH_2 \cdot O \cdot CO$— —NH_2
N

(CLXXIV.)

OCH_3
H CH_2—NH—
N N

(CLXXV.)

References.

1. ADAMS, R.: Über die Chemie der *Senecio*-Alkaloide und verwandter Verbindungen. Angew. Chem. **65**, 433 (1953).

2. ADAMS, R., M. CARMACK and J. E. MAHAN: Structure of Monocrotaline. VII. Structure of Retronecine and Related Bases. J. Amer. Chem. Soc. **64**, 2593 (1942).

3. ADAMS, R., M. CARMACK and E. F. ROGERS: The Alkaloid of *Crotalaria grantiana*. I. Grantianine. J. Amer. Chem. Soc. **64**, 571 (1942).

4. Adams, R. and T. R. Govindachari: *Senecio* Alkaloids: alpha- and beta-Longilobine from *Senecio longilobus*. Festschrift P. Karrer, S. 1 (1949); J. Amer. Chem. Soc. **71**, 1180 (1949).
5. — — *Senecio* Alkaloids: The Isolation of Senecionine from *Senecio cineraria* and Some Observations on the Structure of Senecionine. J. Amer. Chem. Soc. **71**, 1953 (1949).
6. — — *Senecio* Alkaloids: The Alkaloids of *Senecio douglasii, carthamoides, eremophilus, ampullaceus* and *parksii*. J. Amer. Chem. Soc. **71**, 1956 (1949).
7. — — Structure of Monocrotaline. XII. Monocrotalic Acid. J. Amer. Chem. Soc. **72**, 158 (1950).
8. Adams, R., T. R. Govindachari, J. H. Looker and J. D. Edwards, Jr.: *Senecio* Alkaloids: alpha-Longilobine; Structure of alpha-Longinecic Acid. J. Amer. Chem. Soc. **74**, 700 (1952).
9. Adams, R. and K. E. Hamlin, Jr.: Structure of Monocrotaline. VIII. The Proof of Primary and Secondary Hydroxyl Groups in Retronecine. J. Amer. Chem. Soc. **64**, 2597 (1942).
10. Adams, R., K. E. Hamlin, Jr., C. F. Jelinek and R. E. Phillips: The Structure of Riddelliine, the Alkaloid in *Senecio riddellii*. I. J. Amer. Chem. Soc. **64**, 2760 (1942).
11. Adams, R. and F. B. Hauserman: The Total Structure of Monocrotaline. XIII. Synthesis of Dihydroanhydro-monocrotalic Acid. J. Amer. Chem. Soc. **74**, 694 (1952).
12. Adams, R. and W. Herz: Synthesis of Viridifloric Acid. J. Amer. Chem. Soc. **72**, 155 (1950).
13. Adams, R. and N. J. Leonard: Structure of Monocrotaline. XI. Proof of the Structure of Retronecine. J. Amer. Chem. Soc. **66**, 257 (1944).
14. Adams, R. and R. S. Long: Structure of Monocrotaline. IV. Monocrotalic Acid. J. Amer. Chem. Soc. **62**, 2289 (1940).
15. Adams, R. and J. H. Looker: The Identity of alpha-Longilobine and Seneciphylline. J. Amer. Chem. Soc. **73**, 134 (1951).
16. Adams, R. and J. E. Mahan: Basicity Studies of Tertiary Vinyl Amines. J. Amer. Chem. Soc. **64**, 2588 (1942).
17. — — Structure of Monocrotaline. IX. Proof of the Position of the Double Bond in Retronecine. J. Amer. Chem. Soc. **65**, 2009 (1943).
18. Adams, R. and E. F. Rogers: The Structure of Monocrotaline, the Alkaloid in *Crotalaria spectabilis* and *Crotalaria retusa*. I. J. Amer. Chem. Soc. **61**, 2815 (1939).
19. — — Structure of Monocrotaline. V. Retronecine, a Derivative of 1-Methyl-pyrrolizidine. J. Amer. Chem. Soc. **63**, 228 (1941).
20. — — Structure of Monocrotaline. VI. The Structure of Retronecine, Platynecine and Retronecanol. J. Amer. Chem. Soc. **63**, 537 (1941).
21. Adams, R., E. F. Rogers and R. S. Long: The Structure of Monocrotaline. III. Monocrotalic Acid. J. Amer. Chem. Soc. **61**, 2822 (1939).
22. Adams, R., E. F. Rogers and F. J. Sprules: Structure of Monocrotaline. II. Monocrotic Acid obtained by Alkaline Hydrolysis of the Alkaloid. J. Amer. Chem. Soc. **61**, 2819 (1939).
23. Adams, R., P. R. Shafer and B. H. Braun: The Structure of Monocrotaline. XV. J. Amer. Chem. Soc. **74**, 5612 (1952).
24. Adams, R. and B. L. Van Duuren: Dicrotaline. The Structure and Synthesis of Dicrotalic Acid. J. Amer. Chem. Soc. **75**, 2377 (1953).
25. — — Usaramoensine, the Alkaloid in *Crotalaria usaramoensis* E. G. Baker. Integerrimine from *Crotalaria incana* Linn. and Senecionine from *Senecio glabellus* D. C. Stereochemical Relationships. J. Amer. Chem. Soc. **75**, 4631 (1953).

26. ADAMS, R. and B. L. VAN DUUREN: The Structure of Heliotrinic Acid. J. Amer. Chem. Soc. **75**, 4636 (1953).

27. — — Riddelliine, the Alkaloid from *Senecio riddellii*. II. The Structure of Riddellic Acid and the Total Structure of Riddelliine. J. Amer. Chem. Soc. **75**, 4638 (1953).

28. — — Trachelanthicand Viridifloric Acids. J. Amer. Chem. Soc. **74**, 5349 (1952).

28a. — — Stereochemistry of the Pyrrolizidine Bases. J. Amer. Chem. Soc. **76**, 6379 (1954).

29. ADAMS, R., B. L. VAN DUUREN and B. H. BRAUN: The Structure of Monocrotaline. XIV. Synthesis of Monocrotalic Acid. J. Amer. Chem. Soc. **74**, 5608 (1952).

30. ADAMS, R. and J. M. WILKINSON, Jr.: Structure of Monocrotaline. X. Monocrotalic Acid. J. Amer. Chem. Soc. **65**, 2203 (1943).

31. ALLEN, E. Y.: *Senecio platyphyllus* — an Old Native Russian Medicinal Plant and its New Alkaloidal Drug, Platyphylline Bitartarate. Amer. J. Pharm. **117**, 110 (1945) [Chem. Abstr. **39**, 3120 (1945)].

32. Anonymous: Notes on *Senecio* Poisoning. Bull. Imper. Inst. (London) **18**, 435 (1920).

33. — The Causation of Molteno, Pictou or Winton Disease in Cattle and Horses. Bull. Imper. Inst. (London) **9**, 346 (1911).

34. ASAHINA, Y.: Notiz über Seneciosäure. Arch. Pharmaz. **251**, 355 (1913) [Chem. Abstr. **8**, 677 (1914)].

35. BABSKIĬ, E. B.: A new Parasympathicolytic Substance, Platyphyllin. C. R. (Doklady) Acad. Sci. (USSR) **27**, 83 (1940) [Chem. Abstr. **34**, 7011 (1940)].

36. BARGER, G. and J. J. BLACKIE: Alkaloids of *Senecio*. II. Senecionine and Squalidine. J. Chem. Soc. (London) **1936**, 743.

37. — — Alkaloids of *Senecio*. III. Jacobine, Jacodine, and Jaconine. J. Chem. Soc. (London) **1937**, 584.

38. BARGER, G., T. R. SESHADRI, H. E. WATT and T. YABUTA: Alkaloids of *Senecio*. I. Retrorsine. J. Chem. Soc. (London) **1935**, 11.

39. BECKER, R. B., W. M. NEAL, P. T. D. ARNOLD and A. L. SHEALY: A Study of the Palatability and Possible Toxicity of Eleven Species of *Crotalaria*, Especially of *C. spectabilis* ROTH. J. Agricult. Res. **50**, 911 (1935).

40. BLACKIE, J. J.: The Alkaloids of the Genus *Senecio*. Pharmac. J. **138**, 102 (1937) [Chem. Abstr. **31**, 7195 (1937)].

41. BRADBURY, R. B.: The Relationship between Jacobine and Jaconine and the Structure of Jaconecic Acid. Chem. and Ind. **1954**, 1022.

42. BRADBURY, R. B. and C. C. J. CULVENOR: The Alkaloids of *Senecio jacobaea* L. Chem. and Ind. **1954**, 1021; Austral. J. Chem. **7**, 378 (1954).

43. BRIGGS, L. H., J. L. MANGAN and W. E. RUSSELL: Alkaloids of New Zealand *Senecio* Species. I. The Alkaloid from *Senecio kirkii*. J. Chem. Soc. (London) **1948**, 1891.

44. BURG, A. B.: The Behavior of Trimethylamine, Trimethylammino-sulfur Trioxide and Trimethylamine Oxide toward Sulfur Dioxide. J. Amer. Chem. Soc. **65**, 1629 (1943).

45. CHEN, K. K., A. L. CHEN and C. L. ROSE: Action and Toxicity of Retrorsine. J. Pharmacol. exp. Therapeut. **54**, 299 (1935) [Chem. Abstr. **29**, 7497 (1935)].

46. CHEN, K. K., P. N. HARRIS and C. L. ROSE: Action and Toxicity of Platyphylline and Seneciphylline. J. Pharmacol. exp. Therapeut. **68**, 130 (1940) [Chem. Abstr. **34**, 1739 (1940)].

47. CHEN, K. K., P. N. HARRIS and H. A. SCHULZE: Toxicity of Lasiocarpine. J. Pharmacol. exp. Therapeut. **68**, 123 (1940) [Chem. Abstr. **34**, 1738 (1940)].

48. CHRISTIE, S. M. H., M. KROPMAN, E. C. LEISEGANG and F. L. WARREN: The *Senecio* Alkaloids. III. The Structure of Retrorsine and Isatidine, and the Isomerism of Retronecic Acid and Isatinecic Acid. J. Chem. Soc. (London) **1949**, 1700.

49. CHRISTIE, S. M. H., M. KROPMAN, L. NOVELLIE and F. L. WARREN: The *Senecio* Alkaloids. IV. The Structure of Retronecic and Isatinecic Acids. J. Chem. Soc. (London) **1949**, 1703.

50. CLEMO, G. R. and T. A. MELROSE: The Synthesis of 2-Methylpyrrolizidine. J. Chem. Soc. (London) **1942**, 424.

51. CLEMO, G. R. and T. P. METCALFE: The Lupin Alkaloids. IX. The Synthesis of 5 : 5'-Dimethyldi(1 : 2)pyrrolidine. J. Chem. Soc. (London) **1936**, 606.

52. COOK, J. W., E. DUFFY and R. SCHOENTAL: Primary Liver Tumours in Rats Following Feeding with Alkaloids of *Senecio jacobea*. Brit. J. Cancer **4**, 405 (1950) [Chem. Abstr. **45**, 5306 (1951)].

53. CULVENOR, C. C. J.: The Alkaloids of *Heliotropium europaeum* L. II. Isolation and Structures of the Third Major Alkaloid and Two Minor Alkaloids, and Isolation of the Principal *N*-Oxides. Austral. J. Chem. **7**, 287 (1954).

54. CULVENOR, C. C. J., L. J. DRUMMOND and J. R. PRICE: The Alkaloids of *Heliotropium europaeum* L. I. Heliotrine and Lasiocarpine. Austral. J. Chem. **7**, 277 (1954).

55. CULVENOR, C. C. J. and L. W. SMITH: The Separation of *Senecio* Alkaloids and the Nature of Hieracifoline and Pterophine. Chem. and Ind. **1954**, 1386.

56. CUSHNY, A. R.: On the Action of *Senecio* Alkaloids and the Causation of Hepatic Cirrhosis in Cattle (Preliminary Note). Proc. Roy. Soc. (London), Ser. B **84**, 188 (1911).

57. — On the Action of *Senecio* Alkaloids and the Causation of Hepatic Cirrhosis of Cattle (Pictou, Molteno or Winton Desease). J. Pharmacol. exp. Therapeut. **2**, 531 (1910–1911).

58. CUSHNY, A. R. and H. E. WATT: *Senecio* Poisoning. Lancet **199**, 1089 (1920).

59. DANILOVA, A. V. and R. A. KONOVALOVA: Alkaloids of *Senecio* Species. VII. Alkaloids from *Senecio renardi*. J. Gen. Chem. (USSR) **20**, 1921 (1950) [Chem. Abstr. **45**, 2960 (1951)].

60. — — Structure of Platynecic and Senecionic Acids. C. R. (Doklady). Acad. Sci. (USSR) **73**, 315 (1950) [Chem. Abstr. **45**, 2870 (1951)].

61. DANILOVA, A. V., R. A. KONOVALOVA, P. MASSAGETOV and M. GARINA: The Alkaloids of *Senecio sarracenius*. C. R. (Doklady) Acad. Sci. (USSR) **89**, 865 (1953) [Chem. Abstr. **48**, 5875 (1954)].

62. DAVIDSON, J.: The Action of Retrorsine on Rat Liver. J. Pathol. Bacteriol. **40**, 285 (1935) [Chem. Abstr. **29**, 3733 (1935)].

63. DRUMMOND, L. J.: Structure of Lasiocarpic Acid. Nature (London) **167**, 41 (1951).

64. DRY, L. J., M. J. KOEKEMOER and F. L. WARREN: The *Senecio* Alkaloids. X. The Structure of Rosmarinecine and its Synthesis from Retronecine. J. Chem. Soc. (London) **1955**, 59.

65. DRY, L. J. and F. L. WARREN: The *Senecio* Alkaloids. IX. The Synthesis and Resolution of 3 : 4-Dihydroxy-2-methylpentane-3-carboxylic Acids: Viridifloric and Trachelantic Acids. J. Chem. Soc. (London) **1952**, 3445.

65a. — — The Synthesis of (±)-*erythro*-α Hydroxy-β-methoxybutyric Acid. J. South Afr. Chem. Inst. N. S. **6**, 14 (1953).

66. — — The *Senecio* Alkaloids. XII. The Synthesis of (±)-*erythro*-3-Hydroxy-2-methoxy-4-methylpentane-3-carboxylic Acid. J. Chem. Soc. (London) **1955**, 65.

67. DUNSTAN, W. R.: Work on *Senecio latifolius*. Lancet **199**, 1266 (1920); cf. also 56.

68. ELI LILLY and Co.: *Senecio* and Related Alkaloids. Research Today **5** (1949).

69. FISCHER, E. und G. ZEMPLÉN: Synthese der beiden optisch-aktiven Proline. Ber. dtsch. chem. Ges. **42**, 2989 (1909).

70. FONSECA, E. DE CAMARGO: Chemical and Toxicological Study of *Senecio brasiliensis*. Anais faculd. farm. odontol. Univ. São Paulo **9**, 85 (1951) [Chem. Abstr. **47**, 4888 (1953)].

71. FUSON, R. C. and C. L. ZIRKLE: Ring Enlargement by Rearrangement of the 1,2-Aminochloroalkyl Group; Rearrangement of 1-Ethyl-2-chloromethyl-pyrrolidine to 1-Ethyl-3-chloropiperidine. J. Amer. Chem. Soc. **70**, 2760 (1948).

72. GALINOVSKY, F., H. GOLDBERGER und M. PÖHM: Über das Laburnin, ein Alkaloid aus *Cytisus laburnum*. Monatsh. Chem. **80**, 550 (1949).

73. GALINOVSKY, F. und A. REICHARD: Eine Synthese des Pyrrolizidins. Ber. dtsch. chem. Ges. **77**, 138 (1944).

74. GOLDENHERSHEL, I. I.: A New Atropine-like Substance Called Platyphyllin. Klinicheskaya meditsina **31** (3), 56 (1943); Amer. Rev. Soviet Med. **1**, 155 (1943).

75. GRANDVAL, A. et H. LAJOUX: Étude de la sénécionine et de la sénécine. C. R. hebd. Séances Acad. Sci. **120**, 1120 (1895); Bull. soc. chim. France **13**, 942 (1895).

76. GRESHOFF, M.: Mitteilungen aus dem chemisch-pharmakologischen Laboratorium des Botanischen Gartens zu Buitenzorg (Java). Ber. dtsch. chem. Ges. **23**, 3537 (1890).

77. GROVES, K.: Poisonous Constituents of *Amsinckia intermedia*. Washington Agric. Expt. Stat. Bull. **410**, 35 (1941) [Chem. Abstr. **36**, 5251 (1942)].

78. HARRIS, P. N., R. C. ANDERSON and K. K. CHEN: The Action of Senecionine, Integerrimine, Jacobine, Longilobine, and Spartiodine, especially on the Liver. J. Pharmacol. exp. Therapeut. **75**, 69 (1942).

79. — — — The Action of Monocrotaline and Retronecine. J. Pharmacol. exp. Therapeut. **75**, 78 (1942).

80. — — — The Action of Isatidine, Pterophine and Sceleratine. J. Pharmacol. exp. Therapeut. **75**, 83 (1942).

81. — — — The Action of Carthamoidine. J. Pharmacol. exp. Therapeut. **79**, 133 (1943).

82. — — — The Action of Riddelline. J. Pharmacol. exp. Therapeut. **78**, 372 (1943).

83. — — — The Action of Monocrotaline, Isatidine, Pterophine, Sceleratine, and the Alkaloid of *Senecio carthamoides*. Federat. Proc. (Amer. Soc. exp. Biol.) **1**, 37 (1942).

84. — — — The Action of Alloxan, Senecionine, Sulfadiazine and Thiouracil in the Hamster. J. Pharmacol. exp. Therapeut. **87**, 382 (1946).

85. HENDERSON, F. G., P. N. HARRIS and K. K. CHEN: Liver Injury Following Administration of α- and β-Longilobine. Proc. Soc. exper. Biol. Med. **76**, 530 (1951) [Chem. Abstr. **45**, 6296 (1951)].

86. — — — Hepatotoxic Action of *Senecio longilobus*. J. Pharmacol. exp. Therapeut. **110**, 26 (1954).

87. HENRY, T. A.: The Plant Alkaloids, p. 601. 4th ed. London: J. and A. Churchill Ltd. 1949.

88. HOSKING, J. R. and C. W. BRANDT: The Toxic Principle of Ragwort. New Zealand J. Sci. Tech. **17**, 638 (1936) [Chem. Abstr. **30**, 4170 (1936)].

89. KELLY, E. A. and E. V. LYNN: Chemical Study of *Senecio aureus*. J. Amer. Pharmaceut. Assoc. **20**, 755 (1931) [Chem. Abstr. **25**, 5956 (1931)].

90. KERWIN, J. F., G. E. ULLYOT, R. C. FUSON and C. L. ZIRKLE: Rearrangement of 1,2-Aminochloroalkanes. J. Amer. Chem. Soc. **69**, 2961 (1947).
91. KHAYAT, S. M. and A. A. GILDER: Ovine Piroplasmoses in Iraq. Trans. Roy. Soc. trop. Med. **41**, 119 (1947).
92. KOEKEMOER, M. J. and F. L. WARREN: The *Senecio* Alkaloids. VIII. The Occurrence and Preparation of the *N*-Oxides. An Improved Method of Extraction of the *Senecio* Alkaloids. J. Chem. Soc. (London) **1951**, 66.
93. — — The *Senecio* Alkaloids. XI. The Conversion of Rosmarinine to Senecionine and the General Structure of the *Senecio* Alkaloids. J. Chem. Soc. (London) **1955**, 63.
94. KONOVALOV, V. S. and G. P. MEN'SHIKOV: Alkaloids of *Cacalia hastata*. J. Gen. Chem. (USSR) **15**, 328 (1945) [Chem. Abstr. **40**, 3760 (1946)].
95. KONOVALOVA, R. A.: Alkaloids of Species of *Senecio*. Bull. acad. Sci. (USSR), Classe sci. math. nat., Sér. chim. **4**, 961 (1936) [Chem. Abstr. **31**, 5365 (1937)].
96. — Separation of Platyphylline from Seneciphylline. Russian Patent 65708, Jan. 31, 1946 [Chem. Abstr. **40**, 7531 (1946)].
97. — Platyphylline. Russian Patent 69881, Dec. 31, 1947 [Chem. Abstr. **44**, 2042 (1950)].
98. KONOVALOVA, R. A. and A. DANILOVA: Alkaloids of *Senecio* Species. VI. Structure of Seneciphylline. J. Gen. Chem. (USSR) **18**, 1198 (1948) [Chem. Abstr. **43**, 1427 (1949)].
99. KONOVALOVA, R. A. (KONOWALOWA, R.) und A. OREKHOV (ORECHOFF, ORÉKHOFF): Über *Senecio*-Alkaloide, III. Mitt.: Abbau des Platynecins zum Heliotridan. Ber. dtsch. chem. Ges. **69**, 1908 (1936).
100. — — Sur les alcaloïdes du seneçon. 4e mémoire. Alcaloïdes de *senecio vulgaris*. Dégradation de la sénécionine. Bull. soc. chim. France (5) **4**, 1285 (1937).
101. — — Sur les alcaloïdes du seneçon. 5e mémoire. Sur la constitution de la sénéciphylline. Bull. soc. chim. France (5) **4**, 2037 (1937).
102. KONOVALOVA, R. A., A. P. OREKHOV and V. TIDEBEL: Alkaloids of *Senecio platyphyllus*. J. Gen. Chem. (USSR) **8**, 273 (1938) [Chem. Abstr. **32**, 5403 (1938)].
103. KOVYREV, I. G.: Concerning the Parasympathetic Action of Platyphylline. Bull. exptl. Biol. Med. **11**, 92 (1941) [Chem. Zbl. **1943** II, 925; Chem. Abstr. **38**, 5597 (1944)].
104. KROPMAN, M. and F. L. WARREN: The *Senecio* Alkaloids. V. The Structure of Senecic Acid. J. Chem. Soc. (London) **1949**, 2852.
105. — — The *Senecio* Alkaloids. VI. The Isomerisation of Senecic Acid to *trans*-Senecic (Integerrinecic) Acid, and the General Structure of the 'Necic' Acids. J. Chem. Soc. (London) **1950**, 700.
106. LABENSKIĬ, A. S. and G. P. MEN'SHIKOV: Alkaloids of *Lindelofia anchusoides*. I. New Alkaloids, Lindelofine and Lindelofamine, and their Strctures. J. Gen. Chem. (USSR) **18**, 1836 (1948) [Chem. Abstr. **43**, 3827 (1949)].
107. LECHER, H. Z. and W. B. HARDY: Compounds of Trialkylamine Oxides with Sulfur Dioxide and Sulfur Trioxide. J. Amer. Chem. Soc. **70**, 3789 (1948).
108. LEISEGANG, E. C.: *Senecio* Alkaloids. The Ultraviolet Extinction Curves of the Alkaloids and the 'Necic' Acids. J. South African Chem. Inst. **3** (2), 73 (1950).
109. LEISEGANG, E. C. and F. L. WARREN: The *Senecio* Alkaloids. II. Isatinecine. J. Chem. Soc. (London) **1949**, 486.
110. — — The *Senecio* Alkaloids. VII. The Structure of Retrorsine and Isatidine. The Ester Groupings. J. Chem. Soc. (London) **1950**, 702.

111. Leonard, N. J.: The *Senecio* Alkaloids. In: The Alkaloids, Vol. I, p. 107 (R. H. F. Manske and H. L. Holmes, Editors). New York: Academic Press 1949.

112. Leonard, N. J. and K. M. Beck: The Synthesis of Pyrrolizidines. II. Basicities of 8-Alkylpyrrolizidines. J. Amer. Chem. Soc. **70**, 2504 (1948).

113. Leonard, N. J. and D. L. Felley: The Synthesis of Pyrrolizidines. III. Condensation of Nitromethane with Ethyl Crotonate and Subsequent Formation of 1-Methylpyrrolizidine (Heliotridane and Pseudoheliotridane). J. Amer. Chem. Soc. **71**, 1758 (1949).

114. — — The Synthesis of Pyrrolizidines. VI. Stereochemical Correlation of 1-Methyl- and 1-Hydroxymethylpyrrolizidine Isomers with Certain Alkaloid Products. J. Amer. Chem. Soc. **72**, 2537 (1950).

115. Leonard, N. J., L. R. Hruda and F. W. Long: The Synthesis of Pyrrolizidines. J. Amer. Chem. Soc. **69**, 690 (1947).

116. Leonard, N. J. and G. L. Shoemaker: The Synthesis of Pyrrolizidines. IV. Condensation of Nitroparaffins with Methyl Methacrylate and Subsequent Formation of 2-, 6-, and 8-Alkyl-Substituted Pyrrolizidines. J. Amer. Chem. Soc. **71**, 1762 (1949).

117. — — The Synthesis of Pyrrolizidines. V. 8-Hydroxymethylpyrrolizidine and 8-Chloromethylpyrrolizidine. J. Amer. Chem. Soc. **71**, 1762 (1949).

118. MacKay, A. H.: The Alkaloids of *Senecio jacobaea*. Nature (London) **106**, 503 (1920).

119. Manske, R. H. F.: The Alkaloids of *Senecio* Species. I. The Necines and Necic Acids from *S. retrorsus* and *S. jacobaea*. Canad. J. Res. 5, 651 (1931) [Chem. Abstr. **26**, 1609 (1932)].

120. — The Alkaloids of *Senecio* Species. II. Some Miscellaneous Observations. Canad. J. Res. **14** B, 6 (1936) [Chem. Abstr. **30**, 2571 (1936)].

121. — The Alkaloids of *Senecio* Species. III. *Senecio integerrimus, S. longilobus, S. spartioides* and *S. riddellii*. Canad. J. Res. **17** B, 1 (1939) [Chem. Abstr. **33**, 6321 (1939)].

122. — The Alkaloids of *Senecio* Species. IV. *Erechtites hieracifolia* (L) RAF. Canad. J. Res. **17** B, 8 (1939) [Chem. Abstr. **33**, 6321 (1939)].

123. Marais, J. S. C.: Dicrotaline: The toxic Alkaloids from *Crotalaria dura* (Wood and Evans) and *Crotalaria globifera* (E. Mey). Onderstepoort J. Vet. Sci. Animal Ind. **20**, 61 (1944).

124. McCulloch, E. C.: The Experimental Production of Hepatic Cirrhosis by the Seed of *Amsinckia intermedia*. Science (Washington) **91**, 95 (1940).

125. Men'shikov, G. P. (Menschikoff, G.): Über die Alkaloide von *Heliotropium lasiocarpum*. I. Mitt. Ber. dtsch. chem. Ges. **65**, 974 (1932).

126. — Über die Alkaloide von *Heliotropium lasiocarpum*. II. Mitt.: Abbau des Heliotridins zum Heliotridan. Ber. dtsch. chem. Ges. **66**, 875 (1933).

127. — Über die Alkaloide von *Heliotropium lasiocarpum*. III. Mitt.: Über Oxyheliotridan. Ber. dtsch. chem. Ges. **68**, 1051 (1935).

128. — Über die Alkaloide von *Heliotropium lasiocarpum*. IV. Mitt. Abbau des Heliotridans zu einer Pyrrol-Base. Ber. dtsch. chem. Ges. **68**, 1555 (1935).

129. — Alkaloids of *Heliotropium lasiocarpum* and *Trichodesma incanum* fam. Boraginaceae. Bull. Acad. Sci. USSR, Classe sci. math. nat., Sér. chim. **4**, 969 (1936) [Chem. Abstr. **31**, 5365 (1937)].

130. — Über die Synthese eines neuen heterocyclischen, aus zwei kondensierten Pyrrolidinringen aufgebauten Ringsystems. Ber. dtsch. chem. Ges. **69**, 1802 (1936).

131. MEN'SHIKOV, G. P. (MENSCHIKOFF, G.): The Synthesis of 1-Methyl-2-*sec*-butylpyrrolidine. J. Gen. Chem. (USSR) **7**, 1632 (1937) [Chem. Abstr. **31**, 8529 (1937)].

132. — Alkaloids of *Heliotropium lasiocarpum*. The Synthesis and Structure of Heliotridane. Bull. acad. sci. (USSR), Classe sci. math. nat., Sér. chim. **5**, 1035 (1937) [Chem. Abstr. **32**, 2944 (1938)].

133. — Alkaloids of *Heliotropium lasiocarpum*. The Structure of Heliotrinic Acid. J. Gen. Chem. (USSR) **9**, 1851 (1939) [Chem. Abstr. **34**, 4071 (1940)].

134. — Alkaloids of *Trachelanthus korolkovi*. III. Structure of Trachelanthamidine, the Amino Alcohol formed in the Hydrolysis of the Alkaloid Trachelanthamine. J. Gen. Chem. (USSR) **16**, 1311 (1946) [Chem. Abstr. **41**, 3092 (1947)].

135. — Alkaloids of *Trachelanthus korolkovi*. IV. Structure of Trachelanthamine. J. Gen. Chem. (USSR) **17**, 343 (1947) [Chem. Abstr. **42**, 556 (1948)].

136. — Alkaloids of *Cynoglossum viridiflorum*. I. New Alkaloid, Viridiflorine, and its Structure. J. Gen. Chem. (USSR) **18**, 1736 (1948) [Chem. Abstr. **43**, 2625 (1949)].

137. MEN'SHIKOV, G. P. and G. M. BORODINA: Alkaloids of *Trachelanthus korolkovi*. J. Gen. Chem. (USSR) **11**, 209 (1941) [Chem. Abstr. **35**, 7111 (1941)].

138. — — Alkaloids of *Trachelanthus korolkovi*. II. J. Gen. Chem. (USSR) **15**, 225 (1945) [Chem. Abstr. **40**, 2141 (1946)].

139. MEN'SHIKOV, G. P., S. O. DENISOVA and P. S. MASSAGETOV: Alkaloids of *Turneforcia sibirica*. I. New Alkaloid, Turneforcine. J. Gen. Chem. (USSR) **22**, 1465 (1952) [Chem. Abstr. **47**, 7512 (1953)].

140. MEN'SHIKOV, G. P. and E. L. GUREVICH: Alkaloids of *Trachelanthus korolkovi*. V. Synthesis of some Derivatives of Trachelantamidine. J. Gen. Chem. (USSR) **17**, 1714 (1947) [Chem. Abstr. **42**, 2598 (1948)].

141. — — Alkaloids of *Heliotropium supinum*. I. Supinine and its Structure. J. Gen. Chem. (USSR) **19**, 1382 (1949) [Chem. Abstr. **44**, 3486 (1950)].

142. MEN'SHIKOV, G. P. and A. D. KURZOKOV: Alkaloids of *Heliotropium lasiocarpum*. Structure of Heliotrin. J. Gen. Chem. (USSR) **19**, 1702 (1949) [Chem. Abstr. **44**, 1113 (1950)].

143. — — Synthesis in the Pseudoheliotridane Series. J. Gen. Chem. (USSR) **21**, 2515 (1951) [Chem. Abstr. **47**, 5947 (1953)].

144. MEN'SHIKOV, G. P. and M. F. PETROVA: Alkaloids of *Makrotomia echoides*. I. New Alkaloid Makrotomine and its Structure. J. Gen. Chem. (USSR) **22**, 1457 (1952) [Chem. Abstr. **47**, 7512 (1953)].

145. MEN'SHIKOV, G. P. und W. RUBINSTEIN: Über ein Alkaloid aus *Trichodesma incanum*. D. C., I. Mitt. Ber. dtsch. chem. Ges. **68**, 2039 (1935).

146. MEN'SHIKOV, G. P. und E. S. ZHDANOVICH (J. SCHDANOWITSCH): Über die Alkaloide von *Heliotropium lasiocarpum*. V. Mitt. Über Lasiocarpin. Ber. dtsch. chem. Ges. **69**, 1110 (1936).

147. — — Über die Alkaloide von *Heliotropium lasiocarpum*. VI. Mitt. Synthese einiger Pyrrolidin-Derivate. Ber. dtsch. chem. Ges. **69**, 1799 (1936).

148. MÜLLER, A.: Zur Kenntnis der *Senecio*arten in botanisch-medizinischer und pflanzenchemischer Hinsicht mit besonderer Berücksichtigung der Alkaloide. Heil- und Gewürzpflanzen **1924**, 58 Seiten [Chem. Zbl. **1925**, II, 1049].

149. NEAL, W. M., L. L. RUSOFF and C. F. AHMANN: The Isolation and Some Properties of an Alkaloid from *Crotalaria spectabilis* ROTH. J. Amer. Chem. Soc. **57**, 2560 (1935).

150. NOVELLI, A. and A. P. G. DE VARELA: The Alkaloids of *Senecio* from Argentina. I. *Senecio brasiliensis*. An. Asoc. quím. argentina **33**, 176 (1945) [Chem. Abstr. **41**, 2418 (1947)].

151. OREKHOV, A. (ORECHOFF, A.) and R. A. KONOVALOVA (R. KONOWALOWA): Über *Senecio*-Alkaloide, III. Mitt.: Abbau des Platynecins zum Heliotridan. Ber. dtsch. chem. Ges. **69**, 1908 (1936).

152. OREKHOV, A., R. A. KONOVALOVA and V. TIDEBEL (W. TIEDEBEL): Über *Senecio*-Alkaloide. II. Mitt.: Zur Kenntnis des Platyphyllins. Ber. dtsch. chem. Ges. **68**, 1886 (1935).

153. OREKHOV, A. and V. TIDEBBL: Über *Senecio*-Alkaloide, I. Mitt.: Die Alkaloide von *Senecio platyphyllus* D. C. Ber. dtsch. chem. Ges. **68**, 650 (1935).

154. POLONOVSKI, MAX et MICHEL POLONOVSKI: Sur les aminoxydes des alcaloïdes (I). Bull. soc. chim. France (4) **39**, 1147 (1926).

155. PRELOG, V.: Über die zweifache intramolekulare Alkylierung. Liebigs Ann. Chem. **545**, 229 (1940).

156. PRELOG, V. und S. HEIMBACH: Über Pyrrolizidin (1-Aza-bicyclo-(0.3.3.)-octan). Ber. dtsch. chem. Ges. **72**, 1101 (1939).

157. PRELOG, V. und E. ZALÁN: Über die Synthese von *d,l*-Heliotridan (1-Methyl-pyrrolizidin). Helv. Chim. Acta **27**, 531 (1944).

158. RATNOFF, O. D. and G. S. MIRICK: Influence of Sex on the Lethal Effects of a Hepatotoxic Alkaloid, Monocrotaline. Bull. Johns Hopkins Hosp. **84**, 507 (1949) [Chem. Abstr. **43**, 8555 (1949)].

159. RICHARDSON, M. F. and F. L. WARREN: The *Senecio* Alkaloids. I. Rosmarinine. J. Chem. Soc. (London) **1943**, 452.

160. ROSE, C. L., R. D. FINK, P. N. HARRIS and K. K. CHEN: Effect of Hepatotoxic Alkaloids on Prothrombin Time of Rats. J. Pharmacol. exp. Therapeut. **83**, 265 (1945) [Chem. Abstr. **39**, 3839 (1945)].

161. ROSENFELD, I. and O. A. BEATH: Pharmacological Action of *Senecio riddellii*. J. Amer. Pharmaceut. Assoc. **36**, 331 (1947).

162. SAPIRO, M. L.: The Alkaloids of *Senecio ruwenzoriensis*. J. Chem. Soc. (London) **1953**, 1942.

163. SCHOENTAL, R.: *Senecio* Alkaloids and Cancer. British Med. J. **1**, 335 (1954).

164. SCHOENTAL, R., M. A. HEAD and P. R. PEACOCK: *Senecio* Alkaloids. Primary Liver Tumours in Rats as a Result of Treatment with (1) A Mixture of Alkaloids from *S. jacobea* LINN., (2) Retrorsine, (3) Isatidine. Brit. J. Cancer (1954) (in press).

165. SCHOENTAL, R. and B. D. PULLINGER: On the Alleged Oestrogenic and Other Medicinal Properties of *Senecio* Alkaloids (private communication).

166. SHIMOYANA, Y.: Zur Kenntnis von einer neuen, ungesättigten Fettsäure. Apotheker Ztg., **7** 453 (1892).

167. STEYN, D. G.: Toxicology of Plants in South Africa. Central News Agency (S. Africa) p. 457 (1934).

168. — Recent Investigations into the Toxicity of Known and Unknown Poisonous Plants in the Union of South Africa. Report Dir. Vet. Res. Onderstepoort (South Africa) **15** (2), 789 (1929); **17** (2), 715 (1931); Onderstepoort J. Vet. Sci. Animal Ind. **1**, 173 (1933); **3**, 127 (1934); **7**, 172 (1936); **9**, 116–120 (1937).

169. — Crotalariosis in Sheep. Report Dir. Vet. Res. Onderstepoort (South Africa) **18** (2), 947 (1932).

170. STEYN, D. G., G. DE KOCK and P. J. DU TOIT: Studies on the Aetiology of Dunsiekte or Enzootic Liver Disease of Equines in South Africa. Report Dir. Vet. Res. Onderstepoort (South Africa) **17** (2), 617 (1931).

171. STEYN, D. G. and S. J. VAN DER WALT: Recent Investigations into the Toxicity of Known and Unknown Poisonous Plants in the Union of South Africa. XI. Onderstepoort J. Vet. Sci. animal Ind. **16**, 122 (1941).

172. STEYN, D. G. and S. J. VAN DER WALT: Recent Investigations into the Toxicity of Known and Unknown Poisonous Plants in the Union of South Africa. XV. Onderstepoort J. Vet. Sci. Animal Ind. **21**, 48 (1946).
173. SYRNEVA, YU. I.: Comparative Pharmacology of Platyphylline and Atropine. Farmakol. i Toksikol. **9**, 15 (1946) [Chem. Abstr. **41**, 6987 (1947)].
174. THOROLD, P. W. and M. L. SAPIRO: Preliminary Note on the Alkaloids and Toxicity of *Senecio ruwenzoriensis* S. MOORE. East African Agric. J. **19**, (2) October (1953).
175. TRAUTNER, E. M. and O. E. NEUFELD: Alkaloids of *Heliotropium europaeum* Growing in Australia. Austral. J. Sci. **11**, 211 (1949) [Chem. Abstr. **44**, 1230 (1950)].
176. VARDIMAN, P. H.: Experimental Feeding of *Senecio* Silage to Calves. J. Amer. Vet. Med. Assoc. **121**, 397 (1952) [Chem. Abstr. **47**, 4009 (1953)].
177. WAAL, H. L. DE: The South African *Senecio* Alkaloids. Nature (London) **146**, 777 (1940).
178. — Toxic Principles of *Senecio* Species. The Cause of Dunsiekte in Animals and Bread-poisoning in Human Beings. Farming in South Africa **16**, 69 (1941).
179. — The *Senecio* Alkaloids. 1. The Isolation of Isatidine from *Senecio retrosus* and *Senecio isatideus*. Onderstepoort J. Vet. Sci. Animal Ind. **12**, 155 (1939).
180. — The *Senecio* Alkaloids. 2. Hydrogenation, Hydrolysis and Structural Results of Isatidine. Onderstepoort J. Vet. Sci. Animal Ind. **14**, 433 (1940).
181. — Senecio Alkaloids. 3. Chemical Investigations upon the Species Responsible for "Bread-poisoning". The Isolation of Senecionine from *Senecio ilicifolius* THUNB. and a new Alkaloid "Rosmarinine" from *Senecio rosmarinifolius* LINN. Onderstepoort J. Vet. Sci. Animal. Ind. **15**, 241 (1940).
182. — South African *Senecio* Alkaloids. 5. Notes on Isatidine, Rosmarinine and Pterophine, and on the Structure of their Necines and Necic Acids. Onderstepoort J. Vet. Sci. Animal Ind. **16**, 149 (1941).
183. — Suid-Afrikaanse *Senecio*-Alkaloiede (Oorsake van Dunsiekte by diere en Broodvergiftiging by Mense). Tydskr. Wet. en Kuns (South Africa) **2**, 85 (1941).
184. WAAL, H. L. DE and A. CROUS: South African *Senecio* Alkaloids. VII. The Structure of Sceleranecic Acid. J. South African Chem. Inst. N. S. I. **1**, 23 (1948).
185. WAAL, H. L. DE and D. F. LOUW: Suid-Afrikaanse *Senecio*-Alkaloiede. VIII. 'n Aantekening en 'n Verbetering oor die Identiteit van Retronesiensuur; Nuwe Isatinesiensuur en Dewaliensuur. Tydskr. Wet. en Kuns (South Africa) N. R. X. **2**, 171 (1950).
186. — — Suid-Afrikaanse *Senecio*-Alkaloiede. IX. Sceleratiensuur en sy verwantskap met Sceleranesiensuur. Tydskr. Wet. en Kuns (South Africa) N. R. X. **2**, 174 (1950).
187. WAAL, H. L. DE and T. P. PRETORIUS: South African *Senecio* Alkaloids. 6. The Toxic Alkaloids of *Senecio sceleratus* sp. nov. SCHWEIKERDT. Onderstepoort. J. Vet. Sci. Animal Ind. **17**, 181 (1941).
188. WAAL, H. L. DE, W. J. SERFONTEIN and C. F. GARBERS: South African *Senecio* Alkaloids. X. The Structure of Sceleranecic and Sceleratinic Acids. J. South African Chem. Inst. **14**, (1) 115 (1951).
189. WAAL, H. L. DE and J. TIEDT: The *Senecio* Alkaloids. 4. Platyphylline, the Active Principle of *Senecio adnatus*, D. C. Onderstepoort J. Vet. Sci. Animal Ind. **15**, 251 (1940).
190. WAKIM, K. G., P. N. HARRIS and K. K. CHEN: The Effects of Senecionine on the Monkey. J. Pharmacol. exp. Therapeut. **87**, 38 (1946).

191. WARREN, F. L., M. KROPMAN, R. ADAMS, T. R. GOVINDACHARI and J. H. LOOKER: The Identity of beta-Longilobine with Retrorsine. J. Amer. Chem. Soc. **72**, 1421 (1950).

192. WATT, H. E.: The Alkaloids of *Senecio latifolius*. J. Chem. Soc. (London) **95**, 466 (1909).

193. WATT, J. M. and M. G. BREYER-BRANDWYK: Medicinal and Poisonous Plants of South Africa, p. 198. Edinburgh: E. and S. Livingstone. 1932.

194. WILLMOT, F. C. and G. W. ROBERTSON: *Senecio* Disease in Humans, George C. P. 1918. Lancet **199**, 848 (1920); cf. Nature (London) **106**, 321 (1920).

195. ZHDANOVICH, E. S. and G. P. MEN'SHIKOV: The Alkaloids of *Senecio othonnae*. J. Gen. Chem. (USSR) **11**, 835 (1941) [Chem. Abstr. **36**, 4123 (1942)].

(*Received, March 10, 1955.*)

Paper Chromatography in the Study of the Structure of Peptides and Proteins.

By E. O. P. Thompson and A. R. Thompson, Melbourne, Australia.

With 6 Figures.

Contents.

Introduction.

In a review on terminal amino acids in peptides and proteins published in 1944 FOX (*132*) was able to give the position of only one amino acid, phenylalanine, in one protein, insulin, with some degree of certainty [JENSEN and EVANS (*185*)]. In the intervening period great advances in our knowledge of amino acid sequences in peptides and proteins have been made. It is now possible to write the complete structural formula for insulin and for many naturally occurring, short-chain polypeptides, two of which, oxytocin and vasopressin, have been synthesised. Much progress has also been made in elucidating the structure of proteins larger than insulin.

This outstanding progress is the result of the development of techniques for

1. the separation, on a micro scale, of all the amino acids which are known to occur in proteins, of peptides produced by partial hydrolysis of a protein, and also of members of many series of chemical derivatives of amino acids and peptides;

2. the identification, on a micro scale, of peptides produced by degradation of a polypeptide chain.

The identification of peptides may involve characterisation of the "N-terminal" and "C-terminal" residues [SANGER (*325*)] and perhaps further partial hydrolysis, or it may utilise stepwise degradative procedures which work along the peptide chain from the N-terminal or C-terminal residue. Whatever method of attack is chosen, it is essential to employ separation techniques such as paper chromatography. This procedure arose out of the development by MARTIN and SYNGE (*241*) of a new type of chromatography, originally called "partition" chromatography, which made use of the fact that closely related compounds may partition themselves very differently between two liquid phases. By using countercurrent procedures in a chromatogram, separations of a high order of efficiency were achieved. The idea was applied to the amino acids, using filter paper as the support for an aqueous phase, by GORDON, MARTIN, and SYNGE (*152*) and CONSDEN, GORDON, and MARTIN (*77*).

Probably no scientific technique has spread as rapidly and to so many laboratories as paper chromatography. Previous to the introduction of this technique for protein hydrolysates it was only with considerable difficulty that the amino acids present in a protein could be analysed even qualitatively. The separation and qualitative detection of all the amino acids in a protein hydrolysate on a single sheet of filter paper was an inspiring achievement. The simplicity of this technique, the meagre apparatus required for its use, and the micro scale to which it is applicable have resulted in its universal adoption for the analysis of most types of chemical compounds. It has stimulated the development of other techniques of high resolving power, such a zone electrophoresis, ion-exchange chromatography and countercurrent distribution, which have also contributed enormously to the determination of amino acid arrangement in peptides and proteins.

In their classic paper, CONSDEN, GORDON, and MARTIN (*77*) gave a wealth of detail relating to the paper chromatography of amino acids but stated that they developed the paper chromatographic technique "as a preliminary to the study of the products of the partial hydrolysis of proteins". Later, they extended their techniques to the fractionation of peptides and described methods for their elution and identification (*79*). CONSDEN, GORDON, MARTIN, and SYNGE (*83*) applied these techniques to a study of the partial hydrolysis products of *gramicidin S*. After separation on two-dimensional chromatograms, they identified four dipeptides and two tripeptides. From their structures they were able to deduce the pentapeptide sequence of amino acids in this cyclic decapeptide.

CONSDEN, GORDON, and MARTIN (*78, 80, 81, 76*) also used paper chromatography, in conjunction with ionophoretic and ion-exchange methods, to isolate and identify a considerable number of lower acidic peptides from partial acid hydrolysates of *wool*.

The complexity of the hydrolysates from even such a simple peptide as *gramicidin S* emphasised how tremendously complex the mixture of peptides formed on partial hydrolysis of any protein must be, even a protein of low molecular weight such as *insulin*. The elucidation of the sequence of amino acid residues in such a protein seemed at that time a virtually impossible task. Thus CONSDEN (*74*), after reviewing the methods of CONSDEN, GORDON, MARTIN, and SYNGE, expressed the opinion that "it is too much to expect that the structure of a protein could be pieced together from the peptides identified". Similar statements were made about this time by BUTLER, DODDS, PHILLIPS, and STEPHEN (*61*), and by

SANGER (*324*). Nevertheless, SANGER and TUPPY (*334, 335*) commenced an investigation of the peptides obtained by partial hydrolysis of the phenylalanyl chain of oxidised insulin.

SANGER (*321*) had previously characterised insulin by his DNP technique as consisting of only two types of polypeptide chain held together by disulphide bridges. After oxidation with performic acid to split these —S—S— linkages, SANGER (*322*) succeded in isolating two fractions representative of these chains. The A chain, with the N-terminal sequence gly-ileu-val-glu-glu, contained 21 amino acid residues, and the B chain, with the N-terminal sequence phe-val-asp-glu, contained 30 residues* (*323*). SANGER and TUPPY (*334*) used the DNP method in conjunction with fractionation methods including those developed by CONSDEN, GORDON, MARTIN, and SYNGE. The partial hydrolysates were separated into a number of simpler fractions by adsorption on charcoal and ion-exchange resins, and by ionophoresis in a compartment cell. The resulting peptide mixtures were then fractionated on two-dimensional paper chromatograms and their structures determined. From the peptides identified, they were able to deduce five sequences which were present in Fraction B, thereby accounting for all the amino acid residues except one leucine, one tyrosine and two phenylalanine residues. The action of the proteolytic enzymes, pepsin, trypsin and chymotrypsin was next investigated (*335*).

Knowing the structures of the lower peptides, they were able to deduce the structure of many of the peptides obtained in enzymic hydrolysates from their amino acid composition and N-terminal residues. From the results, SANGER and TUPPY found it possible to work out a unique sequence for Fraction B and, consequently, for the "phenylalanyl" chain of insulin.

This outstanding achievement showed that methods were available which enable the amino acid sequence of large portions of protein molecules, at least, to be elucidated. Their methods have set the pattern followed subsequently in all successful investigations of amino acid sequences, and paper chromatography has played, and will continue to play, a large part in investigations of this type.

The present survey is particularly concerned with the techniques of paper chromatography which have application in the determination of amino acid sequences. In these *Fortschritte* SCHROEDER (*343*) has recently published a useful review on column chromatography in sequence studies, and comprehensive reviews have been made by SANGER (*325*), and DESNUELLE (*103*). Attention is also drawn to the published pro-

* For abbreviations cf. p. 292.

ceedings, including informative discussions, of several recent *symposia* embracing this field (*73*, *70*, *266*).

Surveys of progress in the field of paper chromatography have been made by Martin (*236–239*), by Moore and Stein (*257*) and in several *symposia* (*273*, *69*, *192a*) where references to further reviews will be found. Recent *monographs* dealing with paper chromatography are those by Block, LeStrange, and Zweig (*43*), Balston and Talbot (*25*), Cramer (*89*), and Lederer and Lederer (*218*).

I. Principles of Paper Chromatography.

A theory for paper chromatography was developed by Consden, Gordon, and Martin (*77*) as an extension of that of Martin and Synge (*241*) for silica gel columns. They derived a relation between the partition coefficient (α) for a solute and the R_f value for a substance on the paper where

$$\alpha = \frac{\text{concentration of solute in stationary phase}}{\text{concentration of solute in developing phase}}$$

$$R_f = \frac{\text{distance moved by solute}}{\text{distance moved by front of developing phase}}$$

The relation between these terms is,

$$R_f = \frac{A_L}{A_L + \alpha A_S} \quad \text{and} \quad \alpha = \frac{A_L}{A_S}\left(\frac{1}{R_f} - 1\right),$$

where A_L is the cross-sectional area of the developing phase and A_S is the cross-sectional area of the stationary phase. A_L/A_S may be considered constant for a particular solvent and paper so that α should be inversely related to the R_f. In measuring R_f values, the centre of the applied zone is taken as the origin and the distances to the centre of the developed zone and to the front of the developer are used. Provided the applied solution is not too concentrated, the distribution of the solute in the developed zone approximates closely to the normal curve of error.

Consden, Gordon, and Martin (*77*) found good agreement between partition coefficients calculated from the movement of six amino acids on a paper strip, using water-saturated butanol as the developing phase, and those measured directly by England and Cohn (*124*). They therefore considered the filter paper as an inert support for the aqueous phase.

From further theoretical considerations Martin (*238*) developed certain rules for the dependence of partition coefficients on the nature and number of substituent groups in a molecule. These rules suggested, for example, that isomers containing the same functional groups should have identical partition coefficients. However, in fact, two peptides containing the same amino acids but a different sequence can often be separated, as CONSDEN and GORDON (*76*) demonstrated in the case of some peptides containing cysteic acid. The rules enable the R_f values of peptides to be calculated from a knowledge of the R_f values of the amino acids and of a few peptides so that the constants may be determined.

Calculations of this type have been made by PARDEE (*272*) from the data of CONSDEN, GORDON, and MARTIN (*77, 79, 81*), CONSDEN, GORDON, MARTIN, and SYNGE (*83*), KNIGHT (*206*), and COOK and LEVY (*84, 85*) with surprising agreement, using the relationship,

$$RT \log \left(\frac{1}{R_f} - 1 \right)_{\text{peptide}} = (n-1)\,A + B + \Sigma\, RT \log \left(\frac{1}{R_f} - 1 \right)_{\text{amino acids}}$$

where $n-1$ is the number of peptide bonds, A and B are constants for a given phase pair.

SANGER (*325*) has considered the particular case of a dipeptide $A-B$, the R_f of which may be calculated from the following formula, where K is a constant.

$$R_f^{A \cdot B} = \frac{R_f^A\, R_f^B}{K\left(1-R_f^A\right)\left(1-R_f^B\right) + R_f^A\, R_f^B}$$

Although regularities between the R_f values for members of homologous series have also been observed for many series of compounds, the calculated values of α are not always similar to those directly measured between water and the organic solvent. In fact, the concept of liquid-liquid distribution of a solute in paper chromatograms has been questioned, because it appears that other factors, such as adsorption, operate in effecting certain separations. CRAIG (*87*), for example, found the results of "partition" chromatography of very little use in selecting systems for countercurrent distribution. MARTIN (*239*) defended the partition concept but made the important point that the aqueous phase is not just water but a *complex gel* system analogous to a strong polysaccharide solution [cf. HANES and ISHERWOOD (*157*), HORNER, EMRICH, and KIRSCHNER (*175*)]. Filter paper in an atmosphere saturated with water vapour will absorb approximately 22% of water [CONSDEN, GORDON, and MARTIN (*77*)] but the first water absorbed by dry filter paper (approximately 6%) is bound chemically to the amorphous regions of the fibres and is characterised, in particular, by a low diffusion rate compared with the water subsequently bound by capillary sorption, which approaches liquid water in its properties. The R_f of a solute will depend on its hydrophilic character and ability to be incorporated in the water-cellulose complex.

MOORE and STEIN (*257*) have preferred to designate the process by the probable nature of the mobile and stationary phases, as liquid-liquid, liquid-gel or liquid-solid chromatography, although, as they point out, there may be overlapping cases. Paper chromatography is therefore regarded as an example of *liquid-gel* chromatography. The behaviour of solutes will be influenced by the degree of cross-linking and swelling of the gel and their ability to penetrate it.

The abbreviation "papergram" was suggested by TOMARELLI and FLOREY (*387*) and will be used in this survey.

Paper chromatography has found application whenever there is a need for the separation and identification of amino acids, as in the examination of amino acids from natural sources and in the separation, degradation and synthesis of peptides. An early triumph of the method was the demonstration by CONSDEN, GORDON, MARTIN, ROSENHEIM, and SYNGE (*82*) that *norleucine* was not a naturally occurring amino acid as previously suggested. Similarly, DAKIN's "*β-hydroxyglutamic acid*" specimens have been shown by DENT and FOWLER (*102*), using papergrams, to be mixtures of several amino acids, as suggested by BAILEY, CHIBNALL, REES, and WILLIAMS (*24*). The observation of new zones on papergrams of amino acids obtained from natural sources has led to the identification of several new "uncommon" amino acids. A compilation of these new substances has been given by DESNUELLE (*104*). The identification of unknown substances with known compounds has been made by running mixed chromatograms of the two. Provided a variety of solvents, buffered papers, etc., are used, this is good evidence of identity. However, MARTIN (*238*) suggested that isomers containing the same functional groups may have identical partition coefficients and be more readily distinguished by liquid-solid chromatography. SMITH (*354*) has discussed the problem of multiple zone formation by a pure substance. The possibility that a new zone on amino acid papergrams may be a resistant peptide (p. 293) or artifact (p. 293) must also be considered.

II. Paper Chromatographic Techniques.

It is of great importance in the investigation of amino acid sequences to have methods for the successful paper chromatography of amino acids and peptides. Some of the factors involved are discussed in the following sections.

1. Choice of Paper.

Surveys of commercial papers have been made by BULL, HAHN, and BAPTIST (*56*), KOWKABANY and CASSIDY (*208*), ROCKLAND, BLATT, and DUNN (*307*), and BALSTON and TALBOT (*25*).

For qualitative work WHATMAN Nos. 1 and 4 have been most commonly used. No. 4 has a much faster flow rate than No. 1, though resolution is often inferior. WHATMAN Nos. 3 and 3 MM papers give very good resolution of amino acids and peptides and handle at least three times the quantity of material.

ROCKLAND et al. found that the p_H of aqueous extracts of thirteen commercial filter papers varied from 4.8 to 6.5 and that there were marked differences in the ninhydrin colours of chromatograms on these filter papers. The order of the amino acids was usually independant of the type of paper, and most variations in the R_f values involved the basic and dicarboxylic amino acids, with phenol as solvent. Many other workers have noted that R_f values differ from paper to paper. For comparable results, strips should be cut in the same machine direction of the paper.

Although commercial filter papers are comparatively pure, the micro quantities of material which are handled are sometimes subject to interference by trace impurities in the paper, and special procedures have been devised for dealing with them. WYNN (*427*) has described a contaminant which, if not removed, may give rise to "artifact" amino acids on hydrolysis of peptides eluted from the filter paper. Washing with distilled water will remove most impurities and is best effected chromatographically [cf. ISHERWOOD and HANES (*179*)].

HANES, HIRD, and ISHERWOOD (*156*) used WHATMAN No. 3 papers, purified by an elaborate washing procedure, for the isolation of peptides. After elution, the peptides were re-chromatographed on a narrower strip and recovered by elution with 50% ethanol. Amino acid contaminating material from the paper was completely absent. Washing of the paper usually leads to increased flow rates due to the swelling of the paper by the solvent, which is not fully reversible on re-drying. THOMPSON and STEWARD (*384*) found washed WHATMAN No. 1 paper 50% faster.

Other difficulties arise from traces of metal ions in the filter paper which complex with the amino acids to give faster running "beards". CONSDEN, GORDON, and MARTIN (*77*) showed that additional copper increased the proportion of "beard" at the expense of the amino acid. Acid or ammonia washing of the paper did not remove all the copper. By developing papergrams in the presence of reagents that precipitate or form complexes with copper, or other metal impurities, these "beards" were avoided. Reagents commonly used include HCN, H_2S or NH_3 in the atmosphere, or addition of complexing agents (0.1%) such as "cupron" (α-benzoinoxime) or "oxine" (8-hydroxyquinoline) to the solvent.

Buffered papers have been used extensively, and, for certain derivatives of amino acids, reversed phase methods have proved useful.

2. Preparation of the Material for Chromatography.

Successful paper chromatography of amino acid and peptide mixtures often depends on their freedom from salts and long-chain polypeptides or proteins. For this reason, hydrochloric acid is preferred where acids are required since most of the acid can be removed by evaporation *in vacuo*. SMITH and PAGE (*357*) have described the use of long-chain aliphatic tertiary amines in organic solvents for the extraction of mineral acids from mixtures of amino acids such as protein hydrolysates. For alkaline conditions, the volatile trimethylamine may be useful. Baryta is often used because of its convenient removal as an insoluble sulphate or carbonate. When salts are necessary in reaction mixtures, samples of which are subsequently to be isolated pure or analysed chromatographically, ammonium formate and acetate buffers may often be used and removed by volatilisation [SANGER (*322*), HIRS, MOORE, and STEIN (*172*)].

A simple electrolytic method of removing anions and cations was introduced by CONSDEN, GORDON, and MARTIN (*79*). STEIN and MOORE (*364*) have reported conversion of arginine to ornithine by this method, but this can be minimised by the modification in the apparatus made by ACHER, JUTISZ, and FROMAGEOT (*5*), and SIMMONDS (*349*) to reduce the temperature. This method is less suitable for micro amounts.

Ion-exchange procedures, however, are the most versatile methods of freeing amino acids and peptides from contaminants. They are adaptable to small or large quantities. PIEZ, TOOPER, and FOSDICK (*283*) employed a strong base resin (Nalcit SAR), but arginine was not recovered and lysine not quantitatively. An alternative procedure [BOULANGER and BISERTÉ (*50*), PARTRIDGE and BRIMLEY (*276*)] utilises a sulphonated polystyrene in the H^+ form (*e. g.* Dowex 50) to absorb the inorganic cations and amino acids (except cysteic acid); and, after washing free of anions, the amino acids may be displaced by ammonia. Arginine can be eluted with solutions more concentrated than 1 *N*, although quantitative recoveries have not been reported. STEIN (*362*) desalted neutral and acidic amino acids on columns of a strong base resin (Dowex 2) in the hydroxide form and displaced the amino acids with *N*-acetic acid which does not displace absorbed anions. For the basic amino acids, a column of sulphonated polystyrene in the H^+ form was used. After washing with *N*-HCl to remove inorganic cations, the basic amino acids were eluted with 4 *N*-HCl [cf. STEIN and MOORE (*363*)].

Long-chain polypeptide and protein contaminants often cause difficulties in the paper chromatography of amino acids and short-chain peptides, for they tend to streak in many solvents and distort the positions of the amino acids. For their removal a highly cross-linked sulphonated polystyrene resin was used by THOMPSON (*373*) and PARTRIDGE (*275*) to act as a "molecular sieve". The amino acids could penetrate the resin and were readily absorbed, whereas larger molecules were limited to the surface of the resin, which was reduced to a minimum by the use of large beads. After washing away the protein, the amino acids were quantitatively recovered by elution with ammonia. This procedure has been extremely useful for quantitative studies of reactions involving the liberation of amino acids from proteins and polypeptides, *e. g.* with carboxypeptidase.

Alternative methods, such as dialysis, or precipitation of the protein with concentrated alcohol solutions, trichloracetic acid and other protein precipitants, may involve considerable losses due to adsorption on the coagulum, or necessitate the treatment of large volumes of solution.

3. Application of the Material to the Paper.

The smaller the spot applied to the paper, the shorter is the distance which the solvent must run to achieve complete separation. For quantitative experiments, repeated application of a sample often gives unsatisfactory results. Where large volumes of material must be applied, special techniques have been devised to limit the rate of application to the paper or the spot size [URBACH (*399*), MOORE and BOYLEN (*254*), and VON EULER and ELIASSON (*403*)]. It is also possible first to concentrate an aqueous solution on a sheet of polythene in a vacuum desiccator and then "print" the moistened material onto the paper, a procedure devised by CONSDEN, GORDON, and MARTIN (*79*) which is essentially quantitative. Material slightly soluble in water may often be applied in formic acid solution [TOENNIES and KOLB (*386*)] or in ammonia. For long-chain peptide mixtures, SANGER and THOMPSON (*329*) obtained better fractionation if the applied spot was not too small and the paper was placed in the developer while the spot was still moist.

A spot of material, usually applied in 5 μl. or less to WHATMAN No. 1 or No. 4 paper, may contain 1–50 μg., and this may be increased for thicker papers. Larger amounts may be fractionated as a band on the sheet and the substance located by reference to marker strips or possibly by their fluorescence. DATTA, DENT, and HARRIS (*95*) have built a frame for developing many sheets simultaneously. WHATMAN "Seed Testing" paper which is a thick absorbent board can be very heavily loaded. THOMPSON (*378*) has found that chromatographic washing improves these boards. WHATMAN "Accelerator" paper has been used by JONES (*192*), with upward development, to give satisfactory separations of more than 20 mg. of each amino acid per sheet. MITCHELL and HASKINS (*249*) used large numbers of discs, and PORTER (*294*) strips of filter paper clamped together while DANIELSSON (*94*) used a continuous roll of paper, to increase the capacity of papergrams.

The development of automatic fraction collectors has stimulated the use of columns of cellulose powder. SYNGE (*369*) found that separations were slightly inferior to those obtainable with paper sheets. Careful attention to packing is essential; and for this operation, CAMPBELL, WORK, and MELLANBY (*62*) used acetone, subsequently displacing it with butanol-acetic acid-water. Such columns may be used for long periods without repacking.

4. Effect of p_H.

Where samples for chromatography contain acid, *e. g.* hydrolysates of peptides and proteins, CONSDEN, GORDON, and MARTIN (*77*) recommended removal of most of the acid and neutralisation of any bound acid with ammonia vapour, preferably after application to the paper. Without neutralisation, alterations of R_f occurred [LANDUA, FUERST, and AWAPARA (*215*)]. Even with paper buffered in the p_H range 2–12, MCFARREN (*242*) found the p_H of the applied solution could markedly influence the R_f values, and the samples were always applied at p_H 6.2. At some p_H values, multiple spots have been observed for lysine and arginine with phenol or cresol [ARONOFF (*17*, cf. *215*, *242*)]. With acidic solvents, such as butanol-acetic acid-water, variation of the p_H of the applied solution between 1 and 7 has little effect on R_f values [LEVY and CHUNG (*224*)]. The differences in the behaviour of the basic amino acids in an atmosphere of ammonia or acetic acid was used by DENT (*101*) as a test of basicity.

5. Apparatus.

Any method which causes liquid to flow through paper can be used to produce a papergram. BLOCK, LESTRANGE, and ZWEIG (*43*), and LEDERER and LEDERER (*218*) have described the many types of apparatus that have been used. The rate of liquid flow with *descending* development is more rapid than alternative arrangements, and development equivalent to several lengths of paper is readily achieved. This is necessary for difficultly separable mixtures which show optimum separation at low R_f values.

JERMYN and ISHERWOOD (*187*) cut the paper to a series of points and allowed the solvent to drip off, but the solvent flow was diminished by about 50% by this method. A uniform rate of development is obtained by attaching to the bottom of the paper folded sheets of the same paper, sufficient to give runs of two or more lengths [SANGER and TUPPY (*334*), MIETTINEN and VIRTANEN (*245*)]. This avoids liquid accumulation at the bottom and modified distribution of solvent throughout the sheet. WILLIAMS and KIRBY (*419*) used capillary flow *upwards* in the paper, which gives slower development but requires only the simplest of apparatus. Large sheets may be rolled into a cylinder and used in smaller containers where equilibration is faster and saturation easier to maintain, of particular value with very volatile solvents. With ascending flow, substances cannot flow off the paper. R_f values are usually the same by ascending and descending techniques.

The advantages of *circular arcs* on chromatograms were described by RUTTER (*315*) and this technique has been modified by PROOM and WOIWOD (*296*), GIRI and RAO (*148*), ZIMMERMANN and NEHRING (*428*), MARCHAL and MITTWER (*235*), and LAKSHMINARAYANAN (*211*).

The simple apparatus required and the reproducibility of results using RUTTER's technique was found useful by MÜLLER and CLEGG (*261*), LESTRANGE and MÜLLER (*222*), and SAIFER and ORESKES (*318*) for investigation of variables in paper chromatography such as factors affecting liquid flow. The development time with circular paper chromatography is much shorter and resolution better than in other methods of development, for the bands are much sharper. LESTRANGE and MÜLLER claim that, since the R_f values of CONSDEN, GORDON, and MARTIN (*77*) represent a comparison of areas, the R_f of a substance on a strip is equivalent to the square of the R_r value on a circular chromatogram where R_r is the ratio of the radii of solute and solvent. REINDEL and HOPPE (*302*) have shown that suitable "tongues" on strip chromatograms will give as sharp bands as in circular paper chromatography.

It seems unlikely that circular development can compare with the two-dimensional sheet for the resolution of the complex peptide mixtures from partial hydrolysates of proteins, nor is it possible to chromatograph 20 or more peptide hydrolysates simultaneously on one sheet after "printing" the spots from polythene, which is often desirable.

There has been wide variation in the procedures adopted for the establishment of *equilibrium* in chromatographic systems.

For accurate measurement of R_f values, a 24 hour period of equilibration was recommended by BATE-SMITH and WESTALL (*27*). For some solvents longer periods are required [KOWKABANY and CASSIDY (*209*), ISHERWOOD and JERMYN (*180*)]. HANES and ISHERWOOD (*157*) decreased the equilibration period to a few hours by attaching a piece of iron to the paper to keep it in continuous movement by means of a magnet. With buffered papers MCFARREN (*242*) found variations in R_f values for amino acids depending on the equilibration conditions. Complex solvent systems and highly volatile solvents necessitate more attention to equilibration, and many workers have reported improved results with smaller apparatus. It is general practice to equilibrate in the presence of the stationary phase, and the tanks may be lined with a suitable absorbent saturated with the stationary phase to assist equilibration during the running of the chromatogram. The developing phase is added with as little disturbance to the system as possible.

Solvents such as phenol lose water to the paper as they move forward so that there is a marked concentration gradient. Equilibration increases the water content of the advancing solvent and hence the R_f [cf. TOENNIES and KOLB (*386*)]. Volatile components of miscible solvents may evaporate when equilibration is slow and cause waterlogging of the paper [BENTLEY and WHITEHEAD (*29*)]. LEVY and CHUNG

(*224*), however, found no improvement with butanol-acetic acid mixtures after two hour periods of equilibration. With tanks that were continuously in operation for the solvents phenol, collidine, butanol-acetic acid and tertiary amyl alcohol, used in the experiments of SANGER and TUPPY (*334, 335*), SANGER and THOMPSON (*328, 329*), and THOMPSON (*376, 377*) with peptide mixtures and amino acids, no equilibration was carried out and the fractionations were satisfactory. Long equilibration times would have markedly slowed the work. Altogether, amino acids seem less sensitive to poor equilibration than other classes of compounds but with derivatives of amino acids equilibration may be more important.

6. Solvents.

The same solvent systems can be used for chromatographing both amino acids and peptides. A wide variety of mixtures has been employed and the choice of a suitable solvent will depend on the particular mixture being analysed. CONSDEN, GORDON, and MARTIN (*77*) studied the behaviour of amino acids with organic solvents, both miscible and partly miscible with water. For a range of partly miscible solvents, *e. g.* phenol, cresols, alcohols and collidine, they have tabulated the R_f values of amino acids and certain peptides [CONSDEN, GORDON, and MARTIN (*77, 79, 81*)]. KNIGHT (*206*) has also listed R_f values, for a range of peptides from the collection of EMIL FISCHER, with the solvent systems phenol-water and pyridine-amyl alcohol [EDMAN (*116*), BRYANT (*54*)].

The latter solvent was suggested as a substitute for collidine, which has disadvantages such as non-reproducibility with different batches, difficulties of supply, unpleasant odour, and expense. DENT (*101*) found a mixture of 2 : 4-lutidine, 2:4:6-collidine and water (1:1:2) gave resolution similar to that obtained by CONSDEN, GORDON, and MARTIN (*77*). Usually R_f values are low in collidine and high in lutidine. THOMPSON and STEWARD (*384*) used a lutidine : collidine ratio of 2 : 1. Other workers have substituted 2:6-lutidine-water (65:35) [REED (*299*)], 2:4-lutidine-water [McFARREN (*242*), LANDUA, FUERST, and AWAPARA (*215*)] or pyridine-collidine-water mixtures (3:1:1) [ACHER, THAUREAUX, CROCKER, JUTISZ, and FROMAGEOT (*8*)]. Further substitutes for collidine are, acetone-water (60:40 containing 0.5% w/v urea) [BENTLEY and WHITEHEAD (*29*)], methyl cellosolve-water (9:1) [BENDER (*28*), ROCKLAND, BLATT and DUNN (*307*)], mesityl oxide-formic acid-water (1:1:2) [BRYANT and OVERELL (*55*)], and *n*-butanol-acetic acid-water mixtures [WOIWOD (*424*), PHILLIPS (*282*), JONES (*192*)]. The latter solvent mixture is particularly useful. The proportions 4:1:5 used by PARTRIDGE (*274*) for sugars have been widely applied for the chromatography of amino acids and peptides. It has been extensively used by SANGER and TUPPY (*334, 335, 391*) as well as SANGER and THOMPSON (*328, 329*) in the work on insulin, by TUPPY and MICHL (*392*) with oxytocin and by THOMPSON (*376, 377*) with lysozyme hydrolysates. RESSLER, TRIPPETT, and DU VIGNEAUD (*304*), DU VIGNEAUD, RESSLER, and TRIPPETT (*115*), and POPENOE and DU VIGNEAUD (*286, 287*) used the proportions 5:1:5 for peptides from oxytocin and vasopressin.

Three-component systems of this type, in which the third component is readily soluble in both of the partly miscible phases, can be used to increase the amount of water in the organic solvent phase, and thus R_f values may be varied within wide limits.

A mixture, in which the volume of the aqueous phase is much reduced, and which gives similar resolution to the mixture of PARTRIDGE has been used by ROCHE, JUTISZ, LISSITZKY, and MICHEL (*305*) in the proportions 78:5:17. CAMPBELL, WORK, and MELLANBY (*62*) used a 63:10:27 mixture and allowed it to stand for two days for *esterification* to approach equilibrium, when a small aqueous layer was obtained. Some workers always use fresh solvent, and, for this purpose, one phase mixtures 200:30:75 [FRAENKEL-CONRAT (*135*), GLADNER and NEURATH (*150*)] or 40:10:10 [BERRY, SUTTON, CAIN, and BERRY (*31*), BURMA (*58*)], are convenient. Butanol-acetic acid-water mixtures which have stood for long periods give lower R_f values for slower moving zones, and higher R_f values for the aromatic amino acids. In freshly prepared mixtures tyrosine may overlap proline, tryptophan moves slightly faster than tyrosine, and phenylalanine only slightly faster than valine. Resolution is superior with aged mixtures (see *Fig. 3*, p. 285).

n-Butanol-formic acid-water mixtures also give good resolutions and have frequently been applied. ACHER, THAUREAUX, CROCKER, JUTISZ, and FROMAGEOT (*8*) use the proportions 75:15:10. HAUSMANN (*166*) has used the same proportions, but with 2-butanol, to separate all but the basic and acidic amino acids. For these, STEIN (*362*) has used the proportions 65:20:15. WIGGINS and WILLIAMS (*418*) hastened the approach to equilibrium by refluxing the mixture. Tertiary alcohols or ketones reduce esterification [CLAYTON and STRONG (*71*)].

For cysteic acid, which has an extremely low R_f value in most solvents [SIBATANI and FUKUDA (*348*)] and peptides containing cysteic acid [CONSDEN and GORDON (*76*), SANGER and THOMPSON (*328*)], pyridine derivatives are the most useful solvents. The aromatic and basic amino acids also have higher R_f values in these systems than in alcohol mixtures.

Phenol is an excellent solvent giving a wide spread of R_f values and, although it has some disadvantages, it is difficult to find another to replace it.

Many workers purify the phenol before use by distillation from zinc dust [WILLIAMS and KIRBY (*419*)] or aluminium turnings [DRAPER and POLLARD (*112*)]. No time is required for settling out if the phenol is incompletely saturated with water. With phenols in an atmosphere of ammonia, oxidation of the solvent is catalysed by metal impurities, *e. g.* copper, in the paper, giving a brown front which distorts the faster moving amino acids. CONSDEN, GORDON, and MARTIN (*77*) showed that this could be avoided by running the papergrams in an atmosphere of hydrogen cyanide. Cupron or other metal complexing agents added to the phenol are less satisfactory.

With increasing size, polypeptides tend to have high R_f values in phenol and low R_f values in butanol-acetic acid, collidine and most other solvents. GRASSMANN (*153*) mixed phenol and butanol-acetic acid-water phases in various proportions to get a range of R_f values.

SANGER and TUPPY (*334, 335*) observed that *m*-cresol-0.3% NH_3 gave good fractionation of basic peptides and also of large peptides formed by enzymic digestion of the phenylalanyl chain of oxidised insulin. SANGER and THOMPSON (*328*) found *m*-cresol useful for peptides moving fast in phenol. Peptides containing the dicarboxylic amino acids were particularly retarded.

The realisation that *water-miscible* solvents can also be used has led to the use of a much wider range of alcohols, ketones etc., mixed with various quantities of water, acids or bases. R_f values usually increase with increasing water concentration in the solvent but definition of the spots decreases [BENTLEY and WHITEHEAD (*29*)]. With too low a water content streaking occurs [JIRGENSONS (*188*)], and usually no

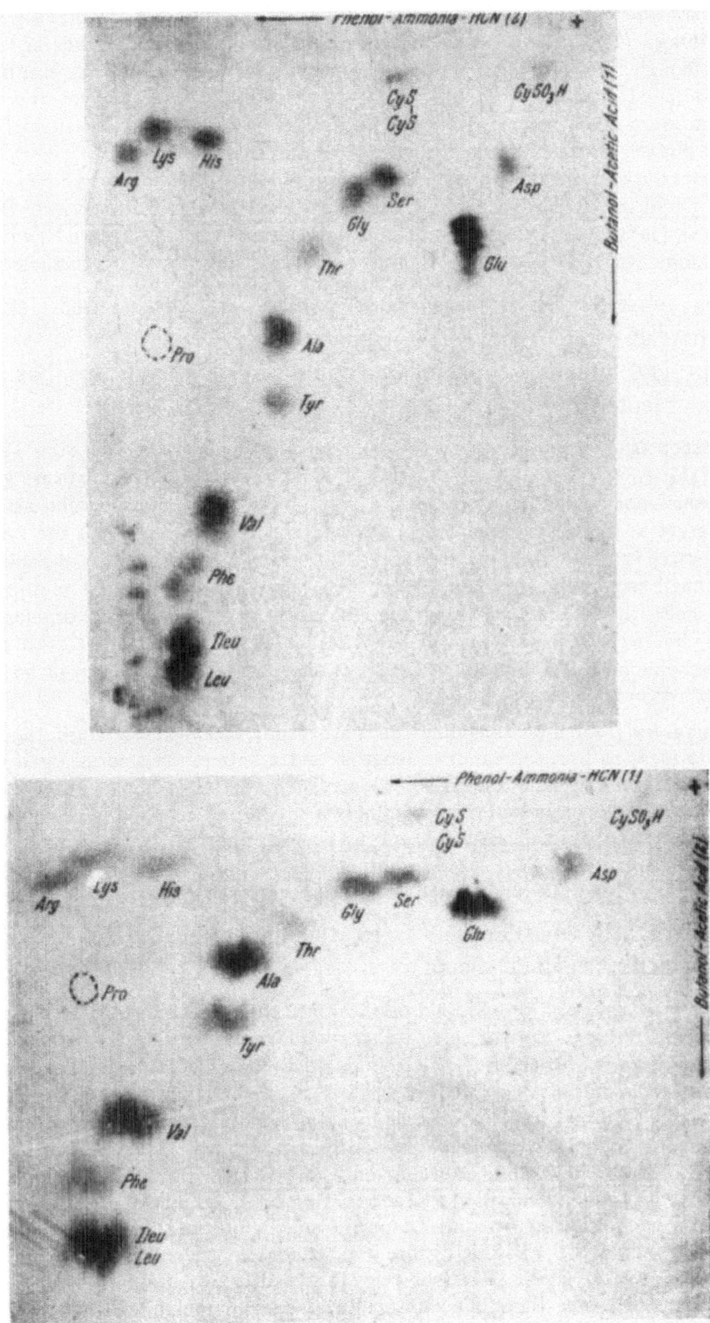

Figs. 1 and 2. Fig. 1 shows the improved resolution of amino acids in an insulin hydrolysate when the papergram is developed in the first dimension with butanol-acetic acid-water (4 : 1 : 5, aged three days) and in the second dimension with phenol-ammonia-HCN, compared with the reverse order as in Fig. 2. (Abbreviations: p. 292.)

movement takes place in the presence of anhydrous solvents [SIBATANI and FUKUDA (*348*), BURMA (*60*)]. There is thus an optimum water content for particular separations, though the order of the amino acids seldom varies. BULL, HAHN, and BAPTIST (*56*) introduced 80% phenol-water as a non-saturated solvent and found little variation in R_f values with water content from 65% to 85% phenol. In this respect phenol contrasts markedly with other solvents.

Useful data on the R_f values of amino acids in many different one phase systems will be found in publications of BENTLEY and WHITEHEAD (*29*), ROCKLAND, BLATT, and DUNN (*307*), SIBATANI and FUKUDA (*348*), BERRY, SUTTON, CAIN, and BERRY (*31*), JIRGENSONS (*188*), BURMA (*58*, *60*) and REDFIELD, and GUZMAN BARRON (*298*).

For *two-dimensional papergrams*, which are often used for the resolution of complex mixtures of amino acids, CONSDEN, GORDON, and MARTIN (*77*) applied a phenol-"collidine" system and this has been widely accepted by other workers.

BOISSONNAS (*45*) introduced two systems, *tert.*-butyl alcohol-methyl ethyl ketone-water (4:4:2), *tert.*-butyl alcohol-methanol-water (4:5:1) for the leucines, valine, methionine, phenylalanine, tyrosine and tryptophan. The other amino acids do not migrate at all under these conditions and can be separated with the systems phenol-water (7:3) *n*-butanol-water (7:3). REDFIELD (*297*) used sheets of Schleicher and Schuell paper No. 507 only 14 cm. square, with ascending chromatography for 2–3 hours in methanol-water-pyridine (80:20:4) and, after drying, development for 2.5–5 hours in *tert.*-butanol-methyl ethyl ketone-water-diethylamine (40:40:20:4). The diethylamine must be removed in an atmosphere of steam for 5–8 hours to avoid background colour.

HAUSMANN (*166*) used ascending chromatography and the system 2-butanol-3% ammonia (5:2) as first solvent (two lengths) and 2-butanol-88% formic acid-water (75:15:10) in the second dimension. This system was also applied by PALADINI and CRAIG (*271*) in a preliminary examination of the peptides obtained by partial acid hydrolysis of tyrocidin A. LANDMANN, DRAKE, and WHITE (*213*), and WHITE and LANDMANN (*416*) have used 2-butanol-3% ammonia (3:1) extensively for fractionation of peptides obtained by enzymic hydrolysis of corticotropin A.

Probably the most widely used two-dimensional system involves butanol-acetic acid and phenol.

For *amino acids* and protein hydrolysates it is the experience of the writers and others [JONES (*192*), WOIWOD (*424*)] that *butanol-acetic acid in the first direction followed by phenol in the second direction gives better resolution and spread of amino acids than does the reverse order* (*Figs. 1–2*, p. 283). Many workers prefer phenol-NH_3-HCN but JONES (*192*) finds ammonia of no advantage with this combination. THOMPSON and STEWARD (*384*) found that, with unwashed papers, histidine did not give a well-defined spot in the absence of ammonia. Moreover, with reagent grade phenol, the elimination of ammonia produced effects due to the acidity of the phenol-water phase, p_H 1 to 2, resulting from acid impurities in the phenol. The p_H was adjusted to 5–5.5 or the phenol redistilled before use. Otherwise aspartic and glutamic acids streaked badly. *A map of spots* showing the positions of common amino acids chromatographed with butanol-acetic acid and phenol-ammonia-HCN is shown in *Fig. 3*. For *peptide mixtures*, however, SANGER and TUPPY (*334*, *335*), SANGER and THOMPSON (*328*, *329*), and THOMPSON (*376*, *377*) obtained better resolution if the phenol was the first solvent followed by butanol-acetic acid. [This was not the experience of BOULANGER and BISERTE (*51*).] Such arrangement is also preferable if the spots are to be located in ultraviolet light. LEVY and CHUNG (*224*) obtained more satisfactory two-dimensional papergrams of amino acids by using butanol-acetic acid in the first dimension and a 1 : 1 *m*-cresol-phenol mixture for the second dimension on buffered paper.

Maximum separation between two substances of similar R_f is best obtained with solvents giving R_f values of approximately 0.2 by developing the paper for a

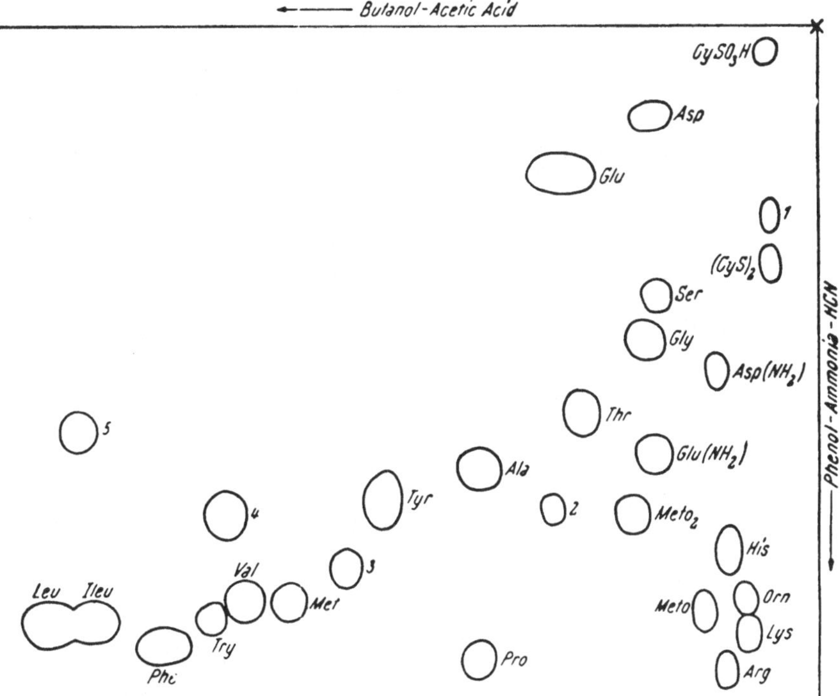

Fig. 3. Diagram showing positions of amino acids on a papergram run with butanol-acetic acid water (4 : 1 : 5, aged three days) and phenol-ammonia-HCN on Whatman No. 1 paper. The amino acids indicated by numbers are, 1. lanthionine; 2. hydroxyproline; 3. α-aminobutyric acid; 4. 3-chlorotyrosine; and 5. 3: 5-dichlorotyrosine. (Abbreviations: p. 292.)

long period [cf. JERMYN and ISHERWOOD (*187*)]. The leucine isomers are a good example of separations achieved in this way by extended development in butanol-benzyl alcohol (1 : 1), in an atmosphere of HCN [CONSDEN, GORDON, and MARTIN (*77*)], or on paper buffered at p_H 8.4 [McFARREN (*242*)], in *tert.*-amyl alcohol-water in the presence of diethylamine [WORK (*426*)] or 8-hydroxyquinoline [THOMPSON (*376*)], in pyridine-amyl alcohol-water [EDMAN (*116*), HEYNS and WALTER (*169*)], in *tert.* butanol-methyl ethyl ketone-water-88% formic acid (160 : 150 : 49 : 1) [BLOCK and VAN DYKE (*42*)] or in 2-butanol-3% ammonia [HAUSMANN (*166*), ROLAND and GROSS (*309*)].

These systems also give good resolution of many other amino acids but lengthy development times are necessary. The system of ROLAND and GROSS, however, gives complete resolution in 48 hours of lysine,

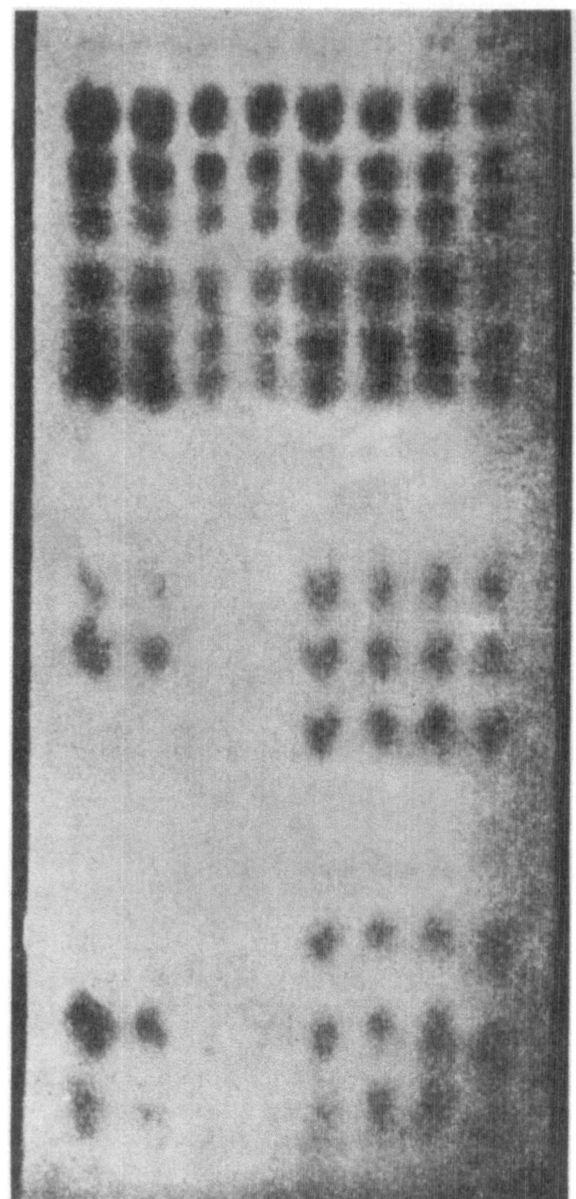

Fig. 4. Papergram of amino acids run in 2-butanol-3% ammonia (3 : 1) at 25° for 48 hours, according to ROLAND and GROSS. Spots (in descending order) are: aspartic acid, glutamic acid, cystine; lysine; arginine; glycine, serine; histidine, threonine; alanine; proline (not visible); tyrosine; valine; methionine; isoleucine; leucine; phenylalanine. [From: Analyt. Chemistry 26, 502 (1954).]

arginine, alanine, proline, tyrosine, valine, methionine, isoleucine, leucine and phenylalanine, as illustrated in *Fig. 4*.

Sometimes it is preferable to identify the sulphur-containing amino acids after *oxidation* to cysteic acid in the case of cystine and cysteine, and to methionine sulphone in that of methionine. DENT (*101*) effected the oxidation with hydrogen peroxide in the presence of catalytic amounts of ammonium molybdate. For peptide hydrolysates, oxidation usually takes place in the presence of chloride ions, and conversion of tyrosine to 3-chloro- and 3,5-dichlorotyrosine, which have different R_f values from tyrosine, is likely [THOMPSON (*382*)].

In investigating the amino acid composition of large numbers of peptides isolated from partial hydrolysates of proteins it is usually not possible to run two-dimensional papergrams on each peptide because of the limitations of apparatus, time, material and labour. Two uni-dimensional papergrams, in conjunction with multiple detection, will usually resolve all of them, for example, in phenol and 2-butanol-ammonia.

Since spraying with 0.1% ninhydrin solution does not destroy all the amino acids on paper (see p. 288) it is also possible to cut out any spot which may contain two amino acids and rerun with a second solvent system which will distinguish them. For this purpose, the spot may be eluted from the paper or used as the "wick" in a circular papergram [GIRI and RAO (*149*)]. For zones overlapping in butanol-acetic acid, the latter authors found that solvents containing pyridine gave good separations. Paper ionophoresis using the "strip transfer" method of DURRUM (*113*) may also be useful for separating overlapping spots. The chromatography may be combined with *ionophoresis* [HAUGAARD and KRONER (*165*), KICKHÖFEN and WESTPHAL (*202*), DURRUM (*113*), BLASS, LECOMTE, and POLONOVSKI (*39*)] but these methods utilise large sheets and have disadvantages similar to those discussed for two-dimensional papergrams.

The effect of *temperature* changes on partly miscible solvent systems may be considerable, as in the case of collidine and lutidine, where the amount of water in the organic phase varies markedly with temperature. However, under-saturated mixtures are now more commonly used to avoid separation of two phases with temperature variation and erratic R_f values. For water-miscible systems where the phase composition does not alter, JIRGENSONS (*188*) and BURMA (*59*) found that in general R_f values increase with the temperature.

7. Drying of Chromatograms.

To obtain quantitative recoveries of amino acids, and presumably peptides, from papergrams, some precautions are necessary in removing the solvent.

This is expecially true in the case of *phenols* at high temperatures [BERRY and CAIN (*30*), FOWDEN and PENNEY (*131*), BRUSH, BOUTWELL, BARTON, and HEIDELBERGER (*53*)]. By radioautographs, BRUSH and coworkers showed that the destruction of amino acids on phenol chromatograms gave compounds moving fast in other solvents. These compounds gave no ninhydrin colour but reformed the amino acids on hydrolysis, suggesting a dehydration reaction. The speed of the air current during drying will, no doubt, influence the recoveries. By drying at room temperature or by washing with ether FOWDEN (*129*) obtained quantitative recoveries. The former method requires very long drying times for thick papers, and a combination of the two methods is often used. Traces of phenol do not prevent the formation of a colour with ninhydrin; in fact, BULL, HAHN, and BAPTIST (*56*) added phenol to their spray to increase the sensitivity. However, SALANDER, PIANO, and PATTON (*320*) remove as much of their solvents as possible because of the importance of obtaining day-to-day reproducibility for quantitative work. Traces of phenolic solvents also mask the fluorescence of the amino acids and give a high background colour with some of the specific colorimetric reagents. Ether and acetone washing is not completely effective in removing traces of solvents and even long drying at 100° will not eliminate them. However, ISHERWOOD and CRUICKSHANK (*178*) report that water vapour readily displaces traces of phenol, acetic and propionic acids during a short steaming.

8. Detection of Amino Acids, Peptides and Proteins on Paper.

The need for reagents more specific than ninhydrin for the detection of amino acids and peptides on papergrams has led to the introduction of several useful new colour reactions. There has been a tendency to prefer *dipping* the papergrams in acetone solutions, rather than spraying them, for more uniform results. JEPSON and SMITH (*186*) have emphasised the usefulness of *successive* specific reagents for detecting the composite nature of a single zone, or two or more components which do not respond to a single general reagent. They recommend that, for maximum information, the reactions be used in the sequence: isatin and ninhydrin, followed by EHRLICH reagent, followed by SAKAGUCHI or diazo reagent.

Radioactive materials are easy to detect and estimate on papergrams. Applications of this specialised technique have been reviewed by LISSITZKY and MICHEL (*228*). Methods particularly developed for the detection of proteins on papergrams have been reviewed by BOULANGER and BISERTE (*51*) and LEDERER and LEDERER (*218*).

a) General Reagents.

Ninhydrin. Ninhydrin in butanol was the reagent originally used by CONSDEN, GORDON, and MARTIN (77) for the detection of amino acids and peptides on papergrams. A wide variety of solvents has now been recommended but experience with quantitative colour development (p. 297) indicated that 95% ethanol or acetone is a preferable solvent.

A concentration of 0.1% is satisfactory for qualitative detection, and only a fraction of the N-terminal amino acid is destroyed [WOIWOD (*423*)]. Higher concentrations give increased sensitivity but background colour is increased unless the colour is allowed to develop at room temperature [DENT (*101*)]. TOENNIES and KOLB (*386*) dip the papers in a 0.25% solution in anhydrous acetone. For quantitative reaction on the paper, ninhydrin concentrations of 1–2% are required.

Amino acids on chromatograms irrigated with acidic solvents give a reddish colour with ninhydrin while basic solvents produce a blue colour. In general, the red colour is converted to the blue by treatment with aqueous pyridine or lutidine [ATKINSON, STUART, and STUCKEY (*18*)]. WOIWOD (*424*) added 0.1% "collidine" to the ninhydrin solution in chloroform to obtain colour differences between the amino acids [cf. HARRIS (*160*)]. LEVY and CHUNG (*224*) used 3–5% collidine and obtained colours which varied with the time and temperature, with the paper and the purity of the collidine. Acetic acid was added to the ninhydrin spray by CONSDEN and GORDON (*75*) to overcome localised alkaline regions when salts were present on the papergram. This was also necessary for paper buffered above p_H 7 [MCFARREN (*242*); cf. (*224*)]. Both these authors developed their colours by heating; otherwise weak colours, if any, are obtained [THOMPSON (*383*)].

The *minimum* detectable amount of different amino acids varies with the concentration of ninhydrin and the conditions used. For certain conditions PRATT and AUCLAIR (*295*), AUCLAIR and DUBREUIL (*19*), and DENT (*101*) found they range from 0.05 to 0.5 μg. for most amino acids but the reaction is less sensitive for histidine, arginine, tryptophan, lysine, hydroxyproline, proline, phenylalanine and tyrosine. KAWERAU and WIELAND (*195*) *preserved* the colour by spraying with a copper nitrate solution and dipping in a solution of methyl methacrylate polymer.

With *peptides*, 0.01% ninhydrin has been used by CONSDEN, GORDON, MARTIN, and SYNGE (*83*) and 0.025% by SANGER and TUPPY (*334*) to reveal the zones with a minimum of destruction of the N-terminal amino acids. Washing the cut out zones with acetone removes any unreacted ninhydrin. Many peptides give yellow colours with ninhydrin, some of which may subsequently turn purple. N-terminal residues of glycine, serine, threonine, proline, hydroxyproline, asparagine and cysteic acid may behave in this way. Other peptides, particularly those with N-terminal valine or isoleucine residues, may be extremely insensitive to the reagent. After cutting out the zones revealed by 0.025% ninhydrin, it is advisable to heat the chromatogram and re-spray with a stronger ninhydrin solution to detect less sensitive peptides.

D- and L-Amino Acids. SYNGE (*368*), JONES (*190*), and BONETTI and DENT (*49*) destroyed *D*-amino acids by spraying the paper with *D*-amino acid oxidase. The keto acids formed were detected by AUCLAIR and PATTON (*20*) by spraying with 2,4-dinitrophenylhydrazine.

α-Amino Acids. CRUMPLER and DENT (*90*) distinguished α-amino acids from other amino acids since the former did not appear on papergrams dusted with basic copper carbonate prior to development with neutral solvents.

Compounds with Amino or Imino Groupings. RYDON and SMITH (*316*) developed a sensitive test for compounds which form an =NCl derivative in which the chlorine atom is readily reduced. After removal of excess chlorine, a starch-KI spray reveals these compounds as blue-black spots. The background colour is reduced in the modified procedure of REINDEL and HOPPE (*301*). They obtained stable spots by exposing the paper to a chlorine dioxide-chlorine atmosphere and developing the colour with a 1% benzidine solution in 10% acetic acid. This test is especially useful for long peptides and proteins where the sensitivity is high, since there are many peptide bonds to react, in contrast to reagents for the free amino groups such as ninhydrin. It is also valuable for substances lacking a free amino group such as pyrrolidone carboxylic acid and pyrrolidonyl peptides, which may be formed by cyclisation of glutamine peptides [SANGER, THOMPSON, and KITAI (*332*)].

b) Non-Specific Reagents.

KEMBLE and MACPHERSON (*196*) used an indicator plus formalin to detect the acid derivatives formed by interaction of the amino groups with the formalin.

The buffering power of amino acids was utilised by PORATH and FLODIN (*289*) to prevent hydrolysis of filter paper sprayed with dilute acid. The sugars produced by hydrolysis of the cellulose react with orcinol giving a reddish background with white spots.

Amino acids may be located with a reagent containing 1,2-naphthoquinone-β-sulphonate (FOLIN's reagent) [MÜTING (*262*), GIRI and NAGABHUSHANAM (*146*)]. MÜTING claims this to be nearly as sensitive as ninhydrin with better colour differentiation. HARRIS and POLLOCK (*161*) found MÜTING's reagent gave relatively weak colours even at a concentration ten times that recommended.

1:2:4-Dinitrofluorobenzene was used by ISHERWOOD and CRUICKSHANK (*178*) to detect amino acids as the yellow dinitrophenyl derivatives. This method (see p. 298) will also be suitable for peptides and requires less than 1 μg. terminal amino acid.

Alloxan, applied as a fresh 0.2% (w/v) methanolic solution, followed by 5–10 minutes heating at 100°, has been recommended by HARRIS and POLLOCK (*161*) as a reagent which gives strong colours with most amino acids, except proline and hydroxyproline, and particularly with cystine and histidine, which give weak ninhydrin colours. Identification is aided by examination in ultraviolet light, and isatin can be included in the reagent, or applied subsequently, as a test for proline and hydroxyproline.

Amino acid and peptide spots *fluoresce* in ultraviolet light [PHILLIPS (*281*)]. This property depends on a reaction between the amino group and the paper on heating at 100° for 15–30 minutes. JONES (*192*) observed that the degree of fluorescence varied with the different batches of paper used, while SANGER and TUPPY (*334*) indicated the deleterious effect of phenol and cresol on the fluorescence. Fluorescent spots which do not contain amino acids or peptides are frequently observed.

c) Specific Colour Reactions for Some Amino Acids and Peptides.

Arginine. The SAKAGUCHI test, involving reaction with α-naphthol and alkaline hypobromite to give a red colour, has been used by CONSDEN, GORDON, and MARTIN (*78*), ACHER and CROCKER (*3*), and ROCHE, THOAI, and HATT (*306*) as a test for arginine, monosubstituted guanidine derivatives, and their peptides. Thus 0.2 μg. of arginine can be detected. JEPSON and SMITH (*186*) have used a dip reagent of 0.1% 8-hydroxyquinoline in acetone followed by a dip in a solution of 0.2 ml. bromine in 100 ml. of

0.5-*N*-NaOH. TUPPY (*390*) sprayed first with alkaline 1% α-naphthol followed by 0.1% diacetyl in alcohol.

Cysteine, Cystine and Methionine. TOENNIES and KOLB (*386*) detected cysteine by the orange colour produced with ammoniacal sodium nitroprusside. This reaction is also given by cystine and cystine peptides after reduction with potassium cyanide.

Both reagents are in methanolic solution suitable for dipping and will detect about 0.5 μg. of cystine. Cysteine, cystine, methionine and methionine sulphoxide have been detected by CONSDEN, GORDON, and MARTIN (*78*), WINEGARD, TOENNIES, and BLOCK (*422*), and TOENNIES and KOLB (*386*) using a platinic iodide reagent. The substances appear as white to yellow spots on a pink background. This reagent may be followed by ninhydrin or nitroprusside treatment.

Histidine. The PAULY reaction, involving treatment with a diazotised amine and alkali to give a red colour, has been widely used because its sensitivity with histidine (0.2 μg.) is greater than that of ninhydrin.

CONSDEN, GORDON, and MARTIN (*78*), DENT (*100*), BERRY, SUTTON, CAIN and BERRY (*31*), and AMES and MITCHELL (*14*) used diazotised sulphanilic acid in aqueous solutions. SANGER and TUPPY (*334*) recommended alcoholic solutions of *p*-anisidine-HCl and amyl nitrite which JEPSON and SMITH (*186*) found suitable for dipping. The reagent can be applied after the acidic EHRLICH reagent for tryptophan provided 2 *N*-NaOH is used to develop the colour. Tyrosine also interacts with diazo reagents to give an orange colour but is much less sensitive than histidine. This reaction is also given by peptides of these amino acids.

Proline and Hydroxyproline. The isatin test adopted by ACHER, FROMAGEOT, and JUTISZ (*4*) for papergrams is sensitive to less than 1 μg. of proline but only 2 μg. of hydroxyproline. It is also given by peptides with these residues N-terminal [cf. MONIER and JUTISZ (*252*)]. SMITH (*360*), and SAIFER and ORESKES (*319*) found that other amino acids also give blue or blue-green colourations at the 1–5 μg. level, although these fade more rapidly. Isatin may be included in the ninhydrin reagent or applied subsequently. CURZON and GILTROW (*91*) have described a test using vanillin, sensitive to 0.5 μg. for ornithine and sarcosine and 1 μg. for proline and hydroxyproline. JEPSON and SMITH (*186*) have described a highly sensitive (0.1 μg./cm².) and specific test for hydroxyproline, involving the formation of a purplish-red colour when the EHRLICH reagent was applied subsequent to the isatin test.

Serine, Threonine and Hydroxylysine. The β-hydroxyamino acids were shown by CONSDEN, GORDON, and MARTIN (*78*) to react with a mixture of periodic acid and NESSLER reagent to give brown spots. In the writers' experience, a preliminary spray with 10% periodic acid followed by NESSLER reagent secures greatly improved sensitivity. The test is also given by peptides where these residues are N-terminal.

Tryptophan. SYNGE and TISELIUS (*370a*) as well as TABONE et al. (*371*) made use of the sensitive EHRLICH reaction with acidified *p*-dimethylaminobenzaldehyde for the detection of tryptophan and its peptides. Smith (*360*) used 1.0% dimethylaminobenzaldehyde in acetone-11 *N*-HCl (9:1) as a dip reagent.

Tyrosine. The Pauly reagents used for histidine will also react with tyrosine to give an orange colour. The reaction is much less sensitive than for histidine, and the aqueous reagents are best. When phenolic solvents have been applied, the background colour may be sufficiently intense to mask the tyrosine. Heating or washing the paper with ether or acetone is not completely satisfactory in removing the phenol (see p. 288).

The red colouration produced with *x*-nitroso-*β*-naphthol followed by nitric acid is specific for tyrosine and its peptides, including 3-mono-halogenated tyrosines [Acher and Crocker (*3*)].

Jepson and Smith (*186*) used a 1% solution in acetone mixed with 5% nitric acid as a dip reagent. This reagent may follow the ninhydrin or isatin tests, but its sensitivity is considerably diminished.

III. The Determination of the Amino Acid Composition of Peptides and Proteins*.

1. Qualitative Analysis.

The amino acid composition of peptides and proteins is determined after complete hydrolysis.

Complete *hydrolysis* of some peptide bonds requires longer periods than the 24 hours refluxing with 6 *N*-HCl, which has long been regarded as a standard condition. The most stable bonds are those involving the carboxyl groups of isoleucine and valine. Thus, Christensen (*67*) isolated val-val from gramicidin after 24 hours boiling with 16% HCl, and the quantitative analyses of Harfenist (*158*), Smith and Stockell (*358*), Smith, Stockell, and Kimmel (*359*), and Hirs (*171*) have emphasised the difficulties of obtaining complete hydrolysis of certain bonds. The ileu-val bond in insulin, for example, required 96 hours for quantitative splitting [Harfenist (*158*)]. It is interesting to note that only 36 hours are required for complete hydrolysis of DNP-ileu-val, in which there is no charged amino group [Rovery and Fabre (*311*)]. Low recoveries

* The *abbreviations* summarised by Sanger (*325*) are followed in this review. Amino acid residues are indicated by the first three letters of the name and cysteic acid residues as $cySO_3H$. Where the sequence of amino acids in a peptide is established, the symbols are separated by hyphens; where the sequence is in doubt, symbols are enclosed in brackets and separated by commas.

ala = alanine, *arg* = arginine, *asp* = aspartic acid, *asp*-(NH_2) = asparagine, *cyS-Scy* = cystine, *glu* = glutamic acid, *glu*-(NH_2) = glutamine, *gly* = glycine, *his* = histidine, *hypro* = hydroxyproline, *ileu* = isoleucine, *leu* = leucine, *lys* = lysine, *met* = methionine, *orn* = ornithine, *phe* = phenylalanine, *pro* = proline, *ser* = serine, *thr* = threonine, and *val* = valine.

of isoleucine and valine after short periods of hydrolysis must also result in low recoveries of amino acids to which they are linked in resistant peptides.

Long periods of hydrolysis before amino acid analysis are not entirely satisfactory since some amino acids are liable to progressive destruction. It is therefore necessary to use different periods of hydrolysis, so that a plateau is reached for those difficult to liberate, and so that extrapolation to zero time can be carried out in order to estimate the unstable amino acids. The recoveries of serine, threonine, cystine, tyrosine, aspartic acid, glutamic acid, arginine, lysine and methionine have been found to decrease with time of hydrolysis [cf. SMITH et al. (*358*, *359*, *356*), HIRS (*171*)]. Destruction factors vary from protein to protein, as do the times for complete hydrolysis.

It is not generally known what products are formed from the amino acids destroyed. Serine and threonine give rise to ammonia [REES (*300*)]. At 150° HEYNS and WALTER (*170*) have noted the formation of α-aminobutyric acid and some glycine from threonine, and alanine from serine [cf. INGRAM (*177*)]. Tryptophan when pure is reasonably stable to hydrolysis, particularly if trace metals present in the HCl are removed by distillation. However, in proteins, the tryptophan residue is less stable, due to reaction with other amino acids and traces of carbohydrates. MONIER and JUTISZ (*251*) found that hydrolysis with purified HCl in an evacuated sealed tube at 75–85° for 15 hours gave no destruction of tryptophan in lysozyme. However, tryptophan in peptides eluted from papergrams and hydrolysed with HCl was found by CONSDEN, GORDON, and MARTIN (*79*), and WHITE and LANDMANN (*416*) to be completely destroyed by contaminating carbohydrate material.

Tryptophan is more stable to alkaline hydrolysis although other amino acids are decomposed or racemised. Serine gives appreciable amounts of glycine and alanine, threonine yields glycine and α-aminobutyric acid, cysteine and cystine produce alanine [WIELAND and WIRTH (*417*)], and arginine gives rise to ornithine and citrulline [WARNER (*406*)].

Artifacts may also be formed either during the preparation of the material for chromatography, or during the run.

Arginine may give ornithine during electrolytic desalting (see p. 277). Lanthionine may be formed from cystine in proteins treated under mild alkaline conditions [HORN, JONES, and RINGEL (*174*)]. BREYHAN (*52*) has detected some decomposition products of glutamic acid and ornithine formed under hydrolysis conditions. Esterification of glutamic acid hydrochloride in butanol-acetic acid has been noted by HEYNS, KOCH, and KÖNIGSDORF (*168*). Cysteine and cystine are decomposed in some solvent systems [CONSDEN, GORDON, and MARTIN (*77*), DENT (*101*)] to give cysteic acid, and methionine may give methionine sulphoxide, according to DENT (*101*). Oxidation of an amino acid mixture with hydrogen peroxide or bromine water, to convert the sulphur-containing amino acids to stable products, was found by VAN HALTEREN (*400*) to yield ninhydrin-positive artifacts, although GALE and VAN HALTEREN (*145*) did not detect such products after autoxidation. In the presence of chloride ions, hydrogen peroxide or performic acid has been shown by THOMPSON (*382*) to form 3-chloro- and 3,5-dichloro-tyrosine. This is the reason for the loss of tyrosine noted by WELLINGTON (*408*).

The qualitative amino acid composition of proteins or peptides is easily determined on papergrams using the solvents and techniques previously discussed. Only micro amounts of material are required and more than enough peptide for this purpose may be isolated on papergrams.

Fig. 5. Method of elution of material from paper "cuts", suitable for large numbers of samples.

The method of elution described by SANGER and TUPPY (*334*) and illustrated in *Fig. 5* is very convenient for large number of "cuts". The use by CONSDEN, GORDON, and MARTIN (*79*) of polythene strips for handling small amounts of material has been discussed (p. 278).

2. Quantitative Analysis.

The simplicity of the paper chromatographic procedure and the comparative ease of separation of the amino acids have resulted in many attempts to use the method for quantitative estimations. The degree of accuracy and reproducibility required of a quantitative procedure is largely dependent on the mixture being analysed and the purpose for which the data are required.

TRISTRAM (*388*) has discussed in some detail the accuracy of various standard analytical procedures for amino acids and the accuracy required for an estimate that n rather than $n \pm 1$ residues occur in a peptide or protein. This requires a mean determination between $n + 0.4$ and $n - 0.4$

residues and hence an accuracy of $\pm 40/n$ %. For short-chain peptides, isolated in large numbers from partial hydrolysates of proteins, n has a low value, and a method which is quick and reliable to an accuracy no greater than $\pm$ 10% is adequate. Where n has a high value, the accuracy of paper procedures may not be sufficient to define the exact number of residues of each amino acid in the molecule. Such information is highly desirable for the determination of amino acid sequences.

Some workers have used paper chromatographic methods to determine the complete amino acid composition of some proteins, and found that their results compare favourably with those of others techniques. While this may be true, it should be realised that, when several solvent systems are required to separate all amino acids, and with sufficient replication and analyses of standard mixtures, the work and time involved may exceed that required by the MOORE and STEIN (*256*) procedure on ion-exchange resins, which has almost excluded other methods for reliable amino acid analyses, with an accuracy of $\pm$ 3%.

Paper methods have been useful in many investigations, such as the analysis of amino acids liberated by stepwise enzymatic attack on a peptide chain, *e. g.* with carboxypeptidase, where a limited range of amino acids is to be determined in samples taken after various periods of time. Thus, many samples can be analysed simultaneously and the accuracy is usually sufficient.

a) Measurements Applied Directly on the Paper.

Reactions applied directly to the paper have an obvious advantage over those involving elution of the material, although the range of reactions is more restricted.

Ninhydrin has been most widely used, since it reacts in the same way with all α-amino acids. For proline, the sensitive reaction with isatin may be employed. Comparison of the *intensity* of ninhydrin colour given by a series of standard amounts of amino acids with several dilutions of the unknown samples chromatographed on the same sheet has been extensively used [POLSON (*284*), POLSON, MOSLEY and WYCKOFF (*285*), BERRY and CAIN (*30*)]. This "spot dilution" method is only suited for one-dimensional runs, and comparison of intensities is more accurate at lower levels. For a wide range of amino acids, the preparation of standard spots may become tedious. Good chromatograms with regular spots are essential, with application of the material to the paper in the same volume of solution.

The possibilities of this method are shown by the accurate analysis of the A chain of oxidised insulin carried out by SANGER (*322*). With the exception of leucine, for which three residues were estimated, the correct composition, $(cySO_3H)_4$, asp_2, glu_4, ser_2, gly_1, ala_1, tyr_2, val_2, leu_2, $ileu_1$, was found [SANGER and THOMPSON (*329*)]. The method has been widely used for estimation of the yields of amino acids liberated by the action of carboxypeptidase on proteins [SANGER and THOMPSON (*329*), GLADNER and NEURATH (*150*, *151*)] and for the estimation of amino acids in DNP-peptides [SANGER (*323*), PORTER (*292*)]. A variation of this method has been proposed by AUCLAIR and DUBREUIL (*19*), based on the minimum detectable amount of ninhydrin colour on a two-dimensional chromatogram.

FISHER, PARSONS, and MORRISON (*127*) reported a quantitative method based on the fact that *spot areas*, or spot lengths for regular shaped spots, are proportional to the logarithm of the amino acid concentrations. The outline of the spot must be clear and may be measured with a planimeter, by counting squares, or by weighing tracings on uniform paper. BLOCK (*40*), and FROMAGEOT and PRIVAT DE GARILHE (*144*) have obtained an accuracy of ± 5% using this method. For two-dimensional chromatograms, the relationship has been amended by ÅKERFELDT (*12*).

BULL, HAHN, and BAPTIST (*56*) plotted the per cent transmission of the spots at different points along a paper strip against the distance from the origin, and integrated the area under the peaks for an estimate of quantity. This approach has also been used by BLOCK (*40*), REDFIELD and GUZMAN BARRON (*298*), and by ROCKLAND and DUNN (*308*). Large numbers of chromatograms may be necessary for accurate results and distinct spots are essential. The symmetrical nature of the curves obtained with this technique led to the observation that the *maximum colour density* of each spot was proportional to the concentration of material in the spot [BLOCK (*41*)]. BLOCK included taurine as an internal standard with each two-dimensional run, to allow for variations between different chromatograms. Unknown solutions can be analysed with an accuracy of ± 10% by reference to standard curves prepared from pure amino acids. Similar methods have also been used by PATTON and CHISM (*278*), ROLAND and GROSS (*309*), and by MCFARREN and MILLS (*243*). SALANDER, PIANO, and PATTON (*320*) stressed the importance of uniform colours from day to day, and found that application of the material in volumes larger than 1 or 2 µl. gave less satisfactory results. The filter paper blank is minimised in this technique, since a base line is found where no amino acid is present. Colour reactions for some of the amino acids may be used, provided that they obey BEER's law over the concentration range employed. The more uniform the conditions, such as size of initial spot, and sharpness of the spots on the finished chromatogram, the more accurate the procedure.

This scanning method has much to recommend it for routine analysis of amino acids. It is less tedious than plotting a peak and measuring the area under the curve. Regular checking of standards is simple. With care, the method should be accurate to ± 5–10% and may be improved using the methods discussed below for full colour development on the chromatogram.

Where material is limited, it is desirable to estimate quantitatively the amino acids in a peptide on the same paper used for qualitative identification. The maximum density method is well suited to this purpose, and, for cases where two peptides overlap on a chromatogram, a quantitative estimation of the hydrolysis products may reveal the components in each of the peptides. WHITE (*413–415*), LANDMANN, DRAKE, and WHITE (*213*), and WHITE and LANDMANN (*416*) have used this technique to analyse peptides isolated from enzymic digests of corticotropin A.

A special method, capable of considerable accuracy and applicable to very small amounts of material, has been developed by KESTON, UDENFRIEND, and LEVY (*201*), and VELICK and UDENFRIEND (*402*) as an *isotope* derivative analysis employing the *p*-iodophenylsulphonyl ("pipsyl") derivatives of the amino acids labelled with I^{131}. The addition, as indicators, of known amounts of one or more pipsylamino acids, labelled with S^{35} instead of I^{131}, is followed by the isolation of a particular derivative in pure form by countercurrent and paper chromatographic methods. The coincidence of S^{35} and I^{131} radiation along the direction of band migration is an excellent criterion of purity [UDENFRIEND (*397*)], and the measured value of the $I^{131}:S^{35}$ ratio in the band, together with the known specific activities and amounts of indicator used, are sufficient for a quantitative estimation. This method is capable of accuracies as low as ± 1%, and is ideally suited for amino acid analyses of protein hydrolysates on a micro scale, although the specialised nature of the technique has limited its use by other workers.

b) Methods Involving Elution from the Paper after Reaction with a Reagent.

Considerable progress has been made in understanding the variables involved in the reaction of *ninhydrin* with amino acids on paper. PATTON and CHISM (*278*) found that 95% ethanol was definitely superior to water-saturated *n*-butanol as a ninhydrin solvent; the colours formed more rapidly, and were destroyed at a slower rate. THOMPSON, ZACHARIUS, and STEWARD (*385*) devised controlled conditions which would give reproducible and maximum colour, comparable in many cases with that found by MOORE and STEIN (*255*) for the reaction with ninhydrin in a test tube. As with the test tube reaction, oxygen was found deleterious to reproducibility. The concentration of ninhydrin must exceed a value of 1% for normal levels of amino nitrogen. High temperatures gave decreased colour yields in the presence, but not in the absence, of air. Humidity also played an important role in the colour development.

Using an alcohol-saturated atmosphere for the development in the presence of carbon dioxide at 60° for 30 minutes, no difference was found between water-saturated butanol and 95% ethanol, but the addition of 2% of a 2:4:6-collidine, 2:4-lutidine (1:3) mixture to buffer the reaction was an advantage. WELLINGTON (*408–410*) also emphasised the importance of humidity, and found a maximum sensitivity in the range 35–40% for all amino acids except aspartic acid. The minimum concentration of ninhydrin recommended was 2% in absolute ethanol. Colour was developed during 30 to 36 hours in a darkened room at 20 ± 1° and 38 ± 2% relative humidity. Both procedures involve about 30 minutes extraction of the colour in 50% aqueous ethanol or isopropyl alcohol. Blanks were cut from the paper and corrections applied based on the area of the spots. The absorption was measured at 570 mμ for all amino acids except proline, hydroxyproline and asparagine, which were measured at 330 mμ. The corrected optical density was then proportional to the amount of amino acid. WELLINGTON has noted a deleterious effect from other chemicals in the atmosphere, particularly hydrochloric acid, and has devised a chamber for controlling the humidity in a stream of compressed air, which reduces the possibility of contamination with other chemicals (*409*). It is not clear what effect the presence of air has on the variability of results under the conditions used (*410*).

GIRI, RADHAKRISHNAN, and VAIDYANATHAN (*147*) have adapted circular paper chromatography for the quantitative estimation of amino acids. Best results were obtained when the concentration of amino acid was less than 15 μg., and a concentration of 0.5% ninhydrin was sufficient. Acetone was preferred as ninhydrin solvent because it dried more rapidly. The paper was heated at 65° for 30 minutes.

These workers and others (BODE, HÜBENER, BRÜCKNER, and HOERES (*44*), MORTREUIL and KHOUVINE (*258*), FISCHER and DÖRFEL (*126*), MEYER and RIKLIS (*244*)) observed that increased sensitivity is obtained by eluting the colour into alcohol solutions containing metal ions such as Cu^{++}, since according to KAWERAU and WIELAND (*195*), with metal ions such as Ni^{++}, Cu^{++}, Cd^{++}, and Zn^{++}, red pigments are formed from the blue ninhydrin colour with an absorption maximum at 510 mμ. The accuracy of this method was assessed at ± 5—10%.

MEYER and RIKLIS (*244*) found that as little as 1 μg. metal ion lessens the normal ninhydrin colour with absorption maximum at 570 mμ, and 20 μg. completely eliminates it. The effect of metal ions may be prevented by the use of versene as a complexing agent. The incorporation of versene in all ninhydrin reagents which require an absorption maximum at 570 mμ seems desirable.

A different approach to the estimation of amino acids on papergrams is that of ISHERWOOD and CRUICKSHANK (*178*) who detected the spots by spraying with a buffered solution of *2:4-dinitrofluorobenzene* under conditions where the reaction was quantitative. The spots then show up as the yellow DNP-derivatives, which can be easily extracted from the paper and estimated colorimetrically.

Reaction was quantitative in 30 minutes at 80° or overnight at room temperature for solutions applied in concentrations less than 0.02 M. Spots containing less than 1 μg. are easily detected by this method if the background due to dinitrophenol is bleached by exposure to hydrogen chloride fumes. Before estimation, the dinitrophenol must be removed, since this absorbs in the same region under the conditions used for the estimation. The DNP-amino acid was isolated by an involved extraction procedure. Blank estimations were low, and never more than 2–4% of those for the DNP-amino acids. The method has been applied only to monoamino monocarboxylic acids and to aspartic and glutamic acids. The latter required some modifications in the extraction conditions. The estimated accuracy was ± 5% and the method was applicable in the range 0.5–20 mg.

An attractive feature of this procedure is the fact that amino acids are easier to chromatograph on paper than the DNP-amino acids. The method should be applicable to peptides as well as amino acids, and may be useful for their estimation, particularly since many peptides give weak colours with ninhydrin. Moreover, the peptides are converted in the process to a derivative useful for their identification. The sublimation of dinitrophenol described by MILLS (*247*) should be valuable for preparative isolations.

c) Methods Involving Elution from the Paper before Reaction.

In this type of approach it is necessary first to locate the zones accurately on the paper. Marker strips have been used; and also ***ninhydrin*** with colour development in two stages. The first spray is usually a weak one which serves to locate the spots, and the colour development is completed in the test tube. The methods of NAFTALIN (*263*), and AWAPARA (*21*) suffered from a lack of reproducibility, common to all the early colorimetric methods using ninhydrin, which MOORE and STEIN (*255*) controlled by incorporating a reducing agent to overcome the deleterious effect of dissolved oxygen. Their reagent is very sensitive and reproducible and has been used by later groups of workers.

LANDUA and AWAPARA (*214*) found that the blanks varied in different batches of paper but were fairly reproducible on the same sheet. BOISSONNAS (*47*) and FOWDEN (*129*) showed that the high blanks were due to ammonia present on the paper and employed alkali to remove it prior to reaction. Both these workers claimed quantitative recoveries of amino acids. FOWDEN (*129*) located the amino acids by their fluorescence in ultraviolet light, which, as previously discussed, is not always a reliable procedure [cf. FOWDEN (*130*)]. BOISSONNAS (*45*, *46*) used ammonia-free solvents for chromatography and a viscous ninhydrin reagent applied on a set of fine metal points to locate the spot. PORATH and FLODIN (*289*) have recommeuded a method (see p. 290) for location of spots without destruction of amino acids for subsequent determination with MOORE-STEIN reagent. PERNIS and WUNDERLY (*280*) stained with a 2% solution of ninhydrin to locate the spots; 38% of a leucine spot reacted during this stage. Final colour development occurred in the test tube. The accuracy with known mixtures was estimated at ± 3–10%.

When the ninhydrin reagent of MOORE and STEIN is used, it is desirable to avoid contamination with ammonia as much as possible, since as little as 0.1 μg is detectable.

TROLL and CANNAN (*389*) as well as COCKING and YEMM (*72*) have recently described ninhydrin reagents which give equal colour intensities with equimolar amounts of all amino acids (except tryptophan and lysine), yet show a much reduced sensitivity to ammonia. These reagents should be of value in reducing blank corrections. Both PERNIS and WUNDERLY (*280*), and KLATZKIN (*205*) locate the amino acids with ninhydrin reagent sensitive to ammonia before treatment with alkali to remove ammonia; this must increase the blank value and the variability in the results.

Methods relying on full colour development of the amino acids have the advantage that, if reproducible, simultaneous chromatography of a range of standards is not necessary. Variation from paper to paper does, however, occur and it is necessary to have rigid conditions regarding pre-treatment of the paper, size of the initial spot, length of time of irrigation, method of drying, colour development, etc. Some measure of control of day-to-day variations is obtained by the inclusion of an internal standard with each two-dimensional chromatogram. No single investigation using ninhydrin has controlled all the variables so far discussed.

Several attempts have been made to avoid the uncertainty of the ninhydrin colorimetric procedures. MARTIN and MITTELMANN (*240*) used the *micro-Kjeldahl* method for estimating amino acids but found that contamination from the solvents introduced large variations [cf. KLATZKIN (*205*)]. SLOTTA and PRIMOSIGH (*351*) used the VAN SLYKE *ninhydrin-CO_2* procedure to estimate the amino acids, after locating them in ultraviolet light. KEMBLE and MACPHERSON (*197*) locate the amino acids with an indicator spray (see p. 290), and estimate the neutral amino acids by a manometric method involving their reaction with *chloramine-T* to liberate carbon dioxide. Ammonia does not interfere in this method and the procedure seems capable of high accuracy.

Amino acids react with *copper phosphate*, and the copper complex which goes into solution has been estimated polarographically [MARTIN and MITTELMANN (*240*), JONES (*191*)] or colorimetrically [WOIWOD (*425*)]. Proline and hydroxyproline react with copper phosphate in the same way as other amino acids, in contrast to their reaction with ninhydrin.

The filtration strip in the colorimetric method may introduce errors [LANDUCCI and PIMONT (*216*)]. BLACKBURN and ROBSON (*38*) have increased the sensitivity of the method by using radioactive copper phosphate. One μg. of α-amino nitrogen could then be determined with an accuracy of $\pm$ 3%; 2 μg. with $\pm$ 2%.

WOIWOD (*425*) focussed attention on *losses* of amino acids depending on the length of travel on a papergram. THOMPSON and STEWARD (*384*) confirmed this, and reported higher losses in the case of the hydroxyamino acids. However, BLACKBURN and ROBSON (*38*) found no losses using a similar technique. Neither did FOWDEN (*129*), when the amino acid was estimated by his ninhydrin procedure, nor ISHERWOOD and CRUICKSHANK whose method has been discussed (p. 298).

An important contribution to the quantitative methods for the determination of amino acids on papergrams was introduced recently by LEVY (*223*). The amino acids were quantitatively converted to their DNP-derivatives by reaction with DNFB *(2:4-dinitrofluorobenzene)* in aqueous solution at p_H 9. All the DNP-derivatives were separated on one sheet of WHATMAN No. 1 paper. The water-soluble DNP-amino acids, DNP-arginine and α-DNP-histidine, were separated during a single-dimensional run in the "toluene" system of BISERTE and OSTEUX (35), whereas the ether soluble derivatives required a second dimensional run with 1.5 *M* phosphate buffer pH 6 (see *Fig.* 6). This process is

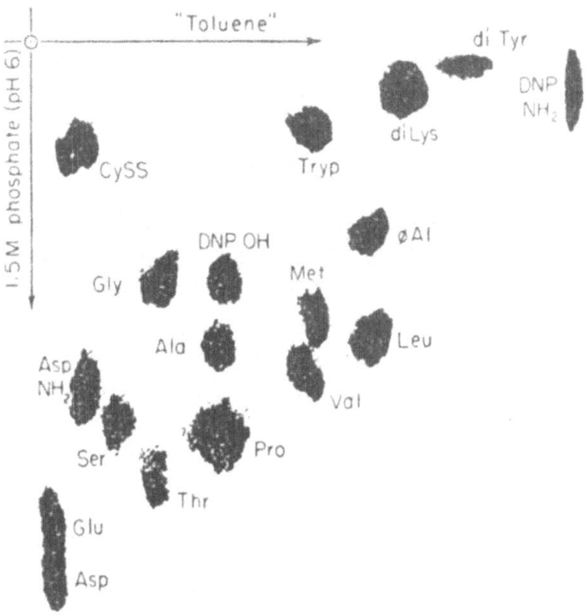

Fig. 6. Separation of the ether-soluble DNP-amino acids on a two-dimensional papergram, according to LEVY. [From: Nature (London) *174*, 126 (1954).]

applicable to mixtures of amino acids with salts, proteins or ammonia, which is an important advantage. The DNP-amino acids are easily eluted and estimated.

Correction factors have been determined, and are independent of the composition of the mixture. The greatest variations from theoretical recovery occurred with tyrosine, histidine and methionine, with correction factors 1.54, 1.62 and 1.21, respectively, but molar ratios were reproducible to within 2–3%, except for tyrosine, where wider variations were unaccountably found.

The accuracy of the method is ± 4% and has already given satisfactory analyses of α-corticotropin (39 residues), ribonuclease (123 residues), growth hormone (361 residues), and insulin (48 residues). It is important in this method to follow accurately the prescribed directions for the DNP reaction, since other procedures may not give quantitative recoveries (see p. 303).

IV. Identification of Terminal Sequences in Peptides and Proteins.

At present, the determination of amino acid sequences in high-molecular polypeptides is essentially a question of identifying terminal sequences and the sequences of peptides formed by partial hydrolysis. The interpretation of results is much simpler with proteins if there is a single peptide chain, and methods for determining the number of free amino or α-imino groups and free carboxyl groups are essential to gain an insight into the gross structural features of the molecules, and as a means of following the separation of component chains.

Proteins may consist of open or branched polypeptide chains or they may be cyclic. Insulin has been shown to have an equal number of N-terminal and C-terminal residues, the molecule consisting of two *open* polypeptide chains held together by disulphide bridges. Chymotrypsinogen has neither N-terminal (DESNUELLE, ROVERY, and FABRE (*110*)] nor C-terminal (GLADNER and NEURATH (*150*)] residues and is thus a *cyclic* chain. With tropomyosin and myosin, however, N-terminal residues have not been detected [BAILEY (*22*)], yet C-terminal residues are present [LOCKER (*229*)]. For these proteins, *branched* polypeptide chains must therefore be postulated. Theoretically, this branching may occur in several ways, but few of these possibilities have received experimental verification for proteins. Branching of the peptide chain of bacitracin A involves the β-carboxyl group of aspartic acid and the ε-amino group of lysine [CRAIG, HAUSMANN, and WEISIGER (*88*), PORATH (*288*), NEWTON and ABRAHAM (*267*)]. These findings, together with the γ-peptide linkage of glutamic acid in glutathione and other natural peptides, suggest that these linkages may also occur in proteins. While it is necessary to keep these possible linkages in mind, there is no doubt that the peptide bond as conceived by HOFMEISTER and FISCHER is the predominant link in proteins.

For proteins, the ability to detect and estimate quantitatively the N- or C-terminal residues is one of the most searching tests of protein purity, since non-stoichiometric amounts of terminal residues will be indicative of adsorbed impurities or contaminating proteins, provided the reagents do not cause peptide bond splitting. Similar techniques can also be used to study the yield of particular end groups set free by enzymatic and chemical hydrolysis.

For peptides, identification of a single N-terminal or C-terminal residue gives valuable evidence of purity and contributes to their identification. Thus, for a tripeptide, the knowledge of both N-terminal and C-terminal residues gives the sequence. A determination of the first two residues from either end would also yield it.

The present section is concerned with methods for the identification of N- and C-terminal residues and sequences. It should be emphasised that *subtractive* methods of end group analysis are less decisive than the positive identification of a derivative of the terminal residue which may be confirmed, in most instances, by the non-appearance of the terminal amino acid in a hydrolysate of the peptide derivative. Quantitative analysis of the amino acids before and after treatment of a peptide with a reagent has been used by DAVOLL, TURNER, PIERCE, and DU VIGNEAUD (*98*), DU VIGNEAUD, RESSLER, and TRIPPETT (*115*), RESSLER, TRIPPETT, and DU VIGNEAUD (*304*), and POPENOE and DU VIGNEAUD (*286*) to identify end groups in the work on oxytocin and vasopressin where only eight amino acids were present to be analysed. FOX, HURST, and ITSCHNER (*133*) have also used this approach. While satisfactory for peptides with small number of residues, the results are necessarily less conclusive for proteins containing large numbers of residues, where the accuracy of the analytical method employed is the limiting factor [cf. FOX, HURST, and WARNER (*134*); DE-FONTAINE and FOX (*99*)].

1. N-Terminal Sequences.

a) Dinitrofluorobenzene Method.

The reagent 2:4-dinitrofluorobenzene (DNFB) introduced by SANGER (*321*) has proved extremely versatile as a means of identifying the free amino groups of proteins and peptides.* It also reacts with the phenolic group of tyrosine, the sulphydryl group of cysteine, and, depending on the conditions, with the imidazole ring of histidine. SANGER (*321*) employed a 66% ethanolic solution saturated with sodium bicarbonate and DNFB. Under these conditions, substitution of histidine occurred [PORTER (*291*)]. SCHROEDER and LEGETTE (*346*) observed that, under these conditions, the yields of DNP-amino acids are far from quantitative, being particularly low for the dicarboxylic amino acids. With 2% or less sodium bicarbonate, almost quantitative yields were found.

LEVY (*223*) has developed a procedure, suitable for peptides or protein hydrolysates, using an autotitrator to maintain the p_H constant (at 9) during stirring of DNFB with an aqueous solution of the amino acid or peptide; the end point was easily determined from the rate of alkali uptake. The yields are nearly quantitative with most amino acids, the exceptions being tyrosine, histidine, and methionine (see p. 301). However, LEVY reported that α-DNP-histidine is formed under his conditions. Aspartic and glutamic acids which are the most difficult to react with DNFB [MILLS (*247*); SCHROEDER and LEGETTE (*346*)] have been coupled quantitatively with DNFB on a papergram by ISHER-

* For formulas cf SCHROEDER (*343*).

WOOD and CRUICKSHANK (see p. 298). SANGER and THOMPSON (*328*) used 1% aqueous trimethylamine in their experiments on peptides. Excess base could then be removed by volatilisation to avoid the presence of large amounts of salt during chromatography of the residual amino acids in hydrolysates of the DNP-peptides.

A comprehensive list of N-terminal residues identified in proteins by the use of DNFB has been published recently by SCHROEDER (*343*). To this list may be added the following data. Three N-terminal histidine residues in *gliadin* (M. W. 27000) [KÖRÖS (*207*)]. A *papaya lysozyme* (M. W. 25000) crystallised by SMITH, KIMMEL, BROWN and THOMPSON (*356*) contains N-terminal glycine residues. *Bacitracin A* gave larger amounts of N-terminal isoleucine after performic acid oxidation [CRAIG, HAUSMANN, and WEISIGER (*88*)] which is interesting in view of the small amounts detected previously by NEWTON and ABRAHAM (*267*); however, PORATH (*288*), and INGRAM (*176*) were unable to find any N-terminal residues in bacitracin A. SCHMID (*339*) could detect no N-terminal residue in acid *glycoprotein*, and isolated only DNP-glucosamine and artifacts derived from it. *Angiotonin* has been purified further and gave DNP-aspartic acid as an N-terminal residue [BUMPUS and PAGE (*57*)]. WILLIAMSON and PASSMANN (*420*) reported N-terminal leucine, and BISERTE and DAUTREVAUX (*34*) the N-terminal sequence, leu-leu-ala-val, for *pepsin*. However, VAN VUNAKIS and HERRIOT (*401*) found an N-terminal sequence ileu-gly for pepsin and leu-leu or ileu for *pepsinogen*. Since the DNP-leucines cannot be distinguished chromatographically as yet, it was necessary to regenerate the amino acids by heating with barium hydroxide [MILLS (*246*)] or ammonia [LOWTHER (*233*)].

The hydrolysis of DNP-proteins or DNP-peptides is usually effected with 6 *N*-hydrochloric acid, which should be redistilled from glass before use. The majority of the DNP-amino acids are stable to long periods of boiling [PORTER and SANGER (*293*)] but DNP-proline is particularly unstable, re-forming proline and dinitrophenol [ACHER and LAURILA (*6*), SCHRAMM and BRAUNITZER (*340*)]. DNP-glycine and DNP-cystine have only a limited stability. DNP-tryptophan decomposes to give a yellow compound running at the same rate on columns [JAMES and SYNGE (*182*)]. DNP-proline was reported to be more stable to concentrated HCl [PORTER and SANGER (*293*)]. However, FRAENKEL-CONRAT and SINGER (*141*) found that DNP-proline after 2 hours heating to 103° yields a yellow artifact with a similar R_f to DNP-proline but with a different absorption spectrum [cf. MONIER and JUTISZ (*252*)]. Concentrated HCl is preferably prepared by saturating redistilled water or 6 *N* acid with purified HCl.

The absorption spectra of DNP-proline and DNP-hydroxyproline and their peptides is different from those of the other DNP-amino acids and DNP-peptides, and may be used to determine whether these residues are N-terminal [SCHROEDER, HONNEN, and GREEN (*344*)].

The *destruction* of DNP-amino acids liberated during hydrolysis of DNP-proteins varies with the composition of the protein concerned.

The determination of destruction factors gives only an approximate correction, since the decomposition is probably different while the DNP-amino acid residue is in peptide linkage, and may also vary with the amino acid sequence near the DNP group. Tryptophan [THOMPSON (*372*)] and cysteine [DESNUELLE, ROVERY, and FABRE (*109*)] cause destruction of DNP-amino acids during hydrolysis; in the case of DNP-papain THOMPSON (*380*) found that tryptophan was chiefly responsible. The determination of correction factors should be made in the presence of DNP-protein rather than protein since all the DNP groups present in the DNP protein share in the destruction, whereas a DNP-amino acid in the presence of protein will suffer the total destructive effect. HANES, HIRD, and ISHERWOOD (*156*) have recently reported greatly improved recoveries of DNP-glycine in DNP-peptides hydrolysed with almost anhydrous acidic mixtures.

Thus, after 8 hours hydrolysis at 100° with a 1:9 mixture of 60% (w/v) perchloric acid and glacial acetic acid, no breakdown of the DNP-glycine in DNP-glycylglycine was detected. More rapid hydrolysis was achieved in 2 hours at 100° with a 20:11:3 mixture of 90% (w/v) formic acid, acetic anhydride and 60% (w/v) perchloric acid and this mixture gave better recoveries of DNP-prolyl-arginine from DNP-salmine than did 11 *N*-HCl [MONIER and JUTISZ (*252*)].

DNP-amino acids are usually extracted into ether, in which they are readily soluble. DNP-asparagine is less soluble in ether than most other DNP-amino acids. DNP-cysteic acid, DNP-arginine and ε-DNP-lysine remain in the aqueous phase together with bis-DNP-histidine. MILLS (*247*) found that continuous ether extraction transferred bis-DNP histidine to the ether phase. Both O-DNP-tyrosine and *im*-DNP histidine are colourless and remain in the aqueous layer. Ethyl acetate is an excellent solvent for DNP-amino acids, but it also extracts small amounts of water-soluble DNP-derivatives. It is more often used for extracting DNP-peptides [SANGER (*323*)].

The *identification* of the DNP-amino acids in the ether and aqueous layers has been made with both columns and papergrams. SCHROEDER (*343*) has discussed the various column procedures which have been widely used for quantitative work. Some DNP-derivatives will travel on unbuffered papers with solvents normally used for amino acids. Butanol-acetic acid is particularly useful for the identification of the colourless water-soluble DNP-derivatives [SANGER and TUPPY (*334*)].

The *solvent systems* most commonly used are *tert.*-amyl alcohol on paper buffered at p_H 6 [BLACKBURN and LOWTHER (*37*)] and the "toluene" system of BISERTE and OSTEUX (*35*). These latter workers made a careful study of the conditions for reproducible results, and particularly recommended that the chromatograms be developed at a constant temperature and in the dark. *Tert.*-amyl alcohol will usually separate clearly the ether-soluble DNP-derivatives of cysteic acid; aspartic acid; glutamic acid; serine and glycine; threonine; alanine and proline; valine, methionine and possibly bis-DNP-lysine; phenylalanine; bis-DNP-tyrosine and the leucines. With the toluene-pyridine-ethylene chlorhydrin-

0.8 N-ammonia (5 : 1 : 3 : 3) system (*35*), DNP-serine and DNP-glycine will separate, but DNP-aspartic acid and DNP-glutamic acid will not; DNP-leucines and DNP-phenylalanine overlap, but are separated from bis-DNP-tyrosine and bis-DNP-lysine.

It must be emphasised that equilibration with these solvent systems and the amounts of DNP-amino acids on the paper may considerably alter the R_f values.

LI and ASH (*226*) used these two systems in their investigations of N-terminal residues in growth hormone. They reported the "toluene" system as unsuitable for water-soluble DNP-amino acids; but the proportions of BISERTE and OSTEUX (*35*) were altered by LEVY (*223*) to 5:1.5:3:3 to obtain satisfactory resolution. BISERTE and OSTEUX have separated DNP-proline and DNP-alanine in a phenol-isoamyl alcohol-water (1:1:1) system. To distinguish between DNP-valine and DNP-methionine, the mixed band was chromatographed in cyclohexane-acetic acid (1:1) or in the other systems, after oxidation of DNP-methionine to the sulphone.

The faster moving DNP-amino acids are among the most difficult to detect, since streaking is more pronounced for them and the spots may be fainter. BISERTE and OSTEUX used the system decalin-acetic acid (1:1) for the separation of DNP-phenylalanine ($R_f = 0.17$), DNP-valine ($R_f = 0.25$) and DNP-leucine ($R_f = 0.53$), all other amino acids moving extremely slowly. Dinitrophenol moves rapidly in this solvent, which was used to eliminate it from DNP-amino acids. The sublimation procedure of MILLS (*247*) is especially convenient for this purpose, and "artifact" spots including dinitrophenol can be distinguished qualitatively by their decolourisation on exposure to hydrochloric acid vapours [SANGER and THOMPSON (*328*)]. The latter authors described a reversed phase system, using silicone-treated paper, for the DNP-amino acids moving fast in the usual systems. A somewhat similar pattern of spots, entirely different from that in the normal systems, is obtained using concentrated aqueous buffer solutions as developers. ROVERY and FABRE (*311*) used a citrate buffer p_H 6.2 system and LEVY (*223*) 1.5 M-phosphate, p_H 6, in the second dimension to separate those DNP-amino acids, (except the DNP-leucines), which were not resolved in the modified "toluene" solvent of BISERTE and OSTEUX. Other solvent mixtures for the separation of particular DNP-amino acids mixtures have been described by LANDMANN, DRAKE, and WHITE (*213*), IWAINSKY (*181*), KENT, LAWSON, and SENIOR (*200*), MONIER and PÉNASSE (*253*), and FELIX and KREKELS (*125*).

The possibility of obtaining clean separations of mixtures of DNP-amino acids on papergrams has resulted in their greatly increased use for *quantitative* estimations after elution from the paper. WEIBULL (*407*) required only 10 mg. protein for an examination of the free amino groups of *Proteus flagella* protein, using the *tert.*-amyl alcohol system of BLACKBURN and LOWTHER (*37*), which gave less spreading of the spots for quantitative work than that of BISERTE and OSTEAUX (*35*). WILLIAMSON and PASSMANN (*420*), ANFINSEN, REDFIELD, CHOATE, PAGE and CARROL (*15*), JOLLÈS and FROMAGEOT (*189*), and LEVY (*223*) have also used paper chromatography of DNP-amino acids for the quantitative estimation of N-terminal residues in proteins. The smaller scale of paper experiments is often advantageous, and LEVY has reported an accuracy of ± 4%, which is better than that usually claimed for column procedures. The separation of a large number of DNP-amino acids on the same sheet of paper is very convenient, as columns may require several transfers, with increased incidental losses. It seems likely that quantitative estimations of amino acids as their DNP-derivatives, separated on papergrams, will be widely used in future. Desalting of peptide hydrolysates is not necessary if the amino acids are converted to their DNP-derivatives which can be extracted into ether.

SANGER (*323*) has shown how the identification of a series of DNP-peptides, isolated from partial acid hydrolysates of a DNP-protein, can be used to deduce the *amino acid sequence* around residues bearing a free amino group. This method has been extensively used to determine N-terminal sequences in proteins; and the early results were discussed by SCHROEDER (*343*). DESNUELLE and FABRE (*107*) have recently reported the N-terminal sequence ileu-val-gly in trypsin, and val-(asp)-lys in trypsinogen.

Concentrated hydrochloric acid at 37° for long periods gives better yields of the higher DNP-peptides in some cases [cf. THOMPSON (*380*)]. However, with DNP-glycylpeptides, SANGER (*323*) obtained better yields using 0.1 *N*-HCl at 100° and extracting the shorter DNP-peptides periodically to avoid their further hydrolysis. ANFINSEN et al. (*15*) have found the N-terminal sequence lys-glu-thr-ala for ribonuclease, although it was necessary to use enzymatic hydrolysis of DNP-ribonuclease to obtain evidence for the sequences beyond the glutamic acid residue. This constitutes a limitation of the method involving partial acid hydrolysis of a DNP-protein, since, if an acid-labile peptide bond involving serine, threonine or tryptophan is close to the N-terminal residue, few or possibly no DNP-peptides will be obtained. Moreover, DNP-peptides containing four or more residues are difficult to work with, since there is probably less of them, they are difficult to extract, and are less soluble in the solvents used for chromatography. Hydrolysis with proteolytic enzymes may be applied to DNP-proteins if the latter are sufficiently soluble.

ANFINSEN and coworkers (*15*) have hydrolysed DNP-ribonuclease and MONIER and JUTISZ (*252*) DNP-salmine with trypsin but DNP-papain was not split [THOMPSON (*380*)].

By coupling the protein with DNFB in aqueous solution, water-soluble DNP-proteins are obtained [ŠORM, KÖRBL, and MATOUŠEK (*361*), LEVY (*223*), FRAENKEL-CONRAT and SINGER (*141*)]. Dinitrobenzene sulphonic acid has been used by EISEN, BELMAN, and CARSTEN (*120*). Oxidation of the protein or DNP-protein with performic acid, to convert the disulphide bonds to sulphonic acid groups, will increase the solubility.

DNP-peptide mixtures from partial hydrolysates may also be fractionated on papergrams. With each additional amino acid residue, the

DNP-peptide will usually have a lower R_f value; and it is preferable to choose a system in which the DNP-amino acid has a high R_f. Columns have been mostly used for this purpose and can fractionate large amounts of DNP-peptides; however, a band on filter paper will give sufficient DNP-peptide for further identification. For some peptides, partial hydrolysis of the DNP-derivative yields a series of smaller DNP-peptides which are not extractable into organic solvent. Chromatography of such a mixture on paper is unsatisfactory, as overlapping with the amino acids and peptides also present occurs. By adsorption of the DNP-compounds on talc [SANGER (*323*)], or 1 : 1 Celite-talc (SCHROEDER, KAY, LEGETTE, HONNEN, and GREEN (*345*)], the amino acids and peptides can be washed away, the DNP-derivatives eluted and then chromatographed [THOMPSON (*377*)]. Methods of determining the molar ratio, amino acids: DNP-peptides contribute to the identification of DNP-peptides and help in the detection of impurities or artifacts [cf. SCHROEDER and coworkers (*345*)]. Duplicate zones of DNP-peptides isolated from partial hydrolysates are sometimes obtained due to incomplete hydrolysis of amide groups.

b) Isothiocyanate Method.

The method of EDMAN (*117*) is receiving increasing use in the determination of N-terminal sequences in peptides and proteins. It has the advantage that the derivatives of proline and glycine are as stable as those of the other amino acids. For detailed formulas cf. SCHROEDER (*343*). The coupling of phenylisothiocyanate with amino or α-imino groups proceeds smoothly at p_H 8–9 and is usually carried out in 1 : 1 pyridine-water [EDMAN (*117*)] or dioxan-water [OTTESEN and WOLLENBERGER (*270*)] at 40°. The desired p_H is maintained by addition of small amounts of sodium hydroxide or triethylamine. By measuring the alkali consumption as a function of time, the end point of the reaction is readily determined, and useful information gathered regarding the number of reacting amino groups [(*270*); CHRISTENSEN (*68*)].

After the reaction is completed, it is necessary to extract exhaustively with benzene or organic solvents to remove excess reagent and material absorbing strongly in the region 260–280 mμ used for estimation of the phenylthiohydantoins [FRAENKEL-CONRAT (*136*)]. Unless this step is carried out carefully, the yield of phenylthiohydantoins will be erroneously high.

For the *quantitative* estimation of N-terminal residues, the phenylthiocarbamyl (PTC) derivative has usually been cyclised with 1 *N*-hydrochloric acid at 100° for 1–2 hours [LANDMANN, DRAKE, and DILLAHA (*212*), FRAENKEL-CONRAT and SINGER (*141*), ROVERY, FABRE, and DESNUELLE (*313*)] and the PTH-derivatives subsequently extracted with ether or ethyl acetate. All thiohydantoin derivatives show an absorption maximum between 267 and 272 mμ, with the exception of proline which has a maximum closer to 273 mμ (*141*). The recoveries of phenylthiohydantoins calculated from a molar extinction coefficient of 16000 are usually about 80% [Fraenkel-Conrat and Fraenkel-Conrat (*138*)]. Under these cyclisation conditions the thiohydantoins of serine, threonine and cysteine are destroyed [Edman (*117*), Christensen (*68*)] and those of histidine and arginine are not extracted from aqueous solution. PTH-lysine was non-extractable from aqueous solution when coupling with phenylisothiocyanate had been carried out at p_H 8 [Christensen (*68*)]. Schramm and Schneider (*341*) have recently reported the formation of thiohydantoins of glycine and alanine, when they were not N-terminal, from both peptides and proteins, due to coupling of the isothiocyanate with their -NH- groups.

Fraenkel-Conrat and Singer (*141*) have shown that some PTH-amino acids are split off rapidly from PTC-peptides in *N*-HCl at 100° and undergo subsequent decomposition, although protein had a stabilising effect. Dahlerup-Petersen, Linderström-Lang and Ottesen (*92*, *93*) reported excellent recoveries from PTC-peptides, using continuous extraction of the phenylthiohydantoin as it is formed in aqueous buffer solutions. Turba (*393*) also achieved high yields using an ion-exchange resin to promote the cyclisation of small, soluble PTC-peptides.

Ingram (*177*) has described thiohydantoins of serine and threonine prepared by cyclisation at p_H 1 and room temperature. Difficulties have been encountered with the derivatives of these amino acids and cysteine due to the formation of dehydro derivatives and polymers. The dehydrothreonine derivative can be recognised by a characteristic absorption maximum at 320 mμ [cf. Fraenkel-Conrat (*136*)]. Landmann, Drake and Dillaha (*212*) obtained reproducible derivatives from serine and corticotropin A which has N-terminal serine, although the chemical nature of these derivatives was not definitely established. Fraenkel-Conrat and Harris (*139*) use 3 *N*-HCl at room temperature for PTC-seryl peptides; then the PTH-serine is stable, whereas in 0.1–0.5*N*-HCl it loses water to form a polymer.

Landmann, Drake, and Dillaha (*212*), and Sjöquist (*350*) have recently described paper chromatographic systems for the separation of the phenylthiohydantoins so that they may be *identified* directly.

This is particularly useful since some PTH-derivatives are destroyed or give rise to artifacts during regeneration to the amino acid. Edman (*117*) used barium hydroxide at 140° for 48 hours to recover the amino acids. Those unstable to alkali have been discussed previously (p. 293). Levy (*223a*) used 6 *N*-HCl at 150° for 16 hours for regeneration of amino acids [cf. Fraenkel-Conrat (*136*), Rovery, Fabre and Desnuelle (*313*), Ingram (*177*), Harris and Li (*163*)] but tryptophan, arginine, serine, threonine, cysteine and cystine were destroyed or only partially regenerated. The yields of other amino acids were quantitative.

The methods for revealing the phenylthiohydantoins on papergrams are not completely satisfactory. Although sensitive, they are not specific; and contamination from other sources, *e. g.* rubber, must be avoided [Sjöquist (*350*), Landmann, Drake and Dillaha (*212*), Acher and Chauvet (*1*)]. The use by Reith and Waldron (*303*) of 3:5-dinitro-4-dimethylamino-phenylisothiocyanate which gives coloured PTH-derivatives may overcome this difficulty.

The EDMAN procedure was developed for *stepwise degradation* of peptides. The cyclisation step was effected with reagents which were unlikely to cause splitting of peptide bonds. EDMAN (*117*) used nitromethane saturated with HCl for this purpose and found that the cyclisation was almost instantaneous at room temperature. However, many PTC-peptides have only a limited solubility in this reagent. Nitromethane-HCl was also used by DU VIGNEAUD and coworkers in their structural investigations on oxytocin and vasopressin. POPENOE and DU VIGNEAUD (*286*) measured the amino acid composition of the residue after successive treatments but there was no reduction in the expected N-terminal amino acid during the third and fourth degradation. Poor quantitative results were also obtained by DU VIGNEAUD, RESSLER, and TRIPPETT (*115*).

Dioxane-HCl was used by FOX, HURST, and ITSCHNER (*133*), and LANDMANN, DRAKE, and DILLAHA (*212*) in investigations of N-terminal sequences in PTC-peptides and proteins. FOX et al. used a quantitative substractive method, and have reported no difficulties with disappearance of the expected N-terminal residues [DE FONTAINE and FOX (*99*), FOX, HURST, and WARNER (*134*)]. LANDMANN and coworkers (*212*) have identified the liberated phenylthiohydantoin on papergrams, however, no quantitative data have been recorded. To overcome the limited solubility of higher PTC-peptides, EDMAN (*118*, *119*) has recently proposed acetic acid-HCl mixtures.

It is, of course, an essential requirement with a stepwise procedure that high yields are obtained at each step, and that the cyclisation reagent does not split peptide bonds. THOMPSON (*381*) has used many of these anhydrous reagents on PTC-carboxypeptidase and PTC-serum albumins. Poor yields of the thiohydantoins were obtained, although the penultimate amino acid residue was exposed in good yield, as judged by the DNP-technique. Some splitting of bonds involving the amino groups of serine and threonine was evident, although the reagents used are not suitable for obtaining good yields in the acyl shift reaction which is possible with these residues [cf. ELLIOTT (*122*)]. LANDMANN, DRAKE, and DILLAHA (*212*) claimed serine as the fifth residue in lysozyme; ACHER, LAURILA, THAUREAUX, and FROMAGEOT (*7*) obtained evidence that arginine was, in fact, the fifth residue. After successive treatments, the residue was coupled with DNFB. DNP-arginine was extracted from the hydrolysate of the DNP-protein but smaller amounts of DNP-serine, DNP-threonine and DNP-aspartic acid were also found. These results suggest that LANDMANN and coworkers detected serine which had been formed by non-specific cleavages, and did not extract the phenylthiohydantoin derivative of arginine.

ACHER and CHAUVET (*1*) used acetic acid – 11.2 N-HCl (5 : 1), for $1^1/_2$ hours at 38°, for cyclisation of PTC-oxidised vasopressin. The treatment was carried through three steps to give the sequence, cySO_3H-tyr-phe, and, although yields were not reported, the authors thought that 5 μmoles would be sufficient to identify successively four or five residues.

To overcome the limited solubility of higher PTC-peptides, OTTESEN and WOLLENBERGER (*270*) used 0.1 N-HCl at 75° for the cyclisation. Good yields of the PTH-derivatives were obtained. With the hexapeptide ala-gly-val-asp-ala-ala, the first cyclisation required $1^1/_2$ hours, the second 14 hours, and during this time, appreciable hydrolysis of the rest of the molecule occurred. This method was applied to insulin in the presence of guanidine hydrochloride by CHRISTENSEN (*68*) and, although the expected products were obtained through five successive degradations, some other amino acids were also present, suggesting scission of other peptide bonds, probably those involving the amino groups of the hydroxyamino acids. Guanidine hydrochloride was introduced by FRAENKEL-CONRAT and FRAENKEL-CONRAT (*138*) to keep the otherwise insoluble PTC-insulin in solution. This insolubility of PTC-proteins means that procedures suitable for PTC-peptides are not necessarily applicable to proteins. FRAENKEL-CONRAT (*137*) has overcome this difficulty in an ingenious way by carrying out all reactions on a paper strip, so that a large surface of protein was accessible to the reagents.

The cyclisation is effected in a moist atmosphere of acetic acid and hydrogen chloride at room temperature for 4–16 hours, and all the PTH-derivatives are extractable from the paper, although half-cystine-PTH, when formed, will remain attached to the peptide chain. The method has been successfully applied to many proteins and to oxytocin, for most of which the N-terminal sequence was already known. Some other amino acids have been detected in addition to the expected ones, but data are not yet available regarding the types of bond split or their quantitative significance.

The method has been applied by HARRIS and LI (*163*) to the N-terminal sequence of sheep *α-corticotropin* (39 residues). The heptapeptide sequence ser-tyr-ser-met-glu-his-phe- has been obtained with yields of PTH-amino acids from 0.58 to 0.85 moles. LANDMANN, DRAKE, and WHITE (*213*) had determined a similar sequence ser-tyr- in hog corticotropin-A, but the third step was equivocal. There is apparently a progressive decrease in yields during successive treatments in the paper strip procedure, so that there has been an upper limit of 5–7 residues detectable in insulin, oxytocin and α-corticotropin, after which the chromatograms of the PTH-derivatives became very complex. The method has promise, however, and may supply a long-felt need for the identification of N-terminal sequences in long-chain polypeptides. Only micro amounts of polypeptide, of the order of 0.5 micromoles, are required.

The EDMAN technique is particularly suited for the location of *asparaginyl* and *glutaminyl* residues situated near the N-terminal residue, since the mild conditions used for cyclisation of the PTC-derivative will not completely deamidate amide groups. Thus, ROVERY and DESNUELLE

(*310*), and THOMPSON (*381*) have independantly shown that the N-terminal residues of serum albumins are actually aspartyl rather than asparaginyl groups. The location of these aspartyl groups assumes that amidised residues would have a stability similar to the free amides. This was the case with carboxypeptidase [THOMPSON (*379*)].

FRAENKEL-CONRAT (*137*) has examined the A and B fractions of oxidised insulin with his paper strip procedure. The results on the B chain were in agreement with those of SANGER, THOMPSON, and KITAI (*332*) [cf. SANGER, THOMPSON, and TUPPY (*333*)] but for the fourth residue of the A chain, a glutaminyl residue was found, compared with the glutamyl residue of these workers.

c) *Other Methods.*

The ***deamination*** method of CONSDEN, GORDON, and MARTIN (*79*) was used to identify many of the peptides in partial hydrolysates of gramicidin S [CONSDEN, GORDON, MARTIN, and SYNGE (*83*)] and wool [CONSDEN, GORDON, and MARTIN (*81*, *76*)]. The N-terminal amino acid fails to appear in a hydrolysate of the degraded peptide. More than one amino acid may disappear after deamination, and it is often uncertain whether this indicates too severe reaction conditions, or contamination with another peptide or amino acid. If the amino acid which is N-terminal occurs again in a peptide, a decrease in intensity is observed which is not conclusive. SANGER and TUPPY (*334*), SANGER and THOMPSON (*328*), and ACHER, CHAUVET, CROCKER, LAURILA, THAUREAUX, and FROMAGEOT (*2*) have also used the method for short peptides with reasonably satisfactory results, although peptides containing tyrosine, arginine or proline created difficulties. The DNP and isothiocyanate methods require more material but the results are much more conclusive.

INGRAM (*176*) has used ***reductive alkylation*** as an end group determination method for peptides. In the presence of formaldehyde and a palladium-charcoal catalyst, hydrogenation of the peptides gave N-dimethyl-derivatives. The dimethylamino acids in hydrolysates were identified on papergrams by indicator solutions or acid-orcinol [PORATH and FLODIN (*289*)]. The amino acids unaffected by this reagent were identified subsequently on the same chromatogram. This reaction took place under mild conditions but required from 0.2–1.0 μmoles of dimethylamino acids although recently KIESSLING and PORATH (*203*) developed a method, sensitive to 5–10 μg., depending on the detection of iodide ions formed by reaction with methyl iodide.

The ***pipsyl*** method (see p. 297) has been adapted as a method of amino end-group analysis by UDENFRIEND and VELICK (*398*). The pipsylamino acids are markedly resistant to acid hydrolysis, and, in the case of proline and glycine, are more stable than the corresponding DNP-derivatives.

The method was applied to several proteins, for some of which comparative data from the DNP technique were available. Whereas the two methods were in excellent agreement as regards ε-amino groups, the yields of N-terminal amino acids were much lower. For horse hemoglobin, two N-terminal valine residues were found, compared with the six reported by PORTER and SANGER (*293*), and one residue each of N-terminal glycine and phenylalanine per 12000 g. of insulin, where SANGER (*321*) found two of each. That this discrepancy is due to incomplete substitution of the N-terminal amino groups is suggested by the finding of a molecular

$$NH_2\text{—}\underset{|}{\overset{R_1}{CH}}\text{—}CO\text{—}NH\text{—}\overset{R_2}{CH}\text{—}CO\ldots + CS_2 + OH^- \qquad (A)$$

$$\Big\downarrow H_2O$$

$$S{=}\underset{S^-}{C}\text{—}NH\text{—}\overset{R_1}{CH}\text{—}CO\text{—}NH\text{—}\overset{R_2}{CH}\text{—}CO\ldots$$

$$\Big\downarrow H^+$$

$$\begin{array}{ccc} HN & \text{——} & \overset{R_1}{CH} \\ | & & | \\ SC & & CO \\ & \diagdown S \diagup & \end{array} + NH_2\text{—}\overset{R_2}{CH}\text{—}CO\ldots$$

$$CH_3O\text{—}\underset{\overset{\|}{S}}{C}\text{—}SCH_3 + NH_2\text{—}\overset{R_1}{CH}\text{—}CO\text{—}NH\text{—}\overset{R_2}{CH}\text{—}CO\ldots \qquad (B)$$

$$\text{PH 8}\Big\downarrow -CH_3SH$$

$$CH_3O\text{—}\underset{\overset{\|}{S}}{C}\text{—}NH\text{—}\overset{R_1}{CH}\text{—}CO\text{—}NH\text{—}\overset{R_2}{CH}\text{—}CO\ldots$$

$$\Big\downarrow CH_3NO_2\text{—}HCl$$

$$^+NH_3\text{—}\overset{R_1}{CH}\text{—}COO^- \xleftarrow{OH^-,\ \text{then}\ H^+} \begin{array}{ccc} HN & \text{——} & \overset{R_1}{CH} \\ | & & | \\ OC & & CO \\ & \diagdown S \diagup & \end{array} + NH_2\text{—}\overset{R_1}{CH}\text{—}CO\ldots$$

weight of 6000 for insulin [HARFENIST and CRAIG (*159*)] and by the low recoveries of pipsylated peptides obtained by LEVY and SLOBODIAN (*225*).

A micro method on paper strips has been investigated by LEONIS (*221*) involving the reaction of amino groups with CS_2 in alkaline solution, with subsequent cyclisation to the *2-thio-5-thiazalidone* of the terminal amino acid (*see the formulae A on p. 313*).

Unfortunately, side reactions take place during the cyclisation and the original peptide is reformed [cf. LINDERSTRÖM-LANG (*227*)]. The chromatograms of the successive terminal residues become increasingly complex and the method is therefore limited to short peptides.

KENNER and KHORANA (*199*) have used *dialkylxanthates* (*p. 313, B*) and ELMORE and TOSELAND (*123*) *methyl acyl dithiocarbamates* to couple smoothly with amino groups of peptides at room temperature. Subsequent treatment with acidic reagents causes cyclisation and selective removal of the N-terminal residue, which may be extracted, and the amino acid regenerated.

$$R_3\text{—CO—NH—}\underset{\underset{\displaystyle S}{\|}}{C}\text{—SCH}_3 + \text{NH}_2\text{—}\overset{\overset{\displaystyle R_1}{|}}{\text{CH}}\text{—CO—NH—}\overset{\overset{\displaystyle R_2}{|}}{\text{CH}}\text{—CO}\ldots$$

$$\big\downarrow \text{PH 8} \quad -\text{CH}_3\text{SH}$$

$$R_3\text{—CO—NH—}\underset{\underset{\displaystyle S}{\|}}{C}\text{—NH—}\overset{\overset{\displaystyle R_1}{|}}{\text{CH}}\text{—CO—NH—}\overset{\overset{\displaystyle R_2}{|}}{\text{CH}}\text{—CO}\ldots$$

$$\big\downarrow \text{H}^+$$

$$\begin{array}{ccc} & & R_1 \\ & & | \\ \text{NH} & \text{———} & \text{CH} \\ | & & | \\ \text{S=C} & & \text{C=O} \\ & \diagdown \quad \diagup & \\ & \text{NH} & \end{array}$$

These derivatives are slightly more soluble than the PTC-peptides obtained in the EDMAN method. The same reagents have been used for the cyclisation but only macro amounts of simple peptides have been investigated so far.

HOLLEY and HOLLEY (*173*) have coupled *1-fluoro-2-nitro-4-carbomethoxybenzene* with the amino group. Reduction of the nitro group allows easy cyclisation, under mild conditions, of the amino compound to the lactam of the 4-carbomethoxy-2-aminophenyl N-terminal amino acid.

So far, these derivatives have only been identified by their melting points, but chromatographic methods on a smaller scale are no doubt possible. This reaction sequence employs the mildest conditions of any stepwise procedure yet described, and it will be interesting to see if it can be adapted to a micro scale.

$$CH_3O{-}\underset{\displaystyle O}{\overset{\|}{C}}{-}C_6H_3(NO_2){-}F + NH_2{-}\underset{R_1}{CH}{-}CO{-}NH{-}\underset{R_2}{CH}{-}CO\ldots + OH^-$$

$$\downarrow \text{PH } 8\text{--}9$$

$$CH_3O{-}\underset{\displaystyle O}{\overset{\|}{C}}{-}C_6H_3(NO_2){-}NH{-}\underset{R_1}{CH}{-}CO{-}NH{-}\underset{R_2}{CH}{-}CO\ldots + F^- + H_2O$$

$$\downarrow H_2,\ Pt$$

$$CH_3O{-}\underset{\displaystyle O}{\overset{\|}{C}}{-}C_6H_3(NH_2){-}NH{-}\underset{R_1}{CH}{-}CO{-}NH{-}\underset{R_2}{CH}{-}CO\ldots$$

$$H_2O \downarrow \begin{array}{l}5\text{ hr. }25^\circ\\ 15\text{ min. }70^\circ\end{array}$$

$$CH_3O{-}\underset{\displaystyle O}{\overset{\|}{C}}{-}C_6H_3\begin{array}{l}{-}NH\\{-}NH{-}CO\end{array}\!\!>\!CH{-}R_1 + NH_2{-}\underset{R_2}{CH}{-}CO\ldots$$

The method of WESSELY, SCHLÖGL, and KORGER (*411*), involving splitting out the first two amino acid residues in a peptide in the form of a *thiohydantoin* derivative, has been applied to a variety of small peptides [cf. WESSELY et al. (*412*)], and recently to lysozyme and synthetic peptides with sequences similar to the known N-terminal sequence of this protein [SCHLÖGL and FABITSCHOWITZ (*337*)].

2. C-Terminal Sequences.

a) Carboxypeptidase Method.

The pancreatic enzyme carboxypeptidase has so far given most information regarding C-terminal residues in peptides and proteins. The C-terminal amino acid is split off the peptide chain, but is contaminated with other amino acids which become C-terminal and are then, in turn, cleaved. In suitable cases they may indicate the sequence of amino acids. Methods for isolating the amino acids free of protein have been discussed (p. 278). SCHROEDER (*343*) has reviewed many of the results obtained with the carboxypeptidase method, mostly referring to paper chromatographic procedures. Special precautions are necessary to ensure that the enzyme is not *contaminated* with amino acids, peptides or other enzymes. SANGER and THOMPSON (*329*) as well as GLADNER and NEURATH (*150*) emphasised the importance of inhibiting any contaminating chymotrypsin and trypsin by previous incubation of the carboxypeptidase with di-isopropylfluorophosphate (DFP) [JANSEN, FELLOWS-NUTTING, JANG, and BALLS (*183*)], which has no effect on carboxypeptidase. Otherwise internal peptide bonds would be split, followed by liberation of many amino acids by the carboxypeptidase. STEINBERG (*365*) has encountered difficulties of this type with *ovalbumin* (which has no C-terminal residue) [cf. LOCKER (*229*)], due to a contaminating *Bac. subtilis* enzyme which is also inhibited by DFP.

Carboxypeptidase is more stable when stored as a suspension in water [ANSON (*16*)], and contaminating amino acids may then be washed away. For enzymatic work, at p_H 7.5–8, the enzyme should be dissolved, *e. g.* in 10% lithium chloride and treated with DFP. The final concentration of salt when added to the substrate should be 0.5% to avoid precipitation of the enzyme [GLADNER and NEURATH (*150*)]. Moreover, LUMRY, SMITH, and GLANTZ (*234*) found 0.3 or higher ionic strength necessary for maximal activity.

It is also essential to use substrates free of adventitious impurities. For enzyme substrates, crystallisation and dialysis may not be adequate to remove peptides or amino acids firmly bound to the active centre [cf. DESNUELLE, ROVERY, and FABRE (*110*)], or released by autolysis, unless an inhibited enzyme preparation, *e. g.* DFP-chymotrypsin, is used. Denatured proteins usually react more readily, provided they remain soluble [cf. DAVIE and NEURATH (*96*)]. THOMPSON (*374, 376*) found carboxypeptidase useful for the study of peptides isolated from papergrams, and for which the DNP-method gave evidence of purity. DNP-peptides, purified if necessary, may also be used as substrates [WALDSCHMIDT-LEITZ and GAUSS (*404*)]. JOLLÈS and FROMAGEOT (*189*) obtained larger amounts of C-terminal leucine from rabbit spleen lysozyme when the insoluble DNP-protein was hydrolysed.

Reaction with carboxypeptidase may be stopped at appropriate intervals by adjusting to p_H 3, or by adding sulphonated polystyrene, which is also applied for the separation of the amino acids from salt and protein. Paper chromatography has invariably been used for identification of the liberated amino acids.

Although native carboxypeptidase does not undergo appreciable *autolysis* at p_H 7.5 and 25° [DAVIE and NEURATH (*96*)], denatured carboxypeptidase is attacked by the active enzyme, releasing a mixture of amino acids [THOMPSON (*383*)], including leucine, valine, lysine, serine, threonine, asparagine, glutamic acid, tyrosine and methionine. This will only become appreciable during repeated treatment of a partially degraded protein substrate contaminated with denatured carboxypeptidase. Treatment of the degraded protein has recently been reported by LOCKER (*229*), GLADNER and NEURATH (*151*), and HARRIS, LI, CONDLIFFE, and PON (*164*) to gain information regarding the sequence of amino acids near the C-terminal residue. Although GLADNER and NEURATH (*150*) found the enzyme fully active after 6 hours in the presence of substrate, multiple treatment was necessary, because non-stoichiometric quantities of amino acids are otherwise liberated, due to substrate or product inhibition. GLADNER and NEURATH (*151*) showed that additions of enzyme only result in further hydrolysis if the products of hydrolysis are removed.

Quantitative analyses are essential to gain insight into the rate of liberation of amino acids, necessary for the deduction of sequences; and a variety of pertinent paper chromatographic methods have been proposed. The different rates of liberation of C-terminal residues [SMITH (*352*)] introduce difficulties in the interpretation when two rapidly liberated residues, *e. g.* phenylalanine and leucine, are adjoining, particularly if the presence of more than one polypeptide chain is suggested by other evidence. Difficulties are caused by resistant amino acids, *e. g.* proline, hydroxyproline, cysteic acid or an amide group, or by residues which are liberated more slowly than expected, due to adjacent proline or acidic amino acids; *e. g.*, the C-terminal asparagine in the glycyl chain of oxidised insulin is preceded by a cysteic acid residue and is liberated slowly by carboxypeptidase. It is set free more readily from insulin but only acetyl insulin gave stoichiometric amounts [SANGER and THOMPSON (*329*)]. Half-cystine residues present in a chain may be released from peptide linkages but remain attached to the polypeptide chain by the disulphide bond.

Other results obtained with this procedure since the review of SCHROEDER (*343*) had appeared include the C-terminal sequences -ala-ileu-met-thr-ser-ileu for ***tropomyosin***, -his-ileu-phe for ***actin***, and one isoleucine residue / 300000 part. wt. of ***myosin***, with significant amounts of alanine, valine and leucine [LOCKER (*229*)]. Isoleucine and phenylalanine were liberated at the same rate from actin, but hydrazinolysis confirmed that phenylalanine was the C-terminal residue. Interpretation difficulties were also encountered by HARRIS, LI, CONDLIFFE, and PON (*164*) with *hypophyseal growth hormone*, which had previously been shown to possess two N-terminal residues. Two moles of phenylalanine were rapidly liberated followed by alanine, leucine and serine. Many possible structures could satisfy these data. This problem emphasises the desirability of the presence of single chains for sequence studies and the need for alternative techniques.

ANFINSEN, REDFIELD, CHOATE, PAGE, and CARROL (*15*) found a C-terminal sequence -met-tyr-ala-(leu or ileu, phe)-val for ***ribonuclease***, and BISERTE and DAUTREVAUX (*34*) the sequence, -gly·leu·ala for ***horse serum albumin***. NEURATH, GLADNER, and DAVIE (*264*) found a total of only two equivalents of isoleucine and histidine per molecule of ***β-lactoglobulin*** which has three N-terminal leucine residues [PORTER (*290*)]. According to NEURATH et al. *horse* and *swine insulins* have similar C-terminal groups to beef insulin. GLADNER and NEURATH (*151*) have established that glycine, serine and leucine are adjacent to the C-terminal leucine and tyrosine residues in ***α-chymotrypsin***.

The interpretation of the results in this case will be aided by an examination of δ-chymotrypsin, recently crystallised as the DIP-derivative

by ROVERY, POILROUX, and DESNUELLE (*314*). Preliminary data indicate that it has an N-terminal isoleucine and a C-terminal leucine residue. The absence of a C-terminal basic amino acid, and the well known specificity of trypsin suggest that the activation of chymotrypsinogen results in the splitting of a bond, involving the carboxyl group of a basic amino acid and an isoleucine residue, to give unstable π-chymotrypsin. Autolysis of π-chymotrypsin liberates a peptide, presumably with the C-terminal basic residue [GLADNER and NEURATH (*150*)], to form δ-chymotrypsin. Further autolysis of δ-chymotrypsin results in the splitting of a –try · ala– bond and in the formation of α-chymotrypsin.

The crystalline β- and γ-chymotrypsins described by KUNITZ (*210*) are formed under conditions favouring autolysis of α-chymotrypsin. The N-terminal residues have been shown to be identical [ROVERY, FABRE, and DESNUELLE (*312*)], and both contain the same C-terminal residues as α-chymotrypsin, but these are liberated more rapidly than is the case for the α compound. The pattern of amino acids subsequently liberated was similar for the β- and γ-forms, but different from that of α-chymotrypsin [GLADNER and NEURATH (*151*)], suggesting that the β- and γ-forms are different crystalline modifications of otherwise identical proteins. The relationship between the β- and γ-chymotrypsins and the parent α-form remains an intriguing one.

HARRIS and LI (*163*) have determined the C-terminal sequence -leu-glu-phe in sheep *α-corticotropin*, identical with that in hog corticotropin-A [WHITE (*414*)]. A resistant group, proline, has been identified as the next residue for both these polypeptides.

There is agreement between VAN VUNAKIS and HERRIOT (*401*), BISERTE and DAUTREVAUX (*34*) and WILLIAMSON and PASSMANN (*421*) regarding the C-terminal residue of pepsin, viz. -ala-(phe, val)-leu-ala (*34*) and -val-leu or ileu-ala (*401*); however WILLIAMSON and PASSMANN found 2 residues of alanine per mole. The different findings regarding the N-terminal sequence of pepsin have been discussed on p. 304. These differences may be due to the several different pepsins and pepsinogens known to exist, and the peculiar p_H stability range of pepsin and pepsinogen [cf. HERRIOT (*167*)].

WHITE (*415*) and WHITE and LANDMANN (*416*) have used carboxypeptidase on peptides originating from partial hydrolysates of corticotropin A isolated on papergrams. With the high enzyme concentration of 1% at 37° some dipeptides are split appreciably, in contrast to earlier studies, with relatively minute enzyme concentrations, on the influence of polar groups on synthetic substrates [SMITH (*352*)]. The dipeptides hydrolysed include arg-tyr, leu-tyr, ser-tyr, and ala-phe whereas glu-phe and leu-ala were not split appreciably. HARRIS (*162*) has found that ser-tyr and pro-phe are hydrolysed by carboxypeptidase.

b) Reduction Methods.

The methods of FROMAGEOT et al. (*143*), and CHIBNALL and REES (*64, 65, 66*) have recently been discussed by JUTISZ (*193*), and CHIBNALL (*63*). It is now clear that, in the case of *insulin*, difficulties have been encountered due to the labile -gly-ser- bond in the phenylalanyl chain, and care must be taken in assessing results on other proteins. Applications subsequent to those reviewed by SCHROEDER (*343*) include the confirmation of a C-terminal leucine residue in *lysozyme* by PÉNASSE, JUTISZ and FROMAGEOT (*279*). The reduction with $LiAlH_4$ was only possible with esterified lysozyme, in contrast to the reduction of insulin and ovomucoid. This may be due to a penultimate basic residue, since amino acids and dipeptides can only be reduced after esterification. A yield of 0.6 mole leucinol was obtained. WILLIAMSON and PASSMANN (*421*) have reported the reduction of *pepsin* with $LiAlH_4$ to give 2 residues of alaninol, in agreement with the results obtained with carboxypeptidase.

JUTISZ, PRIVAT DE GARILHE, SUQUET, and FROMAGEOT (*194*), GRASSMANN, HÖRMANN, and ENDRES (*154*), and JATZKEWITZ and TAM (*184*) have coupled the amino alcohols with DNFB to give the DNP-derivatives, which have been separated with normal and reversed phase systems similar to those used for the DNP-amino acids. The separation and estimation of the DNP-amino alcohols is simpler and more precise than of the amino alcohols themselves, which may also be separated on papergrams with a variety of solvent mixtures [cf. FROMAGEOT and JUTISZ (*142*)]. JUTISZ and coworkers (*194*) have studied the reduction of short acylated peptide esters with $LiAlH_4$ at 0° from 10 to 120 minutes. Yields, estimated as DNP-amino alcohols, varied from 41 to 74% for approximately 50 mg. of material. Less than 3% of peptide bond cleavage was detected. Using these DNP-procedures, it may be possible to examine partial hydrolysates of reduced proteins, and, in suitable cases, to determine C-terminal sequences.

LEGGETT BAILEY (*219*) has had some success in *stepwise degradation* of simple peptides after reduction of the esterified carboxyl group with aluminium hydride. An acyl shift with phosphorus halides was followed by reduction with aluminium hydride to release the terminal amino alcohol and leave the next amino acid residue in the reduced form necessary for further rearrangement. Yields of about 80% have been achieved [cf. ELLIOTT (*122*)].

c) Other Methods.

There have been few attempts made to use the *thiohydantoin* method since SCHROEDER's review (*343*).

TURNER and SCHMERZLER (*395*) have reinvestigated insulin and confirmed the formation of small amounts of aspartic acid as well as alanine after regeneration from the thiohydantoins [WALEY and WATSON (*405*)]. Aspartic and glutamic acids do not form thiohydantoins [SWAN (*366*)], and the detection of aspartic acid is therefore indicative of terminal asparagine. The yields obtained have not been given (*395*) but the aspartic acid spot was weak compared to the alanine spot. The lack of yields is unfortunate, for this reaction is clearly not suitable for many amino acids on this basis. The work of BAPTIST and BULL (*26*) suggests the method is unsuited to C-terminal arginine, lysine, aspartic acid, glutamic acid, serine, and threonine. Proline does not form a thiohydantoin and, consequently, this method cannot be applied even qualitatively to proteins or peptides resistant to carboxypeptidase, suggesting a C-terminal proline group.

Hydrazinolysis, introduced by AKABORI, OHNO, and NARITA (*10*), was reviewed by SCHROEDER (*343*), and further applications have been reported. The reaction does not proceed smoothly, since some amino acids which are not C-terminal have been detected. AKABORI et al. (*9, 11*) have reported carboxyl end groups in ovalbumin, chymotrypsinogen, and α-chymotrypsin, which are not in agreement with the results obtained with carboxypeptidase. LOCKER (*229*) has observed complete destruction of arginine, cysteine, cystine and asparagine after heating for 8 hours at 100° with hydrazine. Other amino acids were recovered in yields exceeding 85%, except cysteic acid (55%), aspartic acid (20%), glutamic acid (40%), and tryptophan (40%). In experiments with tropomyosin, myosin and actin, papergrams revealed a number of amino acids, particularly serine, glutamic acid and alanine, although the expected C-terminal residues from actin and tropomyosin were obtained in larger amounts (25%).

BLACKBURN and LEE (*36*) isolated glycine, alanine, serine and threonine from hydrazinolysed *wool*, using Amberlite IRC-50 to separate the basic hydrazides. This carboxylic resin was also applied by LOCKER (*229*) who showed that arginine, lysine, histidine, and probably tryptophan, were retained on the column.

BUMPUS and PAGE (*57*) carried out hydrazinolysis to detect a leucine or isoleucine residue in *angiotonin*, and SCHLÖGL and WAWERSICH (*338*) have used the method on short-chain peptides. OHNO (*268, 269*) stressed the importance of the purity of hydrazine, and has obtained yields of leucine from *lysozyme* equivalent to 0.9 moles, in 10 hours at 100°, after separation as DNP-leucine from DNP-hydrazides. He has extended this method to estimate the asparaginyl and glutaminyl residues in lysozyme. The total dicarboxylic residues accounted for, however, were considerably lower than the accepted values [THOMPSON (*376*)]. No evidence for glutamyl residues was obtained, which agrees with the observations of SMITH, KIMMEL, and BROWN (*355*).

BOISSONNAS (*48*) described a method suitable for short-chain acylated peptides, which involved *anodic oxidation* in anhydrous methanol. The C-terminal carboxyl group is converted to a methoxyl group and the C-terminal amino acid does not appear on a papergram of the hydrolysate. The method suffers the disadvantages of all subtractive methods, and insufficient results have been reported to assess its value. THOMPSON (*375*) has found that phenylalanine and tyrosine residues are destroyed, even when not in the C-terminal position. TURNER and SCHMERZLER (*396*) have briefly described another subtractive method for identifying the C-terminal residue after heating with pyridine and acetic anhydride at 150° for 2–3 hours. THOMPSON (*378*) has had no success with this method in trial experiments with 0.1 and 1 μmoles ala-gly.

3. Conclusion.

There are several methods available for the determination of N-terminal sequences in peptides and proteins which require only a few micromoles of material. C-terminal methods, however, with the exception of the carboxypeptidase procedure, are of doubtful value even for short peptides, if only micromoles of material are available. Carboxypeptidase is not always satisfactory because of its specificity requirements.

The examination of terminal sequences has revealed differences between related proteins of various *species*, such as haemoglobins [PORTER and SANGER (*293*)], myoglobins [KENDREW, PARRISH, MARRACK, and ORLANS (*198*)], human and bovine serum albumins [THOMPSON (*381*)], and fibrinogens [LORAND and MIDDLEBROOK (*232*)].

The use of these techniques has also thrown considerable light on the enzymatic processes involved in many of the transformations of inactive precursors to biologically active proteins, as already outlined for chymotrypsinogen.

HERRIOT (*167*) has discussed the activation of *pepsinogen*, which is a single peptide chain. A peptide (or peptides) is released from the N-terminal end during conversion to pepsin.

The *trypsinogen*-trypsin conversion has been shown to involve the liberation of a peptide val-(asp)$_{4-6}$-lys [DAVIE and NEURATH (*97*)], from the N-terminal sequence of trypsinogen, thereby exposing the ileu-val-N-terminal sequence of trypsin [DESNUELLE and FABRE (*107*)].

The conversion of *fibrinogen* to fibrin by thrombin has been shown by BAILEY, BETTELHEIM, LORAND, and MIDDLEBROOK (*23*) to involve a loss of two N-terminal glutamyl or glutaminyl residues and the appearance of four N-terminal glycine residues. The peptide material liberated during this transformation has been called "fibrino-peptide" by LORAND (*230*), but has been clearly demonstrated by BETTELHEIM and BAILEY (*33*) to consist of two peptides, one of which had N-terminal glutamic acid, as shown by hydrolysis of the DNP-peptide. The other peptide had no reactive α-amino groups and contained tyrosine in the O-sulphate form [BETTELHEIM (*32*)]. This accounts for the failure of LORAND and MIDDLEBROOK (*231*) to detect it by ultraviolet spectroscopy. Both peptides contained arginine; and SHERRY and TROLL (*347*) have reported that thrombin hydrolyses synthetic substrates containing arginine as the specific amino acid residue.

V. The Determination of Amino Acid Sequences in Long-Chain Polypeptides and Proteins.

1. Separation of Component Chains.

Where the disulphide type of cross-linking occurs in proteins, as is definitely the case with insulin, the component chains are easily released by oxidation. Performic acid, used by SANGER (*322*), is a convenient reagent for this purpose. Other oxidising agents which have been used are the water-stable peracetic acid by ALEXANDER, HUDSON, and FOX (*13*) for oxidation of insoluble substances such as wool, and bromine water [CONSDEN and GORDON (*76*)], which has recently created much interest since it resulted in the specific splitting of the peptide bonds tyrosyl-isoleucine in performic acid-oxidised oxytocin [MUELLER, PIERCE, and DU VIGNEAUD (*260*)], and tyrosyl-phenylalanine in oxidised vasopressin [POPENOE and DU VIGNEAUD (*286*)]. The same result was achieved by treating oxytocin itself in aqueous methanol at — 10° with bromine water [MUELLER, PIERCE, and DU VIGNEAUD (*260*), DU VIGNEAUD et al. (*114*)].

For most proteins, sufficient cystine is indicated by amino acid analyses to allow for disulphide bridges between the known number of open peptide chains. However, for horse hemoglobin, with 5 or 6 N-terminal residues [PORTER and SANGER (*293*), DESNUELLE, ROVERY, and BONJOUR (*108*)], and avidin, with 3 N-terminal residues [FRAENKEL-CONRAT and PORTER (*140*)], only 2–3 and 1 cystine residues, respectively, have been listed by TRISTRAM (*388*); here other types of cross-linking or branching are probably involved. Complete reduction of –S–S–linkages in proteins is more difficult to achieve than complete oxidation, and to prevent reoxidation the resultant –SH groups must be blocked by substituents, *e. g.* benzyl or acetamide groups. This approach may be useful in certain cases, since no extra ionic groups are introduced, in contrast to oxidation.

A treatment with RANEY nickel may also be used to remove sulphur from proteins. Cystine and cysteine residues will thereby be converted to alanine residues, while methionine will give α-aminobutyric acid [MOZINGO, WOLF, HARRIS, and FOLKERS (*259*), COOLEY and WOOD (*86*), TURNER, PIERCE, and DU VIGNEAUD (*394*)].

The separation of the component chains is much more difficult than for short-chain peptides [cf. SYNGE (*370*)].

2. Methods for the Partial Hydrolysis of Peptides and Proteins.

SANGER (*325*) has presented an excellent discussion of the various methods for degrading peptide chains into smaller fragments. *Non-specific* methods of breakdown, such as acid or alkaline hydrolysis, give a peptide mixture, the complexity of which gradually decreases as the more susceptible bonds are split, until finally only amino acids remain. Shortening the time of hydrolysis does not necessarily give higher yields of larger peptides, and is far more likely to give low yields of a wider variety of long-chain material. The most suitable hydrolysates for investigation on papergrams are undoubtedly those containing mainly di- and tri-peptides, which accumulate toward the end of the hydrolysis. This is due to electrostatic effects of the charged amino or carboxyl groups, and to the steric effects of certain amino acid side-chains, particularly those of isoleucine and valine residues.

In partial *acid hydrolysis*, the lability of bonds involving the amino groups of serine and threonine [cf. SANGER (*325*)] and the instability of tryptophan residues considerably reduce the number of peptides that may be formed. The lability of the linkages involving the hydroxyamino acids is due to the ease of hydrolysis of ester bonds formed in the reaction,

$$\ldots NH-\underset{R}{CH}-CO-NH-\underset{\underset{R'}{CHOH}}{CH}-CO\ldots \xrightarrow{H^+} \ldots NH-\underset{R}{CH}-C\overset{N\;-\;CH-CO\ldots}{\underset{O\;-\;CHR'}{\diagdown\quad\diagup}}$$

$$\xrightarrow{H^+,\;H_2O}\ \ldots NH-\underset{R}{CH}-CO-O-\underset{R'}{CH}-\underset{\overset{+}{N}H_3}{CH}-CO\ldots$$

Attempts to use this reaction for a specific fission of peptide chains have so far met with only limited success, since the specificity is not absolute [ELLIOTT (*121*), DESNUELLE and BONJOUR (*105*), MILLS (*248*)] and other peptide bonds are also broken. Tryptophan is likewise destroyed [ELLIOTT (*122*)].

The results of this type of degradation are of great importance to those engaged in obtaining longer peptides for structural studies; and the absence of serine and threonine residues within the peptides may be of advantage for stepwise degradative procedures on the long peptides obtained, since these residues are usually involved in non-specific cleavages. LEGGETT BAILEY (*220*) used $POCl_2F$ to cause the acyl shift from N to

O of the serine and threonine bonds in insulin, with 50% yield. More quantitative rearrangement occurred with esterified insulin.

Partial hydrolysis of polypeptides with concentrated HCl at 37° for 3–10 days is still the most useful set of conditions for obtaining a mixture of small peptides. SYNGE (*367*) used a 1 : 1 mixture of glacial acetic acid and 10 *N*-HCl for partial hydrolysis of gramicidin which was insoluble in 10 *N*-HCl. The remarkable stability of di- and tri-peptides under these conditions is evident from the work of CONSDEN, GORDON, MARTIN, and SYNGE (*83*) who used a 58-day hydrolysate for the isolation of peptides from gramicidin S. No rearrangement of peptide bonds has ever been demonstrated with concentrated HCl. PALADINI and CRAIG (*271*) used papergrams to study various conditions of partial acid hydrolysis for the degradation of tyrocidin A, and selected 6 *N*-HCl at 80° for 6 hours [cf. MONIER and FROMAGEOT (*250*)].

Alkaline hydrolysis is not often used because of the destruction of certain amino acid residues (see p. 293), which occurs even more readily if the amino acids are in peptide linkage [SANGER and TUPPY (*334*)]. However, tryptophan is more stable to alkali, and this may be useful in studying its peptides in sequence investigation. Another approach is to couple the proteins with xanthydrol which then forms acid stable xanthyl-tryptophan [DICKMAN and ASPLUND (*111*)].

Dilute acid hydrolysis has not been widely used, although it shows a different specificity f om that of concentrated acids. The interesting observation of PARTRIDGE and DAVIS (*277*) that very dilute acids cause preferential release of aspartic acid from proteins has not so far proved generally useful for specific cleavages. SANGER and THOMPSON (*331*) found by paper chromatography that, while aspartic acid was the first free amino acid to appear during hydrolysis of the glycyl and phenylalanyl chains of oxidised insulin, the bonds involving cysteic acid residues and the amino groups of serine and threonine residues were also broken; thus low yields of a large variety of peptides were obtained. From studies with model compounds, DESNUELLE and BONJOUR (*106*) concluded that serine and threonine bonds would be labile; and SCHROEDER (*342*) has reported low yields of peptides under these conditions. Moreover, all results with dilute acid hydrolysis must be interpreted with caution, since SANGER and THOMPSON (*330*) were able to demonstrate an *inversion* of peptide sequence with glycyl-valine, to form the more stable valyl-glycine, after boiling with 0.1 *N*-HCl for 24 hours. Similar results have also been reported by SCHAFFER, HARSHMAN and ENGLE (*336*) who isolated 12% glycyl-serine phosphoric acid and 57% phosphoseryl-glycine from P^{32}-labelled DIP-chymotrypsin after $2^1/_2$ hours hydrolysis with 2 *N*-HCl, at 100°. With 12 *N*-HCl at 37°, only phosphoseryl-glycine was formed, and this represents the normal dipeptide sequence in the molecule. The increased stability of the inverted peptide when the serine hydroxyl group is protected is interesting. A similar observation has been made by FLAVIN (*128*) for aspartyl-phosphoserine. Substitution of the hydroxyl group prevents the oxazoline ring formation necessary for the acyl shift and acid lability of serine and threonine bonds (cf. p. 323).

For the isolation of *cystine peptides*, special conditions of hydrolysis are required to avoid interchange reactions. SANGER (*326*) found, for example, that mixtures of bis-DNP-cystine and cystine gave mono-DNP-cystine. RYLE and SANGER (*317*) have observed that HCl more concentrated than 9 N gives the maximum interchange at 37° for this particular reaction. Sulphydryl compounds, *e. g.* cysteine, inhibited the reaction, and with 5 N-HCl rearrangement was negligible. The reaction was also slower with sulphuric acid than with hydrochloric acid. For the preparation of cystine peptides from insulin, SANGER, SMITH, and KITAI (*327*) used a mixture of sulphuric acid and acetic acid containing traces of thioglycolic acid, at 100°, for 30–60 minutes. The mechanism of the reaction in neutral solution is different from that in acid solution and the reaction is slower. It is catalysed by traces of -SH compounds, and, by adding -SH inhibitors such as *p*-chloromercuribenzoate or N-ethyl maleimide, RYLE and SANGER (*317*) prevented rearrangement.

In the absence of more suitable reagents, *proteolytic enzymes* have been used to obtain a more specific type of hydrolysis and peptides larger than those obtainable by acid cleavage. However, these results must be interpreted with caution because of possible rearrangements or transpeptidation reactions. For the glycyl and phenylalanyl chains of oxidised insulin, crystalline pepsin, trypsin and chymotrypsin were used, together with a crystalline preparation of mould protease and crude papain. Only in the latter case was there any evidence of rearrangement which, resulted in the formation of one peptide that did not fit the known sequence in the glycyl chain [SANGER, THOMPSON, and KITAI (*332*)]. Crystalline papain may now be easily prepared from dried latex [KIMMEL and SMITH (*204*)]. By using a variety of enzymes, with different specificities, to obtain evidence for bonds preferentially split by acid or alkali, the chances of the same bond being synthesised by all enzymes is very small.

The only enzyme which has so far shown the exacting specificity revealed by work with synthetic substrates is trypsin [NEURATH and SCHWERT (*265*), SMITH (*353*)], that attacks bonds in which the carboxyl groups of arginine and lysine are involved. All other enzymes so far used on long-chain peptides of known structure attack a diversity of bonds.

The enzyme from *B. subtilis*, crystallised by GÜNTELBERG and OTTESEN (*155*), which gave only a limited proteolysis of ovalbumin to form plakalbumin, had a much lower degree of specificity on the phenylalanyl chain of oxidised insulin [TUPPY (*391*)].

One disadvantage in enzymic degradation is the incomplete splitting of many bonds, which increases the complexity of the peptide mixtures formed. In some cases this may be due to the splitting of one molecule to produce a free amino or carboxyl group which stabilises the adjoining peptide bond, whereas another molecule may split first at the second peptide bond. For example, from the C-terminal amino acid sequence of the A chain of oxidised insulin,

-tyr-glu (NH_2) -- leu + glu + asp (NH_2)-tyr-cySO_3H-asp (NH_2),

SANGER and THOMPSON (*329*) detected 7 peptides in peptic digests because the three susceptible bonds were not completely attacked. The enzyme may also be inhibited by the cleavage products.

3. Isolation and Purification of Peptides.

The mixtures of amino acids and peptides in partial hydrolysates of proteins are usually far too complex for direct fractionation on papergrams, and hence some form of *preliminary* fractionation is essential. The variety of methods available for this purpose has been discussed by SANGER (*325*), and SCHROEDER (*343*). Most useful, in conjunction with paper chromatography for final fractionation, are undoubtedly those methods making use of charge and size differences between the peptides. *Zone electrophoresis* and *ion-exchange chromatography* are receiving increasing use for this purpose. Fractions from a partial acid hydrolysate of lysozyme, separated by elution chromatography on ion-exchange resins, were satisfactorily resolved by THOMPSON (*374, 377*) on two-dimensional papergrams. About 130 peptides were identified and used to establish sequences in the molecule. Valuable evidence regarding the *purity* of separated peptides [cf. SANGER and TUPPY (*335*)] may be given by 1. the presence of only one terminal residue; 2. a quantitative analysis which yields approximately stoichiometric amounts of the amino acid constituents; and 3. a comparison of the known R_f value with that calculated from the amino acid composition (see p. 275).

An enormous advantage of paper chromatography for fractionating and identifying peptides is the small amounts of material required. However, this may limit the number of purification steps and identification tests applicable. It is encouraging that higher peptides from enzymic digests can be fractionated on paper by chromatography or ionophoresis. SANGER and TUPPY (*335*) isolated such peptides on papergrams, and where necessary, used other enzymes for further degradation.

4. Deduction of Polypeptide Sequence from the Structure of Lower Peptides.

The composition and structure of lower peptides may be readily determined by the techniques previously discussed, with further partial hydrolysis if necessary. When a sufficient number of lower peptides has been identified, it is possible to deduce the structure of large portions of the parent molecule. An exact knowledge of the amino acid composition is useful at this stage. For example, all peptides containing an amino acid which occurs only once in the chain must be derived from the sequence around that amino acid, provided that no rearrangement has taken place. It may not be possible to deduce the complete structure from fragments obtained with one type of hydrolysis, because of the preferential splitting of certain bonds; thus, other methods of hydrolysis showing a different type of specificity must be examined.

Once the polypeptide sequence is known, the remaining problems are to locate the amide groups on particular dicarboxylic amino acid residues and the particular half-cystine residues linked by disulphide bridges.

The *amide distribution* in insulin has been determined by SANGER, THOMPSON, and KITAI (*332*) by studying the peptides isolated from enzymic digests where the amide groups remained intact. Where peptides of the same sequence contain a dicarboxylic amino acid residue and were obtained from both acid and enzymic hydrolysates, a comparison of their R_f values, or preferably, their mobility during paper ionophoresis, indicated whether the peptide from enzymic digests was amidised. Comparison of mobilities of two peptides isolated from enzymic digests, which differed by one of them having an extra dicarboxylic amino acid residue, also indicated whether the extra residue was charged or in the amide form. In addition, direct amide estimations on the peptides isolated from enzymic digests were used. The possibility of using stepwise degradative procedures for locating amide groups has been discussed above (p. 311).

The reduction procedures of CHIBNALL and REES (*64*), and FROMAGEOT et al. (*143*) give the total number of asparaginyl and glutaminyl residues in the chain, and, if they could be adapted to micro quantities, they could be used to detect both the C-terminal residue and amide groups in small peptides. C-terminal isoasparaginyl or isoglutaminyl residues, but not those that occur within the chain, can be differentiated from asparaginyl or glutamyl residues by these reduction procedures. SANGER, THOMPSON, and KITAI (*332*) found no evidence for isoglutamine groups in insulin. LAWLER, TAYLOR, SWAN, and DU VIGNEAUD (*217*) used crystalline papain to prepare the free amides from oxytocin and vasopressin. They obtained no evidence for isoglutamine or isoasparagine, which, if present, would have been distinguished from glutamine and asparagine by starch column chromatography or zone ionophoresis.

The *disulphide bridges* in proteins or peptides of known amino acid sequence can be deduced from the structures of cystine peptides, provided precautions are taken to prevent the formation of mixed disulphides between them (see p. 325). SANGER, SMITH, and KITAI (*327*) found papergrams less effective for fractionating cystine peptides obtained from insulin, because they tended to oxidise and interchange during chromatography. Paper ionophoresis was more suitable for their isolation. The cystine peptides were then oxidised to give two cysteic acid peptides which were separated by paper ionophoresis. From the structures determined and the known sequence around the half-cystine residues it was possible to allocate the disulphide bridges—the final requirement for the complete structure of insulin.

VI. Concluding Remarks.

Paper chromatography is only one of a variety of techniques which are now being employed in the investigation of protein structure. In conjunction with other preliminary fractionation procedures, it is possible to deduce much of the amino acid arrangement in polypeptide chains of approximately 100 residues from a study of partial hydrolysates. Proteins which have been investigated so far have been those of low-molecular weight, for which there is good evidence of purity; but experience with these small proteins will enable an extension of these studies to larger molecules. With the limited information at present available no rules governing the arrangement of amino acid sequences have emerged. It may safely be predicted that when our knowledge of protein structure reveals the pattern by which amino acids in proteins are assembled, and suggest structural requirements of enzymes, hormones and other specific biological proteins, paper chromatography will have made great contributions to this understanding.

Much effort has been expended in developing methods for the separation and identification of peptides. Ultimately these methods will not only serve in the elucidation of protein structures but will also contribute to their synthesis and the synthesis of similar molecules with modified properties.

References.

1. ACHER, R. et J. CHAUVET: La structure de la vasopressine de Boeuf. Biochim. Biophys. Acta **14**, 421 (1954).

2. ACHER, R., J. CHAUVET, C. CROCKER, U.R. LAURILA, J. THAUREAUX et C. FROMAGEOT: Isolement et caractérisation des peptides courts obtenus par hydrolyse acide. Étude de la structure du lysozyme et de la vasopressine. Bull. soc. chim. biol. (Paris) **36**, 167 (1954).

3. ACHER, R. et C. CROCKER: Réactions colorées spécifiques de l'arginine et de la tyrosine realisées après chromatographie sur papier. Biochim. Biophys. Acta **9**, 704 (1952).

4. ACHER, R., C. FROMAGEOT et M. JUTISZ: Séparations chromatographiques d'acides aminés et de peptides. V. Les acides aminés individuels de l'insuline avec une note sur le dosage de la proline. Biochim. Biophys. Acta **5**, 81 (1950).

5. ACHER, R., M. JUTISZ et C. FROMAGEOT: Sur la structure du lysozyme. Étude de peptides basiques résultant d'une hydrolyse acide ménagée. Biochim. Biophys. Acta **8**, 442 (1952).

6. ACHER, R. et U. R. LAURILA: Sur la résistance de certains dinitrophénylamino-acides à l'hydrolyse chlorhydrique. Bull. soc. chim. biol. (Paris) **35**, 413 (1953).

7. ACHER, R., U. R. LAURILA, J. THAUREAUX et C. FROMAGEOT: Étude des peptides de la phénylalanine résultant de l'hydrolyse acide et enzymatique du lysozyme. Biochim. Biophys. Acta **14**, 151 (1954).

8. ACHER, R., J. THAUREAUX, C. CROCKER, M. JUTISZ et C. FROMAGEOT: Sur la structure du lysozyme. Quelques peptides de l'histidine et de la tyrosine résultant d'une hydrolyse menagée. Biochim. Biophys. Acta **9**, 339 (1952).

9. AKABORI, S., K. OHNO, T. IKENAKA, A. NAGATA, and I. HARUNA: An Improved Method for the Characterisation and Quantitative Estimation of Carboxyl-terminal Amino Acids in Proteins. Proc. Japan. Acad. **29**, 561 (1953).

10. AKABORI, S., K. OHNO, and K. NARITA: On the Hydrazinolysis of Proteins and Peptides: A Method for the Characterisation of Carboxyl-terminal Amino Acids in Proteins. Bull. Chem. Soc. Japan **25**, 214 (1952).

11. AKABORI, S., N. SAKOTA, H. ONO, Y. OKADA, and H. HANABUSA: The Structure of α-Chymotrypsinogen and α-Chymotrypsin. Symposia Enzyme Chem. (Japan) **9**, 25 (1954) (Chem. Abstr. **48**, 7082 (1954)).

12. ÅKERFELDT, S.: A Spot Area Method for Quantitative Determination of Amino Acids on Two-Dimensional Paper Chromatograms. Acta Chem. Scand. **8**, 521 (1954).

13. ALEXANDER, P., R. F. HUDSON, and M. FOX: The Reaction of Oxidising Agents with Wool. 1. The Division of Cystine into Two Fractions of Widely Differing Reactivities. Biochemic. J. **46**, 27 (1950).

14. AMES, B. N. and H. K. MITCHELL: Paper Chromatography of Imidazoles. J. Amer. Chem. Soc. **74**, 252 (1952).

15. ANFINSEN, C. B., R. R. REDFIELD, W. L. CHOATE, J. PAGE, and W. R. CARROL: Studies on the Gross Structure, Cross Linkages and Terminal Sequences in Ribonuclease. J. Biol. Chem. **207**, 201 (1954).

16. ANSON, M. L.: The Denaturation of Proteins. 9ᵉ conseil de chimie, Institut Internat. Chim. Solvay "Les protéines", p. 201 (1953).

17. ARONOFF, S.: Separation of the Ionic Species of Lysine by Means of Partition Chromatography. Science (Washington) **110**, 590 (1949).

18. ATKINSON, R. O., R. G. STUART, and R. E. STUCKEY: Colours Developed in the Paper Chromatography of Amino Acids. Analyst **75**, 447 (1950).

19. AUCLAIR, J. L. and R. DUBREUIL: Simple Ultramicromethod for the Quantitative Estimation of Amino Acids by Paper Partition Chromatography. Canad. J. Zool. **30**, 109 (1952).

20. AUCLAIR, J. L. and R. L. PATTON: On the Occurrence of D-Alanine in the Haemolymph of the Milkweed Bug, *Oncopeltus fasciatus*. Revue Canad. Biol. **9**, 3 (1950).

21. AWAPARA, J.: Application of Paper Chromatography to the Estimation of Some Free Amino Acids in Tissues of the Rat. J. Biol. Chem. **178**, 113 (1949).

22. BAILEY, K.: End-Group Assay in some Proteins of the Keratin-Myosin Group. Biochemic. J. **49**, 23 (1951).

23. BAILEY, K., F. R. BETTELHEIM, L. LORAND, and W. R. MIDDLEBROOK: Action of Thrombin in the Clotting of Fibrinogen. Nature (London) **167**, 233 (1951).

24. BAILEY, K., A. C. CHIBNALL, M. W. REES, and E. F. WILLIAMS: Critique of the Foreman Method for the Estimation of the Dicarboxylic Acids in Protein Hydrolysates. Appendix 2. On the Possible Presence of Hydroxyglutamic Acid in Protein Hydrolysates. Biochemic. J. **37**, 360 (1943).

25. BALSTON, J. N. and B. E. TALBOT: A Guide to Filter Paper and Cellulose Powder Chromatography (Ed., T. S. G. JONES). London: Reeve Angel and Balston. 1952.

26. BAPTIST, V. H. and H. B. BULL: Determination of the Terminal Carboxyl Residues of Peptides and Proteins. J. Amer. Chem. Soc. **75**, 1727 (1953).

27. BATE-SMITH, E. C. and R. G. WESTALL: Chromatographic Behaviour and Chemical Structure. 1. Some Naturally Occurring Phenolic Substances. Biochim. Biophys. Acta **4**, 427 (1950).

28. BENDER, A. E.: Methyl Cellosolve as a Developing Solvent in Paper Partition Chromatography. Biochemic. J. **48**, xv (1951).

29. BENTLEY, H. R. and J. K. WHITEHEAD: Water-miscible Solvents in the Separation of Amino-acids by Paper Chromatography. Biochemic. J. **46**, 341 (1950).

30. BERRY, H. K. and L. CAIN: Biochemical Individuality. IV. A Paper Chromatographic Technique for Determining Excretion of Amino Acids in the Presence of Interfering Substances. Arch. Biochemistry **24**, 179 (1949).

31. BERRY, H. K., H. E. SUTTON, L. CAIN, and J. S. BERRY: Development of Paper Chromatography for Use in the Study of Metabolic Patterns. Univ. Texas Public. No. 5109, 22 (1951).

32. BETTELHEIM, F. R.: Tyrosine-O-Sulphate in a Peptide from Fibrinogen. J. Amer. Chem. Soc. **76**, 2838 (1954).

33. BETTELHEIM, F. R. and K. BAILEY: The Products of the Action of Thrombin on Fibrinogen. Biochim. Biophys. Acta **9**, 578 (1952).

34. BISERTE, G. et DAUTREVAUX: Enchaînements "NH_2-terminaux" et "COOH-terminaux" de quelques protéines: 1. Cristalbumine de Cheval et pepsine. Bull. soc. chim. biol. (Paris) **36**, 204 (1954).

35. BISERTE, G. et R. OSTEUX: La chromatographie de partage sur papier des dinitrophénylaminoacides. Bull. soc. chim. biol. (Paris) **33**, 50 (1951).

36. BLACKBURN, S. and G. R. LEE: The Terminal Carboxyl Groups of Wool Keratin. J. Textile Inst. **45**, T 487 (1954).

37. BLACKBURN, S. and A. G. LOWTHER: The Separation of N-2:4-Dinitrophenyl Amino-Acids on Paper Chromatograms. Biochemic. J. **48**, 126 (1951).

38. BLACKBURN, S. and A. ROBSON: A Radiochemical Method for the Micro-estimation of α-Amino Acids Separated on Paper Partition Chromatograms. Biochemic. J. **54**, 295 (1953).

39. BLASS, J., O. LECOMTE et J. POLONOVSKI: Sur une technique d'électrophorèse associée à la chromatographie sur papier. "Chromatoionophorèse" appliquée aux aminoacides et bases aminées. Bull. soc. chim. biol. (Paris) **36**, 627 (1954).

40. BLOCK, R. J.: Quantitative Paper Chromatography: A Simplified Procedure. Proc. Soc. exp. Biol. Med. **72**, 337 (1949).

41. — Estimation of Amino Acids and Amines on Paper Chromatograms. Analyt. Chemistry **22**, 1327 (1950).

42. BLOCK, R. J. and H. B. VAN DYKE: Amino Acids in Posterior Pituitary Protein. Arch. Biochem. Biophys. **36**, 1 (1952).

43. BLOCK, R. J., R. LESTRANGE, and G. ZWEIG: Paper Chromatography. A Laboratory Manual. New York: Academic Press. 1952.

44. BODE, F., H. J. HÜBENER, H. BRÜCKNER und K. HOERES: Eine einfache quantitative Bestimmung von Aminosäuren im Papierchromatogramm. Naturwiss. **39**, 524 (1952).

45. BOISSONNAS, R. A.: Séparation rapide des acides aminés par chromatographie ascendante bidimensionnelle sur papier. Helv. Chim. Acta **33**, 1966 (1950).

46. — Révélation ponctuelle des acides aminés, des polypeptides et des sucres séparés par chromatographie sur papier. Helv. Chim. Acta **33**, 1972 (1950).

47. — Dosage colorimétrique des acides aminés separés par chromatographie sur papier. Helv. Chim. Acta **33**, 1975 (1950).

48. — Dégradation des peptides neutres à partir de leur extrémité carboxylique. Helv. Chim. Acta **35**, 2226 (1952).

49. BONETTI, E. and C. E. DENT: The Determination of the Optical Configuration of Naturally Occurring Amino Acids using Specific Enzymes and Paper Chromatography. Biochemic. J. **57**, 77 (1954).

50. BOULANGER, P. et G. BISERTE: Chromatographie sur papier des acides aminés et polypeptides des liquides biologiques. II. Modifications techniques et résultats nouveaux (plasma sanguin). Bull. soc. chim. biol. (Paris) **33**, 1930 (1951).

51. — — Peptides et protéines. Bull. soc. chim. France **1952**, 830.

52. BREYHAN, T.: Über Spaltungsreaktionen von Glutaminsäure und Ornithin unter den Bedingungen der sauren Peptidhydrolyse. Naturwiss. **40**, 271 (1953).

53. BRUSH, M. K., R. K. BOUTWELL, A. D. BARTON, and C. HEIDELBERGER: Destruction of Amino Acids during Filter Paper Chromatography. Science (Washington) **113**, 4 (1951).

54. BRYANT, F.: Pyridine—Amyl Alcohol as a Paper Chromatographic Solvent for Amino Acids. Austral. J. Sci. **13**, 83 (1950).

55. BRYANT, F. and B. T. OVERELL: A New Paper Chromatographic Solvent for Amino Acids. Nature (London) **168**, 167 (1951).

56. BULL, H. B., J. W. HAHN, and V. H. BAPTIST: Filter Paper Chromatography. J. Amer. Chem. Soc. **71**, 550 (1949).

57. BUMPUS, F. M. and I. H. PAGE: Preliminary Studies on the Structure of Angiotonin. Science (Washington) **119**, 849 (1954).

58. BURMA, D. P.: On the Role of Water Contained in the Solvent Used in Filter Paper Chromatography. II. J. Indian Chem. Soc. **28**, 555 (1951).

59. — Effect of Temperature on the R_f Values of Amino Acids during Paper Chromatography with Solvents Completely Miscible with Water. Nature (London) **168**, 565 (1951).

60. — Partition Mechanism of Paper Chromatography. Adsorption of Chromatographed Substances. Analyt. Chemistry **25**, 549 (1953).

61. BUTLER, J. A. V., E. C. DODDS, D. M. P. PHILLIPS, and J. M. L. STEPHEN: The Action of Chymotrypsin and Trypsin on Insulin. Biochemic. J. **42**, 116 (1948).

62. CAMPBELL, P. N., T. S. WORK and E. MELLANBY: The Isolation of a Toxic Substance from Agenized Wheat Flour. Biochemic. J. **48**, 106 (1951).

63. CHIBNALL, A. C.: Chemical Constitution of the Proteins. 9e conseil de chimie. Institut Internat. Chim. Solvay. "Les protéines", p. 119 (1953).

64. CHIBNALL, A. C. and M. W. REES: The Amide and Free Carboxyl Groups of Insulin. Biochemic. J. **48**, XLVII (1951).

65. — — Further Observations on the Amide and Free Carboxyl Groups of Insulin. Biochemic. J. **52**, III (1952).

66. CHIBNALL, A. C. and M. W. REES: Identification and Estimation of the Amide and C-Terminal Residues in Insulin by Reduction of the Ester with Lithium Borohydride. Ciba Found. Symp. "The Chemical Structure of Proteins", p. 70 (1953).

67. CHRISTENSEN, H. N.: The Isolation of Valylvaline from Gramicidin Hydrolysates. J. Biol. Chem. **151**, 319 (1943).

68. — Attempted Successive Applications of the Edman Degradation to Insulin. C. R. Trav. Lab. Carlsberg, Sér. chim. **28**, 265 (1952).

69. Chromatographic Analysis. Discuss. Faraday Soc. **7** (1949).

70. Ciba Symposium, "The Chemical Structure of Proteins". London: Churchill. 1953.

71. CLAYTON, R. A. and F. M. STRONG: New Solvent System for Separation of Amino Acids by Paper Chromatography. Analyt. Chemistry **26**, 1362 (1954).

72. COCKING, E. C. and E. W. YEMM: Estimation of Amino Acids by Ninhydrin. Biochemic. J. **58**, XII (1954).

73. Colloque national sur la structure chimique des protéines. Bull. soc. chim. biol. (Paris) **36**, 11 (1954).

74. CONSDEN, R.: Partition Chromatography on Paper, Its Scope and Application, Nature (London) **162**, 359 (1948).

75. CONSDEN, R. and A. H. GORDON: Effect of Salt on Partition Chromatograms. Nature (London) **162**, 181 (1948).

76. — — A Study of the Peptides of Cystine in Partial Hydrolysates of Wool. Biochemic. J. **46**, 8 (1950).

77. CONSDEN, R., A. H. GORDON, and A. J. P. MARTIN: Qualitative Analysis of Proteins: a Partition Chromatographic Method using Paper. Biochemic. J. **38**, 224 (1944).

78. — — — Ionophoresis in Silica Jelly. A Method for the Separation of Amino Acids and Peptides. Biochemic. J. **40**, 33 (1946).

79. — — — The Identification of Lower Peptides in Complex Mixtures. Biochemic. J. **41**, 590 (1947).

80. — — — Separation of Acidic Amino-acids by Means of a Synthetic Anion Exchange Resin. Biochemic. J. **42**, 443 (1948).

81. — — — A Study of the Acidic Peptides Formed on the Partial Acid Hydrolysis of Wool. Biochemic J. **44**, 548 (1949).

82. CONSDEN, R., A. H. GORDON, A. J. P. MARTIN, O. ROSENHEIM, and R. L. M. SYNGE: The Non-Identity of Thudichum's "Glycoleucine" and nor-Leucine. Biochemic. J. **39**, 251 (1945).

83. CONSDEN, R., A. H. GORDON, A. J. P. MARTIN, and R. L. M. SYNGE: Gramicidin S: The Sequence of Amino-acid Residues. Biochemic. J. **41**, 596 (1947).

84. COOK, A. H. and A. L. LEVY: Studies in the Azole Series. XXVII. A New Method of Peptide Synthesis: Glycyl Peptides. J. Chem. Soc. (London) **1950**, 646.

85. — — Studies in the Azole Series. XXVIII. A New Method of Peptide Synthesis: Alanyl-peptides. J. Chem. Soc. (London) **1950**, 651.

86. COOLEY, S. L. and J. L. WOOD: Desulfurization of Proteins by Raney Nickel. Arch. Biochem. Biophys. **34**, 372 (1951).

87. CRAIG, L. C.: Partition Chromatography and Countercurrent Distribution. Analyt. Chemistry **22**, 1346 (1950).

88. CRAIG, L. C., W. HAUSMANN, and J. R. WEISIGER: Structural Studies with Bacitracin A. J. Amer. Chem. Soc. **76**, 2839 (1954).

89. CRAMER, F.: Paper Chromatography. 2nd ed. London: McMillan, 1954.

90. CRUMPLER, H. R. and C. E. DENT: Distinctive Test for α-Amino Acids in Paper Chromatography. Nature (London) **164**, 441 (1949).

91. CURZON, G. and J. GILTROW: A Chromatographic Colour Reagent for a Group of Amino Acids. Nature (London) **172**, 356 (1953).

92. DAHLERUP-PETERSEN, B.: Rate of Ringclosure in the Edman Method. Acta Chem. Scand. **7**, 1013 (1953).

93. DAHLERUP-PETERSEN, B., K. LINDERSTRÖM-LANG, and M. OTTESEN: Stepwise Degradation of Peptides. Acta Chem. Scand. **6**, 1135 (1952).

94. DANIELSSON, C. E.: A New Type of Preparative Chromatographic Paper Column. Ark. Kemi **5**, 173 (1952).

95. DATTA, S. P., C. E. DENT, and H. HARRIS: Apparatus for Mass-Production Two-Way Paper Chromatography. Biochemic. J. **46**, XLII (1950).

96. DAVIE, E. W. and H. NEURATH: C-Terminal Groups of Trypsinogen, DFP-Trypsin and Carboxypeptidase. J. Amer. Chem. Soc. **74**, 6305 (1952).

97. — — Identification of the Peptide Split from Trypsinogen during Autocatalytic Activation. Biochim. Biophys. Acta **11**, 442 (1953).

98. DAVOLL, H., R. A. TURNER, J. G. PIERCE, and V. DU VIGNEAUD: An Investigation of the Free Amino Groups in Oxytocin and Desulfurized Oxytocin Preparations. J. Biol. Chem. **193**, 363 (1951).

99. DE FONTAINE, D. and S. W. FOX: Application of a Quantitative Method of Peptide Analysis to the N-Terminal Sequence of Lysozyme. J. Amer. Chem. Soc. **76**, 3701 (1954).

100. DENT, C. E.: The Amino-aciduria in Fanconi Syndrome. A Study Making Extensive Use of Techniques Based on Paper Partition Chromatography. Biochemic J. **41**, 240 (1947).

101. — A Study of the Behaviour of Some Sixty Amino Acids and other Ninhydrin Reacting Substances on Phenol-"Collidine" Filter-paper Chromatograms, with Notes on the Occurrence of Some of them in Biological Fluids. Biochemic. J. **43**, 169 (1948).

102. DENT, C. E. and D. I. FOWLER: Paper-chromatographic Analysis of Dakin's Samples of "β-Hydroxyglutamic Acid". Biochemic. J. **56**, 54 (1954).

103. DESNUELLE, P.: Quelques techniques nouvelles pour l'étude de la structure des protéines. Adv. Enzymology **14**, 261 (1953).

104. — The General Chemistry of Amino Acids and Peptides. The Proteins. Vol. 1 A, p. 87 (Ed., H. NEURATH and K. BAILEY). New York: Academic Press. 1953.

105. DESNUELLE, P. et G. BONJOUR: Nouvelles recherches sur l'hydrolyse préférentielle des liaisons sérine et thréonine dans les protéines. Biochim. Biophys. Acta **7**, 451 (1951).

106. — — Quelques modèles simples pour l'étude de l'hydrolyse protéique en milieu faiblement acid. Biochim. Biophys. Acta **9**, 356 (1952).

107. DESNUELLE, P. et C. FABRE: Étude des séquences N-terminales de la trypsine et du trypsinogène. Bull. soc. chim. biol. (Paris) **36**, 181 (1954).

108. DESNUELLE, P., M. ROVERY et G. BONJOUR: Quelques nouvelles observations sur le mode d'attaque de la globine de Cheval et de l'albumine d'oeuf par la pepsine. Biochim. Biophys. Acta **5**, 116 (1950).

109. DESNUELLE, P., M. ROVERY et C. FABRE: Étude des restes N-terminaux dans les sérumalbumines de diverses espèces (suivie d'une remarque sur la stabilité des dinitrophénylaminoacides pendant l'hydrolyse). C. R. hebd. Séances Acad. Sci. **233**, 987 (1951).

110. — — — Extrémités N-terminales de la protéine de l'α-chymotrypsine. Biochim. Biophys. Acta **9**, 109 (1952).

111. DICKMAN, S. R. and R. O. ASPLUND: Effect of Xanthylation on the Recovery of DNP-Amino Acids from Acid Protein Hydrolysates. J. Amer. Chem. Soc. **74**, 5208 (1952).

112. DRAPER, O. J. and A. L. POLLARD: The Purification of Phenol for Paper Partition Chromatography. Science (Washington) **109**, 448 (1949).

113. DURRUM, E. L.: Two-dimensional Electrophoresis and Ionophoresis. J. Coll. Sci. **6**, 274 (1951).

114. DU VIGNEAUD, V., C. RESSLER, J. M. SWAN, C. W. ROBERTS, and P. G. KATSOYANNIS: The Synthesis of Oxytocin. J. Amer. Chem. Soc. **76**, 3115 (1954).

115. DU VIGNEAUD, V., C. RESSLER, and S. TRIPPETT: The Sequence of Amino Acids in Oxytocin, with a Proposal for the Structure of Oxytocin. J. Biol. Chem. **205**, 949 (1953).

116. EDMAN, P.: On the Purification and Chemical Composition of Hypertensin (Angiotonin). Ark. Kemi, Mineral. Geol. **22** A, No. 3 (1946).

117. — Method of Determination of the Amino Acid Sequence in Peptides. Acta Chem. Scand. **4**, 283 (1950).

118. — Note on the Stepwise Degradation of Peptides *via* Phenyl Thiohydantoins. Acta Chem. Scand. **7**, 700 (1953).

119. — Selective Cleavage of Peptides. Ciba Found. Symp. "The Chemical Structure of Proteins", p. 98 (1953).

120. EISEN, H. N., S. BELMAN, and M. E. CARSTEN: The Reaction of 2,4-Dinitrobenzenesulfonic Acid with Free Amino Groups of Proteins. J. Amer. Chem. Soc. **75**, 4583 (1953).

121. ELLIOTT, D. F.: A Search for Specific Chemical Methods for Fission of Peptide Bonds. I. The N-Acyl to O-Acyl Transformation in the Degradation of Silk Fibroin. Biochemic. J. **50**, 542 (1952).

122. — Acyl Migration in the Study of Protein Structure. Ciba Found. Symp. "The Chemical Structure of Proteins", p. 129 (1953).

123. ELMORE, D. T. and P. A. TOSELAND: A New Method of Stepwise Degradation of Peptides. Chem. and Ind. **1953**, 1227.

124. ENGLAND, A. and E. J. COHN: Studies in the Physical Chemistry of Amino Acids, Peptides and Related Substances. IV. The Distribution Coefficients of Amino Acids between Water and Butyl Alcohol. J. Amer. Chem. Soc. **57**, 634 (1935).

125. FELIX, K. und A. KREKELS: Trennung von Dinitrophenylamino-säuren in der Papierchromatographie. Z. physiol. Chem. (Hoppe-Seyler) **290**, 78 (1925).

126. FISCHER, F. G. und H. DÖRFEL: Zur quantitativen Auswertung der Papierchromatogramme von Eiweiß-Hydrolysaten. Biochem. Z. **324**, 544 (1953).

127. FISHER, R. B., D. S. PARSONS, and G. A. MORRISON: Quantitative Paper Chromatography. Nature (London) **161**, 764 (1948).

128. FLAVIN, M.: The Linkage of Phosphate to Protein in Pepsin and Ovalbumin. J. Biol. Chem. **210**, 771 (1954).

129. FOWDEN, L.: The Quantitative Recovery and Colorimetric Estimation of Amino-acids Separated by Paper Chromatography. Biochemic. J. **48**, 327 (1951).

130. — The Effect of Age on the Bulk Protein Composition of *Chlorella vulgaris*. Biochemic. J. **52**, 310 (1952).

131. FOWDEN, L. and J. R. PENNEY: Elimination of Losses in the Quantitative Estimation of Amino Acids by Paper Chromatography. Nature (London) **165**, 846 (1950).

132. FOX, S. W.: Terminal Amino Acids in Peptides and Proteins. Adv. Protein Chem. **2**, 155 (1945).

133. Fox, S. W., T. L. Hurst, and K. F. Itschner: A Microbiological Method for the Determination of Sequences of Amino Acid Residues. J. Amer. Chem. Soc. **73**, 3573 (1951).

134. Fox, S. W., T. L. Hurst, and C. Warner: Sequential and Amino Acid-Residue Compositions of Adrenocorticotropic Hormone Preparations of Various Levels of Activity. J. Amer. Chem. Soc. **76**, 1154 (1954).

135. Fraenkel-Conrat, H.: The Essential Groups of Lysozyme, with Particular Reference to its Reaction with Iodine. Arch. Biochemistry **27**, 109 (1950).

136. — Phenylisothiocyanate as a Reagent for the Identification of the Terminal Amino-Acids. Ciba Found. Symp. "The Chemical Structure of Proteins", p. 102 (1953).

137. — A Technique for Stepwise Degradation of Proteins from the Amino End. J. Amer. Chem. Soc. **76**, 3606 (1954).

138. Fraenkel-Conrat, H. and J. Fraenkel-Conrat: A Method for Determination of the Amino Acid Sequence of Proteins. Acta Chem. Scand. **5**, 1409 (1951).

139. Fraenkel-Conrat, H. and J. I. Harris: private communication.

140. Fraenkel-Conrat, H. and R. R. Porter: The Terminal Amino Groups of Conalbumin, Ovomucoid, and Avidin. Biochim. Biophys. Acta **9**, 557 (1952).

141. Fraenkel-Conrat, H. and B. Singer: The Peptide Chains of Tobacco Mosaic Virus. J. Amer. Chem. Soc. **76**, 180 (1954).

142. Fromageot, C. and M. Jutisz: Identification of C-End Groups in Proteins by Reduction with Lithium Aluminium Hydride. Ciba Found. Symp. "The Chemical Structure of Proteins", p. 82 (1953).

143. Fromageot, C., M. Jutisz, D. Meyer et L. Pénasse: Méthode pour la caractérisation des groupes carboxyliques terminaux des les protéines. Application à l'insuline. Biochim. Biophys. Acta **6**, 283 (1950).

144. Fromageot, C. et M. Privat de Garilhe: La composition du lysozyme en acides aminés. II. Acides aminés totaux. Biochim. Biophys. Acta **4**, 509 (1950).

145. Gale, E. F. and M. B. van Halteren: The Assimilation of Amino Acids by Bacteria. 13. The Effect of Certain Amino Acids on the Accumulation of Free Glutamic Acid by *Staphylococcus aureus*: Extracellular Peptide Formation. Biochemic. J. **50**, 34 (1951).

146. Giri, K. V. and A. Nagabhushanam: Sodium 1,2-Naphthoquinone-4-sulfonate as a Reagent for Identification of Amino Acids and Peptides and for Quantitative Estimation of Proline and Hydroxyproline Separated on Paper Chromatograms. Naturwiss. **39**, 548 (1952).

147. Giri, K. V., A. N. Radhakrishnan, and C. S. Vaidyanathan: Circular Paper Chromatography. VI. The Quantitative Determination of Amino Acids. J. Indian Inst. Sci. **35**, 145 (1953).

148. Giri, K. V. and N. A. N. Rao: A Technique for the Identification of Amino Acids Separated by Circular Paper Chromatography. Nature (London) **169**, 923 (1952).

149. — — Separation and Estimation of Overlapping Amino Acids by Circular Paper Chromatography using Different Solvent Mixtures. Current Sci. (India) **22**, 114 (1953).

150. Gladner, J. A. and H. Neurath: Carboxyl Terminal Groups of Proteolytic Enzymes. I. The Activation of Chymotrypsinogen to α-Chymotrypsin. J. Biol. Chem. **205**, 345 (1953).

151. — — Carboxyl Terminal Groups of Proteolytic Enzymes. II. Chymotrypsins. J. Biol. Chem. **206**, 911 (1954).

152. GORDON, A. H., A. J. P. MARTIN, and R. L. M. SYNGE: Partition Chromatography of Free Amino Acids and Peptides. Biochemic J. **37**, XIII (1943).

153. GRASSMANN, W.: Electron Optical, and Chemical Studies on the Structure of Collagen. Ciba Found. Symp. "The Chemical Structure of Proteins", p. 195 (1953).

154. GRASSMANN, W., H. HÖRMANN und H. ENDRES: Eine Verbesserung der Bestimmung von Aminosäuren am Carboxylende von Peptiden durch Reduktion der Carboxylgruppe. Ber. dtsch. chem. Ges. **86**, 1477 (1953).

155. GÜNTELBERG, A. V. and M OTTESEN: Preparation of Crystals containing the Plakalbumin-forming Enzyme from *Bacillus subtilis*. Nature (London) **170**, 802 (1952).

156. HANES, C. S., F. J. R. HIRD, and F. A. ISHERWOOD: Enzymic Transpeptidation Reactions Involving γ-Glutamyl Peptides and α-Amino-acyl Peptides. Biochemic. J. **51**, 25 (1952).

157. HANES, C. S. and F. A. ISHERWOOD: Separation of the Phosphoric Esters on the Filter Paper Chromatogram. Nature (London) **164**, 1107 (1949).

158. HARFENIST, E. J.: The Amino Acid Compositions of Insulins Isolated from Beef, Pork and Sheep Glands. J. Amer. Chem. Soc. **75**, 5528 (1953).

159. HARFENIST, E. J. and L. C. CRAIG: The Molecular Weight of Insulin. J. Amer. Chem. Soc. **74**, 3087 (1952).

160. HARRIS, G.: Amino Acids and Peptides of Hops and Wort. I. Techniques and New Compounds. J. Inst. Brewing **58**, 417 (1952).

161. HARRIS, G. and J. R. A. POLLOCK: Amino Acids and Peptides of Hops and Wort. II. Pipecolinic Acid, A New Amino Acid in Barley and Hops. J. Inst. Brewing **59**, 28 (1953).

162. HARRIS, J. I.: private communication.

163. HARRIS, J. I. and C. H. LI: N- and C-Terminal Amino Acid Sequences of α-Corticotropin. J. Amer. Chem. Soc. **76**, 3607 (1954).

164. HARRIS, J. I., C. H. LI, P. G. CONDLIFFE, and N. G. PON: Action of Carboxypeptidase on Hypophyseal Growth Hormone. J. Biol. Chem. **209**, 133 (1954).

165. HAUGAARD, G. and T. D. KRONER: Partition Chromatography of Amino Acids with Applied Voltage. J. Amer. Chem. Soc. **70**, 2135 (1948).

166. HAUSMANN, W.: Amino Acid Composition of Crystalline Inorganic Pyrophosphatase Isolated from Bakers Yeast. J. Amer. Chem. Soc. **74**, 3181 (1952).

167. HERRIOT, R. M.: Essential Chemical Structures of Chymotrypsin and Pepsin. In "Mechanism of Enzyme Action", p. 24 (Ed. W. D. MCELROY and B. GLASS). Baltimore: John Hopkins Press. 1954.

168. HEYNS, K., W. KOCH und W. KÖNIGSDORF: Über das papierchromatographische Verhalten einer salzsauren Lösung von Glutaminsäure. Naturwiss. **39**, 381 (1952).

169. HEYNS, K. und W. WALTER: Über Proteine und deren Abbauprodukte. V. Papierchromatographische Trennung der isomeren Leucine. Z. physiol. Chem. (Hoppe-Seyler) **287**, 15 (1951).

170. — — Die Bildung von α-Aminobuttersäure aus Threonin bei der Hydrolyse mit Salzsäure. Naturwiss. **39**, 507 (1952).

171. HIRS, C. H. W.: Structural Studies on Ribonuclease. Federat. Proc. (Amer. Soc. exp. Biol.) **13**, 230 (1954).

172. HIRS, C. H. W., S. MOORE. and W. H. STEIN: Isolation of Amino Acids by Chromatography on Ion Exchange Columns; Use of Volatile Buffers. J. Biol. Chem. **195**, 669 (1952).

173. HOLLEY, R. W. and A. D. HOLLEY: A New Stepwise Degradation of Peptides. J. Amer. Chem. Soc. **74**, 5445 (1952).

174. HORN, M. J., D. B. JONES, and S. J. RINGEL: Isolation of Mesolanthionine from Various Alkali-treated Proteins. J. Biol. Chem. **144**, 87 (1952).

175. HORNER, L., W. EMRICH und A. KIRSCHNER: Experimenteller Beitrag zum Mechanismus der Papierchromatographie. Z. Elektrochem. **56**, 987 (1952).

176. INGRAM, V. M.: The Identification of Peptide End-Groups as Dimethylamino Acids. J. Biol. Chem. **202**, 193 (1953).

177. — Phenylthiohydantoins from Serine and Threonine. J. Chem. Soc. (London) **1953**, 3717.

178. ISHERWOOD, F. A. and D. H. CRUICKSHANK: A New Method for the Colorimetric Estimation of Amino Acids on Paper Chromatograms. Nature (London) **174**, 123 (1954).

179. ISHERWOOD, F. A. and C. S. HANES: Separation and Estimation of Organic Acids on Paper Chromatograms. Biochemic. J. **55**, 824 (1953).

180. ISHERWOOD, F. A. and M. A. JERMYN: Relationship between the Structure of the Simple Sugars and their Behaviour on the Paper Chromatogram. Biochemic. J. **48**, 515 (1951).

181. IWAINSKY, H.: Über den Einfluß von Puffergemischen auf die Trennung von DNP-Aminosäuren mit Hilfe der Papierchromatographie. Z. physiol. Chem. (Hoppe-Seyler) **297**, 195 (1954).

182. JAMES, A. T. and R. L. M. SYNGE: Non-peptide Linkages in Gramicidin. Biochemic. J. **50**, 109 (1951).

183. JANSEN, E. F., M. D. FELLOWS-NUTTING, R. JANG, and A. K. BALLS: Inhibition of the Proteinase and Esterase Activities of Trypsin and Chymotrypsin by Diisopropylfluorophosphate: Crystallisation of Inhibited Chymotrypsin. J. Biol. Chem. **179**, 189 (1949).

184. JATZKEWITZ, H. und N.-D. TAM: Darstellung und papierchromatographische Trennung einiger Dinitrophenyl-aminoalkohole. Z. physiol. Chem. (Hoppe-Seyler) **296**, 188 (1954).

185. JENSEN, H. and E. A. EVANS: Studies on Crystalline Insulin. XVIII. The Nature of the Free Amino Groups in Insulin and the Isolation of Phenylalanine and Proline from Crystalline Insulin. J. Biol. Chem. **108** 1 (1935).

186. JEPSON, J. B. and I. SMITH: "Multiple Dipping" Procedures in Paper Chromatography: A Specific Test for Hydroxyproline. Nature (London) **172**, 1100 (1953).

187. JERMYN, M. A. and F. A. ISHERWOOD: Improved Separation of Sugars on the Paper Partition Chromatogram. Biochemic. J. **44**, 402 (1949).

188. JIRGENSONS, B.: The Influence of Solvent Composition, Temperature and Some Other Factors on the R_f Values of Amino Acids in Paper Chromatography. Univ. Texas Publ. No. 5109, 56 (1951).

189. JOLLÈS, P. et C. FROMAGEOT: La protéine lysante II de la rate du Lapin. III. Détermination des aminoacides N- et C-terminaux. Biochim. Biophys. Acta **14**, 228 (1954).

190. JONES, T. S. G.: The Chemical Nature of Aerosporin. III. The Optical Configuration of the Leucine and Threonine Components. Biochemic. J. **42**, LIX (1948).

191. — The Determination of Amino-acids by Polarographic Reduction of the Copper Complexes. Biochemic. J. **42**, LIX (1948).

192. — The Application of Chromatography to Amino Acids and Peptides. Discuss. Faraday Soc. 7, 285 (1949).

192a. Journée de la chromatographie sur papier. Bull. Soc. chim. France **1952**, 815.

193. JUTI.. M.: Identification des résidus C-terminaux des peptides et des protéines. Bull. soc. chim. biol. (Paris) **36**, 109 (1954).

194. JUTISZ, M., M. PRIVAT DE GARILHE, M. SUQUET et C. FROMAGEOT: Micro-méthode de séparation des N-dinitrophénylaminoalcools par chromatographie en "phase inversée". Bull. soc. chim. biol. (Paris) **36**, 117 (1954).

195. KAWERAU, E. and T. WIELAND: Conservation of Amino-acid Chromatograms. Nature (London) **168**, 77 (1951).

196. KEMBLE, A. R. and H. T. MACPHERSON: Indicator Spray for Amino Acids. Nature (London) **170**, 664 (1952).

197. — — Determination of Monoamino Monocarboxylic Acids by Quantitative Paper Chromatography. Biochemic. J. **56**, 548 (1954).

198. KENDREW, J. C., R. G. PARRISH, J. R. MARRACK, and E. S. ORLANS: The Species Specificity of Myoglobin. Nature (London) **174**, 946 (1954).

199. KENNER, G. W. and H. G. KHORANA: Peptides. II. Selective Degradation by Removal of the Terminal Amino-acid bearing a Free Amino Group. The Use of Alkyl Alkoxydithioformates (Dialkyl Xanthates). J. Chem. Soc. (London) **1952**, 2076.

200. KENT, P. W., G. LAWSON, and A. SENIOR: Chromatographic Separation of Amino Sugars and Amino Acids, Using the N-(2:4-dinitrophenyl) Derivatives. Science (Washington) **113**, 354 (1951).

201. KESTON, A. S., S. UDENFRIEND, and M. LEVY: Determination of Organic Compounds as Isotopic Derivatives. II. Amino Acids by Paper Chromatographic and Indicator Techniques. J. Amer. Chem. Soc. **72**, 748 (1950).

202. KICKHÖFEN, B. und O. WESTPHAL: Über eine einfache Kombination von Papierelektrophorese und Papierchromatographie. Z. Naturforsch. **7 B**, 650 (1952).

203. KIESSLING, H. and J. PORATH: A Method of Detection of N-Dimethylamino Acids on Paper. Acta Chem. Scand. **8**, 859 (1954).

204. KIMMEL, J. R. and E. L. SMITH: Crystalline Papain. I. Preparation, Specificity and Activation. J. Biol. Chem. **207**, 515 (1954).

205. KLATZKIN, C.: Quantitative Determination of Amino Acids Separated by Paper Partition Chromatography. Nature (London) **169**, 422 (1952).

206. KNIGHT, C. A.: Paper Chromatography of Some Lower Peptides. J. Biol. Chem. **190**, 753 (1951).

207. KÖRÖS, Z.: Free Amino Groups of Gliadin. Magyar Kém. Folyóirat **56**, 131 (1950).

208. KOWKABANY, G. N. and H. G. CASSIDY: Investigation of Paper Strip Chromatography. Analyt. Chemistry **22**, 817 (1950).

209. — — Investigation of Paper Chromatography. Factors that may Affect R_f Values in Paper Chromatography. Analyt. Chemistry **24**, 643 (1952).

210. KUNITZ, M.: Formation of New Crystalline Enzymes from Chymotrypsin. Isolation of Beta and Gamma Chymotrypsin. J. Gen. Physiol. **22**, 207 (1938).

211. LAKSHMINARAYANAN, K.: A Simple Technique in Paper Disc Chromatography. Arch. Biochem. Biophys. **49**, 396 (1954).

212. LANDMANN, W. A., M. P. DRAKE, and J. DILLAHA: Paper Chromatography of the 3-Phenyl-2-thiohydantoin Derivatives of Amino Acids with Application to End Group and Sequence Studies. J. Amer. Chem. Soc. **75**, 3638 (1953).

213. LANDMANN, W. A., M. P. DRAKE, and W. F. WHITE: Studies on Pituitary Adrenocorticotropin. VI. An N-Terminal Sequence of Corticotropin-A. J. Amer. Chem. Soc. **75**, 4370 (1953).

214. LANDUA, A. J. and J. AWAPARA: Use of a Modified Ninhydrin Reagent in Quantitative Determination of Amino Acids by Paper Chromatography. Science (Washington) **109**, 385 (1949).

215. LANDUA, A. J., R. FUERST, and J. AWAPARA: Paper Chromatography of Amino Acids. Effect of p_H of Sample. Analyt. Chemistry **23**, 162 (1951).

216. LANDUCCI, J. M. et M. PIMONT: Sur le micro-dosage des acides aminés par la méthode de Woiwod. Application aux spots chromatographiques. Bull. Soc. chim. biol. (Paris) **35**, 1041 (1953).

217. LAWLER, H. C., S. P. TAYLOR, A. M. SWAN, and V. DU VIGNEAUD: Presence of Glutamine and Asparagine in Enzymatic Hydrolysates of Oxytocin and Vasopressin. Proc. Soc. exp. Biol. Med. **87**, 550 (1954).

218. LEDERER, E. and M. LEDERER: Chromatography. A Review of Principles and Applications. New York: Elsevier. 1953.

219. LEGGETT BAILEY, J.: Determination of the Amino-acid Sequence in Peptides. Biochemic. J. **52**, IV (1952).

220. — unpublished results. Quoted by A. C. CHIBNALL (*63*).

221. LEONIS, J.: Structures peptidiques. I. Microdétermination de la séquence des acides aminés dans les petits peptides. Bull. soc. chim. Belgique **61**, 524 (1952).

222. LESTRANGE, R. J. and R. H. MÜLLER: Circular Filter Paper Chromatography. Analyt. Chemistry **26**, 953 (1954).

223. LEVY, A. L.: A Paper Chromatographic Method for the Quantitative Estimation of Amino Acids. Nature (London) **174**, 126 (1954).

223a. — Acid Hydrolysis of 3-Phenyl-2-thiohydantoins. Biochim. Biophys. Acta. **15**, 589 (1954).

224. LEVY, A. L. and D. CHUNG: Two-Dimensional Chromatography of Amino Acids on Buffered Papers. Analyt. Chemistry **25**, 396 (1953).

225. LEVY, M. and E. SLOBODIAN: Sequences of Amino Acid Residues in Silk Fibroin. J. Biol. Chem. **199**, 563 (1952).

226. LI, C. H. and L. ASH: The Nitrogen Terminal End-Groups of Hypophyseal Growth Hormone. J. Biol. Chem. **203**, 419 (1953).

227. LINDERSTRÖM-LANG, K.: Proteins and Enzymes. Lane Medical Lectures. Stanford, Calif.: Univ. Press. 1952.

228. LISSITSKY, S. et R. MICHEL: Substances organiques marquées par les radio-isotopes. Bull. soc. chim. France **1952**, 891.

229. LOCKER, R. H.: C-Terminal Groups in Myosin, Tropomyosin and Actin. Biochim. Biophys. Acta **14**, 533 (1954).

230. LORAND, L.: Fibrino-peptide. Biochemic. J. **52**, 200 (1952).

231. LORAND, L. and W. R. MIDDLEBROOK: Studies on Fibrinopeptide. Biochim. Biophys. Acta **9**, 581 (1952).

232. — — Species Specificity of Fibrinogen as Revealed by End-Group Studies. Science (Washington) **118**, 515 (1953).

233. LOWTHER, A. G.: Identification of N-(2,4-Dinitrophenyl) Amino Acids. Nature (London) **167**, 767 (1951).

234. LUMRY, R., E. L. SMITH, and R. R. GLANTZ: Kinetics of Carboxypeptidase Action. I. Effect of Various Extrinsic Factors on Kinetic Parameters. J. Amer. Chem. Soc. **73**, 4330 (1951).

235. MARCHAL, J. G. et T. MITTWER: Modification apportée à la technique de chromatographie sur papier. Chromatographie en arcs de cercle. C. R. Séances Soc. Biol. **145**, 417 (1951).

236. MARTIN, A. J. P.: The Principles of Chromatography. Endeavour **6**, 21 (1947).

237. — Partition Chromatography. Annu. Rep. Chem. Soc. (London) **45**, 267 (1948).

238. — Some Theoretical Aspects of Partition Chromatography. Biochem. Soc. Symp. **3**, 4 (1949).

239. — Partition Chromatography. Ann. Rev. Biochem. **19**, 517 (1950).

240. MARTIN, A. J. P. and R. MITTELMANN: Quantitative Micro-analysis of Amino Acid Mixtures on Paper Partition Chromatograms. Biochemic. J. **43**, 353 (1948).

241. MARTIN, A. J. P. and R. L. M. SYNGE: A New Form of Chromatogram Employing Two Liquid Phases. I. A Theory of Chromatography. II. Application to the Micro-Determination of the Higher Monoamino Acids in Proteins. Biochemic. J. **35**, 1358 (1941).

242. MCFARREN, E. F.: Buffered Filter-Paper Chromatography of the Amino Acids. Analyt. Chemistry **23**, 168 (1951).

243. MCFARREN, E. F. and J. A. MILLS: Quantitative Determination of Amino Acids on Filter Paper Chromatograms by Direct Photometry. Analyt. Chemistry **24**, 650 (1952).

244. MEYER, H. and E. RIKLIS: Influence of Cations on the Ninhydrin Reaction for the Determination of Amino Acids. Nature (London) **172**, 543 (1950).

245. MIETTINEN, J. K. and A. I. VIRTANEN: A New Technique in Paper Chromatography. Acta Chem. Scand. **3**, 459 (1949).

246. MILLS, G. L.: Identification of Dinitrophenylamino Acids. Nature (London) **165**, 403 (1950).

247. — Observations on the Application of Fluorodinitrobenzene to the Quantitative Analysis of Proteins. Biochemic. J. **50**, 707 (1952).

248. — Specificity of Bond Fission during the Acid Hydrolysis of Insulin. Biochemic. J. **56**, 230 (1954).

249. MITCHELL, H. K. and F. A. HASKINS: A Filter Paper "Chromatopile". Science (Washington) **110**, 278 (1949).

250. MONIER, R. et C. FROMAGEOT: Quelques peptides résultant de l'hydrolyse partielle du lysozyme. Biochim. Biophys. Acta **5**, 224 (1950).

251. MONIER, R. et M. JUTISZ: Étude de l'hydrolyse acide du lysozyme. Possibilité d'éviter la déstruction du tryptophane au cours de cette hydrolyse. Bull. soc. chim. biol. (Paris) **32**, 228 (1950).

252. — — Contribution à l'étude de la structure de la salmine d'*Oncorhynchus*. I. Enchaînement des aminoacides au voisinage du résidu N-terminal et étude de quelques peptides résultant de l'hydrolyse acide partielle. Biochim. Biophys. Acta **14**, 551 (1954).

253. MONIER, R. et L. PÉNASSE: La séparation des dinitrophénylaminoacides par chromatographie sur papier. C. R. hebd. Séances Acad. Sci. **230**, 1176 (1950).

254. MOORE, A. M. and J. B. BOYLEN: Simple Method for Making Transfers in Paper Chromatography. Science (Washington) **118**, 19 (1953).

255. MOORE, S. and W. H. STEIN: Photometric Ninhydrin Method for Use in the Chromatography of Amino Acids. J. Biol. Chem. **176**, 367 (1948).

256. — — Chromatography of Amino Acids on Sulfonated Polystyrene Resins. J. Biol. Chem. **192**, 663 (1951).

257. — — Chromatography. Annu. Rev. Biochem. **21**, 521 (1952).

258. MORTREUIL, M. et Y. KHOUVINE: Dosage des acides aminés par les complexes colorés dicétohydrindylidène-hydrindamine-sels de cadmium. Bull. soc. chim. biol. (Paris) **36**, 425 (1954).

259. MOZINGO, R., D. E. WOLF, S. A. HARRIS, and K. FOLKERS: Hydrogenolysis of Sulfur Compounds by Raney Nickel Catalyst. J. Amer. Chem. Soc. **65**, 1013 (1943).

260. MUELLER, J. M., J. G. PIERCE, and V. DU VIGNEAUD: Treatment of Performic Acid-Oxidized Oxytocin with Bromine Water. J. Biol. Chem. **204**, 857 (1953).

261. MÜLLER, R. H. and D. L. CLEGG: Paper Chromatography. Instruments and Techniques. Physical and Geometric Factors. Kinetic Studies. Analyt. Chemistry **23**, 396, 403, 408 (1951).

262. MÜTING, D.: Über eine qualitative Farbreaktion zur Differenzierung von Aminosäuren auf Filtrierpapier. Naturwiss. **39**, 303 (1952).

263. NAFTALIN, L.: Quantitative Chromatographic Estimation of α-Amino Acids. Nature (London) 161, 763 (1948).

264. NEURATH, H., J. A. GLADNER, and E. W. DAVIE: The Activation of Chymotrypsinogen and Trypsinogen as Viewed by Enzymatic End-Group Analysis. In: "Mechanism of Enzyme Action", p. 50 (Ed., W. D. McELROY and B. GLASS). Baltimore: John Hopkins Press. 1954.

265. NEURATH, H. and G. W. SCHWERT: The Mode of Action of the Crystalline Pancreatic Proteolytic Enzymes. Chem. Rev. 46, 69 (1950).

266. Neuvième conseil de chimie. Institut international de chimie Solvay. "Les Protéines". Bruxelles (1953).

267. NEWTON, G. G. F. and E. P. ABRAHAM: The Nature of Bacitracin A. Biochemic. J. 53, 604 (1953).

268. OHNO, K.: On the Structure of Lysozyme. I. Quantitative Estimation of Carboxyl-Terminal Amino Acid by Improved Hydrazinolysis Method. J. Biochem. (Japan) 40, 621 (1953).

269. — On the Structure of Lysozyme. II. Characterisation of Aspartyl, Asparaginyl and Glutaminyl Residues in Lysozyme. J. Biochem. (Japan) 41, 345 (1954).

270. OTTESEN, M. and A. WOLLENBERGER: Stepwise Degradation of the Peptides Liberated in the Transformation of Ovalbumin to Plakalbumin. C. R. Trav. Lab. Carlsberg, Sér. chim. 28, 463 (1953).

271. PALADINI, A. and L. C. CRAIG: The Chemistry of Tyrocidine. III. The Structure of Tyrocidine A. J. Amer. Chem. Soc. 76, 688 (1954).

272. PARDEE, A. B.: Calculations on Paper Chromatography of Peptides. J. Biol. Chem. 190, 757 (1951).

273. Partition Chromatography. Biochem. Soc. Symp. 3 (1949).

274. PARTRIDGE, S. M.: Filter-paper Partition Chromatography of Sugars. I. General Description and Application to the Qualitative Analysis of Sugars in Apple Juice, Egg White and Foetal Blood of Sheep. Biochemic. J. 42, 238 (1948).

275. — Ion-Exchange Resins as Molecular Sieves. Nature (London) 169, 496 (1952).

276. PARTRIDGE, S. M. and R. C. BRIMLEY: Displacement Chromatography on Synthetic Ion-Exchange Resins. VIII. A Systematic Method for the Separation of Amino Acids. Biochemic. J. 51, 628 (1952).

277. PARTRIDGE, S. M. and H. F. DAVIS: Preferential Release of Aspartic Acid during the Hydrolysis of Proteins. Nature (London) 165, 62 (1950).

278. PATTON, A. R. and P. CHISM: Quantitative Paper Chromatography of Amino Acids. An Evaluation of Techniques. Analyt. Chemistry 23, 1683 (1951).

279. PÉNASSE, L., M. JUTISZ et C. FROMAGEOT: La caractérisation des groupes carboxyles terminaux dans les protéines. III. La leucine comme acide aminé C-terminal du lysozyme. Bull. soc. chim. biol. (Paris) 35, 376 (1953).

280. PERNIS, B. and CH. WUNDERLY: Quantitative Determination of Amino Acids on Filter Paper. Staining in Two Stages. Biochim. Biophys. Acta 11, 209 (1953).

281. PHILLIPS, D. M. P.: Ultraviolet Fluorescence in Paper Chromatography. Nature (London) 161, 53 (1948).

282. — Partition Chromatography of Enzymic Digests of Insulin. Biochem. Biophys. Acta 3, 341 (1949).

283. PIEZ, K. A., E. B. TOOPER, and L. S. FOSDICK: Desalting of Amino Acid Solutions by Ion Exchange. J. Biol. Chem. 194, 669 (1952).

284. POLSON, A.: Quantitative Partition Chromatography and the Composition of *E. Coli*. Biochim. Biophys. Acta 2, 575 (1948).

285. POLSON, A., V. M. MOSLEY, and R. W. G. WYCKOFF: The Quantitative Chromatography of Silk Hydrolysate. Science (Washington) 105, 603 (1947).

286. POPENOE, E. A. and V. DU VIGNEAUD: Degradative Studies on Vasopressin and Performic Acid-Oxidised Vasopressin. J. Biol. Chem. **205**, 133 (1953).

287. — — A Partial Sequence of Amino Acids in Performic Acid-Oxidised Vasopressin. J. Biol. Chem. **206**, 353 (1954).

288. PORATH, J.: Structure of Bacitracin A. Nature (London) **172**, 871 (1953).

289. PORATH, J. and P. FLODIN: A New Method for the Detection of Amino-Acids, Peptides, Proteins and Other Buffering Substances on Paper. Nature (London) **168**, 202 (1951).

290. PORTER, R. R.: The Unreactive Amino Groups of Proteins. Biochim. Biophys. Acta **2**, 105 (1948).

291. — The Reactivity of the Iminazole Ring in Proteins. Biochemic. J. **46**, 304 (1950).

292. — A Chemical Study of Rabbit Antiovalbumin. Biochemic. J. **46**, 473 (1950).

293. PORTER, R. R. and F. SANGER: The Free Amino Groups of Haemoglobins. Biochemic. J. **42**, 287 (1948).

294. PORTER, W. L.: Multiple-Paper Chromatogram. Analyt. Chemistry **23**, 412 (1951).

295. PRATT, J. J. and J. L. AUCLAIR: Sensitivity of the Ninhydrin Reaction in Paper Partition Chromatography. Science (Washington) **108**, 213 (1948).

296. PROOM, H. and A. J. WOIWOD: The Distribution of Glutamic Acid Decarboxylase in the Family *Enterobacteriaceae*, Examined by a Simple Chromatographic Method. J. Gen. Microbiol. **5**, 681 (1951).

297. REDFIELD, R. R.: Two-Dimensional Paper Chromatographic Systems with High Resolving Power for Amino Acids. Biochim. Biophys. Acta **10**, 344 (1953).

298. REDFIELD, R. R. and E. S. GUZMAN BARRON: Quantitative Determination of Amino Acids in Protein Hydrolyzates by Paper Chromatography. Arch. Biochem. Biophys. **35**, 443 (1952).

299. REED, L. J.: The Occurrence of α-Aminobutyric Acid in Yeast Extract. Its Isolation and Identification. J. Biol. Chem. **183**, 451 (1950).

300. REES, M. W.: The Estimation of Threonine and Serine in Proteins. Biochemic. J. **40**, 632 (1946).

301. REINDEL, F. und W. HOPPE: Über eine neue Färbemethode zum Nachweis von Aminosäuren, Peptiden und Eiweißkörpern auf Papierchromatogrammen und Elektropherogrammen. Naturwiss. **40**, 221 (1953).

302. — — Verbesserung des Trennungseffektes bei der Papierchromatographie durch die Formgebung des Papierstreifens. Naturwiss. **40**, 245 (1953).

303. REITH, W. S. and N. M. WALDRON: Studies on the Determination of the Sequence of Amino Acids in Peptides and Proteins. IV. The Synthesis of 3-(4'-Dimethylamino-3' : 5'-dinitrophenyl)-2-thiohydantoin Derivates of Various Amino Acids and their Use for Amino Acid Sequence Determinations. Biochemic. J. **56**, 116 (1954).

304. RESSLER, C., S. TRIPPETT, and V. DU VIGNEAUD: Free Amino Groups of Performic Acid-Oxidised Oxytocin and of its Cleavage Products Formed by Treatment with Bromine Water. J. Biol. Chem. **204**, 861 (1953).

305. ROCHE, J., M. JUTISZ, S. LISSITZKY et R. MICHEL: Sur la chromatographie quantitative des acides aminés iodés radioactifs de la thyroglobuline marquée. Biochim. Biophys. Acta **7**, 257 (1951).

306. ROCHE, J., N. V. THOAI et J. L. HATT: Métabolisme des dérivées guanidylés. III. Analyse chromatographique des dérivées guanidylés. Biochim. Biophys. Acta **14**, 71 (1954).

307. ROCKLAND, L. B., J. L. BLATT, and M. S. DUNN: Small Scale Filter Paper Chromatography. Filter Papers and Solvents. Analyt. Chemistry **23**, 1142 (1951).

308. ROCKLAND, L. B. and M. S. DUNN: Quantitative Determination of Amino Acids on Filter Paper Chromatograms by Direct Photometry. J. Amer. Chem. Soc. **71**, 4121 (1949).

309. ROLAND, J. F., Jr. and A. M. GROSS: Quantitative Determination of Amino Acids Using Monodimensional Paper Chromatography. Analyt. Chemistry **26**, 502 (1954).

310. ROVERY, M. et P. DESNUELLE: Application comparée de la technique de Sanger et de la technique d'Edman-Fraenkel-Conrat à la détermination des résidus N-terminaux des protéines. Bull. soc. chim. biol. (Paris) **36**, 95 (1954).

311. ROVERY, M. et C. FABRE: Chromatographie sur papier en phase aqueuse pour l'étude des dinitrophénylaminoacides et peptides. Bull. soc. chim. biol. (Paris) **35**, 541 (1953).

312. ROVERY, M., C. FABRE et P. DESNUELLE: Extrémités N-terminales de la β- et de la γ-chymotrypsines de Boeuf. Biochim. Biophys. Acta **10**, 481 (1953).

313. — — — Étude de l'activation du chymotrypsinogène et du trypsinogène de Boeuf par détermination des résidus N-terminaux dans les protéines et les enzymes correspondants. Biochim. Biophys. Acta **12**, 547 (1953).

314. ROVERY, M., M. POILROUX et P. DESNUELLE: Cristallisation d'une nouvelle chymotrypsine formée par activation rapide du chymotrypsinogène de Boeuf Biochim. Biophys. Acta **14**, 145 (1954).

315. RUTTER, L.: Some Applications of a Modified Technique in Paper Chromatography. Analyst **75**, 37 (1950).

316. RYDON, H. N. and P. W. G. SMITH: A New Method for the Detection of Peptides and Similar Compounds on Paper Chromatograms. Nature (London) **169**, 922 (1952).

317. RYLE, A. P., and F. SANGER: Disulphide Interchange Reactions. Biochemic. J. **58**, v (1954).

318. SAIFER, A. and I. ORESKES: Circular Paper Chromatography. I. Studies of Physical Factors that May Influence R_f Values. Analyt. Chemistry **25**, 1539 (1953)

319. — — Circular Paper Chromatography. II. Isatin as a Color Reagent for Amino Acids. Science (Washington) **119**, 124 (1954).

320. SALANDER, R. C., M. PIANO, and A. R. PATTON: Accuracy of Quantitative Paper Chromatography in Amino Acid Analysis Addendum. Analyt. Chemistry **25**, 1252 (1953).

321. SANGER, F.: The Free Amino Groups of Insulin. Biochemic. J. **39**, 507 (1945).

322. — Fractionation of Oxidised Insulin. Biochemic. J. **44**, 126 (1949).

323. — The Terminal Peptides of Insulin. Biochemic. J. **45**, 563 (1949).

324. — Application of Partition Chromatography to the Study of Protein Structure. Biochem Soc. Symp. **3**, 21 (1949)

325. — The Arrangement of Amino Acids in Proteins. Adv. Protein Chem. **7**, 1 (1952).

326. — A Disulphide Interchange Reaction. Nature (London) **171**, 1025 (1953).

327. SANGER, F., L. F. SMITH, and R. KITAI: The Disulphide Bridges of Insulin. Biochemic. J. **58**, VI (1954).

328. SANGER, F. and E. O. P. THOMPSON: The Amino-acid Sequence in the Glycyl Chain of Insulin. I. The Identification of Lower Peptides from Partial Hydrolysates. Biochemic. J. **53**, 353 (1953).

329. — — The Amino-acid Sequence in the Glycyl Chain of Insulin. II. The Investigation of Peptides from Enzymic Hydrolysates. Biochemic. J. **53**, 366 (1953).

330. SANGER F. and E. O. P. THOMPSON: The Inversion of a Dipeptide Sequence during Hydrolysis in Dilute Acid. Biochim. Biophys. Acta **9**, 225 (1952).

331. — — unpublished results.

332. SANGER, F., E. O. P. THOMPSON, and R. KITAI: The Amide Groups of Insulin. Biochemic. J. **59**, 509 (1955).

333. SANGER, F., E. O. P. THOMPSON, and H. TUPPY: Amino Acid Sequences in Insulin. Symp. Hormones Protéiques et Derivées des Protéines, p. 26. 11e Congrès Internat. Biochim., Paris (1952).

334. SANGER, F. and H. TUPPY: The Amino-acid Sequence in the Phenylalanyl Chain of Insulin. I. The Identification of Lower Peptides from Partial Hydrolysates. Biochemic. J. **49**, 463 (1951).

335. — — The Amino-acid Sequence in the Phenylalanyl Chain of Insulin. II. The Investigation of Peptides from Enzymic Hydrolysates. Biochemic. J. **59**, 481 (1951).

336. SCHAFFER, N. K., S. HARSHMAN, and R. R. ENGLE: Phosphoserylglycine from Diisopropylphosphoryl Chymotrypsin and Inversion of its Peptide Sequenc by Acid. Federat. Proc. (Amer. Soc. exp. Biol.) **13**, 289 (1954).

337. SCHLÖGL, K. und H. FABITSCHOWITZ: Konstitutionsermittlung von Peptiden. VI. Lysyl-peptide. 11. Mitt. über Peptide. Monatsh. Chem. **84**, 937 (1953).

338. SCHLÖGL, K. und E. WAWERSICH: Eine Methode zur gleichzeitigen Bestimmung der Amino- und Carboxyl-endständigen Aminosäure in Peptiden. Naturwiss. **41**, 38 (1954).

339. SCHMID, K.: The N-Terminal Residue in Acid Glycoprotein. Biochim. Biophys. Acta **14**, 437 (1954).

340. SCHRAMM, G. und G. BRAUNITZER: Prolin als Endgruppe des Tabakmosaikvirus. Z. Naturforsch. **8 B**, 61 (1953).

341. SCHRAMM, G. und J. W. SCHNEIDER: Zum Abbau von Peptiden mit Phenylisothiocyanat. Z. Naturforsch. **9 B**, 209 (1954).

342. SCHROEDER, W. A.: Some Experiments on the Chromatographic Separation and Identification of Peptides in Partial Hydrolysates of Gelatin. Ciba Found. Symp. "The Chemical Structure of Proteins", p. 184 (1953).

343. — Column Chromatography in the Study of the Structure of Peptides and Proteins. Fortschr. Chem. organ. Naturstoffe **11**, 240 (1954).

344. SCHROEDER, W. A., L. R. HONNEN, and F. C. GREEN: Chromatographic Separation and Identification of Some Peptides in Partial Hydrolysates of Gelatin. Proc. Nat. Acad. Sci. (U. S. A.) **39**, 23 (1953).

345. SCHROEDER, W. A., L. M. KAY, J. LEGETTE, L. R. HONNEN, and F. C. GREEN: The Constitution of Gelatin. Separation and Estimation of Peptides in Partial Hydrolysates. J. Amer. Chem. Soc. **76**, 3556 (1954).

346. SCHROEDER, W. A. and J. LEGETTE: A Study of the Quantitative Dinitrophenylation of Amino Acids and Peptides. J. Amer. Chem. Soc. **75**, 4612 (1953).

347. SHERRY, S. and W. TROLL: The Action of Thrombin on Synthetic Substrates. J. Biol. Chem. **208**, 95 (1954).

348. SIBATANI, A. and M. FUKUDA: Importance of Controlling Water Content of the Solvents for Paper Chromatography. J. Biochem. (Japan) **38**, 181 (1951).

349. SIMMONDS, D. H.: Improved Electrolytic Desalter. Analyt. Chemistry **26**, 1253 (1954).

350. SJÖQUIST, J.: Paper Strip Identification of Phenylthiohydantoins. Acta Chem. Scand. **7**, 447 (1953).

351. SLOTTA, K. H. and J. PRIMOSIGH: A New Method of Quantitative Paper Chromatography (of Protein Hydrolysates). Mem. Inst. Butantan **24**, 85 (1952).

352. SMITH, E. L.: The Specificity of Certain Peptidases. Adv. Enzymology **12**, 191 (1951).

353. — Proteolytic Enzymes. In: The Enzymes, Vol. 1, Part. 2, p. 802. New York: Academic Press. 1951.

354. — Pitfalls in Partition Chromatography. Nature (London) **169**, 60 (1952).

355. SMITH, E. L., J. R. KIMMEL, and D. M. BROWN: Crystalline Papain. II. Physical Studies; the Mercury Complex. J. Biol. Chem. **207**, 533 (1954).

356. SMITH, E. L., J. R. KIMMEL, D. M. BROWN, and E. O. P. THOMPSON: Isolation and Properties of a Crystalline Mercury Derivative of a Lysozyme from Papaya Latex. J. Biol. Chem. (in press).

357. SMITH, E. L. and J. E. PAGE: The Acid Binding Properties of Long-Chain Aliphatic Amines. J. Soc. Chem. Ind. **67**, 48 (1948).

358. SMITH, E. L. and A. STOCKELL: Amino Acid Composition of Crystalline Carboxypeptidase. J. Biol. Chem. **207**, 501 (1954).

359. SMITH, E. L., A. STOCKELL, and J. R. KIMMEL: Crystalline Papain. III. Amino Acid Composition. J. Biol. Chem. **207**, 551 (1954).

360. SMITH, I.: Colour Reactions on Paper Chromatograms by a Dipping Technique. Nature (London) **171**, 43 (1953).

361. ŠORM, F., J. KÖRBL, and L. MATOUŠEK: Water-soluble Dinitrophenyl Derivatives of Proteins. Collect. Czechoslov. Chem. Communs. **15**, 295 (1950).

362. STEIN, W. H.: A Chromatographic Investigation of the Amino Acid Constituents of Normal Urine. J. Biol. Chem. **201**, 45 (1953).

363. STEIN, W. H. and S. MOORE: Chromatographic Determination of the Amino Acid Composition of Proteins. Cold Spring Harbor Sympos. Quant. Biol. **14**, 179 (1949).

364. — — Electrolytic Desalting of Amino Acids. Conversion of Arginine to Ornithine. J. Biol. Chem. **190**, 103 (1951).

365. STEINBERG, D.: The Combined Action of Carboxypeptidase and *B. subtilis* Enzyme on Ovalbumin. J. Amer. Chem. Soc. **75**, 4875 (1953).

366. SWAN, J. M.: Thiohydantoins. I. The Preparation of Some 2-Thiohydantoins from Amino Acids and Acylamino Acids. Austral. J. sci. Research A **5**, 711 (1952).

367. SYNGE, R. L. M.: Analysis of a Partial Hydrolysate of Gramicidin by Partition Chromatography with Starch. Biochemic. J. **38**, 285 (1944).

368. — A Further Study of Hydrolysates of Gramicidin. Biochemic. J. **44**, 542 (1949).

369. — Methods for Isolating ω-Amino-acids: γ-Aminobutyric Acid from Rye Grass. Biochemic. J. **48**, 429 (1951).

370. — Electrophoresis, Chromatography and Related Physical Methods in Application to Future Requirements of Protein Chemistry. 9e conseil de chimie. Institut Internat. Chim. Solvay. "Les protéines", p. 153 (1953).

370a. SYNGE, R. L. M. and A. TISELIUS: Some Adsorption Experiments with Amino Acids and Peptides, Especially Compounds of Tryptophan. Acta Chem. Scand. **3**, 231 (1949).

371. TABONE, J., D. ROBERT, S. THOMASSEY et N. MAMOUNAS: Microcaractérisation de la cynurénine en présence de tryptophane. Application à l'étude des hydrolysats alcalins de tryptophane. Bull. soc. chim. biol. (Paris) **32**, 529 (1950).

372. THOMPSON, A. R.: Destruction of DNP-Amino Acids by Tryptophane. Nature (London) **168**, 390 (1951).

373. — The C-Terminal Residue of Lysozyme. Nature (London) **169**, 495 (1952).

374. — Amino Acid Sequences in Lysozyme. Biochim. Biophys. Acta **14**, 581 (1954).

375. THOMPSON, A. R.: Destruction of Phenylalanine and Tyrosine During the Anodic Oxidation of C-Terminal Residues in Acyl Peptides. Biochim. Biophys. Acta **15**, 299 (1954).

376. — Amino Acid Sequence in Lysozyme. I. Displacement Chromatography of Peptides on Ion Exchange Resins. Biochemic. J. (in press).

377. — Amino Acid Sequence in Lysozyme. II. Elution Chromatography of Peptides on Ion Exchange Resins. Biochemic. J. (in press).

378. — unpublished results.

379. THOMPSON, E. O. P.: The N-Terminal Sequence of Carboxypeptidase. Biochim. Biophys. Acta **10**, 633 (1953).

380. — Crystalline Papain. IV. Free Amino Groups and N-Terminal Sequence. J. Biol. Chem. **207**, 563 (1954).

381. — The N-Terminal Sequence of Serum Albumins; Observations on the Thiohydantoin Method. J. Biol. Chem. **208**, 565 (1954).

382. — Modification of Tyrosine during Performic Acid Oxidation. Biochim. Biophys. Acta **15**, 440 (1954).

383. — unpublished results.

384. THOMPSON, J. F. and F. C. STEWARD: Investigation on Nitrogen Compounds and Nitrogen Metabolites in Plants. II. Variables in Two-dimensional Paper Chromatography of Nitrogen Compounds. Plant Physiol. **26**, 421 (1951).

385. THOMPSON, J. F., R. M. ZACHARIUS, and F. C. STEWARD: Investigations of Nitrogen Compounds and Nitrogen Metabolism in Plants. I. The Reaction of Nitrogen Compounds with Ninhydrin on Paper and a Quantitative Procedure. Plant Physiol. **26**, 375 (1951).

386. TOENNIES, G. and J. J. KOLB: Techniques and Reagents for Paper Chromatography. Analyt. Chemistry **23**, 823 (1951).

387. TOMARELLI, R. M. and K. FLOREY: Use of Papergrams in the Study of the Urinary Excretion of Radioactive Sulfur Compounds. Science (Washington) **107**, 630 (1948).

388. TRISTRAM, G. R.: The Amino Acid Composition of Proteins. In: The Proteins, Vol. 1 A, p. 181 (Ed. H. NEURATH and K. BAILEY). New York: Academic Press. 1953.

389. TROLL, W. and R. K. CANNAN: A Modified Photometric Ninhydrin Method for the Analysis of Amino and Imino Acids. J. Biol. Chem. **200**, 803 (1953).

390. TUPPY, H.: Zum Nachweis und zur papierchromatographischen Trennung von Guanidinoverbindungen. Monatsh. Chem. **84**, 342 (1953).

391. — Über die enzymatische Spezifität der bakteriellen Proteinase, die Ovalbumin in Plakalbumin verwandelt. Monatsh. Chem. **84**, 996 (1953).

392. TUPPY, H. and H. MICHL: Über die chemische Struktur des Oxytocins. Monatsh. Chem. **84**, 1011 (1953).

393. TURBA, F.: Degradation of Peptides from the Amino End. Ciba Found. Symp. "The Chemical Structure of Proteins", p. 142 (1953).

394. TURNER, R. A., J. G. PIERCE, and V. DU VIGNEAUD: The Desulphurisation of Oxytocin. J. Biol. Chem. **193**, 359 (1951).

395. TURNER, R. A. and G. SCHMERZLER: C-Terminal Residues in Peptides and Proteins Through Formation of Thiohydantoins. Biochim. Biophys. Acta **13**, 553 (1954)

396. — — A New Method for Identifying C-Terminal Residues in Peptides. J. Amer. Chem. Soc. **76**, 949 (1954).

397. UDENFRIEND, S.: Identification of γ-Aminobutyric Acid in Brain by the Isotope Derivative Method. J. Biol. Chem. **187**, 65 (1950).

398. UDENFRIEND, S. and S. F. VELICK: The Isotope Derivative Method of Protein Amino End-Group Analysis. J. Biol. Chem. **190**, 733 (1951).

399. URBACH, K. F.: Deposition and Simultaneous Concentration of Dilute Solutions in Paper Partition Chromatography. Science (Washington) **109**, 259 (1949).
400. VAN HALTEREN, M. B.: Artefacts in the Chromatography of Mixtures of Amino Acids containing Cysteine. Nature (London) **168**, 1090 (1951).
401. VAN VUNAKIS, H. and R. M. HERRIOT: unpublished results. Quoted by R. M. HERRIOT (*167*).
402. VELICK, S. F. and S. UDENFRIEND: Isotope Derivative Analysis for Proline, Valine, Methionine and Phenylalanine. J. Biol. Chem. **190**, 721 (1951).
403. VON EULER, U. S. and R. ELIASSON: Application of Material in Filter Paper Chromatography. Nature (London) **170**, 664 (1952).
404. WALDSCHMIDT-LEITZ, E. und K. GAUSS: Verfahren zur Bestimmung der Carboxyl-endständigen Aminosäuren in Peptiden. Chem. Ber. **85**, 352 (1952).
405. WALEY, S. G. and J. WATSON: The Stepwise Degradation of Peptides. J. Chem. Soc. (London) **1951**, 2394.
406. WARNER, R. C.: The Kinetics of the Hydrolysis of Urea and of Arginine. J. Biol. Chem. **142**, 705 (1942).
407. WEIBULL, C.: The Free Amino Groups of the Proteus Flagella Protein. Quantitative Determination of Dinitrophenyl Amino Acids using Paper Chromatography. Acta Chem. Scand. **7**, 335 (1953).
408. WELLINGTON, E. F.: An Ultramicro Method for Quantitative Determination of Amino Acids. Canad. J. Chem. **30**, 581 (1952).
409. — The Effect of Relative Humidity on the Reaction of Ninhydrin with Amino Acids on Paper Chromatograms. Canad. J. Chem. **31**, 484 (1953).
410. — The Amino Acid Composition of Some Insect Viruses and their Characteristic Inclusion-body Proteins. Biochemic. J. **57**, 334 (1954).
411. WESSELY, F., K. SCHLÖGL, and G. KORGER: A New Method for the Degradation of Peptides. Nature (London) **169**, 708 (1952).
412. WESSELY, F., K. SCHLÖGL und E. WAWERSICH: Konstitutionsermittlung von Peptiden. II. Die Bestimmung der *Amino*säure, der ihr benachbarten und der Aminos*äure* in Tri- und Tetrapeptiden. 4. Mitt. über Peptide. Monatsh. Chem. **83**, 1426 (1952).
413. WHITE, W. F.: Studies on Adrenocorticotropin. V. The Isolation of Corticotropin-A. J. Amer. Chem. Soc. **75**, 503 (1953).
414. — Studies on Pituitary Adrenocorticotropin. VII. A C-Terminal Sequence of Corticotropin-A. J. Amer. Chem. Soc. **75**, 4877 (1953).
415. — Studies on Pituitary Adrenocorticotropin. IX. Further Sequences near the C-Terminus. J. Amer. Chem. Soc. **76**, 4194 (1954).
416. WHITE, W. F. and W. A. LANDMANN: Studies on Pituitary Adrenocorticotropin. VIII. Synthetic Confirmation of Three Dipeptides from Corticotropin-A. J. Amer. Chem. Soc. **76**, 4193 (1954).
417. WIELAND, T. und L. WIRTH: Papierchromatographische Analyse der durch Erhitzen mit Alkali gebildeten Zersetzungsprodukte von Serin, Threonin und Cystein. Chem. Ber. **82**, 468 (1949).
418. WIGGINS, L. F. and J. H. WILLIAMS: Use of *n*-Butanol-Formic Acid-Water Mixture in the Paper Chromatography of Amino Acids and Sugars. Nature (London) **170**, 279 (1952).
419. WILLIAMS, R. J. and H. KIRBY: Paper Chromatography using Capillary Ascent. Science (Washinhton) **107**, 481 (1948).
420. WILLIAMSON, M. B. and J. M. PASSMANN: The Terminal Amino Group of Pepsin. J. Biol. Chem. **199**, 121 (1952).
421. — — The Terminal Free Carboxyl Groups of Pepsin. Biochim. Biophys. Acta **15**, 246 (1954).

422. Winegard, H. M., G. Toennies, and R. J. Block: Detection of Sulphur-containing Amino Acids on Paper Chromatograms. Science (Washington) **108**, 506 (1948).

423. Woiwod, A. J.: Micro Estimation of Amino-Nitrogen and its Application to Paper Partition Chromatography. Nature (London) **161**, 169 (1948).

424. — A Technique for Examining Large Numbers of Bacterial Culture Filtrates by Partition Chromatography. J. Gen. Microbiol. **3**, 312 (1949).

425. — A Method for the Estimation of Micro Amounts of Amino Nitrogen and its Application to Paper Partition Chromatography. Biochem. J. **45**, 412 (1949).

426. Work, E.: Chromatographic Investigations of Amino Acids from Micro-organisms. I. The Amino-acids of *Corynebacterium Diphtheriae*. Biochim. Biophys. Acta **3**, 400 (1949).

427. Wynn, V.: A Peptide-like Contaminant of Filter Paper. Nature (London) **164**, 445 (1949).

428. Zimmermann, G. und K. Nehring: Über Ring-Papierchromatographie nach der Tropfmethode. Angew. Chem. **63**, 556 (1951).

(*Received, December 28, 1954.*)

Acides aminés iodés et iodoprotéines.

Par JEAN ROCHE et RAYMOND MICHEL, Paris.

Avec 2 figures.

Sommaire.

I. Introduction.

La biochimie des acides aminés iodés et des iodoprotéines connaît depuis quinze ans un développement relevant de causes multiples. L'intérêt qui s'y attache demeure, avant tout, lié au fait que les deux produits hormonaux thyroïdiens connus: la *L*-thyroxine et la *L*-3 : 5 : 3'-triiodothyronine, sont des acides aminés. L'étude des dérivés halogénés de ceux-ci se poursuit de plus en plus dans le cadre de celle des iodoprotéines, considérée tout d'abord comme secondaire, en sorte qu'un chapitre nouveau de la biochimie des protéines s'est peu à peu constitué à partir de travaux ayant conduit à l'identification des deux premiers acides aminés iodés naturels connus. La formation des hormones iodées par action directe de l'halogène sur les protéines et leur biosynthèse au sein de la thyroglobuline justifie l'intérêt de cette orientation des recherches, auxquelles l'emploi des isotopes radioactifs de l'iode, surtout d'I^{131}, a ouvert de vastes possibilités techniques. Une importante revue (*60*) a été consacrée, en 1939, par Sir Charles Harington, à la chimie des composés iodés du corps thyroïde dans cette collection; elle y a été suivie en 1950 d'un article général sur la thyroxine et les combinaisons qui lui sont apparentées, dû à C. Niemann (*112*). Aussi n'y a-t-il pas lieu de décrire à nouveau de nombreux faits déjà présentés dans les volumes antérieurs, mais plutôt d'exposer le cadre actuel de l'étude des acides aminés iodés et des iodoprotéines. C'est pourquoi l'isolement et la synthèse des premiers ne feront l'objet que de brefs rappels, sauf dans le cas de corps récemment étudiés, tandis qu'une plus large place sera réservée à la présentation de divers problèmes non traités dans les deux revues antérieures, dont celle-ci est complémentaire sur le plan de la documentation.

Il est à certains égards artificiel de séparer l'étude d'acides aminés de celle des protéines qui les renferment lorsqu'ils prennent naissance au cours de remaniements structuraux de celles-ci, comme tel est le cas pour les iodothyronines. Nous avons néanmoins divisé cet article en deux parties traitant, l'une des acides aminés iodés, l'autre des iodoprotéines.

Il a, en effet, paru nécessaire de n'aborder la biochimie de ces dernières qu'après avoir présenté celle des premiers considérés en tant que corps organiques.

II. Les acides aminés iodés.

Deux acides aminés présents dans la plupart des protéines, la *L*-tyrosine et la *L*-histidine, existent à l'état de dérivés iodés naturels dans certaines de celles-ci et sont susceptibles de s'halogéner dans leurs cycles respectifs par action directe de l'iode. La substitution nucléaire ($RH + I_2 = RI + IH$) donnant naissance aux dérivés iodés de ces corps n'a pas lieu en présence d'autres acides aminés cycliques ou hétérocycliques, tels que la phénylalanine, la proline, le tryptophane. Certains auteurs ont admis (*110*), pour des raisons de simple analogie, l'existence d'un iodotryptophane; en fait, l'isolement de celui-ci (*118*) n'a pas pu être confirmé et l'iode paraît se borner à oxyder l'indylalanine (*136*). L'action de l'halogène sur les protéines ne comporte pas qu'un processus de substitution; elle conduit en outre à la formation de dérivés d'un acide aminé jamais rencontré à l'état non substitué en dehors des iodoprotéines, la *L*-thyronine. De ce fait, les réactions de l'iode avec les protéines relèvent de deux types divers, dont le mécanisme propre sera étudié avec chacun des groupes de corps auxquels il donne naissance. Enfin, étant donnée l'importance pratique des acides aminés iodés marqués par I^{131} pour des recherches d'ordre métabolique, nous leur consacrerons un bref chapitre.

1. Caractères analytiques généraux des acides aminés iodés.

Nous n'avons retenu parmi ces caractères que ceux faciles à mettre en oeuvre pour identifier les corps et les séparer par chromatographie sur papier. Les acides aminés iodés donnent les mêmes réactions colorées que leurs homologues non substitués, mais dans la plupart des cas avec des modalités particulières, soit dans les solutions, soit sur les taches chromatographiques les renfermant. 15 à 20 μg. sont alors nécessaires

Tableau 1. Colorations des taches des acides aminés iodés développées dans les réactions à la ninhydrine ou de Pauly*.

Réaction	Acide aminé iodé			
	MIT et DIT	DITh	TRITh et Tx	MIH et DIH
Ninhydrine	violet rosé	gris rosé	gris rosé	brun rosé
Réaction de Pauly	rose	rose	rouge orangé	rouge orangé

* *Abbréviations.* MIT = monoiodotyrosine, DIT = diiodotyrosine, DITh = = 3 : 5-diiothyronine, TRITh = 3 : 5 : 3'-triiodothyronine, Tx = thyroxine MIH = monoiodohistidine, DIH = diiodohistidine.

pour une révélation des dérivés iodés de la thyronine et 5 à 10 μg. pour ceux de la tyrosine ou de l'histidine après révélation par la ninhydrine; 10 à 20 fois moins par le réactif au sulfate cérique (*25*). Les solvants chromatographiques les plus appropriés, les valeurs des R_f et les réactions colorées sur papier de ces divers corps, ont été récemment décrits dans un article d'ensemble (*132*), auquel nous empruntons les tableaux 1 et 2.

Parmi les solvants chromatographiques, le mélange de *n*-butanol: acide acétique : eau (78 : 5 : 17) convient à la séparation de la *L*-3-mono-iodotyrosine et de la *L*-3 : 5-diiodotyrosine, mais non à celle des iodo-thyronines, lesquelles demeurent fusionnées en une même tache. Des solvants alcalins conviennent pour résoudre le mélange des derniers corps; l'isopentanol saturé d'ammoniaque 6*N* est à cet égard très utile; le *n*-butanol saturé d'ammoniaque 2*N* et la collidine aqueuse peuvent l'être dans certains cas (voir *tableau 2*).

Tableau 2. Valeurs des R_f des iodures et des acides aminés iodés dans divers solvants (10°). (Pour les abbréviations voir Tableau 1; I^- = iodures.)

Solvant	Valeurs de R_f de							
	MIH	I⁻	DIH	MIT	DIT	DITh	TRITh	Tx
n-Butanol acétique ...	0,12	0,21	0,40	0,43	0,59	0,80	0,83	0,86
n-Butanol ammoniacal	0,21	0,37	0,13	0,13	0,08	0,73	0,70	0,58
Isopentanol-NH_4OH 6*N*	—	0,08	—	—	—	0,38	0,27	0,15
Collidine aqueuse	—	0,70	—	0,30	0,22	0,65	0,50	0,50

Diverses réactions colorées sont particulières à certains acides aminés iodés et seront signalées à propos de chacun. C'est ainsi que la réaction de MILLON et celle au réactif à l'α-nitroso-β-naphtol (*3*), communes à de nombreux phénols, sont positives avec la monoiodotyrosine, mais non avec son homologue diiodé (*135*) et que la réaction d'INGVALDSEN et CAMERON aux ions nitreux n'est présentée que par les acides aminés iodés de structure ortho-diiodophénolique (thyroxine, 3 : 5-diiodotyrosine) (*100*).

2. Iodotyrosines et iodohistidines.

Nous examinerons successivement les caractères des dérivés iodés des deux acides aminés et les modalités de leur formation.

a) *Iodotyrosines*.

1. L-3-Monoiodotyrosine. La substitution d'un seul atome d'iode sur le cycle benzénique de la *L*-tyrosine (formule I) s'opère au voisinage immédiat du radical hydroxylé, en 3 (ou 5). La *L*-3-monoiodotyrosine (II), $C_9H_{10}O_3NI$ (iode = 41,4%) a été d'abord isolée à l'état impur des produits de l'hydrolyse alcaline de l'iodocaséine (*108*) et de l'iodopepsine (*77*). Elle a été ensuite caractérisée comme acide aminé naturel dans les gorgonines (*48*), où elle est parfois très abondante — jusqu'à 8–9% —, puis dans les thyroglobulines (*46*, *126*) et dans les spongines (*151*). La technique chromatographique et son application aux constituants marqués des protéines thyroïdiennes a seule permis de déceler dans celles-ci des quantités de *L*-3-monoiodotyrosine trop faibles pour être accessibles à d'autres moyens d'analyse (0,1—0,2%).

OH	OH	OH
(cycle)	(cycle)—I	I—(cycle)—I
CH_2	CH_2	CH_2
CH—NH_2	CH—NH_2	CH—NH_2
COOH	COOH	COOH
(I.) Tyrosine.	(II.) 3-Monoiodotyrosine.	(III.) 3:5-Diiodotyrosine.

Les dérivés *DL* et *L* ont été synthétisés (*70*) à partir des 3-mononitrotyrosines correspondantes (réduction en amine, diazotation, traitement par l'iode). F = 204—206° (décomp.) et $[\alpha]_D^{20} = -4{,}4°$ dans HCl N; $pK_{OH} = 8{,}2$ au lieu de 10,07 pour la tyrosine (*78*). La *L*-3-monoiodotyrosine est extraite de ses solutions acides par le *n*-butanol, quantitativement aux p_H inférieurs à 2; elle repasse en milieu aqueux par lavage des extraits au moyen de solutions alcalines. Relativement stable dans celles-ci, même à chaud, elle se désiode en présence d'acides minéraux à ébullition. La réaction de MILLON, rouge écarlate avec la *L*-tyrosine, présente une coloration rouge pourpre avec le dérivé monosubstitué, ce qui a permis d'établir une méthode de dosage spectrophotométrique des deux corps dans leurs mélanges (*135*). Cette méthode s'applique aux gorgonines, aux iodoprotéines artificielles, mais non aux thyroglobulines, dont la teneur en *L*-3-monoiodotyrosine est trop faible.

Cet acide aminé est décomposé avec formation de *L*-tyrosine et d'iodures par la déshalogénase thyroïdienne (*143*). Il en est de même de la *L*-3-monobromotyrosine, qui l'accompagne en petite quantité dans les gorgonines (*152*), mais non dans les protéines thyroïdiennes. Il ne paraît pas présenter de propriétés pharmacodynamiques caractéristiques et ne constitue qu'une source d'iode pour les animaux qui l'ingèrent.

2. *L-3:5-Diiodotyrosine.* Le traitement de la *L*-tyrosine par un excès d'halogène conduit à la substitution de deux atomes de celui-ci en position 3:5 sur le cycle benzénique; il donne naissance à la *L*-3:5-diiodotyrosine (III), $C_9H_9O_3NI_2$ (iode = 58,7%).

Celle-ci a été isolée par DRECHSEL (*43*) dès 1895 de l'axe corné d'une gorgone provenant d'une variété mal individualisée d'*Eunicella verrucosa* PALLAS, désignée à cette époque sous le nom de *Gorgonia cavolini* par VON KOCH. Ainsi s'explique le nom d'iodogorgonine qui lui avait tout d'abord été attribué. On l'a extraite par la suite de diverses gorgonines et spongines (*124*), des thyroglobulines (*71*) et de nombreuses iodoprotéines artificielles (*115*). Elle est présente en quantité minime dans les protéines des algues du genre *Laminaria* (*127*), dans celles du jaune de l'oeuf des oiseaux (*134*), dans diverses scléroprotéines d'Invertébrés (*150*).

La constitution de l'acide aminé naturel, longtemps mal définie, a été établie par synthèse (*122*). La technique la plus simple pour réaliser celle-ci est basée sur l'action du chlorure d'iode sur la *L*-tyrosine à 60° et en milieu acétique, les deux corps étant mis en présence en proportions stoechiométriques. L'iode réagit également avec la *L*-tyrosine en milieu ammoniacal, en donnant naissance au dérivé 3,5-diiodé; mais il convient alors qu'il soit en excès (6 atomes d'iode par molécule d'acide aminé au lieu de 4).

La *L*-3 : 5-diiodotyrosine cristallise de ses solutions aqueuses avec deux molécules d'eau, qu'elle perd quand on la dessèche sur P_2O_5 (*68*).

Elle est stable en milieu alcalin et ne se désiode en partie que par ébullition prolongée en présence de NaOH 6*N* et, surtout, de $Ba(OH)_2$ 2*N*, ce qui permet son dosage, comme celui de la 3-monoiodo-*L*-tyrosine dans les produits de l'hydrolyse alcaline des protéines. Elle est, par contre, désiodée par chauffage en milieu acide (SO_4H_2 6*N*) à ébullition et même en 1 heure à 100° à $p_H = 6{,}0$ et au-dessus (maximum de stabilité entre $p_H = 2{,}0$ et 6,0). Son point isoélectrique est inférieur à celui de la *L*-tyrosine ($p_{H_i} = 4{,}29$ au lieu de 5,66 à 25°), principalement en raison de la plus forte dissociation du groupement phénolique. Les valeurs des pK des deux acides aminés sont respectivement: pour la tyrosine 2,20 (—COOH), 9,11 ($—NH_3^+$) et 10,07 (—OH) et, pour la 3 : 5-diiodotyrosine 2,11 (—COOH), 7,82 ($—NH_3^+$) et 6,48 (—OH) (*174*). Cette dernière est extractible par le *n*-butanol de ses solutions acides, facilement aux p_H inférieurs à 2; elle ne l'est pas en revanche en milieu sodique *N* ou aux plus fortes concentrations en soude. On a pu baser sur ce fait la séparation des iodothyronines et des iodotyrosines dans les hydrolysats d'iodoprotéines (*18*). En effet, l'extraction de ceux-ci par le *n*-butanol à $p_H = 1$ entraîne tous les acides aminés et les iodures présents et le lavage de la phase *n*-butanolique au moyen de solutions de soude 2*N*, non miscibles à la première, élimine sélectivement et quantitativement de celle-ci les iodures et tous les acides aminés iodés autres que les iodothyronines.

Le dosage de la *L*-3 : 5-diiodotyrosine après pareil fractionnement des hydrolysats d'iodoprotéines a été longtemps opéré par dosage de l'iode total de la fraction alcalino-insoluble de l'extrait *n*-butanolique (*15*). Cette méthode est dépourvue de sélectivité et ses résultats sont entachés d'une erreur par excès due à la présence de *L*-3-monoiodotyrosine et, éventuellement, d'iodures dans la fraction séparée par lavage avec NaOH 2*N*. La colorimétrie de la réaction brun orangé des *o*-diiodophénols (virage à l'orangé en milieu ammoniacal) (*135*) a pu, en revanche, être mise en oeuvre dans des conditions de spécificité satisfaisantes, car elle est négative avec les bromotyrosines et avec la *L*-3-monoiodotyrosine; elle exige toutefois la présence de 50 à 200 μg. de l'acide aminé (*135*). Dans des conditions appropriées, le dérivé diiodé se déshalogène quantitativement (*144*) par HCl 6*N* et, en milieu alcalin, par action de réducteurs tels que l'hydrate d'hydrazine (*99*) et le stannite de sodium (*99*). La tyrosine libérée peut alors être dosée colorimétriquement par la réaction de MILLON, négative avec l'acide aminé disubstitüé (*144*). Celui-ci perd également la totalité de son iode par traitement à l'acide nitrique en présence de nitrate d'argent. De toute manière, lorsqu'il s'agit de solutions pures, les dosages directs sont plus satisfaisants que ceux basés sur les réactions propres aux produits de dégradation de la diiodotyrosine, qui s'appliquent également au dérivé monoiodé. La séparation chromatographique de ces deux corps permet de les doser sélectivement à l'aide de ces dernières.

L'activité physiologique de la *L*-3 : 5-diiodotyrosine reflète un faible antagonisme avec les hormones thyroïdiennes. Toutefois, il est probable que celui-ci n'est qu'indirect, pour la raison suivante. L'ingestion de la *L*-3 : 5-diiodotyrosine est suivie du rejet d'iodures par l'urine (*169*), ce qui traduit sa dégradation. Or, une augmentation, même légère, de l'iodurémie, provoque une inhibition de la thyroxynogénèse en raison de la diminution de la fixation des iodures par le corps thyroïde, qui en est la conséquence physiologique (*31*). Son effet favorable sur le métabolisme des Batraciens doit sans doute être attribué à l'utilisation par ceux-ci de l'iode apporté par l'acide aminé. La déshalogénase thyroïdienne dégrade la *L*-3 : 5-diiodotyrosine en iodure et tyrosine et les ions I^- formés, réutilisés par la glande, entrent alors à nouveau dans le cycle de la thyroxinogénèse (*143*).

La *L-3 : 5-dibromotyrosine* est également un corps naturel. On la rencontre associée à son homologue iodé dans les gorgonines et les spongines (*106*). L'absence de dérivés bromés de la tyrosine dans le corps thyroïde est probablement due au fait que le système enzymatique oxydant les iodures y est inefficace vis à vis des bromures, faiblement concentrés par l'organe (*177*).

b) Iodohistidines.

L'histidine (IV) peut fixer dans son cycle imidazolique deux atomes d'iode en donnant des dérivés 2 (ou 4) monohalogéné (V) et 2 : 4-dihalogéné (VI, p. 350). Une *L*-2 (ou 4)-monoiodohistidine a été préparée par action directe de l'iode sur la *L*-histidine et isolée à l'état cristallisé (*27*). Le dérivé diiodé prenant naissance en présence d'un excès d'halogène est une huile (*27*); il peut être facilement séparé du précédent par chromatographie (*128*).

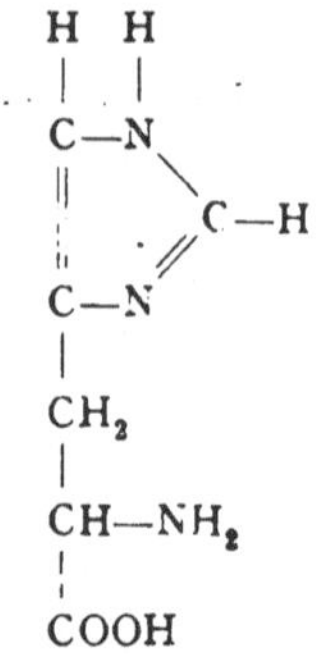

```
H  H              I  H              I  H
|  |              |  |              |  |
C—N               C—N               C—N
||   \            ||   \            ||   \
||    C—H         ||    C—H         ||    C—I
||   //           ||   //           ||   //
C—N               C—N               C—N
|                 |                 |
CH2               CH2               CH2
|                 |                 |
CH—NH2            CH—NH2            CH—NH2
|                 |                 |
COOH              COOH              COOH
```

(IV.) Histidine. 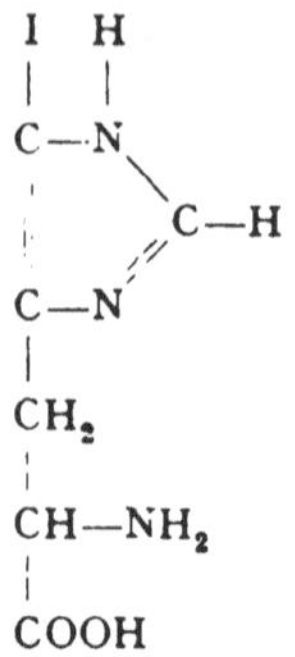 (V.) 2-Monoiodohistidine. 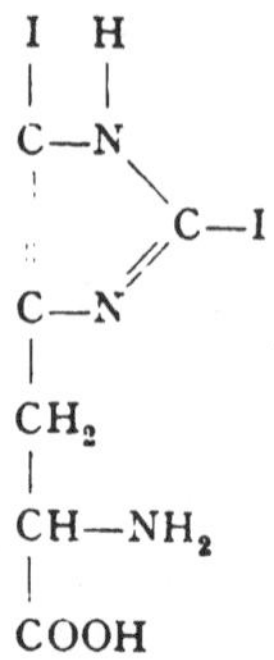 (VI.) 2 : 4-Diiodohistidine.

La *L*-2 (ou 4)-*monoiodohistidine* est un constituant des thyroglobulines et existe à l'état de traces dans l'extrait thyroïdien; elle renferme 2 à 3% de l'iode total de la glande. Elle peut être présente en quantité sensiblement plus grande dans certaines protéines artificiellement iodées, où elle est accompagnée de son isologue 2-4-diiodé lorsque l'halogénation a été opérée en présence d'un excès d'iode (*129*). Les protéines riches en *L*-histidine, comme les globines, constituent alors les meilleures sources de ce dernier produit. La séparation des 2 (ou 4) mono- et 2 : 4-diiodohistidines a été réalisée par chromatographie sur papier (*128*). Les deux dérivés donnent avec la ninhydrine une coloration brun-violet clair et avec l'acide sulfanilique diazoté une teinte jaune orangé pour le produit monosubstitué et rose orangée pour l'isologue disubstitué. Tous deux paraissent dépourvus d'activité physiologique. Comme ils se désiodent spontanément en solution aqueuse, il n'a pas été possible d'étudier l'action des déshalogénases sur eux.

c) *Réaction d'iodur*a*tion de la L-tyrosine, de la L-histidine et de leurs dérivés.*

La formation des iodotyrosines et des iodohistidines est due à des réactions de substitution dont l'étude cinétique avait déjà permis de penser qu'un dérivé monohalogéné prend naissance dans un premier temps, suivi de l'apparition immédiate du dérivé dihalogéné (*128*). On avait tout d'abord admis que l'évolution de l'étape initiale du processus ($RH_2 + I_2 = RIH + HI$) détermine la vitesse de l'halogénation, car «le second atome d'iode réagit instantanément ($RHI + I_2 = RI_2 + HI$) dès que le premier s'est fixé au cycle» (*88, 129*). En fait la séparation chromatographique des produits de la réaction de la *L*-tyrosine ou de la *L*-histidine avec des quantités d'iode comprises entre 0,5 et 10 atomes d'halogène par molécule d'acide aminé a montré qu'il n'en est pas ainsi, comme en rend compte l'examen de la *figure 1*, établie à partir de l'ensemble de nos résultats.

L'ioduration de la *L*-tyrosine conduit à la formation des dérivés mono- et diiodé en proportions diverses selon la quantité d'halogène réagissant. La *L*-3 : 5-diiodotyrosine est seule présente lorsque 6 atomes d'iode au moins ont été mis en oeuvre par molécule de *L*-tyrosine. Les carbones 3 et 5 sont alors substitués, tandis que le carbone 3 l'est seul auparavant dans une partie de l'acide aminé mis en expérience. Dans le cas de la *L*-histidine, le dérivé 2 (ou 4)-monoiodé prédomine jusqu'à ce que 4 atomes d'iode au moins participent à la réaction; le dérivé 2 : 4 diiodé se forme par la suite; il existe pratiquement seul lorsque 7 atomes

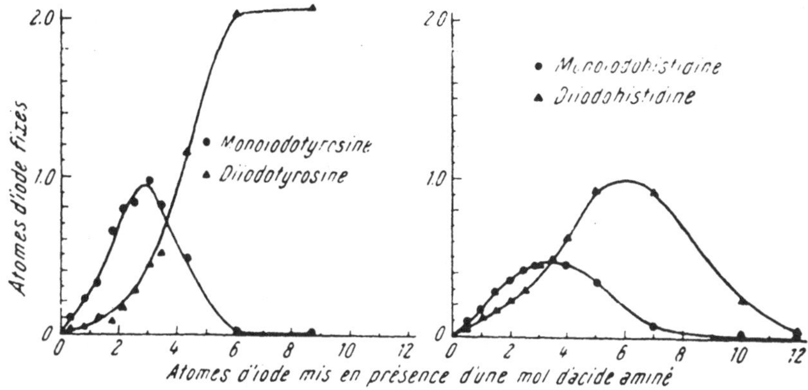

Fig. 1. Quantités d'iode fixées à l'état de dérivé mono- et diiodé de la *L*-tyrosine et de la *L*-histidine en fonction du nombre d'atomes d'iode mis en présence d'une molécule de l'acide aminé. Abscisses: nombre d'atomes d'iode réagissant. Ordonnées: nombre d'atomes d'iode fixés dans les dérivés mono- et disubstitués formés à partir d'une molécule de l'acide aminé.

d'iode réagissent. Des différences importantes se manifestent par ailleurs dans le comportement des deux acides aminés. Comme il est depuis longtemps établi (90), la réactivité du cycle benzénique de la *L*-tyrosine est beaucoup plus grande que celle du cycle imidazolique de la *L*-histidine. Il en résulte que, dans les conditions où la première est saturée en 3 : 5, la seconde est loin de l'être sur ses positions 2 : 4. Le cycle benzénique est stable en présence d'un excès d'halogène, en sorte que la transformation de la *L*-tyrosine en dérivé 3 : 5-diiodé constitue le terme ultime de la réaction. Par contre, le cycle imidazolique, plus labile, subit une oxydation qui le détruit, et cela d'autant plus que la quantité d'halogène mise en oeuvre est plus grande.

Ce fait est à rapprocher de la dégradation de la *L*-histidine dans d'autres conditions où la *L*-tyrosine est stable, en particulier lors de l ur chauffage à ébullition en solution alcaline. On peut admettre que la *L*-histidine est partiellement détruite même en présence de 3 à 4 atomes d'iode par molécule, car on retrouve alors à l'état d'iodures une fraction de l'halogène plus importante que celle pouvant provenir de la réaction nucléaire de substitution. Dans le cas du *L*-tryptophane, la totalité de l'iode réagissant peut être retrouvée à l'état d'iodures, l'oxydation du noyau indolique évoluant alors en dehors de toute substitution dans les conditions où celle-ci a lieu avec les cycles de la *L*-tyrosine et de la *L*-histidine (*136*).

Il y a lieu de signaler une possibilité réactionnelle des noyaux hétérocycliques, à savoir la substitution à certains de leurs atomes d'azote ($—NH^- + I_2 = —NI^- + IH$). Les méthylimidazoles, la 3-N-benzoylhistidine, l'anhydride d'histidine donnent naissance à des composés iodés définis de deux types (*117*). Une première étape de l'halogénation conduit à des combinaisons dans lesquelles l'iode est fixé au carbone 3 avec la N-benzoyl-2 : 4-diiodohistidine, 2 : 2'-diiodohistidine anhydride et 2 : 2' : 4 : 4'-tetraiodohistidine anhydride. Une seconde étape, plus difficile à réaliser, donne naissance à des dérivés iodés au maximum, dans lesquels les =CH— du cycle sont transformés en =CI— et le groupement —NH— en —NI— (VII). La différence de stabilité de ces deux groupements en présence de S_3H_2 permet de les distinguer; le premier demeure alors inaltéré tandis que le second est détruit.

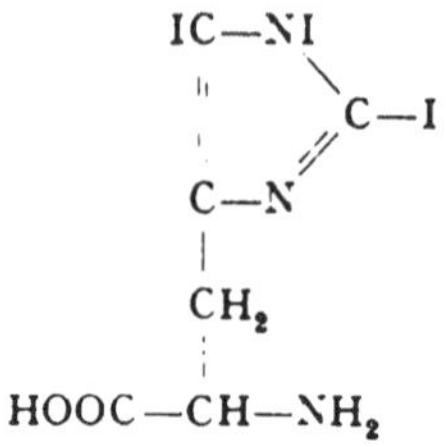

```
IC—NI
‖     \
‖      C—I
‖     //
C—N
|
CH₂
|
HOOC—CH—NH₂
```

(VII.) 2 : 3 : 4-Triiodohistidine.

Le mécanisme biologique de la formation des iodotyrosines et, accessoirement, des iodohistidines, repose sur le processus de substitution étudié plus haut. Celui-ci ne peut s'exercer que lorsque de l'iode libre apparaît au contact des molécules protéiques le fixant sur leurs cycles, ce qui exige l'oxydation enzymatique des iodures présents. Cette oxydation une fois réalisée, il ne semble pas qu'un processus biocatalytique intervienne pour accélérer la fixation de l'halogène. L'individualité d'une tyrosineiodase présumée (*45*) mérite d'être mieux établie, et son rôle dissocié de celui d'une oxydase.

3. Iodothyronines et dérivés.

Pendant longtemps, la 3 : 5 : 3' : 5'-tétraiodothyronine ou thyroxine a été le seul dérivé iodé isolé de la glande thyroïde et possédant toutes les activités biologiques des préparations de celle-ci. Au cours des dernières années, un autre composé de la même série, la *L*-3 : 5 : 3'-triiodothyronine, a été identifié dans la thyroglobuline et synthétisé. Ces deux hormones thyroïdiennes renferment le squelette carboné de la thyronine; elles en diffèrent seulement par le nombre d'atomes d'iode fixés sur les cycles et l'activité biologique du dérivé triiodé est notablement supérieure à celle de son homologue tétrahalogéné.

Des produits synthétiques dérivés de la thyronine ont été préparés et l'un d'entre eux au moins, la *L*-3 : 3' : 5'-triiodothyronine, est présent dans la thyroïde; certains sont biologiquement actifs, d'autre inactifs; par ailleurs, des produits de synthèse d'une structure voisine de celle de la thyronine ont été obtenus.

L'étude des principaux dérivés de la thyronine sera présentée dans l'ordre suivant:

Iodothyronines:

Monoiodothyronines.
Diiodothyronines.
Triiodothyronines.
Thyroxine.

Dérivés de la thyronine:

Modifications portant sur les noyaux et leurs substituants.
Modifications de la chaîne latérale.

a) *Iodothyronines.*

La structure de la thyronine, 4(4'-hydroxyphénoxy)phényl-alanine (VIII) a été établie par HARINGTON et BARGER, et sa synthèse réalisée par les mêmes auteurs.

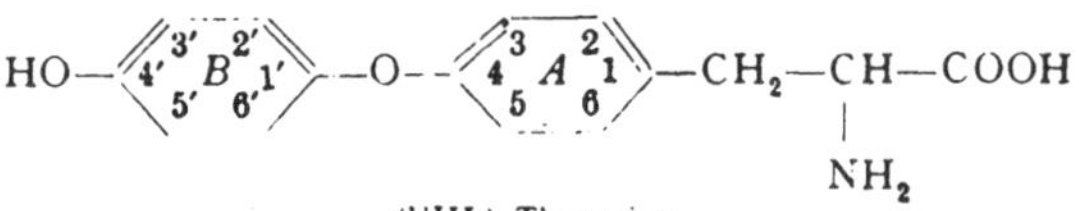

$$HO{-}\underset{(3'\ 2'\ 4'\ B\ 1'\ 5'\ 6')}{C_6H_4}{-}O{-}\underset{(3\ 2\ 4\ A\ 1\ 5\ 6)}{C_6H_4}{-}CH_2{-}\underset{\displaystyle NH_2}{\underset{|}{CH}}{-}COOH$$

(VIII.) Thyronine.

Le cycle porteur de la chaîne alanine est désigné par *A*, le cycle *B* porte la fonction phénol. Les atomes de carbone sont numérotés 1, 2, 3, 4, 5, 6 (cycle *A*) et 1', 2', 3', 4', 5', 6' (cycle *B*).

Les principaux dérivés halogénés sont substitués sur l'un des carbones 3, 5, 3' et 5' ou sur plusieurs de ceux-ci.

Deux produits, la thyronine (Th) et la 3 : 5-diiodothyronine (3 : 5-DITh) permettent de préparer l'ensemble des dérivés iodés sur ces positions. La préparation de la thyronine racémique a été décrite par différents auteurs (67, 59). L'isomère *L* a été obtenu par réduction catalytique de la *L*-thyroxine (32). L'ioduration ménagée et progressive de la thyronine conduit à la 3'-monoiodothyronine (3'-MITh), puis à la 3' : 5'-diiodothyronine (3' : 5'-DITh). L'ioduration ménagée de la 3 : 5-diiodothyronine (3 : 5-DITh) conduit à la 3 : 5 : 3'-triiodothyronine (3 : : 3'-TRITh) et l'on passe à la 3 : 5 : 3' : 5'-tétraiodothyronine ou thyroxine (Tx), par ioduration plus poussée. La réduction catalytique partielle de la 3 : 5-DITh conduit à la 3-monoiodothyronine (3-MITh) et, par hydrogénation plus poussée, à la thyronine. La 3-MITh, produit intermédiaire, donne naissance par ioduration ménagée à la 3 : 3'-diiodothyronine (3 : 3'-DITh) qui fournit la 3 : 3' : 5'-triiodothyronine (3 : 3' : 5'-TRITh) avec un excès d'iode.

En fait, la 3 : 5-DITh permet d'obtenir l'ensemble de ces dérivés, c'est-à-dire : deux monoiodothyronines, trois diiodothyronines, deux triiodothyronines et la thyroxine. La *figure 2* résume les voies par lesquelles ont été synthétisées les différentes iodothyronines.

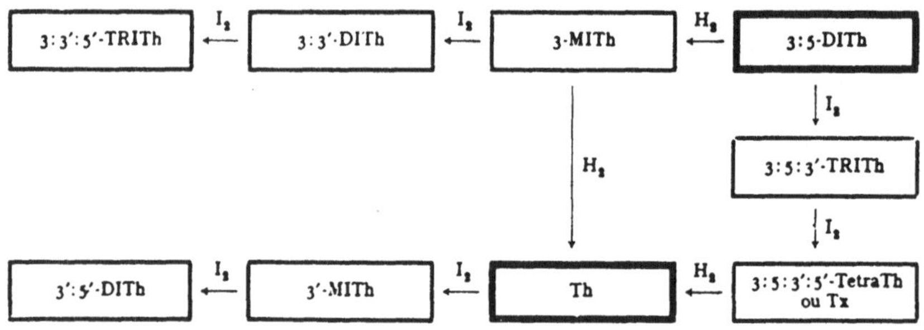

Fig. 2. Synthèses des iodothyronines.

1. Monoiodothyronines. La 3-MITh (IX) est séparée des produits de l'hydrogénation catalytique partielle par cristallisation de son chlorhydrate, dont la neutralisation provoque la précipitation du dérivé monoiodé pur de F. = 206° (*147*). L'ioduration ménagée de la thyronine conduit, comme celle de la tyrosine, au dérivé 3'-monoiodé. La 3'-MITh (X) a pu être préparée à l'état pur et cristallisé, F. = 207°, contrairement à ce que l'on a admis pendant longtemps, à savoir que l'ioduration de la thyronine conduisait à un mélange de produits mal définis (*17*).

I

HO—⟨ ⟩—O—⟨ ⟩—CH_2—CH—COOH

NH_2

(IX.) 3-Monoiodothyronine.

I

HO—⟨ ⟩—O—⟨ ⟩—CH_2—CH—COOH

NH_2

(X.) 3'-Monoiodothyronine.

Les caractères chromatographiques des deux isomères sont différents : leurs R_f sont respectivement de 0,43 pour la 3'- et 0,60 pour la 3-MITh en présence de *n*-butanol saturé de NH_4OH 3*N*.

2. *Diiodothyronines*. Trois isomères. La *DL*-3 : 5-DITh (XI) a été synthétisée pour la première fois par HARINGTON et BARGER (*59*), l'isomère naturel *L* longtemps après par HEMS et ses collaborateurs (*32*). Par déshalogénation catalytique totale, la *L*-thyronine a été obtenue et cristallisée, F. = 245° avec décomposition.

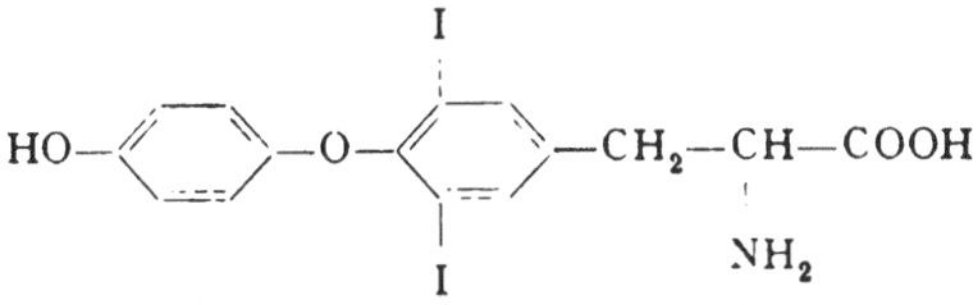

(XI.) 3 : 5-Diiodothyronine.

Le chlorhydrate de la 3 : 5-DITh est peu soluble dans HCl 2*N* à chaud, alors que le chlorhydrate de la 3 : 5 : 3'-TRITh est soluble à chaud et cristallise par refroidissement. Cette propriété est mise à profit pour la purification de la TRITh (*58*).

La 3' : 5'-DITh (XII) a été synthétisée par BLOCK et POWELL (*17*) en condensant la 3 : 5-diiodo-4-méthoxyphénol avec le *p*-nitrochlorobenzène, puis la chaîne alanine est obtenue par la même suite de réactions que celle mise en oeuvre par HARINGTON et BARGER pour préparer la 3 : 5-DITh.

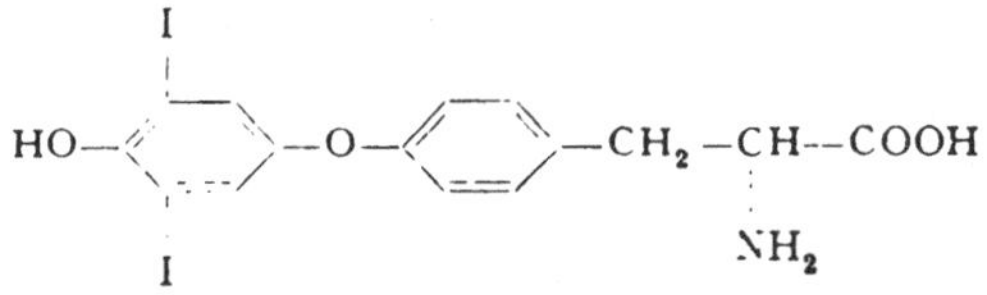

(XII.) 3' : 5'-Diiodothyronine.

Plus récemment, nous avons préparé avec WOLF (*147*) cette DITh par ioduration ménagée de Th en deux étapes successives, F. = 207° avec décomposition. Elle donne la réaction des *o*-diiodophénols avec les ions nitreux, mais elle fournit une réaction négative avec l'α-nitroso-β-naphtol.

La 3 : 3'-DITh (XIII) a été préparée par déshalogénation partielle de la 3 : 5-DITh suivie d'une halogénation partielle, F. = 198° (*147*). Elle ne donne pas la réaction aux ions nitreux, mais elle présente une réaction positive avec l'α-nitroso-β-naphtol. Outre les différences de comportement des 3-diiodothyronines envers des réactions colorées spécifiques

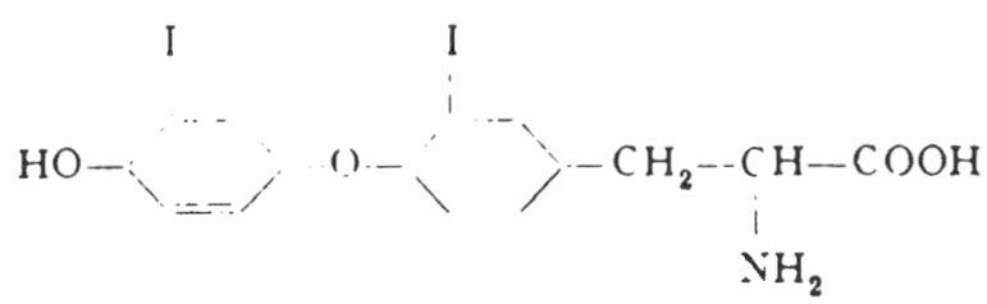

(XIII.) 3 : 3'-Diiodothyronine.

dues à leur différence de structure, des caractères chromatographiques particuliers permettent de les distinguer. C'est ainsi, que par chromatographie sur papier en *n*-butanol saturé par NH_4OH 3*N*, les R_f de la 3 : 5-, de la 3' : 5'- et de la 3 : 3'-DITh sont respectivement de 0,66, 0,43 et 0,53. La 3 : 3'-diiodothyronine a été très récemment caractérisée dans le corps thyroïde (recherches non publiées de J. ROCHE, R. MICHEL, J. NUNEZ et W. WOLF). Son activité antigoîtrogène est voisine de celle de la thyroxine.

3. Triiodothyronines. Deux isomères sont actuellement connus, la 3 : 5 : 3'- et la 3 : 3' : 5'-TRITh. La première (XIV) a été préparée par ioduration ménagée de la 3 : 5-DITh dans des conditions analogues à

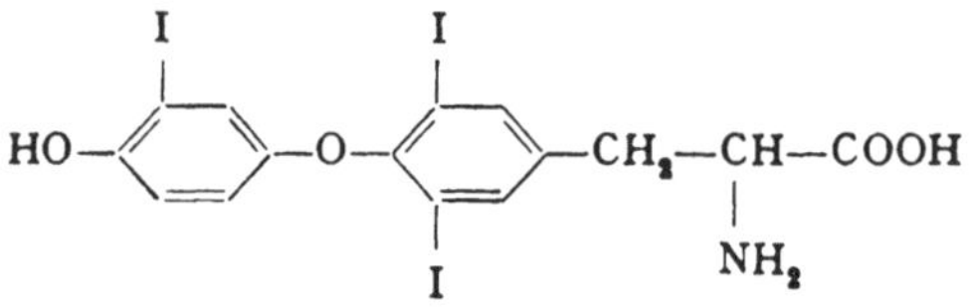

I I
HO—⟨ ⟩—O—⟨ ⟩—CH_2—CH—COOH
I NH_2

(XIV.) 3 : 5 : 3'-Triiodothyronine.

celles utilisées dans la synthèse de la MITh (*131*). Au cours de l'halogénation, il se forme en même temps que la TRITh un peu de thyroxine; celle-ci est partiellement séparée sous forme de chlorhydrate insoluble à chaud (*58*). Une purification ultérieure par chromatographie sur colonne a permis de l'éliminer complétement.

TRITh cristallise en longues aiguilles de F. = 233° (avec décomposition). Elle donne une réaction colorée positive avec l'α-nitroso-β-naphtol et négative avec les ions nitreux. Son spectre ultraviolet montre un maximum d'absorption à 3173 Å. Ses activités biologiques reproduisent celles de l'extrait thyroïdien total, ce qui permet de la considérer comme répondant à la définition d'une hormone. Selon le test envisagé, elle est de 5 à 10 fois plus active que la thyroxine et jusqu'à présent elle paraît présenter toutes les propriétés de la thyroxine. Les seules différences dans son comportement tiennent à l'inégalité de leur vitesse d'utilisation, la TRITh étant plus rapidement et plus intensément métabolisée que la thyroxine.

La 3 : 3' : 5'-TRITh (XV) est aussi un constituant de la thyroglobuline, où elle existe à l'état de traces (*147*). Elle a été repérée par chromatographie en utilisant comme substance de référence un produit obtenu

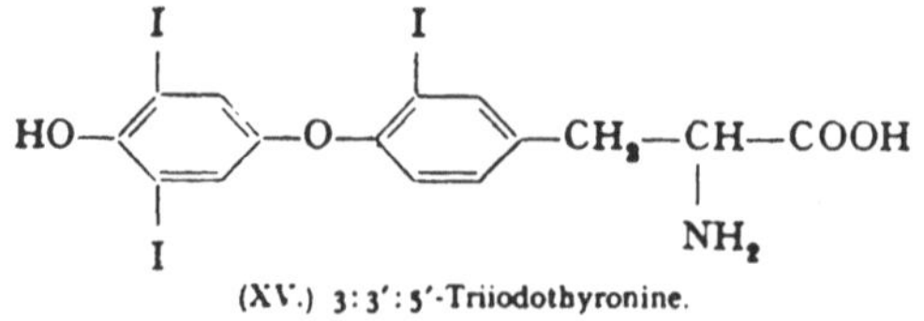

I I
HO—⟨ ⟩—O—⟨ ⟩—CH_2—CH—COOH
I NH_2

(XV.) 3 : 3' : 5'-Triiodothyronine.

par synthèse. Le corps de départ de celle-ci est la 3-MITh. Après ioduration partielle, la 3 : 3'-DITH est isolée et une halogénation ultérieure conduit à la 3 : 3' : 5'-TRITh, F. = 207°.

Les deux TRITh possèdent des caractères chimiques différents. L'isomère 3 : 5 : 3' donne une réaction positive avec l'α-nitroso-β-naphtol, la réaction est négative avec les ions nitreux; c'est l'inverse avec l'autre isomère. Chromatographiquement, ils peuvent être distingués. Le dérivé triiodé en 3 : 5 : 3' a un R_f de 0,64 et celui halogéné en 3 : 3' : 5' un R_f de 0,43 dans le *n*-butanol saturé de NH_4OH 3*N*. Ils sont biologiquement très différents: le dernier ne possède que le centième de l'activité du premier comme agent antigoîtrogène (*148*).

4. Thyroxine (XVI). L'acide β 3 : 5-diiodo-4(3' : 5'-diiodo-4'-hydroxy-phénoxy-phényl)α-aminopropionique, ou thyroxine, dont la formule brute est $C_{15}H_{11}O_4N\ I_4$ (iode = 65,4%), a été isolé sous la forme racémique

$$HO-C_6H_2I_2-O-C_6H_2I_2-CH_2-CH(NH_2)-COOH$$

(XVI.) Thyroxine.

en 1919 par KENDALL et en 1926 par HARINGTON à partir des produits de l'hydrolysat alcalin de poudre de thyroïde (*64*). L'isomère optique naturel *L* a été séparé par la suite de l'hydrolysat enzymatique de la glande. La Tx fond avec décomposition à 235°. Pouvoir rotatoire $[\alpha]_{5461} = -5{,}7°$ en solution éthanol-NaOH.

La constitution de la thyroxine a été établie et sa structure a été confirmée par la réalisation de sa synthèse. Elle a été en outre isolée des iodoprotéines artificielles, où elle se trouve sous forme de dérivé *L*. Elle est vraisemblablement présente à l'état de traces dans les scléroprotéines riches en iode. La thyroxine cristallise en aiguilles microscopiques réunies en gerbes ou en rosettes. Remarquablement peu soluble à $p_H = 4{,}5$ elle précipite quantitativement de ses solutions à ce p_H. Extractible par le *n*-butanol et l'acétate d'éthyle, elle est maintenue en solution dans le premier solvant après lavage par les alcalis, même concentrés, ce qui n'est pas le cas pour les iodotyrosines, lesquelles sont réextraites par les solutions basiques. Cette propriété a été mise à profit pour le dosage de la thyroxine dans les hydrolysats d'iodoprotéines naturelles ou artificielles, lesquelles renferment toujours à côté de la tétraiodothyronine des quantités plus ou moins importantes d'iodotyrosines.

La stabilité de la thyroxine est différente en présence des divers agents chimiques; elle est assez stable en présence des bases à chaud et plus particulièrement de la baryte; cependant une décomposition partielle se produit souvent en milieu sodique, où l'on observe une déshalogénation, et surtout la dégradation oxydative de la chaîne alanine. Les acides minéraux désiodent rapidement la thyroxine à chaud; les atomes d'iode en 3' : 5', plus labiles, sont alors éliminés les premiers. Les agents réducteurs comme l'hydrazine provoquent à chaud une déshalogénation portant principalement sur les atomes d'iode en 3' et 5', tandis que l'hydrogène en présence de catalyseur (palladium) désiode totalement et quantitativement la thyroxine, laissant intact le squelette de la thyronine (*60*).

La thyroxine donne deux séries de sels; les monosels alcalins sont peu solubles jusqu'à 5% dans les mélanges hydroalcooliques; les sels d'argent et de baryum sont insolubles. Elle présente deux principales réactions colorées, l'une caractéristique des *o*-diiodophénols, déjà indiquée à propos de la diiodothyronine, et une coloration pourpre clair avec la N[1]-diéthylsulfanilamide en solution alcaline. Ces deux réactions colorées ont été utilisées pour son dosage; celui-ci exige alors une élimination préalable de la diiodotyrosine.

b) Dérivés de la thyronine.

Les substances naturelles douées d'activité thyroïdienne sont toutes des dérivés iodés de la thyronine, laquelle paraît être totalement dépourvue de la même activité. De nombreux produits de synthèse ont été préparés dans le but de rechercher si la structure de la thyronine est indispensable dans sa totalité aux actions hormonales, ou si des modifications structurales n'apportent simplement que des variations dans l'intensité de celles-ci. D'autres travaux ont été entrepris afin d'établir si l'iode peut être remplacé par d'autres halogènes. Un article publié en 1950 par Niemann (*112*), dans le volume 7 de cette collection a rassemblé les principales données sur les produits synthétiques connus à cette date doués d'activité thyroïdienne. Depuis lors, de nouveaux composés ont été préparés et étudiés; nous en signalerons les principaux, que nous rangerons en deux classes: dérivés de substitution sur les noyaux de la thyronine et dérivés de la thyronine avec modifications de la chaîne latérale.

1. Dérivés substitués sur les noyaux. Plusieurs dérivés de la thyronine ont été préparés; la liste des corps tetrahalogénés est citée par Niemann. L'activité des dérivés 3 : 5-diiodés est supérieure à celle des corps 3 : 5-dibromés et à celle des 3 : 5-dichlorés; il en est d'ailleurs de même pour les dérivés en 3' : 5'. Récemment, Pitt-Rivers (*107*) a étudié l'activité antigoîtrogène des dérivés trihalogénés. La 3 : 5 : 3'-tribromothyronine est environ 12 fois moins active que son isologue triiodé, mais son activité est près de 100 fois plus grande que celle de la 3 : 5 : 3' : 5'-tétrabromothyronine. La comparaison entre les bromoiodothyronines montre que le produit 3 : 5-diiodé et 3'-bromé est plus actif que la 3 : 5-dibromo-3'-

iodothyronine, laquelle est déjà 30% plus active que la thyroxine. Les activités relatives de ces composés montrent l'importance d'un substituant halogéné en ortho par rapport à la fonction phénol.

Bruice et ses collaborateurs (*26*) ont préparé la 3 : 5-diiodo-3' : 5'-diméthyl *DL*-thyronine (XVII) par condensation du N-acétyl-3 : 5-

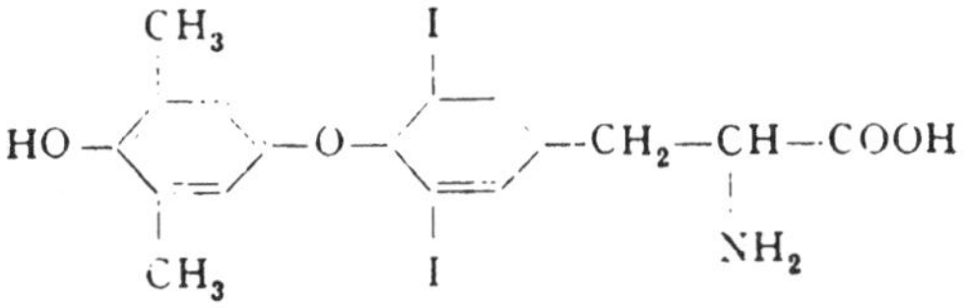

(XVII.) 3 : 5-Diiodo-3' : 5'-diméthylthyronine.

dinitrotyrosine-éthylester avec le 3 : 5-diméthyl-4-méthoxyphénol, puis réduction, diazotation, remplacement des groupements diazonium par l'iode et enfin hydrolyse. Le point de fusion de ce produit est 230°—235°. Il possède une activité 1,6 fois supérieure à celle de la *DL*-thyroxine sur la métamorphose des larves de Batraciens.

La 3 : 5-diiodo-3' : 5'-dinitro-*DL*-thyronine (XVIII) et son isomère, la 3 : 5-dinitro-3' : 5'-diiodo-*LD*-thyronine (XIX) ont été étudiées comme antithyroxiniens, mais leur activité prothyroïdienne ne l'a pas été (*10*).

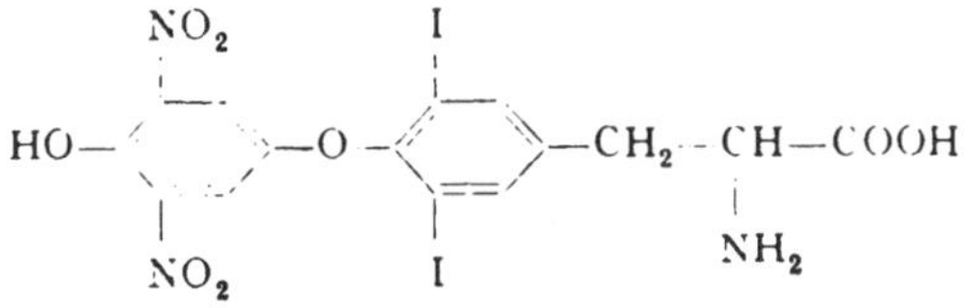

(XVIII.) 3 : 5-Diiodo-3' : 5'-dinitrothyronine.

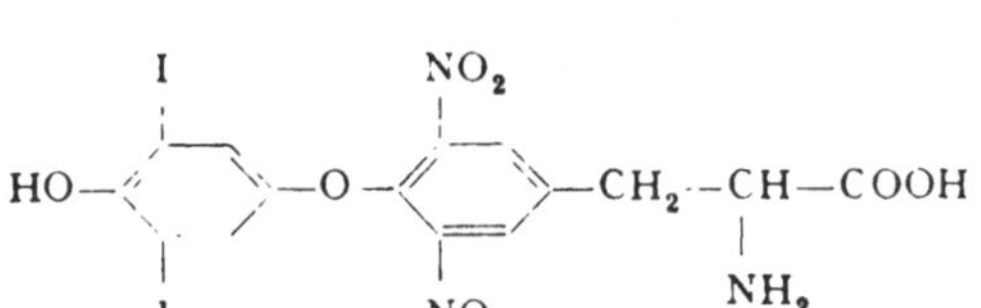

(XIX.) 3 : 5-Dinitro-3' : 5'-diiodothyronine.

La 3 : 5 : 3' : 5'-tétranitrothyronine (XX) est obtenue par nitration de la 3 : 5-dinitrothyronine (F. 222°) (*34*). Selon Lipmann et Dutoit (v. *121*), elle posséderait une activité correspondant à 0,7 p. 1000 celle de la thyroxine

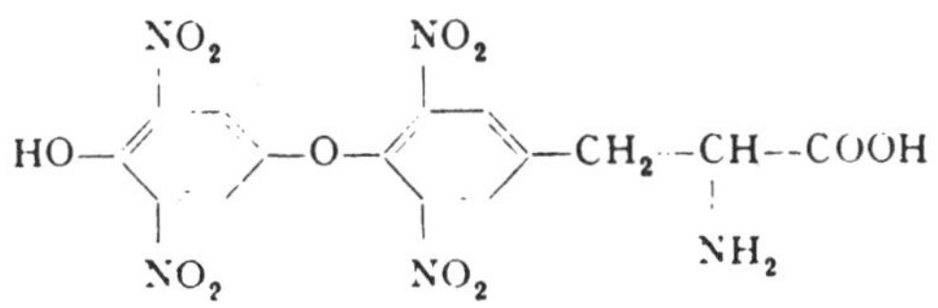

(XX.) 3 : 5 : 3' : 5'-Tétranitrothyronine.

sur la consommation d'oxygène, mais son activité serait nulle sur la métamorphose des Batraciens.

2. *Dérivés substitués sur le phénol.* La thyroxine-O-méthyléther (XXI) est obtenue par méthylation ménagée de la thyroxine-éthylester dans des conditions telles que le groupement azoté n'est pas substitué (*94*). Son activité est beaucoup plus faible que celle de la thyroxine.

(XXI.) Thyroxine-O-méthyléther.

Une partie de la thyroxine et de la 3 : 5 : 3'-triiodothyronine est éliminée par voie biliaire sous forme de dérivés glycuroconjugués sur la fonction phénol (XXII et XXIII) (*167*, *176*). L'éluat d'une bande radioactive d'un chromatogramme provenant de la bile de rats traités à la 3 : 5 : 3'-radio-triiodothyronine a permis de caractériser, après hydrolyse acide, l'acide glycuronique, par le réactif de PARTRIDGE au phtalate acide d'aniline. L'acide glycuronique est également libéré par la β-glycuronidase de la rate et par celle de *Escherichia coli*. Les produits de la glycuroconjugaison des hormones thyroïdiennes donnent une réaction positive avec la ninhydrine et négative avec le réactif de PAULY (v. *146*).

(XXII.) Glycurothyroxine.

(XXIII.) Glycuro-3 : 5 : 3'-triiodothyronine.

Taurog, Abraham et Chaikoff (*167*) ont essayé de synthétiser la glycurothyroxine mais n'y sont point parvenus; ils ont réussi à obtenir le dérivé glycuroconjugué de la 3 : 5-diiodotyrosine et de la tyrosine.

3. Modification de la liaison entre les deux noyaux. Harington (*63*) a synthétisé l'analogue soufré de la thyroxine (thiothyroxine) (XXIV) par

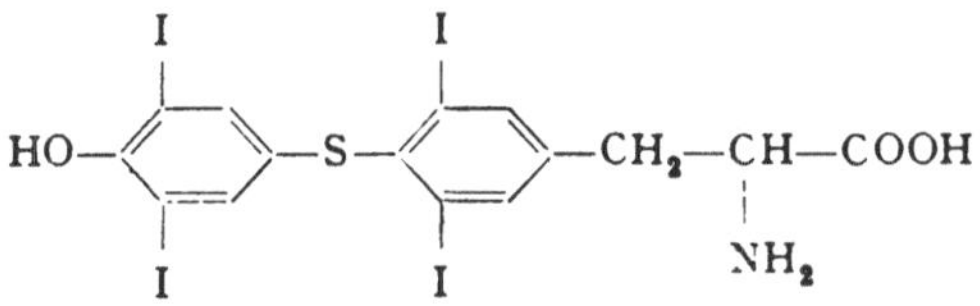

(XXIV.) Thiothyroxine.

une méthode voisine de celle utilisée pour la synthèse de la tétraiodothyronine. Des précautions particulières doivent néanmoins être prises pour l'ioduration de la 3 : 5-diiodothiothyronine.

Hems et ses collaborateurs (*32*) ont préparé le dérivé tétraiodé du diphényle (XXV), donc un composé analogue à la thyroxine, mais sans pont oxydique. La synthèse comporte la condensation du 4-méthoxy-iodo-

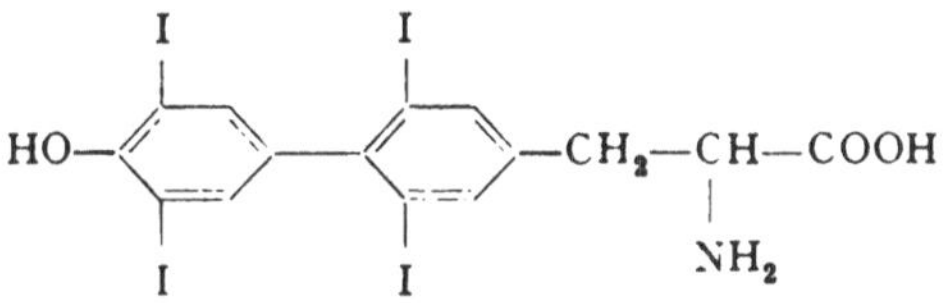

(XXV). Tétraiodo-diphénylthyronine.

benzène avec l'acide 3 : 5-dinitro-4-chlorobenzoïque-méthylester, suivie d'une réduction et d'une diazotation, ce qui conduit au 2 : 6-diiodo-4'-méthoxy-diphényl-4-carboxylate de méthyle. Ce produit est réduit en alcool correspondant, puis transformé en dérivé chloré, lequel est traité par l'acét-amido-malonate d'éthyle, puis hydrolysé et iodé, F. = 202°. L'étude de l'activité thyromimétique de ce corps n'a pas été entreprise; il est doué d'une faible activité antithyroxinienne.

c) Dérivés de la thyrònine avec modifications de la chaîne latérale.

1. Acides aminés et dérivés. L'acide aminé possédant un carbone de moins sur la chaîne alanine de la thyroxine a été préparé (*47*). Ce produit,

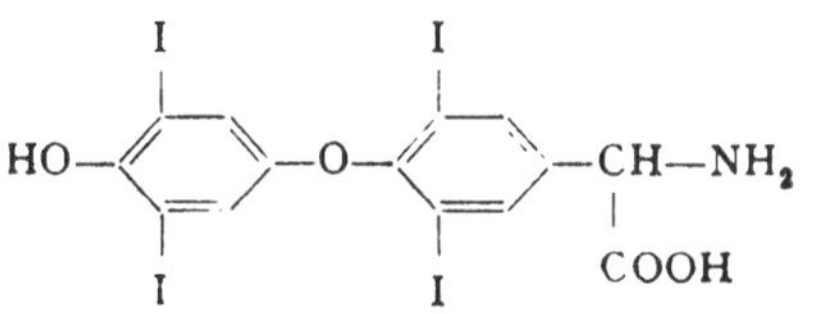

(XXVI.) 3 : 5-Diiodo-4(3' : 5'-diiodo-4'-hydroxyphénoxy)phénylglycine.

qui est donc le dérivé correspondant à la glycine (XXVI), est obtenu à partir du 3 : 5-diiodo-4'-méthoxy-4-phénoxybenzaldéhyde par action de HCN, puis chloruration, amination par H_3N suivie d'hydrolyse et ioduration. Ce composé possède le tiers de l'activité de la thyroxine sur la métamorphose des têtards.

La β-thyroxine (XXVII) a été synthétisée (*37*) par action de NH_2OH sur l'acide diiodothyrocinnamique suivie d'ioduration, F. = 211°. L'activité prothyroïdienne de ce corps n'a pas été étudiée et il ne possède aucune action antithyroxinienne.

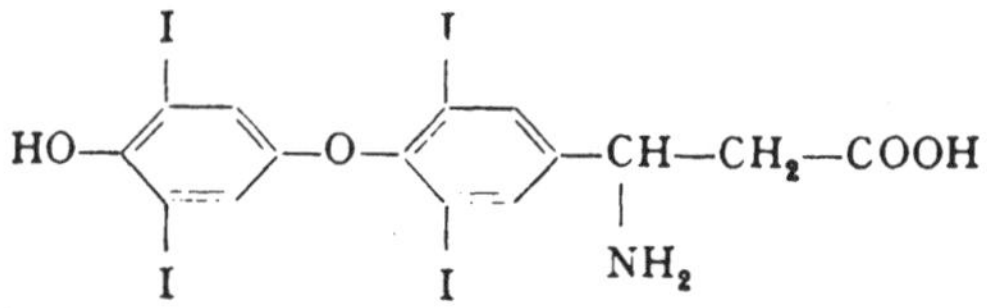

(XXVII.) β-Thyroxine.

La N-acétyl-thyroxine et l'acide N-acétyl-*DL*-thyroxyl-glutamique ont été essayés sur des myxoedémateux; le premier possède le cinquième d'activité de la thyroxine, le deuxième seulement le vingtième (*86*).

Les esters méthylique et éthylique préparés par la méthode de Fischer ont des activités prothyroïdiennes variant de la moitié aux deux tiers de celle de la thyroxine.

2. *Acides.* Les acides 3 : 5 : 3'- et 3 : 5 : 3' : 5'-iodothyropropionique (XXVIII et XXIX) ont été obtenus par réduction de l'acide 3 : 5-diiodo-

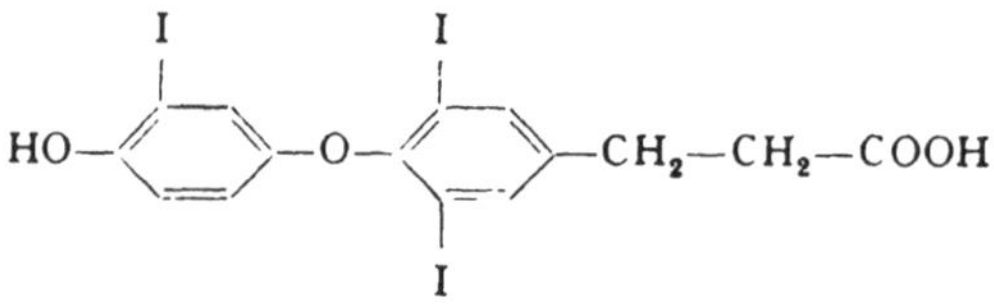

(XXVIII.) Acide 3: 5: 3'-triiodothyropropionique.

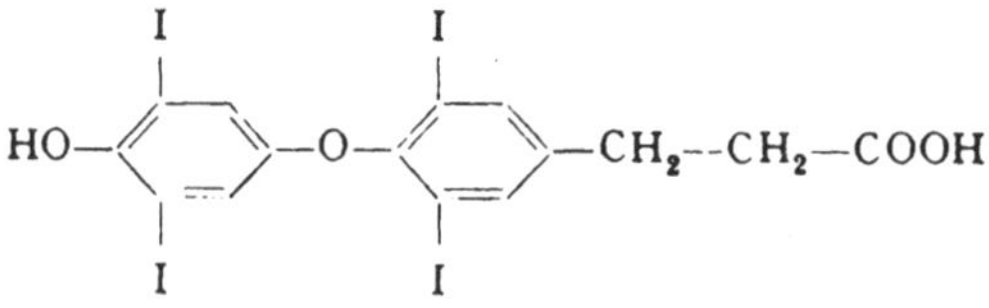

(XXIX.) Acide 3: 5: 3': 5'-tétraiodothyropropionique.

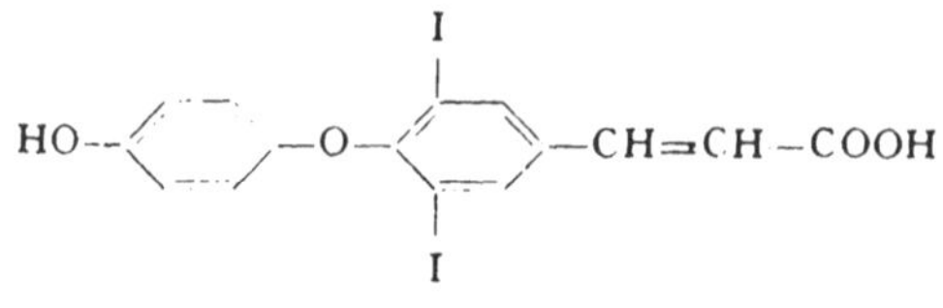

(XXX.) Acide 3: 5-diiodothyrocinnamique.

thyrocinnamique (XXX), suivie d'une ioduration plus ou moins complète. Le dérivé triiodé a un point de fusion de 188°, le tétraiodé de 205°. Ils possèdent une activité antigoîtrogène marquée, voisine de celle de la thyroxine pour le premier, légèrement plus faible pour le second (*148*). L'acide 3 : 5 : 3' : 5'-tétraiodothyropropionique est 130 fois plus actif que la thyroxine sur la métamorphose des pétards de Batraciens (*26*). Cette grande différence de comportement selon le test employé reste jusqu'à présent inexpliquée. On peut néanmoins en rapprocher le fait que la thyroxamine (XXXI), inactive sur les différents tests thyroïdiens (*50*),

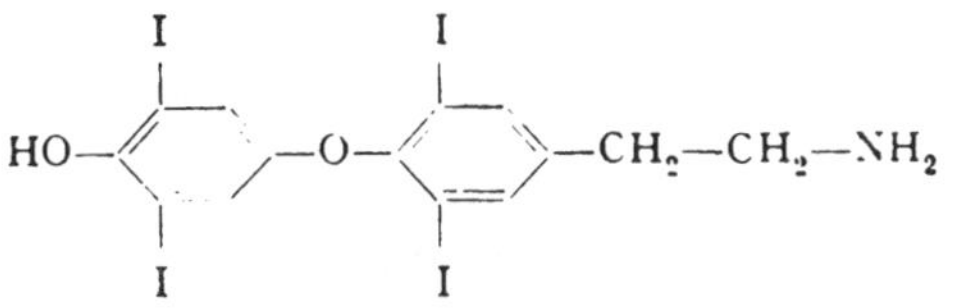

(XXXI.) Thyroxamine.

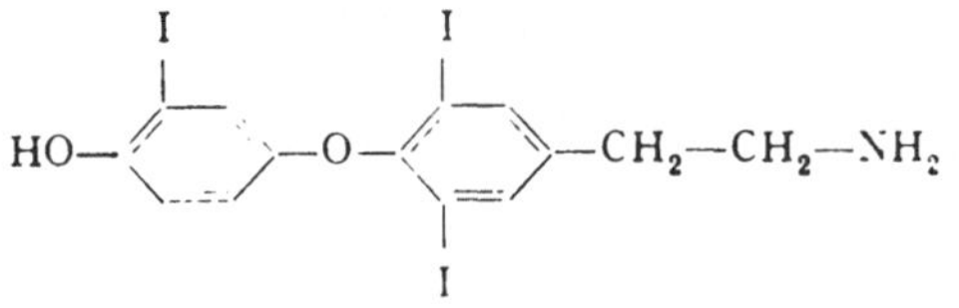

(XXXII.) 3 : 5 : 3'-Triiodothyronamine.

potentialise de façon considérable l'action de l'adrénaline sur l'intestin isolé (*72*); la triiodothyronamine (XXXII) est plus active sur ce même test que la thyroxine.

Les acides 3 : 5 : 3' : 5'-tétraiodo- et 3 : 5 : 3'-triiodo-thyroacétique (XXXIII—XXXIV) ont été synthétisés par Harington et Pitt-Rivers (*69*, *120*). Le produit de départ est le 3 : 5-diiodo-4'-méthoxy-4-phénoxybenzaldéhyde, lequel est transformé en alcool correspondant, puis en dérivé chloré, puis en cyanhydrine, qui, après saponification, conduit à l'acide

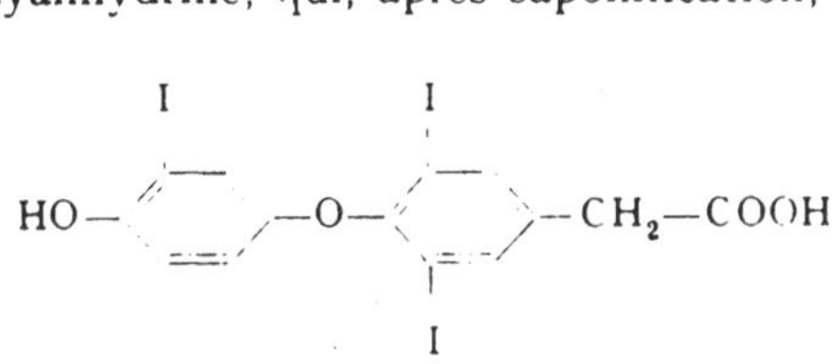

(XXXIII.) Acide 3 : 5 : 3'-triiodothyroacétique.

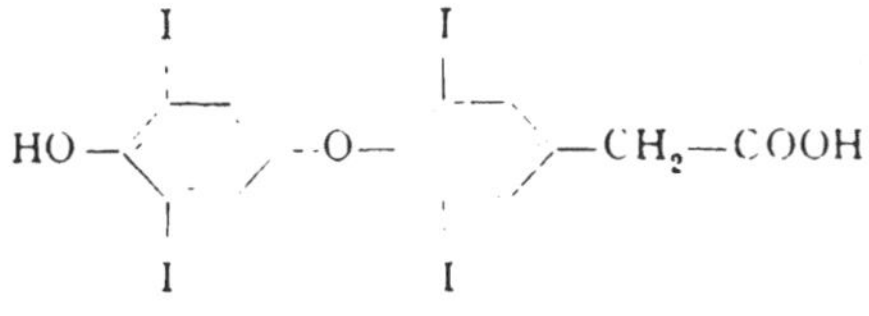

(XXXIV.) Acide 3 : 5 : 3' : 5'-tétraiodothyroacétique.

3 : 5-diiodothyroacétique; une ioduration ménagée conduit au dérivé triiodé, F. = 100° (*120*), une ioduration plus poussée au tétraiodé (*69*). L'activité antigoîtrogène de ces produits est voisine de celle de la thyroxine.

Les acides α-cétoniques dérivés de la *L*- et de la *D*-thyroxine (acide 3 : 5 : 3′ : 5′-tétraiodothyropyruvique) (*146*) et de la *L*-3 : 5 : 3′-triiodothyronine (acide 3 : 5 : 3′-triiodothyropyruvique) ont été caractérisés, le premier dans l'urine, le deuxième dans la bile de rats ayant reçu des injections de leurs précurseurs hormonaux respectifs. Les réactions, qui ont permis leur identification à partir d'éluats de chromatogrammes, sont rassemblées dans le *tableau 3*.

Tableau 3. Réactions de caractérisation des hormones thyroïdiennes et de leurs dérivés α-cétoniques (corps urinaire = acide 3 : 5 : 3′ : 5′-tétraiodothyropyruvique; corps biliaire = acide 3 : 5 : 3′-triiodothyropyruvique).

Réactions étudiées	*L*-thyroxine	Corps urinaire	*L*-3 : 5 : 3′-triiodothyronine	Corps biliaire
I. Réaction à la ninhydrine	+	—	+	—
II. Réactions du groupement phénol				
R. de Pauly	+	+	+	+
R. à l'α-nitroso-β-naphtol	—	—	+	+
III. Réactions des acides α-cétoniques				
R. à la 2 : 4-dinitrophénylhydrazine	—	+	—	+
R. à la semicarbazide	—	+	—	+
R. à l'*o*-phénylènediamine	—	+	—	+
R. de Jaffé	—	+	—	+

La présence d'un groupement phénolique libre sur un cycle dont les deux positions ortho (3′ et 5′) sont substituées, est illustrée par le fait que la réaction de Pauly est positive et celle à l'α-nitroso-β-naphtol négative. Celle d'un groupement α-cétonique l'est par les réactions à la 2 : 4-dinitrophénylhydrazine, à la semicarbazide, à l'*o*-phénylènediamine et de Jaffé; la perte de la réaction à la ninhydrine témoigne de la disparition du groupement α-aminé. Une confusion avec un dérivé cétonique de désioduration partielle pourrait théoriquement être envisagée, mais elle doit être écartée pour deux raisons. D'une part, les résultats obtenus sont identiques avec les deux radiothyroxines marquées, l'une en 3 : 5, l'autre en 3′ : 5′. D'autre part, dans le cas de la dernière, une désioduration respectant l'un des atomes marqués ne pourrait porter que sur l'autre fixé au même cycle. Or, le dérivé cétonique de la *L*-thyroxine étudié diffère, par certains de ses caractères (*tableau 3*), de celui de l'hormone triiodée. Aussi le produit extrait de l'urine d'animaux traités à la *L*-thyroxine paraît-il être l'acide 3 : 5 : 3' : 5'-tétraiodothyropyruvique. Les relations structurales des deux corps biologiquement nouveaux ressortent de leurs formules respectives (XXXV—XXXVI).

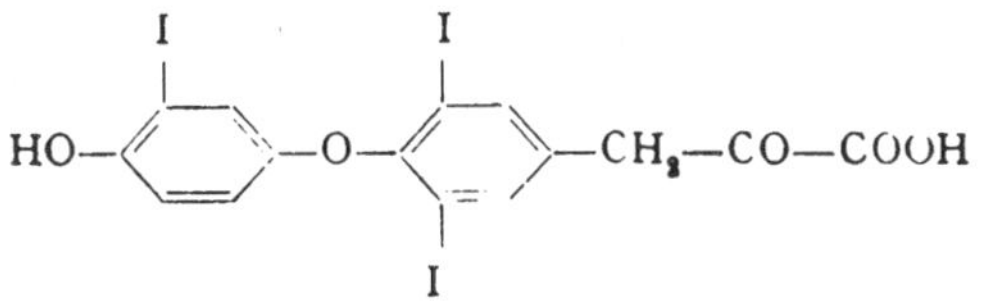

I I

HO—⟨ ⟩—O—⟨ ⟩—CH_2—CO—COOH

I

(XXXV.) Acide 3 : 5 · 3'-triiodothyropyruvique.

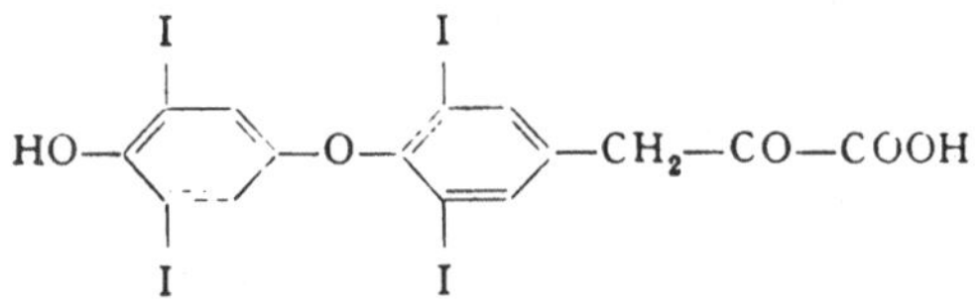

I I

HO—⟨ ⟩—O—⟨ ⟩—CH_2—CO—COOH

I I

(XXXVI.) Acide 3 : 5 : 3' : 5'-tétraiodothyropyruvique.

4. Acides aminés marqués par I^{131}.

La préparation d'acides aminés marqués par I^{131} a seule permis l'étude du métabolisme de certains acides aminés iodés, en particulier de ceux de caractère hormonal. Il a été nécessaire de disposer dans ce but de produits doués d'une radioactivité spécifique très élevée, afin de les administrer aux doses physiologiques minimes qui, seules, ne dérèglent pas le fonctionnement du corps thyroïde et l'utilisation cellulaire de ces produits (quelques microgrammes ou moins pour les animaux d'expérience). Aussi des techniques portant sur de petites quantités de substance ont-elles dû être élaborées et il a été fait un large appel dans ce but à des séparations chromatographiques (*132*).

a) Iodotyrosines marquées.

L'étude radiochromatographique de l'ioduration de la *L*-tyrosine par I^{127} additionné d'I^{131} a permis, comme on l'a vu, de préciser les conditions optima d'obtention de la *L*-3-mono- et de la *L*-3 : 5-diiodotyrosine. Chacune des réactions de substitution (a) et (b) (a: $RH_2 + I_2 = RHI + HI$ et b: $RH_2 + 2\,I_2 = RI_2 + 2\,HI$) évolue en effet préférentiellement selon la quantité relative d'halogène présente. Les produits obtenus peuvent facilement être séparés par chromatographie sur papier et identifiés par leur R_f. Des mesures de la répartition de la radioactivité entre les taches permettent, connaissant la quantité totale d'$I^{131} + I^{127}$ mise en oeuvre par molécule d'acide aminé, de calculer le nombre d'atomes d'iode fixés par celle-ci (*128*).

La *L*-3-monoiodotyrosine est le produit principal de la réaction d'une molécule de *L*-tyrosine avec trois atomes d'iode et la *L*-3 : 5-diiodotyrosine prédomine avec six atomes de l'halogène. La séparation chromatographique sur papier Whatman No 1 est satisfaisante en présence du mélange *n*-butanol : acide acétique : eau (78 : 5 : 17). Elle permet d'obtenir facilement quelques milligrammes ou plus de chacun des dérivés dont la radioactivité est fonction de celle du mélange iodurant. Deux remarques techniques importantes sont à retenir. D'une part, l'emploi de chlorure d'iode à chaud ne permet pas d'obtenir le dérivé monosubstitué. D'autre part, la mise en oeuvre de solutions riches en I_2^{131} est réalisée de la manière la plus satisfaisante par l'emploi d'éther comme solvant; il est ainsi facile de manipuler de 25 à 50 millicuries I_2^{131}.

b) Iodothyronines marquées.

Des problèmes particuliers se sont posés à propos de la synthèse de ces corps, car les atomes d'iode qu'ils renferment peuvent être fixés en différentes positions au squelette carboné de la thyronine. La 3 : 5-diiodo-*L*-thyronine marquée en 3 : 5 a été synthétisée à partir du 3 : 5-diamino-*p*-méthoxyphénoxy-N-acétyl-*L*-phénylalanine-éthylester (*32*) (diazotation, décomposition du diazoïque par I_2^{131} et hydrolyse du 3 : 5-diiodo-*p*-méthoxyphénoxy-N-acétyl-*L*-phénylalanine-éthylester (*schéma 1*). Le dérivé 3 : 5-diiodé homologue (*103*) est surtout important comme précurseur d'iodothyronines. Traité par l'acide iodhydrique en milieu acétique, il donne naissance à la 3* : 5*-diiodo-*L*-thyronine (*103*) (F. = = 245°, iode = 48,4%) avec un rendement de 10% en poids du produit de départ. A partir de ce corps, on peut réaliser l'ioduration en 3' ou en 3' : 5' par action directe de l'halogène ($I_2^{127} + I_2^{131}$) en proportion convenable. La proportion la plus élevée en dérivé triiodé est obtenue avec deux atomes d'iode par molécule d'acide aminé et la saturation en 3' : 5' avec 6 à 8 atomes d'halogène. La séparation chromatographique sert à l'isolement de ces corps; elle exige l'emploi d'un solvant approprié, tel que l'isopentanol saturé d'ammoniaque 6*N* et l'élution en milieu ammoniacal. Il a été possible de synthétiser aussi les *L*-triiodothyronines marquées en 3' ou en 3 : 5, en partant, dans le premier cas, d'un précurseur diiodé non marqué traité par I_2^{131} et, dans le second cas, d'un précurseur marqué traité par I_2^{127}. La préparation des *L*-thyroxines radioactives est relativement plus simple, parce qu'elles peuvent être directement cristallisées, sans séparation chromatographique préalable. Celle du radioisomère marqué en 3' : 5', la première décrite, a été réalisée à partir de la 3 : 5-diiodo-*L*-thyronine non radioactive (*80*) et celle du radioisomère marqué dans tous ses atomes d'iode, à partir du dérivé diiodé marqué (*103*).

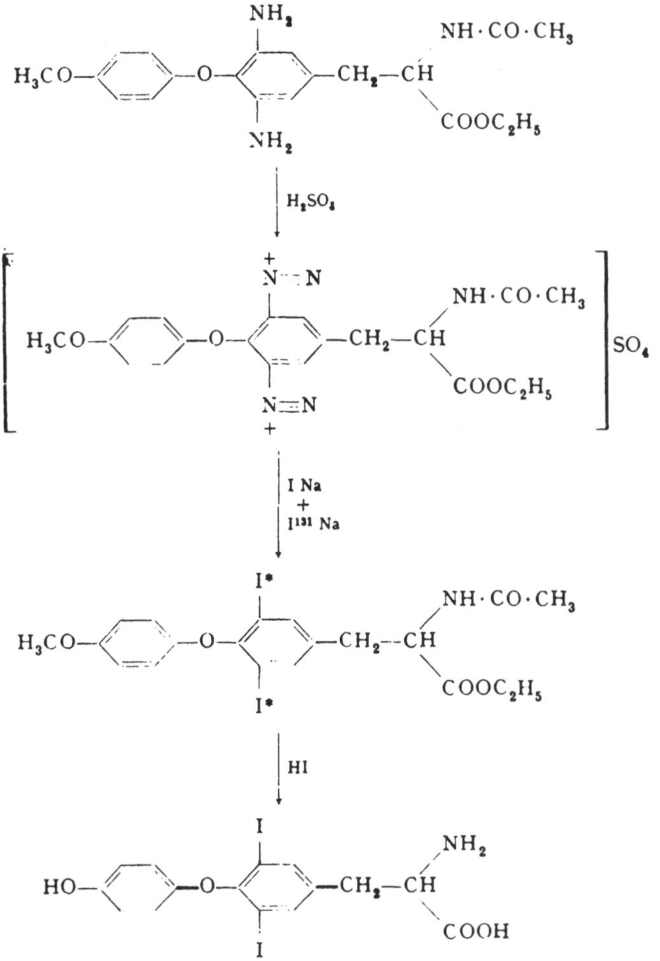

NH_2 / $NH\cdot CO\cdot CH_3$

H_3CO—⟨benzène⟩—O—⟨benzène⟩—CH_2—CH

NH_2 / $COOC_2H_5$

↓ H_2SO_4

[$\overset{+}{N}\equiv N$ / $NH\cdot CO\cdot CH_3$

H_3CO—⟨benzène⟩—O—⟨benzène⟩—CH_2—CH

$N\equiv\overset{+}{N}$ / $COOC_2H_5$] SO_4

↓ I Na + I^{131} Na

I* / $NH\cdot CO\cdot CH_3$

H_3CO—⟨benzène⟩—O—⟨benzène⟩—CH_2—CH

I* / $COOC_2H_5$

↓ HI

I / NH_2

HO—⟨benzène⟩—O—⟨benzène⟩—CH_2—CH

I / COOH

Schéma 1.

III. Iodoprotéines.

Certaines iodoprotéines, les thyroglobulines et diverses scléroprotéines d'Invertébrés, sont des corps naturels, et d'autres des produits artificiellement halogénés. Bien que toutes renferment les mêmes acides aminés iodés, en particulier des iodotyrosines et des iodothyronines, l'identité des problèmes posés par l'étude de toutes les protéines n'est apparue que récemment, depuis que LUDWIG et VON MUTZENBECHER (96) ont établi la formation de thyroxine par action de l'iode sur la caséine et sur diverses protéines. Les iodoprotéines artificielles peuvent être considérées comme des «modèles» de leurs homologues naturels et, de ce fait, l'étude du mécanisme de leur halogénation a présenté un grand intérêt à la fois pour la chimie des protéines et pour celle de la formation des hormones thyroïdiennes et de leurs précurseurs. Il a paru légitime de décrire tout d'abord les iodoprotéines artificielles et les réactions leur donnant naissance dans des conditions chimiquement définies, car l'halogénation biologique des iodoprotéines naturelles paraît relever des mêmes réactions.

1. Protéines artificiellement iodées.

L'action de l'iode sur les protéines a tout d'abord été réalisée dans le but de préparer des produits doués d'une activité antigoîtrogène. Par la suite, la mise en évidence d'iodotyrosines dans les gorgonines et les spongines a orienté les recherches vers le mécanisme de la fixation des halogènes sur les protéines; enfin, l'intérêt de celui-ci s'est beaucoup étendu par la suite, lorsque la formation de thyroxine au cours de l'ioduration de diverses protéines est apparue comme l'une de ses modalités. Jusque vers 1930, les biochimistes ont eu comme principal objet des préoccupations d'ordre quantitatif, tendant à préciser les quantités d'iode que pourrait fixer une protéine dans une ou dans plusieurs réactions successives définies avec celui-ci. Ce point de vue apparaît aujourd'hui comme secondaire, et nous ne le retiendrons que dans la mesure où il a conduit à l'étude de la thyroxinogénèse dans les iodoprotéines artificielles, dont nous exposerons plus longuement les résultats.

a) Ioduration des protéines et réactions de substitution.

L'action de l'halogène comporte des processus de substitution, d'oxydation et, éventuellement, d'addition. Les premiers correspondent au schéma général: $R\mathrm{H} + \mathrm{I}_2 = \mathrm{RI} + \mathrm{IH}$ et portent préférentiellement sur la *L*-tyrosine. Les seconds comportent la transformation des groupements thiol en disulfure en présence d'un excès d'iode, et, dans divers milieux, des phénomènes plus complexes modifiant la structure de la protéine par rupture de certaines liaisons ou par dégradation de certains acides aminés (transformation de méthionine en sulfoxyde; ouverture du cycle indolique du tryptophane, du cycle imidazolique de l'histidine). Les réactions d'addition sont assez mal définies, car, bien que la fixation de l'iode à la liaison thioéther de la méthionine soit hors de doute, celle à l'azote des liaisons peptidiques demeure hypothétique. En outre, il est probable que les teneurs élevées en iode de certaines préparations (près de 20% dans des «periodocaséines») (*93*) est due à une adsorption d'iodures. Les réactions de substitution et d'oxydation ont lieu en présence d'iode libre et leur intensité est sous la dépendance des conditions dans lesquelles elles ont lieu. Un choix judicieux de celles-ci permet de réduire les processus d'oxydation en les limitant à des phénomènes réversibles, en particulier à la transformation de groupements thiol en disulfure, et d'éviter la dénaturation.

L'ioduration sans oxydation autre que celle des groupements thiol et sans dénaturation est réalisée par l'addition de petites quantités de l'halogène (solution iodo-iodurée) à $p_H = 6{,}0$—$7{,}0$ à 0—15°, à de la pepsine (77), à de la sérumalbumine (*24*, *109*), à de la thyroglobuline (*98*). La teneur en iode des produits obtenus est alors fonction de la quantité de réactif mise en oeuvre; elle varie de 1 à 4% d'iode, alors qu'elle est beaucoup plus élevée (de 6 à 14% d'iode) en présence d'un excès d'halogène, dont l'action dénaturante se traduit par une diminution de la solubilité du produit soumis à l'ioduration. Les protéines naturelles présentent une réactivité vis-à-vis de l'halogène moindre que la *L*-tyrosine, mais un comportement inverse après dénaturation par l'urée (*89*). Les protéines végétales solubles dans des mélanges hydroalcooliques sont à cet égard plus stables; l'iodozéine préparée par addition de la quantité d'iode théoriquement nécessaire à la transformation complète de la *L*-tyrosine en *L*-3 : 5-diiodotyrosine ne semble pas dénaturée après avoir fixé de 6,9 à 7,5% d'iode (*111*).

Des iodosérumalbumines cristallisées (*24*) et des iodopepsines cristallisées (77, *78*), de teneurs en iode diverses, ont été obtenues; celles dont on peut préciser qu'elles n'étaient pas dénaturées ne renfermaient pas plus de 3% d'iode. L'étude roentgenographique de la structure de ces iodoprotéines cristallisées mériterait d'être entreprise, afin de préciser la position des restes de *L*-tyrosine substitués et, éventuellement, sa modification en fonction de leur dénaturation progressive. De premiers résultats intéressants ont été obtenus dans ce domaine sur l'iodofibroïne et ils pourraient être utilement étendus.

Les protéines douées d'une activité biologique mériteraient à cet égard de retenir l'attention. L'iodoinsuline décrite par Harington et Neuberger (*65*) renferme 15,4% d'iode, ce qui correspond à la transformation de la totalité de la *L*-tyrosine en dérivé diiodé; elle conserve de 5 a 10% de son pouvoir hypoglycémiant et en récupère de 30 à 50% par une hydrogénation catalytique qui la désiode en partie. Il serait important de préciser dans quelle mesure les modifications de l'activité de cette protéine sont ou non liées à l'halogénation de certains restes de tyrosine. Par ailleurs, l'étude des courbes de titration des iodoprotéines permet de définir le pouvoir tampon dû aux —OH phénoliques de la tyrosine (*89*), en raison de la modification de pK_{OH} provoquée par l'halogénation de cet acide aminé.

Les recherches sur les protéines iodées sans dénaturation sont encore peu nombreuses et il serait souhaitable qu'elles soient entreprises systématiquement, car on peut en attendre des renseignements importants sur les particularités de structure de chacune.

Les documents dont on dispose actuellement ont été presque tous établis sur des produits halogénés dans des conditions où la structure de la protéine a été modifiée, car le but poursuivi était avant tout l'obtention de produits doués d'activité thyroxinienne. La quantité d'iode fixée lorsqu'aucune précaution n'est prise pour éviter la dénaturation de la protéine est fonction de diverses variables. A concentration en halogène maintenue constante, la quantité de celui-ci combinée est plus faible en milieu acide (mélange d'un iodure et d'un iodate en solution sulfurique) qu'en milieu neutre ou faiblement alcalin ou en présence d'iodure d'azote, en milieu ammoniacal. Le dernier de ces réactifs permet d'atteindre les taux d'halogène les plus élevés dans les iodoprotéines. La concentration de l'agent d'ioduration, le p_H et la température interviennent également. Le rendement en thyroxine est le plus fort à $p_H = 8{,}0$ environ, tandis que la formation de 3 : 5-diiodotyrosine est d'autant plus facile que le p_H s'élève.

Sans passer ici en revue les multiples techniques de préparation proposées, il convient de signaler que, la réaction de substitution étant génératrice d'acide iodhydrique, il y a lieu de neutraliser celui à l'aide d'un sel tampon, tel que le bicarbonate de sodium. Par ailleurs, la formation d'iodates à partir d'iodures n'est pas négligeable aux p_H supérieurs à 9,5. En l'état actuel des techniques, les conditions les plus favorables pour préparer des produits actifs est celle élaborée par Reineke et Turner (*156*) et mettant en oeuvre l'action de l'iode en poudre sur des solutions protéiques de $p_H = 7{,}5$—$8{,}0$ renfermant du bicarbonate de sodium.

Chaque protéine s'halogène, dans des conditions expérimentales déterminées, à un taux qui lui est propre et que l'on a cherché à relier à sa composition et à sa structure. De premières recherches (*21*, *22*) ont conduit à différencier deux types de combinaisons iodées, grâce à leur réactivité vis-à-vis de l'acide sulfureux. Il a été signalé plus haut que les produits obtenus en milieu ammoniacal sont les plus riches en halogène; toutefois, après traitement par l'acide sulfureux, leur teneur en iode devient identique à celle des préparations halogénées en milieu acide ou neutre. Le réactif sulfureux décompose les combinaisons du type =N—I, mais non celles du type $>$C—I, en sorte que l'on a admis que les premières existent seules dans les produits iodés en solution acide ou neutre. Bien que de nombreuses données aient été publiées à cet égard (*11*, *12*), on ne peut pas préciser la quantité maxima d'iode qu'une protéine peut renfermer sous l'une ou l'autre forme, en raison de la multiplicité des processus ayant simultanément lieu en présence des réactifs et de la quasi-impossibilité de contrôler le terme de leur évolution.

La disparition de l'aptitude à donner certaines réactions colorées et la formation de produits de dégradation a néanmoins été utilement mise à profit (*21*, *22*, *92*). A partir d'un certain degré d'ioduration, les protéines ne présentent plus la réaction de Millon, positive avec la tyrosine et la

3-monoiodotyrosine, mais négative avec la 3 : 5-diiodotyrosine. De même on assiste à la disparition progressive des réactions de VOISENET et d'ADAMKIEWICZ, propres au cycle indolique du tryptophane, et de celle de PAULY, caractéristique de la tyrosine et de l'histidine. Le chauffage en milieu alcalin ne provoque plus la formation de sulfures, ce qui témoigne d'une dégradation de la cystéine. La réaction du biuret s'atténue puis disparaît, et de l'iodure de méthyle se forme en petite quantité, aux dépens du groupement —CH_3 de la méthionine.

Cette énumération suffit à montrer la complexité des modalités de l'action de l'halogène et nous ne retiendrons ici que celles conduisant à la fixation de celui-ci, à l'exclusion de celles comportant l'oxydation des groupements thiol, disulfure, thiométhyl et la rupture de liaisons peptidiques ou de cycles. Il y a toutefois lieu d'insister sur le fait que l'action d'un excès d'iode va toujours de pair avec des altérations de la structure de la protéine, en dehors du processus de substitution, et que celles-ci même peuvent ne constituer qu'une étape. C'est ainsi que l'étude des courbes de titration de certaines préparations (*6*, *79*) et le dosage de la tyrosine et de ses dérivés (*137*) a montré la destruction d'une partie de cet acide aminé dans certaines conditions opératoires.

La tyrosine fixe préférentiellement l'iode en donnant naissance à la 3-monoiodotyrosine, à la 3 : 5-diiodotyrosine et, éventuellement, par un mécanisme qui sera discuté plus bas, à la thyroxine (*96*) et à la 3 : 5 : 3'-tri-iodothyronine (VOLPERT, communication personnelle). La formation d'iodohistidines (2 et 2 : 4), admise depuis longtemps par de nombreux auteurs (*6*, *11*, *12*) a été récemment établie (*129*). Celle de dérivés du tryptophane a été envisagée (*11*, *92*) en raison de la disparition des réactions colorées de ce corps; en fait, celle-ci traduit la dégradation du cycle indolique de l'acide aminé (*136*). Il découle de ces faits qu'il existe une certaine proportionnalité entre la teneur en tyrosine et, probablement, en histidine, d'une protéine et son aptitude à fixer l'iode. Toutefois, la réactivité de la première demeure beaucoup plus forte à cet égard, même dans des protéines comme les hémoglobines qui en sont relativement pauvres (3,5—4,0%) et sont par ailleurs riches en histidine (8,0—9,0%). L'histidine ne paraît fixer de quantités appréciables d'iode qu'après dihalogénation de la tyrosine. La réaction de substitution affectant les positions 3 et 5, les quantités d'iode qui y participent sont réduites par la nitration de l'une d'elles dans les restes de tyrosine des hémoglobines (*12*). Quant à la phénylalanine, elle ne se combine pas à l'halogène. Dans certains cas, il a été possible de relier de manière assez précise le taux de l'iode fixé à celui de la tyrosine, en particulier, dans les protéines pauvres en histidine, telles que la zéine (*111*, *139*) (I%: 6,9—7,5, tyrosine % = 5,8); l'insuline (*130*, *102*) (I%: 14,8—15,4; tyrosine % = 12,2) et la fibroïne (*102*) (I%: 12,7; tyrosine % = 12,0).

De nombreuses protéines pures ont été iodées à des degrés divers, selon leur composition et les modalités de leur réaction avec l'halogène; aussi nous bornerons-nous à les mentionner, sans indiquer de taux d'iode auxquels on ne peut qu'exceptionnellement accorder une signification absolue. La caséine a fait dans ce domaine l'objet de très nombreux travaux (*19, 23*, entre autres). Aucune protéine ne s'est révélée inapte à s'halogéner et l'on a décrit, en particulier, des dérivés iodés de la sérumalbumine (*24, 109*), de la sérumglobuline (*20*), de la globine (*11, 20*), de la thyroglobuline (*20*), de la gliadine, de la fibroïne (*20, 102, 96*), de la pepsine (*77*), de l'hormone galactogène hypophysaire (*91*) et de peptones (*139*). Tous ces produits présentent des modifications de leur spectre ultraviolet (*51, 113*) et de leur courbe de titration (*68, 79, 111*) caractéristiques de la transformation de leur tyrosine en dérivé disubstitué.

b) Formation de thyroxine au cours de l'ioduration des protéines.

Les iodoprotéines renferment des acides aminés iodés qui, en dehors des produits de substitution de la tyrosine et de l'histidine, dérivent de la thyronine. Le plus abondant de ceux-ci est la thyroxine (*96*), qu'accompagnent de petites quantités de 3:5:3'-triiodothyronine. La formation de la première a été étudiée par de nombreux auteurs après avoir été établie par Ludwig et von Mutzenbecher (*96*), et rapidement confirmée par Harington et Pitt-Rivers (*66*).

On avait tout d'abord pu considérer que les protéines artificiellement iodées renferment un homologue de la thyroxine, dit homothyroxine (*1, 2*). En fait, l'activité biologique de ces produits a pu être rapportée principalement à la thyroxine, isolée à l'état racémique des hydrolysats alcalins d'iodocaséine (*96*), puis à l'état d'isomère *L* de leurs hydrolysats acides soumis à l'extraction *n*-butanolique continue pendant toute la durée de la protéolyse (*155*). Deux problèmes importants, inséparables l'un de l'autre, se sont posés à propos de la formation de la thyroxine et, accessoirement de la triiodothyronine, à savoir la nature des réactions donnant naissance à ces corps et leur rendement.

Muus, Coons et Salter (*109*) ont été les premiers à aborder l'étude quantitative de la synthèse de la thyroxine. L'action de quantités croissantes d'iode sur la sérumalbumine leur a permis d'obtenir des produits stimulant de plus en plus fortement les échanges respiratoires des myxoedémateux à partir de teneurs en halogène de 6,0 jusqu'à 10,0% d'iode. Reineke et Turner (*156*), Reineke, Williamson et Turner (*159*) ont abordé le problème de la formation de produits actifs sur le plan chimique, en reliant les quantités d'iode mises en oeuvre à celles de tyrosine présente dans la protéine.

Les conditions les plus favorables à la formation de dérivés actifs sont: l'addition à + 37° d'iode en poudre à des solutions protéiques renfermant 1,4% CO_3NaH, par

petites portions et sous agitation en 2 h 30, puis l'incubation du milieu en vase clos à 70° pendant 18 h.

La plus forte activité biologique est obtenue lors de l'emploi de 4,5 (*158*) à 6,0 (*159*) atomes d'iode par molécule de tyrosine présente dans la protéine et la dernière de ces valeurs a été en général retenue pour la préparation industrielle d'iodoprotéines.

Des difficultés se sont manifestées au sujet du dosage des produits actifs et le seul critère absolu demeure leur activité biologique. En effet, d'une part la proportion de *L*-thyroxine et de *L*-3 : 5 : 3'-triiodothyronine qu'ils renferment est pratiquement impossible à définir, d'autre part la présence de produits d'oxydation des acides aminés iodés est susceptible d'introduire des erreurs dans l'application des méthodes chimiques de dosage (*122*, *157*, *135*, *164*). C'est ainsi qu'un corps comme le 3 : 5-diiodo-4-hydroxybenzaldéhyde isolé de produits d'hydrolyse des iodoprotéines (*119*) présente la même solubilité dans le *n*-butanol que la *L*-thyroxine. De même, les caractères chromatographiques de certains produits de dégradation de la 3 : 5-diiodotyrosine sont identiques à ceux de la thyroxine dans de nombreux solvants (*165*). On peut admettre que la teneur maxima en thyroxine des iodocaséines est de 1,2—1,5%; la triiodothyronine a été identifiée dans celles-ci mais son taux paraît y être très faible et son existence ne saurait être admise que dans des préparations actives relativement pauvres en halogène*.

L'étude quantitative de la formation des 3-monoiodotyrosine, 3 : 5-diiodotyrosine et thyroxine en présence de quantités croissantes d'iode (*138*, *139*) a été poursuivie sur diverses protéines dans les conditions adoptées par REINEKE et ses collaborateurs ou en milieu ammoniacal (NH_4OH à 5%). Dans tous les cas étudiés (caséine, insuline, thyroglobuline, zéine), la 3-monoiodotyrosine est présente en abondance dans les produits faiblement iodés et la 3 : 5-diiodotyrosine prédomine au fur et à mesure que le degré d'halogénation augmente. La thyroxine et son homologue trisubstitué n'apparaissent qu'à partir d'une certaine teneur en iode (6% pour la caséine, 3% pour la thyroglobuline et 4,5% pour la zéine, entre autres). Lorsque l'halogénation est réalisée en milieu ammoniacal, par l'iodure d'azote, une teneur maxima en thyroxine paraît être atteinte à partir d'un certain taux d'I total fixé. En milieu bicarbonaté ou sodique, on observe, au contraire, un maximum sensiblement plus élevé du rendement en hormone, laquelle serait alors en partie détruite par un excès de réactif, comme la 3 : 5-diiodotyrosine, ce qui explique la présence de 3 : 4-diiodo-4-hydroxybenzaldéhyde dans le milieu (*90*). Ce maximum est atteint lorsque six atomes d'I réagissent avec un reste de tyrosine.

* Les activités biologiques des iodoprotéines ayant été examinées en détail dans une précédente revue (*153*), il ne nous a pas paru utile de l'exposer à nouveau dans cet article, dont l'orientation doit demeurer plus strictement chimique.

L'influence de multiples conditions opératoires, entre autres de l'agitation, de l'oxygénation, de la température, du p_H, des oxydes de manganèse, a été étudiée par Reineke, Turner et leurs collaborateurs (*156*, *158*, *159*). Dans l'ensemble, la formation de l'hormone paraît favorisée par divers facteurs d'oxydation, dans la mesure où ceux-ci ne la détruisent pas au fur et à mesure qu'elle prend naissance.

Des valeurs traduisant le rendement maximum approximatif en thyroxine de l'ioduration de diverses protéines et peptones ont été rassemblées dans le *tableau 4* (*139*).

Tableau 4. Rendement maximum en thyroxine de l'ioduration de protéines et de peptones pepsiques traitées par six atomes d'iode par reste de tyrosine, selon la technique de Reineke et Turner.

Produit étudié	Teneur initiale en tyrosine % A	Teneur en thyroxine % après ioduration B	Rapport $\frac{B \times 100}{A}$
Caséine	7,20	1,65	22,9
Fibroïne	12,00	0,40	3,3
Insuline	12,20	1,35	12,7
Thyroglobuline	3,30	0,68	20,6
Zéine	5,84	1,61	27,5
Peptone I de caséine	1,07	0,04	3,7
Peptone II de caséine	2,70	0,14	5,2

Bien que les teneurs en thyroxine indiquées ne correspondent qu'à un ordre de grandeur, elles traduisent nettement deux faits. Seule une fraction de la tyrosine est susceptible de donner naissance au produit hormonal; on retrouve d'ailleurs la plus grande partie de l'autre à l'état de 3 : 5-diiodotyrosine dans les produits riches en halogène. En outre, le taux de transformation de la tyrosine en thyroxine est très différent d'une protéine à l'autre, ce qui est probablement dû à leur diversité de structure. La teneur en tyrosine ne saurait, en effet, régir à elle seule la réaction et, dans le cas de la fibroïne de la soie, l'étude roentgenographique a montré que les distances séparant les restes iodés de l'acide aminé portés par les fibres protéiques rigides sont trop éloignés pour qu'ils puissent se condenser. L'influence de facteurs structuraux est, d'ailleurs, à coup sûr complexe, car en dehors de la proximité plus ou moins grande des restes, leur réactivité vis-à-vis de l'iode ne saurait être négligée, comme le montre l'augmentation d'affinité des protéines pour celui-ci après une dénaturation (*89*, *90*).

Il est remarquable qu'à cet égard la thyroglobuline, substrat naturel de la thyroxinogénèse, ne paraisse pas présenter une structure se prêtant particulièrement à celle-ci; la spécificité de la fonction thyroïdienne paraît liée à la coordination, dans les cellules épithéliales de la glande, des divers éléments physiologiques du mécanisme de l'ioduration de cette protéine, plutôt qu'à la nature de la thyroglobuline.

c) Mécanisme chimique de formation de la thyroxine et de la 3 : 5 : 3'-triiodothyronine.

La présence simultanée de 3 : 5-diiodotyrosine et de thyroxine dans les protéines halogénées a suggéré l'hypothèse que la première est la substance-mère de la seconde et le *schéma 2* suivant rend compte du mécanisme global de celle-ci (*62*).

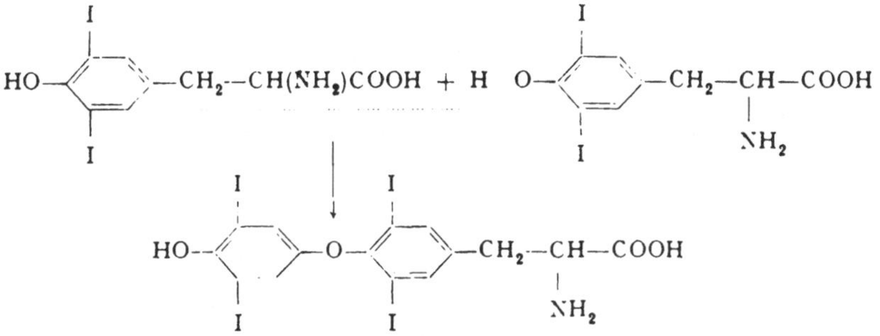

HO–$C_6H_2I_2$–CH_2–CH(NH_2)COOH + H O–$C_6H_2I_2$–CH_2–CH(NH_2)–COOH

↓

HO–$C_6H_2I_2$–O–$C_6H_2I_2$–CH_2–CH(NH_2)–COOH

Schéma 2.

La réaction, réalisant la condensation d'une molécule de 3-mono-iodotyrosine et d'une molécule de 3 : 5-diiodotyrosine, conduirait à la formation des deux triiodothyronines isomères (3 : 5 : 3' et 3 : 3' : 5' selon que l'élimination du reste d'alanine porte sur le dérivé mono- ou diiodé). D'autres hypothèses ont envisagé la formation de 3 : 5-diiodothyronine ou du 4-hydroxyphényl éther de la tyrosine comme produits intermédiaires; elles ne peuvent plus être considérées comme plausibles.

Une première démonstration expérimentale de la synthèse de la thyroxine à partir de 3 : 5-diiodotyrosine a été apportée par VON MUTZENBECHER (*108*). L'incubation à 37° de solutions faiblement alcalines de ce dernier acide aminé conduit en deux semaines à la formation de thyroxine et le sulfite de sodium inhibe cette réaction. Cette importante observation a été confirmée par divers auteurs (*8*, *16*). Comme la présence d'eau oxygénée améliore sensiblement le rendement en thyroxine, on a envisagé qu'un processus d'oxydation participe à la formation de celle-ci. HARINGTON et PITT-RIVERS (*68*) ont réussi à porter à 4,1% la quantité d'hormone recueillie, en l'extrayant par le *n*-butanol au fur et à mesure de son apparition dans le milieu porté à 100°. Dans les conditions adoptées, la thyroxine ne prend naissance qu'en présence d'oxygène ou, mieux, d'eau oxygénée, à un p_H optimum voisin de 10, auquel la 3 : 5-diiodotyrosine est pratiquement présente en totalité à l'état d'ion phénoxyde. Il convient de remarquer que la plus grande partie de la 3 : 5-diiodotyrosine est désiodée au cours des opérations, le milieu s'enrichissant progressivement en iodures. L'agent catalytique de l'oxydation serait pour certains l'iode (*68*), pour d'autres l'hypoiodite alcalin (*83*) formés dans le milieu par dégradation de la 3 : 5-diiodotyrosine; l'addition du premier exerce seule un effet marqué à cet égard.

D'autres oxydants que l'eau oxygénée se sont montrés actifs; tel est le cas de Mn_3O_4 (*153*) et la réaction a été étendue à la condensation de deux molécules d'acide *DL*-3 : 5-diiodo-4-hydroxyphényllactique en un isologue de la thyroxine, l'acide *DL*-3 : 5-diiodo-4(3' : 5'-diiodo-4'-hydroxy)-phénoxyphényllactique (*163*). La constitution du corps participant à ce processus exerce un effet important sur l'intensité de celui-ci. Il a été possible d'obtenir des dérivés de la thyroxine à partir de divers amides ou peptides de la 3 : 5-diiodotyrosine.

L'incubation de la N-acétyl-3 : 5-diiodotyrosine à $p_H = 7{,}5$ et à 38° pendant 11 jours conduit à la N-acétylthyroxine avec un rendement de 7%, celle de l'acide N-acétyl-3 : 5-diiodotyrosylglutamique à l'acide N-acétyl-thyroxylglutamique avec un rendement de 36% et celle de la N-benzoyl-3 : 5-diiodotyrosine n'est pas suivie d'un résultat analogue (*119*). De même, 26% de la *L*-leucyl-*L* tyrosine incubée à 37° à $p_H = 8{,}5$ pendant 22 heures en présence de quatre atomes d'iode par molécule sont condensés en dérivé thyroxinien, le même processus n'ayant pas lieu à partir de divers autres peptides de tyrosine étudiés (*137*). L'incubation de N-acétyl-3 : 5-diiodotyrosylgélatine conduit à la formation de thyroxine avec un rendement d'environ 2%, avec désioduration d'une partie des restes de N-acétyl-3 : 5-diiodotyrosine fixés à la protéine (*140*). Le p_H optimum de la condensation, de 10 avec l'acide aminé diiodé, est de 7,4 avec ses peptides, ce qui peut présenter un intérêt pour la biosynthèse de la thyroxine.

La complexité du phénomène et l'évolution de réactions d'oxydation dans les milieux soumis à l'incubation est illustrée par l'isolement de ceux-ci du 3 : 5-diiodo-4-hydroxybenzaldéhyde, du triiodophénol, des acides oxalique et pyruvique (*119*, *83*). Toutefois certains de ces corps peuvent provenir de réactions indépendantes de la thyroxinogénèse et le mécanisme intime de celle-ci demeure obscur. A partir de travaux de Pummerer et collaborateurs (*123*) sur l'oxydation du *p*-crésol et de divers phénols, on a pu envisager (*83*, *161*) que la condensation initiale de deux molécules de 3 : 5-diiodotyrosine conduit à un produit intermédiaire labile dont la dégradation libère la déhydroalanine, $CH_2{=}C(NH_2){-}COOH$, laquelle serait alors soit hydratée en sérine, soit oxydée en acide pyruvique et ammoniac. En fait, on a bien isolé l'acide pyruvique (*83*) et la sérine (Ohno, communication personnelle), mais de l'alanine a été également obtenue à partir des hydrolysats de milieux d'incubation de la N-acétyl-3 : 5-diiodotyrosine (*119*). Aussi peut-on remarquer avec Pitt-Rivers (*119*) que, si de l'alanine est directement formée au cours de la condensation, celle-ci pourrait reposer sur une dismutation et non sur une oxydation. En l'état actuel de nos connaissances, tous les schémas réactionnels proposés demeurent hypothétiques.

Ainsi, l'étude des protéines artificiellement iodées a fourni aux biochimistes la possibilité de dissocier divers processus et l'extrapolation de ses résultats aux iodoprotéines naturelles s'est montrée féconde. Elle a perdu une partie de son intérêt lorsque l'utilisation du radioiode a permis de suivre directement les étapes de l'halogénation de celles-ci, mais comporte néanmoins d'importantes possibilités de recherches, tant en elle-même que dans l'étude du mécanisme de la condensation des iodotyrosines et de leurs dérivés.

2. Thyroglobuline.

L'intérêt de cette protéine tient principalement à deux faits. D'une part, le corps thyroïde des Vertébrés renferme une fraction importante de l'iode total des organismes, et ceci presque entièrement dans la thyroglobuline de ses vésicules colloïdes. C'est ainsi que, chez l'Homme, on peut estimer à 15—20% de l'halogène total la proportion présente dans la glande, dont le poids correspond à environ 0,03% du poids corporel. D'autre part, deux des acides aminés iodés de la thyroglobuline, à savoir: la *L*-3 : 5 : 3'-triiodothyronine et la *L*-thyroxine constituent des hormones thyroïdiennes.

L'étude autoradiographique de coupes de glandes provenant d'animaux traités par des doses traceuses d'iodures marqués a permis de constater qu'I^{131} est d'abord concentré par les cellules épithéliales (*84*). Il s'y fixe sur la thyroglobuline qu'élaborent celles-ci avant d'être sécrété avec elle dans les vésicules colloïdes que bordent les cellules (*166, 170*). Le cycle de réactions donnant naissance aux hormones dérivées de la thyronine évolue principalement dans les cellules épithéliales, mais il n'est pas certain qu'il ne se poursuive pas dans les vésicules colloïdes. L'iodoprotéine mise en dépôt dans celles-ci n'est pas mobilisée en vue de la sécrétion hormonale. Celle-ci porte exclusivement sur certains de ses acides aminés iodés que libère l'hydrolyse catheptasique préalable de la thyroglobuline. Le rôle biologique de cette protéine est donc d'être le substrat de l'hormonogénèse et la source des produits de la sécrétion interne de la glande.

a) *Préparation.*

L'extrait thyroïdien brut obtenu par extraction au moyen de solutions salines isotoniques (NaCl à 0,9%) de coupes minces de glande congelée est la meilleure matière première (*73*). Il renferme le contenu des vésicules colloïdes dans sa quasi-totalité et seulement des quantités assez faibles de protéines cellulaires; on peut y recueillir plus de 90% de l'iode total de l'organe. L'isolement de l'iodoprotéine thyroïdienne a été tenté par diverses techniques dont les deux objets principaux ont été les suivants: obtenir des produits homogènes pouvant être considérés comme purs et essayer de séparer des protéines de degrés d'halogénation divers susceptibles d'être simultanément présents.

Les recherches préliminaires de BAUMANN et ROOS (*13*), de BUBNOW (v. *14*), de GOURLAY (*57*), d'HUTCHISON (*81*, *82*) ont montré le caractère globulinique de l'iodoprotéine thyroïdienne. OSWALD (*114*) a réalisé la première étude précise de celle-ci, sur des préparations obtenues par précipitation au moyen de $SO_4(NH_4)_2$ à demi saturation de l'extrait thyroïdien (NaCl 0,9%) débarrassé des protéines précipitables par addition d'acide acétique. Le produit obtenu renferme la plus grande partie de la thyroglobuline de l'extrait, en grande partie dénaturée, comme tel est également le cas de celle résultant de la précipitation alcoolique de la même protéine (*10*). La méthode décrite par HEIDELBERGER et PALMER (*73*) conduit à un produit plus pur, qui se dénature partiellement au cours de sa précipitation à $p_H = 4{,}8$, réalisée dans le but d'éliminer le nucléoprotéide qui accompagne l'iodoprotéine. CAVETT et SELJESKOG (*30*) ont réussi à éviter la solubilisation du premier en opérant l'extraction de fines coupes de la glande congelée, débitée au microtome, à l'aide d'une solution isotonique de NaCl à $p_H = 7{,}4$ et à 0° et en séparant la thyroglobuline des fractions protéiques précipitant entre 35 et 45% de la saturation en sulfate d'ammonium.

DERRIEN, MICHEL et ROCHE (*40*) ont entrepris une étude systématique des protéines extraites selon le même procédé et réalisé leur fractionnement en présence de sels neutres (sulfate d'ammonium ou de sodium, mélange équimoléculaire de phosphates mono- et dipotassique). Ils ont ainsi obtenu les préparations que l'on peut considérer actuellement comme les plus pures de cette protéine, électrophorétiquement homogènes, dont les propriétés seront étudiées plus bas.

Les solubilités des protéines présentes dans les extraits bruts des glandes (Boeuf, Cheval, Chien, Homme, Porc), additionnés de divers sels neutres, présentent des différences permettant de les fractionner. Comme par ailleurs, la thyroglobuline est la seule iodoprotéine existant dans les milieux, il a été possible de la localiser parmi ces fractions grâce à sa teneur en iode, ou, lorsque les préparations provenaient d'animaux traités par I^{131}, grâce à sa radioactivité. La thyroglobuline présente des caractères de solubilité constants en présence d'un sel neutre donné, à p_H et température définis, ce qui permet de la séparer des protéines qui l'accompagnent, en quantités ne dépassant pas 10% des protéines totales chez les animaux normaux, plus fortes dans les organes pathologiques. Elle se dénature rapidement aux p_H inférieurs à 5,0, ce qui se traduit par une diminution de sa solubilité. L'étude de la teneur en phosphore et en iode précipitant à diverses concentrations croissantes en un sel neutre a permis de définir les conditions dans lesquelles on peut séparer des extraits glandulaires une thyroglobuline non dénaturée, ne renfermant pas d'impuretés nucléoprotidiques ou phospholipidiques.

Les préparations présentent des courbes de précipitation rigoureusement identiques de l'iode et de l'azote qu'elles renferment; le rapport: N/I des précipités protéiques recueillis au fur et à mesure que l'on augmente la concentration en sel du milieu est alors constant, ce qui traduit l'homogénéité du matériel précipité en ce qui concerne sa teneur en iode.

Le procédé de préparation adopté repose sur le fait que la thyroglobuline en solution diluée (de 1,5 à 2 mg. N protéique par ml.) précipite dans une marge très étroite de concentration en sels neutres à $p_H = 6,5$ et à 20°: de 43 à 48%, en volume, d'une solution équimoléculaire 3,5 *M* de phosphates mono- et dipotassique, de 36 à 41% de la saturation en sulfate d'ammonium par exemple. Les opérations sont conduites sur des solutions délipidées au toluène et dans des conditions qui éliminent au maximum les impuretés. Même après un certain nombre de reprécipitations et remises en solution successives (*39*, *40*), les préparations obtenues, dépourvues de nucléoprotéides, ne renferment plus de phosphore; mais il n'a pas été possible d'en éliminer totalement des glucides, entre autres de la glucosamine. Comme la teneur en celle-ci s'abaisse progressivement jusqu'à environ 2% au cours de la purification, il est probable qu'une petite quantité d'un glycoprotéide d'accompagnement ne peut pas être séparé de la thyroglobuline, laquelle adsorbe par ailleurs très énergiquement la cathepsase des extraits au cours de la précipitation.

La préparation de thyroglobuline radioactive peut être réalisée en utilisant la même technique, mais en suivant les opérations de fractionnement par la mesure de l'intensité du rayonnement des fractions recueillies, au moyen d'un compteur de GEIGER-MÜLLER (*142*). Lorsqu'il est possible d'opérer sur un corps thyroïde dont l'ablation peut être prévue à l'avance, il suffit d'injecter à l'animal d'expérience des iodures marquées en quantité suffisante 24 heures avant l'abattage (1 mc. pour un boeuf ou un porc, 0,2—0,5 mc. pour un chien) (*142*). La thyroglobuline de sujets atteints de la maladie de Basedow a été également préparée (*132*). Dans le cas où pareille disposition ne peut pas être prise, des coupes d'une glande fraîchement prélevée placées dans une solution isotonique additionnée de doses traceuses d'iodures marquées incorporent en quelques heures l'halogène de ceux-ci à leur thyroglobuline, qui peut en être isolée par la suite. L'iodoprotéine marquée de l'homme normal a été ainsi préparée à partir d'organes de sujets sains morts accidentellement (*133*). L'incorporation d'iode marqué à des protéines thyroïdiennes a permis de mettre en évidence des anomalies de la biosynthèse de la thyroglobuline. Dans les extraits thyroïdiens de chiens normaux, traités par des iodures radioactifs 24 heures avant le prélèvement de l'organe, la précipitation de cette protéine par les sels entraîne pratiquement la totalité de la radioactivité présente. Il n'en est plus de même de ceux des animaux porteurs d'un goître expérimental au 6-N-propylthiouracile, où l'on peut mettre en évidence, en dehors de la thyroglobuline normale, une autre iodoprotéine marquée au moyen de sa solubilité (*142*) et de ses caractères électrophorétiques (*38*).

b) Propriétés et composition en acides aminés.

Les caractères de solubilité sommairement indiqués ci-dessus sont ceux d'une globuline animale typique.

Les thyroglobulines hautement purifiées sont homogènes à l'électrophorèse et présentent une mobilité électrophorétique moyenne de 4,54 × 10^{-5} $cm.^2$/volt sec. à p_H = 7,68 dans une solution tampon (phosphates alcalins de /2 = 0,2 (*39*). Aucune hétérogénéité ne se manifeste entre p_H = 6,2 et 7,7 (*98*). La constante de sédimentation de l'iodoprotéine de porc S^0_{20} est 19,4 S et son poids moléculaire M_s 650.000 (*46, 74*). Sa zone de stabilité est comprise entre p_H = 4,8 et 11,3 (*74, 75*) et son point isoélectrique est à p_H = 4,48.

La thyroglobuline de diverses origines animales présente, dans les milieux de forte concentration en sels neutres, une hétérogénéité qui se manifeste identiquement si l'on étudie sa solubilité selon le test de Northrop (solubilité à concentration croissante en protéine dans des milieux de même teneur en sel) ou au moyen de courbes de relargage (solubilité dans des milieux à concentration en sel et à teneur constante en protéine).

Les préparations hautement purifiées, homogènes à l'électrophorèse, présentent alors trois fractions bien individualisées. Toutefois, comme le rapport: N/I est identique dans ces trois constituants, il n'est pas impossible que ceux-ci correspondent à trois états d'une même protéine. Lundgren (97) a, en effet, observé que la thyroglobuline peut se dissocier réversiblement sous l'influence de multiples facteurs, tels que des variations de concentration de ses solutions, des modifications du p_H et de la constante diélectrique du milieu et la présence de certains sels à des molarités faibles. La formation d'agrégats de tailles définies qui se produit alors peut également avoir lieu dans des solutions fortement concentrées en sels neutres.

Tableau 5. Teneur en acides aminés de la thyroglobuline de porc.

Acide aminé	% de la protéine	Acide aminé	% de la protéine
Arginine	12,7	Cystine	3,6
Histidine	2,2	Méthionine	1,3
Lysine	3,4	Alanine	7,4
Phénylalanine	3,1	Glycine	3,7
Tryptophane	2,1	Leucine	12,8
Tyrosine	3,1	Valine	1,4
Diiodotyrosine*	0,5	Sérine	10,8
Thyroxine*	0,2	Proline	14

* Les teneurs en thyroxine et en diiodotyrosine n'ont pas de signification absolue, au même titre que celle en iode total (0,48% dans le produit étudié), car elles varient d'une préparation à l'autre. Celle en monoiodotyrosine ne peut être indiquée que dans son ordre de grandeur, voisin de 0,10%. Les écarts entre les teneurs en ces acides aminés, d'une préparation à l'autre, sont trop faibles pour se refléter de manière appréciable sur le taux de *L*-tyrosine.

La teneur en acides aminés de la thyroglobuline a été étudiée par divers auteurs (*44, 173*) sur des produits de pureté souvent insuffisante et avec des méthodes d'une précision inférieure à celle actuellement en usage. Des données obtenues sur des préparations hautement purifiées et que l'on peut considérer comme correctes dans leur ordre de grandeur, sinon comme très rigoureuses (*39*), ont été rassemblées dans le *tableau* 5; elles mériteraient d'être complétées.

Les caractères les plus remarquables de composition de la thyroglobuline sont sa teneur élevée en arginine, en sérine, en proline et la présence d'acides aminés iodés. En ce qui concerne l'arginine, il a été établi que certaines des liaisons peptidiques auxquelles participe cet acide aminé sont exceptionnellement résistantes à l'hydrolyse acide, contrairement à ce qui a lieu avec diverses autres protéines de taux voisins en arginine, comme la thymohistone et l'édestine (*149*).

Divers problèmes se posent dans le cadre de la biosynthèse des protéines thyroïdiennes et dans celui de son ioduration.

Des recherches ont été entreprises pour préciser, d'une part si l'on trouve dans un corps thyroïde normal plusieurs thyroglobulines et, d'autre part, si l'iodoprotéine thyroïdienne présente des différences notables d'une espèce à l'autre et au cours d'états pathologiques. Il est très peu probable qu'il existe de multiples thyroglobulines dans la glande d'un même sujet; toutes les fractions que l'on peut en isoler ont une composition identique en acides aminés et il en est de même, d'un lot d'organes à l'autre, chez les animaux de même espèce. En revanche, de petites différences ont été mises en évidence entre les préparations provenant d'espèces diverses ou d'organes anormaux. Par exemple, la thyroglobuline de Porc renferme 3,12% de tyrosine et 2,08% de tryptophane, et celle du Chien respectivement 2,42% et 2,28% des mêmes acides aminés (*39*).

Les protéines de glandes fortement hypertrophiées par des goîtres de divers types présentent des compositions légèrement aberrantes par rapport à celle des organes normaux. Chez l'Homme, la thyroglobuline du goître colloïdal est nettement plus riche en tyrosine que celle du goître hyperplasique, alors que la teneur en d'autres acides aminés est à peu près identique (*28, 29*). Chez le Porc et le Boeuf, des glandes du premier type de goître renferment une protéine sensiblement plus riche en tyrosine et plus pauvre en cystine que le corps thyroïde non hypertrophié, comme en témoignent les quelques exemples suivants: 3,65% de tyrosine au lieu de 3,12, et 2,20% de cystine au lieu des 3,60% dans une glande de Porc pesant 110 g.; 5,52% de tyrosine au lieu de 3,52% et 1,60% de cystine au lieu de 3,52% dans une glande de Boeuf pesant 80 g.

Il en découle que la biosynthèse de la protéine thyroïdienne est modifiée par certaines conditions pathologiques, ce qui, comme on le verra plus bas, n'est pas sans influence sur la formation des hormones dans sa molécule.

Bien que l'on ne possède que très peu d'indices sur les particularités de structure de la thyroglobuline, on peut penser qu'elles jouent un rôle important dans deux processus: l'ioduration de cette protéine et sa protéolyse, donc à la fois dans l'hormonogénèse et dans la libération enzymatique des produits actifs, seuls déversés dans le sang circulant. La thyroglobuline est relativement pauvre en *L*-tyrosine et les hormones qu'elle renferme se forment pourtant à partir de cet acide aminé, car on n'a jamais pu mettre en évidence de la *L*-thyronine préformée, dont l'ioduration conduirait directement à la thyroxine ou à la *L*-3 : 5 : 3'-triiodothyronine, sans passer par l'intermédiaire d'iodotyrosines. L'action de l'iode sur la thyroglobuline ne conduit pas à la formation des hormones avec un rendement particulièrement élevé par ailleurs. Il en découle que, si cette protéine présente des caractères la rendant particulièrement apte à remplir sa fonction physiologique, ceux-ci doivent sans doute porter sur la position de restes de tyrosine leur permettant de participer très activement aux réactions conduisant à la formation des hormones ou à leur libération par protéolyse.

L'exceptionnelle stabilité des liaisons peptidiques de l'arginine dans la thyroglobuline traduit un élément de l'architecture moléculaire encore trop imprécis pour faire l'objet de commentaires. En revanche, d'autres faits plus directement liés à la fonction de cette protéine ont été établis dans le même domaine. L'existence en bout de chaîne, avec groupement —NH_2 libre, de restes de tyrosine, de *L*-3 : 5-diiodotyrosine et de *L*-thyroxine a été constatée au moyen des techniques de SANGER et d'EDMAN (*145*). Le fait que de la gélatine, à laquelle des groupes N-carbobenzyloxythyroxylés ont été fixés, est un antigène dont les anticorps précipitent la thyroxine ne suffit pas à démontrer que la thyroglobuline contient l'hormone uniquement dans des restes thyroxylés (*35*). Par ailleurs, les acides aminés iodés de la protéine thyroïdienne sont très facilement libérés par la pepsine et la trypsine pures (*141*). C'est ainsi que 95% de ces acides aminés apparaissent dans les extraits butanoliques de protéolysats pepsiques lorsque 23% des liaisons peptidiques seulement sont rompues. De même, 30% de ces corps sont libérés par la trypsine, laquelle ne rompt aucune liaison peptidique de la tyrosine dans d'autres protéines (*141*).

On ne peut pas se faire encore une représentation de la structure de la thyroglobuline permettant d'interpréter l'évolution des processus d'ioduration et celle de l'hydrolyse physiologique de cette protéine, dont le poids moléculaire élevé rend inopérants les moyens d'étude dont les biochimistes disposent actuellement dans ce domaine.

c) *Iode et acides aminés iodés.*

Le caractère le plus spécifique de la thyroglobuline est qu'elle renferme des acides aminés iodés. Bien que des preuves expérimentales formelles n'aient pas été apportées à ce sujet, l'opinion de SALTER (*161*), selon laquelle les cellules épithéliales de la glande synthétisent une protéine non iodée qu'elles halogènent ensuite à un degré plus ou moins grand, apparaît comme très probablement exacte. Alors que le rapport: I hormonal/I total, est très peu variable d'un organe à l'autre, la teneur en halogène total présente des écarts très importants dans les diverses préparations. Il n'y a plus lieu d'admettre, comme le font de nombreux ouvrages classiques, que le taux de l'iode est une constante de composition de la thyroglobuline au même titre que le taux d'azote total (16,85% N) et que cette protéine est d'autant plus pure qu'elle est plus riche en halogène.

Les différences les plus grandes observées jusqu'ici oscillent entre 0,18 et 0,87% d'I chez des porcs normaux (*100*), 0,16 et 1,12% chez des boeufs normaux (*55*), cette dernière valeur ayant été relevée chez des foetus au voisinage du terme. Le rapport: I hormonal/I total, est égal à 0,30—0,34 chez divers mammifères normaux (*101*) et varie de 0,24 à 0,32 dans les thyroglobulines de onze espèces de Vertébrés zoologiquement aussi éloignés que les Mammifères, les Batraciens, les Reptiles et les Poissons (*176*).

Les données antérieures en désaccord avec celles-ci (*85, 175*) doivent être considérées comme ayant été déterminées à l'aide de techniques non satisfaisantes ou sur des produits anormaux. En effet, des organes pathologiques présentant des troubles de l'hormonogénèse peuvent être très pauvres en iode fixé à des dérivés de la thyronine malgré une certaine richesse en halogène total; le rapport indiqué plus haut est alors inférieur à 0,3 (*29, 39*). L'uniformité de la valeur de ce rapport dans la thyroglobuline des glandes normales, malgré la diversité du degré de l'ioduration de celle-ci, est nécessairement lié à la structure de la protéine. Elle constitue un argument défavorable à la conception, récemment développée par divers auteurs, selon laquelle des acides aminés iodés circulant dans les humeurs seraient directement fixés par la thyroglobuline (*41*), au lieu de se former au sein de celle-ci.

La nature des acides aminés iodés présents dans la thyroglobuline a été examinée plus haut. La *L*-3 : 5-diiodotyrosine est le plus abondant et renferme de 50 à 60% de l'iode total, la *L*-3-monoiodotyrosine de 10 à 15%, la *L*-thyroxine environ 20%, la *L*-3 : 5 : 3'-triiodothyronine de 2 à 10% et la *L*-monoiodohistidine de 1 à 5%. La teneur en hormone triiodée est soumise à des variations assez importantes. Celles-ci peuvent être en liaison avec les teneurs respectives de la protéine en dérivés mono- et diiodé de la *L*-tyrosine, si le dérivé triiodé se forme bien par condensation de ces deux corps molécule à molécule. Les différences saisonnières observées dans les taux des deux hormones suggèrent l'hypothèse que plus l'ioduration de la protéine est rapide, plus la réaction de substitution de l'iode à la tyrosine conduit à la formation du dérivé dihalogéné et va de pair avec la formation de *L*-thyroxine plutôt qu'avec celle de *L*-3 : 5 : 3'-triiodothyronine. La biosynthèse de ces corps paraît calquée sur le mécanisme de leur formation dans les protéines artificiellement iodées.

La thyroglobuline n'est déversée dans le sang circulant que dans des conditions particulières, lorsque le corps thyroïde est soumis à l'action d'un excès d'iodures radioactifs (*168*); sa recherche au moyen de tests immunochimiques d'une très grande sensibilité a toujours été négative chez les sujets normaux (*87*). Il en est ainsi en raison de la protéolyse intrathyroïdienne de la protéine qu'opère une catheptase récemment purifiée (*56*). Aussi trouve-t-on dans l'extrait *n*-butanolique de la glande l'ensemble des acides aminés iodés de la protéine, dont les *L*-3-mono- et *L*-3 : 5-diiodotyrosine sont rapidement désiodées par une déshalogénase (*143*), alors qu'il n'en est pas de même des deux hormones, lesquelles sont, de ce fait, seules déversées dans les vaisseaux sanguins.

3. Scléroprotéines iodées.

De très anciennes observations ont établi que le tissu de soutien (squelette corné) des éponges (*49*) et celui (axe corné) des Gorgones et de divers Anthozoaires (*7*) renferment de l'iode*. Des recherches ultérieures ont montré que cet élément y est souvent associé à de petites quantités de brome et à des traces de chlore (*104*, *105*), les deux premiers halogènes et, éventuellement, le troisième, étant compris dans des scléroprotéines. Plus récemment, un ensemble de recherches, pour lesquelles il a été fait un très large appel à l'emploi d'I^{131} comme indicateur, a mis en évidence la présence de scléroprotéines iodées dans les tissus de nombreux Invertébrés de divers groupes (*105*). Tel est le cas pour l'hypoderme de *Drosophile* (*171*), les soies de vers polychètes, le byssus des moules, le septum stolonique d'ascidies (*52*—*54*) entre autres.

La formation de scléroprotéines, dont les constituants iodés ne sont pas susceptibles de jouer un rôle hormonal, chez des animaux ne possédant pas de corps thyroïde a retenu l'attention en ce qui concerne la biochimie de l'évolution. Elle montre, en effet, que la différenciation des processus d'ioduration des protéines se manifeste phylogénétiquement avant celle du corps thyroïde.

Trois types de scléroprotéines iodées: les antipathines, les gorgonines et les spongines, ont été distingués. Chacun est particulier à un groupe

* Les cendres des éponges sont utilisées depuis des temps reculés par la médecine chinoise pour une thérapeutique empirique du goître, basée sans doute sur leur teneur élevée en iodures; cette pratique subsiste encore dans certains pays du proche Orient.

d'animaux: le premier à des Hexacoralliaires, les Antipathaires, le second aux Gorgonaires et à divers Octocoralliaires, les spongines aux Spongiaires. Nous étudierons leur composition sans négliger les problèmes que pose leur biochimie comparée.

a) Gorgonines et antipathines.

Le résidu corné de décalcification de l'organe de soutien (axe) des Gorgonaires, des Pennatulacées (Octocoralliaires), des Gérardiidés et des Antipathaires (Hexacoralliaires), lavé abondamment à l'eau et extrait par divers solvants organiques, renferme ces protéines. En dehors de données d'orientation, l'étude systématique de la gorgonine de près de cinquante espèces de Gorgonaires, appartenant à six familles, de la protéine de l'axe corné de *Gerardia savaglia* BERTOLINI et d'un petit nombre d'Antipathaires, a été entreprise (*104*, *105*, *124*, *125*).

Elles présentent une extrême variabilité d'un genre à l'autre: de 9,30% I dans certains échantillons provenant d'*Eunicella verrucosa* PALLAS (var. *typica*) à 0,15% I dans d'autres de *Leptogorgia chevallieri* STIASNY. Leur teneur en brome est, en général beaucoup plus faible que celle en iode; elle dépasse rarement 1%, sauf chez les Primnoïdés (4,3% Br chez *Primnoa lepidifera*) et le taux du chlore dans les mêmes protéines est compris entre 0,4 et 0,04%.

Ces données n'ont aucune signification en ce qui concerne la caractérisation des protéines étudiées, car leur teneur en halogènes n'est pas constante d'une préparation à l'autre provenant d'une même espèce et de fragments d'un même axe corné. Comme on le discutera plus bas, il en est ainsi parce que le processus d'halogénation présente une intensité variable et parce que la spécificité de ces diverses scléroprotéines repose sur leur composition en acides aminées, indépendamment du degré d'ioduration ou de bromuration de certaines.

La *L*-3: 5-diiodotyrosine a été le premier acide aminé halogéné isolé des produits de l'hydrolyse alcaline des gorgonines (*43*, *76*), ce qui lui a longtemps valu le nom d'acide iodogorgonique. Elle y est accompagnée de *L*-3: 5-dibromotyrosine (*106*). La *L*-3-monoiodotyrosine est très abondante dans certains (*152*), qui renferment également à un faible taux la *L*-3-monobromotyrosine (*152*), et la teneur en *L*-thyroxine de ces protéines est toujours très faible (0,1—0,2%), sinon nulle. A cet égard, les gorgonines, dont les roentgenogrammes ont montré la structure fibreuse (*33*), méritent d'être rapprochées des fibroïnes artificiellement iodées. Aussi peut-on préciser que, dans un cas comme dans l'autre, la distance séparant les résidus de tyrosine sur des fibres proté ques rigides fait obstacle à leur condensation. Il est possible qu'une monoiodohistidine soit présente dans des antipathines, exceptionnellement riches en histidine et assez pauvres en tyrosine, mais sa destruction au cours de l'hydrolyse acide ou alcaline de ces protéines a jusqu'ici empêché de la mettre en évidence. Il en est de même des chlorotyrosines, dont l'existence a été envisagée.

Les gorgonines ont longtemps été considérées comme des kératines. Les premières analyses de la composition en acides aminés de deux gorgonines ont toutefois permis d'établir qu'elles sont beaucoup plus riches en glycocolle et plus pauvres en acides aminés basiques que la fibroïne de la soie, ce qui a permis de les rapprocher des pseudokératines. La composition de dix huit gorgonines provenant de diverses familles de Gorgonaires et d'une Pennatulacée a été déterminée ensuite par des méthodes chimiques plus précises (*125*); les résultats de ce travail ont été étendus à d'autres acides aminés et confirmés sur onze préparations de même origine, à l'aide de méthodes microbiologiques. Le *tableau 6* reproduit quelques-unes des données obtenues.

Tableau 6. Teneurs en halogènes et en acides aminés de gorgonines de divers Gorgonaires (Octocoralliaires) (% de protéine sèche), d'après DUCHÂTEAU et FLORKIN+ pour les valeurs ne portant pas de mention particulière; d'après ROCHE et EYSSERIC-LAFON (*125*) pour celles marquées du signe*.

Halogènes et acides aminés	Plexauridés			Gorgonellidés	Muricéidés	Gorgonidés
	Eunicella clenocelloides VERILL	*Euplexaura maghrebensis* STIASNY	*Plexaura kükenthalii* MOSER	*Ellisella elongata*	*Paramuricea placomus* L.	*Gorgonia adamsii* VERILL
Iode %	8,9	0,23	2,54	1,80	1,98	0,54
Brome %	1,59	0,30	0,30	2,15	0,34	0,02
Alanine	13,2	11,7	10,2	14,3	9,5	8,6
Acide aspartique	4,5	7,9	5,4	6,9	6,7	7,5
Acide glutamique	4,5	6,8	3,6	3,5	4,1	5,1
Arginine	6,9	6,6	7,2	7,0	8,6	7,2
Glycocolle	20,6	15,0	17,0	16,5	18,8	17,6
Histidine	1,2	1,3	1,0	1,7	2,0	1,3
Isoleucine	2,5	2,8	3,3	3,0	3,1	2,2
Leucine	3,0	3,4	2,5	2,3	2,3	3,1
Lysine	3,7	4,1	3,9	3,6	6,8	4,1
Méthionine	0,5	0,8	0,2	0,2	0,3	0,5
Phénylalanine	1,1	2,4	0,5	1,4	1,1	2,1
Proline	3,5	3,5	5,1	3,9	3,0	4,1
Thréonine	2,6	3,5	3,7	3,7	3,6	3,5
Valine	3,4	3,7	3,0	—	3,9	3,8
Cystine*	2,1	2,8	2,8	3,4	3,6	2,8
Sérine*	2,9	5,6	3,5	3,4	3,9	3,7
Tyrosine* non halog.	4,6	1,1	5,5	2,9	4,7	3,2
Tryptophane*	0,9	—	—	1,2	1,1	0,9

+ Résultats en cours de publication, obtenus par dosages microbiologiques des acides aminés. Nous remercions les auteurs de nous les avoir communiqués.

Aucune différence notable de composition n'a été enregistrée dans des echantillons provenant de divers individus d'une même espèce. En revanche, des écarts sensibles de composition peuvent exister entre des gorgonines de diverses expèces, mais celles des animaux d'un même genre sont relativement caractéristiques de celui-ci. Il existe donc de très nombreuses gorgonines, dont la composition traduit une spécificité d'origine. Dans l'ensemble, ces protéines sont très riches en glycocolle et en alanine, relativement riches en arginine, mais pauvres en histidine, en lysine et en cystine, ce qui permet de les distinguer des autres scléroprotéines. Leur spécificité de genre a été mise à profit pour préciser certains points obscurs de la classification zoologique. C'est ainsi qu'une espèce décrite par LINNÉ sous le nom de *Rhipidigorgia elegans* L. s'est révélée renfermer une gorgonine différente de celle d'autres *Rhipidigorgia* et voisine de celle des *Gorgonia*; un examen minutieux des caractères morphologiques de cette espèce joint à l'étude biochimique de sa protéine de soutien a permis de la ranger dans le genre *Gorgonia* sous le nom de *G. elegans*.

Les scléroprotéines iodées des Hexacoralliaires constituent un groupe de protéines distinct des gorgonines des Octocoralliaires par leur composition en acides aminés, dont rend compte l'examen du *tableau* 7.

Tableau 7. Teneurs en halogènes et en acides aminés d'antipathines d'un Gérardiidé et de divers Antipathaires (% de protéine sèche) (*125*).

Halogènes et acides aminés	Gerardiidé	Antipathaires			
	Gerardia savaglia BERTOLINI	*Antipathes myriophylla* PALLAS	*Antipathes subpinnata* ELLIS et SOLANDER	*Cirripathes anguina* DANA	*Cirripathes spiralis* BLAINV.
Iode total	0,80	2,18	1,29	2,90	4,07
Brome total	0,08	—	—	—	0,20
Monoiodotyrosine	traces	0,50	0,50	4,40	7,14
Diiodotyrosine	1,40	1,40	0,95	2,10	2,20
Tyrosine non halogénée	7,80	4,45	3,50	5,90	6,73
Tyrosine totale	8,43	5,31*	4,41*	9,30*	12,00
Arginine	4,10	3,05	3,05	3,30	3,25
Histidine	13,10	12,45	15,40	15,35	17,60
Lysine	6,00	12,75	10,95	6,85	6,60
Glycocolle	23,00	14,60	13,90	16,40	12,36
Leucine	0,80	—	—	—	0
Valine	0,60	—	—	—	1,68
Tryptophane	1,30	0,80	0,00	1,20	1,60
Cystine	2,22	2,40	2,50	1,50	1,68
Méthionine	indos.	0,30	0,35	0,35	0,90
Sérine	2,50	3,50	4,30	2,50	2,30

* Compte non tenu de la présence éventuelle de bromotyrosines.

Tableau 8. Teneurs en iode et en brome total, en tyrosine non halogénée et en tyrosine totale (halogénée et non halogénée) de gorgonines de diverses origines (% de protéine sèche).

Origine de la protéine	Iode %	Brome %	Tyrosine non halogénée %*	Monoiodo-tyrosine %	Diiodo-tyrosine %	Thyroxine %	Tyrosine totale %
Gorgonellidés:							
Ellisella elongata (Pallas)	1,80	2,15	3,20	2,50	1,20	0	7,23
Ellisella paraplexauroides Stiasny	1,61	1,85	3,00	2,25	2,01	0	8,18
Scirpearia flagellum Johnson	1,82	2,20	2,97	traces	1,30	0	8,34
Gorgonidés:							
Gorgonia adamsii Verrill	0,54	0,02	3,18	traces	0,70	0	3,49
Leptogorgia chevallieri Stiasny	0,15	0,02	0,96	traces	traces	0	0,98
Leptogorgia petechizans (Pallas)	0,60	0,02	2,48	traces	traces	0	2,74
Gorgonia elegans (Duch. et Mich.)	0,08	0,04	1,71	traces	0	0	1,75
Rhipidigorgia flabellum L.	0,62	0,73	7,96	traces	traces	0	9,47
Muricéidés:							
Muriceides chuni Kükenthal	1,95	—	2,48	traces	2,02	0	3,31
Paramuricea placomus L.	1,98	0,34	4,70	traces	1,00	0	5,84
Plexauridés:							
Eunicella ctenocelloides Verrill	8,90	1,59	4,58	8,79	9,54	0,18	14,66
Eunicella verrucosa Pallas (var. *stricta*)	6,78	1,70	4,88	9,25	5,09	0,12	13,24
Eunicella verrucosa Pallas (var. *typica*)	9,30	2,20	5,25	7,13	12,65	0,20	17,14
Euplexaura maghrebensis Stiasny	0,23	0,30	1,10	0	traces	0	1,59
Euplexaura pseudobutikofferi Stiasny	0,49	0,10	2,37	traces	traces	0	2,81
Plexaura kükenthalii Moser	2,54	0,30	5,50	2,35	3,23	0	8,60

* Calculée à partir des résultats de ce tableau en ce qui concerne les iodotyrosines et en considérant la totalité du brome comme compris dans la *L*-3 : 5-dibromotyrosine.

Tableau 9. Teneurs en halogènes et en acides aminés de spongines de divers Spongiaires (% de protéine sèche) (*124*).

Halogènes et acides aminés	Origine				
	Aplysina crassa HYATT	*Aplysina holda* LIND	*Euspongia officinalis*	*Siphonocalina pruvoti* TOPSENT	*Verungia fistularis*
Iode total	2,35	0,47	0,51	0,20	0,50
Brome total	1,20	0,09	0,41	indos.	0,12
Monoiodotyrosine	traces	traces	traces	0	traces
Diiodotyrosine	4,22	traces	0,77	traces	traces
Tyrosine non halogénée	2,07	2,30	0,32	0,77	4,23
Tyrosine totale	3,82	2,63	1,17	0,85	4,70
Arginine	3,00	—	5,75	4,92	3,20
Histidine	0,74	—	0,28	0	1,12
Lysine	9,65	—	5,15	3,50	8,75
Glycocolle	9,50	7,68	15,76	12,00	10,40
Cystine	1,80	0,26	1,50	1,04	2,25
Méthionine	0	0,27	0	0,43	—
Sérine	2,97	6,20	8,00	4,70	6,52
Tryptophane	0,86	traces	0,10	traces	2,30
Leucine	—	1,66	0	—	—
Valine	—	1,33	2,20	—	—

Les antipathines diffèrent de toutes les scléroprotéines actuellement étudiées par une teneur exceptionnellement élevée en histidine, beaucoup plus forte que celle de toutes les autres protéines analysées jusqu'ici; elles sont, en outre, riches en glycocolle et plus pauvres en arginine que les gorgonines. Un caractère biochimique très marqué du groupe d'animaux dont elles proviennent repose donc sur leur composition; l'étude de celle-ci a permis de mieux définir la position des Gérardiidés dans la classification zoologique. En effet, ceux-ci ont été longtemps considéré par certains comme une famille d'Antipathaires (Hexacoralliaires), et par d'autres comme une famille rattachée à l'ordre des Zoanthaires. Comme les arguments morphologiques en faveur de cette dernière opinion ne sauraient être négligés, il est probable que les Gérardiidés constituent un groupe voisin des Antipathaires intermédiaire à ceux-ci et aux Zoanthaires.

En dehors de la différenciation des antipathines et des Gorgonines, on a cherché à définir les facteurs biochimiques limitant l'halogénation de ces protéines. La tyrosine est le seul acide aminé fixant facilement l'iode, aussi la diversité de teneur des gorgonines en celui-ci paraît-elle liée à celle en tyrosine dans la plupart des cas, comme en témoigne l'examen du *tableau 8*.

Seules les gorgonines et les antipathines riches en tyrosine le sont aussi en iode, à ceci près que l'intensité du mécanisme physiologique de l'halogénation est le facteur régissant celle-ci dans les organismes. Ainsi s'explique, d'une part que les gorgonines à teneur en tyrosine élevée de certaines espèces comme *Rhipidigorgia flabellum* L. demeurent pauvres en iode et, d'autre part, que la relation entre les teneurs en halogènes et en tyrosine totale ne soit pas très rigoureuse d'un genre à l'autre. La proportion de cet acide aminé existant à l'état monoiodé étant toujours élevée, il est probable que le processus physiologique de l'halogénation des gorgonines est lent et progressif.

b) Spongines.

La biochimie de ces protéines pose les mêmes problèmes que celle des gorgonines, dont le rôle biologique paraît identique. La présence d'iode dans le squelette corné de nombreux Ceratospongiaires marins ou d'eau douce a été signalée par de nombreux auteurs (*49, 162*), avant qu'elle ait été localisée sur la spongine, protéine fibreuse constituant le résidu d'extraction de ce tissu par des solutions diluées d'acide chlorhydrique, par l'eau, puis par divers solvants organiques. Les teneurs en iode et en brome des spongines sont rarement supérieures à 1%, comme en témoigne l'analyse récente d'échantillons provenant de cinquante-neuf espèces des Bermudes (*95*).

Toutefois, on a pu relever exceptionnellement des teneurs plus élevées en halogènes, surtout en brome, dans *Verungia tenuissima* (1,31% I et 6,25% Br) et dans une espèce indéterminée de *Luffaria* (2,66% I et 4,36% Br) (*4*). Des données antérieures indiquant des valeurs beaucoup plus élevées témoignent sans doute d'erreurs analytiques. De même, l'hypothèse que les spongines sont d'autant plus riches en halogènes qu'elles proviennent d'espèces vivant dans des mers plus chaudes (*42*) doit être abandonnée à la suite de l'étude récente d'un grand nombre d'échantillons de diverses origines (*125, 95*). Des écarts de la teneur en iode de préparations provenant de divers individus d'un même espèce ont été enregistrés; ils atteignent, par exemple, de 0,15 à 0,57% pour la protéine d'*Euspongia officinalis*.

Aussi, la caractérisation des spongines ne peut-elle pas comme celle des gorgonines, reposer sur leur degré d'ioduration.

Le premier acide aminé iodé isolé des hydrolysats alcalins de ces protéines est la *L*-3:5-diiodotyrosine (*5, 116, 172*), qui accompagne la *L*-3:5-dibromotyrosine (*5*). La présence de *L*-3-monoiodotyrosine a été caractérisée (*151*) et celle de *L*-3-monobromotyrosine est probable, comme celle de très faibles traces de *L*-thyroxine.

La composition en acides aminés de ces protéines a fait l'objet d'un nombre restreint de travaux dont les plus anciens conservent un intérêt historique. La spongine d'*Euspongia officinalis* a tout d'abord été dissociée des kératines, en raison de sa teneur plus élevée en glycocolle, et rangée avec les gorgonines, dans les pseudokératines. Les documents les plus récents établis au sujet des spongines à l'aide de méthodes chimiques de dosage, ont été reproduits dans le *tableau 9*, p. 395.

Comme celle des gorgonines, la composition en acides aminés des spongines traduit leur spécificité d'origine. Les écarts observés entre les protéines de deux espèces d'*Aplysina* sont faibles; ils sont importants entre celles-ci et les spongines de *Syphonocalina* et d'*Euspongia*, tandis que la protéine de *Verungia fistularis* en est plus voisine. Le type général de composition de ces protéines diffère sensiblement de celui des gorgonines et des antipathines, et aucune d'elles n'est très riche en tyrosine. On a pu, de ce fait, se demander si d'autres acides aminés que celle-ci ne fixaient pas les halogènes lorsque ceux-ci étaient présents en abondance (5); mais les méthodes actuelles de dosage des quatre mono- et dihalogénotyrosines identifiées et leur séparation quantitative ne donnent pas des résultats assez précis pour formuler à cet égard de conclusion ferme. De toute manière, l'ioduration et la bromuration des spongines sont, en première analyse, fonction de leur teneur en tyrosine et de l'intensité du mécanisme biologique de concentration et d'oxydation des iodures à leur niveau. L'intensité de celui-ci apparaît comme variable, mais son efficacité peut à coup sûr être considérable, puisque les spongines d'organismes d'eau douce, comme *Spongilla lacustris* ESPER de la Seine, peuvent être aussi riches en iode que celles provenant d'éponges marines baignant dans des eaux de mer de beaucoup plus forte teneur en iodures (20 à 40 μg. d'iode par litre).

Bibliographie.

1. ABELIN, I.: Nichtschilddrüsenstoffe mit Schilddrüsenwirkung. II. Mitt.: Einfluß der Abbauprodukte des künstlich jodierten Eiweißes (Homothyroxin) auf das Vogelgefieder und auf die Körpertemperatur des Meerschweinchens. Arch. exp. Pathol. Pharmakol. **175**, 146 (1934).
2. ABELIN, I. und A. FLORIN: Nichtschilddrüsenstoffe mit Schilddrüsenwirkung. Arch. exp. Pathol. Pharmakol. **171**, 443 (1933).
3. ACHER, R. et C. CROCKER: Réactions colorées spécifiques de l'arginine et de la tyrosine réalisées après chromatographie sur papier. Biochim. Biophys. Acta **9**, 704 (1952).
4. ACKERMANN, D. und C. BURCHARD: Zur Kenntnis der Spongine. Z. physiol. Chem. (Hoppe-Seyler) **271**, 183 (1941).
5. ACKERMANN, D. und E. MÜLLER: Über das Vorkommen von Dibromtyrosin neben Dijodtyrosin im Spongin. Z. physiol. Chem. (Hoppe-Seyler) **269**, 146 (1941).
6. ANSON, M. L.: The Sulphydryl Groups of Egg Albumin. J. Gen. Physiol. **24**, 399 (1940).
7. BALARD, M.: Note pour servir à l'histoire naturelle de l'iode. Ann. Chim. Phys., Série 2, **28**, 178 (1825).
8. BARKDOLL, A. E. and W. F. ROSS: Some Experiments on the *in vitro* Formation of Thyroxine from Diiodotyrosine. J. Amer. Chem. Soc. **66**, 898 (1944).
9. BARNES, B. O. and M. JONES: Studies on Thyroglobulin. III. The Thyroglobulin Content of the Thyroid Gland. Amer. J. Physiol. **105**, 556 (1933).
10. BARNES, J. H., R. C. COOKSON, G. T. DICKSON, J. ELKS and V. D. POOLE: The Synthesis of Thyroxine and Related Substances. XIII. Some Further Analogues of Thyroxine. J. Chem. Soc. (London) **1953**, 1448.

11. BAUER, H. und E. STRAUSS: Beiträge zur Kenntnis substituierter Proteine: Nitrierung und Jodierung des Globins. Biochem. Z. **211**, 163 (1929).

12. — — Beiträge zur Kenntnis substituierter Proteine. II. Mitt.: Jodierungen des Hämoglobins und des Globins. Biochem. Z. **284**, 197 (1936).

13. BAUMANN, E. und E. ROOS: Über das normale Vorkommen des Jods im Tierkörper. II. Z. physiol. Chem. (Hoppe-Seyler) **21**, 481 (1895).

14. BLAIZOT-GUÉNOT, S. et J. BLAIZOT: Titrage biologique de la thyroxine et de l'activité thyroxinienne des protéines iodées par dosage de la créatine urinaire. C. R. hebd. Séances Acad. Sci. **223**, 759 (1946).

15. BLAU, N. F.: The Determination of Thyroxine in Thyroid Substance. J. Biol. Chem. **110**, 351 (1935).

16. BLOCK, P., Jr.: A Note on the Conversion of Diiodotyrosine into Thyroxine. J. Biol. Chem. **135**, 51 (1940).

17. BLOCK, P., Jr. and G. POWELL: The Synthesis of 3',5'-Diiodothyronine. J. Amer. Chem. Soc. **64**, 1070 (1942).

18. BLOCK, R. J. and H. B. VICKERY: The Basic Amino Acids of Proteins. A Chemical Relationship Between the Various Keratins. J. Biol. Chem. **93**, 113 (1931).

19. BLUM, F.: Zur Chemie und Physiologie der Jodsubstanz der Schilddrüse. Pflügers Arch. ges. Physiol. **77**, 70 (1899).

20. BLUM, F., LEHMANN et LEISTNER: cité par S. RAUCH, Thèse Doct. Med., Berne. 1945.

21. BLUM, F. und E. STRAUSS: Mitteilungen aus dem Gebiete der Eiweißchemie. I. Über Jodbindungsfähigkeit und Konstitution der Proteine. Z. physiol. Chem. (Hoppe-Seyler) **112**, 111 (1921).

22. — — Mitteilungen aus dem Gebiete der Eiweißchemie. III. Über Jodierung von Proteinen mit Jodstickstoff. Z. physiol. Chem. (Hoppe-Seyler) **127**, 199 (1923).

23. BLUM, F. und W. VAUBEL: Über Halogeneiweißderivate. J. prakt. Chem. **57**, 365 (1898).

24. BONOT, A.: Étude physicochimique de la réaction des précipitines. Cas de la sérumalbumine iodée. Fraction non précipitable d'un système d'anticorps. Bull. soc. chim. biol. (Paris) **21**, 1417 (1939).

25. BOWDEN, C. H., N. F. MACLAGAN and J. H. WILKINSON: The Application of the Ceric Sulphate-Arsenious Acid Reaction to the Detection of Thyroxine and Related Substances. Biochemic. J. **59**, 93 (1955).

26. BRUICE, T. C., R. J. WINZLER and N. KHARASCH: The Thyroxine-like Activity of some New Thyroxine Analogues in Amphibia. J. Biol. Chem. **210**, 1 (1954).

27. BRUNINGS, K. J.: Preparation and Properties of the Iodohistidines. J. Amer. Chem. Soc. **69**, 205 (1947).

28. CAVETT, J. W.: Thyroglobulin Studies. II. The van Slyke Nitrogen Distribution and Tyrosine and Tryptophane Analyses for Normal and Goitrous Human Thyroglobulin. J. Biol. Chem. **114**, 65 (1936).

29. CAVETT, J. W., C. O. RICE and J. F. MCCLENDON: Thyroglobulin Studies. I. Thyroxine and Iodine Content of Normal and Goitrous Human Thyroglobulin. J. Biol. Chem. **110**, 673 (1935).

30. CAVETT, J. W. and S. R. SELJESKOG: The Preparation of Thyroglobulin. J. Biol. Chem. **100**, XXVI (1933).

31. CHAIKOFF, I. L. and A. TAUROG: Application of Radioactive Iodine to Studies in Iodine Metabolism and Thyroid Function. In: The Use of Isotopes in Biology and Medicine, p. 292. Madison: Univ. of Wisconsin Press. 1948.

32. CHALMERS, J. R., G. T. DICKSON, J. ELKS and B. A. HEMS: The Synthesis of Thyroxine and Related Substances. V. A Synthesis of *L*-Thyroxine from *L*-Tyrosine. J. Chem. Soc. (London) **1949**, 3424.

33. CHAMPETIER, G. et E. FAURÉ-FRÉMIET: Étude roentgenographique des kératines sécrétées. C. R. hebd. Séances Acad. Sci. **207**, 1133 (1938).

34. CLAYTON, J. C., G. F. H. GREEN and B. A. HEMS: The Synthesis of Thyroxine and Related Substances. VIII. The Preparation of Some Halogeno- and Nitrodiphenyl Ethers. J. Chem. Soc. (London) **1951**, 2467.

35. CLUTTON, R. F., C. R. HARINGTON and M. E. YUILL: Studies in Synthetic Immunochemistry. III. Preparation and Antigenic Properties of Thyroxyl Derivatives of Proteins, and Physiological Effects of their Antisera. Biochemic. J. **32**, 1119 (1938).

36. COHN, E. J. and J. T. EDSALL: Proteins, Amino Acids and Peptides. New York: Reinhold Publ. 1943.

37. COOKSON, R. C. and G. F. H. GREEN: The Synthesis of Thyroxine and Related Substances. IX. Analogues of Thyroxine with Modified Side Chains. J. Chem. Soc. (London) **1952**, 827.

38. DELTOUR, G. H. et J. BEKAERT: Electrophorèse sur papier et radioautographie de la thyroglobuline marquée de rats normaux ou traités par divers modificateurs de l'activité thyroïdienne. C. R. Séances Soc. Biol. **147**, 388 (1953).

39. DERRIEN, Y., R. MICHEL, K. O. PEDERSEN et J. ROCHE: Recherches sur la préparation et les propriétés de la thyroglobuline pure. II. Biochim. Biophys. Acta **3**, 436 (1949).

40. DERRIEN, Y., R. MICHEL et J. ROCHE: Recherches sur la préparation et les propriétés de la thyroglobuline pure. I. Biochim. Biophys. Acta **2**, 454 (1948).

41. DOBYNS, B.: Discussion to: F. GROSS and R. PITT-RIVERS: Triiodothyronine in Relation to Thyroid Physiology. Recent Progr. Hormone Res. **10**, 120 (1954).

42. DOHRN, A.: Der Ursprung der Chordaten und das Prinzip des Funktionswechsels. Leipzig. 1875.

43. DRECHSEL, E.: Beiträge zur Chemie einiger Seetiere. II. Über das Achsenskelett der *Gorgonia cavolini*. Z. Biol. **33**, 85 (1896).

44. ECKSTEIN, H. C.: The Distribution of some of the more Important Amino Acids in the Globulin of the Thyroid Gland. J. Biol. Chem. **67**, 601 (1926).

45. FAWCETT, D. M. and S. KIRKWOOD: The Synthesis of Organically Bound Iodine by Cell-free Preparations of Thyroid Tissue. J. Biol. Chem. **205**, 795 (1953).

46. FINK, K. and R. M. FINK: The Formation of Monoiodotyrosine from Radioiodine in the Thyroid of Rat and Man. Science (Washington) **108**, 358 (1948).

47. FRIEDEN, E. and R. J. WINZLER: The Synthesis of the Glycine Homolog of Thyroxine. J. Amer. Chem. Soc. **70**, 3511 (1948).

48. FROMAGEOT, C., M. JUTISZ, M. LAFON et J. ROCHE: Sur la présence de monoiodotyrosine dans la gorgonine *(Gorgonia verrucosa)*. C: R. Séances Soc. Biol. **142**, 785 (1948).

49. FYFE, A.: Account of some Experiments, made with the View of Ascertaining the Different Substances from which Iodine can be Procured. Edinbg. philos. Journ. **1**, 254 (1819).

50. GADDUM, J. H.: Quantitative Observations on Thyroxine and Allied Substances. II. Effects on the Oxygen Consumption of Rats. J. Physiol. **68**, 383 (1929—30).

51. GINSEL, L. A.: The Ultraviolet Absorption of Sheep Thyroglobulin. Biochemic. J. **33**, 428 (1939).

52. Gorbman, A.: Identity of an Iodine Storing Tissue in an Ascidian. Science (Washington) **94**, 192 (1941).
53. — Some Aspects of the Comparative Biochemistry of Iodine Utilization and the Evolution of Thyroidal Functions. Physiol. Rev. (in press) (1955).
54. Gorbman, A., M. Clements and R. O'Brien: Utilization of Radioiodine by Invertebrates, with Special Study of Several Annelida and Mollusca. J. exp. Zoology **127**, 75 (1954).
55. Gorbman, A., S. Lissitzky, O. Michel, R. Michel et J. Roche: Métabolisme du radioiode par le foetus de boeuf au voisinage du terme. C. R. Séances Soc. Biol. **145**, 1642 (1951).
56. Gordon, A. W., M. T. McQuillan, P. G. Stanley and V. M. Trikojus: Use of Radioiodine (I^{131}) in the Study of Enzyme Reactions in the Thyroid Gland. Radioisotope Techniques (Oxford) **1**, 37 (1951).
57. Gourlay, F.: The Proteids of the Thyroid and the Spleen. J. Physiol. **16**, 23 (1894).
58. Gross, J. and R. Pitt-Rivers: 3:5:3'-Triiodothyronine. 1. Isolation from Thyroid Gland and Synthesis. Biochemic. J. **53**, 645 (1953).
59. Harington, C. R.: Chemistry of Thyroxine. III. Constitution and Synthesis of Desiodo-thyroxine. Biochemic. J. **20**, 300 (1926).
60. — Chemistry of the Iodine Compounds of the Thyroid. Fortschr. Chem. organ. Naturstoffe **2**, 103 (1939).
61. — Newer Knowledge of the Biochemistry of the Thyroid Gland. J. Chem. Soc. (London) **1944**, 193.
62. — Thyroxine: its Biosynthesis and its Immunochemistry. Proc. Roy. Soc. (London), Ser. B. **132**, 223 (1944).
63. — Synthesis of a Sulfur-containing Analogue of Thyroxine. Biochemic. J. **43**, 434 (1948).
64. Harington, C. R. and G. Barger: Chemistry of Thyroxine. III. Constitution and Synthesis of Thyroxine. Biochemic. J. **21**, 169 (1927).
65. Harington, C. R. and A. Neuberger: Electrometric Titration of Insulin. Preparation and Properties of Iodinated Insulin. Biochemic. J. **30**, 809 (1936).
66. Harington, C. R. and R. Pitt-Rivers: Preparation of Thyroxine from Casein Treated with Iodine. Nature (London) **144**, 205 (1939).
67. — — A New Synthesis of Thyronine. J. Chem. Soc. (London) **1940**, 1101.
68. — — The Chemical Conversion of Di-iodotyrosine into Thyroxine. Biochemic. J. **39**, 157 (1945).
69. — — Note on the Synthesis of the Acetic Acid of Thyroxine. Biochemic. J. **50**, 438 (1952).
70. — — Mono-iodotyrosine. Biochemic. J. **38**, 320 (1944).
71. Harington, C. R. and S. S. Randall: Observations on the Iodine-containing Compounds of the Thyroid Gland. Isolation of *DL*-3:5-Diiodotyrosine. Biochemic. J. **23**, 373 (1929).
72. Harington, C. R. et O. Thibault: Comparaison de l'activité sensibilisatrice de la triiodothyronine et de la thyroxine ainsi que des amines correspondantes. C. R. Séances Soc. Biol. **147**, 78 (1953).
73. Heidelberger, M. and W. W. Palmer: The Preparation and Properties of Thyroglobulin. J. Biol. Chem. **101**, 433 (1933).
74. Heidelberger, M. and K. O. Pedersen: The Molecular Weight and Isoelectric Point of Thyroglobulin. J. Gen. Physiol. **19**, 95 (1935).
75. Heidelberger, M. and T. Svedberg: The Molecular Weight of Thyroglobulin. Science (Washington) **80**, 414 (1934).

76. HENZE, M.: Zur Kenntnis der jodbindenden Gruppe der natürlich vorkommenden Jodeiweißkörper. Die Konstitution der Jodgorgosäure. Z. physiol. Chem. (Hoppe-Seyler) **51**, 64 (1907).
77. HERRIOTT, R. M.: Isolation of Crystalline *L*-Monoiodotyrosine from Partially Iodinated Pepsin. J. Gen. Physiol. **25**, 185 (1941).
78. — Identification of Monoiodotyrosine from Iodinated Pepsin. J. Gen. Physiol. **31**, 19 (1947).
79. HITCHCOCK, D. I.: The Combination of a Standard Gelatin Preparation with Hydrochloric Acid and with Sodium Hydroxide. J. Gen. Physiol. **15**, 125 (1931).
80. HOREAU, A. et P. SÜE: Synthèse de la thyroxine marquée par le radioiode. Séparation dans les organes des divers produits iodés résultant de sa décomposition. Bull. soc. chim. biol. (Paris) **27**, 483 (1945).
81. HUTCHISON, R.: The Chemistry of the Thyroid Gland and the Nature of its Active Constituent. J. Physiol. **20**, 474 (1896).
82. — Further Observations on the Chemistry and Action of the Thyroid Gland. J. Physiol. **23**, 178 (1898—99).
83. JOHNSON, T. B. and L. B. TEWKESBURY, Jr.: The Oxidation of 3, 5-Diiodotyrosine to Thyroxine. Proc. Nat. Acad. Sci. (USA) **28**, 73 (1942).
84. LEBLOND, C. P.: Histological Localization of Radioactive Compounds in Tissue as Illustrated with the Help of Radioiodine. Recent Progr. Hormone Res. **3**, 159 (1948).
85. LELAND, J. P. and G. L. FOSTER: A Method for the Determination of Thyroxine in the Thyroid. J. Biol. Chem. **95**, 165 (1932).
86. LERMAN, J., C. R. HARINGTON and J. H. MEANS: The Physiologic Activity of some Analogues of Thyroxine. J. Clin. Endocrinology **12**, 1306 (1952).
87. LERMAN, J. and H. D. STEBBINS: The Pituitary Type of Myxedema. Further Observations. J. Amer. Med. Assoc. **119**, 391 (1942).
88. LI, C. H.: Catalytic Effect of Acetate and Phosphate Buffers on the Iodination of Thyrosine. J. Amer. Chem. Soc. **66**, 228 (1944).
89. — Kinetics and Mechanism of 2,6-Di-iodotyrosine Formation. J. Amer. Chem. Soc. **64**, 1147 (1942).
90. — Iodination of Tyrosine Groups in Serum Albumin and Pepsin. J. Amer. Chem. Soc. **67**, 1065 (1945).
91. LI, C. H., W. R. LYONS and H. M. EVANS: Studies on Pituitary Lactogenic Hormone. V. Reactions with Iodine. J. Biol. Chem. **139**, 43 (1941).
92. LIEBEN, F. und D. LÁSZLÓ: Über die Jodaufnahme von Caseïn. Biochem. Z. **159**, 110 (1925).
93. LIEBRECHT, A.: Über Jodderivate von Eiweißkörpern (Caseïn). Ber. dtsch. chem. Ges. **30**, 1824 (1897).
94. LOESER, A., H. RULAND und V. M. TRIKOJUS: Darstellung, Eigenschaften und biologische Wirkungen von Derivaten (Äthern) des Thyroxins, Dijodthyronins und Dijodtyrosins. Arch. exp. Pathol. Pharmakol. **189**, 664 (1938).
95. LOW, E. M.: Iodine and Bromine in Sponges. J. Marine Research. **8**, 97 (1949).
96. LUDWIG, W. and P. v. MUTZENBECHER: Die Darstellung von Thyroxin, Monojodtyrosin und Dijodtyrosin aus jodiertem Eiweiß. Z. physiol. Chem. (Hoppe-Seyler) **258**, 195 (1939).
97. LUNDGREN, H. P.: Equilibria in the Thyroid Protein Found by Ultracentrifugal Sedimentation. J. Chem. Physics **6**, 177 (1938).
98. MICHEL, O., R. MICHEL et G. H. DELTOUR: Sur l'ioduration de la thyroglobuline sans dénaturation. Bull. soc. chim. biol. (Paris) **32**, 8 (1950).
99. MICHEL, R.: Sur le microdosage de la diiodotyrosine. Bull. soc. chim. France **12**, 546 (1945).

100. Michel, R.: Contribution à la biochimie des iodoprotéines et des acides aminés iodés. Thèse Doct. Sci. 1948.

101. Michel, R. et M. Lafon: Sur les teneurs en iode et en thyroxine de la thyroglobuline. C. R. Séances Soc. Biol. **140**, 634 (1946).

102. Michel, R. and R. Pitt-Rivers: The Iodination of Silk Fibroin. Biochim. Biophys. Acta **2**, 223 (1948).

103. Michel, R., J. Roche et J. Tata: Synthèse de la L-thyroxine marquée en différentes positions par l'iode radioactif I^{131}. Bull. soc. chim. biol. (Paris) **34**, 466 (1952).

104. Mörner, C. Th.: Zur Kenntnis der organischen Gerüstsubstanz des Anthozoënskeletts. I. Mitt. Z. physiol. Chem. (Hoppe-Seyler) **51**, 33 (1903).

105. — Zur Kenntnis der organischen Gerüstsubstanz des Anthozoënskeletts. II. Mitt. Z. physiol. Chem. (Hoppe-Seyler) **55**, 77 (1908).

106. — Zur Kenntnis der organischen Gerüstsubstanz des Anthozoënskeletts. IV. Mitt. Isolierung und Identifizierung der Bromgorgosäure. Z. physiol. Chem. (Hoppe-Seyler) **88**, 138 (1913).

107. Mussett, M. V. and R. Pitt-Rivers: The Thyroid-like Activity of Triiodothyronine Analogues. Lancet **267**, 1212 (1954).

108. Mutzenbecher, P. v.: Über die Bildung von Thyroxin aus Dijodtyrosin. Z. physiol. Chem. (Hoppe-Seyler) **261**, 253 (1939).

109. Muus, J., A. H. Coons and W. T. Salter: Thyroidal Activity of Iodinated Serum Albumin. IV. The Effect of Progressive Iodination. J. Biol. Chem. **139**, 135 (1941).

110. Neuberg, C.: Beobachtungen an Jodproteinen. Biochem. Z. **27**, 261 (1910).

111. Neuberger, A.: Electrometric Titration of Zein and Iodozein. Biochemic. J. **28**, 1982 (1934).

112. Niemann, C.: Thyroxine and Related Compounds. Fortschr. Chem. organ. Naturstoffe **7**, 167 (1950).

113. Oster, G. and S. Malament: Crystallization and Ultraviolet Spectra of Iodinated Insulin. J. Amer. Chem. Soc. **76**, 3441 (1954).

114. Oswald, A.: Die Eiweißkörper der Schilddrüse. Z. physiol. Chem. (Hoppe-Seyler) **27**, 14 (1899).

115. — Gewinnung von 3,5-Dijodtyrosin aus Jodeiweiß. Z. physiol. Chem. (Hoppe-Seyler) **70**, 310 (1911).

116. — Gewinnung von 3.5-Dijodtyrosin aus Jodeiweiß. IV. Mitt. Die Verhältnisse beim Gorgonin und Spongin. Z. physiol. Chem. (Hoppe-Seyler) **75**, 353 (1911).

117. Pauly, H.: Über jodierte Abkömmlinge des Imidazols und des Histidins. Ber. dtsch. chem. Ges. **43**, 2243 (1910).

118. — Zur Jodierung von Verbindungen des Eiweißgebietes. Z. physiol. Chem. (Hoppe-Seyler) **76**, 291 (1911).

119. Pitt-Rivers, R.: The Oxidation of Diiodotyrosine and Derivatives. Biochemic. J. **43**, 223 (1948).

120. — Physiological Activity of the Acetic-acid Analogues of some Iodinated Thyronines. Lancet **265**, 234 (1953).

121. — Metabolic Effects of Compounds Structurally Related to Thyroxine *in vivo*. Thyronine Derivatives. J. Clin. Endocrinology **14**, 1444 (1954).

122. Pitt-Rivers, R. and S. S. Randall: Preparation and Properties of Physiologically-active Iodinated Proteins. J. Clin. Endocrinology **4**, 221 (1945).

123. Pummerer, R. und A. Rieche: Über die Oxydation der Phenole. IX. Über aromatische Peroxyde und einwertigen Sauerstoff. Ber. dtsch. chem. Ges. **59**, 2161 (1926).

124. ROCHE, J.: Biochimie comparée des scléroprotéines iodées des Anthozoaires et des Spongiaires. Experientia **8**, 45 (1952).

125. ROCHE, J. et M. EYSSERIC-LAFON: Biochimie comparée des scléroprotéines iodées des Anthozoaires (gorgonaires, antipathaires, gérardiidés) et spécificités des gorgonines. Bull. soc. chim. biol. (Paris) **33**, 1437 (1951).

126. ROCHE, J., M. JUTISZ, S. LISSITZKY et R. MICHEL: Chromatographie quantitative des acides aminés radioactifs de la thyroglobuline marquée. Biochim. Biophys. Acta **7**, 257 (1951).

127. ROCHE, J. and M. LAFON: Sur la présence de diiodotyrosine dans les Laminaires. C. R. hebd. Séances Acad. Sci. **229**, 481 (1949).

128. ROCHE, J., S. LISSITZKY, O. MICHEL et R. MICHEL: Étude radiochromatographique des étapes de l'ioduration de la tyrosine et de l'histidine. Biochim. Biophys. Acta **7**, 439 (1951).

129. ROCHE, J., S. LISSITZKY et R. MICHEL: Caractérisation des iodohistidines dans les protéines iodées (thyroglobuline et iodoglobuline). Biochim. Biophys. Acta **8**, 339 (1952).

130. — — — Sur la triiodothyronine et sa formation par ioduration de la diiodothyronine. Biochim. Biophys. Acta **11**, 215 (1953).

131. — — — Sur la présence de la triiodothyronine dans la thyroglobuline et sur sa biosynthèse. Biochim. Biophys. Acta **11**, 220 (1953).

132. — — — Chromatographic Analysis of Radioactive Iodine Compounds from the Thyroid Gland and Body Fluids. Methods Biochem. Analysis **1**, 243 (1954).

133. ROCHE, J., O. MICHEL, G. H. DELTOUR et R. MICHEL: Sur la thyroglobuline humaine marquée par l'iode radioactif, à l'état normal et dans la maladie de Basedow. Ann. Endocrinol. **13**, 1 (1952).

134. ROCHE, J., O. MICHEL, R. MICHEL et M. MAROIS: Sur les modalités de la concentration du radioiode I^{131} dans le vitellus de l'oeuf de poule et sur la formation ovarienne de combinaisons iodées. C. R. Séances Soc. Biol. **145**, 1833 (1951).

135. ROCHE, J. et R. MICHEL: Dosage colorimétrique de la thyroxine, de la diiodotyrosine et de la monoiodotyrosine. Application aux protéines iodées. Biochim. Biophys. Acta **1**, 335 (1947).

136. — — L'action de l'iode sur le tryptophane dans les protéines et à l'état libre. Biochim. Biophys. Acta **2**, 97 (1948).

137. — — Sur la formation de la thyroxine par action de l'iode sur divers peptides renfermant de la tyrosine. Bull. soc. chim. biol. (Paris) **31**. 144 (1949).

138. ROCHE, J., R. MICHEL et M. LAFON: Sur la formation de la thyroxine et de ses précurseurs dans les iodoprotéines. Biochim. Biophys. Acta **1**, 453 (1947).

139. ROCHE, J., R. MICHEL, M. LAFON et D. P. SADHU: Sur la formation de la thyroxine et de ses précurseurs dans les iodoprotéines et les iodopeptones. II. Biochim. Biophys. Acta **3**, 648 (1949).

140. ROCHE, J., R. MICHEL, S. LISSITZKY et S. MAYER: Sur la formation de la thyroxine et de ses précurseurs dans les iodoprotéines. III. Biochim. Biophys. Acta **7**, 446 (1951).

141. ROCHE, J., R. MICHEL, S. LISSITZKY et Y. YAGI: Action de la pepsine et de la trypsine sur la thyroglobuline et les protéines artificiellement iodées marquées par I^{131}. Bull. soc. chim. biol. (Paris) **36**, 143 (1954).

142. ROCHE, J., R. MICHEL, O. MICHEL, G. H. DELTOUR et S. LISSITZKY: Thyroglobuline marquée ou artificiellement iodée sans dénaturation. Formation dans des conditions expérimentales diverses et propriétés. Biochim. Biophys. Acta **6**, 572 (1951).

143. ROCHE, J., R. MICHEL, O. MICHEL et S. LISSITZKY: Sur la déshalogénation enzymatique des iodotyrosines par le corps thyroïde et sur son rôle physiologique. Biochim. Biophys. Acta 9, 161 (1952).
144. ROCHE, J., R. MICHEL et M. MOUTTE: Dosage de la diiodotyrosine. Application aux protéines. Bull. soc. chim. France 11, 490 (1944).
145. ROCHE, J., R. MICHEL, J. NUNEZ et G. LACOMBE: Sur la présence d'acides aminés iodés N-terminaux dans la thyroglobuline de porc. C. R. hebd. Séances Acad. Sci. 240, 464 (1955).
146. ROCHE, J., R. MICHEL et J. TATA: Sur la nature des combinaisons iodées excrétées par le foie et le rein après administration de L-thyroxine et de L-3:5:3'-triiodothyronine. Biochim. Biophys. Acta 15, 500 (1954).
147. ROCHE, J., R. MICHEL et W. WOLF: Préparation de nouveaux dérivés de la DL-thyronine. C. R. hebd. Séances Acad. Sci. 239, 597 (1954).
148. ROCHE, J., R. MICHEL, W. WOLF et N. ETLING: Activité biologique (antigoïtrogène) de quelques nouvelles iodothyronines et iododesaminothyronines. C. R. Séances Soc. Biol. 148, 1738 (1954).
149. ROCHE, J., M. MOURGUE et R. BARET: Structure des protéines et action de l'arginase sur les produits de leur hydrolyse partielle. Bull. soc. chim. biol. (Paris) 34, 337 (1952).
150. ROCHE, J., G. RANSON et M. EYSSERIC-LAFON: Sur la composition des scléroprotéines des coquilles des mollusques (conchiolines). C. R. Séances Soc. Biol. 145, 1474 (1951).
151. ROCHE, J. et Y. YAGI: Sur la présence de la monoiodothyronine dans la spongine. C. R. Séances Soc. Biol. 146, 288 (1952).
152. ROCHE, J., Y. YAGI, R. MICHEL, S. LISSITZKY et M. EYSSERIC-LAFON: Sur la caractérisation de la monobromotyrosine et de la thyroxine dans les gorgonines. Bull. soc. chim. biol. (Paris) 33, 526 (1951).
153. REINEKE, E. P. Thyroactive Iodinated Proteins. Vitamins and Horm. 4, 207 (1946).
154. REINEKE, E. P. and C. W. TURNER: The Effect of Certain Experimental Conditions on the Formation of Thyroxine from Diiodotyrosine. J. Biol. Chem. 162, 369 (1946).
155. — — The Recovery of *L*-Thyroxine from Iodinated Casein by Direct Hydrolysis with Acid. J. Biol. Chem. 149, 563 (1943).
156. — — The Effect of Manganese Compounds and Certain other Factors on the Formation of Thyroxine in Iodinated Casein. J. Biol. Chem. 161, 613 (1945).
157. REINEKE, E. P., C. W. TURNER, G. O. KOHLER, R. D. HOOVER and M. B. BEEZLEY: The Quantitative Determination of Thyroxine in Iodinated Casein having Thyroidal Activity. J. Biol. Chem. 161, 599 (1945).
158. REINEKE, E. P., M. B. WILLIAMSON and C. W. TURNER: The Effect of Progressive Iodination on the Thyroidal Activity of Iodinated Casein. J. Biol. Chem. 143, 285 (1942).
159. — — — The Effect of Progressive Iodination followed by Incubation at High Temperature on the Thyroidal Activity of Iodinated Proteins. J. Biol. Chem. 147, 115 (1943).
160. RIVIÈRE, C., G. GAUTRON et M. THÉLY: Chromatographie et ioduration artificielle de la thyroglobuline. Bull. soc. chim. biol. (Paris) 29, 600 (1947).
161. SALTER, W. T.: Euthyroidism and Thyroid Dysfunction. In: The Chemistry and Physiology of Hormones, p. 104. Washington: Amer. Assoc. Adv. Science. 1944.

162. SARPHATI, E. S.: Samuelis Emanuelis Sarphat. amstelodamensis, etc. Commentatio de Iodie in Certamine litterario civium academiarum etc. praemio ornata. Lugdumi Batavorum. 1835. [Report f. d. Pharmacie **9**, 303 (1837)].

163. SAUL, J. A. and V. M. TRIKOJUS: The Conversion of *DL*-3 : 5-Diiodo-4-hydroxyphenyllactic Acid into an Analogue of Thyroxine. Biochemic. J. **42**, 80 (1948).

164. SIMPSON, G. K., A. G. JOHNSTON and D. TRAILL: A Polarographic Study of Acid-insoluble Fractions from Iodo-proteins. Biochemic. J. **41**, 181 (1947).

165. STANLEY, P. G.: Artefacts Produced from 3-Iodotyrosine and 3 : 5-Diiodotyrosine by Heating with Alkali. Nature (London) **171**, 933 (1953).

166. TATUM, A. L.: A Study of the Distribution of Iodine Between Cells and Colloid in the Thyroid Gland. I. Methods and Results of Study of Beef, Sheep and Pig Thyroid Glands. J. Biol. Chem. **42**, 47 (1920).

167. TAUROG, A., S. ABRAHAM and I. L. CHAIKOFF: Synthesis of Some O-Glucuronides and O-Glucosides of Phenolic Amino Acids. J. Amer. Chem. Soc. **75**, 3473 (1953).

168. TONG, W., A. TAUROG and I. L. CHAIKOFF: Nature of Plasma Iodine Following Destruction of the Rat Thyroid with I^{131}. J. Biol. Chem. **195**, 407 (1952).

169. — — — The Metabolism of I^{131}-Labeled Diiodotyrosine. J. Biol. Chem. **207**, 59 (1954).

170. VAN DYKE, H. B.: A Study of the Distribution of Iodine Between Cells and Colloid in the Thyroid Gland. II. Results of Study of Dog and Human Thyroid Glands. J. Biol. Chem. **45**, 325 (1920).

171. WHEELER, B. M.: Halogen Metabolism of *Drosophila gibberosa*. I. Iodine Metabolism Studied by Means of I^{131}. J. exp. Zoology **115**, 83 (1950).

172. WHEELER, H. L. and L. B. MENDEL: The Iodine Complex in Sponges (3,5-Diiodotyrosine). J. Biol. Chem. **7**, 1 (1909—10).

173. WHITE, A.: Some Analyses of Thyroglobulin. Proc. Soc. exp. Biol. Med. **32**, 1558 (1935).

174. WINNEK, P. S. and C. L. A. SCHMIDT: The Solubilities, Apparent Dissociation Constants, and Thermodynamic Data of the Dihalogenated Tyrosine Compounds. J. Gen. Physiol. **18**, 889 (1935).

175. — — The Solubilities of the *l*-Dihalogenated Tyrosines in Ethanol-Water Mixtures and Certain Related Data. J. Gen. Physiol. **19**, 773 (1936).

176. WOLFF, J. and I. L. CHAIKOFF: The Relation of Thyroxine to Total Iodine in the Thyroid Gland. Endocrinology **41**, 295 (1947).

177. YAGI, Y., R. MICHEL et J. ROCHE: Sur le métabolisme des bromures radioactifs (Br^{82}). Bull. soc. chim. biol. (Paris) **35**, 289 (1953).

(Reçu le 18 février 1955.)

Chemistry and Biochemistry of Snake Venoms.

By KARL SLOTTA, São Paulo, Brazil.

With 4 Figures.

Contents.

I. Introduction.

Snake venoms are viscous liquids which contain poisonous and non-poisonous proteins as well as other organic and inorganic substances. If we could isolate each of these components, particularly the toxic ones, and clarify their constitution, chemistry's task would be fulfilled in this field. Then, after determination of the biological and pharmacological actions of each of these pure substances, the total activity of the approximately 400 different snake venoms could be explained. Unfortunately, we are as yet far from this goal.

Up to now, we do not even know which substances in these poisonous secretions are the toxins. Decidedly, the real toxin, *i. e.* the substance causing the lethal action, is only a very small part of the venom. Of course, the snake venoms contain also substances which only damage the prey, acting slowly and possibly causing its death after several days; but these are not what we consider the real toxins.

In 1938, a toxin from the venomous secretion of a snake was obtained in crystalline form [SLOTTA and FRAENKEL-CONRAT (*207–209, 194*)]. Since this "crotoxin" showed, besides the toxic, also an enzymic action, it was assumed that the toxic effect might be identical with the enzymic one. Since then, several enzymes from snake venoms have been purified and one, "hemolysin", has even been crystallized [DE (*42*)]. But these compounds proved to be non-poisonous or at least much less poisonous than the crude venom.

Furthermore, the most poisonous substance ever isolated from a snake venom, viz. "neurotoxin" ex *Naja naja*, does not possess any enzymic activity [Ghosh (*90*)]. Even though crotoxin represents a uniform and crystalline protein, it is most likely constituted by several sub-units, of which some have enzymic, and some toxic properties. We must therefore distinguish in the snake venoms between toxins and enzymes, and the toxicity cannot be explained any more as the action of one enzyme or the joint action of some of the known enzymes. If, in future, the really toxic groups of the venoms should be found to be special, active centers in the protein molecules, the latter may then be classified as enzymes or enzyme-like toxins. In the meantime, however, they must be considered as "toxins" of unknown chemical character and action mechanism, and in any case as different from the enzymes described up to now.

There are two reasons why we know comparatively much about these enzymes: in the first place, their very definite effects on simple substrates facilitate their enrichment, even though they have rarely been isolated in pure form. On the other hand, snake venoms constitute an excellent source for enzyme research, such as can hardly be found in similar concentration in other vegetal or animal material. This explains why several hundred papers on the enzymes of snake venoms have been published within the past 25 years.

Much less is known about the toxins. Until 25 years ago, the toxins of snake venoms were believed to be nitrogen-free substances (*63*, *64*, *139*). When finally proteins were proved to be the active factors, a search for the prosthetic groups of these toxic proteins began (*156*). After the isolation of crotoxin, it could be shown that the enzymic and toxic activities of such a compound can be due exclusively to its protein structure, without any prosthetic groups.

Only three snake toxins have been isolated so far, viz. neurotoxin [Ghosh *et al.* (*106*)] and cardiotoxin [Sarkar (*182*)] from *Naja naja*, and crotoxin (*207*) from the *Crotalus t. terrificus* venom. The two former toxins have not yet been crystallized but probably they will be further purified; crotoxin is apparently the toxin of rattle-snake venom which, together with an enzyme, forms uniform crystals. Thus, up to the present time we have no snake toxins really separated and pure in our hands: neurotoxin and cardiotoxin are not absolutely pure; and crotoxin has not been separated from phospholipase.

According to the actual status of our knowledge, we can therefore only describe the activity of the snake venoms as a whole, and that of their enzymes and toxins.

Since the venoms of related species are sometimes much alike in their enzymic and toxic effects, the chemist working on snake venoms needs a certain knowledge of the zoological systematics. The value of an otherwise most interesting chemical statement is often diminished by an inexact zoological denomination of the snake.

II. Snakes and Venoms.

1. Zoological Classification of Poisonous Snakes (*18, 1, 169, 3, 79, 53*).

Snakes are one of the principal groups of reptiles of the present day, of which at least 2300 different species are known, that are grouped into thirteen families, based largely on characters of the skull. Poisonous snakes belong to three of these families only:

1. ELAPIDAE (56 genera with 260 species).
2. VIPERIDAE ("true vipers", 15 genera with 69 species).
3. CROTALIDAE ("pit vipers", 6 genera with 77 species).

It is important to note that the usual distinction between poisonous and non-poisonous snakes depends only on the fact whether the snake is able or not to inject the poison directly into the wound of its prey (79). Even the blood and the serum of both poisonous and non-poisonous snakes are toxic to a certain degree. The poisonous substances are concentrated and stored in the parotid glands of the animal and from there find their way directly into the mouth of the "non-poisonous" species. The "poisonous" snakes possess fangs connected to the glands, which enable them to inject the venom into their prey.

The ELAPIDAE have fixed poison fangs which have only a simple groove through which the poison flows down into the wound; whereas the VIPERIDAE and COLUBRIDAE have fangs, whose edges meet and thus produce a tube with openings at the base and at the tip of the fang, forming an injection needle-like device; the maxillary is movable so that the fangs can be erected for use, or folded back on the roof of the mouth.

For our purposes it is not necessary to know the approximately 400 different species of poisonous snakes; they will be listed here only insofar as any chemical or biochemical information on their venoms is available.

The ten pertinent genera of the ELAPIDAE are *Bungarus*, *Naja*, *Sepedon*, *Dendraspis*, *Micrucurus*, *Acanthophis*, *Demansia*, *Denisonia*, *Notechis* and *Pseudechis*. Genus *Naja*, which is found in Asia and Africa, is the most poisonous of all. The best known is *Naja naja* with a length of up to 6 feet (Indian cobra, also called hooded or spectacle cobra, and formerly named *N. tripudians*). It is the most dangerous snake in India and causes most of the 20000 yearly deaths. *Naja hannah* (king cobra), with a length of up to 18 feet, is also found in India and on the Philippines. Its bite is lethal even for a huge Indian elephant, owing to the large volume and the very strong neurotoxic effect of the secretion. In Africa there are mainly three *Naja* species: *N. haje* (aspis or Cleopatra snake), in Egypt, used in ancient times for execution or suicide purposes because of the unfailing effect of its venom. Further, *N. nigricolis* (spitting cobra), which spits its venom to a distance of 3 yards, up to six times consecutively; and *N. flava* (Cape cobra), whose venom is very similar to that of *N. naja*.

Besides the *Naja* species, the krait snakes are particularly important in Asia, especially in India: *Bungarus coeruleus* (common krait) and *B. fasciatus* (banded krait); and in Africa the species *Sepedon haemachates* (black ringhals) and *Dendraspis angusticeps* (mamba).

In America, only three ELAPIDAE genera are to be found, the most noteworthy being *Micrucurus*, with many species and subspecies. These "coral snakes" are famous for their beautiful coloring. Bites and deaths occur rarely.

In Australia there are relatively few non-poisonous, but as many as 76 poisonous species, all of them ELAPIDAE. They occupy a special position as to aspect and composition of their venom. *Acanthophis antarcticus* (death adder) with its broad head, small scales and stocky body is the most viper-like of the ELAPIDAE; it is, together with *Notechis scutatus* (tiger snake), the most dreaded snake of Australia; though small snakes, both of them cause most of the occurring deaths owing to their powerful venom and their aggressiveness. The most frequent species in Australia is *Pseudechis porphyriacus* (black snake); its venom, as well as that of *Denisonia superba* (Australian copperhead) and *Demansia textilis* (brown snake), has also been investigated.

The snakes of both the VIPERIDAE and CROTALIDAE families have movable maxillaries and canalized teeth, but only the CROTALIDAE ("pit vipers") have deep depressions (pits) midway between the eye and the nostrils. It is assumed that this curious organ serves to sense the temperature of objects, so that the snakes may detect the proximity of prey; for they feed on warmblooded animals.

The VIPERIDAE ("true vipers"), which do not possess this organ, are limited to the Old World. The species *Vipera russellii* (daboia) is the one most dreaded in India, China, Japan, and many other Asiatic countries. It invades human habitations, and its bite is usually fatal. In Africa, the vipers belong principally to genus *Bitis*: *B. gabonica* (gaboon viper) in Central Africa and *B. arietans* (puff adder) in South Africa.

All European poisonous snakes are VIPERIDAE and exclusively genus *Vipera*, with seven species. The most widely spread is *Vipera berus* (adder, Kreuzotter, common viper or peliade), the only poisonous snake in England. *V. aspis* (Jura viper) is to be found in Germany, France, Yugoslavia, and Sicily, while *V. ammodytes* (sand viper) is more at home in Eastern Europe, from Hungary to Turkey. Whereas the other European vipers are comparatively small, *V. ammodytes* grows to a length of 3 feet and is the most dangerous of them.

A closer look at the CROTALIDAE (pit vipers) reveals that they are to be found only in Japan, Java, Borneo, Sumatra, on the Philippines and Formosa, on the borders of Asia and in the Americas, but not in Africa. This geographical distribution seems to have a deeper meaning. The American Indians are of Mongolic descent, related to the East Asian peoples. The striking similarity of the poisonous snakes of Japan with the most widely spread American species is presumably due to the same reasons as the migration of the yellow race. For example, genus *Trimesurus*, like *T. flavoviridis* (habú), and *Agkistrodon* (*A. blomhoffi* = = *mamushi*), found in Japan and Okinawa, are very closely related to the American CROTALIDAE species, zoologically as well as by the composition and activities of their venoms.

Two of the best-known North American poisonous snakes belong to the genus *Agkistrodon*: *A. piscivorus* (water moccasin), a semi-aquatic snake which infests lagoons and swamps of the southeastern United States, and *A. mokason* (copperhead), the most widely spread of all North American snakes. These moccasin snakes are particularly dangerous because they lack the tell-tale rattle, and are therefore easily confused with some harmless snakes.

All other poisonous snakes of North America are "rattlers": instead of the single scale that covers the tip of the tail, they have a sheath with two constrictions, and shed the horny epidermal covering of their scales periodically. The characteristic noise is produced by rapid vibration of the whole tail.

The "grand rattlers" are relatively small snakes of the genus *Sistrurus*, which, like *Sistrurus catenatus* (massassauga) live mainly in prairie areas. As to length and dangerousness, they are left far behind by the snakes of genus *Crotalus* (true rattlesnake), the most dangerous snakes of the New World. In the U. S. A., the most dreaded are, *Cr. adamanteus* (Florida diamond back) and *Cr. atrox* (Texas diamond back) which grow up to 6 feet and have $^3/_4$ inch long fangs. *Cr. atrox* causes as many deaths as all other species together. *Cr. horridus* (timber rattler), *Cr. viridis* (prairie rattler), and *Cr. oreganus* (Pacific rattler) are other species of this genus, frequently occurring in the U. S. A.

In Latin America, from Mexico to the Argentine, there exists only one of the 44 species of the genus *Crotalus*: *Cr. terrificus* (cascavel), divided into the subspecies *terrificus* (mainly in Brazil), *durissus* (in Mexico, Venezuela, Columbia), and *basiliscus* (in Mexico). *Cr. t. terrificus* is not very aggressive and its rattles act as a warning, so that there are less accidents than in the case of other poisonous snakes. But on the other hand it is very frequent in Brazil, and its poison is very dangerous, so that if it bites, the death rate is comparatively high.

Another genus of the CROTALIDAE family also represented in Brazil by one species only is *Lachesis*. The terrible *L. muta* (bushmaster) is found from Rio de Janeiro to Central America, its principal habitat being the Amazonas valley. This largest poisonous snake of the world grows to a length of 12 feet, is very aggressive, and injects several milliliters of its very powerful venom by means of its inch-long fangs.

Genus *Bothrops* comprehends many species and subspecies which are all found in Latin America, and thus constitute the largest percentage of its poisonous snakes. *B. atrox* (fer de lance in Martinique, caiçara in Brazil) with a length of up to 6 feet, is responsible for most of the serious accidents in the Central American countries, where it has many different common names. It is found everywhere between Mexico and São Paulo, and in Colombia even at a height of nearly 7000 feet. *B. jararaca* (jararáca) is the most common poisonous snake of Brazil, and also causes the largest number of bites there. It has a length of 3 feet, maximal 5 feet and is found south of Bahia. The other important snakes of the *Bothrops* genus are *B. jararacussu* (jararacuçú), *B. alternata* (urutú), *B. cotiara* (cotiara), *B. neuwiedii* (jararaca pintada), and *B. itapetiningae* (cotiarinha).

2. Poison Apparatus.

The poisonous snakes possess on both sides of the upper jaw, near, below and backwards of the eyes, oval poison glands, which correspond to the glandula parotis, the largest cellular gland of man. These are emptied into an excretory duct, and from there the poison finds its way into the fangs by contraction of the strong surrounding muscles, particularly the masseter muscle with muscle temporalis anterior and medialis.

Since the poison glands not only produce, but also store the venom, they are large in comparison to the snake's size and, in many species extend for several inches behind the head. The volume of the poison glands differs widely according to size and type of snake, and their poison content depends mainly on the time since the last discharge. In the case of *Lachesis muta* (bushmaster) one gland contains sometimes 3 ml. or more of crude venom, *i. e.* about 15 times more than *Bothrops jararaca* or 30 times more than *Crotalus t. terrificus*.

The secretion is a milky, often yellowish, viscous liquid containing about 20–30% solids. *Lachesis muta* (bushmaster) can eject an average of 450–700 mg. of solids (*3*), *Bitis gabonica* (gaboon viper) 600–1000 mg. (*115*), while smaller snakes like *Crotalus t. terrificus* eject only 30–60 mg. (*3*), or *Vipera berus* (adder, Kreuzotter), 10 mg. (*150*). However, the poison glands contain considerably more poison, yet the snakes do not eject all of it at one time. Thus, it was possible to extract from a recently caught *Lachesis muta* 6 ccm. of crude venom which yielded nearly 2 gm. dry venom (*3*).

3. Extraction of Venom.

Snake venoms are extracted in the following public and private institutes for serum preparation and research work:

Central Research Institute, Kasauli, India.
Instituto Butantan, São Paulo, Brazil.
Instituto Pinheiros, São Paulo, Brazil.
Instituto Dr. Carlos Malbran, Argentine.
Commonwealth Serum Laboratory, Melbourne, Australia.
Institut Pasteur, Paris and Lille.
Institute for Infectious Diseases, Japan.
South African Institute for Medical Research, Johannesburg, Transvaal.
Serum and Venene Dept. of the Museum of Port Elizabeth, Natal.
Behringwerke, Marburg a. d. Lahn, Germany.

Some private firms extract venom for pharmaceutical purposes. In homeopathy, rattle snake venom is used against hemorrhagic diathesis, jaundice, epilepsy and cancer; coral-snake venom against common cold and otitis; and *Naja* venom against shock, angina pectoris, migraine, and cancer. *Lachesis* venom has been recommended against fever, and *Viper* venoms against rheumatism, phlebitis and thrombosis (*160*). In allopathy, *Naja* venom in particular has been used against pain in cancer cases, and the venoms of *Bothrops* and *Gaboia* are applied to shorten the coagulation time of blood (*91*).

In many countries large amounts of venom go into the manufacture of these drugs of very doubtful value. The serum institutes as well as some commercial firms maintain big terrariums or snake farms with hemispherical concrete houses where the snakes, being night animals, can stay during day time. In the snake institutes of those countries where snakes are abundant and which are therefore constantly provided with fresh animals by the rural population, the animals are left to die after one or two venom extractions. The Instituto Butantan, *e. g.*, received in the years 1901 to 1945 about 500000 snakes, 300000 of which belonged to the two species, *Crotalus t. terrificus* and *Bothrops jararaca* (*79*).

Poisonous snakes can be kept for years, if treated with sufficient care (oil baths against parasites, protection from cold, proper feeding). It is essential to handle the extraction of venom carefully, by squeezing the glands not too hard, so that the animal ejects only as much venom as it would by biting voluntarily into the proferred dish. Every pressure causes damage of the sensitive glands; a snake injured in this way refuses food and dies. Furthermore, brutally extracted venom contains mucus, which makes it less suitable for chemical investigation or for the preparation of serum. From well kept snakes, venom can be taken every 2 to 4 weeks without harming them.

The amount of fresh venom varies according to the technique of extraction, and to the size and condition of the snakes; 100 animals of

Crotalus t. terrificus yield, as a rule, only 10 ml., corresponding to 2 to 3 gm. of dry venom, whereas *Bothrops jararaca* yields twice as much (*3*).

4. Drying and Conservation of Venom.

Formerly, dry venom was prepared by drying the secretion in the incubator at 37° for a long period. By this procedure the venoms are likely to be damaged by oxidation, overheating, and influence of bacteria; this is the reason for many contradictory statements as to the toxicity and enzyme content. Furthermore, the venoms dried in the incubator contain more than 15% very firmly bound water, as proved by experiments with *Crotalus t. terrificus* venom (*201*, *202*).

Nowadays, the fresh venom is first freed of cell fragments and mucous substances by centrifugation, and then freeze-dried.

When lyophilized and afterwards freed of the last traces of moisture, the venoms are perfectly stable. They can be kept in evacuated and sealed ampoules without loss of activity for many years. But these perfectly dry venoms are much more difficult to dissolve than those which still contain traces of water. This fact suggests that within the molecule a certain amount of water is very firmly bound and not simply adhering (*201*, *202*).

III. The Action of Venoms.

1. Pharmacological Activity.

Crude venoms from all snake species contain toxins and enzymes in varying proportions. Even the venom of the same species varies according to season, geographical location and alimentation, aside from differences due to extraction and storage. Therefore, clinical observations on snake poisoning in man, or pharmacological experiments on whole animals or isolated organs cannot be expected to elucidate the mechanism of action of such complex secretions.

On the other hand, it seems understandable that in the past two milleniums many observations of the symptoms of snake bites, as well as reports on alleged cures have been published. An extensive paper, compiled in 1936 (*188*), comprises the literature from about 300 B.C. to 1900 A.D. More recent publications have demonstrated some essential pharmacological differences between the venoms of the three families of poisonous snakes, and even between species of the same family. But generally the activities overlap, and without a strict chemical separation of the individual components the pharmacological statements are bound to remain very vague. This is the reason why only the principal results will be listed here, without going into details reported elsewhere (*18*, *1*, *169*, *65*, *8*, *9*, *187*, *137*, *3*, *60*, *223*, *15*, *231*, *79*, *235*, *4*).

As to the total activity of the snake venoms, we can distinguish, generally speaking, three principal groups:

a) Venoms that act primarily on the *nervous periphery*, with only secondary effects on circulation, and an almost total lack of local activity. These venoms are frequently called "curare-like" and are produced by the ELAPIDAE.

b) Venoms that primarily and severely affect *circulation*, provoking shock phenomena. Such circulatory venoms are found in all three families.

c) Venoms whose most striking activity is a *local* one. They act histolytically and hemolytically, and cause *hemorrhages*, edema and necrosis. Particularly the CROTALIDAE venoms, *Bothrops* and *Lachesis* species foremost, are hemorrhagic.

a) *"Curare" Group.*

The venoms of the ELAPIDAE which all act curare-like, differ in a certain respect, described as typical or atypical "curare" activity. The venoms of the Indian ELAPIDAE possess the typical, those of the Australian ones the atypical "curare" activity (*136*), which resembles that of certain quaternary bases (*37*).

On the isolated nerve-muscle preparation of the frog, the *Naja naja* venom 1 : 100000 is ineffective, while 1 : 10000 paralyzes the sciaticus nerve in 30 minutes (*88*). *Naja haje* venom, even in a concentration of 1 : 100000, paralyzes the striated muscle and motor end plates (*187*). Paralyzation extends only to the nerve endings but not to the nerve fibers (*135*). On gastrocnemic-sciatic preparation of the frog, the conducting power is not much affected. The venom acts both on the muscle and the neuro-muscular junction (*185*). That this action cannot be considered as affecting the central nervous system is made evident by the fact that, even after breathing has stopped, rhythmical action currents can be found in the phrenic nerve (*136*).

These experiments show, that a substance necessary for transmission is either destroyed or used up. The assumption that it is acetylcholine becomes more probable, since curare (*38*) as well as cholinesterase from *Naja* venom (*128*) split acetylcholine. The hypothesis that cholinesterase is responsible for the curare-like action of some snake venoms became even more likely, when closer observation showed that cholinesterase is present only in curare-like ELAPIDAE venoms.

Unfortunately, however, it has not been possible to determine any parallelism between cholinesterase content and curare-like effect in the various species. But the principal objections against the "curare"-acetylcholine hypothesis are other ones, viz. the destruction of the cholinesterase does not affect the neurotoxic activity (*107*). Cholinesterase from *Naja* venom, enriched twentyfold, displayed no curare-like action and proved to be non-toxic [CHAUDHURI (*27*); SARKAR et al. (*186*)]. On the other hand, neurotoxin concentrated to approximately the same extent showed no cholinesterase effect whatsoever (*90*).

Possibly, the "curare" activity of *Naja* venom is due not so much to the splitting of acetylcholine by cholinesterase as to the circumstance that the synthesis of acetylcholine is inhibited by another factor (*109*). The latter is not identical with neurotoxin, but seems to be the same factor that inhibits respiration and glycolysis. It has been purified to a high degree (pp. 431 and 442).

b) Circulation Group.

Circulation activity is found in the venoms of ELAPIDAE, particularly in the Indian and Australian species, as well as in VIPERIDAE and CROTALIDAE, and it seems to be the most frequent principle of snake venom effects.

It has been known for a long time (*8*) that the bite of many snakes provokes a condition similar to anaphylactic shock. Later on, two research groups have found that the venoms of *Crotalus horridus*, *Cr. atrox*, *Denisonia superba*, and *Naja naja* exert a powerful peripheric activity on the vascular system; the blood pressure drops violently, thereby indirectly causing tachycardia. The striking resemblance between the effect of these venoms and that of histamine appeared clearly in their effect on the isolated smooth muscle (*134*, *60*). Perfusion experiments with internal organs, particularly with the lung, proved that these venoms really liberate histamine from the tissues; its appearance explains several distinct pharmacological symptoms (*68*). On the basis of similar experiments, it was at first assumed that the venoms form simply lysocithin from lecithin (p. 420), and that the former is responsible for the liberation of histamine. The lecithinase activity of the venom does not, however, fully explain the whole biological reaction, and it was soon discovered that the mechanism is much more complicated.

Above all, in the perfusion experiments mentioned a substance was obtained which strongly contracts the guinea-pig ileum but whose effect starts only after a certain lag of time, and which was therefore called the "slow reacting substance" (SRS) [FELDBERG, HOLDEN, and KELLAWAY (*67*)]. Under the influence of snake and insect venoms, and also of bacterial toxins, not only SRS, but also other secondary toxic substances are liberated. Amongst these, there is one polypeptide, called bradykinin, which has been investigated more closely; it is formed from serum proteins under the influence of *Bothrops* venom in particular (*177*). Depending on the type of both venom and receptor, different poisonous fragments from tissue and blood proteins will develop; the snake venoms act as "starter toxins", whereas the substances liberated by them develop their activity as "induced toxins", and are able to multiply the original effect.

However, not only protein fragments, but also adenyl compounds are liberated by the venoms (*138*), and these may be responsible for most of the observed heart symptoms. This can be concluded from the fact that the venoms of snakes like *Naja haje* and *Bitis arietans*, which hardly lower the blood pressure or contract the smooth muscles, do not arrest the heart at all, or only after hours in the diastole. In contrast, the venoms of *Vipera berus*, *V. aspis*, *V. ammodytes*, *V. russellii*, *Naja naja* and *N. flava*, that have a strong action on muscles and blood pressure, are also capable of bringing the heart to a systolic stop in dilutions extending from 1 : 10000 to 1 : 50000 (*187*).

These observations do not exclude the possibility that in certain venoms there may exist a cardiotoxin acting specifically on the heart (*182*).

c) *Hemorrhagic Group.*

In the venoms belonging to this group the local effects of enzymes on tissue and blood are striking; however, even in this case the toxic action is by no means a purely enzymic one. The minimal lethal dose of venoms is considerably lower than that of the purest corresponding enzymes; e. g. that of trypsin is 1.5–2.5 mg./kg. (*176*), against 0.02 mg./kg. for *Bothrops* venom (*222*), both assayed in rabbits by intravenous injection. While the toxicity is neutralized by mixing antiserum to the venom solution before application, the local reactions are thus only partly eliminated (*57*). The latter are principally due to the joint action of proteases, phosphatases and other enzymes.

We are comparatively well informed on the function of the potent proteolytic enzymes of the CROTALIDAE venoms, particularly ex *Bothrops* and *Lachesis* species. They have a threefold function: primarily they damage the endothelia and capillaries at the place of the bite, and later those of the whole body; hemorrhages occur in the nose, mouth, stomach, kidney and also in the cerebral cortex, causing convulsions in some cases.

These same enzymes, however, fulfill a second and a third purpose: when present in small amounts they coagulate the blood, while in stronger concentrations they make it incoagulable (*139*). The biological reasons for these apparently contradictory activities of the same enzyme may be given as follows. At the site of the bite, where the largest amount of venom is present, the proteolytic enzymes act according to their high concentration: they dissolve any fibrin fiber that might have been formed and, furthermore, they digest prothrombin and fibrinogen, before these can act as coagulants. Hence, the blood turns uncoagulable, and the constant oozing of blood sets in which is so characteristic for viper bites (*145*).

However, only where proteolytic (or may be rather histo- or cytolytic) enzyme is present in a very weak concentration, does coagulation set in. Examples: *Vipera russellii* acts cytolytically on the blood platelets, accelerating the liberation of thrombokinase (*84*); *Bothrops* venom splits prothrombin, converting it into thrombin. In both cases the coagulation is effected indirectly; in the case of CROTALIDAE venom there is yet another direct effect: fibrinogen is transformed into fibrin, even without the assistance of thrombin (*55*) (see also p. 428).

d) Toxic Value.

To the chemist, trying to isolate the toxic principle from snake venoms, it is of minor importance why the test animals die. The primary, decisive question is, how many kilograms of animal can be killed by one gram of a certain venom, on an average. This toxicity evaluation has the advantage of giving whole numbers, easy to remember, and which become higher with increasing toxicity of a venom.

The choice of the test animal is, of course, very important, as the *sensitivity* of different animals to snake venom varies widely.

1 gm. of cobra venom kills (18):

20000 kg. of horse
10000 kg. of man (165 persons of 60 kg.)
8333 kg. of mouse
2500 kg. of guinea pig
1430 kg. of rat
1250 kg. of dog

For assaying, pigeons and mice are the most adequate animals. As to the manner of application, the difference between subcutaneous and intravenous injection is not as big in the case of snake venoms as in other venoms (*187*). Subcutaneous injection is preferable, because it corresponds more to the bite and also reveals the local actions (*189*).

Practically, the "mean lethal dose" is that which kills 2 out of 3 animals within the first 24 hours.

The Brazilian authors indicate as "toxic value" the gm. mouse killed by 1 mg. of dry venom: TV (mouse) [SLOTTA and SZYSZKA (*214*, *215*)].

The Indian authors designate as "toxic value" the number of lethal doses for whole pigeons of the same weight (300–310 gm.), injected intramuscularly [GHOSH and DE (*101*)]. Calculating these values as the "gm. pigeon killed by 1 mg. of dry venom", comparable values with TV (mouse) can be obtained. Even though TV (mouse) is 10–15% higher than TV (pigeon), these toxic values are, for all practical purposes, interchangeable.

Toxic values range from 30 to 5000. The *Naja* species have the highest toxic values of the ELAPIDAE; *N. naja* and *N. flava* around 2000 (*227*, *203*), and *N. haje* around 5000. Only *Notechis scutatus* with 4000 approaches it (*150*). For VIPERIDAE and CROTALIDAE, toxic values up to 300 are normal. Thus, *Bothrops jararaca* has a value of around 300; the only exception is the tropical rattler, *Crotalus t. terrificus*, with a toxic value of around 2000 (*214*, *215*, *189*).

These data can only give a faint idea of the toxic values, for it may happen that, with two venom samples of the same species and under the same conditions, an investigator finds toxic values in the ratio of 1 to 4 (*189*). Research on any snake venom should begin with a determination of the toxic values; in general, 6–12 mice of 15–20 gm. are sufficient for that purpose.

For current chemical work and for comparisons the indication of this "toxic value" is adequate but for pharmacological determinations it is not exact enough. Then it is necessary to use approximately 100 mice per sample, and determine the mean lethal doses by planimetric evaluation of the mortality curves in their whole length (*189*, *175*), by the probit analysis (*69*), or the graded response method (*35*).

2. Enzymic Activity.

The enormous importance of enzymes for the activity of snake venoms is understandable if we remember that the poison gland secretions were primarily saliva, needed to digest the prey. The history of evolution shows that the former giant snakes that killed their prey by crushing it, became smaller and smaller. Thus, some families developed other weapons, particularly their poison apparatus; the saliva turned into an aggressive weapon against smaller animals, and a defensive one against larger ones (*2*).

During this process, certain enzymes in the secretion of snakes, like hyaluronidase and *L*-amino acid oxidase, remained digestive ones, while mainly the *proteinases* and *phosphatases* were "specialized" to cause the terrible thromboses, necroses, and hemorrhages for which *Bothrops* bites are notorious.

The amount of peptidases in the venoms is not considerable, but this point is not of importance, since the peptidases of the prey itself are used for its destruction. The animal tissues contain enough peptidases to autolyze and putrify quickly. It is only necessary to activate these peptidases, a task undertaken by *L*-amino acid oxidase, apparently present in all snake venoms [ZELLER (*231*)].

The venoms contain different phosphatases capable of splitting the phosphoric acid linkages of adenosine triphosphates, nucleotides, and cozymases, as well as those of glucose (1- or 6-)phosphates. However, it has not yet been proved whether the 5-nucleotidase is responsible for the destruction of cell nuclei, and the adenosine triphosphate for the shock, as might be assumed (*235*).

The mucolytic enzyme, hyaluronidase, is the only carbohydrase found in snake venoms. This is easy to explain as the snakes ingest only animal food, so that no amylases and hexosidases are needed in the saliva. Only hyaluronidase is required since it makes the tissues more easily penetrable for other enzymes and toxins, although it does not possess any toxic activity (*58*).

IV. Enzymes from Snake Venoms.

1. Esterases.

a) *Phospholipase A* (*Lecithinase A*).

It was discovered 50 years ago (*164*) that certain snake venoms destroy erythrocytes. Some years later it was found that lysocithin, a very poisonous, easily crystallizable substance, is formed enzymatically from yolk lecithin by snake venom, which splits off oleic acid from the lecithin molecules [DELEZENNE and LEDEBT (*52*)]. It was, therefore, assumed that the hemolytic effect of a snake venom is due to the fact that one of its enzymes produces lysocithin (II) from serum lecithin (I), the former being the true hemolytic principle.

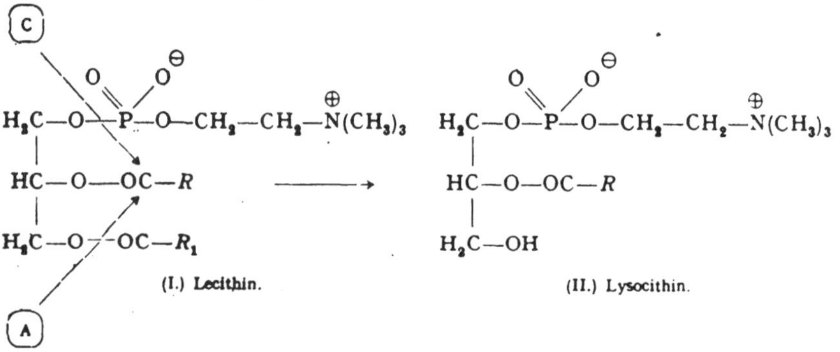

$$\begin{array}{l} \text{(C)} \\ H_2C-O-\overset{O\;\;O^{\ominus}}{\overset{\backslash\backslash\;\,/}{P}}-O-CH_2-CH_2-\overset{\oplus}{N}(CH_3)_3 \\ \;\;| \\ HC-O-OC-R \\ \;\;| \\ H_2C-O-OC-R_1 \\ \text{(A)} \end{array} \longrightarrow \begin{array}{l} H_2C-O-\overset{O\;\;O^{\ominus}}{\overset{\backslash\backslash\;\,/}{P}}-O-CH_2-CH_2-\overset{\oplus}{N}(CH_3)_3 \\ \;\;| \\ HC-O-OC-R \\ \;\;| \\ H_2C-OH \end{array}$$

(I.) Lecithin. (II.) Lysocithin.

(*A* and *C* designate the points of attack of phospholipases *A* and *C* (see p. 416).

This enzyme was first termed phosphatidase (*51*), then lecithinase *A* (*36*) (since meanwhile three further lecithinases had been found which split the lecithin molecule in other ways). The more comprehensive name *phospholipase A* (*167*) is preferable, since this enzyme attacks also other phosphatides, such as cephalin, in the same manner.

Phospholipase A is present in every snake venom as well as practically in all animal organs and tissues. *In vitro* it acts best in phosphate buffer of p_H 7.0 at 38°. In a weak acid solution it is remarkably stable, and not even cyanides prevent its action; at p_H 5.9 it can even withstand boiling.

It is an important fact that 1 mol lysocithin forms with 1 mol cholesterol a complex void of any hemolytic activity. The sera and erythrocytes of different animal species contain lecithin and cholesterol in varying amounts and proportions. There is approximately 0.15% cholesterol in the plasma of man and cattle, while the lecithin value is about 0.2% for man and 0.1% for cattle. Cattle erythrocytes contain roughly the same amount of lecithin as of cholesterol (0.35%), while the human erythrocytes contain about twice as much lecithin (0.4%) as cholesterol (0.17%).

Consequently any test for phospholipase A must refer to blood, serum, or erythrocytes of the same animal species (*40*).

Under equal conditions, the same fraction of *Naja naja* venom gave the following percentages of hemolysis, listed in *Table 1*.

Table 1. Extent of Hemolysis (%) obtained by a *Naja naja* Venom Fraction.

	Guinea pig	Rabbit	Monkey	Human	Horse	Sheep
Whole blood.....	30.6	28.5	13.2	6.9	3.4	nil
Cell suspension...	78.5	62.0	86.9	96.0	79.3	nil

Sheep erythrocytes contain 0.15% lecithin against 0.52% cholesterol (*85*).

For the determination of phospholipase A in snake venoms, it is best to use a suspension of horse erythrocytes (*214*) or guinea pig blood (*39*) with egg lecithin in saline. The Brazilian authors established arbitrary units and found that 1 mg. venom possesses the following lecithinase values ("LV Braz."): crotoxin, 200; venoms from *Crotalus t. terrificus*, 133; *Bothrops jararaca*, 0.7; *Bothrops jararacussu*, 33; *Naja naja*, 15 (*203*). The Indian authors employed other arbitrary units; according to them 1 mg. venom possesses the following lecithinase values ("LV Ind."): *Naja naja*, 100 (*39*); crystallized hemolysin from monocellate and binocellate varieties of *N. naja* venom, 3360 and 3318; and crystallized hemolysin from *Bungarus fasciatus* venom, 1650.

Using another, more quantitative procedure, lysocithin is first formed from lecithin by a certain amount of venom. Then it is allowed to act on erythrocytes, and the amount of the liberated hemoglobin is determined photometrically (*162*, *163*).

By this method, the following comparable values were found:

Bothrops bilineata	0.01	*B. jararacussu*	0.95
B. jararaca	0.07	Bee venom	1.00
Lachesis muta	0.23	*Naja nigricolis*	3.33
B. alternata	0.50	*Crotalus t. terrificus*	0.13
B. neuwiedii	0.70	Crotoxin	0.26
B. atrox	0.80		

Obviously, the difficulties bound to arise by the use of erythrocytes can be avoided: a certain amount of venom acting on pure lecithin splits off some oleic acid which is extracted by means of organic solvents and titrated with phenolphthalein (*62*).

Further, lecithin can be spread out as a monomolecular film on the surface of a saline solution. If 0.01% solutions of snake venom are added, lysocithin is formed which has a totally different surface potential. On measuring the latter, up to 1 γ of phospholipase A can be estimated. Venom concentrations as low as 1 part in 40 millions are thus detected. *Naja naja* venom acts in a particular way: at concentrations of 0.01 to 0.0001 it does not show any detectable hydrolysis of the lecithin film, but on diluting up to 0.00005 a reaction takes place. In higher concentration, other proteins of the venom seem to prevent the phospholipase A from reaching the lecithin layer (*125*).

After electrophoretic or chromatographic separation on filter paper, phospholipase can be easily eluted and determined (*116*). A simpler procedure is to insert the paper strip in a small Plexiglas apparatus, whose cells are charged with an erythrocyte-lecithin suspension; after incubation, the cells containing phospholipase, become clear (*198*). In this or the other manner, it was found that some venoms yield only a single, electrophoretically separable fraction with phospholipase A effect (*Bothrops neuwiedii* (*116*), *Crotalus atrox* (*116*), *Cr. t. terrificus* (*198*), *Trimesurus flavoviridis* (*198*), while some other venoms give two such fractions [*Naja nigricollis* (*116*) and *Bothrops jararaca* (*116*)].

In the venoms of *Denisonia superba* (*123*) and *Naja naja*, a "direct" hemolyzing factor had been found earlier (*40*). It is present in the most basic fractions of *N. naja* and *N. haje* (*116*). The latter venom also contains another "indirect" hemolyzing fraction.

We do not know, how far the "direct" and "indirect" hemolytic effects are different in principle, or whether such differences are only of quantitative nature. It might be possible that a phospholipase A with an extremely strong action attacks the lecithin present in the envelope of the erythrocytes. Whether we assume that the membrane of the erythrocytes consists of mosaic-like protein and lipoid regions, or of an outer lipoid and an inner protein layer, in either case it seems possible, that from the lipoid of the cell envelopes a very effective phospholipase A produces lysocithin which in turn causes hemolysis. Thus, the so-called "direct" hemolysis would be also an "indirect" one, and the two mechanisms would be identical.

It was claimed that the venom's phospholipase A content is responsible for the powerful inhibitory effect on a number of "insoluble" mitochondrial respiratory enzymes, including succinic and pyruvic dehydrogenases; a number of "soluble" enzymes (concerned with glycolysis, fermentation and oxidation) would not be affected (*16*).

b) *Phospholipases B and C.*

The *Bothrops alternata* venom contains besides phospholipase A also phospholipase C which splits lecithin into a diglyceride and phosphocholine, and is known from culture filtrates of *Clostridium welchii* (*224*).

Phospholipase B that removes the saturated fatty acid from lysophosphatides and is present in *Aspergillus oryzae*, *Penicillium notatum* and wasp venom, has not yet been found in snake venoms.

c) *Phosphoesterases.*

All snake venoms, insofar as they have been tested, contain an alkaline phospho*di*esterase (*117*), while *mono*-esterase has only been found in some

ELAPIDAE [*Naja naja*, *N. haje* and *N. melanleuca*, *Sepedon haemachates*, *Notechis scutatus* (*235*)]. The estimation is preferably carried out manometrically at p_H 8.5, using as substrates β-glycerophosphate, 1- or 6-glucosephosphate, inorganic pyrophosphate or diphenylphosphate [ZELLER et. al. (*235*)].

Experiments with *Trimesurus flavoviridis* venom, containing much phosphodiesterase and only traces of mono-esterase, showed that the splitting of diphenylphosphate can be accomplished only with both enzymes present. The ophio-form of the phosphoesterases is quite different from phosphoesterases found in other materials: the p_H optimum of the phosphodiesterase from rice bran is 3–4; from *Aspergillus oryzae*, 5.5; from liver or kidney, 7, whereas the p_H optimum of the ophio-phosphodiesterase is 8.5–9. (*229*).

It is possible to remove the 5-nucleotidase from the venom of *Vipera russellii* or *Crotalus adamanteus* by adsorption on cellulose (*127, 126*) or by precipitation with acetone (*193*) to such an extent that only phosphodiesterase remains. With such preparations, desoxyribonucleic acid can be degraded quantitatively to mono-nucleotides (*118*).

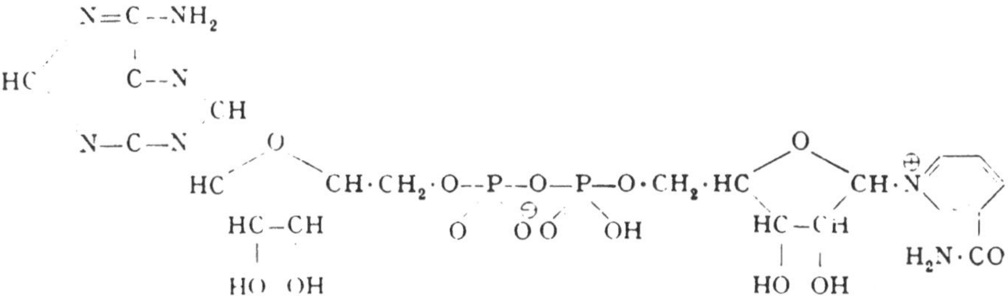

```
      N=C--NH2
         |
HC       C--N
            CH
      N—C—N    O                                    O
          HC    CH·CH2·O--P--O—P—O·CH2·HC             CH·N⊕
          HC—CH          O  OO   OH       HC—CH          H2N·CO
          HO  OH                          HO  OH
```

(III.) Co-enzyme I = DPN (Diphospho-pyridine-nucleotide).

Phosphodiesterase also splits the phosphate linkage of cozymase (= diphospho-pyridine-nucleotide or DPN), thus acting as nucleotide-pyrophosphatase in this case. The inhibition of glycolysis and fermentation, occurring in the presence of venoms, led to the suspicion that the cozymase is damaged by some venom enzymes. This specific action was first observed with *Notechis scutatus* venom (*20*); however, nucleotide-pyrophosphatase occurs in all venoms yet tested, and is particularly abundant in *Bothrops jararaca* (*235*). In this reaction only adenylic acid but no phosphoric acid is formed from cozymase; this became evident when venom solutions acted on cozymase, at first in the presence of zinc ions to inhibit the 5-nucleotidase effect (see below). After the removal of the zinc the 5-nucleotidase became active and phosphoric acid was set free (*235*). Nicotinamide was liberated from coenzyme by *Bungarus fasciatus* but neither by *Naja naja* nor by *Vipera russellii* venom (*11*).

d) 5-Nucleotidase.

Since apparently all snake venoms contain phosphodiesterase as well as 5-nucleotidase (*235*), they are able to split the polynucleotides by way of the oligo- and mono-nucleotides, forming nucleosides [GULLAND and JACKSON (*117*)] (*Chart 1*). Whereas Mg^{++} and Co^{++} activate 5-nucleotidase, Ni^{++}, Zn^{++} and antisera are powerful inhibitors.

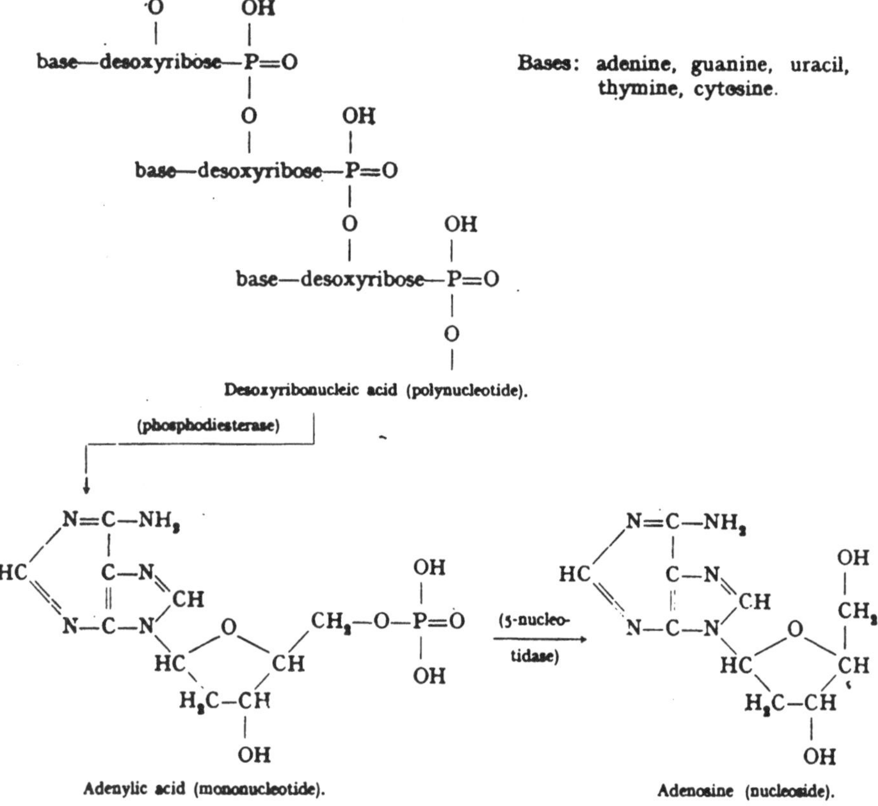

Chart 1. Cleavage of Polynucleotides by some Snake Venom Enzymes.

For the determination of 5-nucleotidase, the millimol phosphates are assayed which are liberated per hour from nucleic acid by 1 mg. of venom, in glycine buffer, at p_H 8.3. *Vipera gabonica* and *Crotalus horridus* gave the highest values (*238*). By another method, the action of some CROTALIDAE venoms on ribonucleic and desoxyribonucleic acids was investigated in bicarbonate solution (*219*). (*Table 2.*)

Table 2. Action of Some CROTALIDAE Venoms on Ribonucleic and Desoxyribonucleic Acids.

CO_2 liberated in 1 hour from		by 1 mg. venom of
ribonucleic acid	desoxyribonucleic acid	
3.2 ml.	7.1 ml.	*Agkistrodon piscivorus*
3.8	—	*Bothrops ammodytoides*
3.7	6.6	*B. atrox*
4.4	8.2	*B. jararaca*
6.6	6.9	*B. jararacussu*
4.0	6.9	*B. neuwiedii*
7.2	10.3	*Crotalus t. terrificus*

Up to now the spectrophotometric method has not been applied for the determination of 5-nucleotidase in snake venoms (*146*, *6*).

e) *Adenosine triphosphatase* (*ATPase*).

All venoms hitherto tested are able to dephosphorylate ATP, but some of them only after Mg^{++} or certain other bivalent ions have been added as activators (*234*). ATP is almost quantitatively hydrolyzed to pyrophosphate and adenylmonophosphate (AMP), which in turn is converted by 5-nucleotidase into adenosine and orthophosphate (*131*) (*Chart 2*).

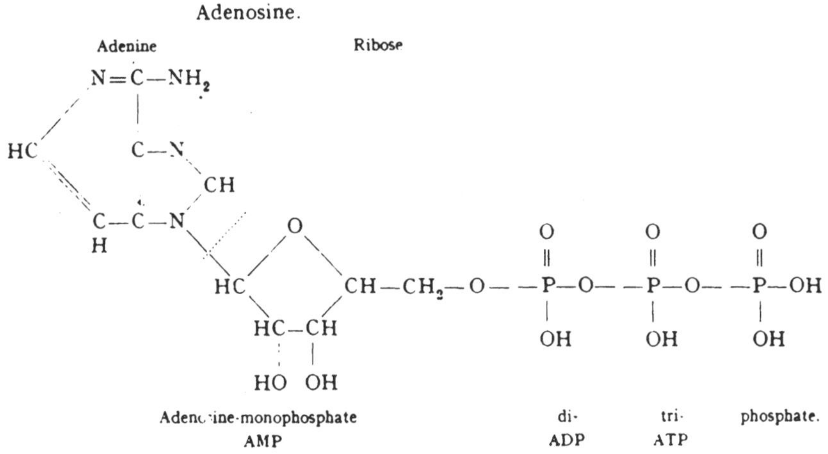

Adenosine.

Adenine Ribose

```
   N=C—NH2
  /  |
HC   C—N
 \\   |   \
  C—C—N    CH
  H     \      O
         \   /   \
          HC       CH—CH2—O— —P—O— —P—O— —P—OH
           \     /             ||      ||      ||
            HC—CH              O       O       O
            |   |              |       |       |
            HO  OH             OH      OH      OH
```

Adenosine-monophosphate AMP · di- ADP · tri- ATP · phosphate.

Chart 2. Action of some Snake Venom Enzymes on Adenosine triphosphate.

The fact that ATPase decomposes energy-rich ATP to adenylic acid may be important for the poison effect, especially if the other phosphatases assist by destroying substances as important for the re-synthesis of ATP as glucosephosphate and cozymase (DPN) (*230*).

The determination of ATPase activity of the venoms was carried out at p_H 8.3 and 38° in glycine buffer [ZELLER (*233*)]. By assaying the liberated phosphoric acid, it became evident that the ELAPIDAE venoms generally display a weeker ATPase activity. It is remarkable that Mg^{++} activates the venoms of different species very differently; thus, *Bitis gabonica* venom is strongly activated, but the *Bothrops jararaca* venom hardly at all.

f) Ophio-cholinesterase (OChEase).

A cholinesterase was discovered in the ELAPIDAE venoms in 1938 [IYENGAR et al. (*128*)]. Later on, it was investigated whether it resembles one of the two cholinesterases formerly obtained from human blood. We know that it is certainly different from both. The cholinesterase, first obtained from human serum and therefore named the *s*-type, acts on benzoylcholine but not on acetyl-β-methylcholine and is not inhibited by excessive substrate concentration — all this in contrast to OChEase. On the other hand, the cholinesterase, first obtained from human erythrocytes and known as the *e*-type, is inactive against methylbutyrate, ethylacetate or tributyrin, whereas OChEase is active in these instances [ZELLER and UTZ (*240*)].

For the determination of OChEase, mainly the manometric method has been used; it consists in liberating CO_2 from a bicarbonate solution by the acid formed in the hydrolysis. All the ELAPIDAE venoms contain OChEase, whereas it was not found in VIPERIDAE or CROTALIDAE. While 1 mg. of *Naja nigricolis* venom produces only 37, *N. naja* gives 4900, *N. melaneuca* 30800, and *Bungarus coeruleus* 35100 microliters of CO_2/hour [ZELLER (*232*)].

An estimation of the OChEase activity can also be performed by colorimetry, using the conversion of acetylcholine into acetohydroxamic acid by means of hydroxylamine (*122*). This procedure makes it possible to follow the total course of the hydrolysis. It is also recommendable for the evaluation of paper chromatograms and paper electrophoresis strips. A new method of potentiometric titration also permits a continuous observation of the hydrolysis (*180*).

2. Carbohydrases: Hyaluronidase.

The snake venoms contain one or more likely several enzymes which enhance the spreading of some substances such as India ink, and therefore this agent was first termed the "spreading factor" [DURAN-REYNALS (*54*)]. It was soon established as being much like or even identical to hyaluronidase from mammalian testicles and leech heads. [Another name for hyaluronidase, invasin, has not been generally accepted (*119*)]. Determination in viscosimetric units showed that 1 mg. of purified testicle extract contains 415 units, against 24 in bee, 27 in *Denisonia superba*, and 111 in *Pseudechis porphyriacus* venoms (*22*). The p_H optimum is at 4.6; and the enzyme is more heat sensitive than phospholipase A [MADINAVEITIA (*151*)]. The hyaluronidase when precipitated from *Crotalus atrox* venom by saturated ammonium sulfate, is six times enriched (*152*). The hyaluronidase of *Bothrops jararaca* and *Cr. t. terrificus* venom is inactivated by the respective antisera (*66, 58*).

Hyaluronidase can be estimated by means of the viscosity and turbidity reduction tests. More satisfactory is the chemical determination of the resulting reducing sugar by the Prussian blue reaction (*168, 196*). Since this method, however, cannot be used in the presence of other reducing substances, the physicochemical assays mentioned are indispensable for the studies on hyaluronidase inhibition by serum. It is important to note that the correlation of the reductometric and turbimetric methods is satisfactory (*173*).

From former determination by means of the India ink method, it can be concluded that *Agkistrodon piscivorus*, and particularly *Crotalus adamanteus*, *Cr. terrificus*, and *Cr. atrox* have the strongest hyaluronidase activity (*231*).

The hyaluronidase (invasin) penetration is prevented or delayed by an inhibitor present in the serum, termed antiinvasin (*119*), apparently a heparin-lipoprotein complex, belonging to the albumin fraction (*132*); it can be split by alcohol into two fractions which act only jointly (*111*). The proteolytic enzymes of the snake venoms split the protein part of this inhibitor, and are thus to be designated as proinvasines (*110*).

3. Proteases.

a) *Proteinases.*

The proteinases of the snake venoms resemble trypsin in certain respects. Venoms of *Agkistrodon piscivorus* as well as the *Bothrops* and *Crotalus* species are rich in proteinases that occur abundantly in the venoms of some vipers, such as *Bitis arietans* but rarely in ELAPIDAE (*55*).

The venoms of *Naja naja* (*89*) and *Vipera russellii* (*97*) act optimally against gelatine or ovalbumin at p_H 7.8–8.0, and against casein at p_H 6.6. The corresponding values for *Echis carinata* and *Bungarus fasciatus* are, 8.0–8.2, and 7.0. Their effect is inhibited by KCN, $HgCl_2$ and H_2S. In all cases 50° is the optimal temperature. *Bothrops jararaca* acts optimally at p_H 10 and between 35–50° (*155*). Since the p_H optimum for the *Naja naja* and *Vipera russellii* venoms is practically identical with that of trypsin, and the critical inactivation temperature for these and other venoms (52–63°) is at least near the 60° given for trypsin, it was claimed that the proteinase of these venoms is identical with trypsin (*95*). However, since the same venoms markedly inhibit the action of added trypsin, this cannot be the case (*97*).

Determination of tyrosine and tryptophane, liberated by eight CROTALIDAE venoms from hemoglobin (*7*), either by means of the phenol reagent (*78*) or spectrophotometrically (*133*), gave the following comparable values (*130*):

Lachesis muta	14.0	*B. jararaca*	3.9
Bothrops jararacussu	6.6	*B. alternata*	1.5
B. atrox	5.8	*B. cotiara*	1.2
B. neuwiedii	5.1	*Crotalus t. terrificus*	1.0

b) *Peptidases.*

In the venoms of *Naja naja, Bungarus fasciatus, Echis carinata,* and *Vipera russellii* not only polypeptidase and carboxypeptidase are present but also dipeptidase (*103*). The peptidase content of snake venoms is, however, usually very small, and slight peptidase activities can be detected only after long incubation.

All snake venoms contain ophio-*L*-amino acid oxidase, which oxidizes any free *L*-amino acid immediately by splitting off the amino group. This was the basis for a simple method to determine the peptidases of the venoms [Zeller and Maritz (*238*)] or peptidases of an other origin, for example from tubercle bacilli (*178*), by a manometric technique using di- or tripeptides as substrates (*178*).

c) *Proteolytic Enzymes and Coagulation.*

It has been suspected for a long time that proteolytic enzymes play a part in blood coagulation (*149*). Trypsin converts prothrombin into thrombin, and is thus the counterpart of the physiologic calcium-platelet system. Papain, however, converts fibrinogen directly into fibrin, and is thus the counterpart of thrombin, as shown in the following diagram (*216*).

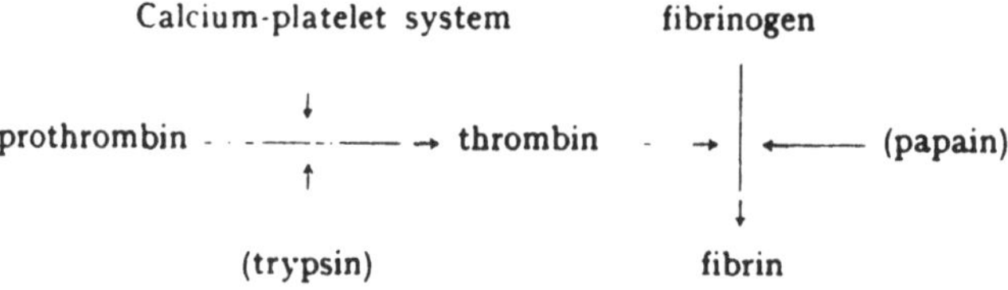

Therefore, several snake venoms were examined as to which act trypsin-like and which papain-like (*55*). The venoms of *Bothrops atrox, B. jararaca, B. nummifera, Crotalus adamanteus, Cr. t. terrificus, Cr. basiliscus,* and *Cr. horridus* interact directly with fibrinogen, like papain or thrombin. The optimum for this coagulation process, at approximately pH 6.5, is the same as for the action of thrombin on fibrinogen. In addition to coagulating fibrinogen directly, three of these venoms, viz., those of *Bothrops atrox*, *B. jararaca*, and, to a lesser extent, *Crotalus t. basiliscus*, convert prothrombin into thrombin, without any intervention of either calcium or platelets. *Notechis scutatus* and *Micrucurus* venom have no effect on fibrinogen but do convert prothrombin into thrombin [EAGLE (55)].

Certain venoms do not convert prothrombin into thrombin, but decompose it. This is the case with *Naja naja*, *N. flava*, *Vipera russellii*, *Bothrops atrox* and *Crotalus horridus*. The venoms of *Vipera ammodytes* and *Agkistrodon piscivorus* decompose not only prothrombin but also fibrinogen (55).

Experiments on the coagulation mechanism of *Bothrops atrox* venom showed that by acetone precipitation and heating the venom solution at 50° for 20 minutes the proteolytic component that destroys prothrombin and fibrinogen can be removed. Venoms treated in this way act only thrombin-like (*129*).

Thus, the action of snake venoms on blood coagulation is such a complex one, that a single standard technique can only yield schematic total values which are the result of many accelerating and inhibiting processes. Frequently, the coagulating activity is indicated in reciprocal minutes of the clotting time on oxalated rabbit plasma. In this manner the coagulating effect of some venoms was determined for the purpose of comparing it with the proteolytic activity. This resulted, on the whole, in a fairly satisfactory correlation (*130*).

Some comparable coagulation values of different venoms follow:

Lachesis muta	6.0	*B. jararaca*	2.6
Bothrops jararacussu	1.5	*B. alternata*	2.4
B. atrox	3.5	*B. cotiara*	1.5
B. neuwiedii	1.7	*Crotalus t. terrificus*	1.0

4. Oxidases.

a) *Catalase.*

Oxidizing enzymes, capable of converting hemoglobin into methemoglobin, have been found in the venoms of *Lachesis muta*, *Crotalus adamanteus*, *Naja bungarus*, *N. naja* but not in *Crotalus t. terrificus* and *Vipera russellii*. The oxidizing enzyme occurring in most of the *Bothrops* venoms is *catalase*, not peroxidase (59). The darkening of the prey's blood after the bite of certain snake species is probably due entirely to this catalase activity and to the resulting conversion of hemoglobin to methemoglobin.

This formation of methemoglobin and metmyoglobin is not arrested by antiserum, therefore even when enough antiserum was immediately applied, lethal accidents did happen, resulting from an intermediate nephron nephrosis (*5*).

b) *Ophio-L-Amino Acid Oxidase (OAAO).*

Nearly all snake venoms have a yellowish color. Spectrographic investigations showed that this color is due to *flavines*, either free or bound to protein (*17*). Determination of the flavin content proved that it not only varies widely in the different snake species, but even in the same species, as much as from 0 to 20 mg.%, depending on the geographical location of the animal.

Probably always riboflavin is present, and its amount shows a remarkable parallelism to the very potent *L*-amino acid oxidase, found in practically all snake venoms as well as in the lungs, livers, kidneys, and erythrocytes (*239*). This *L*-amino acid oxidase is distinctly different from other such oxidases, present in rat kidney or in *Proteus vulgaris* for example. It is not inhibited by Prussic acid or octyl alcohol, in contrast to *Proteus vulgaris* oxidase (*236*). To differentiate it from the other *L*-amino acid oxidases, it is called ophio-*L*-amino acid oxidase (*237*).

It appears that OAAO attacks all *L*-amino acids under the following conditions: Besides the amino and carboxyl group, another organic radical must be linked to the α-carbon atom; thus, glycine is not attacked (*237*). The carboxyl group should be free; therefore, e. g., *L*-tyrosin ethylester is not oxidized (*235*). Another amino group (as present in lysine or arginine) inhibits the enzyme in the same way as does a second carboxyl group (in aspartic or glutamic acid). If such groups are, however, substituted as in benzoyl lysine, or in asparagine or glutamine, the acids are attacked (*235*). Since furthermore the α-amino group must be free, proline is not affected (*236*). The oxidative deamination of the amino acids proceeds according to the equation,

$$R \cdot CH\,(NH_2)\,COOH + O_2 + H_2O = R \cdot CO \cdot COOH + NH_3 + H_2O_2.$$

If the hydrogen peroxide is not decomposed by the catalase present, it will decarboxylate oxidatively the ketocarboxylic acid.

The best analytical method to determine OAAO activity is the manometric one; its results are expressed in mm^3. O_2/hour/mg. dry venom (*231*).

Thanks to the circumstance that the riboflavin content of the venoms corresponds to that of OAAO, it was possible to isolate this enzyme in pure form from *Agkistrodon piscivorus*, and to elucidate its prosthetic group [SINGER and KEARNEY (*191*)]. The purified OAAO is electrophoretically homogeneous at pH 4.9–7.2. Its molecular weight as calculated from the sedimentation constant is 61600, and from the flavin content, 62800. At 38° and the optimal pH (7.0–7.5) 1 mol OAAO is able to oxidize 3100 mol *L*-leucine per minute. The activity rises with the increase of the temperature in the range, 24.5–50°, since in the presence of its substrate OAAO has a marked heat stability (*192*).

In the absence of *L*-amino acids, the prosthetic group of OAAO could be split off by heat denaturation and proved to be flavin-adenine-dinucleotide. This was confirmed by spectral and fluorescence assays as well as by isolation of the cleavage product itself after hydrolysis (*191*).

It seems that in the venom of *Agkistrodon piscivorus* the entire riboflavin content of 4×40^{-4} micromol/gm. is due to the prosthetic group of the OAAO. Probably, flavin is bound in a similar form in the venoms of all other species studied, and thus serves exclusively to build up OAAO. Besides the protein-bound riboflavin, there is always a certain quantity present in free form (*10*).

It is worthwhile recording that amongst the venoms of the *Bothrops* species, which generally contain large amounts of riboflavin and consequently OAAO, there exists a completely white one without any trace of OAAO. Moreover, this *B. itapetiningae* venom does not follow the usual scheme inasmuch as, in contrast to other *Bothrops* venoms, it neither contains any hemaglutinin, nor does it exert any local (dermatotoxic) activity (*56*). A possible connection between OAAO and these effects has so far not been investigated.

Although OAAO is not directly toxic, it seems to be the enzyme which activates the proteases and peptidases of the cells in the bitten animal (*231*). This is shown by the following observation: *Vipera aspis* venom is generally yellow, it contains a considerable amount of OAAO and causes necroses, although it is practically free of proteolytic enzymes. The venom of the same *Vipera aspis*, from the "Département du Gers" in southern France, is colorless and does not display any necrotizing action (*19*). This was confirmed in the case of the Argentine *Crotalus t. terrificus* venom whose shade, from yellow to white, varies with the locality (*10*).

5. Inhibiting Enzymes.

It has been observed for some time that many snake venoms exert a powerful inhibition on glycolysis in cell-free muscle extracts, and on fermentation in yeast maceration juice (*220*). As little as 0.2 mg. per ml. venom of *Notechis scutatus*, *Denisonia superba* or *Bothrops atrox* is sufficient for complete inhibition (*20*). The respective antivenins neutralize this inhibitory effect (*23*).

The enzymic nature of the antifermenting principle is indicated by the fact, that it does not dialyze through cellophane membranes, acts even in very great dilution, and is very sensitive toward pepsin, heat or organic solvents (*23*, *75*).

It was, of course, important to ascertain which stage of the carbohydrate decomposition is inhibited by snake venoms. Certain differences appear when the effect of *Notechis scutatus* venom on several oxidases and dehydrogenases is examined. Those dehydrogenases which require a co-enzyme for their activation, like lactic and malic dehydrogenases, are completely inhibited, while others, like succinic dehydrogenase, are affected only slightly or not at all [CHAIN (*21*)]. As mentioned above, the co-enzyme is decomposed by the phosphodiesterase present in most venoms, and therefore dehydrogenases depending on its presence, such as lactic or malic dehydrogenase, are bound to be weakened in their effect. This alone, however, cannot account for the 90–100% inhibition of glycolysis by the venoms. Evidently, the latter exert a strong action on the cytochrome-cytochrome oxidase system.

Venom samples of *Naja naja*, *Bungarus fasciatus*, and *Vipera russellii* were allowed to act on glucose, succinates, lactates or pyruvates in amounts of 3–30γ/ml. Suspensions of pigeon brain cells have been used as source of enzyme. With air present, the *N. naja* venom showed a marked inhibiting effect on the oxidation of succinic acid. Under anaerobic conditions however, this oxidation was only inhibited to a small extent, while the oxidation of glucose was very strongly affected (*93*).

The oxidation of *p*-phenylene diamine by the cytochrome-cytochrome oxidase system (in form of pigeon brain cells) was inhibited by minute doses of *Naja naja* venom. Even when this venom, by heating to 70° for 30 min., had been deprived of its proteolytic activity, its inhibiting effect on the cytochrome-cytochrome oxidase system was retained to a large extent (*94*).

Among the CROTALIDAE, the inhibitor of *Crotalus t. terrificus* venom has been most closely examined, for it appeared that crotoxin exerts, besides other activities, also the inhibitory one of the crude venom. The inactivation of succinoxidase by crotoxin in rat liver homogenates and mitochondria is by no means due to a lysic effect of lysocithin.

Cytochrome oxidase and succinic dehydrogenase are not considerably affected by crotoxin under conditions which completely inactivate succinoxidase. Succinic dehydrogenase, cytochrome, choline and malic oxidase are inactivated by crotoxin but much more slowly than is succinoxidase. The inactivation of succinoxidase is not overcome by the addition of cytochrome c [NYGAARD (*166*)]. Spectroscopic examination indicated that the bands of the crotoxin-sensitive factor in the reduction of cytochrome c by succinate were located between cytochromes b and c (*165*).

Furthermore, when heating snake venoms (*Naja naja*, *Bungarus fasciatus*, *Agkistrodon piscivorus*, *Vipera russellii*) to 100°, it was found that they still retained their toxicity, and that some of their enzyme inhibitors were not affected (*16*, *92*). It has been suggested that heated venoms exert their inhibitive effects by enzymic degradation of some mitochondrial structures (*16*).

V. Separation, Purification, and Crystallization of Biologically Active Venom Constituents.

1. Non-proteins.

The part played by the non-proteins in snake venom effects is unclear. Of *inorganic substances*, only zinc was found in surprisingly large amounts, and it was claimed that a parallelism exists between the zinc content and the phosphoresterase activity (*50*). The *Naja naja* venom contains 5.27 mg.% zinc, and the zinc content of other ELAPIDAE venoms is not much lower (*174*), while the VIPERIDAE venoms contain only 1–2 mg.% (*50*). As mentioned before, Zn^{++} is a powerful inhibitor of 5-nucleotidase and ATPase. Recently, it was found that Zn^{++} diminishes the toxicity in general. A high zinc content protects the gland against damage by its own secretion. As soon as the zinc concentration decreases, e. g. in the bitten prey, full activation sets in [FLECKENSTEIN et al. (*76, 77*)].

The qualitative *sugar reaction* with naphthol and sulfuric acid is negative for the ELAPIDAE venoms (*Naja haje*, *N. nigricollis*), but positive for the VIPERIDAE venoms (*Vipera russellii*, *Echis carinata*, *V. ammodytes*). As to the CROTALIDAE, all *Bothrops* venoms assayed contain sugar which, however, could not be found in any of the *Crotalus* venoms [FISCHER and DÖRFEL (*71*)]. The importance of carbohydrates for the toxic effects has sometimes been proposed (*106*) but never proved.

Recently, a considerable amount of a *nucleoside* was found in one VIPERIDAE and one CROTALIDAE venom; the venom of *Bitis arietans* contains as much as 12% and that of *Dendraspis viridis* 2.6% adenosine. None of the other venoms examined had any nucleoside. Even though adenosine lowers the blood pressure, it is improbable that it should take any considerable part in the total activity of the venoms (*71*).

2. First Attempts to Separate Active Proteins.

A century ago, it had already been established with the venom of *Vipera berus* (*14*) that snake venoms are of proteinic nature. Later, some authors found that peptones and globulins were present, others believed, according to the dialyzability and solubility, in the presence of albumoses. However, all these reports are only of historical interest. For the first time in 1904, by means of saturated ammonium sulfate, the proteins of a snake venom (rattlesnake) were separated into toxic and non-toxic fractions (*153*). Unfortunately, in the following years, the toxic fractions were described not as proteins, but as nitrogen-free, saponin-like compounds. They were named "ophiotoxin" and "crotalotoxin" (*63, 64*), and even much later a "bothropotoxin" was interpreted as a nitrogen-free substance (*139, 140*).

In 1925, the ammonium sulfate experiments were repeated, and the toxic substance was precipitated from a solution of rattlesnake venom at 46–64% saturation (*225*).

The most important step in the chemical clarification of snake venoms was achieved in the years 1936–39, fairly simultaneously in Germany, India, and Brazil.

The venom of *Bothrops jararaca* was the first to be more closely investigated. By heat coagulation, precipitations at different p_H, alcohol precipitation, dialysis, electrodialysis, and adsorption procedures, a protein fraction was sometimes obtained which was 4.5 times more neurotoxic, and 5 times less coagulating than the crude venom (*140*). These results were subject to variations and the authors were unable to decide whether or not the toxin contained nitrogen (*143*).

The separation of the coagulating component from the neurotoxic one by means of ammonium sulfate precipitation, succeeded only very imperfectly. It seemed, however, certain that the globulin fraction promoted coagulation to a greater extent than the albumin fraction (*141*). The substance thus obtained remained very active even in a concentration of 5×10^{-10}/ g./ml. blood (*142*) but unfortunately, even though some analytical values were obtained (*144*), no real chemical separation of the *Bothrops jararaca* venom components was accomplished. Neither has the protein mixture of this venom since been satisfactorily resolved.

Not even paper electrophoresis provides adequate separation. Thus, two recent papers contradict one another completely. Whereas one report claims that the proteolytic component migrates at p_H 8.6 to the cathode (*116*), according to the other [MICHL (*159*)] all the proteins migrate at p_H 8.8 to the anode, the hardly toxic proteolytic and coagulating fraction as the leading one. Then follows a middle fraction containing the toxin as well as phospholipase and hyaluronidase; and, finally, a fraction with ATPase and desoxyribonuclease activity.

The only well established fact is, that the crude venom contains relatively few sulfur-bearing amino acid units, *i. e.* 5.73% cystine and 1.75% methionine (*201*).

More successful experiments were carried out at the same time with the venom of *Naja naja*, an ELAPIDAE; of *Vipera russellii*, a VIPERIDAE; and of *Crotalus t. terrificus*, a CROTALIDAE. Thus, we do have some information on the venom of at least one principal representative of each of the three most important families of poisonous snakes.

First, it was established that only 14% of the total nitrogen of the *Naja* venom is protein nitrogen, while the rest originates from lecithin. The free fraction of this lecithin is extractible with ether, the rest is bound to protein as lipoprotein. The venom contains 87.56% protein, which can be precipitate with trichloroacetic acid, and from which 36.69% albumin can be separated by heating to 75° for 6 minutes. Surprisingly enough, the toxicity does not suffer by this heating. The heat stability of the *Naja* toxin has recently been confirmed; a solution, heated in a boiling water bath for 15 min., did not lose its toxicity (*16*). By stepwise salting out with ammonium sulfate, 20.31% globulin, 11.31% primary, and 16.81% secondary proteoses were obtained; the latter contained the total toxicity of the venom (*87*), whereas the non-toxic primary proteose and globulin fractions were responsible for the hemolytic activity (*85*).

Daboia venom contains 2.8% lipoids, but in contrast to *Naja* venom, it has no phosphorus in the trichloroacetic acid precipitate, where 15.5% nitrogen (= 96.8% protein) was found. Of the corresponding proteins 23.35% was globulin, 22.12% albumin, and 50.52% consisted of proteoses. The total neurotoxic, coagulating, and hemorrhagic action was present in the secondary proteose fraction [GANGULY and MALKANA (*86*)].

It should be mentioned that the components of *Naja* and *Daboia* venoms were separated by other Indian authors in a similar manner [ROY and CHOPRA (*179*)]. They found in the *Naja* venom 18.0% albumin, 49.0% pseudoglobulin, and 3.6% euglobulin; and in *Daboia* 22.0% albumin, 43.2% pseudoglobulin but no euglobulin. Considering the vagueness of the terms, proteose, euglobulin, etc., the data mentioned can only give a faint idea of the venom composition.

However, in the years 1938–1951 five proteins with different biological activities and a certain degree of purity have been isolated from *Naja naja* venom. It is remarkable that they were 15–20 times more active than the crude venom but hemolysin, which was crystallized, surpassed the activity of the crude venom 33 times.

3. The Active Principles from *Naja* Venoms.

a) *Neurotoxin.*

The so-called neurotoxin of *Naja* venoms, unlike all the other active components of snake venoms, is strongly basic and is a polypeptide. In 1936–1937, simultaneously in Germany and India, these two characteristics of neurotoxin made its isolation possible up to a certain point by salt formation, dialysis, electrodialysis, and cataphoresis. Later, this toxin was much more purified, and now represents the most poisonous compound as yet obtained from snake venoms.

Considering the definitions given on p. 418 the "toxic values" listed in *Table 3* (p. 460) for *Naja* venom and for the purified neurotoxin preparations at different degrees of purity can be calculated.

Table 3. Relative Toxic Values of some *Naja* Venom and Purified Neurotoxin Preparations.

	TV (mouse)	TV (pigeon)
Crude *Naja flava* venom	1000–2000 (*157*)	
Purified neurotoxin from *N. flava* venom	8300 (*157*)	
Crude *N. naja* venom	1300–2600 (*227, 203*)	750 (*98*); 1000 (*101*)–3200 (*96, 106*)
Neurotoxin I from *N. naja*	6700 (*227*)	
Neurotoxin II from *N. naja*	12500 (*156*)	15200 (*101*)
Neurotoxin III from *N. naja*	33000 (*156*)	37200 (*108*)
Highest purified neurotoxin	45400 (*106*)	47000 (*96*) 49200 (*106*)

Evidently, the toxicity of the dry crude venom varies widely. Since only with the highest purified neurotoxin the values were determined in both mouse and pigeon [GHOSH, DE, and CHAUDHURI (*106*)], we can state that 1 mg. of the most poisonous toxin ever obtained from any snake venom may kill about 160 pigeons or 2300 mice, when injected subcutaneously.

From the solution of crude *Naja naja* venom, inactive proteins could be eliminated by successive precipitations with alcohol and acetone, and finally a neurotoxin sample with TV (mouse) of about 6700 was obtained (*227*). It formed a toxic picrate and hydrochloride, and its molecular weight was assayed at 1555 by measuring the dialysis coefficients in comparison to clupein (*227*). By another procedure, a neurotoxin from *N. flava* venom with TV (mouse) of 8300, was isolated by rapid dialysis through cellophane [MICHEEL et al. (*157*)]; by the same method, its molecular weight was determined as 2500–4000 (*156*). When comparing the elementary analysis of the crude venom with that of this neurotoxin preparation, an increase of the sulfur content becomes conspicuous (*157*):

Crude venom: C = 47.06%; H = 6.05%; S = 4.6%; N = 14.64%; (NH_2 = 2.69%); ash = 3.0%.

Neurotoxin: C = 45.2%; H = 7.0%; S = 5.5%; N = 14.7%; ash = 3.0%.

Further purification of the neurotoxin was obtained by cataphoresis. Neurotoxin from *N. naja* venom, pre-purified by ultrafiltration and dialysis, was exposed to a current of 15–20 mamp., 3000–3500 volts, in a 5-liter six-cell cataphoresis apparatus, at 25°. The neurotoxin recovered from the cathode cell after 17 hours, had a TV (mouse) of 12500, and that from another cell, of 33000. However, this solution lost much of its activity on evaporation (*156*).

The Indian authors first tried to separate the neurotoxin by electrodialysis or cataphoresis (*98*) and by adsorption on various adsorbents at different p_H values (*49*). In 1938, they worked out some methods of purification of the neurotoxin (*101*, *105*), by making use of precipitation and adsorption techniques alone. Later these procedures were perfected by using highly toxic crude venom [TV (pigeon) 3200] as starting material (*106*, *96*). First, the principal amount of accessory proteins was eliminated by precipitation with sodium sulfate. The neurotoxin was then adsorbed on tungstic acid and after dissolution of the adsorbent, precipitated by methyl alcohol. It then possessed a TV (pigeon) value of 32000 (*108*). This product was further purified by another precipitation with methyl alcohol [TV (pigeon), 37200 (*108*)] or by fractional precipitation of the accessory proteins with ammonium sulfate, repetition of the tungstic acid adsorption, and methyl alcohol precipitation. This resulted in a dry neurotoxin sample showing the highest toxicity so far attained, viz. TV (mouse) 45500; TV (pigeon) 49200 (*106*), and 47000 (*96*) respectively.

The yields have not been stated but it seems obvious that the total losses during this purifying process were considerable. It cannot be decided at the present time whether the neurotoxin amounts to about 30% of the total protein, and whether it contains half of the sulfur found in the *Naja* venom, as had been formerly claimed (*109*).

b) *Constitution of Neurotoxin.*

According to the available analytical determinations which, however, have been carried out at different stages of purity of the *Naja* neurotoxin, this compound is a basic polypeptide with the isoelectric point of about 11 (*101*), a molecular weight of 2300–4000, and a sulfur content of 5.5% (*157*). By applying various methods to the crude *Naja* venom, all of the sulfur was found in cystine (*203*). Neither methionine, nor any trace of another sulfur-containing amino acid or other component could be detected. Thus, the presence of 3 cystine residues, equivalent to 6 sulfur atoms in a molecule of the molecular weight 3500 can be postulated (*195*).

Against this assumption two hypotheses have been formulated. One of them claims that the sulfur in the *Naja* neurotoxin is present in a thiolactone ring. However, the laborious assays carried out in support of this hypothesis have not been able to provide any conclusive evidence (*158*).

On the other hand, some experiments with various reducing agents made it appear doubtful that the decrease of toxicity caused by them should only be due to their influence on the sulfur bond. Sodium sulfite, cysteine, and ascorbic acid, after having acted for 24 hours on very dilute, neutral solutions, brought about a decrease in toxicity of only 30% (at most) in neurotoxin and less than 10% in crude venoms. By sodium bisulfite, however, the toxicity was nearly completely destroyed within the same time. Hence, it has been proposed that this reagent acts specifically on the carbohydrate contained in neurotoxin, and thereby produces inactivation (*106*). It should be noted at this point that no sugar was found in the venoms of *N. haje* and *N. nigricollis* (*71*).

In contradiction to both hypotheses just mentioned *Naja* venom when inactivated by potassium cyanide acquired a certain reducing power and gave a positive color reaction with sodium nitroprussate (*226*). Inactivation, reducing power, and color reaction thus produced vary, depending on the time and concentration, in such a manner that three clearly defined levels can be observed: a) a 50% inactivation, a 1 ml. consumption of N/500 iodine solution per 25 mg. of venom, and a faint nitroprussate reaction; b) a 90% inactivation, a 2 ml. iodine consumption, and a marked nitroprussate reaction; and c) complete inactivation, a 3 ml. iodine consumption, and a strong nitroprussate reaction [Wellers (*226*)].

We may assume that by means of the cyanide reaction first one, then another, and finally the third (and last) of the -S-S-bridges is broken, and that the molecule becomes accordingly less active. The investigations carried out so far do not, however, enable us to assert this definitely (*195*).

The amount of the three basic amino acids in the highest purified *Naja* neurotoxin was compared to that in highly active crude venom. The latter is rich in lysine, while the neurotoxin is remarkably rich in arginine (*96*). On the basis of the amino acid nitrogen content as per cent of the total nitrogen, the following values can be approximately computed (*Table 4*).

Table 4. Basic Amino Acids in Purified *Naja* Neurotoxin.

	TV (pigeon)	Arginine	Histidine	Lysine
Dry crude *Naja naja* venom	3200	3.5%	2.2%	11.0%
Purified *N.* neurotoxin	47000	12.0%	4.0%	4.0%

According to these figures, the three amino acids are present in the neurotoxin in a proportion similar to that in thymus histone. The greatest difference between the structure of neurotoxin and that of histones or protamines consists in its high cystine content (20.6%). This value was obtained by postulating a sulfur content of 5.5% even for the highest purified neurotoxin preparations (*157*).

c) *Hemolysin.*

In 1936, two research groups have demonstrated that by treating *Naja naja* venom solution with sodium chloride all the hemolysin is precipitated and that this precipitate is free of neurotoxin (*87, 99*). After heating a faintly alkaline solution of this precipitate at 86°, and then by submitting to cataphoresis the solution obtained by heat coagulation, a hemolysin fraction four times more active than the crude venom was recovered (*99*).

On p. 421 two test methods were described which had been employed to control the purification of phospholipase A originating from *Crotalus t. terrificus* and *Naja naja* venoms. The arbitrary units used in Brazil (*214, 215*), are in a proportion of about 15 : 100 to the Indian units (*39*), since the hemolytic activity of 1 mg. of *N. naja* crude venom was determined as 15 units with one method (*203*), and as 100 units with the other one (*39*). Thus, with all due reservations, because of the mentioned variability of crude venom activities, 1 unit (Braz.) can be considered to be equal to 6.6 units (Ind.).

Avoiding heat coagulation and cataphoresis, it was possible to increase the hemolytic component 11-fold by precipitation and adsorption methods (*39, 90*). From a solution of 1 grn. of *N. naja* crude venom (= 100000 units Ind.) 388.8 mg. hemolysin with 92500 units were precipitated by sodium chloride. After adsorption on tungstic acid and removal from the adsorbent (p_H 9.2) 111.7 mg. of hemolysin with 75200 units Ind. were recovered from the evaporated solution (*39*).

A similarly active product was prepared by precipitating inactive proteins from a solution of 100 mg. of venom (= 10000 u. Ind.), first by methanol and sodium chloride, and then by acetone. The thus obtained 10.45 mg. of hemolysin (7550 units) was then precipitated in an acetone concentration of 60–90%. After the adsorption of inert proteins on ferric hydroxide, the yield was 5.88 mg. of hemolysin (6500 units), corresponding to an 11-fold enrichment (*39*).

This product, however, was not pure enough to crystallize. Therefore, the hemolysin was repeatedly precipitated by 15% sodium chloride at p_H 3. Inactive proteins were then removed at p_H 4.2 by precipitation with ammonium sulfate and sodium chloride of a lower concentration or by adsorption on Alumina C_r. When the final solution of hemolysin reached at p_H 5.6 an ammonium sulfate saturation of 60%, a suspension of needles was obtained. After reprecipitation, these crystals contained 2.94% of the protein and 52% of the hemolysin present in the crude cobra venom [De (*42*)].

d) *Properties and Composition of Crystalline Hemolysin.*

Crystallized hemolysin obtained from the monocellate and binocellate varieties of *Naja naja* venom has 3360, resp. 3318 hemolytic units (Ind.) per mg., while crystalline hemolysin *ex Bungarus fasciatus* contained only 1650 such units (*43, 44*). The precipitin reaction indicated that the hemolysins from the two varieties of the *N. naja* venom are much more closely related to each other than to the *Bungarus* hemolysin (*43*). Since 7.2 mg. of this crystallized hemolysin is necessary to kill a 300-gm. pigeon (*44*), this enzyme appears to be practically non-toxic when compared to crude venom of which 0.1–0.4 mg. is the letal dose.

Hemolysin could be inactivated by some oxidizing agents such as iodine, hydrogen peroxide, ferricyanide or benzoquinone, and the inactivated product reactivated with hydrogen sulfide, potassium cyanide, ascorbic acid, cysteine, or reduced glutathione (*41*). A 1% solution of crystallized hemolysin was half inactivated by irradiation with ultraviolet light but this product could not be reactivated. Immunization of rabbits was carried out with hemolysin and its oxidized and irradiated products. Only the irradiated hemolysin did not precipitate with the immun serum (*43*).

The crystalline hemolysin from *Naja naja* venom is most stable at p_H 6.0, and at this p_H the "critical" inactivation temperature, at which the activity is reduced to 50% of the original value within 60 minutes, was observed to be 62°. In contrast, crude *Naja naja* venom was activated by heating at 62°, probably due to the elimination of inhibitors (*45*). The isoelectric point of crystalline hemolysin (from *N. naja*) has been established by the microcataphoretic method as being at p_H 8.55, and by the U-tube method as 8.61 (*46*). The molecular weight has been determined as 31900 (*47*) and 35200 (*48*) by the diffusion method, and as 33200 by a computation based on the chemical composition [De (*48*)].

Crystalline hemolysin was found to contain:

C, 48.66%; H, 6.21%; N, 15.92%; S, 3.88%.

Arginine	18.94%;	Cysteine	11.60%;	Tyrosine	4.34%;		
Histidine	11.25%;	Methionine	3.62%;	Tryptophan	1.81%;		
Lysine	5.24%.						

These data on the cysteine and methionine contents of hemolysin are contradictory to results obtained by other authors who by the same analytical method have found all the sulfur in crude *Naja naja* venom in the form or cystine and methionine (*203*).

Titration of amino groups in 90% acetone and of carboxyl groups in 90% alcohol, before and after hydrolysis, resulted in an average residue weight of 115.2. Calculating with 40 as the minimum possible number of sulfur atoms, the molecular formula was stipulated as being about $C_{1350}H_{2060}O_{525}N_{380}S_{40}$ (*48*).

e) Cholinesterase.

Cholinesterase from horse serum was first purified by ammonium sulfate fractionation and adsorption on aluminum hydroxide (*218*). From two varieties of *Naja naja* venom, the cholinesterase could be isolated with an activity increased 20-fold in relation to the crude venom; and the enzyme was 22 times more active than the one prepared from serum. It was not necessary to use other methods for the isolation than the fractional precipitation with sodium and ammonium sulfate [CHAUDHURI (*26*)]. Cataphoretic and adsorption procedures failed to produce further increase in activity (*27*). Cholinesterase from the venom of *Bungarus fasciatus* was enriched 11-fold by the same process (*27*).

The purified enzyme sample was completely free from the other active principles present in crude venoms (*26*). It was ascertained that it acts specially on choline esters (*29*). The isoelectric point of the cholinesterase adsorbed on quartz was established by the microcataphoretic method as p_H 5.55 and by the U-tube cataphoretic method as p_H 5.9 (*28*).

Experiments to determine the stability of cholinesterase showed that the enzyme lost half of its activity when exposed for an hour to a temperature as low as 44°. Ultraviolet light caused inactivation; in a 5% solution of sodium sulfate, however, considerably less inactivation took place (*30*).

The optimum condition for the activity is p_H 8.4, while the maximum stability was found to be at p_H 7.4 (*32*). The activity increases with increasing acetylcholine concentration to a maximum at about 0.5% acetylcholine but a substrate concentration in excess up to 1% causes no inhibition (*31*).

Crude (*33*) as well as purified (*34*) ophio-cholinesterase can be inhibited by iodine, hydrogen peroxide, ferricyanide, silver nitrate, mercuric chloride, iodoacetic acid, phenylmercuric chloride and *p*-benzoquinone. The original activity is restored by treatment with hydrogen sulfide, cysteine, ascorbic acid or sodium cyanide.

f) Cardiotoxin.

After neurotoxin, hemolysin and cholinesterase had been isolated from the *Naja* venom, it was possible to compare the effect of these purified enzymes and that of crude venom on respiration and heart. Cholinesterase had little effect on blood pressure or respiration of rabbits, and only in large amounts did it stimulate the toad heart slightly (*186*). Non-crystalline, purified hemolysin stopped the perfused toad's heart, but with the same amount of crystalline hemolysin neither irregular beating, nor any shortening of the amplitude took place (*184*). Purified neurotoxin manifested its toxicity by arresting the respiratory movements; when these movements were reinstated by artificial respiration, further quantities of the same neurotoxin could be administered without producing any other toxic effects (*184*). Under similar circumstances crude *Naja* venom finally led to impairing the circulation and stopping the heart, in spite of artificial respiration.

As a result of the discussed evidence, the existence of an as yet unknown cardiac principle in *Naja* venom appears all the more probable That its chemical properties are similar to those of hemolysin could only be suspected, as it was still present in the non-crystalline hemolysin. The isolation was accomplished by precipitating crude venom three times with 22% sodium sulfate at p_H 6.5, and then twice with 17% sodium chloride at p_H 2.8. By further precipitation with 24% sodium chloride at p_H 4.2, followed by precipitation with 30% sodium chloride at p_H 6.0, and removal of the salts by dialysis and precipitation with ice cold alcohol, a purified *cardiotoxin* was obtained. This had still 40% of the cardiotoxic activity in 4% of the original venom and was practically free from neurotoxic and hemolytic activities. The cardiac effect was determined and expressed in arbitrary toad or cat units, indicating the amount of cardiotoxin necessary to bring about the stoppage of perfused toad heart or the arrest of a cat's heart beat (*128*). The cardiac activity decreases with exposure to ultraviolet light (*181*). Using either the microcataphoretic or the U-tube method [SARKAR (*182*)], the isoelectric point of cardiotoxin was established as about p_H 8.2. The molecular weight was determined by a diffusion method, dialyzing 1 and 2% solutions of p_H 7.4 at 5° through a porous alundum diaphragm for two days. The diffusion rate was assayed by the cardiac activity of the diffusate, and corresponded to a molecular weight of approximately 46200 (*183*).

Cardiotoxin acts directly on the muscle only, leaving the nerve and the neuromuscular junction intact, whereas crude *Naja naja* venom acts on both the muscle and the neuromuscular junction (*185*).

g) *Inhibitor.*

It had been verified earlier that the inhibitory activity of the *Naja naja* venom is not due to previously isolated proteins and especially not to the proteolytic enzyme system contained in it. Therefore, attempts were made to separate the inhibitor of the oxidation of glucose, lactic, piruvic, or succinic acid. When a 1% venom solution was treated with 20% sodium chloride and incubated at p_H 9 for 30 minutes, the inhibitor remained in solution, and the precipitated proteins contained the proteolytic factor. On repeating this procedure, the inhibitor, by acidification to p_H 2, could be obtained free from any proteolytic activity (*94*).

Further purification of the cytochrome cytochrome-oxidase inhibitor was accomplished by precipitations with sodium and ammonium sulfates at varying p_H. The final product contained only 2.12% of the initial protein but 33% of the inhibitory activity and was 16 times more active than the crude venom [CHATTERJEE (*24*)]. In solutions of p_H 7.4, its half inactivation temperature is about 80°. The inhibitor is fairly stable within the p_H range of 1.0 to 9.6, the stability optimum being at p_H 7.4. Contrary to several other active principles of *Naja naja* venom (hemolysin, cholinesterase and cardiotoxin), the inhibitor is not affected by ultraviolet irradiation. It was found to be free from neurotoxin and cholinesterase, and only contaminated by traces of cardiotoxin (*24*). The inhibiting effect of this factor on the cytochrome-oxidase system is markedly slower than that of hydrocyanic acid and hydrosulfide but the necessary molar concentration is much lower. The process of inhibition seems to be noncompetitive (*25*).

Table 5 shows the five proteins, with their characteristic biological activities, which have been isolated from *N. naja* venom:

Table 5. Biologically Active Proteins Isolated from the *Naja naja* Venom.

	Isoelectric point	TV (pigeon)	More active than crude venom	Mol. weight
Neurotoxin	11	49200	16.6 times	3500
Hemolysin	8.6	42	33.6 times	33000
Cardiotoxin	8.2	—	15 times	46200
Cholinesterase	5.7	—	20 times	—
Inhibitor	?	22	16 times	—

These proteins have been isolated mainly by precipitation and adsorption methods; experiments to separate them by means of electrophoresis will be discussed on p. 451.

4. Neurotoxin from *Bungarus fasciatus* Venom.

In *Bungarus fasciatus* as in *Naja* venom, inactive protein and hemolysin could be precipitated by means of 24% sodium sulfate. Adsorption on tungstic acid yielded neurotoxin free from hemolysin but the recovery was very small. Better separation of neurotoxin was achieved after the repeated adsorption of inactive proteins on ferric hydroxide in solution of p_H 7.2. In this manner, 75% of the neurotoxin with a 9-fold activity against crude venom was recovered, but it still contained some hemolysin (*104*).

5. The Neurotoxic and Coagulating Principles from *Vipera russellii* Venom.

After several experiments (*103*) which resulted in a 5-fold concentration of the neurotoxin from *Vipera russellii* and the determination of its isoelectric point as being 5.8 (*100*), this neurotoxin could be concentrated 7.8-fold, after the previous precipitation of inactive proteins from a 24% sodium sulfate solution. At a 24–30% sodium sulfate concentration the neurotoxin precipitated and, after redissolution, was more readily adsorbed on Alumina C than the accessory proteins. The toxin could then be eluted by 1% disodium phosphate and 5% glycerol (*104*).

It is important to note that the neurotoxin from *Vipera russellii* is far more sensitive than that *ex Naja* to the action of reducing agents like H_2S, $NaHSO_3$, Na_2SO_3, cysteine, ascorbic acid, and Zn + HCl (*105*).

6. Active Proteins from *Crotalus t. terrificus* Venom.

a) Coagulin.

The CROTALIDAE belong to the solenotodentes (σωλῆνος = tube or canal), *i. e.* to those snakes which, from an evolutionary point of view, possess the most highly developed and complicated venom apparatus. It could, therefore, be assumed that their venom might also be chemically more complex than that of the proteroglyphodentes (γλύφειν = dig in). However, this does not seem to be the case.

At the beginning of this century it was demonstrated that a toxic globulin precipitates from a solution of rattlesnake venom half saturated with ammonium sulfate (*153*). This globulin represents by far the most toxic component of the CROTALIDAE venoms. It displays a number of different biological activities, and has been crystallized from the venom of the South American rattlesnake, *Crotalus t. terrificus*. At one time it was assumed that the *Crotalus t. terrificus* venom contains also a cardiotoxin with direct action on the myocardium (*217*), but this has not been confirmed until now.

Furthermore, the CROTALIDAE venoms contain one, or perhaps two, active proteins that are responsible for their proteolytic, histolytic, hemorrhagic, hemocoagulating, and anti-hemocoagulating effects. Whereas these proteins do not play a considerable part in rattlesnake venoms, they are important as to the quantity and action in the *Bothrops* species. Even though the chemical investigation of the proteolytic and coagulating substances in *Bothrops* venom has not yielded satisfactory results so far (p. 428), it seems probable that they belong to the globulin fraction (*141*), whereas they are bound to an albumin in *Crotalus t. terrificus.*

Many facts point to the possibility that the proteolytic as well as the coagulating activity are due to the same protein. On comparing, *e. g.*, the different CROTALIDAE venoms with respect to these activities, a striking parallelism appears (*130*). When the separation of the proteolytic enzyme from *Crotalus t. terrificus* was attempted, it retained its coagulating activity, even when various preparative methods were applied

(*206*). Thus, it seems likely that the same albumin molecule has a proteolytic as well as a coagulating activity; and as long as these have not been separated, the name *coagulin* should be retained (*194*).

Coagulin was obtained, when a solution of *Crotalus t. terrificus* venom was precipitated by ammonium sulfate to 45% saturation. On removing the ammonium sulfate from the precipitate by dialysis, the globulin fraction separates and the supernatant solution contains the coagulin. The latter was obtained as a colorless substance by lyophilization, and then purified by redissolving in distilled water and centrifuging the undissolved residue. Another portion of the coagulin was isolated in the same manner from the mother liquors of the crotoxin preparation.

Coagulin, which represents only up to 10% of the *Crotalus* crude venom, could be concentrated at best to show a tenfold coagulating and proteolytic activity per mg. as compared with that of the crude venom [SLOTTA and FRAENKEL-CONRAT (*206*)]. The preparation must be carried out rapidly, since coagulin is very sensitive to alcohol, to heat and even to standing at room temperature.

b) *Preparation of Crotoxin* (*207*, *208*).

For the preparation of amorphous crotoxin two methods were used:

From 15 ml. of freshly drawn and centrifuged venom, 1.25 gm., corresponding to 42% of the dry content of the crude venom, was isolated by heat coagulation and precipitation near the isoelectric point.

Fig. 1. Crotoxin crystals. [From: Ber. dtsch. chem. Ges. *71*, 1076 (1938).]

Inactive proteins, some crotoxin, and coagulin were precipitated from 2 gm. of dry venom by ammonium sulfate up to a saturation of approximately 45%. From the centrifuged solution, 312 mg. could be separated by saturation with ammonium sulfate up to 52%, and another 430 mg. of crotoxin by saturation up to 62%.

The solution of amorphous crotoxin in 1% acetic acid at 55° was treated with 1% pyridine to a p_H 4.4. When cooling this solution slowly, crystallization set in after about 30 minutes. Upon five recrystallizations, 74 mg. of crystallized crotoxin was obtained from 140 mg. amorphous crotoxin. Only 2.3 mg. of crotoxin dissolve in 1 ml. of water but it is easily soluble in 4% sodium chloride solution.

c) *Composition, Chemical Properties and Homogeneity of Crotoxin.*

Amorphous crotoxin: N, 15.8%, 16.0% (*207, 208*); S, 4.02%, 4.05% (*202, 203*).

Specific rotation in 0.85% NaCl solution, $[\alpha]_D^{23} = -0.71° \pm 0.09°$ (*207*), (*208*).

Crotoxin (five times recrystallized) (*206*):

C, 50.77%;	H, 6.41% (*202, 203*);	N, 15.86% (*207, 208*);	S, 3.96% (*202, 203*)
50.56	6.47	15.90	4.03

The solubility curve of crotoxin at different ammonium sulfate concentrations shows the straight line typical of pure proteins (*Fig. 2*).

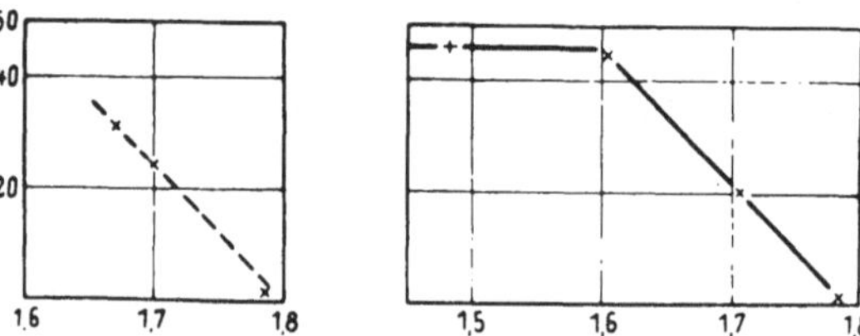

Fig. 2. Solubility curves of crotoxin at different ammonium sulfate concentrations.

[Abscissa: log. of the % saturation of ammonium sulfate. Ordinate: mg. crotoxin in solution (*207, 208*).]

The *molecular weight* of crotoxin was determined by sedimentation and diffusion experiments in the ultracentrifuge [GRALÉN and SVEDBERG (*114*)]. The partial specific volume of 0.704 is comparatively low for a protein. The sedimentation constant was 3.13, independent of the concentration and p_H within the range of 5.5–10.0, and the diffusion constant was found to be 8.59. Crotoxin behaved as a homogeneous substance in these sedimentation and diffusion assays; and its computed molecular weight was 30000; sedimentation equilibrium measurements gave 30500.

Electrophoresis experiments [LI and FRAENKEL-CONRAT (*148, 147*)] proved the homogeneity of crystalline crotoxin. It is worth mentioning at this point that crotoxin does not display the phenomenon of "reversible boundary spreading" to a detectable extent, in contrast to some other proteins studied. The very sharp boundaries appeared after two hours, in a field gradient of 6.40 volts per cm., in phosphate and acetate buffers of pH 7.20, 7.0, 6.3, 4.90, 4.40, 4.23, and 3.91.

The *isoelectric point* was found to be 4.71, and the mobility, 4.65×10^{-5} cm.^{2}/sec./volt, in a solution of ionic strength 0.10 (*148*).

Amino acid composition: Twodimensional paper partition chromatography showed that the crotoxin molecule contains the following 18 amino acids: arginine, lysine, aspartic and glutamic acids, phenylalanine, tryptophan, tyrosine, glycine, alanine, proline, serine, threonine, cystine, and methionine. By special techniques of monodimensional chromatography, it was possible to demonstrate that crotoxin contains also valine, histidine, leucine, and isoleucine but no hydroxyproline (*211, 212*).

Even before crotoxin was isolated and crystallized, it had been shown that the sulfur is not present in *Crotalus* venom as a thiolactone, but as a normal -S-S-linkage (*204, 205*). The average of eight determinations was, 13.2 ± 0.3% cystine (methods of FOLIN, SULLIVAN, and BÄRNSTEIN); and that of three more determinations, 1.36 ± 0.05% methionine (BÄRNSTEIN) (*201, 202*).

The amino nitrogen (VAN SLYKE) was determined at 0.78; 0.89% before (*82*), and at 12.5% after acid hydrolysis (*210*). Titration demonstrated the presence of 0.119 equivalent per cent carboxyl groups (*210*), whereas when using dyes (*81*), 0.142 equivalent per cent total acid and 0.111 equivalent per cent basic groups were found (*82*).

A quantitative estimation (*213*) of some amino acids in the hydrolyzate by means of paper chromatography resulted in: 4.5% phenylalanine; 9.4% tyrosine; 7.4% arginine; 1.9% histidine, and 6.3% lysine.

A complete quantitative amino acid analysis of crotoxin was carried out using the following chromatographic technique: the hydrolyzate was chromatographed monodimensionally on paper, and the separated amino acids were determined colorimetrically in solution as copper salts after development with ninhydrin (*70*). The values listed in *Table 6* were obtained [FISCHER and DÖRFEL (*72*)].

Table 6. Amino Acid Analysis of Crotoxin.

Amino acid	%	Amino acid	%
Alanine	4.62%	Serine	5.03%
Glycine	6.52%	Threonine	4.96%
Valine	1.73%	Cystine/2	12.25%
Leucine	3.62%	Methionine	2.09%
Isoleucine	2.96%	Arginine	8.48%
Proline	5.31%	Histidine	2.25%
Phenylalanine	9.64%	Lysine	6.30%
Tyrosine	9.53%	Aspartic acid	12.22%
Tryptophan	4.18%	Glutamic acid	13.86%

Another recent analysis (*80*) carried out using the dinitrofluorobenzene method has yielded results that are in general accordance with the amino acid composition listed in Table 6, except that the phenylalanine value is appreciably lower and the tyrosine value higher than given; both figures are more in line with earlier determinations [cf. above and (*82*)]. Furthermore, a recent estimation of tryptophan in crotoxin [FRAENKEL-CONRAT (*80*, *82*)] by means of various procedures has given values of about 2.5%, *versus* 4.18% in Table 6, p. 447.

Compared to the amino acid composition of other proteins (*221*), the high content of aromatic amino acids in crotoxin is striking regardless of which will be the final values. From the very accurate elementary analysis it can be deduced that more aspartic and glutamic acid is present than has been found. In summary the following tentative formula may be assumed:

$C_{1230}H_{1780}O_{430}N_{330}S_{36}$; mol. weight about 29000, which would require: C, 50.53%; H, 6.13%; N, 15.73%; S, 3.95%, and which could be written as:

ala_{16}, gly_{24}, val_4, leu_8, iso_6, pro_{16}, phe_8, tyr_{16}, try_6, ser_{16}, thr_{12}, cys $(SS/2)_{12}$, met_4, arg_{12}, his_4, lys_{12}, asp_{32}, glu_{32}, $ammonia_8$. It must be stressed that these formulations are hypothetical.

When the amino acid composition just given is compared with the analytical data obtained for amino-N, total basic, total acidic, and carboxylic groups, the agreement is mostly within the range of the expected experimental error.

d) Biological Properties of Crotoxin.

After intramuscular injection of the approximate lethal dosis of crotoxin, it has been observed that the animals drag their hind-legs when running. This paralysis phenomenon, however, is not the cause of death. The animals die from paralysis of the respiratory muscles, the mean lethal dose (*175*), as determined in a great number of mice, being 0.27 γ/g. mouse [SCHÖTTLER* (*190*)] (TV = 3700), against 0.6 γ/g. mouse [TV = = 1670 for the crude venom (*189*)].

The crotoxin effect can be neutralized by a specific antivenin. In this event, a chemical reaction takes place between the antigen (mol. weight, 30000) and the antitoxin (mol. weight, 150000). The radio in the equivalence zone is near 10, *i. e.* twice that observed between diphtheria toxin (mol. weight, 70000) and equine antitoxin [BIER (*12*, *13*)].

The toxicity of crotoxin is severely diminished or even disappears in almost every kind of chemical conversion. This distinguishes crotoxin from many other biologically active proteins. For example, for the insulin effect phenolic, carboxyl and disulfide groups are essential, but not

* To whom the writer is grateful for this determination.

amino, hydroxyl, amide and guanidyl groups. In crotoxin, esterification of the carboxyl groups or acetylation of most of the amino groups results in extensive detoxication. Furthermore, inactivation is caused by sulfation of the aliphatic hydroxyls, as well as by iodination of most of the phenolic groups, or coupling of the latter with imidazole groupings. Complete inactivation is caused by formaldehyde, particularly in alkaline solution or in the presence of alanine in neutral or acid solutions. This is regarded as a proof that some or all indole, guanidyl, and amide groups are essential for the toxic activity of crotoxin (*82*); but to what degree the hemolytic and other enzymic activities suffer by these chemical reactions, has not yet been investigated. Recent studies have shown (*80*) that the toxicity and hemolytic activity of crotoxin are not impaired when, by means of carboxypeptidase, three amino acids are split off from the carboxyl end of the peptide chains.

The hemolytic and hyaluronidase activities of crotoxin are not responsible for its toxic effect. They are about 50–100% higher in crotoxin than in the crude venom. Taking into consideration the considerable variation of these activities in the crude venom, as well as the instability of the solutions (*214, 215*) and the limit of error of the methods, the lecithinase (LV) and hyaluronidase (HV) values can only be approximately evaluated as follows (*Table 7*).

Table 7. Lecithinase and Hyaluronidase Values.

	LV (*214, 215*)	HV (*196*)
Crude *Crotalus t. terrificus* venom	133	76
Crotoxin	200	113

Crotoxin possesses the total smooth muscle stimulating activity of the crude venom, as was shown by comparative experiments on isolated guineapig ilea. The contractions set in after 30 seconds, and, although the intestine pieces retained their full sensitivity to histamine, they were in both instances, after a single experiment, almost completely desensitized against another dose of the toxin (*199*).

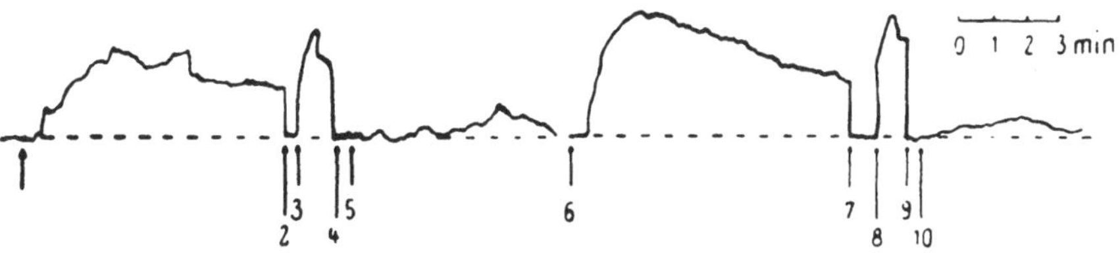

Fig. 3. Stimulating effect of crude *Crotalus* venom and of crotoxin on smooth muscles (unpublished). — At 1 and 5: Crude *Crotalus t. terrificus* venom (1 : 1 000 000). — At 6 and 10: Crystallized crotoxin (1 : 1 000 000). — At 2 and 8: Histamine (1 : 25 000). — At 3, 4, 7, and 9: Washing.

The inhibition or inactivation of several vital enzymes, like succinic dehydrogenase and others, represents another essential property of crotoxin (*165*, *166*, *228*). It has so far not been investigated whether this inhibitory effect is higher in crotoxin than in the crude venom.

Three kinds of enzymic effects present in crude venom do not exist in crotoxin: the proteolytic, coagulating, and 5-nucleotidase activities. The former two are present in the albumin fraction, called "coagulin". The 5-nucleotidase, found in the crude *Crotalus* venom to a higher degree than in many other crude venoms, is missing in crotoxin.

Determinations of desoxyribonuclease activity resulted in, 8.2 (*219*) and 9.85 (*197*) for *Bothrops* venom; 10.3 (*219*) and 13.50 (*197*) for *Crotalus* venom; and 0 for crotoxin (*197*).

e) The Structure of Crotoxin.

A precipitate from a solution of *Crotalus* crude venom, saturated with sodium chloride, contained 9–10% of the initial hemolytic, but only 2–2.5% of the original toxic activity (*102*). This experiment, however, did not decide whether crotoxin is homogeneous or a mixture of two proteins (*209*).

Judging from solubility curve, electrophoresis and ultracentrifuging, crystalline crotoxin is doubtlessly as pure and uniform as any other protein can be at the present stage of protein chemistry. Since even uniform and pure protein molecules can be composed of sub-units with the same or different biological activities, such a complicated internal structure can be assumed also for crotoxin. It might represent an easily crystallizable, salt-like compound of the acid protein, hemolysin, with several basic polypeptide groups, responsible for the toxic effect (*195*). There is much in favor of this theory, although definite proofs are not yet available.

On determination of the molecular weight of crotoxin by means of the ultracentrifuge, it was observed that crotoxin, upon standing for three days at 37° in a solution of p_H 8.7, contained the usual high-molecular component (sedimentation constant, 3.39 and 3.08), but in addition also some decomposition products of low molecular weight (*114*).

Partition chromatography on a Hyflo-Supercel kieselguhr column with ammonium sulfate-cellosolve-water (*154*) gave a curve with a double peak but all the fractions obtained were almost completely non-poisonous [Fraenkel-Conrat (*80*)].

Experiments to determine the end groups of crotoxin by means of dinitrofluorobenzene resulted in two physiologically inactive dinitrobenzene derivatives. One of them, amounting to 2/3 of the initial weight, was water insoluble, the other one water soluble. They were hydrolyzed separately. From the water-insoluble derivative twice as much ε-dinitrophenyl-lysine could be recovered than from the water soluble fraction. A study of the complete amino acid composition of these two fractions by the DNFB method has indicated (*80*) significant differences for almost all amino acid values. Particularly marked were these differences for the acidic amino acids (almost twice higher in the water soluble fraction) and for arginine as well as histidine (2–3 times higher in the insoluble fraction).

These findings are in accordance with the concept that crotoxin represents an ionic complex formed by a strongly basic and a strongly acidic protein [SLOTTA (*195*)] which can be separated when a substantial portion of the basic groups are blocked by dinitrophenyl residues.

As to the N-terminal amino acids, it should be noted that serine and small amounts of unidentified amino acids have been found in both fractions, although in much higher amounts in the soluble one. The C-terminal amino acids, as released by carboxypeptidase, suggested the sequence, -ser-thr-asp for the soluble, and -glu-asp for the insoluble fraction (*80*), and possibly they also indicate the presence of two peptide chains in both instances.

Finally, separation by means of paper electrophoresis was tried without leading to definite results. Whereas crotoxin migrated at p_H 3 — 6 uniformly, slight tendencies of a separation below or above this range could be noted (*197*). Better results were obtained with chromatography which up to then had hardly been used for the separation of the protein fractions of snake venoms (*83*, *170*, *171*).

When crotoxin was chromatographed ascendent for 4 hours in an 8% sodium chloride solution with a mixture of 60 ml. of Na_2HPO_4-citric acid buffer (p_H 7.5) and 30 ml. of *n*-propanol, separation into a quicker and a slower fraction was achieved. Only the latter showed any hemolytic activity (*200*).

VI. Electrophoresis of Snake Venoms.

While it proved comparatively easy to resolve by electrophoresis and paper electrophoresis such complex protein mixtures as plasma and serum into their different albumin and globulin fractions, this has so far not been achieved with snake venoms.

As is well known, for the electrophoretic separation of a protein mixture it is advisable to select a p_H, either above or below the isoelectric point of all components, in order to reduce electrostatic interaction. This is comparatively easy in the case of serum proteins, whose isoelectric points lie between 4.6 and 7.3 and which migrate to the anode at p_H 7.7, or even 8.6, showing considerable mobilities.

The ELAPIDAE venoms represent a mixture of proteins, some with isoelectric points from 5,7 to over 11, as shown in the case of *Naja naja*.

To avoid electrostatic interaction, it seems thus advisable to carry out electrophoresis at a p_H under 5.7.

A number of electrophoretic patterns appear in *Fig. 4*.

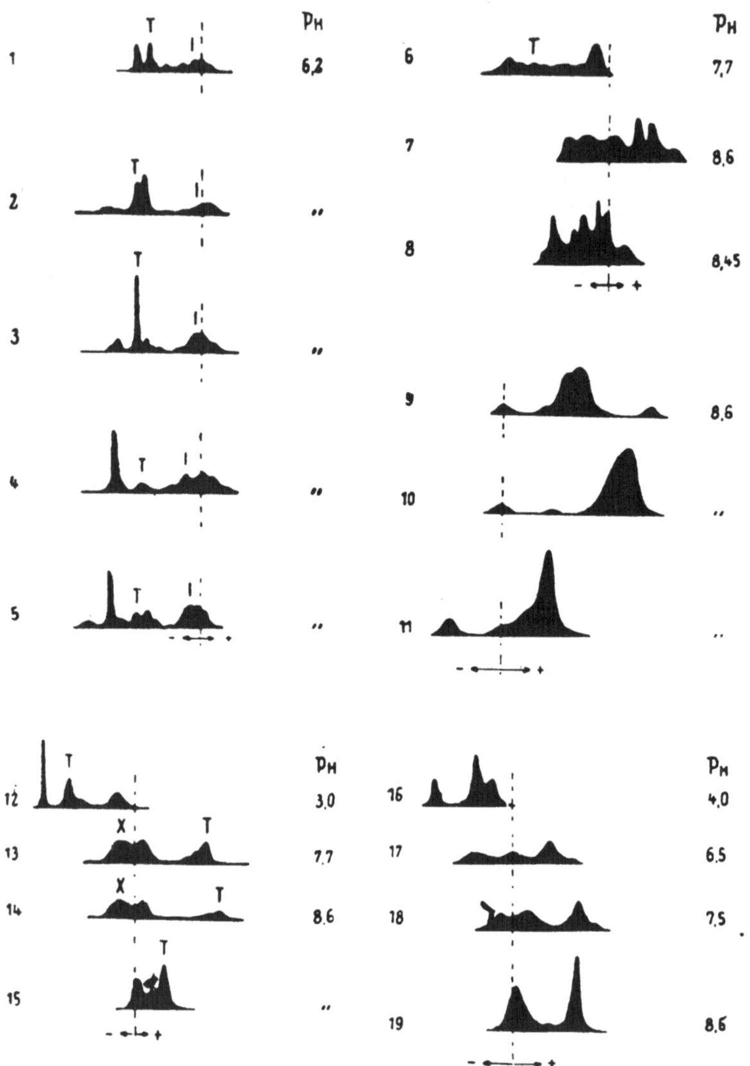

Fig. 4. Electrophoretic patterns of some snake venoms.

1–5. Electrophoretic pattern of 5 ELAPIDAE venoms; phosphate buffer p_H 6.2; ionic strength, 0.2.

Ascending columns: 1, *Naja nigricollis*; 2, *N. flava*; 3, *N. haje*; 4, *Sepedon haemachates*, and 5, *N. naja* [POLSON, JOURBERT, and HAIG (*172*)].

6–8, Electrophoretic patterns of *Bothrops jararacussu* venom:
   6, Phosphate buffer; pH 7,7; i. str. 0.1; 6.6 volts/cm.; 10800 sec. (*112*).
   7, Barbiturate-citrate buffer; pH 8.6; i. str. 0.1; Whatman 1; 110 volts; 11 hours (*116*).
   8, Barbiturate-citrate buffer; pH 8.45; i. str. 0.2; Whatman 1; 140 volts; $12^1/_2$ hours (*161*).

9–11, *Crotalus t. terrificus*: barbiturate-citrate buffer; pH 8.6 (*113*).
   9, *Cr. t. terrificus* (north); i. str. 0,1; 8.5 volts/cm.; 7200 sec.
   10, *Cr. t. terrificus* (center); i. str. 0.1; 8.6 volts/cm.; 7440 sec.
   11, *Cr. t. terrificus* (south); i. str. 0.2; 2.4 volts/cm.; 14400 sec.

12–15, *Cr. t. terrificus* (Argentine) (*112*):
   12, Glycine buffer; pH 3.0; i. str. 0.1; 3.4 volts/cm.; 14400 sec.
   13, Phosphate buffer; pH 7.7; i. str. 0.1; 3.4 volts/cm.; 10800 sec.
   14, Barbiturate-citratė buffer; pH 0.088; i. str. 0.2; 8.6 volts/cm.
   15, Barbiturate-citrate buffer; pH 0.3; i. str. 0.3; 1.6 volts/cm.; 12600 sec.

16–19, *Cr. t terrificus* (Butantan); pH 4–8.6 (*124*).

In 1946, five ELAPIDAE venoms were electrophorized in the TISELIUS apparatus in 1% solution, buffered with phosphate to a pH 6.2. Samples were drawn off at a fixed position in one of the limbs of the U-tube; toxicity and hemolytic activity were assayed by mouse and red cell tests, respectively. Two components are common to all venoms: the nontoxic component *I*, and component *T* with a mobility of about 6.4×10^{-5} cm.²/sec./volt, combining the neurotoxic and hemolytic activities (*172*) (see patterns 1–5, Fig. 4).

Recently, by electrophoresis of *Naja nigricollis* and *N. haje* venoms at pH 4.0; 6.2; 6.8 and 8.5 in hydrochloric acid-veronal-acetate buffer, 8 und 10 fractions were determined, respectwely (*73*). But in contradiction to an earlier report (*172*), all these fractions contained toxin, viz. the most powerful fraction contained twice as much as the crude venom, the least powerful one third or one fourth. The sum of the toxicities of all fractions was approximately equal to that of the crude venom. Thus, a resolution of the neurotoxin by means of electrophoresis was not achieved. Besides the phospholipase A, which remained to a considerable degree at the starting point, a directly hemolyzing, more basic component was found (*120*).

Several ELAPIDAE venoms were also examined by means of paper electrophoresis in a phenobarbital-sodium acetate buffer of pH 8.6. The hemolytic and proteolytic activities were determined, as well as the inhibition of egg-yolk coagulation by different fractions (*116*). At pH 8.6, practically all the proteins migrated to the cathode, but in accordance with what was mentioned above, the separation was very incomplete, since the electrostatic interaction at this pH is particularly strong. Under apparently identical conditions, *e. g.*, the same black cobra venom gave quite different patterns. Notwithstanding the evident defects of the method, the above mentioned activities were assigned to certain fractions of *Naja haje*, *N. nigricollis* and black cobra venoms.

When VIPERIDAE or CROTALIDAE venoms were electrophorized, the strongly basic fractions observed in the ELAPIDAE venoms were missing. The two VIPERIDAE venoms, *Bothrops arietans* and *Echis carinatus*, were separated by electrophoresis, respectively, into seven or five fractions, which all migrated to the cathode in buffer below p_H 5, and to the anode in buffer above p_H 5 (74).

Two studies on CROTALIDAE venoms were carried out with the venoms of three *Bothrops* species (*B. jararacussu*, *B. jararaca* and *B. atrox*) and that of *Crotalus t. terrificus*. Unfortunately, in most of these experiments a buffer of p_H 8.6 was employed, whereby interaction of proteins carrying opposite charges resulted in the formation of protein-protein complexes (*113*). The electrophoresis of *B. jararacussu* venom gave, in phosphate buffer of p_H 7.7, a curve (*112*) which is of special interest, because it can be compared with two others obtained by means of paper electrophoresis (*116*, *161*) (cf. Fig. 4, patterns 6—8). These experiments, carried out independently in three different laboratories, have demonstrated conclusively how carefully electrophoretic and paper electrophoretic separations of protein venoms must be evaluated. Of course, the respective methods and conditions were not identical but very similar. Even so, three completely different patterns were obtained.

Electrophoresis of *Crotalus* venom originating from different parts of the South American continent (p. 452, paterns 9–11) showed that even with the same method and under identical conditions the venom of one and the same snake species can give very different patterns. The toxic component (crotoxin) corresponds always to the highest peak on the right side (*113*). In the venoms from the south of Brazil and from the Argentine, another component was found migrating to the cathode at p_H 8.6. It was called "crotamin" and causes paralysis (*113*) (pattern 11, Fig. 4).

This basic component had been discovered earlier in the *Crotalus* venoms from the Argentine: in these experiments it was trapped in the salt boundaries. In patterns 13 and 14 it is designated by *x*, whereas the crotoxin is marked by *T*.

Liquid, Brazilian, *Crotalus* venom gave similar patterns (see p. 452, 16–19) (*124*).

Paper electrophoretic experiments with the venoms of European VIPERIDAE (*Viper ammodytes*, *V. berus*) and the Indian *V. russellii*, as well as with a CROTALIDAE venom (*Crotalus atrox*) showed that at p_H 8.6 most of the proteins migrated to the anode, while under the same conditions practically the total protein of the ELAPIDAE venoms migrated to the cathode. Thus, in the case of unknown dry venoms, ELAPIDAE can easily be differentiated from VIPERIDAE and CROTALIDAE venoms in a simple paper electrophoretic experiment. Recently, *Bothrops jararaca* venom was separated by a new electrophoresis technique, using wedge-shaped paper strips or thin starch plates. At pH 8.8, protease, *L*-amino acid oxidase, the coagulating component, and the yolk coagulation inhibiting one (*121*) migrated to the anode faster, whereas phosphatase and desoxyribonuclease were slower than the toxin [MICHL (*159*)].

Summarizing, the classical and the paper electrophoresis methods when applied to snake venoms have yielded few analytical or preparative results, considering the amount of time and work employed. Electrophoresis has, however, played an important part in the determination of the homogeneity of crotoxin (*148*, *147*). It is to be hoped that we may soon use this method to test the homogeneity of other isolated proteins with enzymic or toxic activity. For the chemical elucidation of snake venoms, the most important achievement will always be the isolation of homogeneous and if possible crystallized proteins from the still mysterious mixture produced by the poison glands.

References.

1. ACTON, H. W. and R. KNOWLES: Snakes and Snake Poisoning. In: The Practice of Medicine in the Tropics, by W. BYAM and R. B. ARCHIBALD. Vol. 1. London: Frowde and Hodder a. Stoughton. 1921.

2. AMARAL, A. DO: Die Schlangen in der Wissenschaft. Medizinische Welt **9**, 743 (1935).

3. — Animais veníferos, venenos e antivenenos. São Paulo: Caça e pesca, Editora Ltd. 1945.

4. — Snake Venom Poisoning. Textbook of Medicine by R. L. CECIL and R. F. LOEB. 8th ed. Philadelphia: W. B. Saunders Co. 1951.

5. AMORIN, M. F. and R. F. MELLO: Intermediate Nephron Nephrosis from Snake Poisoning in Man. Amer. J. Pathol. **30**, 479 (1954).

6. ANFINSEN, C. B., R. R. REDFIELD, W. L. CHOATE, J. PAGE, and W. R. CARROL: Studies of the Gross Structure, Cross-linkage and Terminal Sequences in Ribonuclease. J. Biol. Chem. **207**, 202 (1954).

7. ANSON, M. L.: The Estimation of Pepsin, Papain and Cathepsin with Hemoglobin. J. Gen. Physiol. **22**, 79 (1939).

8. ARTHUS, M.: De l'anaphylaxie à L'imunité. Paris: Masson et Cie. 1921.

9. — Traité de physiologie normale et pathologique, vol. **1**, p. 1113, ed. G. H. Roger et L. Binet. Paris: Masson et Cie. 1933.

10. BARRIO, A.: Variaciones en el contenido de riboflavina y 1-ofi-amino ácido-oxidasa en CROTALIDAE de la Argentina. Ciencie e Investigación **8**, 36 (1952).

11. BHATTACHARYA, K. L.: Effect of Snake Venoms on Co-enzyme-1. J. Indian Chem. Soc. **30**, 685 (1953).

12. BIER, O. G.: Estudo quantitativo da reação de floculação entre o antiveneno crotálico e uma fração purificada do veneno da cascavel neotrópica (*Crotalus t. terrificus*). Mem. Inst. Butantan **20**, 31 (1947).

13. — Étude quantitative de la neutralisation des toxines par les antitoxines et des venins par les antivenins. Revue Canad. biol. **6**, 729 (19 7).

14. BONAPARTE, L. L., Prince: Analyse du venin de vipère et découverte de la vipérine. Gaz. Tosc. Sci. med.-fis. (Firenze) **1843**, 169.

15. BOQUET, P.: Venins de serpents et antivenins. Paris: Ed. médicales Flameron. 1948.

16. BRAGANÇA, B. M. and J. H. QUASTEL: Amino Acid Oxidations by Snake Venoms. Biochemic. J. **53**, 88 (1953).

17. BROOKS, G.: Étude chimique et spectrographique de la fluorescence des venins de serpents. C. R. hebd. Séances Acad. Sci. **209**, 248 (1939).

18. CALMETTE, A.: Les venins, les animaux vénimeux et la sérathérapie antivénimeuse. Paris: Masson et Cie. 1907.

19. CÉSARI, E., J. BAUCHE et P. BOQUET: Sur une race de vipère aspic (*Vipera aspis*) à venin blanc. C. R. hebd. Séances Acad. Sci. **201**, 683 (1935).

20. CHAIN, E.: Effect of Snake Venoms on Glycolysis and Fermentation in Cell-free Extracts. Quart. J. exp. Physiol. **26**, 300 (1937).

21. — Inhibition of Dehydrogenases by Snake Venom. Biochemic. J. **33**, 407 (1939).

22. CHAIN, E. and E. S. DUTHIE: Identity of Hyaluronidase and Spreading Factor. Brit. J. exp. Pathol. **21**, 324 (1940).

23. CHAIN, E. and L. J. GOLDWORTHY: Studies on the Chemical Nature of the Antifermenting Principle in Black Tiger Snake Venom. Quart. J. exp. Physiol. **27**, 375 (1938).

24. CHATTERJEE, A. K.: Snake Venoms. I. The Isolation of an Inhibitor from Cobra Venom. Indian J. med. Res. **37**, 241 (1949).

25. — Studies on Snake Venoms. II. The Mechanism of Inhibition of the Cytochromeoxydase System Caused by an Active Principle isolated from Cobra Venom. Ann. Biochem. exp. Med. **12**, 79 (1952).

26. CHAUDHURI, D. K.: Isolation of Cholinesterase from Cobra Venom (*Naja tripudians*). Science and Culture (India) **8**, 5 (1942).

27. — Cholinesterase. Ann. Biochem. exp. Med. **4**, 77 (1944).

28. — Studies on Cholinesterase. Ann. Biochem. exp. Med. **6**, 91 (1946).

29. — Cholinesterase. III. Specificity of Cholinesterase. Ann. Biochem. exp. Med. **9**, 67 (1949).

30. — Cholinesterase. IV. Influence of Temperature and Irradiation on the Activity of Cholinesterase. Ann. Biochem. exp. Med. **9**, 73 (1949).

31. — Cholinesterase. V. Effect of Substrate Concentration and Temperature on the Rate of Hydrolysis. Ann. Biochem. exp. Med. **9**, 79 (1949).

32. — Cholinesterase. VI. Optimum p_H and Stability in Relation to p_H. Ann. Biochem. exp. Med. **9**, 85 (1949).

33. — Cholinesterase. VII. Effect of Different Chemicals on the Activity of Crude and Pure Cholinesterase. Ann. Biochem. exp. Med. **10**, 65 (1950).

34. — Cholinesterase. VIII. Reversible Inactivation of Cholinesterase. Ann. Biochem. exp. Med. **10**, 71 (1950).

35. CHRISTENSEN, P. and D. J. FINNEY: Standardization of Cobra (*Naja flava*) Venom using the Graded Response Method. J. Immunology **70**, 7 (1953).

36. CONTARDI, A. und A. ERCOLI: Über die enzymatische Spaltung der Lecithine und Lysocithine. Biochem. Z. **261**, 275 (1933).

37. COWAN, S. L. and H. R. ING: The Effect of Quaternary Ammonium Salts upon Nerve. J. Physiol. **79**, 75 (1933).

38. DALE, H. H., W. FELDBERG, and M. VOGT: Release of Acetylcholine of Voluntary Motor Nerve Endings. J. Physiol. **86**, 353 (1936).

39. DE, S. S.: Studies on Haemolysin of Cobra Venom. I. Investigations on the Isolation of Haemolysin from Cobra (*Naja naja*) Venom. Ind. J. med. Res. **27**, 531 (1939).

40. — Studies on Haemolysin of Cobra Venom. II. Effect of Different Substances on the Activity of Cobra Haemolysin. Ind. J. med. Res. **27**, 793 (1940).

41. De, S. S.: Studies on Haemolysin of Cobra Venom. III. Reversible Inactivation of Haemolysin of Cobra Venom. Ind. J. med. Res. **27**, 807 (1940).

42. — Crystalline Haemolysin from Cobra (*Naja naja*) Venom. Science and Culture (India) **6**, 675 (1941).

43. — Antigenic Properties of Crystalline Haemolysin. Ann. Biochem. exp. Med. **2**, 237 (1942).

44. — Physiochemical Studies on Haemolysin. I. Crystalline Haemolysin. Ann. Biochem. exp. Med. **4**, 45 (1944).

45. — Physicochemical Studies on Haemolysin. II. p_H and Heat Stability of Haemolysin. J. Indian Chem. Soc. **21**, 290 (1944).

46. — Physicochemical Studies on Haemolysin. III. Isoelectric Point of Haemolysin. J. Indian Chem. **21**, 292 (1944).

47. — Physicochemical Studies on Haemolysin. IV. Molecular Weight of Haemolysin. J. Indian Chem. Soc. **21**, 307 (1944).

48. — Physicochemical Studies on Haemolysin. V. On the Composition of Haemolysin. J. Indian Chem. Soc. **22**, 10 (1945).

49. De, S. S. and B. N. Ghosh: Studies in Adsorption of the Neurotoxin and Haemolysin of Cobra (*Naja naja*) Venom by Various Adsorbents at Different p_H, with a View to their Isolation. J. Indian Chem. Soc. **14**, 748 (1937).

50. Delezenne, C.: Le zinc constituant cellulaire de l'organisme animal. Sa présence et son rôle dans le venin des serpents. Ann. Inst. Pasteur **33**, 68 (1919).

51. Delezenne, C. et E. Fourneau: Constitution du phosphatide hémolysant (lysocithine) provenant de l'action du venin de cobra sur le vitellus de l'oeuf de poule. Bull. soc. chim. France **15**, 421 (1914).

52. Delezenne, C. et S. Ledebt: Action du venin de cobra sur le sérum de cheval. Ses rapports avec l'hémolyse. C. R. hebd. Séances Acad. Sci. **152**, 790 (1911); Formation de substances hémolytiques et de substances toxiques aux dépens du vitellus de l'oeuf soumis à l'action du cobra. C. R. hebd. Séances Acad. Sci. **153**, 81 (1911); Nouvelle contribution à l'étude des substances hémolytiques dérivées du sérum et du vitellus de l'oeuf, soumis à l'action des venins. C. R hebd. Séances Acad. Sci. **155**, 1101 (1912).

53. Ditmars, R. L.: Snakes of the World. New York: Macmillan Co. 1951.

54. Duran-Reynals, F.: Spreading Factor in Certain Snake Venoms and its Relation to their Mode of Action. J. exp. Medicine **69**, 69 (1939).

55. Eagle, H.: The Coagulation of Blood by Snake Venoms and its Physiologic Significance. J. exp. Medicine **65**, 613 (1937).

56. Eichbaum, F. W.: Hemaglutininas nos venenos de serpentes sulamericanas. Mem. Inst. Butantan **19**, 229 (1946).

57. — Ação dermatotóxica de venenos ofídicos e sua neutralisação pelos antivenenos. Mem. Inst. Butantan **20**, 79 (1947).

58. — O fator de difusão („spreading factor") dos venenos de *Bothrops jararaca* e *Crotalus t. terrificus*. Mem. Inst. Butantan **20**, 95 (1947).

59. Eichbaum, F. W. and H. Stammreich: Oxidizing Agents in Snake Venoms. An. Acad. brasil. Ciênc. **23**, 91 (1951).

60. Essex, H. E.: Certain Animal Venoms and their Physiologic Action. Physiol. Rev. **25**, 148 (1945).

61. Essex, H. E. and J. Markowitz: The Physiologic Actio of Rattlesnake Venom (Crotalin). I. Effect on Blood Pressure: Symptoms and Postmortem Observations. Amer. J. Physiol. **92**, 317 (1930); II. The Effect of Crotalin on Surviving Organs. Amer. J. Physiol. **92**, 329 (1930); VII. The Similarity of Crotalin Shock and Anaphylactic Shock. Amer. J. Physiol. **92**, 698 (1930);

VIII. A. Comparison of the Physiologic Action of Crotalin and Histamine. Amer. J. Physiol. **92**, 705 (1930).

62. Fairbairn, D.: The Phospholipase of the Venom of the Cottonmouth Mocassin (*Agkistrodon piscivorus* L.). J. Biol. Chem. **157**, 633 (1945).

63. Faust, E. St.: Über das Ophiotoxin aus dem Gifte der ostindischen Brillenschlange, Cobra di Capello (*Naja tripudians*). Arch. exp. Pathol. Pharmakol. **56**, 236 (1907).

64. — Über das Crotalotoxin aus dem Gifte der nordamerikanischen Klapperschlange (*Crotalus adamanteus*). Arch. exp. Pathol. Pharmakol. **64**, 244 (1911).

65. — Tierische Gifte. In: Handbuch der experimentellen Pharmakologie von A. Heffter, 2. Band, 2. Hälfte, S. 1782. Berlin: J. Springer. 1924.

66. Favilli, G.: Mucolytic Effect of Several Diffusing Agents and Diazotized Compounds. Nature (London) **145**, 866 (1940).

67. Feldberg, W., H. F. Holden, and C. H. Kellaway: The Formation of Lysocithin and a Muscle-stimulating Substance by Snake Venom. J. Physiol. **94**, 232 (1938).

68. Feldberg, W. and C. H. Kellaway: Liberation of Histamine from the Perfused Lung by Snake Venoms. J. Physiol. **90**, 257 (1937).

69. Finney, D. J.: Probit Analysis. 2nd ed. London: Cambridge Univ. Press. 1952; Statistical Method in Biological Assay. New York: Hafner Publ. Co. 1952.

70. Fischer, F. G. und H. Dörfel: Zur quantitativen Auswertung der Papierchromatogramme von Eiweiß-Hydrolysaten. Biochem. Z. **324**, 544 (1953).

71. — — Adenosin im Gift der Puffotter *Bitis arietans* (Schlangengifte. I). Z. physiol. Chem. (Hoppe-Seyler) **296**, 232 (1954).

72. — — Die Aminosäuren-Zusammensetzung von Crotoxin (Schlangengifte. IV). Z. physiol. Chem. (Hoppe-Seyler) **297**, 278 (1954).

73. Fischer, F. G. und W. P. Neumann: Die elektrophoretische Analyse der Gifte von *Naja haje* und *Naja nigricollis* (Schlangengifte. II). Z. physiol. Chem. (Hoppe-Seyler) **297**, 92 (1954).

74. — — Die elektrophoretische Analyse der Gifte von *Bitis arietans* und *Echis carinatus* (Schlangengifte. III). Z. physiol. Chem. (Hoppe-Seyler) **297**, 100 (1954).

75. Fleckenstein, A., G. Berg, J. Gayer und S. Schoenig: Über die Dehydrasenhemmung durch Schlangengift und die Inaktivierung des Dehydrasen-hemmenden Prinzips durch antitoxische Sera. Arch. exp. Pathol. Pharmakol. **213**, 265 (1951).

76. Fleckenstein, A. und H. Gerhardt: Über die biologische Bedeutung des hohen Zinkgehalts in Schlangengiften. Zink als Schlangengift-Inhibitor. Arch. exp. Pathol. Pharmakol. **214**, 135 (1952).

77. Fleckenstein, A. und W. Jaeger: Weitere Ergebnisse über die Blockierung der Bienengift- und Schlangengiftwirkung durch Zinksalze. Arch. exp. Pathol. Pharmakol. **215**, 163 (1952).

78. Folin, O. and V. Ciocalteu: On Tyrosine and Tryptophan Determinations in Proteins. J. Biol. Chem. **73**, 627 (1927).

79. Fonseca, F. da: Animais peçonhentos. São Paulo: Inst. Butantan. 1949.

80. Fraenkel-Conrat, H. L.: Intern. Confer. Animal Venoms. Berkeley. 1954.

81. Fraenkel-Conrat, H. and M. Cooper: The Use of Dyes for the Determination of Acid and Basic Groups in Proteins. J. Biol. Chem. **154**, 239 (1944).

82. Fraenkel-Conrat, H. and J. Fraenkel-Conrat: Inactivation of Crotoxin by Group-specific Reagents. Biochim. Biophys. Acta **5**, 98 (1950).

83. Franklin, A. E., J. H. Quastel, and S. F. van Straten: Paper Chromatography of Protein Mixtures and Blood Plasmas. Proc. Soc. exp. Biol. Med. **77**, 783 (1951).

84. Ganguly, S. N.: Mechanism of the Coagulant Action of Daboia Venom on Blood. Indian J. med. Res. **24**, 525 (1936).

85. — Haemolysis by the Venom of the Indian Cobra (*Naja tripudians*). Indian J. med. Res. **24**, 1165 (1937).

86. Ganguly, S. N. and M. T. Malkana: Studies on Indian Snake Venoms. I. Daboia Venom: its Chemical Composition, Protein Fractions and their Physiological Action. Indian J. med. Res. **23**, 997 (1936).

87. — — Studies on Indian Snake Venoms. II. Cobra Venom: its Chemical Composition, Protein Fractions and their Physiological Actions. Indian J. med. Rés. **24** 281 (1936).

88. Gautrelet, J. et N. Halpern: Action du venin de Cobra sur l'excitabilité neuro-musculaire de la Grenouille. C. R. Séances Soc. Biol. **113**, 1486 (1933).

89. Ghosh, B. N.: The Enzymes in Snake Venom. I. Their Action on Haemoglobin and on Protein Solutions of Different p_H. J. Indian chem. Soc. **13**, 450 (1936).

90. — Die Enzyme der Schlangengifte. Österr. Chem.-Ztg. **43**, 158 (1940).

91. — Therapeutic Use of Snake Venom. Indian and Eastern Chemist **22**, 187, 207 (1941).

92. Ghosh, B. N. and K. L. Bhattacharya: Presence of Inhibitor of Pyruvic Dehydrogenase in some Snake Venoms. Science and Culture (India) **18**, 253 (1952).

93. Ghosh, B. N. and A. K. Chatterjee: Effect of Snake Venoms on the Oxidation of Glucose and its Metabolites in Cell Suspensions. J. Indian Chem. Soc. **25**, 360 (1948).

94. Ghosh, B. N., A. K. Chatterjee, and A. C. Sinha: Effect of Snake Venom on the Cytochrome-Cytochrome Oxidase System. J. Indian Chem. Soc. **25**, 384 (1948).

95. Ghosh, B. N. and D. K. Chaudhuri: Enzymes in Snake Venom. IV. J. Indian Chem. Soc. **15**, 566 (1938).

96. — — Estimation of Some of the Amino Acids in Cobra (*Naja naja*) Neurotoxin. J. Indian Chem. Soc. **20**, 22 (1943).

97. Ghosh, B. N. and S. S. De: Effectof Snake Venom on the Action of Trypsin and Pancreatic Juice. Science and Culture (India) **2**, 223 (1936).

98. — — The Migration of the Toxic Constituents of Cobra (*Naja naja*) Venom at Various p_H in an Electric Field. Indian J. med. Res. **24**, 1175 (1937).

99. — — Partial Purification of the Toxic Components of Cobra (*Naja naja*) Venom. Science and Culture (India) **2**, 585 (1937).

100. — — Determination of the Isoelectric Point of the Neurotoxin of Russel Viper Venom. Science and Culture (India) **3**, 297 (1937).

101. — — Investigation on the Isolation of the Neurotoxin and Haemolysin in Cobra (*Naja naja*) Venom. Indian J. med. Res. **25**, 779 (1938).

102. — — Proteins of Rattlesnake Venom. Nature (London) **143**, 380 (1939).

103. Ghosh, B. N., S. S. De, and D. P. Bhattacharya: Partial Separation of the Neurotoxin of Russel Viper Venom. Science and Culture (India) **3**, 298 (1937).

104. — — — Investigations on the Isolation of the Active Principles from the Venoms of *Bungarus fasciatus* and *Vipera russellii*. Indian J. med. Res. **26**, 753 (1939).

105. Ghosh, B. N., S. S. De, and D. K. Chaudhuri: Destruction of the Neurotoxin of Cobra (*Naja naja*) and Daboia (*Vipera russellii*) Venom by Various Reducing Agents. Science and Culture (India) **4**, 198 (1938).

106. — — — Separation of the Neurotoxin from the Crude Cobra Venom and Study of the Action of a Number of Reducing Agents on it. Indian J. med. Res. **29**, 367 (1941).

107. — — — Enzymes in Snake Venom. Ann. Biochem. exp. Med. **1**, 31 (1941); Trop. Diseases Bull. **39**, 718 (1942).

108. Ghosh, B. N., S. S. De, and N. L. Kundu: The Separation of Neurotoxin from the Crude Cobra (*Naja naja*) Venom. Science and Culture (India) **4**, 133 (1938).

109. Ghosh, B. N., S. S. De, and N. K. Sarkar: Effect of Cobra (*Naja naja*) Venom and its Constituents on the Synthesis of Acetylcholine by the Brain Cells of the Rats and Pigeons. J. Indian Chem. Soc. **21**, 93 (1944).

110. Glick, D. and B. Sylvén: Evidence for the Heparin Nature of the Non-specific Hyaluronidase Inhibitor in Tissue Extracts and Blood Serum. Science (Washington) **113**, 388 (1951).

111. Goldberg, A. and E. Haas: Separation of Antinvasin in two Components. J. Biol. Chem. **170**, 757 (1947).

112. Gonçalves, J. M. and A. Polson: The Electrophoretic Analysis of Snake Venoms. Arch. Biochem. **13**, 253 (1947).

113. Gonçalves, J. M. and L. G. Vieira: Estudos sobre venenos de serpentes brasileiras. I. Análise electroforética. An. Acad. bras. Ciênc. **22**, 141 (1950).

114. Gralén, N. and T. Svedberg: The Molecular Weight of Crotoxin. Biochem. J. **32**, 1375 (1938).

115. Grasset, E.: La vipère du Gabon. Acta tropica **3**, 97 (1946).

116. Grassmann, W. und K. Hannig: Elektrophoretische Untersuchungen an Schlangen- und Insektentoxinen. Z. physiol. Chem. (Hoppe-Sevler) **296**, 30 (1954).

117. Gulland, J. M. and E. M. Jackson: Phosphoesterases of Bone and Snake Venoms. Biochemic. J. **32**, 590 (1938); 5-Nucleotidase. Biochemic. J. **32**, 597 (1938); The Constitution of Yeast Nucleic Acid. J. Chem. Soc. (London) **1938**, 1492.

118. Gulland, J. M. and E. O. Walsh: The Constitution of Yeast Ribonucleic Acid. IX. Alkali-labil Linkages. J. Chem. Soc. (London) **1945**, 172.

119. Haas, E.: On the Mechanism of Invasin. I. Antinvasin I, an Enzyme in Plasma; II. Proinvasin I, an Enzyme in Pathogenic Bacteria and in Venoms; III. Antinvasin II, an Enzyme in Plasma. J. Biol. Chem. **163**, 63, 89, 101 (1946).

120. Habermann, E.: Zur pharmakologischen Charakterisierung elektrophoretischer Fraktionen der Gifte von *Naja nigricollis* und *Naja haje*. Z. physiol. Chem. (Hoppe-Seyler) **297**, 104 (1954).

121. Habermann, E. und W. Neumann: Die Hemmung der Hitzekoagulation von Eigelb durch Bienengift — ein Phospholipase-Effekt. Z. physiol. Chem. (Hoppe-Seyler) **297**, 179 (1954).

122. Hestrin, S.: The Reaction of Acetylcholine and other Carboxylic Acid Derivates with Hydroxylamine and its Analytical Application. J. Biol. Chem. **180**, 249 (1949).

123. Holden, H. F.: Haemolysis by Australian Snake Venoms. 3. Some Factors which Influence the Action of the Venom of the Copperhead. Austral. J. exp. Biol. med. Sci. **13**, 103 (1935).

124. Höxter, G.: In: K. Slotta, Zur Chemie der Schlangengifte. Experientia **9**, 81 (1953).

125. HUGHES, A.: The Action of Snake Venoms on Surface Films. Biochemic. J. **29**, 437 (1935).

126. HURST, R. O. and G. C. BUTLER: The Chromatographic Separation of Phosphatases in Snake Venoms. J. Biol. Chem. **193**, 91 (1951).

127. HURST, R. O., J. A. LITTLE, and G. C. BUTLER: The Enzymatic Degradation of Thymonucleic Acid. II. The Hydrolysis of Oligonucleotides. J. Biol. Chem. **188**, 705 (1951).

128. IYENGAR, N. K., K. B. SEHRA, B. MUKERJI, and R. N. CHOPRA: Cholinesterase in Cobra Venom. Current Sci. (India) **7**, 51 (1938).

129. JÁNSZKY, B.: Action of the Venom of *Bothrops atrox* on Fibrinogen. Science (Washington) **110**, 307 (1949).

130. — The Relation Between the Proteolytic and Blood Clotting Activity of Snake Venoms. Arch. Biochemistry **28**, 139 (1950).

131. JOHNSON, M., M. A. G. KAYE, R. HEMS, and H. A. KREBS: Enzymic Hydrolysis of Adenosine Phosphates by Cobra Venom. Biochemic. J. **54**, 625 (1953).

132. KAISER, E.: Untersuchungen über den Trypsin- und Hyaluronidase-Inhibitor des menschlichen Serums. Biochem. Z. **324**, 344 (1953).

133. — Fermentchemische Untersuchungen an Spinnengiften. Monatsh. Chem. **84**, 482 (1953).

134. KELLAWAY, C. H.: The Action of Australian Snake Venoms on Plain Muscle. Brit. J. exp. Pathol. **10**, 281 (1929).

135. — The Venom of the Ornamented Snake *Denisonia maculata*. Austral. J. exp. Biol. med. Sci. **12**, 47 (1934).

136. — Some Peculiarities of Australian Snake Venoms. Trans. Roy. Soc. Trop. Med. **27**, 9 (1933).

137. — Animal Poisons. Annu. Rev. Biochem. **8**, 541 (1939).

138. KELLAWAY, C. H. and E. R. TRETHEWIE: A Note on the Extraction of Adenyl Compounds from Tissues. Austral. J. exp. Biol. med. Sci. **18**, 265 (1940).

139. KLOBUSITZKY, D. v.: Biochemische Studien über die Gifte der Schlangengattung *Bothrops*. I. Mitt. Die blutgerinnungsfördernde Wirkung und die Reinigung des Giftdrüsensekretes der *Bothrops jararaca*. Arch. exp. Pathol. Pharmakol. **179**, 204 (1935).

140. — Biochemische Studien über die Gifte der Schlangengattung *Bothrops*. II. Mitt. Eine verbesserte Methode zur Herstellung von Bothropotoxin. Arch. exp. Pathol. Pharmakol. **180**, 479 (1936).

141. KLOBUSITZKY, D. v. und P. KÖNIG: Biochemische Studien über die Gifte der Schlangengattung *Bothrops*. III. Mitt. Die Trennung der gerinnungsfördernden Substanz von dem Bothropstoxin und den übrigen Sekretbestandteilen. Arch. exp. Pathol. Pharmakol. **181**, 387 (1936).

142. — — Biochemische Studien über die Gifte der Schlangengattung *Bothrops*. IV. Mitt. Die Wirkung der gerinnungsfördernden Substanz *in vivo*. Arch. exp. Pathol. Pharmakol. **182**, 577 (1936).

143. — — Biochemische Studien über die Gifte der Schlangengattung *Bothrops*. V. Mitt. Über den Stickstoff- und Schwefelgehalt des Bothropotoxin. Z. physiol. Chem. (Hoppe-Seyler) **255**, I—III (1938).

144. — — Biochemische Studien über die Gifte der Schlangengattung *Bothrops*. VI. Mitt. Kurzer Bericht über verschiedene, in den Jahren 1936 bis 1937 gewonnene Versuchsergebnisse. Arch. exp. Pathol. Pharmakol. **192**, 271 (1939).

145. KOHLSCHÜTTER, E. und W. MINNING: Ein Fall von Spättod nach Biß der südamerikanischen Viper *Lachesis alternatus*. Dtsch. med. Wschr. **1936**, 2043.

146. KUNITZ, M.: A Spectophotometric Method for the Measurement of Ribonuclease Activity. J. Biol. Chem. **164**, 565 (1946).

147. LI, C. H.: In: K. SLOTTA „Zur Chemie der Schlangengifte". Experientia **9**, 81 (1953).
148. LI, C. H. and H. FRAENKEL-CONRAT: Electrophoresis of Crotoxin. J. Amer. Chem. Soc. **64**, 1586 (1942).
149. LINK, TH.: Der Einfluß der Schlangengifte auf die Blutgerinnung. Z. Immunitätsforschung **85**, 504 (1935).
150. MAASS, TH. A.: Schlangen und Schlangen-Gifte. Tabulae Biologicae Periodicae, vol. 3 (= Tabulae Biologicae vol. IX, p. 105). 1934.
151. MADINAVEITIA, J.: Diffusing Factors. 7. Concentration of the Mucinase from Testicular Extracts and from *Crotalus atrox* Venom. Biochemic. J. **35**, 447 (1941).
152. MADINAVEITIA, J. and T. H. H. QUIBELL: Diffusing Factors. 9. The Effect of Salts on the Action of Testicular Extracts on the Viscosity of Vitreous Humour Preparations. Biochemic. J. **35**, 456 (1941).
153. MARSHALL, J.: The Constituents of the Venom of the Rattlesnake. Science (Washington) **19**, 715 (1904).
154. MARTIN, A. J. P. and R. R. PORTER: The Chromatographic Fractionation of Ribonuclease. Biochemic. J. **49**, 215 (1951).
155. MARTIRANI, J. and M. P. AZEVEDO: Acçao proteolítica do veneno da *Bothrops jararaca* (Wied). Mem. Inst. Butantan **22**, 47 (1950).
156. MICHEEL, F., H. DIETRICH und G. BISCHOFF: III. Mitteilung über Schlangengifte. Über die Neurotoxine aus Giften von Cobra-Arten. Z. physiol. Chem. (Hoppe-Seyler) **249**, 157 (1937).
157. MICHEEL, F. und F. JUNG: Zur Kenntnis der Schlangengifte. II. Z. physiol. Chem. (Hoppe-Seyler) **239**, 217 (1936).
158. MICHEEL, F., H. SCHMITZ, G. BODE und H. EMDE: Zur Kenntnis der Schlangengifte 4. bis 12. Mitt. Ber. dtsch. chem. Ges. **71**, 703, 1302, 1446, 2653 (1938); **72**, 68, 397, 1724 (1939); Naturwiss. **26**, 298 (1938); Z. physiol. Chem. (Hoppe-Seyler) **265**, 266 (1940).
159. MICHL, H.: Elektrophoretische und enzymatische Untersuchungen des Jararaca-Toxins. Naturwiss. **41**, 403 (1954); Über das Gift der *Bothrops jararaca*. Monatsh. Chem. **85**, 1240 (1954).
160. MOSIG, A.: Die Giftschlangen und ihre Bedeutung für die Pharmazie und Therapie im Lichte neuerer Erkenntnisse. Pharmazie **1951**, 381 und 451.
161. NEUMANN, W. und E. HABERMANN: Zur papierelektrophoretischen Fraktionierung tierischer Gifte. Naturwiss. **39**, 286 (1952).
162. — — Über die Phospholipase A des Bienengiftes. Z. physiol. Chem. (Hoppe-Seyler) **296**, 166 (1954).
163. NEUMANN, W., E. HABERMANN und H. HANSEN: Differenzierung von zwei hämolysierenden Faktoren im Bienengift. Arch. exp. Pathol. Pharmakol. **217**, 130 (1953).
164. NOC, F.: Sur quelques propriétés physiologiques des différents venins de serpents. Ann. Inst. Pasteur **18**, 387 (1904).
165. NYGAARD, A. P.: Factors Involved in the Enzymic Reduction of Cytochrome C. J. Biol. Chem. **204**, 655 (1953).
166. NYGAARD, A. P. and J. B. SUMNER: The Effect of Lecithinase A on the Succinoxidase System J. Biol. Chem. **200**, 723 (1953).
167. OGAWA, K.: Über die fermentative Lysolecithinbildung. J. Biochemistry (Japan) **24**, 389 (1936).
168. PARK, J. T. and M. J. JOHNSON: A Submicrodetermination of Glucose. J. Biol. Chem. **181**, 149 (1949).
169. PHISALIX, M.: Animaux venimeux et venins, vols. 1 et 2. Paris: Masson et Cie. 1922.

170. Piantanida, M. and N. Muić: Paper-Strip Chromatography of the Proteinic Components of Ammodytes Viper Venom. Arch. Biochem. Biophys. **46**, 110 (1953).

171. — — Properties of the Toxin of *Vipera ammodytes*. Radovi Yugoslav. Akad. Ananosti i Umjetnosti **298**, 207 (1953).

172. Polson, A., J. F. Jourbert, and D. A. Haig: Electrophoretic Examination of Cobra Venoms. Biochemic. J. **40**, 265 (1946).

173. Rapport, M. M., K. Meyer, and A. Linker: Correlation of Reductimetric and Turbimetric Methods for Hyaluronidase Assay. J. Biol. Chem. **186**, 615 (1950).

174. Ray, P.: Estimation of Zinc in Snake Venoms by Micro-quinaldinate Method. J. Indian Chem. Soc. **17**, 681 (1940).

175. Reed, L. J. and H. Muench: A Simple Method of Estimating Fifty Per Cent End-points. Amer. J. Hyg. **27**, 493 (1938).

176. Rocha e Silva, M.: Beiträge zur Pharmakologie des Trypsins. Arch. exp. Pathol. Pharmakol. **194**, 335 (1940).

177. Rocha e Silva, M., W. T. Beraldo, and G. Rosenfeld: Bradykinin, a Hypotensive and Smooth Muscle Stimulating Factor Released from Plasma Globulin by Snake Venoms and by Trypsin. Amer. J. Physiol. **156**, 261 (1949).

178. Roulet F. und E. A. Zeller: Über die Enzyme des *Mycobacterium tuberculosis* und anderer säurefester Bakterien. 3. Mitt. Über den enzymatischen Abbau von *L*-Peptiden durch säurefeste Bakterien. Helv. Chim. Acta **31**, 1915 (1948).

179. Roy, A. C. and R. N. Chopra: Some Biochemical Characteristics of Snake Venom. Indian J. med. Res. **26**, 241 (1938).

180. Sabine, J. C.: A Continuous Titration for Cholinesterase Determinations. U. S. Atomic Energy Comm., Tech. Inform. Service, Oak Ridge, Tenn. AECU-2575 (1953).

181. Sarkar, N. K.: Effect of Ultraviolet Rays on the Stability of Cobra Venom and Cardiotoxin. Ann. Biochem. exp. Med. **6**, 87 (1946).

182. — Isolation of Cardiotoxin from Cobra Venom (*Naja tripudians*, Monocellate Variety). J. Indian Chem. Soc. **24**, 227 (1947).

183. — Determination of Molecular Weight of Cardiotoxin by Diffusion Method. J. Indian Chem. Soc. **24**, 61 (1947).

184. — Existence of a Cardiotoxin Principle in Cobra Venom. Ann. Biochem. exp. Med. **8**, 11 (1947).

185. Sarkar, N. K. and S. R. Maitre: Action of Cobra Venom and Cardiotoxin on Gastrocnemius-sciatic Preparation of a Frog. Amer. J. Physiol. **163**, 209 (1950).

186. Sarkar, N. K., S. R. Maitre, and B. N. Ghosh: The Effect of Neurotoxin, Hemolysin and Cholinesterase Isolated from Cobra Venom on Heart, Blood Pressure and Respiration. Indian J. med. Res. **30**, 453 (1942).

187. Schaumann, O.: Pharmakologische Versuche mit Schlangengiften und Schlangensera. Behringwerk-Mitt. (Marburg a. d. Lahn) **7**, 33 (1936).

188. Schlossberger, H.: Die Geschichte der Schlangengift Forschung. Behringwerk-Mitt. (Marburg a. d. Lahn) **7**, 1 (1936).

189. Schöttler, W. A.: Toxicity of the Principal Snake Venoms of Brazil. J. Tropical Med. **31**, 489 (1951).

190. — (unpublished).

191. Singer, T. P. and E. B. Kearney: The *L*-Amino Acid Oxidases of Snake Venom. I. Prosthetic Group of the *L*-Amino Acid Oxidase of Mocassin Venom. Arch. Biochemistry **27**, 348 (1950).

192. — — The *L*-Amino Acid Oxidases of Snake Venom. II. Isolation and Characterization of Homogeneous *L*-Amino Acid Oxidase. Arch. Biochemistry **29**, 190 (1950).

193. Sinsheimer, R. L. and J. Koerner: A Purification of Venom Phosphodiesterase. J. Biol. Chem. **198**, 293 (1952).

194. Slotta, K. H.: A crotoxina, primeira substância pura dos venenos ofídicos. An. Acad. brasil. Scienc. **10**, 195 (1938).

195. — Zur Chemie der Schlangengifte. Experientia **9**, 81 (1953).

196. Slotta, K. and A. Ballester: Determinação colorimétrica da hialuronídase dos venenos ofídicos. Mem. Inst. Butantan **26**, 311 (1954).

197. — — (unpublished).

198. Slotta, K. and P. Borchert: Sôbre o fator hemolítico dos venenos ofídicos. Mem. Inst. Butantan **26**, 297 (1954).

199. — — Histamina e toxinas protéicas no veneno de abelha. Mem. Inst. Butantan **26**, 279 (1954).

200. — — (unpublished).

201. Slotta, K. H. und W. Forster: Schlangengifte. IV. Mitt. Quantitative Bestimmung der schwefelhaltigen Bausteine. Ber. dtsch. chem. Ges. **71**, 1082 (1938).

202. — — Estudos químicos sôbre os venenos ofídicos. 5. Determinação quantitativa dos componentes que contêm enxofre. Mem. Inst. Butantan **12**, 513 (1938).

203. Slotta, K. H., W. Forster und H. L. Fraenkel-Conrat: Schlangengifte. V. Mitt. Über die schwefelhaltigen Bausteine des Cobragiftes. Ber. dtsch. chem. Ges. **71**, 1623 (1938).

204. Slotta, C. H. and H. L. Fraenkel-Conrat: Estudos químicos sôbre os venenos ofídicos. 2. Sôbre a forma de ligação do enxofre. Mem. Inst. Butantan **11**, 121 (1937).

205. — — Schlangengifte. II. Mitt. Über die Bindungsart des Schwefels. Ber. dtsch. chem. Ges. **71**, 264 (1938).

206. — — Two Active Proteins from Rattlesnake Venom. Nature (London) **142**, 213 (1938).

207. — — Schlangengifte. III. Mitt. Reinigung und Kristallisation des Klapperschlangengiftes. Ber. dtsch. chem. Ges. **71**, 1076 (1938).

208. — — Estudos químicos sôbre os venenos ofídicos. 4. Purificação e cristalisação do veneno da cobra cascavel. Mem. Inst. Butantan **12**, 505 (1938).

209. — — Crotoxin. Nature (London) **144**, 290 (1939).

210. — — (unpublished).

211. Slotta, K. and J. Primosigh: Estudos químicos sôbre os venenos ofídicos. 6. Composição da crotoxina. Mem. Inst. Butantan **23**, 51. (1951). [English translation by E. R. Hope. Defense Scient. Inform. Serv. Canada (1953)].

212. — — Amino-Acid Composition of Crotoxin. Nature (London) **168**, 696 (1951).

213. — — A New Method of Quantitative Paper Chromatography. Mem. Inst. Butantan **24** (2), 85 (1952).

214. Slotta, K. H. und G. Szyszka: Schlangengifte. I. Mitt. Bestimmung von Gift-, Koagulase- und Lecithinase-Wert und Beeinflussung der Gifte durch physikalische und chemische Mittel. Ber. dtsch. chem. Ges. **71**, 258 (1938).

215. — — Estudos químicos sôbre os venenos ofídicos. 1. Determinação de sua toxicidade em camondongos. Mem. Inst. Butantan **11**, 109 (1937).

216. Slotta, C. H., G. Szyszka, and H. L. Fraenkel-Conrat: Estudos químicos sôbre os venenos ofídicos. 3. Teor da coagulação e da lecithinase. Mem. Inst. Butantan **11**, 133 (1937).

217. Soaje Echagüe, E.: Alteraciones circulatorias producidas por el veneno de *Crotalus terrificus*. Rev. soc. Argent. Biol. **16**, 475 (1940).

218. Stedmann Ed. and Ell. Stedmann: The Purification of Cholinesterase. Biochemic. J. **29**, 2563 (1935).

219. TABORDA, A. R., L. C. TABORDA, J. N. WILLIAMS, Jr., and C. A. ELVEHJEM: A Study of the Ribonuclease Activity of Snake Venoms. J. Biol. Chem. **194**, 227 (1952). A Study of the Desoxyribonuclease Activity of Snake Venoms. J. Biol. Chem. **195**, 207 (1952).

220. TAGUET, C., E. ROUSEAU et R. DUMATRAS: Recherches relatives à l'action du venin de Cobra. C. R. Séances Soc. Biol. **113**, 9 (1933).

221. TRISTRAM, G. R.: The Proteins, vol. I, Part. A, p. 181, Ed. by H. Neurath and K. Bailey. New York: Academic Press. 1953.

222. VELLARD, J.: Propriétés du venin des principales espèces de serpents du Venezuela. Ann. Inst. Pasteur **60**, 511 (1938).

223. — Enfermedades producidas por animales venenosas. Therapeutica clinica, 4. parte. Buenos Aires: E. Ateneu. 1946.

224. VIDAL BREARD, J. J. and V. E. ELIAS: Biochemistry of Snake Venoms. I. Lecithinase C Activity in Venom of *Bothrops alternatus*. Arch. farm. bioquim. Tucumán **5**, 77 (1950).

225. WELKER, W. H.: Fractionation of the Proteins of Rattlesnake Venom. J. Lab. clin. Med. **10**, 298 (1925).

226. WELLERS, G.: Action du cyanure de potassium sur le venin de cobra brut. Revue canad. biol. **8**, 139 (1949).

227. WIELAND, H. und W. KONZ: Einige Beobachtungen am Gift der Brillenschlange (*Naja tripudians*). Sitz.-Ber. math.-nat. Abt. bayr. Akad. Wiss. **1936**, 177.

228. YANG, C. C.: Effects of Crotoxin on the Succinic Oxidase System. J. Formosan Med. Assoc. **53**, 59 (1954).

229. YOSHIDA, S.: Phosphodiesterase. J. Biochem. (Japan) **34**, 23 (1941).

230. ZELLER, E. A.: Über ein Adenosintriphosphorsäure (ATP) spaltendes Enzym der Schlangengifte. Experientia **4**, 194 (1948).

231. — Enzymes of Snake Venoms and their Biological Significance. Adv. Enzymology **8**, 459 (1948).

232. — Über die Cholinesterase der Schlangengifte. 5. Mitt. über die Biochemie tierischer Gifte. Helv. Chim. Acta **32**, 94 (1949).

233. — Über Phosphatasen. II. Über eine neue Adenosintriphosphatase. Helv. Chim. Acta **33**, 821 (1950).

234. — The Formation of Pyrophosphate from Adenosine Triphosphate in the Presence of a Snake Venom. Arch. Biochemistry **28**, 138 (1950).

235. — Enzymes as Essential Components of Toxins. In: "The Enzymes" by J. B. Sumner and K. Myrbäck. Vol. I, Part. 2, p. 986. New York: Academic Press. 1951.

236. ZELLER, E. A., B. ISELIN und A. MARITZ: Über das Vorkommen der Ophio-*l*-aminosäure-oxydase. 4. Mitt. über eine neue *l*-Aminosäure-oxydase. Helv. Physiol. Acta **4**, 233 (1946).

237. ZELLER, E. A. und A. MARITZ: Über eine neue *l*-Aminosäure-oxydase. 1. Mitt. Helv. Chim. Acta **27**, 1888 (1944); 2. Mitt. (Ophio-*l*-Aminosäure-oxydase). Helv. Chim. Acta **28**, 365 (1945).

238. — — Demonstration einer neuen Peptidase-Bestimmungsmethode (Fermentchemische Methoden. III.). Helv. Physiol. Acta **3**, C 6 (1945).

239. ZELLER, E. A., A. MARITZ und B. ISELIN: Über eine neue *l*-Aminosäure-oxydase (Ophio-*l*-aminosäure-oxydase). 3. Mitt. Helv. Chim. Acta **28**, 1615 (1945).

240. ZELLER, E. A. und D. C. UTZ: Über die Spezifität der Cholinesterase der Schlangengifte. 6. Mitt. über die Biochemie tierischer Gifte. Helv. Chim. Acta **32**, 338 (1949).

(Received, November 2, 1954.)

Gene Structure and Gene Action.

By G. W. Beadle, Pasadena, California.

With 1 Figure.

Contents.

Introduction.

During the past several years there have been developments that have profoundly changed current thinking about the structure, reproduction, mutation and function of genes. It is the author's purpose to summarize some of these without systematically reviewing all recent literature. Accordingly, many important contributions to the subject are not referred to. Readers interested in additional examples and further details will have no difficulty in finding the desired references in the papers cited below.

In a sense this review is a continuation of a similar one prepared for Vol. V of this series by the writer seven years ago (*3*).

The Gene as a Biological Unit.

Genes, first postulated by Mendel 90 years ago, are believed to be particulate units of inheritance, transmitted in linearly arranged blocks that correspond to the visible chromosomes in cell nuclei. The many genes of a single chromosome recombine during sexual reproduction by a process of crossing over between homologous chromosomes (corresponding chromosomes contributed by the two parents). Classically, crossing over occurs between but not within genes, and the gene is defined in terms of this irreducibility by crossing over. The probability of crossing over between two genes in a chromosome is a function of their distance apart and it is this relation that makes possible the genetic localization of genes on genetic maps.

To be investigated by genetic methods a gene must exist in at least two forms. This implies mutability. It is presumed that all genes are potentially mutable, although the frequencies of mutation may vary widely, and that such mutation underlies all organic evolution from the time of appearance on earth of the first successful self-duplicating and mutable system.

Many, possibly all, genes are capable of existing in more than two forms, a condition called multiple allelism. True alleles of a single gene cannot by definition be recombined by crossing over, *i. e.*, a given chromosome can carry only one allele of a single gene at any one time. Alleles are also identified by their genetic interaction, that is, first generation hybrids between recessive allelic mutant types show a mutant appearance, whereas such hybrids between non-allelic mutant types are of a non-mutant appearance*. In recent years a number of instances have been found in which genes, formerly thought to be allelic, show rare crossing over between them but at the same time give an interaction characteristic of alleles. This close association of related genes, a condition known as pseudoallelism is not accidental, but is evidently highly significant in terms of gene action (*30*).

Like gene mutations, rearrangements of chromosome segments occur spontaneously with a low frequency. Also like gene mutations, chromosome rearrangements can be increased many fold by high-energy radiations and by treatment with certain chemicals such as mustard gas. Following ionizing radiations of certain characteristics, segments of chromosomes may be lost (deficiency), have their orders reversed (inversion) or be exchanged for segments of non-homologous chromosomes (translocation). Combined genetic and cytological study of such rearrangements permits

* Thus if two recessive dwarf mutant strains of maize give tall plants in the first generation of a hybrid between them, it is concluded that the mutant genes responsible are not allelic. If, on the other hand, the first generation hybrid between the two dwarf strains is made up of dwarf plants, the result is taken as evidence that the two mutant genes are alleles. Representing the normal and mutant alleles of one gene by D_1 and d_1 and of the other by D_2 and d_2, the two conditions are symbolized as follows:

(1) For non-alleles, $d_1 d_1 D_2 D_2$ (dwarf) × $D_1 D_1 d_2 d_2$ (dwarf) gives $D_1 d_1 D_2 d_2$, tall, in the first generation because normal dominant alleles are present for both genes.

(2) For alleles, $d_1 d_1 D_2 D_2$ (dwarf) × $d_1 d_1 D_2 D_2$ (dwarf) gives $d_1 d_1 D_2 D_2$, dwarf, because no normal dominant allele of the d_1 gene is contributed by either parent.

the localization of specific genes in the cytologically visible chromosomes. Cytological maps show the expected correspondence in gene sequence and spacing with genetic maps.

Chromosome rearrangements are often accompanied by genetic changes like those resulting from gene mutation. Thus a deficiency may be so small that it results in the loss of a single gene. Its effect cannot be distinguished by genetic means from a mutation of a gene to a completely inactive form. The distinction is possible if the deficiency can be seen cytologically but in most cases this is difficult or impossible. The difficulty of distinguishing true gene mutations and chromosome rearrangements when the latter are extremely small has lead STADLER (*43*) to question whether ionizing radiations ever produce true gene mutations. Similar considerations have caused GOLDSCHMIDT (*15*) to doubt the correctness of the classical concept of the gene. Although one cannot at the moment categorically answer questions of this kind, most geneticists continue to have confidence in the gene as a discrete unit, the mutation of which differs in a basic way from simple loss or rearrangement of genetic material (*37*).

The Chemistry of the Genetic Material.

From their staining reactions and their absorption of ultraviolet radiation, chromosomes have long been believed to be made up of nucleoproteins. Enzyme digestion studies (*32*) and direct chemical analysis (*35*) confirm this view and further suggest that chromosome continuity is dependent on certain peptide linkages. The effectiveness of ultraviolet radiation of different wave lengths but of equal energy in producing mutations parallels the absorption spectrum of nucleic acid and is thus consistent with the view that nucleic acid is an important gene constituent (*20, 44*).

Recently a method of disrupting and solubilizing chromosome material has been discovered (*32*) that does not depend on breaking peptide bonds. If sea urchin sperm, fruit fly salivary gland cells or grasshopper testis cells are first treated with a metal chelating agent such as citrate or Versene (ethyleneamine tetraacetic acid) and subsequently with distilled water, the desoxyribonucleic acid (DNA) can be extracted in close association with most of the protein. The extracted particles appear in the electron microscope to be about 4000 Å long and 400 Å wide. It is presumed that the chromosome is built up of these macromolecular complexes of protein and DNA linked together linearly by bridges of Ca^{++} and Mg^{++} ions as well as by "the interactions making for 'insolubility' at moderate ionic strengths" (*32*). It is tempting to believe that these units correspond to genes and that crossing over always involves the breakage and reunion of intergenic connections.

Transforming Principles.

The view that DNA has *primary* genetic significance is now widely held. The first line of evidence leading toward the conclusion that DNA may be responsible for primary gene specificity comes from investigations of the transforming principles of bacteria. It is now more than a decade since AVERY and co-workers (*2*) demonstrated that treatment of a strain of Pneumococcus under the right conditions with a DNA preparation from a strain differing in capsular specificity was capable of permanently transforming the treated strain to the capsular specificity of the donor.

This line of work has been greatly extended in recent years by EPHRUSSI-TAYLOR (*13*), HOTCHKISS (*24*, *25*) and others. It has now been shown that many characters can be permanently transferred from donor to recipient strain through DNA as free from contaminants as it is presently possible to prepare this substance.

The proportion of cells of a treated population that are transformed varies with conditions but is usually low, ranging from less than 1 per cent to as high as 17 per cent. There are indications that transformability varies from one stage of the division cycle to another (*24*).

Although unfortunately it is not possible to do classical genetic studies on pneumococci, it does appear that there are produced simultaneously in a single strain of bacteria many DNA transforming principles. These are presumed to correspond to as many non-allelic genes. In each of these a single mutational event is capable of changing the nature of one biological unit of DNA. There are instances of such biological DNA units that correspond to multiple alleles in the sense that a given strain may possess any one, but not more than one, of three such DNA units.

If a DNA preparation from donors is used to transform recipients differing by two transformable characters, it is found that for some such pairs the two characters are transformed independently (*25*). Thus if 1.0 per cent of the recipients is transformed for one character and an equal proportion for the other, the doubly transformed cells are approximately the 0.01 per cent of the population expected by simple coincidence. HOTCHKISS and MARMUR (*25*) have found, however, that when the two characters of the donor are: ability to utilize the sugar mannitol and resistance to streptomycin, the double transformations are approximately ten times more frequent than would be expected on the basis of the product of their separate probabilities. If the two DNA-s are derived from separate donor cultures, double transformations are no higher than expected on a simple coincidence basis. This is also true if the entire DNA extraction was made from a mixed population of two types of cells each carrying one of the transforming principle.

These results with bacterial transformations strongly suggest that the specific properties of the genes of these organisms are represented by DNA-s and that transformation consists in the actual transfer of such gene DNA from one cell to another. They further suggest that the segments of DNA that are introduced into the recipient are often larger than single genes (high incidence of simultaneous transformation for mannitol utilization and streptomycin resistance when these originate from cells carrying both DNA-s which are presumably "linked") but smaller than the whole genic complex (expected coincidence value for double transformation of unlinked DNA units).

Other interpretations of the role of DNA in bacterial transformation are possible. A specific donor DNA might, for example, induce a specific mutation in a particular gene of the recipient without actually replacing its DNA. This is a less simple hypothesis since it would require a different DNA for each gene as well as modified forms of each DNA unit for all of the alleles of a given gene.

The Genetics of Viruses.

Additional evidence that gene specificity resides in DNA comes from recent work on bacteriophage (also known as bacterial viruses or phage), particularly the T-2 phage of *Escherichia coli*.

The life cycle of T-2 phage is as follows:

T-2 phage particles are tadpole shaped with a "head" roughly 800 Å in diameter and a "tail" somewhat longer. They are composed of protein and DNA. By osmotic shock the DNA is separated from the protein leaving the latter as empty head membranes known as "ghosts". The DNA is released in the form of long fibers about 20 Å in diameter. Thus the DNA appears to be inside the protein envelope of the phage head.

If, under suitable conditions, phages are added to a culture of susceptible bacteria, the phages are soon adsorbed by tips of their tails to the bacterial surfaces, one to many phages per cell depending on their concentration relative to cells. The adsorption occurs in several steps, the first one presumably involving union through complementary patterns of electrostatic charges on the protein membrane of the bacteria and on the tails of the phages (*40*).

Shortly after adsorption, phage-DNA is injected into the host cell through the tail. The phage protein coat remains outside. This is known from tracer experiments in which phage-DNA is labeled with P^{32} or in which phage protein is labeled with S^{35}. After infection of the bacteria, the coats can be knocked off in a Waring blendor and separated from the bacteria. If P^{32}-labeled phage are used most of the radioactivity is found in the bacteria. On the other hand, if S^{35} is used as a label, the radioactivity is largely recovered in the coats (*18, 19*).

There follows a period of about 8 to 12 minutes during which no infective phage can be recovered from the hosts after sonic disruption or other types of lysis. The phages are said to be in a "vegetative" stage. During this phase the phage-DNA increases to 50 to 100 phage units. Its phosphorus comes partly from host-DNA and partly from phosphate in the external medium.

After this period of vegetative phage multiplication, phage protein begins to appear and is shortly combined with phage-DNA to give complete infective phage particles. These increase linearly in number for another 10 to 15 minutes at which time lysis occurs with release of 100 to 300 phage particles per cell.

Phages are mutable in several respects, for example, in host range and in rate of lysis. If two mutant phages, one differing from standard in host range and the other in rate of lysis, are allowed to infect a single bacterial cell, the progeny consist of the two parental types of phages plus two recombination types, one non-mutant and one a double mutant (*17*). Such recombination experiments show that there are many genetic loci in a phage. In T-2 three groups of linked "genes" have been reported.

It is supposed that recombination occurs in T-2 through a series of several successive rounds of random pairwise "mating" of genetic units during the multiplication of vegetative phages (*29*, *47*).

The above account indicates that DNA is the sole carrier of genetic information in bacteriophages. It appears to be the only material that enters the host cell. If so, it is evidently capable of carrying directions for the synthesis of corresponding protein coats, coats made of protein that appears to be serologically different from any protein of the bacterial host. It is evident, too, that in the environment provided by its specific host, phage-DNA is replicated. Thus phage-DNA seems to possess the essential properties of genes—ability to replicate in a suitable environment, the ability to mutate, and the ability to transfer specificity to other macromolecules.

The Structure of DNA.

In 1953 Watson and Crick (*48*) proposed a structure of DNA that may well represent one of the most significant advances in biology in recent years.

This structure consists of two polynucleotide chains coiled around a common axis in a right-handed helix 20 Å in diameter. The two chains run in opposite directions around the helix, *i. e.*, the 3′–5′ phosphate di-ester linkages are in that order in one chain and in the 5′–3′ order in the other. The purine and pyrimidine bases are directed inward in planes at right angles to the axis of the helix.

There are two structures, A and B, that are interconvertible with changes in humidity and differ in spacing of the base pairs. It is to the B form that the following details refer.

The base pairs are spaced 3.4 Å apart and the helix repeats, *i. e.*, makes one turn, for every 10 base pairs. The molecular weight per turn of the helix is approximately 6700. As a lower limit of the molecular weight of DNA *in vivo* we may take that of the material isolated under the most favorable conditions. This gives a value of 6700000 or 1000 turns of the helix. If the total DNA in T-2 phage were a continuous single fiber, its length would correspond to approximately 20000 turns.

An especially significant feature of the structure is the complementary nature of the base sequences in the two chains. This results from the fact that the purines of one chain are hydrogen bonded to the pyrimidines of its complement and that structural considerations dictate that adenine be bonded to thymine and guanine to cytosine.

The proposed structure is supported by X-ray diffraction data, by base ratio analyses and by other chemical considerations [Crick (*7*), Todd (*46*)]. In addition it has great appeal from a biological point of view because it provides a reasonable basis for specificity, replication, and mutation of the gene.

Gene Specificity.

Gene specificity seems adequately accounted for by the proportions and sequences of base pairs in the DNA double helix. If the structure is given schematically as in *Chart 1* and *Figure 1*, one sees that the

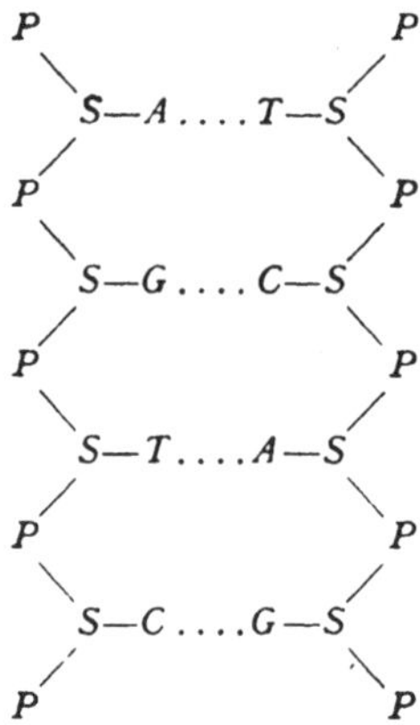

Chart 1. Schematic Structure of Desoxyribonucleic Acid (DNA), according to Watson and Crick (*48*) (*P* = phosphate; *S* = sugar; *A* = adenine; *T* = thymine; *G* = guanine; and *C* = cytosine).

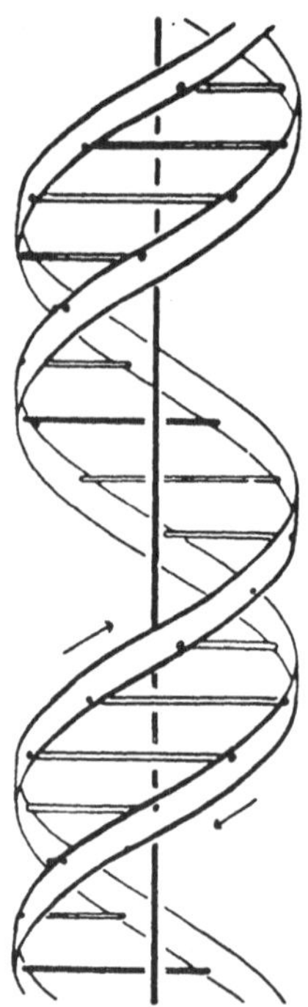

Fig. 1. The structure of DNA proposed by Watson and Crick (*48*) in 1953. The two phosphate-sugar chains are coiled round a common axis and joined together by hydrogen bonds between the nucleotide bases. The phosphate and sugar groups are on the outside of the helix, whilst the bases are on the inside. The figure is diagrammatic. The vertical line marks the fiber axis. [From: Nature (London) *171*, 737 (1953) and Cold Spring Harbor Sympos. Quant. Biol., *18*, 125 (1953)].

base pairs shown represent the four possibilities. Presumably these may be present in varying proportions and sequences. Considering that the DNA in a single haploid cell of a higher plant or animal consists of something like 10^8 base pairs, there is adequate basis for gene specificity.

Gene Reproduction.

In gene replication it is supposed that the two polynucleotide chains of DNA separate by breakage of hydrogen bonds and that each half reconstructs a paired helix by assembling a complementary chain hydrogen bonded to the original. The details of this process are not understood. It is not known, for example, how the paired helices separate, whether by untwisting, by breakage or by some other mechanism (*48 a*). Two kinds of breakage have been suggested. In one of these schemes [Delbrück (*11*)] it is proposed that systematic breakage and reunion occur at every half turn of the helix during the process of replication in such a way that one double helix becomes two double helices after replication. The other suggestion [Dekker and Schachman (*10*)] is that there are breaks in the polynucleotide chains about every five turns of the helix but that continuity of the double structure is maintained because breaks in the two chains do not coincide. Such breaks could of course facilitate separation of the two chains of the helix prior to or during replication. The details by which a single chain builds up its complement likewise remain to be determined (*12*).

Gene Mutation.

Gene mutation might have its explanation in the Watson-Crick structure, as has been suggested (*48 a*), by occasional mistakes in assembling complementary base pairs during replication. In its most probable tautomeric form, adenine forms hydrogen bonds with thymine. In a less probable tautomeric form it will form hydrogen bonds with cytosine in a manner that will allow the pair to fit the double helix. If this happens, one of the halves of the helix will produce a double helix during replication in which an adenine-thymine base pair is replaced by a guanine-cytosine pair. Judging from mutation rates, the frequency of such mistakes in the replication of a specific gene must be very low.

Gene Function.

The Watson-Crick model of DNA enables us to consider gene function in a manner that was not previously possible. If the bacteriophage life cycle outlined above is correct, it is clear that phage-DNA must direct protein synthesis in the host in such a way that a correct phage protein coat is made. Presumably this is a protein that is never made by an uninfected bacterium.

In higher organisms, too, many instances, some of them reviewed below, are known in which genes appear to direct protein synthesis. It seems clear from this and several other lines of evidence that somehow the base sequence of DNA must be translatable into the amino acid sequence of proteins. The fact that DNA is frequently found closely bound to protein as nucleoprotein suggests that this can be a direct transfer of information from DNA to protein. Possibilities as to how this transfer might be made have been suggested by GAMOW (*14*) and others, but the process is far from being clearly understood at present.

The observations that several plant viruses appear to contain only ribose nucleic acid (RNA) and that protein synthesis in higher plants and animals seems to occur in the cytoplasm in close association with RNA-containing mitochondria and microsomes, suggests that RNA also has the property of directing that part of the protein synthesizing mechanism that leads to specificity.

Unfortunately no structure of RNA comparable to the WATSON-CRICK structure of DNA has been proposed. This constitutes an important gap in present knowledge, for in the absence of such a structure it is not at all clear what the structural relation is between DNA and RNA or between RNA and protein. Many possibilities exist. One is that primary genetic information is carried by DNA, that this is transferred to RNA and that, through RNA, protein specificity is arrived at (*42*).

A question of very great importance is whether RNA is capable of replication without direct reference to the information carried in DNA. A simple interpretation of the replication of RNA plant viruses suggests that the answer is yes.

RNA-containing structures of higher organisms, for example plastids and plastid primordia, appear to be able to replicate under conditions in which some of their specific properties are maintained independently of the genetic constitution of the cell nucleus (*41*). Such cases also suggest that RNA may exist in the form of a directly replicating structure. But again our knowledge of details leaves much to be desired.

The relation between DNA as found in bacteriophages and as portrayed by the WATSON-CRICK model seems quite satisfactory. As a result one feels a certain degree of satisfaction in thinking about gene replication, recombination, and function in terms of this model. When one considers the chromosomes of higher organisms, however, the gap is disturbingly great. Chromosomes of these organisms in the most spun out stages of prophase are still visible in the light microscope, which means they are perhaps 500 to a 1000 times greater in cross sectional area than a DNA double helix. Perhaps in the resting stage the discrepancy is less. Considering that such chromosomes contain protein as well as DNA (*35*), it would seem that they must be quite complex structures as compared with the simple DNA fibers of phage. On the other hand the small number of nuclear divisions required for the expression of experimentally induced mutations in corn or *Drosophila* is difficult to reconcile with the notion that the genes of these organisms exist in a much replicated form throughout the division cycle of the cell. Until the relation between chromosomes and DNA fibers are better understood it is not easy to see how a single mechanism of recombination can apply to both.

In the course of the development of biochemical genetics many instances have been found in which particular chemical reactions are subject to genetic control, presumably through enzymes as intermediaries. Unfortunately in most of these cases the enzymes concerned have not themselves been investigated in detail. Nevertheless the hypothesis that many genes act by directing the synthesis of proteins that lend specificity to enzyme systems has been a fruitful one in both genetics and biochemistry. It is likewise a working hypothesis that is readily arrived at from the DNA-RNA-protein interrelations suggested by the WATSON-CRICK model.

It is not the purpose of this paper to catalog the known cases in which there appear to be close and perhaps fairly direct relations between genes and proteins. This has already been done several times (*3, 4, 5, 8, 9, 16, 22, 23, 28, 49, 50*). But there are a number of recently reported examples that for one reason or another are worthy of discussion.

Genetics of Hemoglobin Structure.

The genetic control of the characteristics of the hemoglobin molecule in man was first suggested by the studies of sickle cell trait and anemia by PAULING, ITANO, SINGER, and WELLS (*39*). Here it is found that the normal allele of a particular gene results in the presence of normal hemoglobin (*A*) in red blood cells. If a specific mutant form of the gene is substituted for its normal allele in both chromosomes, mainly abnormal hemoglobin (*S*) is present. This hemoglobin is recognized *in vitro* by its characteristic electrophoretic mobility (*42a*) and in the patient by a serious anemia in which red cells become highly distorted.

If in the two homologous chromosomes of an individual *both* forms of the gene are present, then *both* forms of hemoglobin are likewise present. This is what one would expect if each of the two alleles were directly responsible for its own kind of hemoglobin.

That the sickle cell mutant gene is highly disadvantageous is obvious from the fact that individuals homozygous* for it rarely survive to

* The individual homozygous for the *S* allele receives this mutant allele from *both* parents (symbolically, is *SS*), while an individual heterozygous receives the mutant allele from one parent and its normal counterpart from the other (symbolized *Ss* where *S* is the mutant allele and *s* the normal allele).

reproduce. Yet the incidence of the heterozygote reaches a value 40 per cent in a number of populations. This suggests either an unheard of mutation rate in these populations or some strong selective advantage. Recently this curious puzzle has been solved by Allison (*1*) and others who have found that individuals with sickle cell trait (heterozygotes) are much more resistant to malaria than are normal individuals. In highly malarious areas, heterozygotes evidently have a sufficient advantage to counterbalance the high incidence of fatal sickle cell anemia (homozygotes). Evidently malarial parasites find the red cells of heterozygotes an unfavorable environment for their survival and multiplication.

The discovery of sickle cell hemoglobin has stimulated a successful search for other abnormal hemoglobins. In addition to fetal hemoglobin (*F*), which may be a precursor of normal adult hemoglobin (*A*) and may persist in the adult if its conversion to hemoglobin *A* is genetically blocked as it appears to be in thalassemia (*23*), there have now been reported in man three additional abnormal hemoglobins, *C*, *D*, and *E* (*6*, *26*, *27*), each associated with disease and each having a characteristic electrophoretic mobility as compared with hemoglobins *A*, *S*, and *F*. Each appears to be gene-controlled but there is not yet sufficient evidence to say whether *A*, *S*, *C*, *D*, and *E* hemoglobins are determined by a series of five multiple alleles as one might expect on the basis of a simple assumption that there is a single gene from which the globin moiety of hemoglobin derives its specific properties.

Genes and Enzymes.

Two examples of genetically conditioned thermolability of enzymes have been worked out in some detail. The first involves the synthesis by *E. coli* of pantothenic acid from β-alanine and pantoic acid (*31*). Extracts of normal bacteria contain an enzyme that catalyzes the synthesis of pantothenic acid from these precursors. One mutant type shows no detectable activity for this enzyme. A second mutant type possesses an enzyme that is active at low temperatures (25°) but that is rapidly inactivated at temperatures above 30°. Interaction studies show no evidence of agents that inhibit or protect the enzyme and confirm the conclusion that for genetic reasons the thermolabile form of the enzyme is qualitatively different from its counterpart in the normal strain.

Another case of a genetically controlled qualitative change in a specific enzyme has been reported by Horowitz and Fling (*21*) in *Neurospora*. One allele of a gene permits the formation of a normally thermostable tyrosinase, but when another allele of this gene is present, a thermolabile tyrosinase is produced. That it differs qualitatively from its normal counterpart is confirmed by partial purification, by kinetic analysis of its thermal inactivation, and by studies of mixtures of stable and labile enzyme preparations. In both this case and in that of the pantothenate-synthesizing enzyme it is tempting to believe that the altered enzyme results from a modified DNA structure in a specific controlling gene. As has been pointed out, however, there is an alternative interpretation in which it is postulated that one form of the enzyme is a precursor of the other (*21*).

From the early days of *Neurospora* chemical genetics, there has been known a one-gene mutant with a requirement for both isoleucine and valine. The situation in this and in related mutants of *E. coli* has now been made clear (*38*) and can be simply summarized as follows: Isoleucine (III) is formed by the amination of its keto acid analogue (II) which in turn comes from the α,β-dihydroxy compound (I). In a similar manner valine (VI) is synthesized by way of its keto acid (V) and the α,β-dihydroxyacid analogue (IV). Thus the reactions may be represented

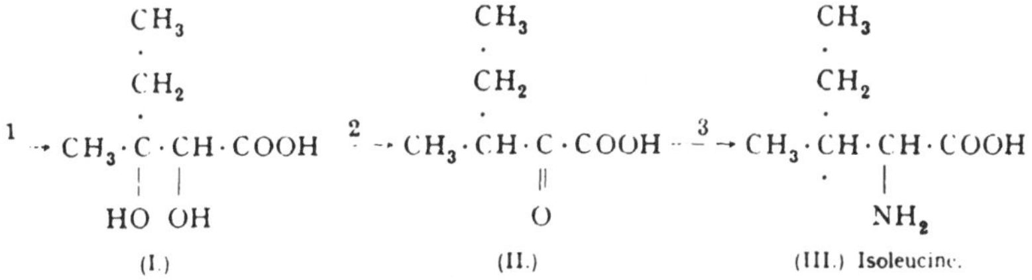

```
           CH3                     CH3                      CH3
            ·                       ·                        ·
           CH2                     CH2                      CH2
            ·                       ·                        ·
1 → CH3·C·CH·COOH    2 → CH3·CH·C·COOH    3 → CH3·CH·CH·COOH
        |  |                        ||                   ·  |
       HO OH                        O                      NH2
        (I.)                      (II.)                (III.) Isoleucine.
```

and

```
           CH3                     CH3                      CH3
            ·                       ·                        ·
1′ → CH3·C·CH·COOH   2′ → CH3·CH·C·COOH   3′ → CH3·CH·CH·COOH
         |  |                       ||                      |
        HO OH                       O                      NH2
        (IV.)                     (V.)                 (VI.) Valine.
```

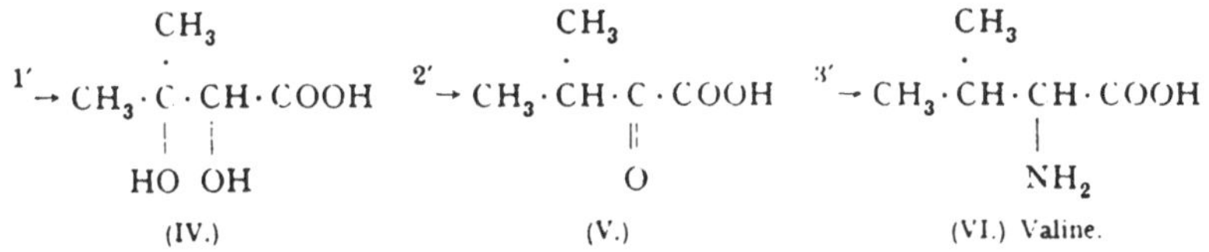

Chart 2. Biosynthesis of Isoleucine and Valine.

as in *Chart 2*. Reactions 3 and 3′ are both catalyzed by a single transaminase which is absent in a mutant strain of *E. coli* requiring both isoleucine and valine. It therefore seems clear that these two closely related reactions are simultaneously blocked by a single mutation because they are both catalyzed by the same enzyme.

Reactions 2 and 2′ are catalyzed by a dihydroxyacid dehydrase which is reduced to about 10% of its normal amount in the valine-isoleucine-requiring *Neurospora* mutant and is essentially absent in a corresponding mutant in *E. coli*. The experimental evidence for a single enzyme in this case is not complete, but the hypothesis that the situation is similar to that in which a single gene controls reactions 3 and 3′ by way of a single enzyme now seems a most reasonable one.

Genes and Aromatic Amino Acids.

Another series of genetically controlled reactions that deserves special mention is that concerned with the biosynthesis of the aromatic amino acids tyrosine, phenylalanine, tryptophan, *p*-aminobenzoic acid, and the related *p*-hydroxybenzoic acid. A few years ago very little could have been said about the biosynthesis of these compounds. Now, as a result of the work of a number of investigators using mutant types of *E. coli* [cf. DAVIS (*8*, *9*)] and *Neurospora* [TATUM *et al.* (*45*)], several precursors and a number of enzymes that catalyze their reactions have been identified and their interrelations made clear.

In addition to the end products and several of the well-known intermediates, the compounds concerned are, (VII)-(XI).

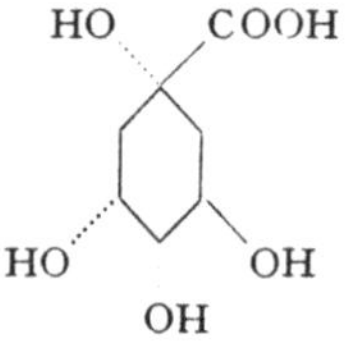

(VII.) Quinic acid. 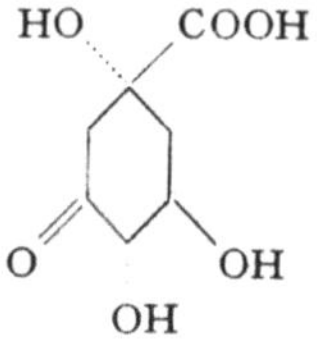 (VIII.) 5-Dehydroquinic acid. 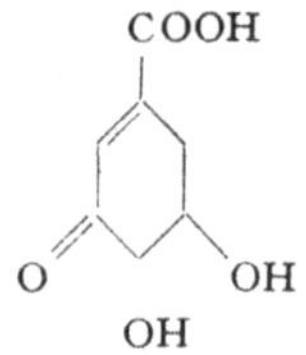 (IX.) 5-Dehydroshikimic acid

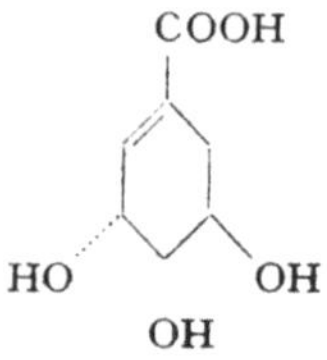

(X.) Shikimic acid. 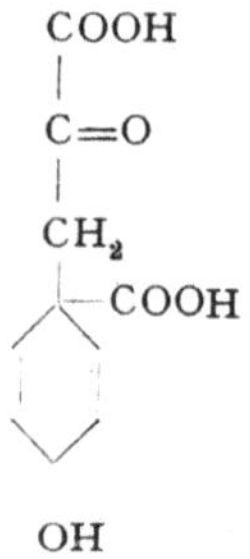 (XI.) Prephenic acid.

The scheme of biosynthesis proposed (*9*) is given in *Chart 3*.

Reaction 2 in *E. coli* is catalyzed by the enzyme 5-dehydroquinase for which several important characteristics have been determined. In mutants blocked in this reaction 5-dehydroquinic acid is found if special steps are taken to favor its accumulation. In mutants with a complete genetic block no enzyme was detected by methods that would measure as little as 1/300 of the normal amount. In mutants in which this block is incomplete, intermediate amounts of enzyme were found.

Reaction 3 is catalyzed by the enzyme 5-dehydroshikimic reductase. Wild-type *E. coli* yields this enzyme in amounts sufficient to account for its growth rate whereas in a mutant strain blocked between 5-dehydroshikimic acid and shikimic acid none could be demonstrated. It is of interest to note that shikimic acid does not serve well as a growth factor

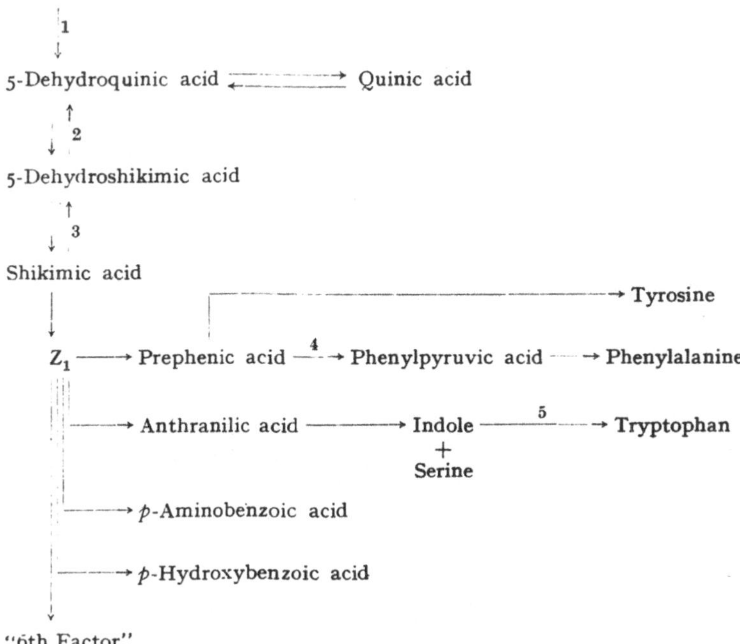

Chart 3. Scheme of Biosynthesis of Some Aromatic Acids in *E. coli.* (Z_1 represents an accumulated intermediate not yet identified.)

for such a mutant because accumulated dehydroshikimic acid interferes with its utilization. If by adding a second mutant dehydroshikimic synthesis is prevented, shikimic acid is readily used.

The evidence indicating that the three compounds 5-dehydroquinic acid, 5-dehydroshikimic acid and shikimic acid are true intermediates is convincing. Any one of these substances will serve as a substitute for the six end products mentioned in Chart 3 for mutants blocked prior to the substance concerned. All are found to be accumulated by appropriate mutants that block subsequent steps in the synthesis. Finally, reactions 1 and 2 seem pretty clearly to be one-step reactions catalyzed by single, well characterized enzymes. Absence of either enzyme leads to the predicted growth factor requirements.

For several reasons quinic acid, on the other hand, is believed not to be in the main line of synthesis (9).

Mutants blocked in step 4 show the curious property of accumulating a growth factor for themselves. This is because such mutants accumulate prephenic acid and excrete it into the culture medium where it is spontaneously converted to phenylpyruvic acid if the culture medium becomes acid, as it normally does. Although it goes spontaneously in acid medium, reaction 4 is enzymatically catalyzed *in vivo*, for wild type extracts contain such an enzyme whereas those of certain phenylalanine-requiring mutants do not (*9*).

Reaction 5 is catalyzed by tryptophan synthetase (desmolase), which can be prepared from wild type strains *Neurospora* but which is absent in mutant strains that cannot carry out the reaction (*36, 51*). The symbol *tp* is used to designate the mutant gene responsible for a tryptophan requirement. Yanofsky (*52*) has isolated a series of twenty-five mutants of independent occurence all of which are blocked in reaction 5. Twenty-two have been examined for tryptophan synthetase activity and all were found to lack it. The first two mutants of the series were found to be alleles (or closely linked genes) but it is not yet reported whether the remaining twenty-three are likewise alleles. A suppressor gene, not at the locus of tp_1 and tp_2, is known that suppresses tp_2—by partially restoring tryptophan synthetase activity—but not tp_1. Two other suppressor genes have been studied each of which suppresses one or two *tp* alleles but not the same ones.

The evidence so far available is consistent with the assumption that all twenty-five *tp* genes are alleles, and that in the presence of any one of them the formation of active tryptophan synthetase does not take place. The fact that a non-allelic suppressor may restore enzyme activity to a strain carrying one mutant allele but not to a strain carrying another mutant allele of the same gene suggests that tryptophan synthetase, like the pantothenic acid enzyme and tyrosinase discussed above, can be genetically altered in qualitatively different ways. In each case it has greatly reduced or no activity in the particular genetic environment in which the mutant strain was selected.

If this interpretation is correct, it should not be expected that suppressors active for one allele would be active for another. A particular suppressor gene may well alter the cellular environment in such a manner that an enzyme altered in one way will have its activity restored while one altered in a different manner will remain unaffected (*23*).

Summarizing Discussion.

While the proof cannot yet be regarded as complete, it now seems probable that the primary specificity of genes resides in DNA.

The evidence is strong that genic DNA has the structure proposed by Watson and Crick (*48*), that is, a double helix of two polynucleotide chains with the bases of one chain hydrogen bonded to those of the other in such a way as to make the base sequence in one chain the complement of that in the other. This structure suggests that gene replication involves simultaneous separation of complementary chains and synthesis of new complementary chains by each single chain. The details of how this happens and the nature of the final precursors from which new polynucleotide chains are constructed remain to be determined.

The WATSON-CRICK structure provides for the determination of gene specificity through base sequence in DNA helices. It is reasonable to suppose that mutation results from a change in base sequency by replacement of one base by another, by loss of one or more bases or by rearrangement of base sequence.

It seems likely that the information present in genes in the form of specific base sequences is transferred to proteins as specific amino acid sequences. It is not known whether this transfer can be made directly from DNA to protein or requires RNA as an intermediate step.

The structure of RNA and its exact role in plant viruses and in higher plants and animals remain to be discovered.

In higher organisms genetic DNA seems always to be associated with protein but in a manner not now clear. There is some evidence to indicate that the chromosomes of higher animals, and presumably plants, are built up of macromolecular complexes of nucleic acid and protein bound together linearly by bridges of divalent ions, presumably Ca^{++} and Mg^{++}. It is possible that these units correspond to individual genes and that normal crossing over involves breakage and reunion only in intergenic connections.

Whether the chromosomal DNA in higher plants and animals is continuous over the length of a chromosome, as it seems to be in phage DNA for at least 1000 turns of the helix and perhaps as many as 20000 turns, or is interrupted at intervals of the order of 100 turns, as MAZIA (*32*) suggests, is a question of the greatest importance. In either case, it seems reasonable to suppose that the information necessary for determining the amino acid sequence of a given protein is represented by the base sequence of a discrete segment of DNA helix. By one definition such a DNA segment would constitute a gene.

On the basis of a continuous DNA helix it is difficult to understand why crossing over should not be intragenic as well as intergenic. Available experimental evidence is, in fact, probably inadequate to exclude intragenic recombination. If it occurs, it could conceivably account for the observations on which the concept of pseudoallelism is based (cf. p. 467). On the MAZIA hypothesis of an interrupted DNA-protein chromosome structure one is inclined to postulate that each discrete segment carries the information essential for the construction of a specific protein (or other macromolecule) and that these units normally remain intact during cell division and throughout the life cycle of the organism.

If the genic material is continuous, the question immediately arises as to what determines the limits of a single gene? If it is that portion of a DNA helix that corresponds to a specific protein, what is it in the DNA helix that determines the two ends of the protein molecule? Is it perhaps a combination of base pairs in the DNA helix for which there are no corresponding amino acids?

It is conceivable that the segments of DNA corresponding to different proteins may overlap. If this were a frequent situation, it would be expected that many single gene mutations would lead to alterations in two different proteins and that such pairs of proteins would have segments with identical amino acid sequences. There is no convincing evidence in favor of this situation nor is the evidence sufficient to exclude it.

In a remarkable series of investigations with *Zea mays* McCLINTOCK (*33, 34*) has recently discovered phenomena that clearly involve the regulation of gene activity in a manner not previously known. She has shown that there exists a special kind of genetic material, possibly heterochromatic in nature, that may be transposed by an unknown mechanism from place to place in the chromosome complement. When placed in close proximity to an active form of a gene such a unit may inactivate the gene. This inactivation is reversible, for, when the unit is removed by transposition, gene activity is restored. Proximity to such units may lead to instability of gene action in a manner that suggests frequent gene mutation. Localized chromosome breakage may also be induced by these extragenic units. Although there can be little doubt about the importance of this new approach, the relation of findings of McCLINTOCK to the facts and interpretations reviewed above is not obvious.

The hypothesis of gene structure and gene action summarized here is no doubt far from complete. It is evident, too, that it may well be incorrect in many details. Whatever its final fate may be, however, there can be little doubt that the WATSON-CRICK structure of DNA represents a great forward step and that it will continue to give impetus to the search for further knowledge about the nature of the gene.

References.

1. ALLISON, A. C.: Protection Afforded by Sickle-cell Trait Against Subtertian Malarial Infection. Brit. Med. J. 1954, 290.

2. AVERY, O. T., C. M. McLEOD and M. McCARTY: Studies on the Nature of the Substance Inducing Transformation of Pneumococcal Types: Induction of Transformation by a Desoxyribonucleic Acid Fraction Isolated from Pneumococcus Type III. J. exp. Medicine 79, 137 (1944).

3. BEADLE, G. W.: Some Recent Developments in Chemical Genetics. Fortschr. Chem. organ. Naturstoffe 5, 300 (1948).

4. BONNER, D. M.: Gene-enzyme Relationships in Neurospora. Cold Spring Harbor Sympos. Quant. Biol. 16, 143 (1951).

5. CATCHESIDE, D. E.: The Genetics of Microorganisms, p. 223. London: Pitman Publ. Corp. 1951.

6. CHERNOFF, A. I., V. MINNICH and S. CHONGCHAREONSUK: Hemoglobin E, a Hereditary Abnormality of Human Hemoglobin. Science (Washington) **120**, 605 (1954).
7. CRICK, F. H. C.: The Complementary Structure of DNA. Proc. Nat. Acad. Sci. (USA) **40**, 756 (1954).
8. DAVIS, B. D.: Recent Applications of Genetic Methods to the Study of Intermediary Metabolism in Microorganisms. Sympos. Microb. Metabolism, Inst. Superiore Sanità (Rome) p. 23 (1953).
9. — Biosynthesis of the Aromatic Amino Acids. In: Amino Acid Metabolism, W. D. McElroy and B. Glass, Eds. Baltimore: Johns Hopkins Univ. Press 1955. p. 799.
10. DEKKER, C. A. and H. K. SCHACHMAN: On the Macromolecular Structure of Desoxyribonucleic Acid: an Interrupted Two-strand Model. Proc. Nat. Acad. Sci. (USA) **40**, 894 (1954).
11. DELBRÜCK, M.: On the Replication of Desoxyribonucleic Acid (DNA). Proc. Nat. Acad. Sci. (USA) **40**, 783 (1954).
12. — Wie vermehrt sich ein Bacteriophage? Angew. Chem. **66**, 391 (1954).
13. EPHRUSSI-TAYLOR, H.: Genetic Aspects of Transformations of Pneumococci. Cold Spring Harbor Sympos. Quant. Biol. **16**, 445 (1951).
14. GAMOW, G.: Possible Mathematical Relation Between Desoxyribonucleic Acid and Proteins. Kong. Danske Vidensk. Selsk., Biol. Medd. **22**, 1 (1954).
15. GOLDSCHMIDT, R. B.: The Theory of the Gene: Chromosomes and Genes. Cold Spring Harbor Sympos. Quant. Biol. **16**, 1 (1951).
16. HALDANE, J. B. S.: The Biochemistry of Genetics, p. 144. New York: Macmillan Co. 1954.
17. HERSHEY, A. D.: Inheritance in Bacteriophage. Adv. Genetics **5**, 89 (1953).
18. — Functional Differentiation Within Particles of a Bacteriophage T 2. Cold Spring Harbor Sympos. Quant. Biol. **18**, 135 (1953).
19. HERSHEY, A. D. and M. CHASE: Independent Functions of Viral Protein and Nucleic Acid in Growth of Bacteriophage. J. Gen. Physiol. **36**, 39 (1952).
20. HOLLAENDER, A. and C. W. EMMONS: Wavelength Dependence of Mutation Production in the Ultraviolet with Special Emphasis on Fungi. Cold Spring Harbor Sympos. Quant. Biol. **9**, 179 (1941).
21. HOROWITZ, N. H. and M. FLING: Genetic Determination of Tyrosinase Thermostability in Neurospora. Genetics **38**, 360 (1953).
22. HOROWITZ, N. H. and U. LEUPOLD: Some Recent Studies Bearing on the One Gene—One Enzyme Hypothesis. Cold Spring Harbor Sympos. Quant. Biol. **16**, 65 (1951).
23. HOROWITZ, N. H. and R. D. OWEN: Physiological Aspects of Genetics. Annu. Rev. Physiology **16**, 81 (1954).
24. HOTCHKISS, R. D.: Cyclical Behavior in Pneumococcal Growth and Transformability Occasioned by Environmental Changes. Proc. Nat. Acad. Sci. (USA) **40**, 49 (1954).
25. HOTCHKISS, R. D. and J. MARMUR: Double Marker Transformations as Evidence of Linked Factors in Desoxyribonucleate Transforming Agents. Proc. Nat. Acad. Sci. (USA) **40**, 55 (1954).
26. ITANO, H. A.: Human Hemoglobin. Science (Washington) **117**, 89 (1953).
27. ITANO, H. A., W. R. BERGREN and P. STURGEON: Identification of a Fourth Abnormal Human Hemoglobin. J. Amer. Chem. Soc. **76**, 2278 (1954).
28. LEDERBERG, J.: Problems in Microbial Genetics. Heredity **2**, 145 (1948).
29. LEVINTHAL, C.: Recombination in Phage: its Relationship to Heterozygosis and Growth. Cold Spring Harbor Sympos. Quant. Biol. **18**, 13 (1953).

30. LEWIS, E. B.: Pseudoallelism and Gene Evolution. Cold Spring Harbor Sympos. Quant. Biol. 16, 159 (1951).
31. MAAS, W. K. and B. D. DAVIS: Production of an Altered Pantothenate-Synthesizing Enzyme by a Temperature-sensitive Mutant of *Escherichia coli*. Proc. Nat. Acad. Sci. (USA) 38, 785 (1952).
32. MAZIA, D.: The Particulate Organization of the Chromosome. Proc. Nat. Acad. Sci. (USA) 40, 521 (1954).
33. MCCLINTOCK, B.: Chromosome Organization and Genic Expression. Cold Spring Harbor Sympos. Quant. Biol. 16, 13 (1951).
34. — Induction of Instability at Selected Loci in Maize. Genetics 38, 579 (1953).
35. MIRSKY, A. E. and H. RIS: The Chemical Composition of Isolated Chromosomes. J. Gen. Physiol. 31, 7 (1947).
36. MITCHELL, H. K. and J. LEIN: A Neurospora Mutant Deficient in the Enzymatic Synthesis of Tryptophan. J. Biol. Chem. 175, 481 (1948).
37. MULLER, H. J.: The Gene. Proc. Roy. Soc. (London), Ser. B 134, 1 (1947).
38. MYERS, J. W. and E. A. ADELBERG: The Biosynthesis of Isoleucine and Valine. I. Enzymatic Transformation of the Dihydroxy Acid Precursors to the Keto Acid Presursors. Proc. Nat. Acad. Sci. (USA) 40, 493 (1954).
39. PAULING, L., H. A. ITANO, S. J. SINGER and I. C. WELLS: Sickle Cell Anemia, a Molecular Disease. Science (Washington) 110, 543 (1949).
40. PUCK, T. T.: The First Steps of Virus Invasion. Cold Spring Harbor Sympos. Quant. Biol. 12, 149 (1953).
41. RENNER, O.: Zur Kenntnis der Plastiden- und Plasmavererbung. Cytologia, Fujii Jubiläumsband 1937, 644.
42. RICH, A. and J. D. WATSON: Some Relations Between DNA and RNA. Proc. Nat. Acad. Sci. (USA) 40, 759 (1954).
42a. SCHEINBERG, I., R. S. HARRIS and J. L. SPITZER: Differential Titration by Means of Paper Electrophoresis and the Structure of Human Hemoglobins. Proc. Nat. Acad. Sci. (USA) 40, 777 (1954).
43. STADLER, L. J.: The Gene. Science (Washington) 120, 811 (1954).
44. STADLER, L. J. and F. M. UBER: Genetic Effects of Ultraviolet Radiation in Maize. IV. Comparisons of Monochromatic Radiations. Genetics 27, 84 (1942).
45. TATUM, E. L., S. R. GROSS, G. EHRENSVÄRD and L GARNJOBST: Synthesis of Aromatic Compounds by Neurospora. Proc. Nat. Acad. Sci. (USA) 40, 271 (1954).
46. TODD, A. R.: Chemical Structure of the Nucleic Acids. Proc. Nat. Acad. Sci. (USA) 40, 748 (1954).
47. VISCONTI, N. and M. DELBRÜCK: The Mechanism of Genetic Recombination in Phage. Genetics 38, 5 (1953).
48. WATSON, J. D. and F. H. C. CRICK: A Structure for Deoxyribose Nucleic Acid. Nature (London) 171, 737 (1953).
48a. — — The Structure of DNA. Cold Spring Harbor Sympos. Quant. Biol. 18, 123 (1953).
49. WRIGHT, S.: The Physiology of the Gene. Physiol. Rev. 21, 487 (1941).
50. — Genes as Physiological Agents: General Considerations. Amer. Naturalist 79, 289 (1945).
51. YANOFSKY, C.: The Effects of Gene Change on Tryptophan Desmolase Formation. Proc. Nat. Acad. Sci. (USA) 38, 215 (1952).
52. — Further Studies With the td Mutants of Neurospora. Genetics 38, 702 (1953).

(Received, December 31, 1954.)

Namenverzeichnis. Index of Names. Index des Auteurs.

Sachverzeichnis. Index of Subjects. Index des Matières.

Manzsche Buchdruckerei, Wien IX.